PENGUIN BOOKS

BEHAVE

Robert M. Sapolsky is the author of several works of nonfiction, including *A Primate's Memoir*, *The Trouble with Testosterone*, and *Why Zebras Don't Get Ulcers*. He is a professor of biology and neurology at Stanford University and the recipient of a MacArthur Foundation genius grant. He lives in San Francisco with his wife, two children, and dogs.

* * *

Praise for *Behave*

One of *The Washington Post*'s 10 Best Books of the Year
One of *The Wall Street Journal*'s Best Books of the Year
One of *The New York Times* Critics' Favorite Books of the Year
Winner of the 2018 Phi Beta Kappa Award in Science

"A masterly cross-disciplinary scientific study of human behavior: What in our glands, our genes, our childhoods explains our species' capacity for both altruism and brutality? This comprehensive and friendly survey of a 'big sprawling mess of a subject' is leavened by an impressive data-to-silly-joke ratio. It has my vote for science book of the year." —Parul Sehgal, *The New York Times*

"Sapolsky has created an immensely readable, often hilarious romp through the multiple worlds of psychology, primatology, sociology, and neurobiology to explain why we behave the way we do. It is hands-down one of the best books I've read in years. I loved it." —Dina Temple-Raston, *The Washington Post*

"It's no exaggeration to say that *Behave* is one of the best nonfiction books I've ever read." —David P. Barash, *The Wall Street Journal*

"A quirky, opinionated, and magisterial synthesis of psychology and neurobiology that integrates this complex subject more accessibly and completely than ever. . . . A wild and mind-opening ride into a better understanding of just where our behavior comes from. Darwin would have been thrilled."
—Richard Wrangham, *The New York Times Book Review*

"[Sapolsky's] new book is his magnum opus. . . . A stunning achievement and an invaluable addition to the canon of scientific literature, certain to kindle debate for years to come." —*Minneapolis Star Tribune*

"As wide as it is deep, this book is colorful, electrifying, and moving. Sapolsky leverages his deep expertise to ask the most fundamental questions about being human—from acts of hate to acts of love, from our compulsion to dehumanize to our capacity to rehumanize."
—David Eagleman, PhD, neuroscientist at Stanford University, author, presenter of PBS's *The Brain*

"A monumental contribution to the scientific understanding of human behavior that belongs on every bookshelf and many a course syllabus . . . It is a magnificent culmination of integrative thinking, on par with similar authoritative works, such as Jared Diamond's *Guns, Germs, and Steel* and Steven Pinker's *The Better Angels of Our Nature*."
—Michael Shermer, *American Scholar*

"*Behave* is the best detective story ever written, and the most important. If you've ever wondered why someone did something—good or bad, vicious or generous—you need to read this book. If you think you already know why people behave as they do, you need to read this book. In other words, everybody needs to read it. It should be available on prescription (side effects: chronic laughter; highly addictive). They should put *Behave* in hotel rooms instead of the Bible: the world would be a much better, wiser place."
—Kate Fox, author of *Watching the English*

"Magisterial . . . In this extraordinary survey of the science of human behavior, [Sapolsky] takes the reader on an epic journey. . . . He makes the book consistently entertaining, with an infectious excitement at the puzzles he explains. . . . A miraculous synthesis of scholarly domains."
—Steven Poole, *The Guardian*

"Rarely does an almost eight-hundred-page book keep my attention from start to finish, but if anyone can save evolutionary biology from TED talkers and pop-science fabulists, it might be Sapolsky. . . . *Behave* ranges at great length from moral philosophy to social science, genetics to Sapolsky's home turf of neurons and hormones—but all of it is aimed squarely at the question of why humans are so awful to each other, and whether the condition is terminal."

—*Vulture*

"Robert Sapolsky's students must love him. In *Behave*, the primatologist, neurologist, and science communicator writes like a teacher: witty, erudite, and passionate about clear communication. You feel like a lucky auditor in a fast-paced undergraduate course, where the implications of fascinating scientific findings are illuminated through topical stories and pop-culture allusions."

—*Nature*

"Sapolsky's book shows in exquisite detail how culture, context, and learning shape everything our genes, brains, hormones, and neurons do."

—*The Times Literary Supplement* (London)

"*Behave* is like a great historical novel, with excellent prose and encyclopedic detail. It traces the most important story that can ever be told."

—Edward O. Wilson

"Truly all-encompassing . . . Detailed, accessible, fascinating."

—*The Telegraph* (UK)

"*Behave* is a beautifully crafted work about the biology of morality. Sapolsky makes multiple passes at the target, using different time scales and systems. He shows you how all the perspectives and systems connect, and he makes you laugh and marvel along the way. Sapolsky is not just a leading primatologist; he's a great writer and a superb guide to human nature."

—Jonathan Haidt, New York University, author of *The Righteous Mind*

"A wide-ranging, learned survey of all the making-us-tick things that, for better or worse, define us as human . . . An exemplary work of popular science, challenging but accessible."

—*Kirkus Reviews* (starred review)

"[Sapolsky] weaves science storytelling with humor. . . . [His] big ideas deserve a wide audience and will likely shape thinking for some time."

—*Publishers Weekly* (starred review)

"[Sapolsky] does an excellent job of bringing together the expansive literature of thousands of fascinating studies with clarity and humor. . . . A tour de force."
—*Library Journal* (starred review)

"Sapolsky finds not the high moral drama of the soul choosing good or evil but rather down-to-earth biology. . . . A remarkably encyclopedic survey of the sciences illuminating human conduct."
—*Booklist* (starred review)

"Read Robert Sapolsky's marvelous book *Behave* and you'll never again be surprised by the range and depth of our own bad behavior. We all carry the potential for unconscious biases, to be damaged by our childhoods and map that damage onto our own loved ones, and to form the tribal 'Us' groups that treat outsiders as lesser 'Thems.' But to read this book is also, marvelously, to be given the hope that we have much more control of those behaviors than we think. And *Behave* gives us more than hope—it gives us the knowledge of how to act on that aspiration, to manifest more of our best selves and less of our worst, individually and as a society. That's very good news indeed."
—Charles Duhigg, author of *The Power of Habit* and *Smarter Faster Better*

"This is a miraculous book, by far the best treatment of violence, aggression, and competition ever. . . . Its depth and breadth of scholarship are amazing, building on Sapolsky's own research and his vast knowledge of the neurobiology, genetic, and behavioral literature. For instance, *Behave* includes fair evaluations of complex debates (like over sociobiology) that I was involved in, and tackles controversial questions such as whether our hunter-gatherer ancestors warred on each other. He even takes on free will with a clarity usually absent from the writings of philosophers on the subject. All this is done brilliantly with a light and funny touch that shows why Sapolsky is recognized as one of the greatest teachers in science today."
—Paul R. Ehrlich, author of *Human Natures*

Behave

THE BIOLOGY *of* HUMANS *at* OUR BEST *and* WORST

Robert M. Sapolsky

PENGUIN BOOKS

PENGUIN BOOKS
An imprint of Penguin Random House LLC
375 Hudson Street
New York, New York 10014
penguin.com

First published in the United States of America by Penguin Press,
an imprint of Penguin Random House LLC, 2017
Published in Penguin Books 2018

Illustration credits appear on page 774.

ISBN 9780143110910 (paperback)

THE LIBRARY OF CONGRESS HAS CATALOGED THE
HARDCOVER EDITION AS FOLLOWS:
Names: Sapolsky, Robert M., author.
Title: Behave: the biology of humans at our best and worst / Robert M. Sapolsky.
Description: New York: Penguin Press, 2017.
Identifiers: LCCN 2016056755 (print) | LCCN 2017006806 (ebook) |
ISBN 9781594205071 (hardback) | ISBN 9780735222786 (ebook)
Subjects: LCSH: Neurophysiology. | Neurobiology. | Animal behavior. |
BISAC: SCIENCE / Life Sciences / Biology / General. | SOCIAL SCIENCE /
Criminology. | SCIENCE / Life Sciences / Neuroscience.
Classification: LCC QP351 .S27 2017 (print) | LCC QP351 (ebook) |
DDC 612.8–dc23
LC record available at https://lccn.loc.gov/2016056755

Printed in the United States of America
5 7 9 10 8 6

Interior Illustrations by Tanya Maiboroda on pages 24, 30, 70, 72, 75, 128, 163, 214, 267, 280, 286, 287, 322, 438, 448, 497, 567, 568, 604, 625, 690, 695, 696, 697, 698, 700

Designed by Amanda Dewey

To Mel Konner, who taught me.

To John Newton, who inspired me.

To Lisa, who saved me.

Contents

Introduction

The fantasy always runs like this: A team of us has fought our way into his secret bunker. Okay, it's a fantasy, let's go whole hog. *I've* single-handedly neutralized his elite guard and have burst into his bunker, my Browning machine gun at the ready. He lunges for his Luger; I knock it out of his hand. He lunges for the cyanide pill he keeps to commit suicide rather than be captured. I knock that out of his hand as well. He snarls in rage, attacks with otherworldly strength. We grapple; I manage to gain the upper hand and pin him down and handcuff him. "Adolf Hitler," I announce, "I arrest you for crimes against humanity."

And this is where the medal-of-honor version of the fantasy ends and the imagery darkens. What would I do with Hitler? The viscera become so raw that I switch to passive voice in my mind, to get some distance. What should be done with Hitler? It's easy to imagine, once I allow myself. Sever his spine at the neck, leave him paralyzed but with sensation. Take out his eyes with a blunt instrument. Puncture his eardrums, rip out his tongue. Keep him alive, tube-fed, on a respirator. Immobile, unable to speak, to see, to hear, only able to feel. Then inject him with something that will give him a cancer that festers and pustulates in every corner of his body, that will grow and grow until every one of his cells shrieks with agony, till every moment feels like an infinity spent in the fires of hell.

That's what should be done with Hitler. That's what I would want done to Hitler. That's what I would do to Hitler.

I've had versions of this fantasy since I was a kid. Still do at times. And when I really immerse myself in it, my heart rate quickens, I flush, my fists clench. All those plans for Hitler, the most evil person in history, the soul most deserving of punishment.

But there is a big problem. I don't believe in souls or evil, think that the word "wicked" is most pertinent to a musical, and doubt that punishment should be relevant to criminal justice. But there's a problem with that, in turn—I sure feel like some people should be put to death, yet I oppose the death penalty. I've enjoyed plenty of violent, schlocky movies, despite being in favor of strict gun control. And I sure had fun when, at some kid's birthday party and against various unformed principles in my mind, I played laser tag, shooting at strangers from hiding places (fun, that is, until some pimply kid zapped me, like, a million times and then snickered at me, which made me feel insecure and unmanly). Yet at the same time, I know most of the lyrics to "Down by the Riverside" ("ain't gonna study war no more") plus when you're supposed to clap your hands.

In other words, I have a confused array of feelings and thoughts about violence, aggression, and competition. Just like most humans.

To preach from an obvious soapbox, our species has problems with violence. We have the means to create thousands of mushroom clouds; shower heads and subway ventilation systems have carried poison gas, letters have carried anthrax, passenger planes have become weapons; mass rapes can constitute a military strategy; bombs go off in markets, schoolchildren with guns massacre other children; there are neighborhoods where everyone from pizza delivery guys to firefighters fears for their safety. And there are the subtler versions of violence—say, a childhood of growing up abused, or the effects on a minority people when the symbols of the majority shout domination and menace. We are always shadowed by the threat of other humans harming us.

If that were solely the way things are, violence would be an easy problem to approach intellectually. AIDS—unambiguously bad news—eradicate. Alzheimer's disease—same thing. Schizophrenia, cancer, malnutrition, flesh-eating bacteria, global warming, comets hitting earth—ditto.

The problem, though, is that violence doesn't go on that list. Sometimes we have no problem with it at all.

This is a central point of this book—we don't hate violence. We hate and fear the *wrong* kind of violence, violence in the wrong context. Because violence in the right context is different. We pay good money to watch it in a stadium, we teach our kids to fight back, we feel proud when, in creaky middle age, we manage a dirty hip-check in a weekend basketball game. Our conversations are filled with military metaphors—we rally the troops after our ideas get shot down. Our sports teams' names celebrate violence—Warriors, Vikings, Lions, Tigers, and Bears. We even think this way about something as cerebral as chess—"Kasparov kept pressing for a murderous attack. Toward the end, Kasparov had to oppose threats of violence with more of the same."[1] We build theologies around violence, elect leaders who excel at it, and in the case of so many women, preferentially mate with champions of human combat. When it's the "right" type of aggression, we love it.

It is the ambiguity of violence, that we can pull a trigger as an act of hideous aggression or of self-sacrificing love, that is so challenging. As a result, violence will always be a part of the human experience that is profoundly hard to understand.

This book explores the biology of violence, aggression, and competition—the behaviors and the impulses behind them, the acts of individuals, groups, and states, and when these are bad or good things. It is a book about the ways in which humans harm one another. But it is also a book about the ways in which people do the opposite. What does biology teach us about cooperation, affiliation, reconciliation, empathy, and altruism?

The book has a number of personal roots. One is that, having had blessedly little personal exposure to violence in my life, the entire phenomenon scares the crap out of me. I think like an academic egghead, believing that if I write enough paragraphs about a scary subject, give enough lectures about it, it will give up and go away quietly. And if everyone took enough classes about the biology of violence and studied hard, we'd all be able to take a nap between the snoozing lion and lamb. Such is the delusional sense of efficacy of a professor.

Then there's the other personal root for this book. I am by nature majorly pessimistic. Give me any topic and I'll find a way in which things will fall apart. Or turn out wonderfully and somehow, because of that, be poignant and sad. It's a pain in the butt, especially to people stuck around me. And when I had kids, I realized that I needed to get ahold of this tendency big time. So I looked for evidence that things weren't quite that bad. I started small, practicing on them—don't cry, a *T. rex* would never come and eat you; of course Nemo's daddy will find him. And as I've learned more about the subject of this book, there's been an unexpected realization—the realms of humans harming one another are neither universal nor inevitable, and we're getting some scientific insights into how to avoid them. My pessimistic self has a hard time admitting this, but there is room for optimism.

THE APPROACH IN THIS BOOK

I make my living as a combination neurobiologist—someone who studies the brain—and primatologist—someone who studies monkeys and apes. Therefore, this is a book that is rooted in science, specifically biology. And out of that come three key points. First, you can't begin to understand things like aggression, competition, cooperation, and empathy without biology; I say this for the benefit of a certain breed of social scientist who finds biology to be irrelevant and a bit ideologically suspect when thinking about human social behavior. But just as important, second, you're just as much up the creek if you rely *only* on biology; this is said for the benefit of a style of molecular fundamentalist who believes that the social sciences are destined to be consumed by "real" science. And as a third point, by the time you finish this book, you'll see that it actually makes no sense to distinguish between aspects of a behavior that are "biological" and those that would be described as, say, "psychological" or "cultural." Utterly intertwined.

Understanding the biology of these human behaviors is obviously important. But unfortunately it is hellishly complicated.[2] Now, if you were interested in the biology of, say, how migrating birds navigate, or in the mating reflex that occurs in female hamsters when they're ovulating, this would be an easier task. But that's not what we're interested in. Instead, it's human behavior, human social behavior, and in many cases abnormal human social behavior. And it is indeed a mess, a subject involving brain chemistry, hormones, sensory cues, prenatal environment, early experience, genes, both biological and cultural evolution, and ecological pressures, among other things.

How are we supposed to make sense of all these factors in thinking about behavior? We tend to use a certain cognitive strategy when dealing with complex, multifaceted phenomena, in that we break down those separate facets into categories, into buckets of explanation. Suppose there's a rooster standing next to you, and there's a chicken across the street. The rooster gives a sexually solicitive gesture that is hot by chicken standards, and she promptly runs over to mate with him (I haven't a clue if this is how it works, but let's just suppose). And thus we have a key behavioral biological question—why did the chicken cross the road? And if you're a psychoneuroendocrinologist, your answer would be "Because circulating estrogen levels in that chicken worked in a certain part of her brain to make her responsive to this male signaling," and if you're a bioengineer, the answer would be "Because the long bone in the leg of the chicken forms a fulcrum for her pelvis (or some such thing), allowing her to move forward rapidly," and if you're an evolutionary biologist, you'd say, "Because over the course of millions of years, chickens that responded to such gestures at a time that they were fertile left more copies of their genes, and thus this is now an innate behavior in chickens," and so on, thinking in categories, in differing scientific disciplines of explanation.

The goal of this book is to avoid such categorical thinking. Putting facts into nice cleanly demarcated buckets of explanation has its advantages—for example, it can help you remember facts better. But it can wreak havoc on your ability to *think* about those facts. This is because the boundaries between different categories are often arbitrary, but once some arbitrary boundary exists, we forget that it is arbitrary and get way too impressed with its importance. For example, the visual spectrum is a continuum of wavelengths from violet to red, and it is arbitrary where boundaries are put for different color names (for example, where we see a transition from "blue" to "green"); as proof of this, different languages arbitrarily split up the visual spectrum at different points in coming up with the words for different colors. Show someone two roughly similar colors. If the color-name boundary in that person's language happens to fall between the two colors, the person will overestimate the difference between the two. If the colors fall in the same category, the opposite happens. In other words, when you think categorically, you have trouble seeing how similar or different two things are. If you pay lots of attention to where boundaries are, you pay less attention to complete pictures.

Thus, the official intellectual goal of this book is to avoid using categorical buckets when thinking about the biology of some of our most complicated behaviors, even more complicated than chickens crossing roads.

What's the replacement?

A behavior has just occurred. Why did it happen? Your first category of explanation is going to be a neurobiological one. What went on in that person's brain a second before the behavior happened? Now pull out to a slightly larger field of vision, your next category of explanation, a little earlier in time. What sight, sound, or smell in the previous seconds to minutes triggered the nervous system to produce that behavior? On to the next explanatory category. What hormones acted hours to days earlier to change how responsive that individual was to the sensory stimuli that trigger the nervous system to produce the behavior? And by now you've increased your field of vision to be thinking about neurobiology and the sensory world of our environment *and* short-term endocrinology in trying to explain what happened.

And you just keep expanding. What features of the environment in the prior weeks to years changed the structure and function of that person's brain and thus changed how it responded to those hormones and environmental stimuli? Then you go further back to the childhood of the individual, their fetal environment, then their genetic makeup. And then you increase the view to encompass factors larger than that one individual—how has culture shaped the behavior of people living in that individual's group?—what ecological factors helped shape that culture—expanding and expanding until considering events umpteen millennia ago and the evolution of that behavior.

Okay, so this represents an improvement—it seems like instead of trying to explain all of behavior with a single discipline (e.g., "Everything can be explained with knowledge about this particular [take your pick:] hormone/gene/childhood event"), we'll be thinking about a bunch of disciplinary buckets. But something subtler will be done, and this is the most important idea in the book: when you explain a behavior with one of these disciplines, you are implicitly invoking all the disciplines—any given type of explanation is the end product of the influences that preceded it. It has to work this way. If you say, "The behavior occurred because of the release of neurochemical Y in the brain," you are also saying, "The behavior occurred because the heavy secretion of hormone X this morning increased the levels of neurochemical Y." You're also saying, "The behavior occurred because the environment in which that person was raised made her brain more likely to release neurochemical Y in response to certain types of stimuli." And you're also saying, ". . . because of the gene that codes for the particular version of neurochemical Y." And if you've so much as whispered the word "gene," you're also saying, ". . . and because of the millennia of factors that shaped the evolution of that particular gene." And so on.

There are not different disciplinary buckets. Instead, each one is the end product of all the biological influences that came before it and will influence all the factors that follow it. Thus, it is impossible to conclude that a behavior is caused by *a* gene, *a* hormone, *a* childhood trauma, because the second you invoke one type of explanation, you are de facto invoking them all. No buckets. A "neurobiological" or "genetic" or "developmental" explanation for a behavior is just shorthand, an expository convenience for temporarily approaching the whole multifactorial arc from a particular perspective.

Pretty impressive, huh? Actually, maybe not. Maybe I'm just pretentiously saying, "You have to think complexly about complex things." Wow, what a revelation. And maybe what I've been tacitly setting up is this full-of-ourselves straw man of "Ooh, we're going to think subtly. We won't get suckered into simplistic answers, not like those chicken-crossing-the-road neurochemists and chicken evolutionary biologists and chicken psychoanalysts, all living in their own limited categorical buckets."

Obviously, scientists aren't like that. They're smart. They understand that they need to take lots of angles into account. Of necessity, their research may focus on a narrow subject, because there are limits to how much one person can obsess over. But of course they know that their particular categorical bucket isn't the whole story.

Maybe yes, maybe no. Consider the following quotes from some card-carrying scientists. The first:

> Give me a dozen healthy infants, well formed, and my own specified world to bring them up in and I'll guarantee to take any one at random and train him to become any type of specialist I might select—doctor, lawyer, artist, merchant-chief and yes, even beggar-man thief, regardless of his talents, penchants, tendencies, abilities, vocations, and race of his ancestors.[3]

This was John Watson, a founder of behaviorism, writing around 1925. Behaviorism, with its notion that behavior is completely malleable, that it can be shaped into anything in the right environment, dominated American psychology in the midtwentieth century; we'll return to behaviorism, and its considerable limitations. The point is that Watson was pathologically caught inside a bucket having to do with the environmental influences on development. "I'll guarantee . . . to train him to become any type." Yet we are not all born the same, with the same potential, regardless of how we are trained.*[4]

The next quote:

> Normal psychic life depends upon the good functioning of brain synapses, and mental disorders appear as a result of synaptic derangements. . . . It is necessary to alter these synaptic adjustments and change the paths chosen by the impulses in their constant passage so as to modify the corresponding ideas and force thought into different channels.[5]

Alter synaptic adjustments. Sounds delicate. Yeah, right. These were the words of the Portuguese neurologist Egas Moniz, around the time he was awarded the Nobel Prize in 1949 for his development of frontal leukotomies. Here was an individual pathologically stuck in a bucket having to do with a crude version of the nervous system. Just tweak those microscopic synapses with a big ol' ice pick (as was done once leukotomies, later renamed frontal lobotomies, became an assembly line operation).

And a final quote:

> The immensely high reproduction rate in the moral imbecile has long been established. . . . Socially inferior human material is enabled . . . to penetrate and finally to annihilate the healthy nation. The selection for toughness, heroism, social utility . . . must be accomplished by some human institution if mankind, in default of selective factors, is not to be ruined by domestication-induced degeneracy. The racial idea as the basis of our state has already accomplished much in this respect. We

* Shortly after making this pronouncement, Watson fled academia amid sexual scandal. He eventually resurfaced as the vice president of an advertising company. You may not be able to shape people into anything you wish, but at least you can often shape them into buying some useless gewgaw.

must—and should—rely on the healthy feelings of our Best and charge them . . . with the extermination of elements of the population loaded with dregs.[6]

This was Konrad Lorenz, animal behaviorist, Nobel laureate, cofounder of the field of ethology (stay tuned), regular on nature TV programs.[7] Grandfatherly Konrad, in his Austrian shorts and suspenders, being followed by his imprinted baby geese, was also a rabid Nazi propagandist. Lorenz joined the Nazi Party the instant Austrians were eligible, and joined the party's Office of Race Policy, working to psychologically screen Poles of mixed Polish/German parentage, helping to determine which were sufficiently Germanized to be spared death. Here was a man pathologically mired in an imaginary bucket related to gross misinterpretations of what genes do.

These were not obscure scientists producing fifth-rate science at Podunk U. These were among the most influential scientists of the twentieth century. They helped shape who and how we educate and our views on what social ills are fixable and when we shouldn't bother. They enabled the destruction of the brains of people against their will. And they helped implement final solutions for problems that didn't exist. It can be far more than a mere academic matter when a scientist thinks that human behavior can be entirely explained from only one perspective.

OUR LIVES AS ANIMALS AND OUR HUMAN VERSATILITY AT BEING AGGRESSIVE

So we have a first intellectual challenge, which is to always think in this interdisciplinary way. The second challenge is to make sense of humans as apes, primates, mammals. Oh, that's right, we're a kind of animal. And it will be a challenge to figure out when we're just like other animals and when we are utterly different.

Some of the time we are indeed just like any other animal. When we're scared, we secrete the same hormone as would some subordinate fish getting hassled by a bully. The biology of pleasure involves the same brain chemicals in us as in a capybara. Neurons from humans and brine shrimp work the same way. House two female rats together, and over the course of weeks they will synchronize their reproductive cycles so that they wind up ovulating within a few hours of each other. Try the same with two human females (as reported in some but not all studies), and something similar occurs. It's called the Wellesley effect, first shown with roommates at all-women's Wellesley College.[8] And when it comes to violence, we can be just like some other apes—we pummel, we cudgel, we throw rocks, we kill with our bare hands.

So some of the time an intellectual challenge is to assimilate how similar we can be to other species. In other cases the challenge is to appreciate how, though human physiology resembles that of other species, we use the physiology in novel ways. We activate the classical physiology of vigilance while watching a scary movie. We activate a stress response when thinking about mortality. We secrete hormones related to nurturing and social bonding, but in response to an adorable baby panda. And this certainly applies to aggression—we use the same muscles as does a male chimp attacking a sexual competitor, but we use them to harm someone because of their ideology.

Finally, sometimes the only way to understand our humanness is to consider solely humans, because the things we do are unique. While a few other species have regular nonreproductive sex, we're the only ones to talk afterward about how it was. We construct cultures premised on beliefs concerning the nature of life and can transmit those beliefs multigenerationally, even between two individuals separated by millennia—just consider that perennial best seller, the Bible. Consonant with that, we can harm by doing things as unprecedented as and no more physically taxing than pulling a trigger, or nodding consent, or looking the other way. We can be passive-aggressive, damn with faint praise, cut with scorn, express contempt with patronizing concern. All species are unique, but we are unique in some pretty unique ways.

Here are two examples of just how strange and unique humans can be when they go about harming one another and caring for one another. The first example involves, well, my wife. So we're in the minivan, our kids in the back, my wife driving. And this complete jerk cuts us off, almost causing an accident, and in a way that makes it clear that it wasn't distractedness on his part, just sheer selfishness. My wife honks at him, and he flips us off. We're livid, incensed. Asshole-where's-the-cops-when-you-need-them, etc. And suddenly my wife announces that we're going to follow him, make him a little nervous. I'm still furious, but this doesn't strike me as the most prudent thing in the world. Nonetheless, my wife starts trailing him, right on his rear.

After a few minutes the guy's driving evasively, but my wife's on him. Finally both cars stop at a red light, one that we know is a long one. Another car is stopped in front of the villain. He's not going anywhere. Suddenly my wife grabs something from the front seat divider, opens her door, and says, "Now he's going to be sorry." I rouse myself feebly—"Uh, honey, do you really think this is such a goo—" But she's out of the car, starts pounding on his window. I hurry over just in time to hear my wife say, "If you could do something that mean to another person, you probably need this," in a venomous voice. She then flings something in the window. She returns to the car triumphant, just glorious.

"What did you throw in there!?"

She's not talking yet. The light turns green, there's no one behind us, and we just sit there. The thug's car starts to blink a very sensible turn indicator, makes a slow turn, and heads down a side street into the dark at, like, five miles an hour. If it's possible for a car to look ashamed, this car was doing it.

"Honey, what did you throw in there, tell me?"

She allows herself a small, malicious grin.

"A grape lollipop." I was awed by her savage passive-aggressiveness—"You're such a mean, awful human that something must have gone really wrong in your childhood, and maybe this lollipop will help correct that just a little." That guy was going to think twice before screwing with us again. I swelled with pride and love.

And the second example: In the mid-1960s, a rightist military coup overthrew the government of Indonesia, instituting the thirty-year dictatorship of Suharto known as the New Order. Following the coup, government-sponsored purges of communists, leftists, intellectuals, unionists, and ethnic Chinese left about a half million dead.[9] Mass executions, torture, villages torched with inhabitants trapped inside. V. S. Naipaul, in his book *Among the Believers: An Islamic Journey*, describes hearing rumors while in Indonesia that when a paramilitary group would arrive to exterminate every person in some village, they would, incongruously, bring along a traditional gamelan orchestra. Eventually Naipaul encountered an unrepentant veteran of a massacre, and he asked him about the rumor. Yes, it is true. We would bring along gamelan musicians, singers, flutes, gongs, the whole shebang. Why? Why would you possibly do that? The man looked puzzled and gave what seemed to him a self-evident answer: "Well, to make it more beautiful."

Bamboo flutes, burning villages, the lollipop ballistics of maternal love. We have our work cut out for us, trying to understand the virtuosity with which we humans harm or care for one another, and how deeply intertwined the biology of the two can be.

One

The Behavior

We have our strategy in place. A behavior has occurred—one that is reprehensible, or wonderful, or floating ambiguously in between. What occurred in the prior second that triggered the behavior? This is the province of the nervous system. What occurred in the prior seconds to minutes that triggered the nervous system to produce that behavior? This is the world of sensory stimuli, much of it sensed unconsciously. What occurred in the prior hours to days to change the sensitivity of the nervous system to such stimuli? Acute actions of hormones. And so on, all the way back to the evolutionary pressures played out over the prior millions of years that started the ball rolling.

So we're set. Except that when approaching this big sprawling mess of a subject, it is kind of incumbent upon you to first define your terms. Which is an unwelcome prospect.

Here are some words of central importance to this book: aggression, violence, compassion, empathy, sympathy, competition, cooperation, altruism, envy, schadenfreude, spite, forgiveness, reconciliation, revenge, reciprocity, and (why not?) love. Flinging us into definitional quagmires.

Why the difficulty? As emphasized in the introduction, one reason is that so many of these terms are the subject of ideological battles over the appropriation and distortions of their meanings.*[1] Words pack power and these definitions are laden with values, often wildly idiosyncratic ones. Here's an example, namely the ways I think about the word "competition": (a) "competition"—your lab team races the Cambridge group to a discovery (exhilarating but embarrassing to admit to); (b) "competition"—playing pickup soccer (fine, as long as the best player shifts sides if the score becomes lopsided); (c) "competition"—your child's teacher announces a prize for the best outlining-your-fingers Thanksgiving turkey drawing (silly and perhaps a red flag—if it keeps happening, maybe complain to the principal); (d) "competition"—whose deity is more worth killing for? (try to avoid).

But the biggest reason for the definitional challenge was emphasized in the introduction—these terms mean different things to scientists living inside different disciplines. Is "aggression" about thought, emotion, or something done with muscles? Is "altruism" something that can be studied mathematically in various species, including bacteria, or are we discussing moral development in kids? And implicit in these different perspectives, disciplines have differing tendencies toward lumping and splitting—these scientists believe that behavior X consists of two different subtypes, whereas those scientists think it comes in seventeen flavors.

Let's examine this with respect to different types of "aggression."[2] Animal behaviorists dichotomize between offensive and defensive aggression, distinguishing between, say, the intruder and the resident of a territory; the biology underlying these two versions differs. Such scientists also distinguish between conspecific aggression (between members of the same species) and fighting off a predator. Meanwhile, criminologists distinguish between impulsive and premeditated aggression. Anthropolo-

* I recently found a startling example of unorthodox defining of terms. This concerned Menachem Begin, one of the surprising architects of the Camp David Peace Accords in 1978 as the prime minister of Israel. In the mid-1940s he headed the Irgun, the Zionist paramilitary group intent on driving Britain out of Palestine in order to facilitate the founding of Israel. The Irgun raised money to buy arms through extortion and robbery, hanged two captive British soldiers and booby-trapped their bodies, and carried out a series of bombings including, most notoriously, an attack on British headquarters at Jerusalem's King David Hotel, an act that killed not only numerous British officials but also scores of Arab and Jewish civilians. And Begin's account of these activities? "Historically we were not 'terrorists.' We were, strictly speaking, *anti-terrorists*" (my emphasis).

gists care about differing levels of organization underlying aggression, distinguishing among warfare, clan vendettas, and homicide.

Moreover, various disciplines distinguish between aggression that occurs reactively (in response to provocation) and spontaneous aggression, as well as between hot-blooded, emotional aggression and cold-blooded, instrumental aggression (e.g., "I want your spot to build my nest, so scram or I'll peck your eyes out; this isn't personal, though").[3] Then there's another version of "This isn't personal"—targeting someone just because they're weak and you're frustrated, stressed, or pained and need to displace some aggression. Such third-party aggression is ubiquitous—shock a rat and it's likely to bite the smaller guy nearby; a beta-ranking male baboon loses a fight to the alpha, and he chases the omega male;* when unemployment rises, so do rates of domestic violence. Depressingly, as will be discussed in chapter 4, displacement aggression can decrease the perpetrator's stress hormone levels; giving ulcers can help you avoid getting them. And of course there is the ghastly world of aggression that is neither reactive nor instrumental but is done for pleasure.

Then there are specialized subtypes of aggression—maternal aggression, which often has a distinctive endocrinology. There's the difference between aggression and ritualistic *threats* of aggression. For example, many primates have lower rates of actual aggression than of ritualized threats (such as displaying their canines). Similarly, aggression in Siamese fighting fish is mostly ritualistic.†

* I've observed a remarkable example of this among the baboons that I've studied in East Africa. Over the thirty-odd years I've watched them, I've seen a handful of instances of what I believe warrants the seemingly human-specific term "rape"—where a male baboon will forcibly vaginally penetrate a female who is not in estrus, who is not sexually receptive, who struggles to prevent it, and who gives every indication of distress and pain when it happens. And each of these instances has been the act of the former alpha male in the hours after he has been toppled from his position.

† There is a great contemporary version of human ritualistic aggression, namely the *haka* ritual performed by rugby teams from New Zealand. Just before the game starts, the Kiwis line up midfield and perform this neo-Maori war dance, complete with rhythmic stamping, menacing gestures, guttural shouting, and histrionically threatening facial expressions. It's cool to see from afar on YouTube (even better is watching the YouTube clip of Robin Williams doing a *haka* display at Charlie Rose on PBS), while up close it typically appears to scare the bejesus out of the other team. However, some opposing teams have come up with ritualistic responses straight out of the baboon playbook—getting in the *haka*-ers' faces and trying to stare them down. Other teams come out with ritualistic responses that are pure human uniqueness—ignoring the *haka*-ers while nonchalantly warming up; using their smartphones to film the display, thereby emasculating it to something vaguely touristy in flavor; tepidly applauding afterward with great condescension. One response initially seems uniquely human but would be understandable to other primates after some translating—the sports newsletter for one Australian team printed a photo of the mortal enemy New Zealanders doing a *haka*, with each player brandishing a Photoshopped woman's handbag.

Getting a definitional handle on the more positive terms isn't easy either. There's empathy versus sympathy, reconciliation versus forgiveness, and altruism versus "pathological altruism."[4] For a psychologist the last term might describe the empathic codependency of enabling a partner's drug use. For a neuroscientist it describes a consequence of a type of damage to the frontal cortex—in economic games of shifting strategies, individuals with such damage fail to switch to less altruistic play when being repeatedly stabbed in the back by the other player, despite being able to verbalize the other player's strategy.

When it comes to the more positive behaviors, the most pervasive issue is one that ultimately transcends semantics—does pure altruism actually exist? Can you ever separate doing good from the expectation of reciprocity, public acclaim, self-esteem, or the promise of paradise?

This plays out in a fascinating realm, as reported in Larissa MacFarquhar's 2009 *New Yorker* piece "The Kindest Cut."[5] It concerns people who donate organs not to family members or close friends but to strangers. An act of seemingly pure altruism. But these Samaritans unnerve everyone, sowing suspicion and skepticism. Is she expecting to get paid secretly for her kidney? Is she that desperate for attention? Will she work her way into the recipient's life and do a *Fatal Attraction*? What's her deal? The piece suggests that these profound acts of goodness unnerve because of their detached, affectless nature.

This speaks to an important point that runs through the book. As noted, we distinguish between hot-blooded and cold-blooded violence. We understand the former more, can see mitigating factors in it—consider the grieving, raging man who kills the killer of his child. And conversely, affectless violence seems horrifying and incomprehensible; this is the sociopathic contract killer, the Hannibal Lecter who kills without his heart rate nudging up a beat.*[6] It's why *cold-blooded* killing is a damning descriptor.

* A fascinating, grotesque example of this is Munchausen Syndrome by Proxy, where a woman (it's overwhelmingly a female disorder) generates illnesses in her child out of a pathological need for the attention, care, and envelopment of the medical system. This is not someone falsely telling the pediatrician that her child had a fever last night. This is giving children emetics to induce vomiting, poisoning them, smothering them to induce symptoms of hypoxia—often with fatal consequences. One feature of the disorder is a stunning lack of affect in the mothers. One would expect an air of spittle-flecked madness to match the actions. Instead, there is cold detachment, as if they could

Similarly, we expect that our best, most prosocial acts be warmhearted, filled with positive affect. Cold-blooded goodness seems oxymoronic, is unsettling. I was once at a conference of neuroscientists and all-star Buddhist monk meditators, the former studying what the brains of the latter did during meditation. One scientist asked one of the monks whether he ever stops meditating because his knees hurt from all that cross-leggedness. He answered, "Sometimes I'll stop sooner than I planned, but not because it hurts; it's not something I notice. It's as an act of kindness to my knees." "Whoa," I thought, "these guys are from another planet." A cool, commendable one, but another planet nonetheless. Crimes of passion and good acts of passion make the most sense to us (nevertheless, as we shall see, dispassionate kindness often has much to recommend it).

Hot-blooded badness, warmhearted goodness, and the unnerving incongruity of the cold-blooded versions raise a key point, encapsulated in a quote of Freud's emphasized by Elie Wiesel, the Nobel Peace Prize winner and concentration camp survivor: "The opposite of love is not hate; its opposite is indifference." The biologies of strong love and strong hate are similar in many ways, as we'll see.

Which reminds us that we don't hate aggression; we hate the wrong kind of aggression but love it in the right context. And conversely, in the wrong context our most laudable behaviors are anything but. The motoric features of our behaviors are less important and challenging to understand than the meaning behind our muscles' actions.

This is shown in a subtle study.[7] Subjects in a brain scanner entered a virtual room where they encountered either an injured person in need of help or a menacing extraterrestrial; subjects could either bandage or shoot the individual. Pulling a trigger and applying a bandage are different behaviors. But they are similar, insofar as bandaging the injured person and shooting the alien are both the "right" things. And contemplating those two different versions of doing the right thing activated the same circuitry in the most context-savvy part of the brain, the prefrontal cortex.

---

simply be lying to a veterinarian about their supposedly sick goldfish or to customer service at Sears about their supposedly broken toaster, if doing so would bring the same psychological benefits. For a lengthy overview of Munchausen Syndrome by Proxy, see R. Sapolsky, "Nursery Crimes," in *Monkeyluv and Other Essays on Our Lives as Animals* (New York: Simon and Schuster/Scribner, 2005).

And thus those key terms that anchor this book are most difficult to define because of their profound context dependency. I will therefore group them in a way that reflects this. I won't frame the behaviors to come as either pro- or antisocial—too cold-blooded for my expository tastes. Nor will they be labeled as "good" and "evil"—too hot-blooded and frothy. Instead, as our convenient shorthand for concepts that truly defy brevity, this book is about the biology of our best and worst behaviors.

Two

One Second Before

Various muscles have moved, and a behavior has happened. Perhaps it is a good act: you've empathically touched the arm of a suffering person. Perhaps it is a foul act: you've pulled a trigger, targeting an innocent person. Perhaps it is a good act: you've pulled a trigger, drawing fire to save others. Perhaps it is a foul act: you've touched the arm of someone, starting a chain of libidinal events that betray a loved one. Acts that, as emphasized, are definable only by context.

Thus, to ask the question that will begin this and the next eight chapters, why did that behavior occur?

As this book's starting point, we know that different disciplines produce different answers—because of some hormone; because of evolution; because of childhood experiences or genes or culture—and as the book's central premise, these are utterly intertwined answers, none standing alone. But on the most proximal level, in this chapter we ask: What happened one second before the behavior that caused it to occur? This puts us in the realm of neurobiology, of understanding the brain that commanded those muscles.

This chapter is one of the book's anchors. The brain is the final common pathway, the conduit that mediates the influences of all the distal factors to be covered in the chapters to come. What happened an hour, a decade, a million years earlier? What happened were factors that impacted the brain and the behavior it produced.

This chapter has two major challenges. The first is its god-awful length. Apologies; I've tried to be succinct and nontechnical, but this is foundational material that needs to be covered. Second, regardless of how nontechnical I've tried to be, the material can overwhelm someone with no background in neuroscience. To help with that, please wade through appendix 1 around now.

Now we ask: What crucial things happened in the second before that pro- or antisocial behavior occurred? Or, translated into neurobiology: What was going on with action potentials, neurotransmitters, and neural circuits in particular brain regions during that second?

THREE METAPHORICAL (BUT NOT LITERAL) LAYERS

We start by considering the brain's macroorganization, using a model proposed in the 1960s by the neuroscientist Paul MacLean.[1] His "triune brain" model conceptualizes the brain as having three functional domains:

Layer 1: An ancient part of the brain, at its base, found in species from humans to geckos. This layer mediates automatic, regulatory functions. If body temperature drops, this brain region senses it and commands muscles to shiver. If blood glucose levels plummet, that's sensed here, generating hunger. If an injury occurs, a different loop initiates a stress response.

Layer 2: A more recently evolved region that has expanded in mammals. MacLean conceptualized this layer as being about emotions, somewhat of a mammalian invention. If you see something gruesome and terrifying, this layer sends commands down to ancient layer 1, making you shiver with emotion. If you're feeling sadly unloved, regions here prompt layer 1 to generate a craving for comfort food. If you're a rodent and smell a cat, neurons here cause layer 1 to initiate a stress response.

Layer 3: The recently evolved layer of neocortex sitting on the upper surface of the brain. Proportionately, primates devote more of their brain to this layer than do other species. Cognition, memory storage, sensory processing, abstractions, philosophy, navel contemplation. Read a scary passage of a book, and layer 3 signals layer 2 to make you feel frightened, prompting layer 1 to initiate shivering. See an ad for Oreos and feel a craving—layer 3 influences layers 2 and 1. Contemplate the fact that loved ones won't live forever, or kids in refugee camps, or how the Na'vis' home tree was destroyed by those jerk humans in *Avatar* (despite the fact that, wait, *Na'vi aren't real!*), and layer 3 pulls layers 2 and 1 into the picture, and you feel sad and have the same sort of stress response that you'd have if you were fleeing a lion.

Thus we've got the brain divided into three functional buckets, with the usual advantages and disadvantages of categorizing a continuum. The biggest disadvantage is how simplistic this is. For example:

- **a.** Anatomically there is considerable overlap among the three layers (for example, one part of the cortex can best be thought of as part of layer 2; stay tuned).
- **b.** The flow of information and commands is not just top down, from layer 3 to 2 to 1. A weird, great example explored in chapter 15: if someone is holding a cold drink (temperature is processed in layer 1), they're more likely to judge someone they meet as having a cold personality (layer 3).
- **c.** Automatic aspects of behavior (simplistically, the purview of layer 1), emotion (layer 2), and thought (layer 3) are not separable.
- **d.** The triune model leads one, erroneously, to think that evolution in effect slapped on each new layer without any changes occurring in the one(s) already there.

Despite these drawbacks, which MacLean himself emphasized, this model will be a good organizing metaphor for us.

THE LIMBIC SYSTEM

To make sense of our best and worst behaviors, automaticity, emotion, and cognition must all be considered; I arbitrarily start with layer 2 and its emphasis on emotion.

Early-twentieth-century neuroscientists thought it obvious what layer 2 did. Take your standard-issue lab animal, a rat, and examine its brain. Right at the front would be these two gigantic lobes, the "olfactory bulbs" (one for each nostril), the primary receptive area for odors.

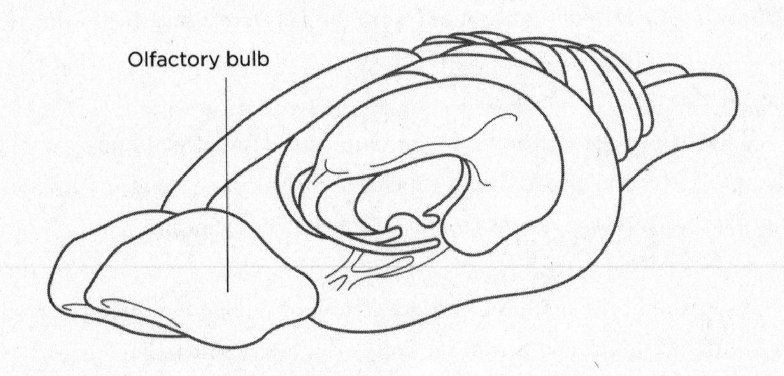

Neuroscientists at the time asked what parts of the brain these gigantic rodent olfactory bulbs talked to (i.e., where they sent their axonal projections). Which brain regions were only a single synapse away from receiving olfactory information, which were two synapses, three, and so on?

And it was layer 2 structures that received the first communiqués. Ah, everyone concluded, this part of the brain must process odors, and so it was termed the rhinencephalon—the nose brain.

Meanwhile, in the thirties and forties, neuroscientists such as the young MacLean, James Papez, Paul Bucy, and Heinrich Klüver were starting to figure out what the layer 2 structures did. For example, if you lesion (i.e., destroy) layer 2 structures, this produces "Klüver-Bucy syndrome," featuring abnormalities in sociality, especially in sexual and aggressive behaviors. They concluded that these structures, soon termed the "limbic system" (for obscure reasons), were about emotion.

Rhinencephalon or limbic system? Olfaction or emotion? Pitched street battles ensued until someone pointed out the obvious—for a rat, emotion and olfaction are nearly synonymous, since nearly all the environmental stimuli that elicit emotions in a rodent are olfactory. Peace in our time. In a rodent, olfactory inputs are what the limbic system most depends on for emotional news of the world. In contrast, the primate limbic system is more informed by visual inputs.

Limbic function is now recognized as central to the emotions that fuel our best and worst behaviors, and extensive research has uncovered the functions of its structures (e.g., the amygdala, hippocampus, septum, habenula, and mammillary bodies).

There really aren't "centers" in the brain "for" particular behaviors. This is particularly the case with the limbic system and emotion. There is indeed a sub-subregion of the motor cortex that approximates being the "center" for making your left pinkie bend; other regions have "center"-ish roles in regulating breathing or body temperature. But there sure aren't centers for feeling pissy or horny, for feeling bittersweet nostalgia or warm protectiveness tinged with contempt, or for that what-is-that-thing-called-love feeling. No surprise, then, that the circuitry connecting various limbic structures is immensely complex.

The Autonomic Nervous System and the Ancient Core Regions of the Brain

The limbic system's regions form complex circuits of excitation and inhibition. It's easier to understand this by appreciating the deeply held desire of every limbic structure—to influence what the hypothalamus does.

Why? Because of its importance. The hypothalamus, a limbic structure, is the interface between layers 1 and 2, between core regulatory and emotional parts of the brain.

Consistent with that, the hypothalamus gets massive inputs from limbic layer 2 structures but disproportionately sends projections to layer 1 regions. These are the evolutionarily ancient midbrain and brain stem, which regulate automatic reactions throughout the body.

For a reptile such automatic regulation is straightforward. If muscles are working hard, this is sensed by neurons throughout the body that send signals up the spine to layer 1 regions, resulting in signals back down the spine that increase heart rate and blood pressure; the result is more oxygen and glucose for the muscles. Gorge on food, and stomach walls distend; neurons embedded there sense this and pass on the news, and soon blood vessels in the gut dilate, increasing blood flow and facilitating digestion. Too warm? Blood is sent to the body's surface to dissipate heat.

All of this is automatic, or "autonomic." And thus the midbrain and brain-stem regions, along with their projections down the spine and out to the body, are collectively termed the "autonomic nervous system."*

And where does the hypothalamus come in? It's the means by which the limbic system influences autonomic function, how layer 2 talks to layer 1. Have a full bladder with its muscle walls distended, and midbrain/brain-stem circuitry votes for urinating. Be exposed to something sufficiently terrifying, and limbic structures, via the hypothalamus, persuade the midbrain and brain stem to do the same. This is how emotions change bodily functions, why limbic roads eventually lead to the hypothalamus.†

The autonomic nervous system has two parts—the sympathetic and parasympathetic nervous systems, with fairly opposite functions.

The sympathetic nervous system (SNS) mediates the body's response to arousing circumstances, for example, producing the famed "fight or flight" stress response. To use the feeble joke told to first-year medical students, the SNS mediates the "four *F*s—fear, fight, flight, and sex."

* It is also often called the "involuntary nervous system," contrasting it with the "voluntary nervous system." The latter is about conscious, voluntary movement and involves neurons in "motor" regions of the brain and their projections down the spine to skeletal muscles.

† Just as a warning of complexities to come, the hypothalamus consists of a bunch of different nuclei, each receiving a unique orchestration of limbic inputs and equivalently distinctive outputs to various midbrain/brain-stem regions. And while each hypothalamic nucleus has a different set of functions, they all fall under the general rubric of autonomic regulation.

Particular midbrain/brain-stem nuclei send long SNS projections down the spine and on to outposts throughout the body, where the axon terminals release the neurotransmitter norepinephrine. There's one exception that makes the SNS more familiar. In the adrenal gland, instead of norepinephrine (aka noradrenaline) being released, it's epinephrine (aka the famous adrenaline).*

Meanwhile, the parasympathetic nervous system (PNS) arises from different midbrain/brain-stem nuclei that project down the spine to the body. In contrast to the SNS and the four *F*s, the PNS is about calm, vegetative states. The SNS speeds up the heart; the PNS slows it down. The PNS promotes digestion; the SNS inhibits it (which makes sense—if you're running for your life, avoiding being someone's lunch, don't waste energy digesting breakfast).† And as we will see chapter 14, if seeing someone in pain activates your SNS, you're likely to be preoccupied with your own distress instead of helping; turn on the PNS, and it's the opposite. Given that the SNS and PNS do opposite things, the PNS is obviously going to be releasing a different neurotransmitter from its axon terminals—acetylcholine.‡

There is a second, equally important way in which emotion influences the body. Specifically, the hypothalamus also regulates the release of many hormones; this is covered in chapter 4.

So the limbic system indirectly regulates autonomic function and hormone release. What does this have to do with behavior? Plenty—because the autonomic and hormonal states of the body feed back to the brain, influencing behavior (typically unconsciously).§ Stay tuned for more in chapters 3 and 4.

---

* And to needlessly complicate things further, thus explaining why this is buried in a footnote, there is actually an intervening synapse between the long spinal projection neurons of the SNS and the SNS neurons reaching target cells. This is the second neuron in the two-step pathway that releases norepinephrine. The first neuron in each pathway releases acetylcholine.

† Nice logical piece of this: Suppose you're stressed, not by running for your life from a lion but by having to give a speech. Your mouth gets dry, the first step of your SNS shutting down digestion until a more auspicious time.

‡ Like the SNS, the PNS gets the brain to target organs via two steps. And as one complication, the SNS and PNS branches aren't always working in complete opposition; in some cases they function in a more cooperative, sequential manner. For example, erection and ejaculation involve coordination between the SNS and PNS that is so complicated that it's a miracle that any of us were conceived.

§ In other words, layers 2 and 3 can influence the autonomic functions of layer 1, which alters events throughout the body, which in turn influences all the parts of the brain. Loops and loops.

The Interface Between the Limbic System and the Cortex

Time to add the cortex. As noted, this is the brain's upper surface (its name comes from the Latin *cortic*, meaning "tree bark") and is the newest part of the brain.

The cortex is the gleaming, logical, analytical crown jewel of layer 3. Most sensory information flows there to be decoded. It's where muscles are commanded to move, where language is comprehended and produced, where memories are stored, where spatial and mathematical skills reside, where executive decisions are made. It floats above the limbic system, supporting philosophers since at least Descartes who have emphasized the dichotomy between thought and emotion.

Of course, that's all wrong, as shown by the temperature of a cup—something processed in the hypothalamus—altering assessment of the coldness of someone's personality. Emotions filter the nature and accuracy of what is remembered. Stroke damage to certain cortical regions blocks the ability to speak; some sufferers reroute the cerebral world of speech through emotive, limbic detours—they can sing what they want to say. The cortex and limbic system are not separate, as scads of axonal projections course between the two. Crucially, those projections are bidirectional—the limbic system talks to the cortex, rather than merely being reined in by it. The false dichotomy between thought and feeling is presented in the classic *Descartes' Error*, by the neurologist Antonio Damasio of the University of Southern California; his work is discussed later.[2]

While the hypothalamus dwells at the interface of layers 1 and 2, it is the incredibly interesting frontal cortex that is the interface between layers 2 and 3.

Key insight into the frontal cortex was provided in the 1960s by a giant of neuroscience, Walle Nauta of MIT.*[3] Nauta studied what brain

* Nauta was not only a towering scientist but also a force of integrity, as well as a renowned teacher who made neuroanatomy, taught in three-hour evening classes, borderline fun. During college I did research in the lab next to his, and I was so in awe of the guy that I'd find every possible autonomic excuse to go to the bathroom whenever I saw him heading in that direction, just for the chance to offhandedly say hello to him by the urinals. (My awe grew further

regions sent axons to the frontal cortex and what regions got axons from it. And the frontal cortex was bidirectionally enmeshed with the limbic system, leading him to propose that the frontal cortex is a quasi member of the limbic system. Naturally, everyone thought him daft. The frontal cortex was the most recently evolved part of the very highbrow cortex—the only reason why the frontal cortex would ever go slumming into the limbic system would be to preach honest labor and Christian temperance to the urchins there.

Naturally, Nauta was right. In different circumstances the frontal cortex and limbic system stimulate or inhibit each other, collaborate and coordinate, or bicker and work at cross-purposes. It really is an honorary member of the limbic system. And the interactions between the frontal cortex and (other) limbic structures are at the core of much of this book.

Two more details. First, the cortex is not a smooth surface but instead is folded into convolutions. The convolutions form a superstructure of four separate lobes: the temporal, parietal, occipital, and frontal, each with different functions.

The Cortex

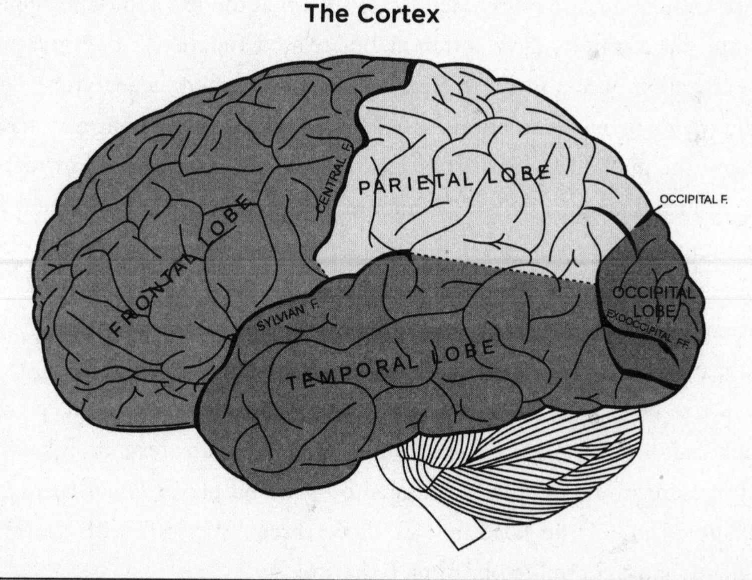

later, with my discovery that he and his family sheltered Jews from Nazis in Holland during World War II and are cited in the Holocaust Museum in Washington, DC.)

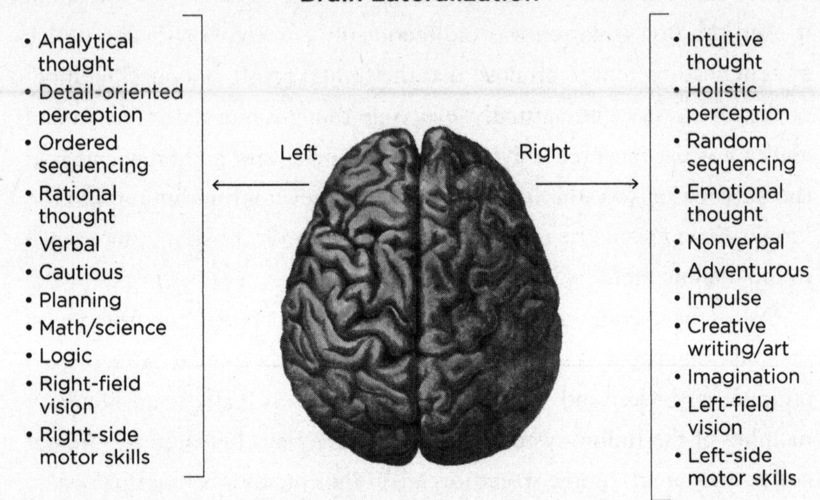

Second, brains obviously have left and right sides, or "hemispheres," that roughly mirror each other.

Thus, except for the relatively few midline structures, brain regions come in pairs (a left and right amygdala, hippocampus, temporal lobe, and so on). Functions are often lateralized, such that the left and right hippocampi, for example, have different but related functions. The greatest lateralization occurs in the cortex; the left hemisphere is analytical, the right more involved in intuition and creativity. These contrasts have caught the public fancy, with cortical lateralization exaggerated by many to an absurd extent, where "left brain"–edness has the connotation of anal-retentive bean counting and "right brain"–edness is about making mandalas or singing with whales. In fact the functional differences between the hemispheres are generally subtle, and I'm mostly ignoring lateralization.

We're now ready to examine the brain regions most central to this book, namely the amygdala, the frontal cortex, and the mesolimbic/mesocortical dopamine system (discussion of other bit-player regions will be subsumed under the headings for these three). We start with the one arguably most central to our worst behaviors.

THE AMYGDALA

The amygdala* is the archetypal limbic structure, sitting under the cortex in the temporal lobe. It is central to mediating aggression, along with other behaviors that tell us tons about aggression.

A First Pass at the Amygdala and Aggression

The evidence for the amygdala's role in aggression is extensive, based on research approaches that will become familiar.

First there's the correlative "recording" approach. Stick recording electrodes into numerous species' amygdalae† and see when neurons there have action potentials; this turns out to be when the animal is being aggressive.‡ In a related approach, determine which brain regions consume extra oxygen or glucose, or synthesize certain activity-related proteins, during aggression—the amygdala tops the list.

Moving beyond mere correlation, if you lesion the amygdala in an animal, rates of aggression decline. The same occurs transiently when you temporarily silence the amygdala by injecting Novocain into it. Conversely, implanting electrodes that stimulate neurons there, or spritzing in excitatory neurotransmitters (stay tuned), triggers aggression.[4]

Show human subjects pictures that provoke anger, and the amygdala activates (as shown with neuroimaging). Sticking an electrode in someone's amygdala and stimulating it (as is done before certain types of neurosurgery) produces rage.

The most convincing data concern rare humans with damage restricted

* The word is from the Greek ἀμυγδαλή (thank you, Wikipedia), which means "almond," which the amygdala very vaguely resembles. Weirdly, it turns out that the word also means "tonsil," which must have produced some major malpractice suits when ancient Greeks went in for tonsillectomies.

† The amygdala is one of those "bilateral" structures, meaning that there are two of them, one in each hemisphere, mirroring each other.

‡ A note about specificity. To be confident that the amygdala really is selectively about aggression, you also have to show that it activates more than do *other* brain regions and doesn't activate as much during a hodgepodge of *other* behaviors.

to the amygdala, either due to a type of encephalitis or a congenital disorder called Urbach-Wiethe disease, or where the amygdala was surgically destroyed to control severe, drug-resistant seizures originating there.[5] Such individuals are impaired in detecting angry facial expressions (while being fine at recognizing other emotional states—stay tuned).

And what does amygdala damage do to aggressive behavior? This was studied in humans where amygdalotomies were done not to control seizures but to control aggression. Such psychosurgery provoked fiery controversy in the 1970s. And I don't mean scientists not saying hello to each other at conferences. I mean a major public shit storm.

The issue raised bioethical lightning rods: What counted as pathological aggression? Who decided? What other interventions had been tried unsuccessfully? Were some types of hyperaggressive individuals more likely to go under the knife than others? What constituted a cure?[6]

Most of these cases concerned rare epileptics where seizure onset was associated with uncontrollable aggression, and where the goal was to contain that behavior (these papers had titles such as "Clinical and physiological effects of stereotaxic bilateral amygdalotomy for intractable aggression"). The fecal hurricane concerned the involuntary lopping out of the amygdala in people without epilepsy but with a history of severe aggression. Well, doing this could be profoundly helpful. Or Orwellian. This is a long, dark story and I will save it for another time.

Did destruction of the human amygdala lessen aggression? Pretty clearly so, when violence was a reflexive, inchoate outburst preceding a seizure. But with surgery done solely to control behavior, the answer is, er, maybe—the heterogeneity of patients and surgical approaches, the lack of modern neuroimaging to pinpoint exactly which parts of the amygdala were destroyed in each individual, and the imprecision in the behavioral data (with papers reporting from 33 to 100 percent "success" rates) make things inconclusive. The procedure has almost entirely fallen out of practice.

The amygdala/aggression link pops up in two notorious cases of violence. The first concerns Ulrike Meinhof, a founder in 1968 of the Red

Army Faction (aka the Baader-Meinhof Gang), a terrorist group responsible for bombings and bank robberies in West Germany. Meinhof had a conventional earlier life as a journalist before becoming violently radicalized. During her 1976 murder trial, she was found hanged in her jail cell (suicide or murder? still unclear). In 1962 Meinhof had had a benign brain tumor surgically removed; the 1976 autopsy showed that remnants of the tumor and surgical scar tissue impinged on her amygdala.[7]

A second case concerns Charles Whitman, the 1966 "Texas Tower" sniper who, after killing his wife and mother, opened fire atop a tower at the University of Texas in Austin, killing sixteen and wounding thirty-two, one of the first school massacres. Whitman was literally an Eagle Scout and childhood choirboy, a happily married engineering major with an IQ in the 99th percentile. In the prior year he had seen doctors, complaining of severe headaches and violent impulses (e.g., to shoot people from the campus tower). He left notes by the bodies of his wife and his mother, proclaiming love and puzzlement at his actions: "I cannot rationaly [*sic*] pinpoint any specific reason for [killing her]," and "let there be no doubt in your mind that I loved this woman with all my heart." His suicide note requested an autopsy of his brain, and that any money he had be given to a mental health foundation. The autopsy proved his intuition correct—Whitman had a glioblastoma tumor pressing on his amygdala. Did Whitman's tumor "cause" his violence? Probably not in a strict "amygdaloid tumor = murderer" sense, as he had risk factors that interacted with his neurological issues. Whitman grew up being beaten by his father and watching his mother and siblings experience the same. This choirboy Eagle Scout had repeatedly physically abused his wife and had been court-martialed as a Marine for physically threatening another soldier.* And, perhaps indicative of a thread running through the family, his brother was murdered at age twenty-four during a bar fight.[8]

* Wait, don't the Marines want you to be physically threatening? Don't they train you to be that way? This is a great example of the big theme of this book, namely the context dependency of our best and worst behaviors: what the Marines train people for is to be majorly physically threatening . . . in certain contexts only.

A Whole Other Domain of Amygdaloid Function to the Center Stage

Thus considerable evidence implicates the amygdala in aggression. But if you asked amygdala experts what behavior their favorite brain structure brings to mind, "aggression" wouldn't top their list. It would be fear and anxiety.[9] Crucially, the brain region most involved in feeling afraid and anxious is most involved in generating aggression.

The amygdala/fear link is based on evidence similar to that supporting the amygdala/aggression link.[10] In lab animals this has involved lesioning the structure, detecting activity in its neurons with "recording electrodes," electrically stimulating it, or manipulating genes in it. All suggest a key role for the amygdala in perceiving fear-provoking stimuli and in expressing fear. Moreover, fear activates the amygdala in humans, with more activation predicting more behavioral signs of fear.

In one study subjects in a brain scanner played a Ms. Pac-Man–from–hell video game where they were pursued in a maze by a dot; if caught, they'd be shocked.[11] When people were evading the dot, the amygdala was silent. However, its activity increased as the dot approached; the stronger the shocks, the farther away the dot would be when first activating the amygdala, the stronger the activation, and the larger the self-reported feeling of panic.

In another study subjects waited an unknown length of time to receive a shock.[12] This lack of predictability and control was so aversive that many *chose* to receive a stronger shock immediately. And in the others the period of anticipatory dread increasingly activated the amygdala.

Thus the human amygdala preferentially responds to fear-evoking stimuli, even stimuli so fleeting as to be below conscious detection.

Powerful support for an amygdaloid role in fear processing comes from post-traumatic stress disorder (PTSD). In PTSD sufferers the amygdala is overreactive to mildly fearful stimuli and is slow in calming down after being activated.[13] Moreover, the amygdala expands in size with long-term PTSD. This role of stress in this expansion will be covered in chapter 4.

The amygdala is also involved in the expression of anxiety.[14] Take a deck of cards—half are black, half are red; how much would you wager that the top card is red? That's about risk. Here's a deck of cards—at least one is black, at least one is red; how much would you wager that the top card is red? That's about ambiguity. The circumstances carry identical probabilities, but people are made more anxious by the second scenario and activate the amygdala more. The amygdala is particularly sensitive to unsettling circumstances that are social. A high-ranking male rhesus monkey is in a sexual consortship with a female; in one condition the female is placed in another room, where the male can see her. In the second she's in the other room along with a rival of the male. No surprise, that situation activates the amygdala. Is that about aggression or anxiety? Seemingly the latter—the extent of activation did not correlate with the amount of aggressive behaviors and vocalizations the male made, or the amount of testosterone secreted. Instead, it correlated with the extent of anxiety displayed (e.g., teeth chattering, or self-scratching).

The amygdala is linked to social uncertainty in other ways. In one neuroimaging study, a subject would participate in a competitive game against a group of other players; outcomes were rigged so that the subject would wind up in the middle of the rankings.[15] Experimenters then manipulated game outcomes so that subjects' rankings either remained stable or fluctuated wildly. Stable rankings activated parts of the frontal cortex that we'll soon consider. Instability activated the frontal cortex plus the amygdala. Being unsure of your place is unsettling.

Another study explored the neurobiology of conforming.[16] To simplify, a subject is part of a group (where, secretly, the rest are confederates); they are shown "X," then asked, "What did you see?" Everyone else says "Y." Does the subject lie and say "Y" also? Often. Subjects who stuck to their guns with "X" showed amygdala activation.

Finally, activating specific circuits within the amygdala in mice turns anxiety on and off; activating others made mice unable to distinguish between safe and anxiety-producing settings.*[17]

* By the way, what does mouse anxiety look like? Mice dislike bright lights and open spaces—go figure, for a nocturnal animal that lots of species like to eat. So one measure of mouse anxiety is how long it takes for a mouse to go into the center of a brightly lit area to get some food.

The amygdala also helps mediate both innate and learned fear.[18] The core of innate fear (aka a phobia) is that you don't have to learn by trial and error that something is aversive. For example, a rat born in a lab, who has interacted only with other rats and grad students, instinctually fears and avoids the smell of cats. While different phobias activate somewhat different brain circuitry (for example, dentist phobia involves the cortex more than does snake phobia), they all activate the amygdala.

Such innate fear contrasts with things we learn to fear—a bad neighborhood, a letter from the IRS. The dichotomy between innate and learned fear is actually a bit fuzzy.[19] Everyone knows that humans are innately afraid of snakes and spiders. But some people keep them as pets, give them cute names.* Instead of inevitable fear, we show "prepared learning"—learning to be afraid of snakes and spiders more readily than of pandas or beagles.

The same occurs in other primates. For example, lab monkeys who have never encountered snakes (or artificial flowers) can be conditioned to fear the former more readily than the latter. As we'll see in the next chapter, humans show prepared learning, being predisposed to be conditioned to fear people with a certain type of appearance.

The fuzzy distinction between innate and learned fear maps nicely onto the amygdala's structure. The evolutionarily ancient central amygdala plays a key role in innate fears. Surrounding it is the basolateral amygdala (BLA), which is more recently evolved and somewhat resembles the fancy, modern cortex. It's the BLA that learns fear and then sends the news to the central amygdala.

Joseph LeDoux at New York University has shown how the BLA learns fear.†[20] Expose a rat to an innate trigger of fear—a shock. When this "unconditioned stimulus" occurs, the central amygdala activates, stress hor-

* We even have that profound renunciation of arachnophobia, namely kids becoming bereft when Charlotte dies in *Charlotte's Web*.

† As an important point, throughout this book, whenever I describe work done by Jane Doe or Joe Smith, I actually mean "work done by Doe and a team of her postdocs, technicians, grad students, and collaborators spread far and wide over the years." I'll be referring solely to Doe or Smith for brevity, not to imply that they did all the work on their own—science is utterly a team process. In addition, as long as we're at it, another point: At endless junctures throughout the book, I'll be reporting the results of a study, along the lines of, "And when you do whatever to this or that brain region/neurotransmitter/hormone/gene/etc., X happens." What I mean is that *on the average* X happens, and at a statistically reliable rate. There is always lots of variability, including individuals in whom nothing happens or even the opposite of X occurs.

mones are secreted, the sympathetic nervous system mobilizes, and, as a clear end point, the rat freezes in place—"What was that? What do I do?" Now do some conditioning. Before each shock, expose the rat to a stimulus that normally does not evoke fear, such as a tone. And with repeated coupling of the tone (the conditioned stimulus) with the shock (the unconditioned one), fear conditioning occurs—the sound of the tone alone elicits freezing, stress hormone release, and so on.*

LeDoux and others have shown how auditory information about the tone stimulates BLA neurons. At first, activation of those neurons is irrelevant to the central amygdala (whose neurons are destined to activate following the shock). But with repeated coupling of tone with shock, there is remapping and those BLA neurons acquire the means to activate the central amygdala.†

BLA neurons that respond to the tone only once conditioning has occurred would also have responded if conditioning instead had been to a light. In other words, these neurons respond to the meaning of the stimulus, rather than to its specific modality. Moreover, if you electrically stimulate them, rats are easier to fear-condition; you've lowered the threshold for this association to be made. And if you electrically stimulate the auditory sensory input at the same time as shocks (i.e., there's no tone, just activation of the pathway that normally carries news of the tone to the amygdala), you cause fear conditioning to a tone. You've engineered the learning of a false fear.

There are synaptic changes as well. Once conditioning to a tone has occurred, the synapses coupling the BLA and central nucleus neurons have become more excitable; how this occurs is understood at the level of changes in the amount of receptors for excitatory neurotransmitters in dendritic spines in these circuits.‡ Furthermore, conditioning increases levels of "growth factors," which prompt the growth of new connections

* This is termed "Pavlovian conditioning" in a nod to Ivan Pavlov; it's the same process by which Pavlov's dogs learned to associate the conditioned stimulus of a bell with the unconditioned stimulus of food, so that the former eventually was able to provoke salivation. Less reliable are "operant conditioning" approaches, in which the degree to which something is scary is assessed by how much an individual will work to avoid being exposed to it.

† As usual in science, things aren't all that clean—some of those "plastic" changes during fear conditioning also occur in the central amygdala.

‡ Just to make things more complicated, the BLA neurons are probably talking to the central amygdala neurons via middlemen called intercalated cells.

between BLA and central amygdala neurons; some of the genes involved have even been identified.

We've now got learning to be afraid under our belts.*[21] Now conditions change—the tone still occurs now and then, but no more shock. Gradually the conditioned fear response abates. How does "fear extinction" occur? How do we learn that this person wasn't so scary after all, that different doesn't necessarily equal frightening? Recall how a subset of BLA neurons respond to the tone only once conditioning has occurred. Another population does the opposite, responding to the tone once it's no longer signaling shock (logically, the two populations of neurons inhibit each other). Where do these "Ohhh, the tone isn't scary anymore" neurons get inputs from? The frontal cortex. When we stop fearing something, it isn't because some amygdaloid neurons have lost their excitability. We don't passively forget that something is scary. We actively learn that it isn't anymore.†

The amygdala also plays a logical role in social and emotional decision making. In the Ultimatum Game, an economic game involving two players, the first makes an offer as to how to divide a pot of money, which the other player either accepts or rejects.[22] If the latter, neither gets anything. Research shows that rejecting an offer is an emotional decision, triggered by anger at a lousy offer and the desire to punish. The more the amygdala activation in the second player after an offer, the more likely the rejection. People with damaged amygdalae are atypically generous in the Ultimatum Game and don't increase rejection rates if they start receiving unfair offers.

---

* I would be remiss not to touch on one issue in this field—when a new fear is learned, where is that memory stored? Next to the amygdala is the hippocampus, which plays a key role in "explicit" learning about straightforward facts (e.g., someone's name). While the hippocampus is where short-term knowledge about the name is turned into long-term memory, the memory trace itself is most likely in the cortex. The hippocampus, to use a metaphor that will probably be obsolete by the time this book sees the light of day, is the keyboard, the conduit, the portal to the cortical hard drive where a memory is stored. Is the amygdala solely the keyboard (with fear memories stored elsewhere), or is it the hard drive as well? This has been an ongoing, unresolved debate in the field, with the "keyboard + hard drive" view championed by LeDoux and the keyboard-alone view espoused by the equally accomplished scientist James McGaugh of the University of California at Irvine.

† An example of the sort of complexity we're up against here: Both fear conditioning and fear extinction involve activation of inhibitory neurons. Hmm, that commonality seems weird, given opposite outcomes. It turns out that extinction involves activation of neurons that inhibit excitatory neurons, while fear conditioning involves activation of inhibitory neurons that inhibit *other* inhibitory neurons that project onto excitatory neurons. A double negative, equaling a positive.

Why? These individuals understand the rules and can give sound, strategic advice to other players. Moreover, they use the same strategies as control subjects in a nonsocial version of the game, when believing the other player is a computer. And they don't have a particularly long view, undistracted by the amygdala's emotional tumult, reasoning that their noncontingent generosity will induce reciprocity and pay off in the long run. When asked, they anticipate the same levels of reciprocity as do controls.

Instead, these findings suggest that the amygdala injects implicit distrust and vigilance into social decision making.[23] All thanks to learning. In the words of the authors of the study, "The generosity in the trust game of our BLA-damaged subjects might be considered pathological altruism, in the sense that inborn altruistic behaviors have not, due to BLA damage, been un-learned through negative social experience." In other words, the default state is to trust, and what the amygdala does is learn vigilance and distrust.

Unexpectedly, the amygdala and one of its hypothalamic targets also play a role in male sexual motivation (other hypothalamic nuclei are central to male sexual performance)* but not female.† What's that about? One neuroimaging study sheds some light. "Young heterosexual men" looked at pictures of attractive women (versus, as a control, of attractive men). Passively observing the pictures activated the reward circuitry just alluded to. In contrast, *working* to see the pictures—by repeatedly pressing a button—also activated the amygdala. Similarly, other studies show that the amygdala is most responsive to positive stimuli when the value of the reward is shifting. Moreover, some BLA neurons that respond in that circumstance also respond when the severity of something aversive is shifting—these neurons are paying attention to change, independent of direction. For them, "the amount of reward is changing" and "the amount of punishment is changing" are the same. Studies like these clarify that

* How might you distinguish between sexual motivation and performance in a male rat? Well, the latter is easy—what's the guy's frequency and latency when with a sexually receptive female? But sexual motivation? This is measured by determining how often a male will press a lever in order to get access to a female.

† I can't resist mentioning a case report of a woman with epileptic seizures initiating in the amygdala. Before seizure onset, she would have the delusion that she was male, including a sense that she had a deep voice and hairy arms.

the amygdala isn't about the pleasure of experiencing pleasure. It's about the uncertain, unsettled yearning for a potential pleasure, the anxiety and fear and anger that the reward may be smaller than anticipated, or may not even happen. It's about how many of our pleasures and our pursuits of them contain a corrosive vein of dis-ease.*[24]

The Amygdala as Part of Networks in the Brain

Now that we know about the subparts of the amygdala, it's informative to consider its extrinsic connections—i.e., what parts of the brain send projection to it, and what parts does it project to?[25]

SOME INPUTS TO THE AMYGDALA

Sensory inputs. For starters, the amygdala, specifically the BLA, gets projections from all the sensory systems.[26] How else can you get terrified by the shark's theme music in *Jaws*? Normally, sensory information from various modalities (eyes, ears, skin . . .) courses into the brain, reaching the appropriate cortical region (visual cortex, auditory cortex, tactile cortex . . .) for processing. For example, the visual cortex would engage layers and layers of neurons to turn pixels of retinal stimulation into recognizable images before it can scream to the amygdala, "It's a gun!" Importantly, some sensory information entering the brain takes a shortcut, bypassing the cortex and going directly to the amygdala. Thus the amygdala can be informed about something scary before the cortex has a clue. Moreover, thanks to the extreme excitability of this pathway, the amygdala can respond to stimuli that are too fleeting or faint for the cortex to note. Additionally, the shortcut projections form stronger, more excitable synapses in the BLA than do the ones from the sensory cortex; emotional arousal enhances fear conditioning through this pathway. This shortcut's power is shown in the case of a man with stroke damage to his visual cortex, producing "cortical blindness." While unable to process

* In contrast to this picture of growing, precarious arousal, the amygdala deactivates in both men and women during orgasm.

most visual information, he still recognized emotional facial expressions via the shortcut.*

Crucially, while sensory information reaches the amygdala rapidly by this shortcut, it isn't terribly accurate (since, after all, accuracy is what the cortex supplies). As we'll see in the next chapter, this produces tragic circumstances where, say, the amygdala decides it's seeing a handgun before the visual cortex can report that it's actually a cell phone.

Information about pain. The amygdala receives news of that reliable trigger of fear and aggression, namely pain.[27] This is mediated by projections from an ancient, core brain structure, the "periaqueductal gray" (PAG); stimulation of the PAG can evoke panic attacks, and it is enlarged in people with chronic panic attacks. Reflecting the amygdala's roles in vigilance, uncertainty, anxiety, and fear, it's unpredictable pain, rather than pain itself, that activates the amygdala. Pain (and the amygdala's response to it) is all about context.

Disgust of all stripes. The amygdala also receives a hugely interesting projection from the "insular cortex," an honorary part of the prefrontal cortex, which we will consider at length in later chapters.[28] If you (or any other mammal) bite into rancid food, the insular cortex lights up, causing you to spit it out, gag, feel nauseated, make a revolted facial expression—the insular cortex processes gustatory disgust. Ditto for disgusting smells.

Remarkably, humans also activate it by thinking about something *morally* disgusting—social norm violations or individuals who are typically stigmatized in society. And in that circumstance its activation drives that of the amygdala. Someone does something lousy and selfish to you in a game, and the extent of insular and amygdaloid activation predicts how much outrage you feel and how much revenge you take. This is all about sociality—the insula and amygdala don't activate if it's a computer that has stabbed you in the back.

The insula activates when we eat a cockroach or imagine doing so. And the insula and amygdala activate when we think of the neighboring

* This shortcut has been most cleanly demonstrated for auditory information, by LeDoux. The evidence for other sensory modalities has been more inferential.

tribe as loathsome cockroaches. As we'll see, this is central to how our brains process "us and them."

And finally, the amygdala gets tons of inputs from the frontal cortex. Much more to come.

SOME OUTPUTS FROM THE AMYGDALA

Bidirectional connections. As we'll see, the amygdala talks to many of the regions that talk to it, including the frontal cortex, insula, periaqueductal gray, and sensory projections, modulating their sensitivity.

The amygdala/hippocampus interface. Naturally, the amygdala talks to other limbic structures, including the hippocampus. As reviewed, typically the amygdala learns fear and the hippocampus learns detached, dispassionate facts. But at times of extreme fear, the amygdala pulls the hippocampus into a type of fear learning.[29]

Back to the rat undergoing fear conditioning. When it's in cage A, a tone is followed by a shock. But in cage B, the tone isn't. This produces context-dependent conditioning—the tone causes fearful freezing in cage A but not in cage B. The amygdala learns the stimulus cue—the tone—while the hippocampus learns about the contexts of cage A versus B. The coupled learning between amygdala and hippocampus is very focalized—we all remember the view of the plane hitting the second World Trade Center tower, but not whether there were clouds in the background. The hippocampus decides whether a factoid is worth filing away, depending on whether the amygdala has gotten worked up over it. Moreover, the coupling can rescale. Suppose someone robs you at gunpoint in an alley in a bad part of town. Afterward, depending on the circumstance, the gun can be the cue and the alley the context, or the alley is the cue and the bad part of town the context.

Motor outputs. There's a second shortcut regarding the amygdala, specifically when it's talking to motor neurons that command movement.[30] Logically, when the amygdala wants to mobilize a behavior—say, fleeing—it talks to the frontal cortex, seeking its executive approval. But if sufficiently aroused, the amygdala talks directly to subcortical, reflexive motor pathways. Again, there's a trade-off—increased speed by bypassing the cortex, but decreased accuracy. Thus the input shortcut may prompt you to see the cell phone as a gun. And the output shortcut may prompt you to pull a trigger before you consciously mean to.

Arousal. Ultimately, amygdala outputs are mostly about setting off alarms throughout the brain and body. As we saw, the core of the amygdala is the central amygdala.[31] Axonal projections from there go to an amygdala-ish structure nearby called the bed nucleus of the stria terminalis (BNST). The BNST, in turn, projects to parts of the hypothalamus that initiate the hormonal stress response (see chapter 4), as well as to midbrain and brain-stem sites that activate the sympathetic nervous system and inhibit the parasympathetic nervous system. Something emotionally arousing occurs, layer 2 limbic amygdala signals layer 1 regions, and heart rate and blood pressure soar.*

The amygdala also activates a brain-stem structure called the locus coeruleus, akin to the brain's own sympathetic nervous system.[32] It sends norepinephrine-releasing projections throughout the brain, particularly the cortex. If the locus coeruleus is drowsy and silent, so are you. If it's moderately activated, you're alert. And if it's firing like gangbusters, thanks to inputs from an aroused amygdala, all neuronal hands are on deck.

The amygdala's projection pattern raises an important point.[33] When is the sympathetic nervous system going full blast? During fear, flight, fight, and sex. Or if you've won the lottery, are happily sprinting down a soccer field, or have just solved Fermat's theorem (if you're that kind of person). Reflecting this, about a quarter of neurons in one hypothalamic nucleus are involved in both sexual behavior and, when stimulated at a higher intensity, aggressive behavior in male mice.

This has two implications. Both sex and aggression activate the sympathetic nervous system, which in turn can influence behavior—people feel differently about things if, say, their heart is racing versus beating slowly. Does this mean that the pattern of your autonomic arousal

---

* Just to bring some specificity into it, the precise pattern of which subregions of the hypothalamus and which autonomic relay nuclei are activated can vary with the type of stimulus—thus the fear and aggression associated with responding to a predator are somewhat different from those in response to the menace of a member of one's own species; similarly, the pattern of response in a rodent to the smell of a cat is a bit different from the response to a cat itself.

influences *what* you feel? Not really. But autonomic feedback influences the *intensity* of what is felt. More on this in the next chapter.

The second consequence reflects a core idea of this book. Your heart does roughly the same thing whether you are in a murderous rage or having an orgasm. Again, the opposite of love is not hate, it's indifference.

This concludes our overview of the amygdala. Amid the jargon and complexity, the most important theme is the amygdala's dual role in both aggression and facets of fear and anxiety. Fear and aggression are not inevitably intertwined—not all fear causes aggression, and not all aggression is rooted in fear. Fear typically increases aggression only in those already prone to it; among the subordinate who lack the option of expressing aggression safely, fear does the opposite.

The dissociation between fear and aggression is evident in violent psychopaths, who are the antithesis of fearful—both physiologically and subjectively they are less reactive to pain; their amygdalae are relatively unresponsive to typical fear-evoking stimuli and are smaller than normal.[34] This fits with the picture of psychopathic violence; it is not done in aroused reaction to provocation. Instead, it is purely instrumental, using others as a means to an end with emotionless, remorseless, reptilian indifference.

Thus, fear and violence are not always connected at the hip. But a connection is likely when the aggression evoked is reactive, frenzied, and flecked with spittle. In a world in which no amygdaloid neuron need be afraid and instead can sit under its vine and fig tree, the world is very likely to be a more peaceful place.*

We now move to the second of the three brain regions we're considering in detail.

* Apologies to Micah 4:4.

THE FRONTAL CORTEX

I've spent decades studying the hippocampus. It's been good to me; I'd like to think I've been the same in return. Yet I think I might have made the wrong choice back then—maybe I should have studied the frontal cortex all these years. Because it's the most interesting part of the brain.

What does the frontal cortex do? Its list of expertise includes working memory, executive function (organizing knowledge strategically, and then initiating an action based on an executive decision), gratification postponement, long-term planning, regulation of emotions, and reining in impulsivity.[35]

This is a sprawling portfolio. I will group these varied functions under a single definition, pertinent to every page of this book: *the frontal cortex makes you do the harder thing when it's the right thing to do.*

To start, here are some important features of the frontal cortex:

> It's the most recently evolved brain region, not approaching full splendor until the emergence of primates; a disproportionate percentage of genes unique to primates are active in the frontal cortex. Moreover, such gene expression patterns are highly individuated, with greater interindividual variability than average levels of whole-brain differences between humans and chimps.

> The human frontal cortex is more complexly wired than in other apes and, by some definitions as to its boundaries, proportionately bigger as well.[36]

> The frontal cortex is the last brain region to fully mature, with the most evolutionarily recent subparts the very last. Amazingly, it's not fully online until people are in their *midtwenties*. You'd better bet this factoid will be relevant to the chapter about adolescence.

> Finally, the frontal cortex has a unique cell type. In general, the human brain isn't unique because we've evolved unique types of neurons, neurotransmitters, enzymes, and so on. Human and fly neurons are remarkably similar; the uniqueness is quantitative—for every fly neuron, we have a gazillion more neurons and a bazillion more connections.[37]

The sole exception is an obscure type of neuron with a distinctive shape and pattern of wiring, called von Economo neurons (aka spindle neurons). At first they seemed to be unique to humans, but we've now found them in other primates, whales, dolphins, and elephants.* That's an all-star team of socially complex species.

Moreover, the few von Economo neurons occur only in two subregions of the frontal cortex, as shown by John Allman at Caltech. One we've heard about already—the insula, with its role in gustatory and moral disgust. The second is an equally interesting area called the anterior cingulate. To give a hint (with more to come), it's central to empathy.

So from the standpoint of evolution, size, complexity, development, genetics, and neuron type, the frontal cortex is distinctive, with the human version the most unique.

The Subregions of the Frontal Cortex

Frontal cortical anatomy is hellishly complicated, and there are debates as to whether some parts of the primate frontal cortex even exist in "simpler" species. Nonetheless, there are some useful broad themes.

In the very front is the *pre*frontal cortex (PFC), the newest part of the frontal cortex. As noted, the frontal cortex is central to executive function. To quote George W. Bush, within the frontal cortex, it's the PFC that is "the decider." Most broadly, the PFC chooses between conflicting options—Coke or Pepsi; blurting out what you really think or restraining yourself; pulling the trigger or not. And often the conflict being resolved

* Strongly suggesting that these neurons independently evolved on three separate occasions, given the evolutionary distances between primates, cetaceans, and elephants. The nearest relatives of elephants, for example, are hyraxes and manatees. The convergent evolution of von Economo neurons from three separate lineages emphasizes that these cells go hand in hand with major sociality.

is between a decision heavily driven by cognition and one driven by emotions.

Once it has decided, the PFC sends orders via projections to the rest of the frontal cortex, sitting just behind it. Those neurons then talk to the "premotor cortex," sitting just behind it, which then passes it to the "motor cortex," which talks to your muscles. And a behavior ensues.*

Before considering how the frontal cortex influences social behavior, let's start with a simpler domain of its function.

The Frontal Cortex and Cognition

What does "doing the harder thing when it's the right thing to do" look like in the realm of cognition (defined by Princeton's Jonathan Cohen as "the ability to orchestrate thought and action in accordance with internal goals")?[38] Suppose you've looked up a phone number in a city where you once lived. The frontal cortex not only remembers it long enough to dial but also considers it strategically. Just before dialing, you consciously recall that it is in that other city and retrieve your memory of the city's area code. And then you remember to dial "1" before the area code.†

The frontal cortex is also concerned with focusing on a task. If you step off the curb planning to jaywalk, you look at traffic, paying attention to motion, calculating whether you can cross safely. If you step off looking for a taxi, you pay attention to whether a car has one of those lit taxicab thingies on top. In a great study, monkeys were trained to look at a screen of dots of various colors moving in particular directions; depending on a signal, a monkey had to pay attention to either color or movement. Each signal indicating a shift in tasks triggered a burst of PFC activity and, coupled with that, suppression of the stream of information (color or movement) that was now irrelevant. This is the PFC getting you to do the

* To give a sense of this, consider someone deciding whether to press a button. The frontal cortex makes its decision; know its neurons' firing patterns, and you can predict the decision with 80 percent accuracy about seven hundred milliseconds before the person is consciously aware of their decision.

† This quaintly obsolete paragraph is written with the recognition that most of this is irrelevant in the age of smartphones and the constant companionship of Siri.

harder thing; remembering that the rule has changed, don't do the previous habitual response.[39]

The frontal cortex also mediates "executive function"—considering bits of information, looking for patterns, and then choosing a strategic action.[40] Consider this truly frontally demanding test. The experimenter tells a masochistic volunteer, "I'm going to the market and I'm going to buy peaches, cornflakes, laundry detergent, cinnamon . . ." Sixteen items recited, the volunteer is asked to repeat the list. Maybe they correctly recall the first few, the last few, list some near misses—say, nutmeg instead of cinnamon. Then the experimenter repeats the same list. This time the volunteer remembers a few more, avoids repeating the nutmeg incident. Now do it again and again.

This is more than a simple memory test. With repetition, subjects notice that four of the items are fruits, four for cleaning, four spices, four carbs. They come in categories. And this changes subjects' encoding strategy as they start clumping by semantic group—"Peaches. Apples. Blueberries—no, I mean blackberries. There was another fruit, can't remember what. Okay, cornflakes, bread, doughnuts, muffins. Cumin, nutmeg—argh, again!—I mean cinnamon, oregano . . ." And throughout, the PFC imposes an overarching executive strategy for remembering these sixteen factoids.*

The PFC is essential for categorical thinking, for organizing and thinking about bits of information with different labels. The PFC groups apples and peaches as closer to each other in a conceptual map than are apples and toilet plungers. In a relevant study, monkeys were trained to differentiate between pictures of a dog and of a cat. The PFC contained individual neurons that responded to "dog" and others that responded to "cat." Now the scientists morphed the pictures together, creating hybrids with varying percentages of dog and cat. "Dog" PFC neurons responded about as much to hybrids that were 80 percent dog and 20 percent cat, or

---

* This test is reminiscent of something called the California Verbal Learning Test (CVLT). My wife, who spent her professional youth as a neuropsychologist, would practice tests on me when she was in grad school; the CVLT was, without question, the worst. It was insanely stressful—I'd be a sopping mess by the time she'd finally call it an evening. But on the other hand, this will pay off handsomely in a few decades when I ace the neuropsych tests out of habit, despite being seriously demented . . . and thus don't get appropriate medical care. Hmm, I may need to rethink this.

60:40, as to 100 percent dog. But not to 40:60—"cat" neurons would kick in there.[41]

The frontal cortex aids the underdog outcome, fueled by thoughts supplied from influences that fill the rest of this book—stop, those aren't your cookies; you'll go to hell; self-discipline is good; you're happier when you're thinner—all giving some lone inhibitory motor neuron more of a fighting chance.

Frontal Metabolism and an Implicit Vulnerability

This raises an important point, pertinent to the social as well as cognitive functions of the frontal cortex.[42] All this "I wouldn't do that if I were you"–ing by the frontal cortex is taxing. Other brain regions respond to instances of some contingency; the frontal cortex tracks rules. Just think how around age three, our frontal cortices learned a rule followed for the rest of our lives—don't pee whenever you feel like it—and gained the means to enact that rule by increasing their influence over neurons regulating the bladder.

Moreover, the frontal mantra of "self-discipline is good" when cookies beckon is also invoked when economizing to increase retirement savings. Frontal cortical neurons are generalists, with broad patterns of projections, which makes for more work.[43]

All this takes energy, and when it is working hard, the frontal cortex has an extremely high metabolic rate and rates of activation of genes related to energy production.[44] Will*power* is more than just a metaphor; self-control is a finite resource. Frontal neurons are expensive cells, and expensive cells are vulnerable cells. Consistent with that, the frontal cortex is atypically vulnerable to various neurological insults.

Pertinent to this is the concept of "cognitive load." Make the frontal cortex work hard—a tough working-memory task, regulating social behavior, or making numerous decisions while shopping. Immediately afterward performance on a different frontally dependent task declines.[45] Likewise during multitasking, where PFC neurons simultaneously participate in multiple activated circuits.

Importantly, increase cognitive load on the frontal cortex, and afterward subjects become less prosocial*—less charitable or helpful, more likely to lie.[46] Or increase cognitive load with a task requiring difficult emotional regulation, and subjects cheat more on their diets afterward.†[47]

So the frontal cortex is awash in Calvinist self-discipline, a superego with its nose to the grindstone.[48] But as an important qualifier, soon after we're potty-trained, doing the harder thing with our bladder muscles becomes automatic. Likewise with other initially demanding frontal tasks. For example, you're learning a piece of music on the piano, there's a difficult trill, and each time as you approach it, you think, "Here it comes. Remember, tuck my elbow in, lead with my thumb." A classic working-memory task. And then one day you realize that you're five measures past the trill, it went fine, and you didn't have to think about it. And that's when doing the trill is transferred from the frontal cortex to more reflexive brain regions (e.g., the cerebellum). This transition to automaticity also happens when you get good at a sport, when metaphorically your body knows what to do without your thinking about it.

The chapter on morality considers automaticity in a more important realm. Is resisting lying a demanding task for your frontal cortex, or is it effortless habit? As we'll see, honesty often comes more easily thanks to automaticity. This helps explain the answer typically given after someone has been profoundly brave. "What were you thinking when you dove into the river to save that drowning child?" "I wasn't thinking—before I knew it, I had jumped in." Often the neurobiology of automaticity mediates doing the hardest moral acts, while the neurobiology of the frontal cortex mediates working hard on a term paper about the subject.

The Frontal Cortex and Social Behavior

Things get interesting when the frontal cortex has to add social factors to a cognitive mix. For example, one part of the monkey PFC contains

* There's a key exception to this, to be covered in chapter 13 on morality.

† There is an ongoing controversy in this field as to whether it is "willpower" or "motivation" being decreased by cognitive load. For our purposes, let's think of them as synonymous.

neurons that activate when the monkey makes a mistake on a cognitive task or observes another monkey doing so; some activate only when it's a particular animal who made the mistake. In a neuroimaging study humans had to choose something, balancing feedback obtained from their own prior choices with advice from another person. Different PFC circuits tracked "reward-driven" and "advice-driven" cogitating.[49]

Findings like these segue into the central role of the frontal cortex in social behavior.[50] This is appreciated when comparing various primates. Across primate species, the bigger the size of the average social group, the larger the relative size of the frontal cortex. This is particularly so with "fission-fusion" species, where there are times when subgroups split up and function independently for a while before regrouping. Such a social structure is demanding, requiring the scaling of appropriate behavior to subgroup size and composition. Logically, primates from fission-fusion species (chimps, bonobos, orangutans, spider monkeys) have better frontocortical inhibitory control over behavior than do non-fission-fusion primates (gorillas, capuchins, macaques).

Among humans, the larger someone's social network (measured by number of different people texted), the larger a particular PFC subregion (stay tuned).[51] That's cool, but we can't tell if the big brain region causes the sociality or the reverse (assuming there's causality). Another study resolves this; if rhesus monkeys are randomly placed into social groups, over the subsequent fifteen months, the bigger the group, the larger the PFC becomes—social complexity expands the frontal cortex.

We utilize the frontal cortex to do the harder thing in social contexts—we praise the hosts for the inedible dinner; refrain from hitting the infuriating coworker; don't make sexual advances to someone, despite our fantasies; don't belch loudly during the eulogy. A great way to appreciate the frontal cortex is to consider what happens when it is damaged.

The first "frontal" patient, the famous Phineas Gage, was identified in 1848 in Vermont. Gage, the foreman on a railroad construction crew, was injured when an accident with blasting powder blew a thirteen-pound iron tamping rod through the left side of his face and out the top front of

The two known pictures of Gage, along with the tamping rod.

his skull. It landed eighty feet away, along with much of his left frontal cortex.[52]

Remarkably, he survived and recovered his health. But the respected, even-keeled Gage was transformed. In the words of the doctor who followed him over the years:

> The equilibrium or balance, so to speak, between his intellectual faculties and animal propensities, seems to have been destroyed. He is fitful, irreverent, indulging at times in the grossest profanity (which was not previously his custom), manifesting but little deference for his fellows, impatient of restraint or advice when it conflicts with his desires, at times pertinaciously obstinate, yet capricious and vacillating, devising many plans of future operations, which are no sooner arranged than they are abandoned in turn for others appearing more feasible.

Gage was described by friends as "no longer Gage," was incapable of resuming his job and was reduced to appearing (with his rod) as an exhibit displayed by P. T. Barnum. Poignant as hell.

Amazingly, Gage got better. Within a few years of his injury, he could resume work (mostly as a stagecoach driver) and was described as being broadly appropriate in his behavior. His remaining right frontal cortical tissue had taken on some of the functions lost in the injury. Such malleability of the brain is the focus of chapter 5.

Another example of what happens when the frontal cortex is damaged is observed in frontotemporal dementia (FTD), which starts by damaging the frontal cortex; intriguingly, the first neurons killed are those mysterious von Economo neurons that are unique to primates, elephants, and cetaceans.[53] What are people with FTD like? They exhibit behavioral disinhibition and socially inappropriate behaviors. There's also an apathy and lack of initiating behavior that reflects the fact that the "decider" is being destroyed.*

Something similar is seen in Huntington's disease, a horrific disorder due to a thoroughly weird mutation. Subcortical circuits that coordinate signaling to muscles are destroyed, and the sufferer is progressively incapacitated by involuntary writhing movements. Except that it turns out that there is frontal damage as well, often before the subcortical damage. In about half the patients there's also behavioral disinhibition—stealing, aggressiveness, hypersexuality, bursts of compulsive, inexplicable gambling.† Social and behavioral disinhibition also occur in individuals with stroke damage in the frontal cortex—for example, sexually assaultive behavior in an octogenarian.

There's another circumstance where the frontal cortex is hypofunctional, producing similar behavioral manifestations—hypersexuality, outbursts of emotion, flamboyantly illogical acts.[54] What disease is this? It

* The apathy is in contrast to early-stage Alzheimer's sufferers, who, after making some horrible social blunder due to memory problems—say, asking after the health of someone's spouse because they didn't remember that the person died years ago—are mortified.

† Ian McEwan's novel *Saturday* pivots around behavioral disinhibition due to Huntington's in a central character. It's brilliant.

isn't. You're dreaming. During REM sleep, when dreaming occurs, the frontal cortex goes off-line, and dream scriptwriters run wild. Moreover, if the frontal cortex is stimulated while people are dreaming, the dreams become less dreamlike, with more self-awareness. And there's another nonpathological circumstance where the PFC silences, producing emotional tsunamis: during orgasm.

One last realm of frontal damage. Adrian Raine of the University of Pennsylvania and Kent Kiehl of the University of New Mexico report that criminal psychopaths have decreased activity in the frontal cortex and less coupling of the PFC to other brain regions (compared with nonpsychopathic criminals and noncriminal controls). Moreover, a shockingly large percentage of people incarcerated for violent crimes have a history of concussive trauma to the frontal cortex.[55] More to come in chapter 16.

The Obligatory Declaration of the Falseness of the Dichotomy Between Cognition and Emotion

The PFC consists of various parts, subparts, and sub-subparts, enough to keep neuroanatomists off the dole. Two regions are crucial. First there is the dorsal part of the PFC, especially the dorsolateral PFC (dlPFC)—don't worry about "dorsal" or "dorsolateral"; it's just jargon.* The dlPFC is the decider of deciders, the most rational, cognitive, utilitarian, unsentimental part of the PFC. It's the most recently evolved part of the PFC and the last part to fully mature. It mostly hears from and talks to other cortical regions.

In contrast to the dlPFC, there's the ventral part of the PFC, particularly the ventromedial PFC (vmPFC). This is the frontocortical region that the visionary neuroanatomist Nauta made an honorary member of the limbic system because of its interconnections with it. Logically, the

* A quick primer for directions in the brain, for anyone who cares. They come in three dimensions: (1) Dorsal/ventral. Dorsal = the top of the brain (in the same way that the fin on the top of a horizontal dolphin is the dorsal fin). Ventral = the bottom. (2) Medial/lateral. Medial = at the midline of the brain, when viewed in cross section. Lateral = as far as possible from the midline, moving left or right. Thus the "dorsolateral" PFC is the part of the PFC that is on top and to the outside. (3) Anterior/posterior. At the front or back of the brain. Lateralized brain structures come in pairs—one in the left hemisphere, one in the right, both at the same place in the dorsal/ventral and anterior/posterior planes, but in opposite locations in the medial/lateral plane.

vmPFC is all about the impact of emotion on decision making. And many of our best and worst behaviors involve interactions of the vmPFC with the limbic system and dlPFC.*

The functions of the cognitive dlPFC are the essence of doing the harder thing.[56] It's the most active frontocortical region when someone forgoes an immediate reward for a bigger one later. Consider a classic moral quandary—is it okay to kill one innocent person to save five? When people ponder the question, greater dlPFC activation predicts a greater likelihood of answering yes (but as we'll see in chapter 13, it also depends on how you ask the question).

Monkeys with dlPFC lesions can't switch strategies in a task when the rewards given for each strategy shift—they perseverate with the strategy offering the most immediate reward.[57] Similarly, humans with dlPFC damage are impaired in planning or gratification postponement, perseverate on strategies that offer immediate reward, and show poor executive control over their behavior.† Remarkably, the technique of transcranial magnetic stimulation can temporarily silence part of someone's cortex, as was done in a fascinating study by Ernst Fehr of the University of Zurich.[58] When the dlPFC was silenced, subjects playing an economic game impulsively accepted lousy offers that they'd normally reject in the hopes of getting better offers in the future. Crucially, this was about sociality—silencing the dlPFC had no effect if subjects thought the other player was a computer. Moreover, controls and subjects with silenced dlPFCs rated lousy offers as being equally unfair; thus, as concluded by the authors, "subjects [with the silenced dlPFC] behave as if they can no longer implement their fairness goals."

What are the functions of the emotional vmPFC?[59] What you'd expect, given its inputs from limbic structures. It activates if the person you're rooting for wins a game, or if you listen to pleasant versus dissonant

* To help keep "dlPFC" and "vmPFC" straight, I'll constantly refer to their falsely dichotomized functions, just as a reminder—"the cognitive dlPFC" and "the emotional vmPFC." Or here's a mnemonic—"dl" of the cognitive dlPFC standing for "deliberative," the "vm" of the emotional vmPFC as "very (e)motional." Lame, but it's saved me on a few occasions.

† Moreover, dlPFC patients are poor at the difficult task of taking someone else's perspective. This is a subtype of something called Theory of Mind, and involves interactions between the dlPFC and a brain region called the temporoparietal juncture. More in a later chapter.

music (particularly if the music provokes a shiver-down-the-spine moment).

What are the effects of vmPFC damage?[60] Lots of things remain normal—intelligence, working memory, making estimates. Individuals can "do the harder thing" with purely cognitive frontal tasks (e.g., puzzles where you have to give up a step of progress in order to gain two more).

The differences appear when it comes to making social/emotional decisions—vmPFC patients just can't decide.* They understand the options and can sagely advise someone else in similar circumstances. But the closer to home and the more emotional the scenario, the more they have problems.

Damasio has produced an influential theory about emotion-laden decision making, rooted in the philosophies of Hume and William James; this will soon be discussed.[61] Briefly, the frontal cortex runs "as if" experiments of gut feelings—"How would I feel if this outcome occurred?"—and makes choices with the answer in mind. Damaging the vmPFC, thus removing limbic input to the PFC, eliminates gut feelings, making decisions harder.

Moreover, eventual decisions are highly utilitarian. vmPFC patients are atypically willing to sacrifice one person, including a family member, to save five strangers.[62] They're more interested in outcomes than in their underlying emotional motives, punishing someone who accidentally kills but not one who tried to kill but failed, because, after all, no one died in the second case.

It's Mr. Spock, running on only the dlPFC. Now for a crucial point. People who dichotomize between thought and emotion often prefer the former, viewing emotion as suspect. It gums up decision making by getting sentimental, sings too loudly, dresses flamboyantly, has unsettling amounts of armpit hair. In this view, get rid of the vmPFC, and we'd be more rational and function better.

But that's not the case, as emphasized eloquently by Damasio. People

* A reminder—as with all good studies of individuals with damage to particular brain regions, not only is there a control comparison group of people with no brain damage, but there's an additional control group of people with damage to other, unrelated parts of the brain.

with vmPFC damage not only have trouble making decisions but also make bad ones.[63] They show poor judgment in choosing friends and partners and don't shift behavior based on negative feedback. For example, consider a gambling task where reward rates for various strategies change without subjects knowing it, and subjects can shift their play strategy. Control subjects shift optimally, even if they can't verbalize how reward rates have changed. Those with vmPFC damage don't, even when they *can* verbalize. Without a vmPFC, you may know the meaning of negative feedback, but you don't know the *feeling* of it in your gut and thus don't shift behavior.

As we saw, without the dlPFC, the metaphorical superego is gone, resulting in individuals who are now hyperaggressive, hypersexual ids. But without a vmPFC, behavior is inappropriate in a detached way. This is the person who, encountering someone after a long time, says, "Hello, I see you've put on some weight." And when castigated later by their mortified spouse, they will say with calm puzzlement, "But it's true." The vmPFC is not the vestigial appendix of the frontal cortex, where emotion is something akin to appendicitis, inflaming a sensible brain. Instead it's essential.[64] It wouldn't be if we had evolved into Vulcans. But as long as the world is filled with humans, evolution would never have made us that way.

Activation of the dlPFC and vmPFC can be inversely correlated. In an inspired study where a keyboard was provided to jazz pianists inside a brain scanner, the vmPFC became more active and the dlPFC less so when subjects improvised. In another study, subjects judged hypothetical harmful acts. Pondering perpetrators' responsibility activated the dlPFC; deciding the amount of punishment activated the vmPFC.* When subjects did a gambling task where reward probabilities for various strategies shifted and they could always change strategies, decision making reflected two factors: (a) the outcome of their most recent action (the better that had turned out, the more vmPFC activation), and (b) reward rates from all the previous rounds, something requiring a long retrospective view (the

* For those who care, some of the strongest responses are found in a subregion of the vmPFC called the orbitofrontal cortex.

better the long-term rewards, the more dlPFC activation). Relative activation between the two regions predicted the decision subjects made.[65]

A simplistic view is that the vmPFC and dlPFC perpetually battle for domination by emotion versus cognition. But while emotion and cognition can be somewhat separable, they're rarely in opposition. Instead they are intertwined in a collaborative relationship needed for normal function, and as tasks with both emotive and cognitive components become more difficult (making an increasingly complex economic decision in a setting that is increasingly unfair), activity in the two structures becomes more synchronized.

The Frontal Cortex and Its Relationship with the Limbic System

We now have a sense of what different subdivisions of the PFC do and how cognition and emotion interact neurobiologically. This leads us to consider how the frontal cortex and limbic system interact.

In landmark studies Joshua Greene of Harvard and Princeton's Cohen showed how the "emotional" and "cognitive" parts of the brain can somewhat dissociate.[66] They used philosophy's famous "runaway trolley" problem, where a trolley is bearing down on five people and you must decide if it's okay to kill one person to save the five. Framing of the problem is key. In one version you pull a lever, diverting the trolley onto a side track. This saves the five, but the trolley kills someone who happened to be on this other track; 70 to 90 percent of people say they would do this. In the second scenario you push the person in front of the trolley with your own hands. This stops the trolley, but the person is killed; 70 to 90 percent say no way. The same numerical trade-off, but utterly different decisions.

Greene and Cohen gave subjects the two versions while neuroimaging them. Contemplating intentionally killing someone with your own hands activates the decider dlPFC, along with emotion-related regions that respond to aversive stimuli (including a cortical region activated by emotionally laden words), the amygdala, and the vmPFC. The more amygdaloid activation and the more negative emotions the participant reported in deciding, the less likely they were to push.

And when people contemplate detachedly pulling a lever that inadvertently kills someone? The dlPFC alone activates. As purely cerebral a decision as choosing which wrench to use to fix a widget. A great study.*

Other studies have examined interactions between "cognitive" and "emotional" parts of the brain. A few examples:

> Chapter 3 discusses some unsettling research—stick your average person in a brain scanner, and show him a picture of someone of another race for only a tenth of a second. This is too fast for him to be aware of what he saw. But thanks to that anatomical shortcut, the amygdala knows . . . and activates. In contrast, show the picture for a longer time. Again the amygdala activates, but then the cognitive dlPFC does as well, inhibiting the amygdala—the effort to control what is for most people an unpalatable initial response.
>
> Chapter 6 discusses experiments where a subject plays a game with two other people and is manipulated into feeling that she is being left out. This activates her amygdala, periaqueductal gray (that ancient brain region that helps process physical pain), anterior cingulate, and insula, an anatomical picture of anger, anxiety, pain, disgust, sadness. Soon afterward her PFC activates as rationalizations kick in—"This is just a stupid game; I have friends; my dog loves me." And the amygdala et al. quiet down. And what if you do the same to someone whose frontal cortex is not fully functional? The amygdala is increasingly activated; the person feels increasingly distressed. What neurological disease is involved? None. This is a typical teenager.
>
> Finally, the PFC mediates fear extinction. Yesterday the rat learned, "That tone is followed by a shock," so the sound of the tone began to

* We will return to Greene's subsequent "trolleyology" work at length in the chapter on morality. Broadly, it shows that the differing decisions pivot around (a) the personal/impersonal contrast between pulling a lever and pushing with your own hands, (b) the means/side effect contrast between the person's death being a necessity and its being an unintentional by-product, (c) the psychological distance to the potential victim.

trigger freezing. Today there are no shocks, and the rat has acquired another truth that takes precedence—"but not today." The first truth is still there; as proof, start coupling tone with shock again, and freezing to tone is "reinstated" faster than the association was initially learned.

Where is "but not today" consolidated? In the PFC, after receiving information from the hippocampus.[67] The medial PFC activates inhibitory circuits in the BLA, and the rat stops freezing to the tone. In a similar vein but reflecting cognition specific to humans, condition people to associate a blue square on a screen with a shock, and the amygdala will activate when seeing that square—but less so in subjects who reappraise the situation, activating the medial PFC by thinking of, say, a beautiful blue sky.

This segues into the subject of regulating emotion through thought.[68] It's hard to regulate thought (try not thinking about a hippo) but even tougher with emotion; research by my Stanford colleague and close friend James Gross has explored this. First off, "thinking differently" about something emotional differs from simply suppressing the expression of the emotions. For example, show someone graphic footage of, say, an amputation. Subjects cringe, activate the amygdala and sympathetic nervous system. Now one group is instructed to hide their emotions ("I'm going to show you another film clip, and I want you to hide your emotional reactions"). How to do so most effectively? Gross distinguishes between "antecedent" and "response"-focused strategies. Response-focused is dragging the emotional horse back to the barn after it's fled—you're watching the next horrific footage, feeling queasy, and you think, "Okay, sit still, breathe slowly." Typically this causes even greater activation of the amygdala and sympathetic nervous system.

Antecedent strategies generally work better, as they keep the barn door closed from the start. These are about thinking/feeling about something else (e.g., that great vacation), or thinking/feeling differently about what you're seeing (reappraisals such as "That isn't real; those are just actors"). And when done right, the PFC, particularly the dlPFC, activates, the amygdala and sympathetic nervous system are damped, and subjective distress decreases.*

Antecedent reappraisal is why placebos work.[69] Thinking, "My finger is about to be pricked by a pin," activates the amygdala along with a circuit of pain-responsive brain regions, and the pin hurts. Be told beforehand that the hand cream being slathered on your finger is a powerful analgesic cream, and you think, "My finger is about to be pricked by a pin, but this cream will block the pain." The PFC activates, blunting activity in the amygdala and pain circuitry, as well as pain perception.

Thought processes like these, writ large, are the core of a particularly effective type of psychotherapy—cognitive behavioral therapy (CBT)—for the treatment of disorders of emotion regulation.[70] Consider someone with a social anxiety disorder caused by a horrible early experience with trauma. To simplify, CBT is about providing the tools to reappraise circumstances that evoke the anxiety—remember that in this social situation those awful feelings you're having are about what happened back then, not what is happening now.†

Controlling emotional responses with thought like this is very top down; the frontal cortex calms the overwrought amygdala. But the PFC/limbic relationship can be bottom up as well, when a decision involves a gut feeling. This is the backbone of Damasio's somatic marker hypothesis. Choosing among options can involve a cerebral cost-benefit analysis. But it also involves "somatic markers," internal simulations of what each outcome would feel like, run in the limbic system and reported to the vmPFC. The process is not a thought experiment; it's an emotion experiment, in effect an emotional memory of a possible future.

A mild somatic marker activates only the limbic system.[71] "Should I do behavior A? Maybe not—the possibility of outcome B feels scary." A more vivid somatic marker activates the sympathetic nervous system as

* Given the circuitry of the PFC, the most probable sequence is activation of the dlPFC, then activation of the vmPFC, then inhibition of the amygdala.

† And this extends to a metalevel of reappraisal, as Gross has shown that one mediator of treatment outcome when using CBT for social anxiety is the *belief* that one can effectively reappraise.

well. "Should I do behavior A? Definitely not—I can feel my skin getting clammy at the possibility of outcome B." Experimentally boosting the strength of that sympathetic signal strengthens the aversion.

This is a picture of normal collaboration between the limbic system and frontal cortex.[72] Naturally, things are not always balanced. Anger, for example, makes people less analytical and more reflexive in decisions about punishment. Stressed people often make hideously bad decisions, marinated in emotion; chapter 4 examines what stress does to the amygdala and frontal cortex.*

The effects of stress on the frontal cortex are dissected by the late Harvard psychologist Daniel Wegner in an aptly titled paper, "How to Think, Say or Do Precisely the Worst Thing on Any Occasion."[73] He considers what Edgar Allan Poe called the "imp of the perverse":

> We see a rut coming up in the road ahead and proceed to steer our bike right into it. We make a mental note not to mention a sore point in conversation and then cringe in horror as we blurt out exactly that thing. We carefully cradle the glass of red wine as we cross the room, all the while thinking "don't spill," and then juggle it onto the carpet under the gaze of our host.

Wegner demonstrated a two-step process of frontocortical regulation: (A) one stream identifies X as being *very* important; (B) the other stream tracks whether the conclusion is "*Do* X" or "*Never* do X." And during stress, distraction, or heavy cognitive load, the two streams can dissociate; the A stream exerts its presence without the B stream saying which fork in the road to take. The chance that you will do precisely the wrong thing rises not despite your best efforts but because of a stress-boggled version of them.

* And then there are circumstances where the limbic system overwhelms the frontal cortex, where there is no such thing as a good decision, where each choice is worse than the other. Think about what is, for a parent, probably the most excruciating scene in all of cinema—in *Sophie's Choice*, when Sophie must make the Choice, when, without warning, she has seconds to choose which of her two children lives, which dies. Making her bludgeoned, unimaginable choice requires her frontocortical neurons to send signals to her prefrontal cortex and on to her motor cortex—after all, she eventually says words and moves her hands, pushing one child forward. And the bidirectionality of the circuitry is shown by the fact that her limbic system, no doubt, was screaming in agony to the frontal cortex.

This concludes our overview of the frontal cortex; the mantra is that it makes you do the harder thing when that is the right thing. Five final points:

- "Doing the harder thing" effectively is not an argument for valuing either emotion or cognition more than the other. For example, as discussed in chapter 11, we are our most prosocial concerning in-group morality when our rapid, implicit emotions and intuitions dominate, but are most prosocial concerning out-group morality when cognition holds sway.
- It's easy to conclude that the PFC is about preventing imprudent behaviors ("Don't do it; you'll regret it"). But that isn't always the case. For example, in chapter 17 we'll consider the surprising amount of frontocortical effort it can take to pull a trigger.
- Like everything about the brain, the structure and function of the frontal cortex vary enormously among individuals; for example, resting metabolic rate in the PFC varies approximately thirtyfold among people.* What causes such individual differences? See the rest of this book.[74]
- "Doing the harder thing when it's the right thing to do." "Right" in this case is used in a neurobiological and instrumental sense, rather than a moral one.
- Consider lying. Obviously, the frontal cortex aids the hard job of resisting the temptation. But it is also a major frontocortical task, particularly a dlPFC task, to lie competently, to control the emotional content of a signal, to generate an abstract distance between message and meaning. Interestingly, pathological liars have atypically large amounts of white matter in the PFC, indicating more complex wiring.[75]

* Consider individuals with "repressive" personalities. Such individuals have highly regimented affect and behavior—they're not emotionally expressive and aren't great at reading emotions in others. They like ordered, structured, predictable lives, can tell you what they're having for dinner a week from Thursday, and complete everything on time. And they have elevated metabolism in the frontal cortex and elevated circulating levels of stress hormones, showing that it can be enormously stressful to construct a world in which nothing stressful ever occurs.

But again, the "right thing," in the setting of the frontal cortically assisted lying, is amoral. An actor lies to an audience about having the feelings of a morose Danish prince. A situationally ethical child lies, telling Grandma how excited she is about her present, concealing the fact that she already has that toy. A leader tells bold-faced lies, starting a war. A financier with Ponzi in his blood defrauds investors. A peasant woman lies to a uniformed thug, telling him she does not know the whereabouts of the refugees she knows are hiding in her attic. As with much about the frontal cortex, it's context, context, context.

Where does the frontal cortex get the metaphorical motivation to do the harder thing? For this we now look at our final branch, the dopaminergic "reward" system in the brain.

THE MESOLIMBIC/MESOCORTICAL DOPAMINE SYSTEM

Reward, pleasure, and happiness are complex, and the motivated pursuit of them occurs in at least a rudimentary form in many species. The neurotransmitter dopamine is central to understanding this.

Nuclei, Inputs, and Outputs

Dopamine is synthesized in multiple brain regions. One such region helps initiate movement; damage there produces Parkinson's disease. Another regulates the release of a pituitary hormone. But the dopaminergic system that concerns us arises from an ancient, evolutionarily conserved region near the brain stem called the ventral tegmental area (henceforth the "tegmentum").

A key target of these dopaminergic neurons is the last multisyllabic brain region to be introduced in this chapter, the nucleus accumbens (henceforth the "accumbens"). There's debate as to whether the accumbens should count as part of the limbic system, but at the least it's highly limbic-ish.

Here's our first pass at the organization of this circuitry:[76]

a. The tegmentum sends projections to the accumbens and (other) limbic areas such as the amygdala and hippocampus. This is collectively called the "mesolimbic dopamine pathway."

b. The tegmentum also projects to the PFC (but, significantly, not other cortical areas). This is called the "mesocortical dopamine pathway." I'll be lumping the mesolimbic plus mesocortical pathways together as the "dopaminergic system," ignoring their not always being activated simultaneously.*

c. The accumbens projects to regions associated with movement.

d. Naturally, most areas getting projections from the tegmentum and/or accumbens project back to them. Most interesting will be the projections from the amygdala and PFC.

Reward

As a first pass, the dopaminergic system is about reward—various pleasurable stimuli activate tegmental neurons, triggering their release of dopamine.[77] Some supporting evidence: (a) drugs like cocaine, heroin, and alcohol release dopamine in the accumbens; (b) if tegmental release of dopamine is blocked, previously rewarding stimuli become aversive; (c) chronic stress or pain depletes dopamine and decreases the sensitivity of dopamine neurons to stimulation, producing the defining symptom of depression—"anhedonia," the inability to feel pleasure.

Some rewards, such as sex, release dopamine in every species

* In humans activation of the dopaminergic system is typically assessed with functional imaging techniques like fMRI, which detect changes in metabolism in different parts of the brain. To be precise, while an increase in metabolic demands in these regions is typically due to neurons there having lots of (dopamine-releasing) action potentials, the two are not synonymous. Nonetheless, for simplicity I'm using "dopaminergic signaling increases," "dopamine pathways activate," "dopamine is released" interchangeably.

examined.[78] For humans, just thinking about sex suffices.*[79] Food evokes dopamine release in hungry individuals of all species, with an added twist in humans. Show a picture of a milkshake to someone after they've consumed one, and there's rarely dopaminergic activation—there's satiation. But with subjects who have been dieting, there's *further* activation. If you're working to restrict your food intake, a milkshake just makes you want another one.

The mesolimbic dopamine system also responds to pleasurable aesthetics.[80] In one study people listened to new music; the more accumbens activation, the more likely subjects were to buy the music afterward. And then there is dopaminergic activation for artificial cultural inventions—for example, when typical males look at pictures of sports cars.

Patterns of dopamine release are most interesting when concerning social interactions.[81] Some findings are downright heartwarming. In one study a subject would play an economic game with someone, where a player is rewarded under two circumstances: (a) if both players cooperate, each receives a moderate reward, and (b) stabbing the other person in the back gets the subject a big reward, while the other person gets nothing. While both outcomes increased dopaminergic activity, the bigger increase occurred after cooperation.†

Other research examined the economic behavior of punishing jerks.[82] In one study subjects played a game where player B could screw over player A for a profit. Depending on the round, player A could either (a) do nothing, (b) punish player B by having some of player B's money taken (at no cost to player A), or (c) pay one unit of money to have two units taken from player B. Punishment activated the dopamine system, especially when subjects had to pay to punish; the greater the dopamine increase during no-cost punishment, the more willing someone was to pay to punish. Punishing norm violations is satisfying.

* And, in a fact that hints at a world of sex differences, dopaminergic responses to sexually arousing visual stimuli are greater in men than in women. Remarkably, this difference isn't specific to humans. Male rhesus monkeys will forgo the chance to drink water when thirsty in order to see pictures of—I'm not quite sure how else to say this—crotch shots of female rhesus monkeys (while not being interested in other rhesus-y pictures).

† As an important point, study subjects were all female.

Another great study, carried out by Elizabeth Phelps of New York University, concerns "overbidding" in auctions, where people bid more money than anticipated.[83] This is interpreted as reflecting the additional reward of besting someone in the competitive aspect of bidding. Thus, "winning" an auction is intrinsically socially competitive, unlike "winning" a lottery. Winning a lottery and winning a bid both activated dopaminergic signaling in subjects; losing a lottery had no effect, while losing a bidding war inhibited dopamine release. Not winning the lottery is bad luck; not winning an auction is social subordination.

This raises the specter of envy. In one neuroimaging study subjects read about a hypothetical person's academic record, popularity, attractiveness, and wealth.[84] Descriptions that evoked self-reported envy activated cortical regions involved in pain perception. Then the hypothetical individual was described as experiencing a misfortune (e.g., they were demoted). More activation of pain pathways at the news of the person's good fortune predicted more dopaminergic activation after learning of their misfortune. Thus there's dopaminergic activation during schadenfreude—gloating over an envied person's fall from grace.

The dopamine system gives insights into jealousy, resentment, and invidiousness, leading to another depressing finding.[85] A monkey has learned that when he presses a lever ten times, he gets a raisin as a reward. That's just happened, and as a result, ten units of dopamine are released in the accumbens. Now—surprise!—the monkey presses the lever ten times and gets *two* raisins. Whoa: twenty units of dopamine are released. And as the monkey continues to get paychecks of two raisins, the size of the dopamine response returns to ten units. Now reward the monkey with only a single raisin, and dopamine levels *decline.*

Why? This is our world of habituation, where nothing is ever as good as that first time.

Unfortunately, things have to work this way because of our range of rewards.[86] After all, reward coding must accommodate the rewarding properties of both solving a math problem and having an orgasm. Dopaminergic responses to reward, rather than being absolute, are relative to the reward value of alternative outcomes. In order to accommodate the pleasures of both mathematics and orgasms, the system must constantly rescale to accommodate the range of intensity offered by particular stimuli. The response to any reward must habituate with repetition, so that the system can respond over its full range to the next new thing.

This was shown in a beautiful study by Wolfram Schultz of Cambridge University.[87] Depending on the circumstance, monkeys were trained to expect either two or twenty units of reward. If they unexpectedly got either four or forty units, respectively, there'd be an identical burst of dopamine release; giving one or ten units produced an identical decrease. It was the relative, not absolute, size of the surprise that mattered over a tenfold range of reward.

These studies show that the dopamine system is bidirectional.[88] It responds with scale-free increases for unexpected good news and decreases for bad. Schultz demonstrated that following a reward, the dopamine system codes for discrepancy from expectation—get what you expected, and there's a steady-state dribble of dopamine. Get more reward and/or get it sooner than expected, and there's a big burst; less and/or later, a decrease. Some tegmental neurons respond to positive discrepancy from expectation, others to negative; appropriately, the latter are local neurons that release the inhibitory neurotransmitter GABA. Those same neurons participate in habituation, where the reward that once elicited a big dopamine response becomes less exciting.*

Logically, these different types of coding neurons in the tegmentum (as well as the accumbens) get projections from the frontal cortex—that's where all the expectancy/discrepancy calculations take place—"Okay, I thought I was going to get 5.0 but got 4.9. How big of a bummer is that?"

Additional cortical regions weigh in. In one study subjects were shown an item to purchase, with the degree of accumbens activation predicting how much a person would pay.[89] Then they were told the price; if it was less than what they were willing to spend, there was activation of the

* Remarkably, in a gambling paradigm where both outcomes result in a shock, after a while, getting the lesser of the two shocks begins to activate dopamine signaling.

emotional vmPFC; more expensive, and there'd be activation of that disgust-related insular cortex. Combine all the neuroimaging data, and you could predict whether the person would buy the item.

Thus, in typical mammals the dopamine system codes in a scale-free manner over a wide range of experience for both good and bad surprises and is constantly habituating to yesterday's news. But humans have something in addition, namely that we invent pleasures far more intense than anything offered by the natural world.

Once, during a concert of cathedral organ music, as I sat getting goose-flesh amid that tsunami of sound, I was struck with a thought: for a medieval peasant, this must have been the loudest human-made sound they ever experienced, awe-inspiring in now-unimaginable ways. No wonder they signed up for the religion being proffered. And now we are constantly pummeled with sounds that dwarf quaint organs. Once, hunter-gatherers might chance upon honey from a beehive and thus briefly satisfy a hardwired food craving. And now we have hundreds of carefully designed commercial foods that supply a burst of sensation unmatched by some lowly natural food. Once, we had lives that, amid considerable privation, also offered numerous subtle, hard-won pleasures. And now we have drugs that cause spasms of pleasure and dopamine release a thousandfold higher than anything stimulated in our old drug-free world.

An emptiness comes from this combination of over-the-top nonnatural sources of reward and the inevitability of habituation; this is because unnaturally strong explosions of synthetic experience and sensation and pleasure evoke unnaturally strong degrees of habituation.[90] This has two consequences. First, soon we barely notice the fleeting whispers of pleasure caused by leaves in autumn, or by the lingering glance of the right person, or by the promise of reward following a difficult, worthy task. And the other consequence is that we eventually habituate to even those artificial deluges of intensity. If we were designed by engineers, as we consumed more, we'd desire less. But our frequent human tragedy is that the more we consume, the hungrier we get. More and faster and stronger.

What was an unexpected pleasure yesterday is what we feel entitled to today, and what won't be enough tomorrow.

The Anticipation of Reward

Thus, dopamine is about invidious, rapidly habituating reward. But dopamine is more interesting than that. Back to our well-trained monkey working for a reward. A light comes on in his room, signaling the start of a reward trial. He goes over to the lever, presses ten times, and gets the raisin reward; this has happened often enough that there's only a small increase in dopamine with each raisin.

However, importantly, lots of dopamine is released when the light first comes on, signaling the start of the reward trial, before the monkey starts lever pressing.

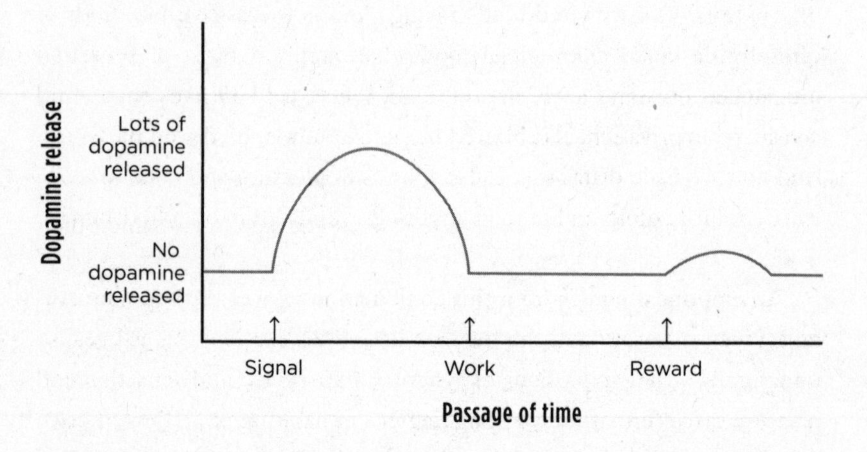

In other words, once reward contingencies are learned, dopamine is less about reward than about its anticipation. Similarly, work by my Stanford colleague Brian Knutson has shown dopamine pathway activation in people in anticipation of a monetary reward.[91] Dopamine is about mastery and expectation and confidence. It's "I know how things work; this is going to be great." In other words, the pleasure is in the anticipation of reward, and the reward itself is nearly an afterthought (unless, of course, the reward fails to arrive, in which case it's the most important thing in the world). If you know your appetite will be sated, pleasure is more about the appetite than about the sating.* This is hugely important.

Anticipation requires learning.[92] Learn Warren G. Harding's middle name, and synapses in the hippocampus become more excitable. Learn that when the light comes on it's reward time, and it's hippocampal amygdaloid and frontal cortical neurons projecting to dopamine neurons that become more excitable.

This explains context-dependent craving in addiction.[93] Suppose an alcoholic has been clean and sober for years. Return him to where the alcohol consumption used to occur (e.g., that rundown street corner, that fancy men's club), and those potentiated synapses, those cues that were learned to be associated with alcohol, come roaring back into action, dopamine surges with anticipation, and the craving inundates.

Can a reliable cue of an impending reward eventually become rewarding itself? This has been shown by Huda Akil of the University of Michigan. A light in the left side of a rat's cage signals that lever pressing will produce a reward from a food chute on the right side. Remarkably, rats eventually will work for the chance to hang around on the left side of the cage, just because it feels so nice to be there. The signal has gained the dopaminergic power of what is being signaled. Similarly, rats will work to be exposed to a cue that signals that *some kind* of reward is likely, without knowing what or when. This is what fetishes are, in both the anthropological and sexual sense.[94]

Schultz's group has shown that the magnitude of an anticipatory dopamine rise reflects two variables. First is the size of the anticipated reward. A monkey has learned that a light means that ten lever presses earns one unit of reward, while a tone means ten presses earns ten units. And soon a tone provokes more anticipatory dopamine than does a light. It's "This is going to be great" versus "This is going to be *great*."

The second variable is extraordinary. The rule is that the light comes

* The phenomenon reminds me of the terribly cynical observation of a dormmate in college, one with a long string of tumultuously disastrous relationships: "A relationship is the price you pay for the anticipation of it."

on, you press the lever, you get the reward. Now things change. Light comes on, press the lever, get the reward . . . only 50 percent of the time. Remarkably, once that new scenario is learned, far more dopamine is released. Why? Because nothing fuels dopamine release like the "maybe" of intermittent reinforcement.[95]

This additional dopamine is released at a distinctive time. The light comes on in the 50 percent scenario, producing the usual anticipatory dopamine rise before the lever pressing starts. Back in the predictable days when lever pressing always earned a reward, once the pressing was finished, dopamine levels remained low until the reward arrived, followed by a little dopamine blip. But in this 50 percent scenario, once the pressing is finished, dopamine levels start rising, driven by the uncertainty of "maybe yes, maybe no."

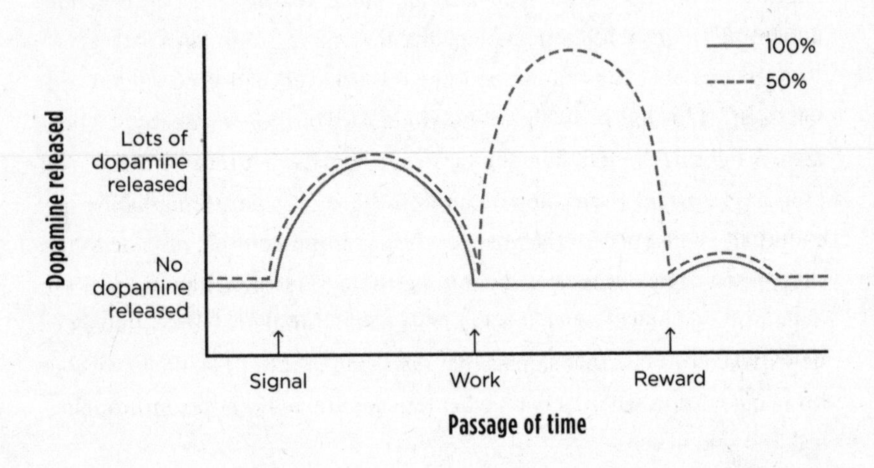

Modify things further; reward now occurs 25 or 75 percent of the time. A shift from 50 to 25 percent and a shift from 50 to 75 percent are exactly opposite, in terms of the likelihood of reward, and work from Knutson's group shows that the greater the probability of reward, the more activation in the medial PFC.[96] But switches from 50 to 25 percent and from 50 to 75 percent both reduce the magnitude of uncertainty. And the secondary rise of dopamine for a 25 or 75 percent likelihood of reward is smaller than for 50 percent. Thus, anticipatory dopamine release peaks with the greatest uncertainty as to whether a reward will occur.* Interestingly, in circumstances of uncertainty, enhanced anticipatory dopamine release is mostly in the mesocortical rather than mesolimbic pathway, implying that uncertainty is a more cognitively complex state than is anticipation of predictable reward.

None of this is news to the honorary psychologists running Las Vegas. Logically, gambling shouldn't evoke much anticipatory dopamine, given the astronomical odds against winning. But the behavioral engineering—the 24-7 activity and lack of time cues, the cheap alcohol pickling frontocortical judgment, the manipulations to make you feel like today is your lucky day—distorts and shifts the perception of the odds into a range where dopamine pours out and, oh, why not, let's try again.

The interaction between "maybe" and the propensity for addictive gambling is seen in a study of "near misses"—when two out of three reels line up in a slot machine. In control subjects there was minimal dopaminergic activation after misses of any sort; among pathological gamblers, a near miss activated the dopamine system like crazy. Another study concerned two betting situations with identical probabilities of reward but different levels of information about reward contingencies. The circumstance with less information (i.e., that was more about ambiguity than risk) activated the amygdala and silenced dopaminergic signaling; what is perceived to be well-calibrated risk is addictive, while ambiguity is just agitating.[97]

Pursuit

So dopamine is more about anticipation of reward than about reward itself. Time for one more piece of the picture. Consider that monkey trained to respond to the light cue with lever pressing, and out comes the reward; as we now know, once that relationship is established, most dopamine release is anticipatory, occurring right after the cue.

* This fact prompted Greene, in a conversation with me, to dryly note how Harvard's budget projections incorporate the expectation that if they work hard enough, approximately 50 percent of junior faculty will receive tenure.

What happens if the post–light cue release of dopamine doesn't occur?[98] Crucially, the monkey doesn't press the lever. Similarly, if you destroy its accumbens, a rat makes impulsive choices, instead of holding out for a delayed larger reward. Conversely, back to the monkey—if instead of flashing the light cue you electrically stimulate the tegmentum to release dopamine, the monkey presses the lever. Dopamine is not just about reward anticipation; it fuels the *goal-directed behavior* needed to gain that reward; dopamine "binds" the value of a reward to the resulting work. It's about the motivation arising from those dopaminergic projections to the PFC that is needed to do the harder thing (i.e., to work).

In other words, dopamine is not about the happiness of reward. It's about the happiness of pursuit of reward that has a decent chance of occurring.*[99]

This is central to understanding the nature of motivation, as well as its failures (e.g., during depression, where there is inhibition of dopamine signaling thanks to stress, or in anxiety, where such inhibition is caused by projections from the amygdala).[100] It also tells us about the source of the frontocortical power behind willpower. In a task where one chooses between an immediate and a (larger) delayed reward, contemplating the immediate reward activates limbic targets of dopamine (i.e., the mesolimbic pathway), whereas contemplating the delayed reward activates frontocortical targets (i.e., the mesocortical pathway). The greater the activation of the latter, the more likely there'll be gratification postponement.

These studies involved scenarios of a short burst of work soon followed by reward.[101] What about when the work required is prolonged, and reward is substantially delayed? In that scenario there is a secondary rise of dopamine, a gradual increase that fuels the sustained work; the extent of the dopamine ramp-up is a function of the length of the delay and the anticipated size of the reward:

* And as a great example of the happiness of pursuit, where the rewarding quality of something is as much in the process as in the end product, the mesolimbic dopamine system plays a key role in motivating maternal care in female rats.

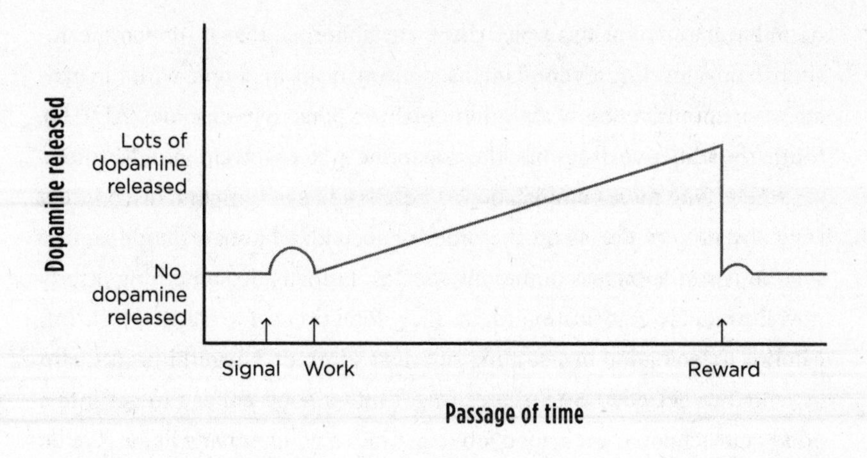

This reveals how dopamine fuels delayed gratification. If waiting X amount of time for a reward has value Z; waiting 2X should logically have value ½Z; instead we "temporally discount"—the value is smaller, e.g., ¼Z. We don't like waiting.

Dopamine and the frontal cortex are in the thick of this phenomenon. Discounting curves—a value of ¼Z instead of ½Z—are coded in the accumbens, while dlPFC and vmPFC neurons code for time delay.[102]

This generates some complex interactions. For example, activate the vmPFC or inactivate the dlPFC, and short-term reward becomes more alluring. And a cool neuroimaging study of Knutson's gives insight into impatient people with steep temporal discounting curves; their accumbens, in effect, underestimates the magnitude of the delayed reward, and their dlPFC overestimates the length of the delay.[103]

Collectively these studies show that our dopaminergic system, frontal cortex, amygdala, insula, and other members of the chorus code for differing aspects of reward magnitude, delay, and probability with varying degrees of accuracy, all influencing whether we manage to do the harder, more correct thing.[104]

Individual differences among people in the capacity for gratification postponement arise from variation in the volume of these individual neural voices.[105] For example, there are abnormalities in dopamine response profiles during temporal discounting tasks in people with the maladaptive impulsiveness of attention-deficit/hyperactivity disorder (ADHD). Similarly, addictive drugs bias the dopamine system toward impulsiveness.

Phew. One more complication: These studies of temporal discounting typically involve delays on the order of seconds. Though the dopamine system is similar across numerous species, humans do something utterly novel: we delay gratification for insanely long times. No warthog restricts calories to look good in a bathing suit next summer. No gerbil works hard at school to get good SAT scores to get into a good college to get into a good grad school to get a good job to get into a good nursing home. We do something even beyond this unprecedented gratification delay: we use the dopaminergic power of the happiness of pursuit to motivate us to work for rewards that come *after we are dead*—depending on your culture, this can be knowing that your nation is closer to winning a war because you've sacrificed yourself in battle, that your kids will inherit money because of your financial sacrifices, or that you will spend eternity in paradise. It is extraordinary neural circuitry that bucks temporal discounting enough to allow (some of) us to care about the temperature of the planet that our great-grandchildren will inherit. Basically, it's unknown how we humans do this. We may merely be a type of animal, mammal, primate, and ape, but we're a profoundly unique one.

A Final Small Topic: Serotonin

This lengthy section has concerned dopamine, but an additional neurotransmitter, serotonin, plays a clear role in some behaviors that concern us.

Starting with a 1979 study, low levels of serotonin in the brain were shown to be associated with elevated levels of human aggression, with end points ranging from psychological measures of hostility to overt violence.[106] A similar serotonin/aggression relationship was observed in other mammals and, remarkably, even crickets, mollusks, and crustaceans.

As work continued, an important qualifier emerged. Low serotonin didn't predict premeditated, instrumental violence. It predicted *impulsive* aggression, as well as cognitive impulsivity (e.g., steep temporal discounting or trouble inhibiting a habitual response). Other studies linked low serotonin to impulsive suicide (independent of severity of the associated psychiatric illness).[107]

Moreover, in both animals and humans pharmacologically decreasing serotonin signaling increases behavioral and cognitive impulsivity (e.g., impulsively torpedoing a stable, cooperative relationship with a player in an economic game).[108] Importantly, while increasing serotonin signaling did not lessen impulsiveness in normal subjects, it did in subjects prone toward impulsivity, such as adolescents with conduct disorder.

How does serotonin do this? Nearly all serotonin is synthesized in one brain region,* which projects to the usual suspects—the tegmentum, accumbens, PFC, and amygdala, where serotonin enhances dopamine's effects on goal-directed behavior.[109]

This is as dependable a finding as you get in this business.[110] Until we get to chapter 8 and look at genes related to serotonin, at which point everything becomes a completely contradictory mess. Just as a hint of what's to come, one gene variant has even been referred to, straight faced, by some scientists as the "warrior gene," and its presence has been used successfully in some courtrooms to lessen sentences for impulsive murders.

CONCLUSIONS

This completes our introduction to the nervous system and its role in pro- and antisocial behaviors. It was organized around three themes: the hub of fear, aggression, and arousal centered in the amygdala; the hub of reward, anticipation, and motivation of the dopaminergic system; and the hub of frontal cortical regulation and restraint of behavior. Additional brain regions and neurotransmitters will be introduced in subsequent chapters. Amid this mountain of information, be assured that the key

* Its name—raphe nucleus—is not essential.

brain regions, circuits, and neurotransmitters will become familiar as the book progresses.

Hang on. So what does this all mean? It's useful to start with three things that this information doesn't mean:

1. First, there's the lure of needing neurobiology to confirm the obvious. Someone claims that, for example, their crappy, violent neighborhood leaves them so anxious that they can't function effectively. Toss them in a brain scanner and flash pictures of various neighborhoods; when their own appears, the amygdala explodes into activity. "Ah," it is tempting to conclude, "we've now *proven* that the person really does feel frightened."

   It shouldn't require neuroscience to validate someone's internal state. An example of this fallacy was reports of atrophy of the hippocampus in combat vets suffering from PTSD; this was in accord with basic research (including from my lab) showing that stress can damage the hippocampus. The hippocampal atrophy in PTSD got a lot of play in Washington, helping to convince skeptics that PTSD is an organic disorder rather than neurotic malingering. It struck me that if it took brain scans to convince legislators that there's something tragically, organically damaged in combat vets with PTSD, then these legislators have some neurological problems of their own. Yet it required precisely this to "prove" to many that PTSD was an organic brain disorder.

   The notion that "if a neuroscientist can demonstrate it, we know that the person's problem is for real" has a corollary—the fancier the neurobiology utilized, the more reliable the verification. That's simply not true; for example, a good neuropsychologist can discern more of what's happening to someone with subtle but pervasive memory problems than can a gazillion-dollar brain scanner.

   It shouldn't take neuroscience to "prove" what we think and feel.

2. There's been a proliferation of "neuro-" fields. Some, like neuroendocrinology and neuroimmunology, are stodgy old institutions by now. Others are relatively new—neuroeconomics, neuromarketing, neuroethics, and, I kid you not, neuroliterature and neuroexistentialism. In other words, a hegemonic neuroscientist might conclude that their field explains everything. And with that comes the danger, raised by the *New Yorker* writer Adam Gopnik under the sardonic banner of "neuroskepticism," that explaining everything leads to forgiving everything.[111] This premise is at the heart of debates in the new field of "neurolaw." In chapter 16 I will argue that it is wrong to think that understanding must lead to forgiveness—mainly because I think that a term like "forgiveness," and others related to criminal justice (e.g., "evil," "soul," "volition," and "blame"), are incompatible with science and should be discarded.
3. Finally, there is the danger of thinking that neuroscience supports a tacit sort of dualism. A guy does something impulsive and awful, and neuroimaging reveals that, unexpectedly, he's missing all his PFC neurons. There's a dualist temptation now to view his behavior as more "biological" or "organic" in some nebulous manner than if he had committed the same act with a normal PFC. However, the guy's awful, impulsive act is equally "biological" with or without a PFC. The sole difference is that the workings of the PFC-less brain are easier to understand with our primitive research tools.

So What Does All of This Tell Us?

Sometimes these studies tell us what different brain regions do. They are getting fancier, telling us about circuits, thanks to the growing time resolution of neuroimaging, transitioning from "This stimulus activates brain regions A, B, C" to "This stimulus activates both A and B, and then C, and C activates only if B does." And identifying what specific regions/circuits do gets harder as studies become subtler. Consider, for example, the fusiform face area. As discussed in the next chapter, it is a cortical region that responds to faces in humans and other primates. We primates sure are social creatures.

But work by Isabel Gauthier of Vanderbilt University demonstrates something more complicated. Show pictures of different cars, and the fusiform activates—in automobile aficionados.[112] Show pictures of birds, and ditto among bird-watchers. The fusiform isn't about faces; it's about recognizing examples of things from categories that are emotionally salient to each individual.

Thus, studying behavior is useful for understanding the nature of the brain—ah, isn't it interesting that behavior A arises from the coupling of brain regions X and Y. And sometimes studying the brain is useful for understanding the nature of behavior—ah, isn't it interesting that brain region A is central to both behavior X and behavior Y. For example, to me the most interesting thing about the amygdala is its dual involvement in both aggression and fear; you can't understand the former without recognizing the relevance of the latter.

A final point related to the core of this book: While this neurobiology is mighty impressive, the brain is not where a behavior "begins." It's merely the final common pathway by which all the factors in the chapters to come converge and create behavior.

Three

Seconds to Minutes Before

Nothing comes from nothing. No brain is an island.

Thanks to messages bouncing around your brain, a command has been sent to your muscles to pull that trigger or touch that arm. Odds are that a short time earlier, something outside your brain prompted this to happen, raising this chapter's key questions: (a) What outside stimulus, acting through what sensory channel and targeting which parts of the brain, prompted this? (b) Were you aware of that environmental stimulus? (c) What stimuli had your brain made you particularly sensitive to? And, of course, (d) what does this tell us about our best and worst behaviors?

Varied sensory information can prompt the brain into action. This can be appreciated by considering this variety in other species. Often we're clueless about this because animals can sense things in ranges that we can't, or with sensory modalities we didn't know exist. Thus, you must think like the animal to learn what is happening. We'll begin by seeing how this pertains to the field of ethology, the science of interviewing an animal in its own language.

UNIVERSAL RULES VERSUS KNOBBY KNEES

Ethology formed in Europe in the early twentieth century in response to an American brand of psychology, "behaviorism." Behaviorism descended from the introduction's John Watson; the field's famed champion was B. F. Skinner. Behaviorists cared about universalities of behavior across species. They worshipped a doozy of a seeming universal concerning stimulus and response: rewarding an organism for a behavior makes the organism more likely to repeat that behavior, while failure to get rewarded or, worse, punishment for it, makes the organism less likely to repeat it. Any behavior can be made more or less common through "operant conditioning" (a term Skinner coined), the process of controlling the rewards and punishments in the organism's environment.

Thus, for behaviorists (or "Skinnerians," a term Skinner labored to make synonymous) virtually any behavior could be "shaped" into greater or lesser frequency or even "extinguished" entirely.

If all behaving organisms obeyed these universal rules, you might as well study a convenient species. Most behaviorist research was done on rats or, Skinner's favorite, pigeons. Behaviorists loved data, no-nonsense hard numbers; these were generated by animals pressing or pecking away at levers in "operant conditioning boxes" (aka "Skinner boxes"). And anything discovered applied to any species. A pigeon is a rat is a boy, Skinner preached. Soulless droid.*

Behaviorists were often right about behavior but wrong in really important ways, as many interesting behaviors don't follow behaviorist rules.†[1] Raise an infant rat or monkey with an abusive mother, and it becomes more attached to her. And behaviorist rules have failed when humans love the wrong abusive person.

---

* An urban legend has persisted forever that Skinner raised his daughter in a giant Skinner box, where she learned to lever press away for all her needs. Naturally, according to the legend, when she grew up she went mad, committed suicide, sued him, tried to murder him, etc. All untrue.

† When I was in college, Skinner came to my dorm once for dinner and gave an extraordinarily dogmatic talk afterward. This produced an odd thought on my part as I listened to him. "Wow, this guy is, like, a *total* Skinnerian."

Meanwhile, ethology was emerging in Europe. In contrast with behaviorism's obsession with uniformity and universality of behavior, ethologists loved behavioral variety. They'd emphasize how every species evolves unique behaviors in response to unique demands, and how one had to open-mindedly observe animals in their natural habitats to understand them ("Studying rat social behavior in a cage is like studying dolphin swimming behavior in a bathtub" is an ethology adage). They'd ask, What, objectively, is the behavior? What triggered it? Did it have to be learned? How did it evolve? What is the behavior's adaptive value? Nineteenth-century parsons went into nature to collect butterflies, revel in the variety of wing colors, and marvel at what God had wrought. Twentieth-century ethologists went into nature to collect behavior, revel in its variety, and marvel at what evolution had wrought. In contrast to lab coat–clad behaviorists, ethologists tromped around fields in hiking shoes and had fetching knobby knees.*

Sensory Triggers of Behavior in Some Other Species

Using an ethological framework, we now consider sensory triggers of behavior in animals.†[2] First there's the auditory channel. Animals vocalize to intimidate, proclaim, and seduce. Birds sing, stags roar, howler monkeys howl, orangutans give territorial calls audible for miles. As a subtle example of information being communicated, when female pandas

* It's obvious which tribe I root for, being some manner of ethologist myself (but just to take this paean down a notch, remember that one of ethology's founders was the odious Konrad Lorenz). In an inspired move, the three founders of ethology—Lorenz, Niko Tinbergen, and Karl von Frisch—were awarded the Nobel Prize in Physiology or Medicine in 1973. The biomedical community was appalled. Giving the prize to guys with foot fungi, whose main research technique was looking through binoculars—what's that got to do with medicine? Of the trio, Lorenz was the energetic self-promoter and flashy popularizer, Tinbergen, one of my heroes, was the deep thinker and amazing experimentalist, and von Frisch played electric bass and didn't say much.

† How would ethologists figure out what sensory information is pertinent to an animal? An example: Among gulls, the mother's beak has a conspicuous red spot. When she brings food to the chicks, they peck at the beak, and Mom regurgitates the food. Here's how Tinbergen proved that the red spot triggers the pecking behavior: A subtraction approach, where he'd paint over the red spot on birds; chicks no longer pecked. A replication approach, where he'd take a two-by-four, paint a red dot on it, and wave it over the nest; chicks would start pecking. Or super stimulation, where he'd paint a *gigantic* red dot on the mother's beak; chicks would go berserk with the pecking. This approach now incorporates robotics, where ethologists have built, for example, robotic bees that infiltrate and fiendishly deceive bee colonies by dancing about nonexistent food sources, which the bees then fly off in search of.

ovulate, their vocalizations get higher, something preferred by males. Remarkably, the same shift and preference happens in humans.

There are also visual triggers of behavior. Dogs crouch to invite play, birds strut their plumage, monkeys display their canines menacingly with "threat yawns." And there are visual cues of cute baby–ness (big eyes, shortened muzzle, round forehead) that drive mammals crazy, motivating them to care for the kid. Stephen Jay Gould noted that the unsung ethologist Walt Disney understood exactly what alterations transformed rodents into Mickey and Minnie.*[3]

Then there are animals signaling in ways we can't detect, requiring creativity to interview an animal in its own language.[4] Scads of mammals scent mark with pheromones—odors that carry information about sex, age, reproductive status, health, and genetic makeup. Some snakes see in infrared, electric eels court with electric songs, bats compete by jamming one another's feeding echolocation signals, and spiders identify intruders by vibration patterns on their webs. How about this: tickle a rat and it chirps ultrasonically as its mesolimbic dopamine system is activated.

Back to the rhinencephalon/limbic system war and the resolution ethologists already knew: for a rodent, emotion is typically triggered by olfaction. Across species the dominant sensory modality—vision, sounds, whichever—has the most direct access to the limbic system.

Under the Radar: Subliminal and Unconscious Cuing

It's easy to see how the sight of a knife, the sound of a voice calling your name, a touch on your hand can rapidly alter your brain.[5] But crucially, tons of subliminal sensory triggers occur—so fleeting or minimal that we don't consciously note them, or of a type that, even if noted, seems irrelevant to a subsequent behavior.

Subliminal cuing and unconscious priming influence numerous behaviors unrelated to this book. People think potato chips taste better

* A great example of interspecies cute responses: A significant factor in how much money people pledge to donate to help an endangered species is the relative size of the animal's eyes. Big googly eyes loosen purse strings.

when hearing crunching sounds. We like a neutral stimulus more if, just before seeing it, a picture of a smiling face is flashed for a twentieth of a second. The more expensive a supposed (placebo) painkiller, the more effective people report the placebo to be. Ask subjects their favorite detergent; if they've just read a paragraph containing the word "ocean," they're more likely to choose Tide—and then explain its cleaning virtues.[6]

Thus, over the course of seconds sensory cues can shape your behavior unconsciously.

A hugely unsettling sensory cue concerns race.[7] Our brains are incredibly attuned to skin color. Flash a face for less than a tenth of a second (one hundred milliseconds), so short a time that people aren't even sure they've seen something. Have them guess the race of the pictured face, and there's a better-than-even chance of accuracy. We may claim to judge someone by the content of their character rather than by the color of their skin. But our brains sure as hell *note* the color, real fast.

By one hundred milliseconds, brain function already differs in two depressing ways, depending on the race of the face (as shown with neuroimaging). First, in a widely replicated finding, the amygdala activates. Moreover, the more racist someone is in an implicit test of race bias (stay tuned), the more activation there is.[8]

Similarly, repeatedly show subjects a picture of a face accompanied by a shock; soon, seeing the face alone activates the amygdala.[9] As shown by Elizabeth Phelps of NYU, such "fear conditioning" occurs faster for other-race than same-race faces. Amygdalae are prepared to learn to associate something bad with Them. Moreover, people judge neutral other-race faces as angrier than neutral same-race faces.

So if whites see a black face shown at a subliminal speed, the amygdala activates.[10] But if the face is shown long enough for conscious processing, the anterior cingulate and the "cognitive" dlPFC then activate and inhibit the amygdala. It's the frontal cortex exerting executive control over the deeper, darker amygdaloid response.

Second depressing finding: subliminal signaling of race also affects the fusiform face area, the cortical region that specializes in facial recognition.[11] Damaging the fusiform, for example, selectively produces "face blindness" (aka prosopagnosia), an inability to recognize faces. Work by John Gabrieli at MIT demonstrates less fusiform activation for other-race faces, with the effect strongest in the most implicitly racist subjects. This isn't about novelty—show a face with purple skin and the fusiform responds as if it's same-race. The fusiform isn't fooled—"That's not an Other; it's just a 'normal' Photoshopped face."

In accord with that, white Americans remember white better than black faces; moreover, mixed-race faces are remembered better if described as being of a white rather than a black person. Remarkably, if mixed-race subjects are told they've been assigned to one of the two races for the study, they show less fusiform response to faces of the arbitrarily designated "other" race.[12]

Our attunement to race is shown in another way, too.[13] Show a video of someone's hand being poked with a needle, and subjects have an "isomorphic sensorimotor" response—hands tense in empathy. Among both whites and blacks, the response is blunted for other-race hands; the more the implicit racism, the more blunting. Similarly, among subjects of both races, there's more activation of the (emotional) medial PFC when considering misfortune befalling a member of their own race than of another race.

This has major implications. In work by Joshua Correll at the University of Colorado, subjects were rapidly shown pictures of people holding either a gun or a cell phone and were told to shoot (only) gun toters. This is painfully reminiscent of the 1999 killing of Amadou Diallo. Diallo, a West African immigrant in New York, matched a description of a rapist. Four white officers questioned him, and when the unarmed Diallo started to pull out his wallet, they decided it was a gun and fired forty-one shots. The underlying neurobiology concerns "event-related potentials" (ERPs), which are stimulus-induced changes in electrical activity of the brain (as assessed by EEG—electroencephalography). Threatening faces produce a distinctive change (called the P200 component) in the ERP waveform in under two hundred milliseconds. Among white subjects, viewing someone black evokes a stronger P200 waveform than viewing someone white, regardless of whether the person is armed. Then, a few milliseconds later, a second, inhibitory waveform (the N200 component) appears, originating from the frontal cortex—"Let's think a sec about what we're seeing before we shoot." Viewing a black individual evokes less of an N200 waveform than does seeing someone white. The greater the P200/N200 ratio (i.e., the greater the ratio of I'm-feeling-threatened to Hold-on-a-sec), the greater the likelihood of shooting an unarmed black individual. In another study subjects had to identify fragmented pictures of objects. Priming white subjects with subliminal views of black (but not white) faces made them better at detecting pictures of weapons (but not cameras or books).[14]

Finally, for the same criminal conviction, the more stereotypically African a black individual's facial features, the longer the sentence.[15] In contrast, juries view black (but not white) male defendants more favorably if they're wearing big, clunky glasses; some defense attorneys even exploit this "nerd defense" by accessorizing their clients with fake glasses, and prosecuting attorneys ask whether those dorky glasses are real. In other words, when blind, impartial justice is supposedly being administered, jurors are unconsciously biased by racial stereotypes of someone's face.

This is so depressing—are we hardwired to fear the face of someone of another race, to process their face less as a face, to feel less empathy? No. For starters, there's tremendous individual variation—not everyone's amygdala activates in response to an other-race face, and those exceptions are informative. Moreover, subtle manipulations rapidly change the amygdaloid response to the face of an Other. This will be covered in chapter 11.

Recall the shortcut to the amygdala discussed in the previous chapter, when sensory information enters the brain. Most is funneled through that sensory way station in the thalamus and then to appropriate cortical region (e.g., the visual or auditory cortex) for the slow, arduous process of decoding light pixels, sound waves, and so on into something identifiable. And finally information about it ("It's Mozart") is passed to the limbic system.

As we saw, there's that shortcut from the thalamus directly to the amygdala, such that while the first few layers of, say, the visual cortex are futzing around with unpacking a complex image, the amygdala is already thinking, "That's a gun!" and reacting. And as we saw, there's the trade-off: information reaches the amygdala fast *but is often inaccurate.*[16] The amygdala thinks it knows what it's seeing before the frontal cortex slams on the brakes; an innocent man reaches for his wallet and dies.

Other types of subliminal visual information influence the brain.[17] For example, the gender of a face is processed within 150 milliseconds. Ditto with social status. Social dominance looks the same across cultures—direct gaze, open posture (e.g., leaning back with arms behind the head), while subordination is signaled with averted gaze, arms sheltering the torso. After a mere 40-millisecond exposure, subjects accurately distinguish high- from low-status presentations. As we'll see in chapter 12, when people are figuring out stable status relations, logical areas of the frontal cortex (the vmPFC and dlPFC) activate; but in the case of unstable, flip-flopping relations, the amygdala also activates. It's unsettling when we're unsure who gets ulcers and who gives them.

There's also subliminal cuing about beauty.[18] From an early age, in both sexes and across cultures, attractive people are judged to be smarter, kinder, and more honest. We're more likely to vote for attractive people or hire them, less likely to convict them of crimes, and, if they are convicted, more likely to dole out shorter sentences. Remarkably, the medial orbitofrontal cortex assesses both the beauty of a face and the goodness of a behavior, and its level of activity during one of those tasks predicts the level during the other. The brain does similar things when contemplating beautiful minds, hearts, and cheekbones. And assumes that cheekbones tell something about minds and hearts. This will be covered in chapter 12.

Though we derive subliminal information from bodily cues, such as posture, we get the most information from faces.[19] Why else evolve the fusiform? The shape of women's faces changes subtly during their ovulatory cycle, and men prefer female faces at the time of ovulation. Subjects guess political affiliation or religion at above-chance levels just by looking at faces. And for the same transgression, people who look embarrassed—blushing, eyes averted, face angled downward and to the side—are more readily forgiven.

Eyes give the most information.[20] Take pictures of two faces with different emotions, and switch different facial parts between the two with cutting and pasting. What emotion is detected? The one in the eyes.*[21]

Eyes often have an implicit censorious power.[22] Post a large picture of a pair of eyes at a bus stop (versus a picture of flowers), and people become more likely to clean up litter. Post a picture of eyes in a workplace coffee room, and the money paid on the honor system triples. Show a pair of eyes on a computer screen and people become more generous in online economic games.

Subliminal auditory cues also alter behavior.[23] Back to amygdaloid activation in whites subliminally viewing black faces. Chad Forbes of the University of Delaware shows that the amygdala activation increases if loud rap music—a genre typically associated more with African Americans than with whites—plays in the background. The opposite occurs when evoking negative white stereotypes with death metal music blaring.

Another example of auditory cuing explains a thoroughly poignant anecdote told by my Stanford colleague Claude Steele, who has done seminal research on stereotyping.[24] Steele recounts how an African American male grad student of his, knowing the stereotypes that a young black man evokes on the genteel streets of Palo Alto, whistled Vivaldi when walking home at night, hoping to evoke instead "Hey, that's not Snoop Dogg. That's a dead white male composer [exhale]."

No discussion of subliminal sensory cuing is complete without considering olfaction, a subject marketing people have salivated over since we were projected to watch Smell-O-Vision someday. The human olfactory system is atrophied; roughly 40 percent of a rat's brain is devoted to olfactory processing, versus 3 percent in us. Nonetheless, we still have

* Unconscious cues aren't always about faces and posture. Among closely matched teams or individual male athletic competitors, wearing a red jersey boosts performance. This has been shown in Olympic boxing, tae kwon do, and wrestling, for rugby and soccer teams, and when playing a virtual gladitorial computer game. It has been speculated to reflect the fact that in many species (e.g., mandrill monkeys and widow birds) male dominance displays involve flashing a red body part, where more testosterone equals more intense red. I'm dubious about the explanation, as it feels like cherry-picking examples from other species.

unconscious olfactory lives, and as in rodents, our olfactory system sends more direct projections to the limbic system than other sensory systems. As noted, rodent pheromones carry information about sex, age, reproductive status, health, and genetic makeup, and they alter physiology and behavior. Similar, if milder, versions of the same are reported in some (but not all) studies of humans, ranging from the Wellesley effect, discussed in the introduction, to heterosexual women preferring the smell of high-testosterone men.

Importantly, pheromones signal fear. In one study researchers got armpit swabs from volunteers under two conditions—either after contentedly sweating during a comfortable run, or after sweating in terror during their first tandem skydive (note—in tandem skydives you're yoked to the instructor, who does the physical work; so if you're sweating, it's from panic, not physical effort). Subjects sniffed each type of sweat and couldn't consciously distinguish between them. However, sniffing terrified sweat (but not contented sweat) caused amygdaloid activation, a bigger startle response, improved detection of subliminal angry faces, and increased odds of interpreting an ambiguous face as looking fearful. If people around you smell scared, your brain tilts toward concluding that you are too.[25]

Finally, nonpheromonal odors influence us as well. As we'll see in chapter 12, if people sit in a room with smelly garbage, they become more conservative about social issues (e.g., gay marriage) without changing their opinions about, say, foreign policy or economics.

Interoceptive Information

In addition to information about the outside world, our brains constantly receive "interoceptive" information about the body's internal state. You feel hungry, your back aches, your gassy intestine twinges, your big toe itches. And such interoceptive information influences our behavior as well.

This brings us to the time-honored James-Lange theory, named for William James, a grand mufti in the history of psychology, and an obscure

Danish physician, Carl Lange. In the 1880s they independently concocted the same screwy idea. How do your feelings and your body's automatic (i.e., "autonomic") function interact? It seems obvious—a lion chases you, you feel terrified, and thus your heart speeds up. James and Lange suggested the opposite: you subliminally note the lion, speeding up your heart; then your conscious brain gets this interoceptive information, concluding, "Wow, my heart is racing; I must be terrified." In other words, you decide what you feel based on signals from your body.

There's support for the idea—three of my favorites are that (a) forcing depressed people to smile makes them feel better; (b) instructing people to take on a more "dominant" posture makes them feel more so (lowers stress hormone levels); and (c) muscle relaxants decrease anxiety ("Things are still awful, but if my muscles are so relaxed that I'm dribbling out of this chair, things must be improving"). Nonetheless, a strict version of James-Lange doesn't work, because of the issue of specificity—hearts race for varying reasons, so how does your brain decide if it's reacting to a lion or an exciting come-hither look? Moreover, many autonomic responses are too slow to precede conscious awareness of an emotion.[26]

Nonetheless, interoceptive information influences, if not determines, our emotions. Some brain regions with starring roles in processing social emotions—the PFC, insular cortex, anterior cingulate cortex, and amygdala—receive lots of interoceptive information. This helps explain a reliable trigger of aggression, namely pain, which activates most of those regions. As a repeating theme, pain does not cause aggression; it amplifies preexisting tendencies toward aggression. In other words, pain makes aggressive people more aggressive, while doing the opposite to unaggressive individuals.[27]

Interoceptive information can alter behavior more subtly than in the pain/aggression link.[28] One example concerns how much the frontal cortex has to do with willpower, harking back to material covered in the last chapter. Various studies, predominantly by Roy Baumeister of Florida State University, show that when the frontal cortex labors hard on some cognitive task, immediately afterward individuals are more aggressive and less empathic, charitable, and honest. Metaphorically, the frontal cortex says, "Screw it. I'm tired and don't feel like thinking about my fellow human."

This seems related to the metabolic costs of the frontal cortex doing the harder thing. During frontally demanding tasks, blood glucose levels drop, and frontal function improves if subjects are given a sugary drink (with control subjects consuming a drink with a nonnutritive sugar substitute). Moreover, when people are hungry, they become less charitable and more aggressive (e.g., choosing more severe punishment for an opponent in a game).* There's debate as to whether the decline in frontal regulation in these circumstances represents impaired capacity for self-control or impaired motivation for it. But either way, over the course of seconds to minutes, the amount of energy reaching the brain and the amount of energy the frontal cortex needs have something to do with whether the harder, more correct thing happens.

Thus, sensory information streaming toward your brain from both the outside world and your body can rapidly, powerfully, and automatically alter behavior. In the minutes before our prototypical behavior occurs, more complex stimuli influence us as well.

Unconscious Language Effects

Words have power. They can save, cure, uplift, devastate, deflate, and kill. And unconscious priming with words influences pro- and antisocial behaviors.

One of my favorite examples concerns the Prisoner's Dilemma, the economic game where participants decide whether to cooperate or compete at various junctures.[29] And behavior is altered by "situational labels"—call the game the "Wall Street Game," and people become less cooperative. Calling it the "Community Game" does the opposite. Similarly, have subjects read seemingly random word lists before playing.

* Findings like this should not be confused with the rationale behind the "Twinkie defense." In 1978 San Francisco mayor George Moscone and city supervisor Harvey Milk—the first openly gay politician in California—were assassinated by Dan White, a disgruntled ex-supervisor. According to the common misconception, during his trial, White's defense attorneys argued that his addiction to sugary junk food somehow impaired his judgment and self-control. In reality, the defense argued that White suffered from diminished capacity because of his depression, and his shift from a healthy diet to one of junk food was merely evidence of his depressed state.

Embedding warm fuzzy prosocial words in the list—"help," "harmony," "fair," "mutual"—fosters cooperation, while words like "rank," "power," "fierce," and "inconsiderate" foster the opposite. Mind you, this isn't subjects reading either Christ's Sermon on the Mount or Ayn Rand. Just an innocuous string of words. Words unconsciously shift thoughts and feelings. One person's "terrorist" is another's "freedom fighter"; politicians jockey to commandeer "family values," and somehow you can't favor both "choice" and "life."*[30]

There are more examples. In Nobel Prize–winning research, Daniel Kahneman and Amos Tversky famously showed word framing altering decision making. Subjects decide whether to administer a hypothetical drug. If they're told, "The drug has a 95 percent survival rate," people, including doctors, are more likely to approve it than when told, "The drug has a 5 percent death rate."†[31] Embed "rude" or "aggressive" (versus "considerate" or "polite") in word strings, and subjects interrupt people more immediately afterward. Subjects primed with "loyalty" (versus "equality") become more biased toward their team in economic games.[32]

Verbal primes also impact moral decision making.[33] As every trial lawyer knows, juries decide differently depending on how colorfully you describe someone's act. Neuroimaging studies show that more colorful wording engages the anterior cingulate more. Moreover, people judge moral transgressions more harshly when they are described as "wrong" or "inappropriate" (versus "forbidden" or "blameworthy").

Even Subtler Types of Unconscious Cuing

In the minutes before a behavior is triggered, subtler things than sights and smells, gas pain, and choice of words unconsciously influence us.

In one study, subjects filling out a questionnaire expressed stronger

* One recent study demonstrates a pointed version of this—describing someone as "African American" evokes associations with higher levels of education and income than does describing them as "black."

† A recent study shows life-and-death consequences of linguistic cuing. For the same storm intensity, hurricanes arbitrarily given female names kill more people than do those with male names (names alternate between the two genders). Why? People unconsciously take male-named hurricanes more seriously and are more likely to comply with evacuation orders. And this despite both male and female names being selected for their innocuousness—this isn't comparing Hurricane Mary Poppins with Hurricane Vlad the Impaler.

egalitarian principles if there was an American flag in the room. In a study of spectators at English football matches, a researcher planted in the crowd slips, seemingly injuring his ankle. Does anyone help him? If the plant wore the home team's sweatshirt, he received more help than when he wore a neutral sweatshirt or one of the opposing team. Another study involved a subtle group-membership manipulation—for a number of days, pairs of conservatively dressed Hispanics stood at train stations during rush hour in predominately white Boston suburbs, conversing quietly in Spanish. The consequence? White commuters expressed more negative, exclusionary attitudes toward Hispanic (but not other) immigrants.[34]

Cuing about group membership is complicated by people belonging to multiple groups. Consider a famous study of Asian American women who took a math test.[35] Everyone knows that women are worse at math than men (we'll see in chapter 9 how that's not really so) and Asian Americans are better at it than other Americans. Subjects primed beforehand to think about their racial identity performed better than did those primed to think about their gender.

Another realm of rapid group influences on behavior is usually known incorrectly. This is the "bystander effect" (aka the "Genovese syndrome").[36] This refers to the notorious 1964 case of Kitty Genovese, the New Yorker who was raped and stabbed to death over the course of an hour outside an apartment building, while thirty-eight people heard her shrieks for help and didn't bother calling the police. Despite that being reported by the *New York Times*, and the collective indifference becoming emblematic of all that's wrong with people, the facts differed: the number was less than thirty-eight, no one witnessed the entire event, apartment windows were closed on that winter's night, and most assumed they were hearing the muffled sounds of a lover's quarrel.*

The mythic elements of the Genovese case prompt the quasi myth that in an emergency requiring brave intervention, the more people

* For a horrifying, documented example of bystanders being at least as callous as the apartment dwellers were reputed to be, Google the case of the death of two-year-old Wang Yue.

present, the less likely anyone is to help—"There's lots of people here; someone else will step forward." The bystander effect does occur in non-dangerous situations, where the price of stepping forward is inconvenience. However, in dangerous situations, the more people present, the *more* likely individuals are to step forward. Why? Perhaps elements of reputation, where a larger crowd equals more witnesses to one's heroics.

Another rapid social-context effect shows men in some of their lamest moments.[37] Specifically, when women are present, or when men are prompted to think about women, they become more risk-taking, show steeper temporal discounting in economic decisions, and spend more on luxury items (but not on mundane expenses).* Moreover, the allure of the opposite sex makes men more aggressive—for example, more likely in a competitive game to punish the opposing guy with loud blasts of noise. Crucially, this is not inevitable—in circumstances where status is achieved through prosocial routes, the presence of women makes men more prosocial. As summarized in the title of one paper demonstrating this, this seems a case of "Male generosity as a mating signal." We'll return to this theme in the next chapter.

Thus, our social environment unconsciously shapes our behavior over the course of minutes. As does our physical environment.

Now we come to the "broken window" theory of crime of James Q. Wilson and George Kelling.[38] They proposed that small signs of urban disarray—litter, graffiti, broken windows, public drunkenness—form a slippery slope leading to larger signs of disarray, leading to increased crime. Why? Because litter and graffiti as the norm mean people don't care or are powerless to do anything, constituting an invitation to litter or worse.

Broken-window thinking shaped Rudy Giuliani's mayoralty in the 1990s, when New York was turning into a Hieronymus Bosch painting. Police commissioner William Bratton instituted a zero-tolerance policy toward minor infractions—targeting subway fare evaders, graffiti artists,

* In these studies the control situation is when subjects are in the presence of another man. And FYI, the presence of men has no such effects on the behavior of women.

vandals, beggars, and the city's maddening infestation of squeegee men. Which was followed by a steep drop in rates of serious crime. Similar results occurred elsewhere; in Lowell, Massachusetts, zero-tolerance measures were experimentally applied in only one part of the city; serious crime dropped only in that area. Critics questioned whether the benefits of broken-window policing were inflated, given that the approach was tested when crime was already declining throughout the United States (in other words, in contrast to the commendable Lowell example, studies often lacked control groups).

In a test of the theory, Kees Keizer of the University of Groningen in the Netherlands asked whether cues of one type of norm violation made people prone to violating other norms.[39] When bicycles were chained to a fence (despite a sign forbidding it), people were more likely to take a shortcut through a gap in the fence (despite a sign forbidding it); people littered more when walls were graffitied; people were more likely to steal a five-euro note when litter was strewn around. These were big effects, with doubling rates of crummy behaviors. A norm violation increasing the odds of that *same* norm being violated is a conscious process. But when the sound of fireworks makes someone more likely to litter, more unconscious processes are at work.

A Wonderfully Complicating Piece of the Story

We've now seen how sensory and interoceptive information influence the brain to produce a behavior within seconds to minutes. But as a complication, the brain can alter the *sensitivity* of those sensory modalities, making some stimuli more influential.

As an obvious one, dogs prick up their ears when they're alert—the brain has stimulated ear muscles in a way that enables the ears to more easily detect sounds, which then influences the brain.[40] During acute stress, all of our sensory systems become more sensitive. More selectively, if you're hungry, you become more sensitive to the smell of food. How does something like this work? A priori, it seems as if all sensory roads lead to the brain. But the brain also sends neuronal projections *to* sensory organs. For example, low blood sugar might activate particular hypothalamic neurons. These, in turn, project to and stimulate receptor neurons in the nose that respond to food smells. The stimulation isn't enough to give those receptor neurons action potentials, but it now takes fewer food odorant molecules to trigger one. Something along these lines explains how the brain alters the selective sensitivity of sensory systems.

This certainly applies to the behaviors that fill this book. Recall how eyes carry lots of information about emotional state. It turns out that the brain biases us toward preferentially looking at eyes. This was shown by Damasio, studying a patient with Urbach-Wiethe disease, which selectively destroys the amygdala. As expected, she was poor at accurately detecting fearful faces. But in addition, while control subjects spent about half their face-gazing time looking at eyes, she spent half that. When instructed to focus on the eyes, she improved at recognizing fearful expressions. Thus, not only does the amygdala detect fearful faces, but it also biases us toward obtaining information about fearful faces.[41]

Psychopaths are typically poor at recognizing fearful expressions (though they accurately recognize other types).[42] They also look less at eyes than normal and improve at fear recognition when directed to focus on eyes. This makes sense, given the amygdaloid abnormalities in psychopaths noted in chapter 2.

Now an example foreshadowing chapter 9's focus on culture. Show subjects a picture of an object embedded in a complex background. Within seconds, people from collectivist cultures (e.g., China) tend to look more at, and remember better, the surrounding "contextual" information, while people from individualistic cultures (e.g., the United States) do the same with the focal object. Instruct subjects to focus on the domain that their culture doesn't gravitate toward, and there's frontal cortical activation—this is a difficult perceptual task. Thus, culture literally shapes how and where you look at the world.*[43]

* As an important point, this is indeed a case of acculturation, rather than a reflection of populational genetic differences—East Asian Americans show the typical American pattern.

CONCLUSIONS

No brain operates in a vacuum, and over the course of seconds to minutes, the wealth of information streaming into the brain influences the likelihood of pro- or antisocial acts. As we've seen, pertinent information ranges from something as simple and unidimensional as shirt color to things as complex and subtle as cues about ideology. Moreover, the brain also constantly receives interoceptive information. And most important, much of these varied types of information is subliminal. Ultimately, the most important point of this chapter is that in the moments just before we decide upon some of our most consequential acts, we are less rational and autonomous decision makers than we like to think.

Four

Hours to Days Before

We now take the next step back in our chronology, considering events from hours to days before a behavior occurs. To do so, we enter the realm of hormones. What are the effects of hormones on the brain and sensory systems that filled the last two chapters? How do hormones influence our best and worst behaviors?

While this chapter examines various hormones, the most attention is paid to one inextricably tied to aggression, namely testosterone. And as the punch line, testosterone is far less relevant to aggression than usually assumed. At the other end of the spectrum, the chapter also considers a hormone with cult status for fostering warm, fuzzy prosociality, namely oxytocin. As we'll see, it's not quite as groovy as assumed.

Those who are unfamiliar with hormones and endocrinology, please see the primer in appendix 2.

TESTOSTERONE'S BUM RAP

Testosterone is secreted by the testes as the final step in the "hypothalamic/pituitary/testicular" axis; it has effects on cells throughout the body (including neurons, of course). And testosterone is everyone's usual suspect when it comes to the hormonal causes of aggression.

Correlation and Causality

Why is it that throughout the animal kingdom, and in every human culture, males account for most aggression and violence? Well, what about testosterone and some related hormones (collectively called "androgens," a term that, unless otherwise noted, I will use simplistically as synonymous with "testosterone")? In nearly all species males have more circulating testosterone than do females (who secrete small amounts of androgens from the adrenal glands). Moreover, male aggression is most prevalent when testosterone levels are highest (adolescence, and during mating season in seasonal breeders).

Thus, testosterone and aggression are linked. Furthermore, there are particularly high levels of testosterone receptors in the amygdala, in the way station by which it projects to the rest of the brain (the bed nucleus of the stria terminalis), and in its major targets (the hypothalamus, the central gray of the midbrain, and the frontal cortex). But these are merely correlative data. Showing that testosterone *causes* aggression requires a "subtraction" plus a "replacement" experiment. Subtraction—castrate a male. Do levels of aggression decrease? Yes (including in humans). This shows that something coming from the testes causes aggression. Is it testosterone? Replacement—give that castrated individual replacement testosterone. Do precastration levels of aggression return? Yes (including in humans).

Thus, testosterone causes aggression. Time to see how wrong that is.

The first hint of a complication comes after castration, when average levels of aggression plummet in every species. But, crucially, not to zero. Well, maybe the castration wasn't perfect, you missed some bits of testes. Or maybe enough of the minor adrenal androgens are secreted to maintain the aggression. But no—even when testosterone and androgens are completely eliminated, some aggression remains. Thus, some male aggression is testosterone independent.*

This point is driven home by castration of some sexual offenders, a

* This is no surprise to fans of the history of eunuchs, who were a mainstay of the military of Imperial China, prized as fierce soldiers.

legal procedure in a few states.[1] This is accomplished with "chemical castration," administration of drugs that either inhibit testosterone production or block testosterone receptors.* Castration decreases sexual urges in the subset of sex offenders with intense, obsessive, and pathological urges. But otherwise castration doesn't decrease recidivism rates; as stated in one meta-analysis, "hostile rapists and those who commit sex crimes motivated by power or anger are not amenable to treatment with [the antiandrogenic drugs]."

This leads to a hugely informative point: the more experience a male had being aggressive prior to castration, the more aggression continues afterward. In other words, the less his being aggressive in the future requires testosterone and the more it's a function of social learning.

On to the next issue that lessens the primacy of testosterone: What do individual levels of testosterone have to do with aggression? If one person has higher testosterone levels than another, or higher levels this week than last, are they more likely to be aggressive?

Initially the answer seemed to be yes, as studies showed correlation between individual differences in testosterone levels and levels of aggression. In a typical study, higher testosterone levels would be observed in those male prisoners with higher rates of aggression. But being aggressive *stimulates* testosterone secretion; no wonder more aggressive individuals had higher levels. Such studies couldn't disentangle chickens and eggs.

Thus, a better question is whether differences in testosterone levels among individuals *predict* who *will be* aggressive. And among birds, fish, mammals, and especially other primates, the answer is generally no. This has been studied extensively in humans, examining a variety of measures of aggression. And the answer is clear. To quote the British endocrinologist John Archer in a definitive 2006 review, "There is a weak and inconsistent association between testosterone levels and aggression in [human] adults, and . . . administration of testosterone to volunteers typically does not increase their aggression." The brain doesn't pay attention to fluctuations of testosterone levels within the normal range.[2]

---

* One exception: Texas, where they still use a knife.

(Things differ when levels are made "supraphysiological"—higher than the body normally generates. This is the world of athletes and bodybuilders abusing high-dose testosterone-like anabolic steroids; in that situation risk of aggression does increase. Two complications: it's not random who would *choose* to take these drugs, and abusers are often already predisposed toward aggression; supraphysiological levels of androgens generate anxiety and paranoia, and increased aggression may be secondary to that.)[3]

Thus, aggression is typically more about social learning than about testosterone, and differing levels of testosterone generally can't explain why some individuals are more aggressive than others. So what does testosterone actually do to behavior?

Subtleties of Testosterone Effects

When looking at faces expressing strong emotions, we tend to make microexpressions that mimic them; testosterone decreases such empathic mimicry.*[4] Moreover, testosterone makes people less adept at identifying emotions by looking at people's eyes, and faces of strangers activate the amygdala more than familiar ones and are rated as less trustworthy.

Testosterone also increases confidence and optimism, while decreasing fear and anxiety.[5] This explains the "winner" effect in lab animals, where winning a fight increases an animal's willingness to participate in, and its success in, another such interaction. Part of the increased success probably reflects the fact that winning stimulates testosterone secretion, which increases glucose delivery and metabolism in the animal's muscles and makes his pheromones smell scarier. Moreover, winning increases the number of testosterone receptors in the bed nucleus of the stria terminalis (the way station through which the amygdala communicates with the rest of the brain), increasing its sensitivity to the hormone. Success in everything from athletics to chess to the stock market boosts testosterone levels.

Confident and optimistic. Well, endless self-help books urge us to be

* Where, as with all these studies, neither the subjects nor the scientist observing them know whether the volunteers received testosterone or placebo, and where the testosterone levels produced are always within the normal range.

precisely that. But testosterone makes people *over*confident and *overly* optimistic, with bad consequences. In one study, pairs of subjects could consult each other before making individual choices in a task. Testosterone made subjects more likely to think their opinion was correct and to ignore input from their partner. Testosterone makes people cocky, egocentric, and narcissistic.[6]

Testosterone boosts impulsivity and risk taking, making people do the easier thing when it's the dumb-ass thing to do.[7] Testosterone does this by decreasing activity in the prefrontal cortex and its functional coupling to the amygdala and increasing amygdaloid coupling with the thalamus—the source of that shortcut path of sensory information into the amygdala. Thus, more influence by split-second, low-accuracy inputs and less by the let's-stop-and-think-about-this frontal cortex.

Being fearless, overconfident, and delusionally optimistic sure feels good. No surprise, then, that testosterone can be pleasurable. Rats will work (by pressing levers) to be infused with testosterone and show "conditioned place preference," returning to a random corner of the cage where infusions occur. "I don't know why, but I feel good whenever I stand there."[8,9]

The underlying neurobiology fits perfectly. Dopamine is needed for place-preference conditioning to occur, and testosterone increases activity in the ventral tegmentum, the source of those mesolimbic and mesocortical dopamine projections. Moreover, conditioned place preference is induced when testosterone is infused directly into the nucleus accumbens, the ventral tegmentum's main projection target. When a rat wins a fight, the number of testosterone receptors increases in the ventral tegmentum and accumbens, increasing sensitivity to the hormone's feel-good effects.[10]

So testosterone does subtle things to behavior. Nonetheless, this doesn't tell us much because everything can be interpreted every which way. Testosterone increases anxiety—you feel threatened and become more reactively aggressive. Testosterone decreases anxiety—you feel cocky and overconfident, become more preemptively aggressive. Testosterone increases risk taking—"Hey, let's gamble and invade." Testosterone increases risk taking—"Hey, let's gamble and make a peace offer."

Testosterone makes you feel good—"Let's start another fight, since the last one went swell." Testosterone makes you feel good—"Let's all hold hands."

It's a crucial unifying concept that testosterone's effects are hugely context dependent.

Contingent Testosterone Effects

This context dependency means that rather than causing X, testosterone amplifies the power of something else to cause X.

A classic example comes from a 1977 study of groups of male talapoin monkeys.[11] Testosterone was administered to the middle-ranking male in each group (say, rank number 3 out of five), increasing their levels of aggression. Does this mean that these guys, stoked on 'roids, started challenging numbers 1 and 2 in the hierarchy? No. They became aggressive jerks to poor numbers 4 and 5. Testosterone did not create new social patterns of aggression; it exaggerated preexisting ones.

In human studies testosterone didn't raise baseline activity in the amygdala; it boosted the amygdala's response and heart-rate reactivity to angry faces (but not to happy or neutral ones). Similarly, testosterone did not make subjects more selfish and uncooperative in an economic game; it made them more punitive when provoked by being treated poorly, enhancing "vengeful reactive aggression."[12]

The context dependency also occurs on the neurobiological level, in that the hormone shortens the refractory period of neurons in the amygdala and amygdaloid targets in the hypothalamus.[13] Recall that the refractory period comes in neurons after action potentials. This is when the neuron's resting potential is hyperpolarized (i.e., when it is more negatively charged than usual), making the neuron less excitable, producing a period of silence after the action potential. Thus, shorter refractory periods mean a higher rate of action potentials. So is testosterone causing action potentials in these neurons? No. It's causing them to fire at a faster rate *if* they are stimulated by something else. Similarly, testosterone increases amygdala response to angry faces, but not to other sorts. Thus, if the amygdala is already responding to some realm of social learning, testosterone ups the volume.

A Key Synthesis: The Challenge Hypothesis

Thus, testosterone's actions are contingent and amplifying, exacerbating preexisting tendencies toward aggression rather than creating aggression out of thin air. This picture inspired the "challenge hypothesis," a wonderfully unifying conceptualization of testosterone's actions.[14] As proposed in 1990 by the superb behavioral endocrinologist John Wingfield of the University of California at Davis, and colleagues, the idea is that rising testosterone levels increase aggression only at the time of a challenge. Which is precisely how things work.

The explains why basal levels of testosterone have little to do with subsequent aggression, and why increases in testosterone due to puberty, sexual stimulation, or the start of mating season don't increase aggression either.[15]

But things are different during challenges.[16] Among various primates, testosterone levels rise when a dominance hierarchy first forms or undergoes reorganization. Testosterone rises in humans in both individual and team sports competition, including basketball, wrestling, tennis, rugby, and judo; there's generally a rise in anticipation of the event and a larger one afterward, especially among winners.* Remarkably, *watching* your favorite team win raises testosterone levels, showing that the rise is less about muscle activity than about the psychology of dominance, identification, and self-esteem.

Most important, the rise in testosterone after a challenge makes aggression more likely.[17] Think about this. Testosterone levels rise, reaching the brain. If this occurs because someone is challenging you, you head in

* This is a rich literature showing the subtleties of the human psyche. The winning effect on testosterone is lessened in circumstances where people feel like they won by luck or where, despite winning, they feel like they underperformed. In contast, the effect is enhanced among people who went into the competition having the strongest psychological motives for domination. Finally, testosterone levels can rise robustly in "losers" who nonetheless performed far better than they anticipated. Thus one might see testosterone levels rise after a marathon in a guy who came in at the back of the pack but is triumphant because he was sure he was going to drop dead halfway through, and may decline in the guy who comes in third but was expected to win. We all belong to numerous hierarchies, but some of the most powerful are the ones in our heads based on our internal standards.

the direction of aggression. If an identical rise occurs because days are lengthening and mating season is approaching, you decide to fly a thousand miles to your breeding grounds. And if the same occurs because of puberty, you get stupid and giggly around that girl who plays clarinet in the band. The context dependency is remarkable.*[18]

The challenge hypothesis has a second part to it. When testosterone rises after a challenge, it doesn't prompt aggression. Instead it prompts *whatever behaviors are needed to maintain status.* This changes things enormously.

Well, maybe not, since maintaining status for, say, male primates consists mostly of aggression or threats of it—from slashing your opponent to giving a "You have no idea who you're screwing with" stare.[19]

And now for some flabbergastingly important research. What happens if defending your status requires you to be nice? This was explored in a study by Christoph Eisenegger and Ernst Fehr of the University of Zurich.[20] Participants played the Ultimatum Game (introduced in chapter 2), where you decide how to split money between you and another player. The other person can accept the split or reject it, in which case neither of you gets anything. Prior research had shown that when someone's offer is rejected, they feel dissed, subordinated, especially if news of that carries into future rounds with other players. In other words, in this scenario status and reputation rest on being fair.

And what happens when subjects were given testosterone beforehand? *People made more generous offers.* What the hormone makes you do depends on what counts as being studly. This requires some fancy neuroendocrine wiring that is sensitive to social learning. You couldn't ask for a finding more counter to testosterone's reputation.

The study contained a slick additional finding that further separated

* All these circumstances of testosterone levels rising raises a question: why not just produce higher levels all the time and save the effort? For one thing, all those androgens are lousy for the cardiovascular system. But more important, they'd get in the way of various prosocial behaviors. For example, among monogamous birds and rodents, if testosterone levels don't drop around the time the female gives birth, the males won't act paternally. And some similar patterns seem to apply to humans: Fathers have lower testosterone levels than age-matched, married men without children, and more involved fathers have lower levels than less involved ones. Moreover, evoking nurturing behavior in men lowers testosterone levels, as does the birth of a man's child. And when compared with high-testosterone fathers, those with lower average testosterone levels are rated as better parents by their partners and have more activation of their reward-related ventral tegmentum when seeing a picture of their child.

testosterone myth from reality. As per usual, subjects got either testosterone or saline, without knowing which. Subjects who believed it was testosterone (independent of whether it actually was) made less generous offers. In other words, testosterone doesn't necessarily make you behave in a crappy manner, but *believing* that it does and that you're drowning in the stuff makes you behave in a crappy manner.

Additional studies show that testosterone promotes prosociality in the right setting. In one, under circumstances where someone's sense of pride rides on honesty, testosterone decreased men's cheating in a game. In another, subjects decided how much of a sum of money they would keep and how much they would publicly contribute to a common pool shared by all the players; testosterone made most subjects more prosocial.[21]

What does this mean? Testosterone makes us more willing to do what it takes to attain and maintain status. And the key point is what it takes. Engineer social circumstances right, and boosting testosterone levels during a challenge would make people compete like crazy to do the most acts of random kindness. In our world riddled with male violence, the problem isn't that testosterone can increase levels of aggression. The problem is the frequency with which we reward aggression.

OXYTOCIN AND VASOPRESSIN: A MARKETING DREAM

If the point of the preceding section is that testosterone has gotten a bum rap, the point of this one is that oxytocin (and the closely related vasopressin) is coasting in a Teflon presidency. According to lore, oxytocin makes organisms less aggressive, more socially attuned, trusting, and empathic. Individuals treated with oxytocin become more faithful partners and more attentive parents. It makes lab rats more charitable and better listeners, makes fruit flies sing like Joan Baez. Naturally, things are more complicated, and oxytocin has an informative dark side.

Basics

Oxytocin and vasopressin are chemically similar hormones; the DNA sequences that constitute their genes are similar, and the two genes occur close to each other on the same chromosome. There was a single ancestral gene that, a few hundred million years ago, was accidentally "duplicated" in the genome, and the DNA sequences in the two copies of the gene drifted independently, evolving into two closely related genes (stay tuned for more in chapter 8). This gene duplication occurred as mammals were emerging; other vertebrates have only the ancestral version, called vasotocin, which is structurally between the two separate mammalian hormones.

For twentieth-century neurobiologists, oxytocin and vasopressin were pretty boring. They were made in hypothalamic neurons that sent axons to the posterior pituitary. There they would be released into circulation, thereby attaining hormone status, and have nothing to do with the brain ever again. Oxytocin stimulated uterine contraction during labor and milk letdown afterward. Vasopressin (aka "antidiuretic hormone") regulated water retention in the kidneys. And reflecting their similar structures, each also had mild versions of the other one's effects. End of story.

Neurobiologists Take Notice

Things became interesting with the discovery that those hypothalamic neurons that made oxytocin and vasopressin also sent projections throughout the brain, including the dopamine-related ventral tegmentum and nucleus accumbens, hippocampus, amygdala, and frontal cortex, all regions with ample levels of receptors for the hormones. Moreover, oxytocin and vasopressin turned out to be synthesized and secreted elsewhere in the brain. These two boring, classical peripheral hormones affected brain function and behavior. They started being called "neuropeptides"—neuroactive messengers with a peptide structure—which is a fancy way of saying they are small proteins (and, to avoid writing "oxytocin and vasopressin" endlessly, I will refer to them as neuropeptides; note though that there are other neuropeptides).

The initial findings about their behavioral effects made sense.[22] Oxytocin prepares the body of a female mammal for birth and lactation; logically, oxytocin also facilitates maternal behavior. The brain boosts oxytocin production when a female rat gives birth, thanks to a hypothalamic circuit with markedly different functions in females and males. Moreover, the ventral tegmentum increases its sensitivity to the neuropeptide by increasing levels of oxytocin receptors. Infuse oxytocin into the brain of a virgin rat, and she'll act maternally—retrieving, grooming, and licking pups. Block the actions of oxytocin in a rodent mother,*[23] and she'll stop maternal behaviors, including nursing. Oxytocin works in the olfactory system, helping a new mom learn the smell of her offspring. Meanwhile, vasopressin has similar but milder effects.

Soon other species were heard from. Oxytocin lets sheep learn the smell of their offspring and facilitates female monkeys grooming their offspring. Spray oxytocin up a woman's nose (a way to get the neuropeptide past the blood-brain barrier and into the brain), and she'll find babies to look more appealing. Moreover, women with variants of genes that produce higher levels of oxytocin or oxytocin receptors average higher levels of touching their infants and more synchronized gazing with them.

So oxytocin is central to female mammals nursing, *wanting* to nurse their child, and remembering which one is their child. Males then got into the act, as vasopressin plays a role in paternal behavior. A female rodent giving birth increases vasopressin and vasopressin receptor levels throughout the body, including the brain, of the nearby father. Among monkeys, experienced fathers have more dendrites in frontal cortical neurons containing vasopressin receptors. Moreover, administering vasopressin enhances paternal behaviors. However, an ethological caveat: this occurs only in species where males are paternal (e.g., prairie voles and marmoset monkeys).[24]†

Then, dozens of millions of years ago, some rodent and primate species independently evolved monogamous pair-bonding, along with the

* In studies like these, this is typically accomplished either with the administration of a drug that blocks oxytocin receptors or with genetic engineering techniques to eliminate the gene for oxytocin or for the oxytocin receptor.
† In other words, a familiar theme: vasopressin doesn't cause paternal behavior; it facilitates it in species that are already predisposed to it.

neuropeptides central to the process.[25] Among marmoset and titi monkeys, which both pair-bond, oxytocin strengthens the bond, increasing a monkey's preference for huddling with her partner over huddling with a stranger. Then there was a study that is embarrassingly similar to stereotypical human couples. Among pair-bonding tamarin monkeys, lots of grooming and physical contact predicted high oxytocin levels in female members of a pair. What predicted high levels of oxytocin in males? Lots of sex.

Beautiful, pioneering work by Thomas Insel of the National Institute of Mental Health, Larry Young of Emory University, and Sue Carter of the University of Illinois has made a species of vole arguably the most celebrated rodent on earth.[26] Most voles (e.g., montane voles) are polygamous. In contrast, *prairie* voles form monogamous mating pairs for life. Naturally, this isn't quite the case—while they are "social pair-bonders" with their permanent relationships, they're not quite perfect "sexual pair-bonders," as males might mess around on the side. Nonetheless, prairie voles pair-bond more than other voles, prompting Insel, Young, and Carter to figure out why.

First finding: sex releases oxytocin and vasopressin in the nucleus accumbens of female and male voles, respectively. Obvious theory: prairie voles release more of the stuff during sex than do polygamous voles, causing a more rewarding buzz, encouraging the individuals to stick with their partner. But prairie voles don't release more neuropeptides than montane voles. Instead, prairie voles have more of the pertinent receptors in the nucleus accumbens than do polygamous voles.* Moreover, male prairie voles with a variant of the vasopressin receptor gene that produced more receptors in the nucleus accumbens were stronger pair-bonders. Then the scientists conducted two tour de force studies. First they engineered the brains of male mice to express the prairie vole version of the vasopressin receptor in their brains, and they groomed and huddled more

* This turned out to be due to a genetic difference between the two species. Interestingly, it is not a difference in the DNA sequence that constitutes the gene for the vasopressin receptor. It is a difference in the sequence that constitutes the on/off switch for the gene. More on this in chapter 8.

with familiar females (but not with strangers). Then the scientists engineered the brains of male montane voles to have more vasopressin receptors in the nucleus accumbens; the males became more socially affiliative with individual females.*

What about versions of vasopressin receptor genes in other species? When compared with chimps, bonobos have a variant associated with more receptor expression and far more social bonding between females and males (although, in contrast to prairie voles, bonobos are anything but monogamous).[27]

How about humans? This is tough to study, because you can't measure these neuropeptides in tiny brain regions in humans and instead have to examine levels in the circulation, a fairly indirect measure.

Nevertheless, these neuropeptides appear to play a role in human pair-bonding.[28] For starters, circulating oxytocin levels are elevated in couples when they've first hooked up. Furthermore, the higher the levels, the more physical affection, the more behaviors are synchronized, the more long-lasting the relationship, and the happier interviewers rate couples to be.

Even more interesting were studies where oxytocin (or a control spray) was administered intranasally. In one fun study, couples had to discuss one of their conflicts; oxytocin up their noses, and they'd be rated as communicating more positively and would secrete less stress hormones. Another study suggests that oxytocin unconsciously strengthens the pair-bond. Heterosexual male volunteers, with or without an oxytocin spritz, interacted with an attractive female researcher, doing some nonsense task. Among men in stable relationships, oxytocin increased their distance from the woman an average of four to six inches. Single guys, no effect. (Why didn't oxytocin make them stand closer? The researchers indicated that they were already about as close as one could get away

* This generated all sorts of caustic discussions at conferences as to whether this constituted a case of "gene transfer" (i.e., the value-neutral process of transferring a novel gene into an individual in order to alter a function) or "gene therapy" (i.e., transferring a gene in order to cure those montane males of the disease of infidelity). It strikes me that if that research had been carried out instead at Berkeley during the Summer of Love in 1967, the gene-therapy goal would have been to make prairie voles transcend their Middle American bourgeois genetics and become polygamous. The times, they are a changin' to quote a recent Nobel Laureate.

with.) If the experimenter was male, no effect. Moreover, oxytocin caused males in relationships to spend less time looking at pictures of attractive women. Importantly, oxytocin didn't make men rate these women as less attractive; they were simply less interested.[29]

Thus, oxytocin and vasopressin facilitate bonding between parent and child and between couples.* Now for something truly charming that evolution has cooked up recently. Sometime in the last fifty thousand years (i.e., less than 0.1 percent of the time that oxytocin has existed), the brains of humans and domesticated wolves evolved a new response to oxytocin: when a dog and its owner (but not a stranger) interact, they secrete oxytocin.[30] The more of that time is spent gazing at each other, the bigger the rise. Give dogs oxytocin, and they gaze longer at their humans . . . which raises the humans' oxytocin levels. So a hormone that evolved for mother-infant bonding plays a role in this bizarre, unprecedented form of bonding between species.

In line with its effects on bonding, oxytocin inhibits the central amygdala, suppresses fear and anxiety, and activates the "calm, vegetative" parasympathetic nervous system. Moreover, people with an oxytocin receptor gene variant associated with more sensitive parenting also have less of a cardiovascular startle response. In the words of Sue Carter, exposure to oxytocin is "a physiological metaphor for safety." Furthermore, oxytocin reduces aggression in rodents, and mice whose oxytocin system was silenced (by deleting the gene for oxytocin or its receptor) were abnormally aggressive.[31]

Other studies showed that people rate faces as more trustworthy, and are more trusting in economic games, when given oxytocin (oxytocin had no effect when someone thought they were playing with a computer, showing that this was about social behavior).[32] This increased trust was interesting. Normally, if the other player does something duplicitous in the game, subjects are less trusting in subsequent rounds; in contrast, oxytocin-treated investors didn't modify their behavior in this way. Stated scientifically, "oxytocin inoculated betrayal aversion among investors";

* Note that the entire romantic coupling literature I just discussed solely concerns heterosexual couples. As far as I know, very little has been studied in this realm with gay or lesbian subjects.

stated caustically, oxytocin makes people irrational dupes; stated more angelically, oxytocin makes people turn the other cheek.

More prosocial effects of oxytocin emerged. It made people better at detecting happy (versus angry, fearful, or neutral) faces or words with positive (versus negative) social connotations, when these were displayed briefly. Moreover, oxytocin made people more charitable. People with the version of the oxytocin receptor gene associated with more sensitive parenting were rated by observers as more prosocial (when discussing a time of personal suffering), as well as more sensitive to social approval. And the neuropeptide made people more responsive to social reinforcement, enhancing performance in a task where correct or wrong answers elicited a smile or frown, respectively (while having no effect when right and wrong answers elicited different-colored lights).[33]

So oxytocin elicits prosocial behavior, and oxytocin is released when we experience prosocial behavior (being trusted in a game, receiving a warm touch, and so on). In other words, a warm and fuzzy positive feedback loop.[34]

Obviously, oxytocin and vasopressin are the grooviest hormones in the universe.* Pour them into the water supply, and people will be more charitable, trusting, and empathic. We'd be better parents and would make love, not war (mostly platonic love, though, since people in relationships would give wide berths to everyone else). Best of all, we'd buy all sorts of useless crap, trusting the promotional banners in stores once oxytocin starts spraying out of the ventilation system.

Okay, time to settle down a bit.

Prosociality Versus Sociality

Are oxytocin and vasopressin about prosociality or social competence? Do these hormones make us see happy faces everywhere or become more interested in gathering accurate social information about faces? The latter

* Discriminating online shoppers can now buy "Liquid Trust," touted as "the world's first oxytocin pheromone product." Perhaps worse, perfectly straitlaced scientific publications have referred to oxytocin as the "love drug" or the "cuddle drug." The "cuddle" part is puzzling, since the literature is about oxytocin-laden prairie voles huddling, not cuddling, and the former doesn't evoke images of luv-fests as much as of tubercular huddled masses yearning to breathe free.

isn't necessarily prosocial; after all, accurate information about someone's emotions makes them easier to manipulate.

The Groovy Neuropeptide School supports the idea of ubiquitous prosociality.[35] But the neuropeptides also foster social interest and competence. They make people look at eyes longer, increasing accuracy in reading emotions. Moreover, oxytocin enhances activity in the temporoparietal juncture (that region involved in Theory of Mind) when people do a social-recognition task. The hormone increases the accuracy of assessments of other people's thoughts, with a gender twist—women improve at detecting kinship relations, while men improve at detecting dominance relations. In addition, oxytocin increases accuracy in remembering faces and their emotional expressions, and people with the "sensitive parenting" oxytocin receptor gene variant are particularly adept at assessing emotions. Similarly, the hormones facilitate rodents' learning of an individual's smell, but not nonsocial odors.

Neuroimaging research shows that these neuropeptides are about social competence, as well as prosociality.[36] For example, variants of a gene related to oxytocin signaling* are associated with differing degrees of activation of the fusiform face area when looking at faces.

Findings like these suggest that abnormalities in these neuropeptides increase the risk of disorders of impaired sociality, namely autism spectrum disorders (ASD) (strikingly, people with ASD show blunted fusiform responses to faces).[37] Remarkably, ASD has been linked to gene variants related to oxytocin and vasopressin, to nongenetic mechanisms for silencing the oxytocin receptor gene, and to lower levels of the receptor itself. Moreover, the neuropeptides improve social skills in some individuals with ASD—e.g., enhancing eye contact.

Thus, sometimes oxytocin and vasopressin make us more prosocial, but sometimes they make us more avid and accurate social information gatherers. Nonetheless, there is a happy-face bias, since accuracy is most enhanced for positive emotions.[38]

Time for more complications.

* For the truly interested, the gene codes for a protein called CD38, which facilitates oxytocin secretion from neurons.

Contingent Effects of Oxytocin and Vasopressin

Recall testosterone's contingent effects (e.g., making a monkey more aggressive, but only toward individuals he already dominates). Naturally, these neuropeptides' effects are also contingent.[39]

One factor already mentioned is gender: oxytocin enhances different aspects of social competence in women and men. Moreover, oxytocin's calming effects on the amygdala are more consistent in men than in women. Predictably, neurons that make these neuropeptides are regulated by both estrogen and testosterone.[40]

As a really interesting contingent effect, oxytocin enhances charitability—but only in people who are already so. This mirrors testosterone's only raising aggression in aggression-prone people. Hormones rarely act outside the context of the individual and his or her environment.[41]

Finally, a fascinating study shows cultural contingencies in oxytocin's actions.[42] During stress, Americans seek emotional support (e.g., telling a friend about their problem) more readily than do East Asians. In one study oxytocin receptor gene variants were identified in American and Korean subjects. Under unstressful circumstances, neither cultural background nor receptor variant affected support-seeking behavior. During stressful periods, support seeking rose among subjects with the receptor variant associated with enhanced sensitivity to social feedback and approval—but only among the Americans (including Korean Americans). What does oxytocin do to support-seeking behavior? It depends on whether you're stressed. And on the genetic variant of your oxytocin receptor. And on your culture. More to come in chapters 8 and 9.

And the Dark Side of These Neuropeptides

As we saw, oxytocin (and vasopressin) decreases aggression in rodent females. Except for aggression in defense of one's pups, which the neuropeptide increases via effects in the central amygdala (with its involvement in instinctual fear).[43]

This readily fits with these neuropeptides enhancing maternalism, including snarling don't-get-one-step-closer maternalism. Similarly, vasopressin enhances aggression in paternal prairie vole males. This finding comes with a familiar additional contingency. The more aggressive the male prairie vole, the less that aggression decreases after blocking of his vasopressin system—just as in the case of testosterone, with increased experience, aggression is maintained by social learning rather than by a hormone/neuropeptide. Moreover, vasopressin increases aggression most in male rodents who are already aggressive—yet another biological effect depending on individual and social context.[44]

And now to really upend our view of these feel-good neuropeptides. For starters, back to oxytocin enhancing trust and cooperation in an economic game—but not if the other player is anonymous and in a different room. When playing against strangers, oxytocin *decreases* cooperation, enhances envy when luck is bad, and enhances gloating when it's good.[45]

Finally, beautiful studies by Carsten de Dreu of the University of Amsterdam showed just how unwarm and unfuzzy oxytocin can be.[46] In the first, male subjects formed two teams; each subject chose how much of his money to put into a pot shared with teammates. As usual, oxytocin increased such generosity. Then participants played the Prisoner's Dilemma with someone from the other team.* When financial stakes were high, making subjects more motivated, oxytocin made them *more* likely to preemptively stab the other player in the back. Thus, oxytocin makes you more prosocial to people like you (i.e., your teammates) but spontaneously lousy to Others who are a threat. As emphasized by De Dreu, perhaps oxytocin evolved to enhance social competence to make us better at identifying who is an Us.

In De Dreu's second study, Dutch student subjects took the Implicit Association Test of unconscious bias.† And oxytocin exaggerated biases against two out-groups, namely Middle Easterners and Germans.[47]

---

* In the Prisoner's Dilemma each of the two players must decide whether to cooperate. If they both cooperate, they each get, say, two units of reward. If they both backstab, they each get one unit. If one cooperates and the other backstabs, the cheater gets three units and the stooge gets none.

† The IAT will be described in detail in a later chapter—briefly, the test takes advantage of the fact that it takes us milliseconds longer to process pairings of information that seem discordant than pairings that make sense; thus if

Then came the study's truly revealing second part. Subjects had to decide whether it was okay to kill one person in order to save five. In the scenario the potential sacrificial lamb's name was either stereotypically Dutch (Dirk or Peter), German (Markus or Helmut), or Middle Eastern (Ahmed or Youssef); the five people in danger were unnamed. Remarkably, oxytocin made subjects *less* likely to sacrifice good ol' Dirk or Peter, rather than Helmut or Ahmed.

Oxytocin, the luv hormone, makes us more prosocial to Us and worse to everyone else. That's not generic prosociality. That's ethnocentrism and xenophobia. In other words, the actions of these neuropeptides depend dramatically on context—who you are, your environment, and who that person is. As we will see in chapter 8, the same applies to the regulation of genes relevant to these neuropeptides.

THE ENDOCRINOLOGY OF AGGRESSION IN FEMALES

Help!

This topic confuses me. Here's why:

- This is a domain where the ratios of two hormones can matter more than their absolute levels, where the brain responds the same way to (a) two units of estrogen plus one unit of progesterone and (b) two gazillion units of estrogen plus one gazillion units of progesterone. This requires some complex neurobiology.
- Hormone levels are extremely dynamic, with hundredfold changes in some within hours—no male's testes ever had to navigate the endocrinology of ovulation or childbirth. Among other things, recreating such endocrine fluctuations in lab animals is tough.
- There's dizzying variability across species. Some breed year-round,

you're prejudiced against group X, it takes you longer to process a pairing of group X with a positive term—for example, "wonderful"—than with a negative term—"dangerous."

others only in particular seasons; nursing inhibits ovulation in some, stimulates it in others.

- Progesterone rarely works in the brain as itself. Instead it's usually converted into various "neurosteroids" with differing actions in different brain regions. And "estrogen" describes a soup of related hormones, none of which work identically.
- Finally, one must debunk the myth that females are always nice and affiliative (unless, of course, they're aggressively protecting their babies, which is cool and inspirational).

Maternal Aggression

Levels of aggression rise in rodents during pregnancy, peaking around parturition.*[48] Appropriately, the highest levels occur in species and breeds with the greatest threat of infanticide.[49]

During late pregnancy, estrogen and progesterone increase maternal aggression by increasing oxytocin release in certain brain regions, bringing us back to oxytocin promoting maternal aggression.[50]

Two complications illustrate some endocrine principles.† Estrogen contributes to maternal aggression. But estrogen can also *reduce* aggression and enhance empathy and emotional recognition. It turns out there are two different types of receptors for estrogen in the brain, mediating these opposing effects and with their levels independently regulated. Thus, same hormone, same levels, different outcome if the brain is set up to respond differently.[51]

The other complication: As noted, progesterone, working with estrogen, promotes maternal aggression. However, on its own it decreases aggression and anxiety. Same hormone, same levels, diametrically opposite outcomes depending on the presence of a second hormone.[52]

Progesterone decreases anxiety through a thoroughly cool route. When

* Maternal aggression involves the amygdala; no surprise there. But (harking back to chapter 1 and its discussion of the heterogeneity of types of aggression) it is also uniquely and crucially dependent on a tiny brain region that hasn't been mentioned before, the ventral premammillary nucleus of the hypothalamus.

† The next two paragraphs can be skipped if you already have enough complications in your life.

it enters neurons, it is converted to another steroid;* this binds to GABA receptors, making them more sensitive to the inhibitory effects of GABA, thereby calming the brain. Thus, direct cross-talk between hormones and neurotransmitters.

Bare-Knuckled Female Aggression

The traditional view is that other than maternal aggression, any female-female competition is passive, covert. As noted by the pioneering primatologist Sarah Blaffer Hrdy of the University of California at Davis, before the 1970s hardly anyone even researched competition among females.[53]

Nevertheless, there is plenty of female-female aggression. This is often dismissed with a psychopathology argument—if, say, a female chimp is murderous, it's because, well, she's crazy. Or female aggression is viewed as endocrine "spillover."[54] Females synthesize small amounts of androgens in the adrenals and ovaries; in the spillover view, the process of synthesizing "real" female steroid hormones is somewhat sloppy, and some androgenic steroids are inadvertently produced; since evolution is lazy and hasn't eliminated androgen receptors in female brains, there's some androgen-driven aggression.

These views are wrong for a number of reasons.

Female brains don't contain androgen receptors simply because they come from a similar blueprint as male brains. Instead, androgen receptors are distributed differently in the brains of females and males, with higher levels in some regions in females. There has been active selection for androgen effects in females.[55]

Even more important, female aggression makes *sense*—females can increase their evolutionary fitness with strategic, instrumental aggression.[56] Depending on the species, females compete aggressively for resources (e.g., food or nesting places), harass lower-ranking reproductive competitors into stress-induced infertility, or kill each other's infants (as in chimps).

---

* Called allopregnanolone.

And in the bird and (rare) primate species where males are actually paternal, females compete aggressively for such princes.

Remarkably, there are even species—primates (bonobos, lemurs, marmosets, and tamarins), rock hyraxes, and rodents (the California mouse, Syrian golden hamsters, and naked mole rats)—where females are socially dominant and more aggressive (and often more muscular) than males.[57] The most celebrated example of a sex-reversal system is the spotted hyena, shown by Laurence Frank of UC Berkeley and colleagues.* Among typical social carnivores (e.g., lions), females do most of the hunting, after which males show up and eat first. Among hyenas it's the socially subordinate males who hunt; they are then booted off the kill by females so that the kids eat first. Get this: In many mammals erections are a sign of dominance, of a guy strutting his stuff. Among hyenas it's reversed—when a female is about to terrorize a male, he gets an erection. ("Please don't hurt me! Look, I'm just a nonthreatening male.")†

What explains female competitive aggression (in sex-reversal species or "normal" animals)? Those androgens in females are obvious suspects, and in some sex-reversal species females have androgen levels that equal or even trump those in males.[58] Among hyenas, where this occurs, spending fetal life awash in Mom's plentiful androgens produces a "pseudohermaphrodite"‡—female hyenas have a fake scrotal sack, no external vagina, and a clitoris that is as large as a penis and gets erect as well.§ Moreover, some of the sex differences in the brain seen in most mammals don't occur in hyenas or naked mole rats, reflecting their fetal androgenization.

---

* Hyenas have a terrible reputation, thanks to outdated zoology that characterizes them, sneeringly, as "scavengers" (snarkiness that makes little sense, given that most of us just scavenge dead stuff in the supermarket). Rather than living off the scraps of lions' kills, they are highly effective hunters. Most often, it is scavenging lions trying to drive hyenas off a kill, rather than the other way around. And real hyenas don't sing inane songs like in *The Lion King.*

† Think about this: Among typical mammals, when males are terrified, they lose erections. Among hyenas, that's when they get one (and when some moth-eaten male has the chance to mate, he's probably terrified out of his mind). This implies some very different wiring of the autonomic nervous system, whereby stress promotes rather than inhibits erections.

‡ More than two thousand years ago Aristotle, for reasons obscure to even the most learned, dissected some dead hyenas, discussing them in his treatise *Historia Animalium*, VI, XXX. He drew the incorrect conclusion that these animals were hermaphrodites, possessing all the machinery of both sexes.

§ This leads us to another great factoid from the hyena world. If a low-ranking female is being menaced by a high-ranking one, the subordinate gets a clitoral erection—"Please don't hurt me; look, I'm just like one of those bedraggled, innocuous males."

This suggests that elevated female aggression in sex-reversal species arises from the elevated androgen exposure and, by extension, that the diminished aggression among females of other species comes from their low androgen levels.

But complications emerge. For starters, there are species (e.g., Brazilian guinea pigs) where females have high androgen levels but aren't particularly aggressive or dominant toward males. Conversely, there are sex-reversal bird species without elevated androgen levels in females. Moreover, as with males, individual levels of androgens in females, whether in conventional or sex-reversal species, do not predict individual levels of aggression. And most broadly, androgen levels don't tend to rise around periods of female aggression.[59]

This makes sense. Female aggression is mostly related to reproduction and infant survival—maternal aggression, obviously, but also female competition for mates, nesting places, and much-needed food during pregnancy or lactation. Androgens disrupt aspects of reproduction and maternal behavior in females. As emphasized by Hrdy, females must balance the proaggression advantages of androgens with their antireproductive disadvantages. Ideally, then, androgens in females should affect the "aggression" parts of the brain but not the "reproduction/maternalism" parts. Which is precisely what has evolved, as it turns out.*[60]

Perimenstrual Aggression and Irritability

Inevitably we turn to premenstrual syndrome (PMS)†—the symptoms of negative mood and irritability that come around the time of menstruation (along with the bloating of water retention, cramps, acne . . .). There's a lot of baggage and misconceptions about PMS (along with PMDD—premenstrual dysphoric disorder, where symptoms are severe enough to impair normal functioning; it affects 2 to 5 percent of women).[61]

---

* This is due to an obscure hormone called DHEA (dehydroepiandrosterone) that is converted to an androgen within only certain neurons, and, even stranger, some of those neurons even synthesize their own androgens.

† Many people more properly think of PMS as *peri*menstrual syndrome, in that the symptoms occur typically not just before the onset of menses, but for a few days after as well.

The topic is mired in two controversies—what causes PMS/PMDD, and how is it relevant to aggression? The first is a doozy. Is PMS/PMDD a biological disease or a social construct?

In the extreme "It's just a social construct" school, PMS is *entirely* culture specific, meaning it occurs only in certain societies. Margaret Mead started this by asserting in 1928 in *Coming of Age in Samoa* that Samoan women don't have mood or behavioral changes when menstruating. Since the Samoans were enshrined by Mead as the coolest, most peaceful and sexually free primates east of bonobos, this started trendy anthropological claims that women in other hip, minimal-clothing cultures had no PMS either.* And naturally, cultures with rampant PMS (e.g., American primates) were anti-Samoans, where symptoms arose from mistreatment and sexual repression of women. This view even had room for a socioeconomic critique, with howlers like "PMS [is] a mode for the expression of women's anger resulting from her oppressed position in American capitalist society."†[62]

An offshoot of this view is the idea that in such repressive societies, it's the most repressed women who have the worst PMS. Thus, depending on the paper, women with bad PMS must be anxious, depressed, neurotic, hypochondriacal, sexually repressed, toadies of religious repression, or more compliant with gender stereotypes and must respond to challenge by withdrawing, rather than by tackling things head on. In other words, not a single cool Samoan among them.

Fortunately, this has mostly subsided. Numerous studies show normal shifts in the brain and behavior over the course of the reproductive cycle, with as many behavioral correlates of ovulation as of menses.‡[63] PMS, then, is simply a disruptively extreme version of those shifts. While PMS

* Mead has been posthumously savaged by some Oceanic anthropologists for supposedly having painted a grossly inaccurate picture of Samoa as the Garden of Eden, in part because of her ideological desire to see Samoa that way. Naturally, others defend her strongly (admitting that she did have a pretty starry-eyed view of sexuality among "her" Samoans).

† This literature also has produced sentences like "Such a symbolic analysis is consistent with the hermeneutic, meaning-centered focus of the 'new cross-cultural psychiatry.'" I haven't a clue what in hell that means.

‡ For example, the "fusiform face area" is more responsive to faces in ovulating women than in those menstruating. Similarly, the "emotional" vmPFC is more responsive to men's faces when women are approaching ovulation than when they are approaching menses; the higher the ratio of estrogen to progesterone in the bloodstream during that preovulatory phase, the more vmPFC responsiveness. Finally, women find faces of men judged to look "aggressive" to be more attractive when they are ovulating.

is real, symptoms vary by culture. For example, perimenstrual women in China report less negative affect than do Western women (raising the issue of whether they experience less and/or report less). Given the more than one hundred symptoms linked to PMS, it's not surprising if different symptoms predominate in different populations.

As strong evidence that perimenstrual mood and behavioral changes are biological, they occur in other primates.[64] Both female baboons and female vervet monkeys become more aggressive and less social before their menses (without, to my knowledge, having issues with American capitalism). Interestingly, the baboon study showed increased aggressiveness only in dominant females; presumably, subordinate females simply couldn't express increased aggressiveness.

All these findings suggest that the mood and behavioral shifts are biologically based. What *is* a social construct is medicalizing and pathologizing these shifts as "symptoms," a "syndrome," or "disorder."

Thus, what is the underlying biology? A leading theory points to the plunging levels of progesterone as menses approaches and thus the loss of its anxiolytic and sedating effects. In this view, PMS arises from too extreme of a decline. However, there's not much actual support for this idea.

Another theory, backed by some evidence, concerns the hormone beta-endorphin, famed for being secreted during exercise and inducing a gauzy, euphoric "runner's high." In this model PMS is about abnormally low levels of beta-endorphin. There are plenty more theories but very little certainty.

Now for the question of how much PMS is associated with aggression. In the 1960s, studies by Katharina Dalton, who coined the term "premenstrual syndrome" in 1953, reported that female criminals committed their crimes disproportionately during their perimenstrual period (which may tell less about committing a crime than about getting caught).[65] Other studies of a boarding school showed a disproportionate share of "bad marks" for behavioral offenses going to perimenstrual students. However, the prison studies didn't distinguish between violent and nonviolent crimes, and the school study didn't distinguish between aggressive acts and infractions like tardiness. Collectively, there is little evidence that women tend toward aggression around their menses or that violent women are more likely to have committed their acts around their menses.

Nevertheless, defense pleas of PMS-related "diminished responsibility" have been successful in courtrooms.[66] A notable 1980 case concerned Sandie Craddock, who murdered a coworker and had a long rap sheet with more than thirty convictions for theft, arson, and assault. Incongruously but fortuitously, Craddock was a meticulous diarist, having years of records of not just when she was having her period but also when she was out about town on a criminal spree. Her criminal acts and times of menses matched so closely that she was put on probation plus progesterone treatment. And making the case stranger, Craddock's doctor later reduced her progesterone dose; by her next period, she had been arrested for attempting to knife someone. Probation again, plus a wee bit more progesterone.

These studies suggest that a small number of women do show perimenstrual behavior that qualifies as psychotic and should be mitigating in a courtroom.* Nevertheless, normal garden-variety perimenstrual shifts in mood and behavior are not particularly associated with increased aggression.

STRESS AND IMPRUDENT BRAIN FUNCTION

The time before some of our most important, consequential behaviors can be filled with stress. Which is too bad, since stress influences the decisions we make, rarely for the better.

* The broad question of what this book's avalanche of information says about criminal justice will be considered in chapter 16. I thank Dylan Alegria, a research assistant of mine, for superb assistance in reviewing the PMS/criminality literature.

The Basic Dichotomy of the Acute and the Chronic Stress Response

We begin with a long-forgotten term from ninth-grade biology. Remember "homeostasis"? It means having an ideal body temperature, heart rate, glucose level, and so on. A "stressor" is anything that disrupts homeostatic balance—say, being chased by a lion if you're a zebra, or chasing after a zebra if you're a hungry lion. The stress response is the array of neural and endocrine changes that occur in that zebra or lion, designed to get them through that crisis and reestablish homeostasis.*[67]

Critical events in the brain mediate the start of the stress response. (Warning: the next two paragraphs are technical and not essential.) The sight of the lion activates the amygdala; amygdaloid neurons stimulate brain-stem neurons, which then inhibit the parasympathetic nervous system and mobilize the sympathetic nervous system, releasing epinephrine and norepinephrine throughout the body.

The amygdala also mediates the other main branch of the stress response, activating the paraventricular nucleus (PVN) in the hypothalamus. And the PVN sends projections to the base of the hypothalamus, where it secretes corticotropin-releasing hormone (CRH); this triggers the pituitary to release adrenocorticotropic hormone (ACTH), which stimulates glucocorticoid secretion from the adrenals.

Glucocorticoids plus the sympathetic nervous system enable an organism to survive a physical stressor by activating the classical "fight or flight" response. Whether you are that zebra or that lion, you'll need energy for your muscles, and the stress response rapidly mobilizes energy into circulation from storage sites in your body. Furthermore, heart rate and blood pressure increase, delivering that circulating energy to exercising muscles faster. Moreover, during stress, long-term building projects—growth, tissue repair, and reproduction—are postponed until after the

* For real aficionados: In recent years "homeostasis" has been expanded and fancified into the new, elegant concept of "allostasis." Most basically, it incorporates the fact that an ideal homeostatic set point in the body varies dramatically depending on the circumstance.

crisis; after all, if a lion is chasing you, you have better things to do with your energy than, say, thicken your uterine walls. Beta-endorphin is secreted, the immune system is stimulated, and blood clotting is enhanced, all useful following painful injury. Moreover, glucocorticoids reach the brain, rapidly enhancing aspects of cognition and sensory acuity.

This is wonderfully adaptive for the zebra or lion; try sprinting without epinephrine and glucocorticoids, and you'll soon be dead. Reflecting its importance, this basic stress response is ancient physiology, found in mammals, birds, fish, and reptiles.

What is not ancient is how stress works in smart, socially sophisticated, recently evolved primates. For primates the definition of a stressor expands beyond merely a physical challenge to homeostasis. In addition, it includes thinking you're *going to be* thrown out of homeostasis. An anticipatory stress response is adaptive if there really is a physical challenge coming. However, if you're constantly but incorrectly convinced that you're about to be thrown out of balance, you're being an anxious, neurotic, paranoid, or hostile primate who is *psychologically* stressed. And the stress response did not evolve for dealing with this recent mammalian innovation.

Mobilizing energy while sprinting for your life helps save you. Do the same thing chronically because of a stressful thirty-year mortgage, and you're at risk for various metabolic problems, including adult-onset diabetes. Likewise with blood pressure: increase it to sprint across the savanna—good thing. Increase it because of chronic psychological stress, and you've got stress-induced hypertension. Chronically impair growth and tissue repair, and you'll pay the price. Ditto for chronically inhibiting reproductive physiology; you'll disrupt ovulatory cycles in women and cause plummeting erections and testosterone levels in men. Finally, while the acute stress response involves enhanced immunity, chronic stress suppresses immunity, increasing vulnerability to some infectious diseases.*

* More info for aficionados: The suppression of immunity and inflammation during chronic stress is caused by glucocorticoids. This is the reason why glucocorticoids are used to suppress the immune system in people with an overactive immune system (i.e., an autoimmune disease), to prevent rejection of a transplanted organ, or to suppress an overactive inflammatory response. This is what happens when people are put on immunosuppressive/anti-inflammatory "steroids" like cortisone or prednisone (two synthetic glucocorticoids).

We have a dichotomy—if you're stressed like a normal mammal in an acute physical crisis, the stress response is lifesaving. But if instead you chronically activate the stress response for reasons of psychological stress, your health suffers. It is a rare human who sickens because they can't activate the stress response when it is needed. Instead, we get sick from activating the stress response too often, too long, and for purely psychological reasons. Crucially, the beneficial effects of the stress response for sprinting zebras and lions play out over the course of seconds to minutes. But once you take stress to the time course of this chapter (henceforth referred to as "sustained" stress), you'll be dealing with adverse consequences. Including some unwelcome effects on the behaviors that fill this book.

A Brief Digression: Stress That We Love

Either running from a lion or dealing with years of traffic jams is a drag. Which contrasts with stress that we love.[68]

We love stress that is mild and transient and occurs in a benevolent context. The stressful menace of a roller-coaster ride is that it will make us queasy, not that it will decapitate us; it lasts for three minutes, not three days. We love that kind of stress, clamor for it, pay to experience it. What do we call that optimal amount of stress? Being engaged, engrossed, and challenged. Being stimulated. Playing. The core of psychological stress is loss of control and predictability. But in benevolent settings we happily relinquish control and predictability to be challenged by the unexpected—a dip in the roller-coaster tracks, a plot twist, a difficult line drive heading our way, an opponent's unexpected chess move. Surprise me—this is fun.

This brings up a key concept, namely the inverted *U*. The complete absence of stress is aversively boring. Moderate, transient stress is wonderful—various aspects of brain function are enhanced; glucocorticoid levels in that range enhance dopamine release; rats work at pressing levers in order to be infused with just the right amount of glucocorticoids. And as stress becomes more severe and prolonged, those good effects

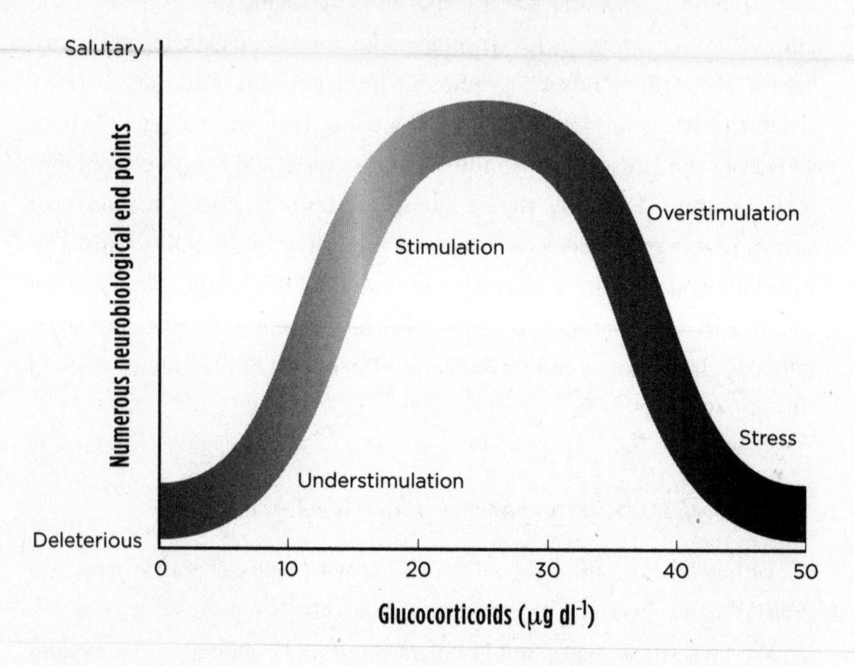

disappear (with, of course, dramatic individual differences as to where the transition from stress as stimulatory to overstimulatory occurs; one person's nightmare is another's hobby).*

We love the right amount of stress, would wither without it. But back now to sustained stress and the right side of the inverted *U*.

Sustained Stress and the Neurobiology of Fear

For starters, sustained stress makes people implicitly (i.e., not consciously) look more at angry faces. Moreover, during stress, that sensory shortcut from the thalamus to the amygdala becomes more active, with

* How does the brain pull off an inverted *U*, where a moderate rise in glucocorticoid levels enhances memory (for example), while a bigger rise does the opposite? One solution that the brain has evolved is to have two receptor systems for glucocorticoids. One (the "MR") is responsive to small increases in glucocorticoid levels above baseline and mediates the stimulatory effects. The other ("GR") receptors respond only to big, prolonged increases and mediate the adverse effects. Predictably, levels of the two types of receptors vary by brain region, person, and circumstance.

more excitable synapses; we know the resulting trade-off between speed and accuracy. Compounding things further, glucocorticoids decrease activation of the (cognitive) medial PFC during processing of emotional faces. Collectively, stress or glucocorticoid administration decreases accuracy when rapidly assessing emotions of faces.[69]

Meanwhile, during stress things aren't going great in the amygdala. The region is highly sensitive to glucocorticoids, with lots of glucocorticoid receptors; stress and glucocorticoids increase excitability of amygdaloid neurons,* particularly in the basolateral amygdala (the BLA), with its role in learning fear. Thus, this is another contingent hormone action—glucocorticoids don't cause action potentials in amygdaloid neurons, don't invent excitation. Instead they amplify preexisting excitation. Stress and glucocorticoids also increase levels of CRH in the BLA, and of a growth factor that builds new dendrites and synapses (brain-derived neurotrophic factor, or BDNF).[70]

Recall from chapter 2 how during a fearful situation the amygdala recruits the hippocampus into remembering contextual information about the event (e.g., the amygdala remembers the thief's knife, whereas the hippocampus remembers where the robbery occurred).[71] Stress strengthens this recruitment, making the hippocampus a temporary fear-laden suburb of the amygdala. Thanks to these glucocorticoid actions in the amygdala,† stress makes it easier to learn a fear association and to consolidate it into a long-term memory.

This sets us up for a positive feedback loop. As noted, with the onset of stress, the amygdala indirectly activates the glucocorticoid stress response. And in turn glucocorticoids increase amygdala excitability.

Stress also makes it harder to *unlearn* fear, to "extinguish" a conditioned fear association. This involves the prefrontal cortex, which causes fear extinction by inhibiting the BLA (as covered in chapter 2); stress weakens the PFC's hold over the amygdala.[72]

---

* As just stated, stress increases overall excitability in the amygdala. This involves the *inhibition* of particular neurons—namely the inhibitory GABA interneurons. Inhibiting the inhibitors in the circuit causes the activity of the big, excitatory glutamate-releasing neurons to increase.

† Plus, more obscurely, via the sympathetic nervous system indirectly activating the amygdala by way of that norepinephrine-releasing projection into it from the locus coeruleus (the brain-stem region that was mentioned briefly in chapter 2—its activation leads to arousal throughout the brain).

Recall what fear extinction is about. You've learned to fearfully associate a light with a shock, but today the light keeps coming on with no shock. Extinction is not passively forgetting that light equals shock. It is the BLA actively learning that light no longer equals shock. Thus stress facilitates learning fear associations but impairs learning fear extinction.

Sustained Stress, Executive Function, and Judgment

Stress compromises other aspects of frontal cortical function. Working memory is disrupted; in one study, prolonged administration of high glucocorticoid levels to healthy subjects impaired working memory into the range seen after frontal cortical damage. Glucocorticoids accomplish this by enhancing norepinephrine signaling in the PFC so much that, instead of causing aroused focus, it induces chicken-with-its-head-cut-off cognitive tumult, and by enhancing disruptive signaling from the amygdala to the PFC. Stress also desynchronizes activation in different frontocortical regions, which impairs the ability to shift attention between tasks.[73]

These stress effects on frontal function also make us perseverative—in a rut, set in our ways, running on automatic, being habitual. We all know this—what do we typically do during a stressful time when something isn't working? The same thing again, many more times, faster and more intensely—it becomes unimaginable that the usual isn't working. This is precisely where the frontal cortex makes you do the harder but more correct thing—recognize that it's time for a change. Except for a stressed frontal cortex, or one that's been exposed to a lot of glucocorticoids. In rats, monkeys, and humans, stress weakens frontal connections with the hippocampus—essential for incorporating the new information that should prompt shifting to a new strategy—while strengthening frontal connections with more habitual brain circuits.[74]

Finally, the decreased frontal function and increased amygdaloid function during stress alter risk-taking behavior. For example, the stress of sleep deprivation or of public speaking, or the administration of high glucocorticoid levels, shifts people from protecting against losses to seeking bigger gains when gambling. This involves an interesting gender difference—in general, major stressors make people of both genders more risk taking. But moderate stressors bias men toward, and women away from, risk taking. In the absence of stress, men tend toward more risk taking than women; thus, once again, hormones enhance a preexisting tendency.[75]

Whether one becomes irrationally risk taking (failing to shift strategy in response to a declining reward rate) or risk averse (failing to respond to the opposite), one is incorporating new information poorly. Stated most broadly, sustained stress impairs risk assessment.[76]

Sustained Stress and Pro- and Antisociality

During sustained stress, the amygdala processes emotional sensory information more rapidly and less accurately, dominates hippocampal function, and disrupts frontocortical function; we're more fearful, our thinking is muddled, and we assess risks poorly and act impulsively out of habit, rather than incorporating new data.[77] This is a prescription for rapid, reactive aggression; stress and acute administration of glucocorticoids increase such aggression in both rodents and humans. We have two familiar qualifications: (a) rather than creating aggression, stress and glucocorticoids increase sensitivity to social triggers of aggression; (b) this occurs most readily in individuals already predisposed toward aggression. As we will see in the next chapter, stress over the course of weeks to months produces a less nuanced picture.

There's an additional depressing reason why stress fosters aggression—because it reduces stress. Shock a rat and its glucocorticoid levels and blood pressure rise; with enough shocks, it's at risk for a "stress" ulcer. Various things can buffer the rat during shocks—running on a running wheel, eating, gnawing on wood in frustration. But a particularly effective buffer is for the rat to bite another rat. Stress-induced (aka frustration-induced) displacement aggression is ubiquitous in various species. Among baboons, for example, nearly half of aggression is this type—a high-ranking male loses a fight and chases a subadult male, who promptly bites a female, who then lunges at an infant. My research shows that within the same dominance rank, the more a baboon tends to displace aggression after losing a fight, the lower his glucocorticoid levels.[78]

Humans excel at stress-induced displacement aggression—consider how economic downturns increase rates of spousal and child abuse. Or consider a study of family violence and pro football. If the local team unexpectedly loses, spousal/partner violence by men increases 10 percent soon afterward (with no increase when the team won or was expected to lose). And as the stakes get higher, the pattern is exacerbated: a 13 percent increase after upsets when the team was in playoff contention, a 20 percent increase when the upset is by a rival.[79]

Little is known concerning the neurobiology of displacement aggression blunting the stress response. I'd guess that lashing out activates dopaminergic reward pathways, a surefire way to inhibit CRH release.*[80] Far too often, giving an ulcer helps avoid getting one.

More bad news: stress biases us toward selfishness. In one study subjects answered questions about moral decision-making scenarios after either a social stressor or a neutral situation.† Some scenarios were of low emotional intensity ("In the supermarket you wait at the meat counter and an elderly man pushes to the front. Would you complain?"), others high intensity ("You meet the love of your life, but you are married and have children. Would you leave your family?"). Stress made people give more egoistic answers about emotionally intense moral decisions (but not milder ones); the more glucocorticoid levels rose, the more egoistic the answers. Moreover, in the same paradigm, stress lessened how altruistic people claimed they'd be concerning personal (but not impersonal) moral decisions.[81]

We have another contingent endocrine effect: stress makes people more egoistic, but only in the most emotionally intense and personal circumstances.‡ This resembles another circumstance of poor frontal

* And the underlying neurobiology is probably similar to what is going on in those other realms of poor decision making during stress, e.g., eating or drinking more.

† The test is a standard in the field called the Trier Social Stress Test, which is a fifteen-minute combination of a mock job interview and a mental arithmetic task, both carried out in front of a panel of stone-faced evaluators.

‡ Note that these are studies about what people *say* they would do, not what they *actually* do. The difference between the two will be considered in chapter 13, when considering moral reasoning versus moral action.

function—recall from chapter 2 how individuals with frontal cortical damage make reasonable judgments about someone else's issues, but the more personal and emotionally potent the issue, the more they are impaired.

Feeling better by abusing someone innocent, or thinking more about your own needs, is not compatible with feeling empathy. Does stress decrease empathy? Seemingly yes, in both mice and humans. A remarkable 2006 paper in *Science* by Jeffrey Mogil of McGill University showed the rudiments of mouse empathy—a mouse's pain threshold is lowered when it is near another mouse in pain, but only if the other mouse is its cagemate.[82]

This prompted a follow-up study that I did with Mogil's group involving the same paradigm. The presence of a strange mouse triggers a stress response. But when glucocorticoid secretion is temporarily blocked, mice show the same "pain empathy" for a strange mouse as for a cagemate. In other words, to personify mice, glucocorticoids narrow who counts as enough of an "Us" to evoke empathy. Likewise in humans—pain empathy was not evoked for a stranger unless glucocorticoid secretion was blocked (either after administration of a short-acting drug or after the subject and stranger interacted socially). Recall from chapter 2 the involvement of the anterior cingulate cortex in pain empathy. I bet that glucocorticoids do some disabling, atrophying things to neurons there.

Thus, sustained stress has some pretty unappealing behavioral effects. Nonetheless there are circumstances where stress brings out the magnificent best in some people. Work by Shelley Taylor of UCLA shows that "fight or flight" is the typical response to stress in males, and naturally, the stress literature is predominantly studies of males by males.[83] Things often differ in females. Showing that she can match the good old boys when it comes to snappy sound bites, Taylor framed the female stress response as being more about "tend and befriend"—caring for your young and seeking social affiliation. This fits with striking sex differences in stress management styles, and tend-and-befriend most likely reflects the female stress response involving a stronger component of oxytocin secretion.

Naturally, things are subtler than "male = fight/flight and female = tend/befriend." There are frequent counterexamples to each; stress elicits prosociality in more males than just pair-bonded male marmosets, and we saw that females are plenty capable of aggression. Then there's Mahatma Gandhi and Sarah Palin.* Why are some people exceptions to these gender stereotypes? That's part of what the rest of this book is about.

Stress can disrupt cognition, impulse control, emotional regulation, decision making, empathy, and prosociality. One final point. Recall from chapter 2 how the frontal cortex making you do the harder thing when it's the right thing is value free—"right thing" is purely instrumental. Same with stress. Its effects on decision making are "adverse" only in a neurobiological sense. During a stressful crisis, an EMT may become perseverative, making her ineffectual at saving lives. A bad thing. During a stressful crisis, a sociopathic warlord may become perseverative, making him ineffectual at ethnically cleansing a village. Not a bad thing.

SOME IMPORTANT DEBUNKING: ALCOHOL

No review of the biological events in the minutes to hours prior to a behavior can omit alcohol. As everyone knows, alcohol lessens inhibitions, making people more aggressive. Wrong, and in a familiar way—alcohol only evokes aggression only in (a) individuals prone to aggression (for example, mice with lower levels of serotonin signaling in the frontal cortex and men with the oxytocin receptor gene variant less responsive to oxytocin are preferentially made aggressive by alcohol) and (b) those who *believe* that alcohol makes you more aggressive, once more showing the power of social learning to shape biology.[84] Alcohol works differently in everyone else—for example, a drunken stupor has caused many a quickie Vegas wedding that doesn't seem like a great idea with the next day's sunrise.

* Okay, that was a juvenile cheap shot thrown in merely to increase the number of moose buying copies of this book.

SUMMARY AND SOME CONCLUSIONS

- Hormones are great; they run circles around neurotransmitters, in terms of the versatility and duration of their effects. And this includes affecting the behaviors pertinent to this book.
- Testosterone has far less to do with aggression than most assume. Within the normal range, individual differences in testosterone levels don't predict who will be aggressive. Moreover, the more an organism has been aggressive, the less testosterone is needed for future aggression. When testosterone does play a role, it's facilitatory—testosterone does not "invent" aggression. It makes us more sensitive to triggers of aggression, particularly in those most prone to aggression. Also, rising testosterone levels foster aggression only during challenges to status. Finally, crucially, the rise in testosterone during a status challenge does not necessarily increase aggression; it increases whatever is needed to maintain status. In a world in which status is awarded for the best of our behaviors, testosterone would be the most prosocial hormone in existence.
- Oxytocin and vasopressin facilitate mother-infant bond formation and monogamous pair-bonding, decrease anxiety and stress, enhance trust and social affiliation, and make people more cooperative and generous. But this comes with a huge caveat—these hormones increase prosociality only toward an Us. When dealing with Thems, they make us more ethnocentric and xenophobic. Oxytocin is not a universal luv hormone. It's a parochial one.
- Female aggression in defense of offspring is typically adaptive and is facilitated by estrogen, progesterone, and oxytocin. Importantly, females are aggressive in many other evolutionarily adaptive circumstances. Such aggression is facilitated by the presence of androgens in females and by complex neuroendocrine tricks for generating androgenic signals in "aggressive," but not "maternal"

or "affiliative," parts of the female brain. Mood and behavioral changes around the time of menses are a biological reality (albeit poorly understood on a nuts-and-bolts level); in contrast, pathologizing these shifts is a social construct. Finally, except for rare, extreme cases, the link between PMS and aggression is minimal.

- Sustained stress has numerous adverse effects. The amygdala becomes overactive and more coupled to pathways of habitual behavior; it is easier to learn fear and harder to unlearn it. We process emotionally salient information more rapidly and automatically, but with less accuracy. Frontal function—working memory, impulse control, executive decision making, risk assessment, and task shifting—is impaired, and the frontal cortex has less control over the amygdala. And we become less empathic and prosocial. Reducing sustained stress is a win-win for us and those stuck around us.
- "I'd been drinking" is no excuse for aggression.
- Over the course of minutes to hours, hormonal effects are predominantly contingent and facilitative. Hormones don't determine, command, cause, or invent behaviors. Instead they make us more sensitive to the social triggers of emotionally laden behaviors and exaggerate our preexisting tendencies in those domains. And where do those preexisting tendencies come from? From the contents of the chapters ahead of us.

Five

Days to Months Before

Our act has occurred—the pulling of a trigger or the touching of an arm that can mean such different things in different contexts. Why did that just happen? We've seen how, seconds before, that behavior was the product of the nervous system, whose actions were shaped by sensory cues minutes to hours before, and how the brain's sensitivity to those cues was shaped by hormonal exposure in the preceding hours to days. What events in the prior days to months shaped that outcome?

Chapter 2 introduced the plasticity of neurons, the fact that things alter in them. The strength of a dendritic input, the axon hillock's set point for initiating an action potential, the duration of the refractory period. The previous chapter showed that, for example, testosterone increases the excitability of amygdaloid neurons, and glucocorticoids decrease excitability of prefrontal cortical neurons. We even saw how progesterone boosts the efficacy with which GABA-ergic neurons decrease the excitability of other neurons.

Those versions of neural plasticity occur over hours. We now examine more dramatic plasticity occurring over days to months. A few months is enough time for an Arab Spring, for a discontented winter, or for STDs to spread a lot during a Summer of Love. As we'll see, this is also sufficient time for enormous changes in the brain's structure.

NONLINEAR EXCITATION

We start small. How can events from months ago produce a synapse with altered excitability today? How do synapses "remember"?

When neuroscientists first approached the mystery of memory at the start of the twentieth century, they asked that question on a more macro level—how does a brain remember? Obviously, a memory was stored in a single neuron, and a new memory required a new neuron.

The discovery that adult brains don't make new neurons trashed that idea. Better microscopes revealed neuronal arborization, the breathtaking complexity of branches of dendrites and axon terminals. Maybe a new memory requires a neuron to grow a new axonal or dendritic branch.

Knowledge emerged about synapses, neurotransmitter-ology was born, and this idea was modified—a new memory requires the formation of a new synapse, a new connection between an axon terminal and a dendritic spine.

These speculations were tossed on the ash heap of history in 1949, because of the work of the Canadian neurobiologist Donald Hebb, a man so visionary that even now, nearly seventy years later, neuroscientists still own bobblehead dolls of him. In his seminal book, *The Organization of Behaviour*, Hebb proposed what became the dominant paradigm. Forming memories doesn't require new synapses (let alone new branches or neurons); it requires the strengthening of *preexisting* synapses.[1]

What does "strengthening" mean? In circuitry terms, if neuron A synapses onto neuron B, it means that an action potential in neuron A more readily triggers one in neuron B. They are more tightly coupled; they "remember." Translated into cellular terms, "strengthening" means that the wave of excitation in a dendritic spine spreads farther, getting closer to the distant axon hillock.

Extensive research shows that experience that causes repeated firing across a synapse "strengthens" it, with a key role played by the neurotransmitter glutamate.

Recall from chapter 2 how an excitatory neurotransmitter binds to its receptor in the postsynaptic dendritic spine, causing a sodium channel to open; some sodium flows in, causing a blip of excitation, which then spreads.

Glutamate signaling works in a fancier way that is essential to learning.[2] To simplify considerably, while dendritic spines typically contain only one type of receptor, those responsive to glutamate contain two. The first (the "non-NMDA") works in a conventional way—for every little smidgen of glutamate binding to these receptors, a smidgen of sodium flows in, causing a smidgen of excitation. The second (the "NMDA") works in a nonlinear, threshold manner. It is usually unresponsive to glutamate. It's not until the non-NMDA has been stimulated over and over by a long train of glutamate release, allowing enough sodium to flow in, that this activates the NMDA receptor. It suddenly responds to all that glutamate, opening its channels, allowing an explosion of excitation.

This is the essence of learning. The lecturer says something, and it goes in one ear and out the other. The factoid is repeated; same thing. It's repeated enough times and—aha!—the lightbulb goes on and suddenly you get it. At a synaptic level, the axon terminal having to repeatedly release glutamate is the lecturer droning on repetitively; the moment when the postsynaptic threshold is passed and the NMDA receptors first activate is the dendritic spine finally getting it.

"AHA" VERSUS ACTUALLY REMEMBERING

But this has only gotten us to first base. The lightbulb going on in the middle of the lecture doesn't mean it'll still be on in an hour, let alone during the final exam. How can we make that burst of excitation persist, so that NMDA receptors "remember," are more easily activated in the future? How does the potentiated excitation become long term?

This is our cue to introduce the iconic concept of LTP—"long-term potentiation." LTP, first demonstrated in 1966 by Terje Lømo at the University of Oslo, is the process by which the first burst of NMDA receptor activation causes a prolonged increase in excitability of the synapse.* Hundreds of productive careers have been spent figuring out how LTP works, and the key is that when NMDA receptors finally activate and open their channels, it is calcium, rather than sodium, that flows in. This causes an array of changes; here are a few:

- The calcium tidal wave causes more copies of glutamate receptors to be inserted into the dendritic spine's membrane, making the neuron more responsive to glutamate thereafter.†
- The calcium also alters glutamate receptors that are already on the front lines of that dendritic spine; each will now be more sensitive to glutamate signals.‡
- The calcium also causes the synthesis of peculiar neurotransmitters in the dendritic spine, which are released and travel *backward* across the synapse; there they increase the amount of glutamate released from the axon terminal after future action potentials.

In other words, LTP arises from a combination of the presynaptic axon terminal yelling "glutamate" more loudly and the postsynaptic dendritic spine listening more attentively.

As I said, additional mechanisms underlie LTP, and neuroscientists debate *which* is most important (the one they study, naturally) in neurons in organisms when they are actually learning. In general, the debate has been whether pre- or the postsynaptic changes are more crucial.

After LTP came a discovery that suggests a universe in balance. This

* Although nothing was known about NMDA and non-NMDA receptors at the time.

† Where do more copies of the receptors come from? Miles away from that dendritic spine, in the center of that neuron, is the nucleus, containing the DNA, which includes genes coding for glutamate receptors. Somehow the nucleus needs to hear that a calcium wave occurred in one dendritic spine in the boondocks. The nucleus then directs the synthesis of more copies of the receptor, which are then shipped to that specific spine, out of the neuron's ten thousand. That's insanely hard. Typically, instead, there are extra glutamate receptors mothballed inside dendritic spines, and the calcium tidal wave is the signal that pulls them onto the spine's membrane.

‡ For aficionados, the non-NMDA receptors are "phosphorylated," which causes their sodium channels to stay open longer.

is LTD—long-term "depression"—experience-dependent, long-term decreases in synaptic excitability (and, interestingly, the mechanisms underlying LTD are not merely the opposite of LTP). LTD is not the functional opposite of LTP either—rather than being the basis of generic forgetting, it sharpens a signal by erasing what's extraneous.

A final point about LTP. There's long term and there's *long* term. As noted, one mechanism underlying LTP is an alteration in glutamate receptors so that they are more responsive to glutamate. That change might persist for the lifetime of the copies of that receptor that were in that synapse at the time of the LTPing. But that's typically only a few *days*, until those copies accumulate bits of oxygen-radical damage and are degraded and replaced with new copies (similar updating of all proteins constantly occurs). Somehow LTP-induced changes in the receptor are transferred to the next generation of copies. How else can octogenarians remember kindergarten? The mechanism is elegant but beyond the scope of this chapter.

All this is cool, but LTP and LDP are what happens in the hippocampus when you learn explicit facts, like someone's phone number. But we're interested in other types of learning—how we learn to be afraid, to control our impulses, to feel empathy, or to feel nothing for someone else.

Synapses utilizing glutamate occur throughout the nervous system, and LTP isn't exclusive to the hippocampus. This was a traumatic discovery for many LTP/hippocampus researchers—after all, LTP is what occurred in Schopenhauer's hippocampus when he read Hegel, not what the spinal cord does to make you more coordinated at twerking.*

Nonetheless, LTP occurs throughout the nervous system.†[3] For example, fear conditioning involves synapses LTPing in the basolateral amygdala. LTP underlies the frontal cortex learning to control the amygdala. It's how dopaminergic systems learn to associate a stimulus with a

* Actually, LTP in the spinal cord has more to do with "neuropathic" pain, syndromes where a severe injury causes all sorts of nonnoxious stimuli to start hurting chronically—in effect, your spine has "learned" to always feel pain. Interestingly, such LTP arises in part from the inflammation that accompanies the initial injury.

† The mechanisms underlying LTP elsewhere in the nervous system often differ from those of hippocampal LTP—some involve a third class of glutamate receptors; some may not even involve glutamate. The LTP old guard has generally coped with the indignity of LTP outside the hippocampus by viewing the hippocampal kind as classical, canonical, textbook, divine, etc., and the rest as chintzy knockoffs.

reward—for example, how addicts come to associate a location with a drug, feeling cravings when in that setting.

Let's add hormones to this, translating some of our stress concepts into the language of neural plasticity. Moderate, transient stress (i.e., the good, stimulatory stress) promotes hippocampal LTP, while prolonged stress disrupts it and promotes LTD—one reason why cognition tanks at such times. This is the inverted-*U* concept of stress writ synaptic.[4]

Moreover, sustained stress and glucocorticoid exposure enhance LTP and suppress LTD in the amygdala, boosting fear conditioning, and suppress LTP in the frontal cortex. Combining these effects—more excitable synapses in the amygdala, fewer ones in the frontal cortex—helps explain stress-induced impulsivity and poor emotional regulation.[5]

Rescued from the Trash

The notion of memory resting on the strengthening of preexisting synapses dominates the field. But ironically, the discarded idea that memory requires the formation of new synapses has been resuscitated. Techniques for counting all of a neuron's synapses show that housing rats in a rich, stimulatory environment increases their number of hippocampal synapses.

Profoundly fancy techniques let you follow one dendritic branch of a neuron over time as a rat learns something. Astonishingly, over minutes to hours a new dendritic spine emerges, followed by an axon terminal hovering nearby; over the next weeks, they form a functioning synapse that stabilizes the new memory (and in other circumstances, dendritic spines retract, eliminating synapses).

Such "activity-dependent synaptogenesis" is coupled to LTP—when a synapse undergoes LTP, the tsunami of calcium rushing into the spine can diffuse and trigger the formation of a new spine in the adjacent stretch of the dendritic branch.

New synapses form throughout the brain—in motor-cortex neurons when you learn a motoric task, or in the visual cortex after lots of visual stimulation. Stimulate a rat's whiskers a lot, and ditto in the "whisker cortex."[6]

Moreover, when enough new synapses form in a neuron, the length and number of branches in its dendritic "tree" often expand as well, increasing the strength and number of the neurons that can talk to it.

Stress and glucocorticoids have inverted-*U* effects here as well. Moderate, transient stress (or exposure to the equivalent glucocorticoid levels) increases spine number in the hippocampus; sustained stress or glucocorticoid exposure does the opposite.[7] Moreover, major depression or anxiety—two disorders associated with elevated glucocorticoid levels—can reduce hippocampal dendrite and spine number. This arises from decreased levels of that key growth factor mentioned earlier this chapter, BDNF.

Sustained stress and glucocorticoids also cause dendritic retraction and synapse loss, lower levels of NCAM (a "neural cell adhesion molecule" that stabilizes synapses), and less glutamate release in the frontal cortex. The more of these changes, the more attentional and decision-making impairments.[8]

Recall from chapter 4 how acute stress strengthens connectivity between the frontal cortex and motoric areas, while weakening frontal-hippocampal connections; the result is decision making that is habitual, rather than incorporating new information. Similarly, chronic stress increases spine number in frontal-motor connections and decreases it in frontal-hippocampal ones.[9]

Continuing the theme of the amygdala differing from the frontal cortex and hippocampus, sustained stress increases BDNF levels and expands dendrites in the BLA, persistently increasing anxiety and fear conditioning.[10] The same occurs in that way station by which the amygdala talks to the rest of the brain (the BNST—bed nucleus of the stria terminalis). Recall that while the BLA mediates fear conditioning, the central amygdala is more involved in innate phobias. Interestingly, stress seems not to increase the force of phobias or spine number in the central amygdala.

There's wonderful context dependency to these effects. When a rat secretes tons of glucocorticoids because it's terrified, dendrites atrophy in the hippocampus. However, if it secretes the same amount by voluntarily running on a running wheel, dendrites expand. Whether the amygdala is also activated seems to determine whether the hippocampus interprets the glucocorticoids as good or bad stress.[11]

Spine number and branch length in the hippocampus and frontal cortex are also increased by estrogen.[12] Remarkably, the size of neurons' dendritic trees in the hippocampus expands and contracts like an accordion throughout a female rat's ovulatory cycle, with the size (and her cognitive skills) peaking when estrogen peaks.*

Thus, neurons can form new dendritic branches and spines, increasing the size of their dendritic tree or, in other circumstances, do the opposite; hormones frequently mediate these effects.

Axonal Plasticity

Meanwhile, there's plasticity at the other end of the neuron, where axons can sprout offshoots that head off in novel directions. As a spectacular example, when a blind person adept at Braille reads in it, there's the same activation of the tactile cortex as in anyone else; but amazingly, uniquely, there is also activation of the *visual* cortex.[13] In other words, neurons that normally send axons to the fingertip-processing part of the cortex instead have gone miles off course, growing projections to the visual cortex. One extraordinary case concerned a congenitally blind woman, adept at Braille, who had a stroke in her visual cortex. And as a result, she lost the ability to read Braille—the bumps on the page felt flattened, imprecise—while other tactile functions remained. In another study, blind subjects were trained to associate letters with distinctive tones, to the point where they could hear a sequence of tones as letters and words. When these individuals would "read with sound," they'd activate the part of the visual cortex activated in sighted individuals when reading. Similarly,

* Equally remarkably, over the course of the menstrual cycle in humans, the amount of myelin in the corpus callosum, the massive bundle of axons that connects the two hemispheres, fluctuates as well.

when a person who is deaf and adept at American Sign Language watches someone signing, there is activation of the part of their auditory cortex normally activated by speech.

The injured nervous system can "remap" in similar ways. Suppose there is stroke damage to the part of your cortex that receives tactile information from your hand. The tactile receptors in your hand work fine but have no neurons to talk to; thus you lose sensation in your hand. In the subsequent months to years, axons from those receptors can sprout off in new directions, shoehorning their way into neighboring parts of the cortex, forming new synapses there. An imprecise sense of touch may slowly return to the hand (along with a less precise sense of touch in the part of the body projecting to the cortical region that accommodated those refugee axon terminals).

Suppose, instead, that tactile receptors in the hand are destroyed, no longer projecting to those sensory cortical neurons. Neurons abhor a vacuum, and tactile neurons in the wrist may sprout collateral axonal branches and expand their territory into that neglected cortical region. Consider blindness due to retinal degeneration, where the projections to the visual cortex are silenced. As described, fingertip tactile neurons involved in reading Braille sprout projections into the visual cortex, setting up camp there. Or suppose there is a pseudoinjury: after merely five days of subjects being blindfolded, auditory projections start to remap into the visual cortex (and retract once the blindfolds come off).[14]

Consider how fingertip tactile neurons carrying information about Braille remap to the visual cortex in someone blind. The sensory cortex and visual cortex are far away from each other. How do those tactile neurons "know" (a) that there's vacant property in the visual cortex; (b) that hooking up with those unoccupied neurons helps turn fingertip information into "reading"; and (c) how to send axonal projections to this new cortical continent? All are matters of ongoing research.

What happens in a blind person when auditory projection neurons expand their target range into the inactive visual cortex? More acute hearing—the brain can respond to deficits in one realm with compensations in another.

So sensory projection neurons can remap. And once, say, visual cortex neurons are processing Braille in a blind person, *those* neurons need to remap where they project to, triggering further downstream remapping. Waves of plasticity.

Remapping occurs regularly throughout the brain in the absence of injury. My favorite examples concern musicians, who have larger auditory cortical representation of musical sounds than do nonmusicians, particularly for the sound of their own instrument, as well as for detecting pitch in speech; the younger the person begins being a musician, the stronger the remapping.[15]

Such remapping does not require decades of practice, as shown in beautiful work by Alvaro Pascual-Leone at Harvard.[16] Nonmusician volunteers learned a five-finger exercise on the piano, which they practiced for two hours a day. Within a few days the amount of motor cortex devoted to the movement of that hand expanded, but the expansion lasted less than a day without further practice. This expansion was probably "Hebbian" in nature, meaning preexisting connections transiently strengthened after repeated use. However, if subjects did the daily exercise for a crazed four weeks, the remapping persisted for many days afterward. This expansion probably involved axonal sprouting and the formation of new connections. Remarkably, remapping also occurred in volunteers who spent two hours a day *imagining* playing the finger exercise.

As another example of remapping, after female rats give birth, there is expansion of the tactile map representing the skin around the nipples. As a rather different example, spend three months learning how to juggle, and there is expansion of the cortical map for visual processing of movement.*[17]

Thus, experience alters the number and strength of synapses, the extent of dendritic arbor, and the projection targets of axons. Time for the biggest revolution in neuroscience in years.

---

* Not all remapping is logical; some is just plain weird. A few years back, during an extremely stressed period, I developed a tic—when I'd be acutely upset about something, the second and third fingers on my left hand would rhythmically contract for a few seconds. What the hell was that about? No idea, but I marvel at the randomness of the remapping, at how unpleasant tumult in limbic circuitry somehow tapped into this motor circuit.

DIGGING DEEPER IN THE ASH HEAP OF HISTORY

Recall the crude, Neanderthal-ish notion that new memories require new neurons, an idea discarded when Hebb was in diapers. The adult brain does not make new neurons. You've got your maximal number of neurons around birth, and it's downhill from there, thanks to aging and imprudence.

You see where we're heading—adult brains, including aged human brains, do make new neurons. The finding is truly revolutionary, its discovery epic.

In 1965 an untenured associate professor at MIT named Joseph Altman (along with a longtime collaborator, Gopal Das) found the first evidence for adult neurogenesis, using a then-novel technique. A newly made cell contains newly made DNA. So, find a molecule unique to DNA. Get a test tube full of the stuff and attach a miniscule radioactive tag to each molecule. Inject it into an adult rat, wait awhile, and examine its brain. If any neurons contain that radioactive tag, it means they were born during the waiting period, with the radioactive marker incorporated into the new DNA.

This is what Altman saw in a series of studies.[18] As even he notes, the work was initially well received, being published in good journals, generating excitement. But within a few years something shifted, and Altman and his findings were rejected by leaders in the field—it couldn't be true. He failed to get tenure, spent his career at Purdue University, lost funding for his adult neurogenesis work.

Silence reigned for a decade until an assistant professor at the University of New Mexico named Michael Kaplan extended Altman's findings with some new techniques. Again this caused mostly crushing rejection by senior figures in the field, including one of the most established men in neuroscience, Pasko Rakic of Yale.[19]

Rakic publicly rejected Kaplan's (and tacitly Altman's) work, saying he had looked for new neurons himself, they weren't there, and Kaplan was mistaking other cell types for neurons. At a conference he notoriously told Kaplan, "Those may look like neurons in New Mexico, but they don't in New Haven." Kaplan soon left research (and a quarter century later, amid the excitement of the rediscovery of adult neurogenesis, wrote a short memoir entitled "Environmental Complexity Stimulates Visual Cortex Neurogenesis: Death of a Dogma and a Research Career").

The field lay dormant for another decade until unexpected evidence of adult neurogenesis emerged from the lab of Fernando Nottebohm of Rockefeller University. Nottebohm, a highly accomplished and esteemed neuroscientist, as good an old boy as you get, studied the neuroethology of birdsong. He demonstrated something remarkable, using new, more sensitive techniques: new neurons are made in the brains of birds that learn a new territorial song each year.

The quality of the science and Nottebohm's prestige silenced those who doubted that neurogenesis occurred. Instead they questioned its relevance—oh, that's nice for Fernando and his birdies, but what about in real species, in mammals?

But this was soon convincingly shown in rats, using newer, fancier techniques. Much of this was the work of two young scientists, Elizabeth Gould of Princeton, and Fred "Rusty" Gage of the Salk Institute.

Soon lots of other people were finding adult neurogenesis with these new techniques, including, lo and behold, Rakic.[20] A new flavor of skepticism emerged, led by Rakic. Yes, the adult brain makes new neurons, but only a few, they don't live long, and it doesn't happen where it really counts (i.e., the cortex); moreover, this has been shown only in rodents, not in primates. Soon it was shown in monkeys.*[21] Yeah, said the skeptics, but not humans, and besides, there's no evidence that these new neurons are integrated into preexisting circuits and actually function.

All of that was eventually shown—there's considerable adult neurogenesis in the hippocampus (where roughly 3 percent of neurons are

* Nottebohm, interviewed in an excellent *New Yorker* piece recounting this history, said, "Pasko has taken on the role of hard-nosed defender of standards. And that's fine—it's even warranted. . . . [But] as much as I hate to say this, I think Pasko Rakic single-handedly held the field of neurogenesis back by at least a decade."

replaced each month) and lesser amounts in the cortex.[22] It happens in humans throughout adult life. Hippocampal neurogenesis, for example, is enhanced by learning, exercise, estrogen, antidepressants, environmental enrichment, and brain injury* and inhibited by various stressors.†[23] Moreover, the new hippocampal neurons integrate into preexisting circuits, with the perky excitability of young neurons in the perinatal brain. Most important, new neurons are essential for integrating new information into preexisting schemas, something called "pattern separation." This is when you learn that two things you previously thought were the same are, in fact, different—dolphins and porpoises, baking soda and baking powder, Zooey Deschanel and Katy Perry.

Adult neurogenesis is the trendiest topic in neuroscience. In the five years after Altman's 1965 paper was published, it was cited (a respectable) twenty-nine times in the literature; in the last five, more than a thousand. Current work examines how exercise stimulates the process (probably by increasing levels of certain growth factors in the brain), how new neurons know where to migrate, whether depression is caused by a failure of hippocampal neurogenesis, and whether the neurogenesis stimulated by antidepressants is required for such medications to work.[24]

Why did it take so long for adult neurogenesis to be accepted? I've interacted with many of the principals and am struck by their differing takes. At one extreme is the view that while skeptics like Rakic were ham-handed, they provided quality control and that, counter to how path-of-the-hero epics go, some early work in the field was not all that

* The fact that brain injury, such as a stroke, triggers neurogenesis created huge excitement—wow, the brain has a means to try to repair itself after an injury, how cool is that? What was obvious from the start is that whatever compensatory neurogenesis there is, there isn't a ton, since so many neurological insults leave the nervous system an irreparable mess afterward. But to add insult to injury, work in that area began to show that sometimes the new neurons actually made things worse, migrating where they shouldn't, integrating into circuits the wrong way, making those circuits seizure prone. To metaphorically appropriate a concept from chapter 1, this seems a case of neuronal pathological altruism—beware when freshly minted neurons that may not yet know feces from Shinola want to lend a helping hand.

† Listing these various factors that "enhance" or "inhibit" neurogenesis glosses over lots of detail. The number of new neurons that are integrated into circuits reflects (a) the number of new cells that are formed from stem cells in the brain; (b) the percentage of new cells that differentiate into neurons (as opposed to glial cells); and (c) the rate at which new neurons survive and form functional synapses. Each of these manipulations—learning, exercise, stress, etc.—targets different steps. Complicating things further is the fact that not all stressors are equal. If a rodent secretes glucocorticoids because it thinks there is a predator around and the fight-or-flight sirens are going off, neurogenesis is inhibited. But if it secretes glucocorticoids while voluntarily running in a running wheel, it enhances neurogenesis (in other words, the contrast between "bad" and "good" stress).

solid. At the other extreme is the view that Rakic et al., having failed to find adult neurogenesis, couldn't accept that it existed. This psychohistorical view, of the old guard clinging to dogma in the face of changing winds, is weakened a bit by Altman's not having been a young anarchist running amok in the archives; in fact, he is a bit older than Rakic and other principal skeptics. All of this needs to be adjudicated by historians, screenwriters, and soon, I hope, by the folks in Stockholm.

Altman, who at the time of this writing is eighty-nine, published a 2011 memoir chapter.[25] Parts of it have a plaintive, confused tone—everyone was so excited at first; what happened? Maybe he spent too much time in the lab and too little marketing the discovery, he suggests. There's the ambivalence of someone who spent a long time as a scorned prophet who at least got to be completely vindicated. He's philosophical about it—hey, I'm a Hungarian Jew who escaped from a Nazi camp; you take things in stride after that.

SOME OTHER DOMAINS OF NEUROPLASTICITY

We've seen how in adults experience can alter the number of synapses and dendritic branches, remap circuitry, and stimulate neurogenesis.[26] Collectively, these effects can be big enough to actually change the size of brain regions. For example, postmenopausal estrogen treatment increases the size of the hippocampus (probably through a combination of more dendritic branches and more neurons). Conversely, the hippocampus atrophies (producing cognitive problems) in prolonged depression, probably reflecting its stressfulness and the typically elevated glucocorticoid levels of the disease. Memory problems and loss of hippocampal volume also occur in individuals with severe chronic pain syndromes, or with Cushing's syndrome (an array of disorders where a tumor causes extremely elevated glucocorticoid levels). Moreover, post-traumatic stress disorder is associated with increased volume (and, as we know, hyperreactivity) of the amygdala. In all of these instances it is unclear how much the stress/glucocorticoid effects are due to changes in neuron number or to changes in amounts of dendritic processes.*

One cool example of the size of a brain region changing with experience concerns the back part of the hippocampus, which plays a role in memory of spatial maps. Cab drivers use spatial maps for a living, and one renowned study showed enlargement of that part of the hippocampus in London taxi drivers. Moreover, a follow-up study imaged the hippocampus in people before and after the grueling multiyear process of working and studying for the London cabbie license test (called the toughest test in the world by the *New York Times*). The hippocampus enlarged over the course of the process—in those who passed the test.[27]

Thus, experience, health, and hormone fluctuations can change the size of parts of the brain in a matter of months. Experience can also cause long-lasting changes in the numbers of receptors for neurotransmitters and hormones, in levels of ion channels, and in the state of on/off switches on genes in the brain (to be covered in chapter 8).[28]

With chronic stress the nucleus accumbens is depleted of dopamine, biasing rats toward social subordination and biasing humans toward depression. As we saw in the last chapter, if a rodent wins a fight on his home territory, there are long-lasting increases in levels of testosterone receptors in the nucleus accumbens and ventral tegmentum, enhancing testosterone's pleasurable effects. There's even a parasite called *Toxoplasma gondii* that can infect the brain; over the course of weeks to months, it makes rats less fearful of the smell of cats and makes humans less fearful and more impulsive in subtle ways. Basically, most anything you can measure in the nervous system can change in response to a sustained stimulus. And importantly, these changes are often reversible in a different environment.†

* As one additional, grim piece of this picture of neural plasticity, extremes of chronic stress and glucocorticoid overexposure can also kill hippocampal neurons. While this is probably pertinent to nightmare extremes of stress, it is unclear how relevant it might be to more garden-variety sustained stress.

† For example, the phenomenon where experience can turn the on/off switch on a gene in a particular direction used to be considered permanent; that is turning out not to be the case. Similarly, the hippocampal atrophy in Cushing's syndrome appears to reverse within a year or so after the tumor is removed. As one disturbing exception to this theme, most studies suggest that the atrophy of the hippocampus with long-term major depression persists after the depression is successfully treated. Moreover, the reversibility of some of these effects (for example, of stress-induced retraction of dendritic processes) decreases with age.

SOME CONCLUSIONS

The discovery of adult neurogenesis is revolutionary, and the general topic of neuroplasticity, in all its guises, is immensely important—as is often the case when something the experts said couldn't be turns out to be.[29] The subject is also fascinating because of the nature of the revisionism—neuroplasticity radiates optimism. Books on the topic are entitled *The Brain That Changes Itself*, *Train Your Mind, Change Your Brain*, and *Rewire Your Brain: Think Your Way to a Better Life*, hinting at the "new neurology" (i.e., no more need for neurology once we can fully harness neuroplasticity). There's can-do Horatio Alger spirit every which way you look.

Amid that, some cautionary points:

- One recalls caveats aired in other chapters—the ability of the brain to change in response to experience is value free. Axonal remapping in blind or deaf individuals is great, exciting, and moving. It's cool that your hippocampus expands if you drive a London cab. Ditto about the size and specialization of the auditory cortex in the triangle player in the orchestra. But at the other end, it's disastrous that trauma enlarges the amygdala and atrophies the hippocampus, crippling those with PTSD. Similarly, expanding the amount of motor cortex devoted to finger dexterity is great in neurosurgeons but probably not a societal plus in safe crackers.
- The extent of neuroplasticity is most definitely finite. Otherwise, grievously injured brains and severed spinal cords would ultimately heal. Moreover, the limits of neuroplasticity are quotidian. Malcolm Gladwell has explored how vastly skilled individuals have put in vast amounts of practice—ten thousand hours is his magic number. Nevertheless, the reverse doesn't hold: ten thousand hours of practice does not guarantee the neuroplasticity needed to make any of us a Yo-Yo Ma or LeBron James.

Manipulating neuroplasticity for recovery of function does have enormous, exciting potential in neurology. But this domain is far from the concerns of this book. Despite neuroplasticity's potential, it's unlikely that we'll ever be able to, say, spritz neuronal growth factors up people's noses to make them more open-minded or empathic, or to target neuroplasticity with gene therapy to blunt some jerk's tendency to displace aggression.

So what's the subject good for in the realm of this book? I think the benefits are mostly psychological. This recalls a point from chapter 2, in the discussion of the neuroimaging studies demonstrating loss of volume in the hippocampus of people with PTSD (certainly an example of the adverse effects of neuroplasticity). I sniped that it was ridiculous that many legislators needed pictures of the brain to believe that there was something desperately, organically wrong with veterans with PTSD.

Similarly, neuroplasticity makes the functional malleability of the brain tangible, makes it "scientifically demonstrated" that brains change. That people change. In the time span considered in this chapter, people throughout the Arab world went from being voiceless to toppling tyrants; Rosa Parks went from victim to catalyst, Sadat and Begin from enemies to architects of peace, Mandela from prisoner to statesman. And you'd better bet that changes along the lines of those presented in this chapter occurred in the brains of anyone transformed by these transformations. A different world makes for a different worldview, which means a different brain. And the more tangible and real the neurobiology underlying such change seems, the easier it is to imagine that it can happen again.

Six

Adolescence; or, Dude, Where's My Frontal Cortex?

This chapter is the first of two focusing on development. We've established our rhythm: a behavior has just occurred; what events in the prior seconds, minutes, hours, and so on helped bring it about? The next chapter extends this into the developmental domain—what happened during that individual's childhood and fetal life that contributed to the behavior?

The present chapter breaks this rhythm in focusing on adolescence. Does the biology introduced in the preceding chapters work differently in an adolescent than in an adult, producing different behaviors? Yes.

One fact dominates this chapter. Chapter 5 did in the dogma that adult brains are set in stone. Another dogma was that brains are pretty much wired up early in childhood—after all, by age two, brains are already about 85 percent of adult volume. But the developmental trajectory is much slower than that. This chapter's key fact is that the final brain region to fully mature (in terms of synapse number, myelination, and metabolism) is the frontal cortex, not going fully online until the mid*twenties*.[1]

This has two screamingly important implications. First, no part of the adult brain is more shaped by adolescence than the frontal cortex. Second, nothing about adolescence can be understood outside the context of delayed frontocortical maturation. If by adolescence limbic, autonomic, and endocrine systems are going full blast while the frontal cortex is still working out the assembly instructions, we've just explained why adolescents are so frustrating, great, asinine, impulsive, inspiring, destructive, self-destructive, selfless, selfish, impossible, and world changing. Think about this—adolescence and early adulthood are the times when someone is most likely to kill, be killed, leave home forever, invent an art form, help overthrow a dictator, ethnically cleanse a village, devote themselves to the needy, become addicted, marry outside their group, transform physics, have hideous fashion taste, break their neck recreationally, commit their life to God, mug an old lady, or be convinced that all of history has converged to make this moment the most consequential, the most fraught with peril and promise, the most demanding that they get involved and make a difference. In other words, it's the time of life of maximal risk taking, novelty seeking, and affiliation with peers. All because of that immature frontal cortex.

THE REALITY OF ADOLESCENCE

Is adolescence real? Is there something qualitatively different distinguishing it from before and after, rather than being part of a smooth progression from childhood to adulthood? Maybe "adolescence" is just a cultural construct—in the West, as better nutrition and health resulted in earlier puberty onset, and the educational and economic forces of modernity pushed for childbearing at later ages, a developmental gap emerged between the two. Voilà! The invention of adolescence.*[2]

---

* The delayed attainment of legal adulthood in the West also sometimes reflected something as mundane as muscle mass. In thirteenth-century England the age of legal majority was raised from fifteen to twenty-one years—protective armor was getting heavier, and it wasn't until the older age that males were typically strong enough to comport themselves in armor on the battlefield. There is no mention as to whether the age of majority of the horses carrying these heavier weights was raised as well. But sometimes technological advancements have made it possible

As we'll see, neurobiology suggests that adolescence is for real, that the adolescent brain is not merely a half-cooked adult brain or a child's brain left unrefrigerated for too long. Moreover, most traditional cultures do recognize adolescence as distinct, i.e., it brings some but not all of the rights and responsibilities of adulthood. Nonetheless, what the West invented is the longest period of adolescence.*

What does seem a construct of individualistic cultures is adolescence as a period of intergenerational conflict; youth of collectivist cultures seem less prone toward eye rolling at the dorkiness of adults, starting with parents. Moreover, even within individualistic cultures adolescence is not universally a time of acne of the psyche, of Sturm und Drang. Most of us get through it just fine.

THE NUTS AND BOLTS OF FRONTAL CORTICAL MATURATION

The delayed maturation of the frontal cortex suggests an obvious scenario, namely that early in adolescence the frontal cortex has fewer neurons, dendritic branches, and synapses than in adulthood, and that levels increase into the midtwenties. Instead, levels *decrease.*

This occurs because of a truly clever thing evolved by mammalian brains. Remarkably, the fetal brain generates far more neurons than are found in the adult. Why? During late fetal development, there is a dramatic competition in much of the brain, with winning neurons being the ones that migrate to the correct location and maximize synaptic connections to other neurons. And neurons that don't make the grade? They undergo "programmed cell death"—genes are activated that cause them to shrivel and die, their materials then recycled. Neuronal overproduction

---

for younger adolescents to join the ranks of adult occupations—it has been pointed out that the development of lightweight automatic weapons has been a boon for the usefulness of the estimated 300,000 child soldiers on earth.

* Not to mention the idea that adults should aspire to still be adolescent in many ways—to retain or regain adolescent tastes for novelty and sociality, adolescent levels of hair on the head and cellulite in the thighs, and adolescent refractory periods. Hunter-gatherers aren't interested in *"Look ten years younger!"* They want to look like elders, so they can boss everyone around.

followed by competitive pruning (which has been termed "neural Darwinism") allowed the evolution of more optimized neural circuitry, a case of less being more.

The same occurs in the adolescent frontal cortex. By the start of adolescence, there's a greater volume of gray matter (an indirect measure of the total number of neurons and dendritic branches) and more synapses than in adults; over the next decade, gray-matter thickness declines as less optimal dendritic processes and connections are pruned away.*[3] Within the frontal cortex, the evolutionarily oldest subregions mature first; the spanking-new (cognitive) dorsolateral PFC doesn't even start losing gray-matter volume until late adolescence. The importance of this developmental pattern was shown in a landmark study in which children were neuroimaged and IQ tested repeatedly into adulthood. The longer the period of packing on gray-matter cortical thickness in early adolescence before the pruning started, the higher the adult IQ.

Thus, frontal cortical maturation during adolescence is about a more efficient brain, not more brain. This is shown in easily misinterpreted neuroimaging studies comparing adolescents and adults.[4] A frequent theme is how adults have more executive control over behavior during some tasks than do adolescents and show more frontal cortical activation at the time. Now find a task where, atypically, adolescents manage a level of executive control equal to that of adults. In those situations adolescents show *more* frontal activation than adults—equivalent regulation takes less effort in a well-pruned adult frontal cortex.

That the adolescent frontal cortex is not yet lean and mean is demonstrable in additional ways. For example, adolescents are not at adult levels of competence at detecting irony and, when trying to do so, activate the dmPFC more than do adults. In contrast, adults show more activation in the fusiform face region. In other words, detecting irony isn't much of a frontal task for an adult; one look at the face is enough.[5]

What about white matter in the frontal cortex (that indirect measure

* Not surprisingly, the peak of frontal cortical gray matter comes earlier in girls than in boys. Beyond that, what is most striking is the lack of major sex differences in the trajectory of adolescent brain development.

of myelination of axons)? Here things differ from the overproduce-then-prune approach to gray matter; instead, axons are myelinated throughout adolescence. As discussed in appendix 1, this allows neurons to communicate in a more rapid, coordinated manner—as adolescence progresses, activity in different parts of the frontal cortex becomes more correlated as the region operates as more of a functional unit.[6]

This is important. When learning neuroscience, it's easy to focus on individual brain regions as functionally distinct (and this tendency worsens if you then spend a career studying just one of them). As a measure of this, there are two high-quality biomedical journals out there, one called *Cortex*, the other *Hippocampus*, each publishing papers about its favorite brain region. At neuroscience meetings attended by tens of thousands, there'll be social functions for all the people studying the same obscure brain region, a place where they can gossip and bond and court. But in reality the brain is about circuits, about the patterns of functional connectivity among regions. The growing myelination of the adolescent brain shows the importance of increased connectivity.

Interestingly, other parts of the adolescent brain seem to help out the underdeveloped frontal cortex, taking on some roles that it's not yet ready for. For example, in adolescents but not adults, the ventral striatum helps regulate emotions; we will return to this.[7]

Something else keeps that tyro frontal cortex off-kilter, namely estrogen and progesterone in females and testosterone in males. As discussed in chapter 4, these hormones alter brain structure and function, including in the frontal cortex, where gonadal hormones change rates of myelination and levels of receptors for various neurotransmitters. Logically, landmarks of adolescent maturation in brain and behavior are less related to chronological age than to the time since puberty onset.[8]

Moreover, puberty is not just about the onslaught of gonadal hormones. It's about *how* they come online.[9] The defining feature of ovarian endocrine function is the cyclicity of hormone release—"It's that time of the month." In adolescent females puberty does not arrive full flower, so to speak, with one's first period. Instead, for the first few years only about half of cycles actually involve ovulation and surges of estrogen and progesterone. Thus, not only are young adolescents experiencing these first ovulatory cycles, but there are also higher-order fluctuations in whether the ovulatory fluctuation occurs. Meanwhile, while adolescent males don't have equivalent hormonal gyrations, it can't help that their frontal cortex keeps getting hypoxic from the priapic blood flow to the crotch.

Thus, as adolescence dawns, frontal cortical efficiency is diluted with extraneous synapses failing to make the grade, sluggish communication thanks to undermyelination, and a jumble of uncoordinated subregions working at cross-purposes; moreover, while the striatum is trying to help, a pinch hitter for the frontal cortex gets you only so far. Finally, the frontal cortex is being pickled in that ebb and flow of gonadal hormones. No wonder they act adolescent.

Frontal Cortical Changes in Cognition in Adolescence

To appreciate what frontal cortical maturation has to do with our best and worst behaviors, it's helpful to first see how such maturation plays out in cognitive realms.

During adolescence there's steady improvement in working memory, flexible rule use, executive organization, and frontal inhibitory regulation (e.g., task shifting). In general, these improvements are accompanied by increasing activity in frontal regions during tasks, with the extent of the increase predicting accuracy.[10]

Adolescents also improve at mentalization tasks (understanding someone else's perspective). By this I don't mean emotional perspective (stay tuned) but purer cognitive challenges, like understanding what objects look like from someone else's perspective. The improvement in detecting irony reflects improvement in abstract cognitive perspective taking.

Frontal Cortical Changes in Emotional Regulation

Older teenagers experience emotions more intensely than do children or adults, something obvious to anyone who ever spent time as a teenager. For example, they are more reactive to faces expressing strong emotions.*[11] In adults, looking at an "affective facial display" activates the amygdala, followed by activation of the emotion-regulating vmPFC as they habituate to the emotional content. In adolescence, though, the vmPFC response is less; thus the amygdaloid response keeps growing.

Chapter 2 introduced "reappraisal," in which responses to strong emotional stimuli are regulated by thinking about them differently.[12] Get a bad grade on an exam, and there's an emotional pull toward "I'm stupid"; reappraisal might lead you instead to focus on your not having studied or having had a cold, to decide that the outcome was situational, rather than a function of your unchangeable constitution.

Reappraisal strategies get better during adolescence, with logical neurobiological underpinnings. Recall how in early adolescence, the ventral striatum, trying to be helpful, takes on some frontal tasks (fairly ineffectively, as it's working above its pay grade). At that age reappraisal engages the ventral striatum; more activation predicts less amygdaloid activation and better emotional regulation. As the adolescent matures, the prefrontal cortex takes over the task, and emotions get steadier.†[13]

Bringing the striatum into the picture brings up dopamine and reward, thus bringing up the predilection of adolescents for bungee jumping.

ADOLESCENT RISK TAKING

In the foothills of the Sierras are California Caverns, a cave system that leads, after an initial narrow, twisting 30-foot descent down a hole, to

* An interesting exception is that adolescents do not have particularly strong responses to disgusting stimuli, either on a subjective level or on the level of insular cortex activation.

† With frontal regulation of emotions emerging later in males than females.

an abrupt 180-foot drop (now navigable by rappelling). The Park Service has found skeletons at the bottom dating back centuries, explorers who took one step too far in the gloom. And the skeletons are always those of adolescents.

As shown experimentally, during risky decision making, adolescents activate the prefrontal cortex less than do adults; the less activity, the poorer the risk assessment. This poor assessment takes a particular form, as shown by Sarah-Jayne Blakemore of University College London.[14] Have subjects estimate the likelihood of some event occurring (winning the lottery, dying in a plane crash); then tell them the actual likelihood. Such feedback can constitute good news (i.e., something good is actually more likely than the person estimated, or something bad is less likely). Conversely, the feedback can constitute bad news. Ask subjects to estimate the likelihood of the same events again. Adults incorporate the feedback into the new estimates. Adolescents update their estimates as adults do for good news, but feedback about bad news barely makes a dent. (Researcher: "How likely are you to have a car accident if you're driving while drunk?" Adolescent: "One chance in a gazillion." Researcher: "Actually, the risk is about 50 percent; what do you think your own chances are now?" Adolescent: "Hey, we're talking about me; one chance in a gazillion.") We've just explained why adolescents have two to four times the rate of pathological gambling as do adults.[15]

So adolescents take more risks and stink at risk assessment. But it's not just that teenagers are more willing to take risks. After all, adolescents and adults don't equally desire to do something risky and the adults simply don't do it because of their frontal cortical maturity. There is an age difference in the sensations sought—adolescents are tempted to bungee jump; adults are tempted to cheat on their low-salt diet. Adolescence is characterized not only by more risking but by more novelty seeking as well.*[16]

Novelty craving permeates adolescence; it is when we usually develop our stable tastes in music, food, and fashion, with openness to novelty declining thereafter.[17] And it's not just a human phenomenon. Across the rodent

* With the peak of sensation seeking coming and going earlier in females than in males.

life span, it's adolescents who are most willing to eat a new food. Adolescent novelty seeking is particularly strong in other primates. Among many social mammals, adolescents of one sex leave their natal group, emigrating into another population, a classic means to avoid inbreeding. Among impalas there are groups of related females and offspring with one breeding male; the other males knock around disconsolately in "bachelor herds," each scheming to usurp the breeding male. When a young male hits puberty, he is driven from the group by the breeding male (and to avoid some Oedipus nonsense, this is unlikely to be his father, who reigned many breeding males ago).

But not among primates. Take baboons. Suppose two troops encounter each other at some natural boundary—say, a stream. The males threaten each other for a while, eventually get bored, and resume whatever they were doing. Except there's an adolescent, standing at the stream's edge, riveted. New baboons, a whole bunch of 'em! He runs five steps toward them, runs back four, nervous, agitated. He gingerly crosses and sits on the other bank, scampering back should any new baboon glance at him.

So begins the slow process of transferring, spending more time each day with the new troop until he breaks the umbilical cord and spends the night. He wasn't pushed out. Instead, if he has to spend one more day with the same monotonous baboons he's known his whole life, he'll scream. Among adolescent chimps it's females who can't get off the farm fast enough. We primates aren't driven out at adolescence. Instead we desperately crave novelty.*

Thus, adolescence is about risk taking and novelty seeking. Where does the dopamine reward system fit in?

Recall from chapter 2 how the ventral tegmentum is the source of the mesolimbic dopamine projection to the nucleus accumbens, and of the mesocortical dopamine projection to the frontal cortex. During adolescence, dopamine projection density and signaling steadily increase in both pathways (although novelty seeking itself peaks at midadolescence, probably reflecting the emerging frontal regulation after that).[18]

* What this doesn't explain is why, for example, it's males who leave among baboons and females who do among chimps, nor does it explain why novelty seeking varies among humans. That will be touched on obliquely in chapter 10.

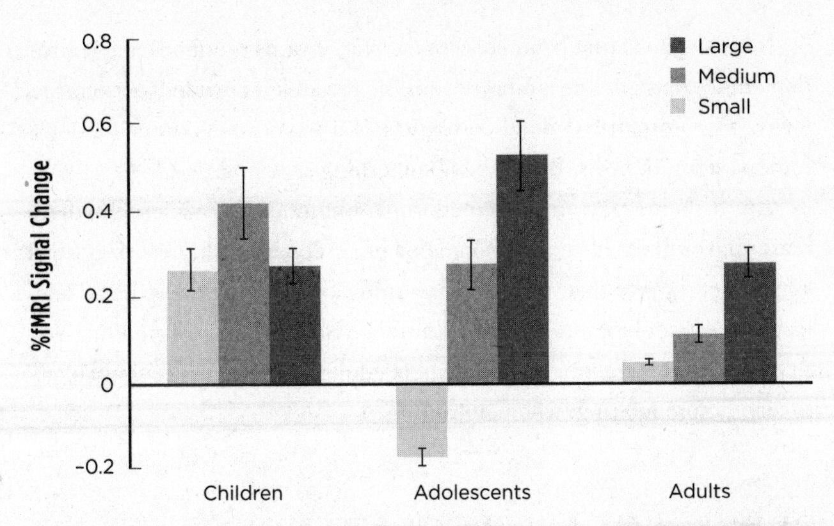

Changes in the amount of dopaminergic activity in the "reward center" of the brain following different magnitudes of reward. For the adolescents, the highs are higher, the lows lower.

It's unclear how much dopamine is released in anticipation of reward. Some studies show more anticipatory activation of reward pathways in adolescents than in adults, while others show the opposite, with the least dopaminergic responsiveness in adolescents who are most risk taking.[19]

Age differences in absolute levels of dopamine are less interesting than differences in patterns of release. In a great study, children, adolescents, and adults in brain scanners did some task where correct responses produced monetary rewards of varying sizes (see figure above).[20] During this, prefrontal activation in both children and adolescents was diffuse and unfocused. However, activation in the nucleus accumbens in adolescents was distinctive. In children, a correct answer produced roughly the same increase in activity regardless of size of reward. In adults, small, medium, and large rewards caused small, medium, and large increases in accumbens activity. And adolescents? After a medium reward things looked the same as in kids and adults. A large reward produced a humongous increase, much bigger than in adults. And the small reward? Accumbens activity *declined.* In other words, adolescents experience bigger-than-expected rewards more positively than do adults and smaller-than-expected rewards as aversive. A gyrating top, nearly skittering out of control.

This suggests that in adolescents strong rewards produce exaggerated dopaminergic signaling, and nice sensible rewards for prudent actions feel lousy. The immature frontal cortex hasn't a prayer to counteract a dopamine system like this. But there is something puzzling.

Amid their crazy, unrestrained dopamine neurons, adolescents have reasoning skills that, in many domains of perceiving risk, match those of adults. Yet despite that, logic and reasoning are often jettisoned, and adolescents act adolescent. Work by Laurence Steinberg of Temple University has identified a key juncture where adolescents are particularly likely to leap before looking: when around peers.

PEERS, SOCIAL ACCEPTANCE, AND SOCIAL EXCLUSION

Adolescent vulnerability to peer pressure from friends, especially peers they want to accept them as friends, is storied. It can also be demonstrated experimentally. In one Steinberg study adolescents and adults took risks at the same rate in a video driving game. Adding two peers to egg them on had no effect on adults but tripled risk taking in adolescents. Moreover, in neuroimaging studies, peers egging subjects on (by intercom) lessens vmPFC activity and enhances ventral striatal activity in adolescents but not adults.[21]

Why do adolescents' peers have such social power? For starters, adolescents are more social and more complexly social than children or adults. For example, a 2013 study showed that teens average more than four hundred Facebook friends, far more than do adults.[22] Moreover, teen sociality is particularly about affect, and responsiveness to emotional signaling—recall the greater limbic and lesser frontal cortical response to emotional faces in adolescents. And teens don't rack up four hundred Facebook friends for data for their sociology doctorates. Instead there is the frantic need to belong.

This produces teen vulnerability to peer pressure and emotional contagion. Moreover, such pressure is typically "deviance training," increasing the odds of violence, substance abuse, crime, unsafe sex, and poor health habits (few teen gangs pressure kids to join them in tooth flossing followed by random acts of kindness). For example, in college dorms the excessive drinker is more likely to influence the teetotaling roommate than the reverse. The incidence of eating disorders in adolescents spreads among peers with a pattern resembling viral contagion. The same occurs with depression among female adolescents, reflecting their tendency to "co-ruminate" on problems, reinforcing one another's negative affect.

Neuroimaging studies show the dramatic sensitivity of adolescents to peers. Ask adults to think about what they imagine others think of them, then about what they think of themselves. Two different, partially overlapping networks of frontal and limbic structures activate for the two tasks. But with adolescents the two profiles are the same. "What do you think about yourself?" is neurally answered with "Whatever everyone else thinks about me."[23]

The frantic adolescent need to belong is shown beautifully in studies of the neurobiology of social exclusion. Naomi Eisenberger of UCLA developed the fiendishly clever "Cyberball" paradigm to make people feel snubbed.[24] The subject lies in a brain scanner, believing she is playing an online game with two other people (naturally, they don't exist—it's a computer program). Each player occupies a spot on the screen, forming a triangle. The players toss a virtual ball among themselves; the subject is picking whom to throw to and believes the other two are doing the same. The ball is tossed for a while; then, unbeknownst to the subject, the experiment begins—the other two players stop throwing the ball to her. She's being excluded by those creeps. In adults there is activation of the periaqueductal gray, anterior cingulate, amygdala, and insular cortex. Perfect—these regions are central to pain perception, anger, and disgust.*

* Studies using this Cyberball paradigm typically have an important control group: The subject is in the same three-way game of virtual catch when they are told, "Oops, there is a computer glitch. We've lost contact with the other two players. Hang out in there while we fix this." While things are being "fixed," the other two players toss the ball. In other words, the person is left out, but because of technical problems, not for social reasons. And none of those brain regions activate. (Mind you, if it were me when in a less secure state of mind, it would certainly cross my mind that by the time that computer glitch was fixed, the other two would have already bonded and realized they were happier without me as part of the game and would continue to exclude me or, if they started throwing the ball back to me, would be doing so out of sheer condescension, thereby causing my mesolimbic dopamine system to instantly atrophy.)

And then, after a delay, the ventrolateral PFC activates; the more activation, the more the cingulate and insula are silenced and the less subjects report being upset afterward. What's this delayed vlPFC activation about? "Why am I getting upset? This is just a stupid game of catch." The frontal cortex comes to the rescue with perspective, rationalization, and emotion regulation.

Now do the study with teenagers. Some show the adult neuroimaging profiles; these are ones who rate themselves as least sensitive to rejection and who spend the most time with friends. But for most teenagers, when social exclusion occurs, the vlPFC barely activates; the other changes are bigger than in adults, and the subjects report feeling lousier—adolescents lack sufficient frontal forcefulness to effectively hand-wave about why it doesn't matter. Rejection *hurts* adolescents more, producing that stronger need to fit in.[25]

One neuroimaging study examined a neural building block of conformity.[26] Watch a hand moving, and neurons in premotor regions that contribute to moving your own hand become a bit active—your brain is on the edge of imitating the movement. In the study, ten-year-olds watched film clips of hand movements or facial expressions; those most vulnerable to peer influence (assessed on a scale developed by Steinberg)* had the most premotor activation—but only for emotional facial expressions. In other words, kids who are more sensitive to peer pressure are more prepared to imitate someone else's emotionality. (Given the age of the subjects, the authors framed their findings as potentially predictive of later teen behavior.)†

This atomistic level of explaining conformity might predict something about which teens are likely to join in a riot. But it doesn't tell much about who chooses not to invite someone to a party because the cool kids think she's a loser.

* The inventory requires the person to indicate the extent to which various statements tapping into social conformity apply to them—"Some people go along with their friends just to keep their friends happy," "Some people say things they don't really believe because they think it will make their friends respect them more," and so on.

† Some readers will recognize that those premotor neurons that start to imitate the movement being observed are "mirror neurons." As will be seen in a later chapter, the mirror neuron system is fascinating, amid enormous amounts of hype.

Another study showed neurobiological correlates of more abstract peer conformity. Recall how the adolescent ventral striatum helps the frontal cortex reappraise social exclusion. In this study, young adolescents most resistant to peer influence had the strongest such ventral striatal responses. And where might a stronger ventral striatum come from? You know the answer by now: you'll see in the remaining chapters.

EMPATHY, SYMPATHY, AND MORAL REASONING

By adolescence, people are typically pretty good at perspective taking, seeing the world as someone else would. That's usually when you'll first hear the likes of "Well, I still disagree, but I can see how he feels that way, given his experience."

Nonetheless, adolescents are not yet adults. Unlike adults, they are still better at first- than third-person perspective taking ("How would *you* feel in her situation?" versus "How does *she* feel in her situation?").[27] Adolescent moral judgments, while growing in sophistication, are still not at adult levels. Adolescents have left behind children's egalitarian tendency to split resources evenly. Instead, adolescents mostly make meritocratic decisions (with a smattering of utilitarian and libertarian viewpoints thrown in); meritocratic thinking is more sophisticated than egalitarian, since the latter is solely about outcomes, while the former incorporates thinking about causes. Nonetheless, adolescents' meritocratic thinking is less complex than adults'—for example, adolescents are as adept as adults at understanding how individual circumstances impact behavior, but not at understanding systemic circumstances.

As adolescents mature, they increasingly distinguish between intentional and accidental harm, viewing the former as worse.[28] When contemplating the latter, there is now less activation of three brain regions related to pain processing, namely the amygdala, the insula, and the premotor areas (the last reflecting the tendency to cringe when hearing about pain being inflicted). Meanwhile, there is increasing dlPFC and vmPFC activation when contemplating intentional harm. In other words, it is a frontal task to appreciate the painfulness of someone's being harmed intentionally.

As adolescents mature, they also increasingly distinguish between harm to people and harm to objects (with the former viewed as worse); harm to people increasingly activates the amygdala, while the opposite occurs for harm to objects. Interestingly, as adolescents age, there is *less* differentiation between recommended punishment for intentional and unintentional damage to objects. In other words, the salient point about the damage becomes that, accidental or otherwise, the damn thing needs to be fixed—even if there is less crying over spilled milk, there is no less cleaning required.*

What about one of the greatest things about adolescents, with respect to this book's concerns—their frenzied, agitated, incandescent ability to feel someone else's pain, to feel everyone's pain, to try to make everything right? A later chapter distinguishes between sympathy and empathy—between feeling *for* someone in pain and feeling *as* that someone. Adolescents are specialists at the latter, where the intensity of feeling *as* the other can border on *being* the other.

This intensity is no surprise, being at the intersection of many facets of adolescence. There are the abundant emotions and limbic gyrations. The highs are higher, the lows lower, empathic pain scalds, and the glow of doing the right thing makes it seem plausible that we are here for a purpose. Another contributing factor is the openness to novelty. An open mind is a prerequisite for an open heart, and the adolescent hunger for new experiences makes possible walking miles in lots of other people's shoes. And there is the egoism of adolescence. During my late adolescence I hung out with Quakers, and they'd occasionally use the aphorism "All God has is thee." This is the God of limited means, not just needing the help of humans to right a wrong, but needing you, you only, to do so.

* I have not seen any studies that look at maturational sophistication about circumstances where harm to an object produces enormous emotional harm to individuals—for example, the destruction of religious relics. As will be covered in a later chapter, there is vast power to such symbolic objects.

The appeal to egoism is tailor-made for adolescents. Throw in inexhaustible adolescent energy plus a feeling of omnipotence, and it seems possible to make the world whole, so why not?

In chapter 13 we consider how neither the most burning emotional capacity for empathy nor the most highfalutin moral reasoning makes someone likely to actually do the brave, difficult thing. This raises a subtle limitation of adolescent empathy.

As will be seen, one instance where empathic responses don't necessarily lead to acts is when we think enough to rationalize ("It's overblown as a problem" or "Someone else will fix it"). But feeling too much has problems as well. Feeling someone else's pain is painful, and people who do so most strongly, with the most pronounced arousal and anxiety, are actually *less* likely to act prosocially. Instead the personal distress induces a self-focus that prompts avoidance—"This is too awful; I can't stay here any longer." As empathic pain increases, your own pain becomes your primary concern.

In contrast, the more individuals can regulate their adverse empathic emotions, the more likely they are to act prosocially. Related to that, if a distressing, empathy-evoking circumstance increases your heart rate, you're less likely to act prosocially than if it decreases it. Thus, one predictor of who actually acts is the ability to gain some detachment, to ride, rather than be submerged, by the wave of empathy.

Where do adolescents fit in, with their hearts on their sleeves, fully charged limbic systems, and frontal cortices straining to catch up? It's obvious. A tendency toward empathic hyperarousal that can disrupt acting effectively.[29]

This adolescent empathy frenzy can seem a bit much for adults. But when I see my best students in that state, I have the same thought—it used to be so much easier to be like that. My adult frontal cortex may enable whatever detached good I do. The trouble, of course, is how that same detachment makes it easy to decide that something is not my problem.

ADOLESCENT VIOLENCE

Obviously, the adolescent years are not just about organizing bake sales to fight global warming. Late adolescence and early adulthood are when violence peaks, whether premeditated or impulsive murder, Victorian fisticuffs or handguns, solitary or organized (in or out of a uniform), focused on a stranger or on an intimate partner. And then rates plummet. As has been said, the greatest crime-fighting tool is a thirtieth birthday.

On a certain level the biology underlying the teenaged mugger is similar to that of the teen who joins the Ecology Club and donates his allowance to help save the mountain gorillas. It's the usual—heightened emotional intensity, craving for peer approval, novelty seeking, and, oh, that frontal cortex. But that's where similarities end.

What underlies the adolescent peak in violence? Neuroimaging shows nothing particularly distinct about it versus adult violence.[30] Adolescent and adult psychopaths both have less sensitivity of the PFC and the dopamine system to negative feedback, less pain sensitivity, and less amygdaloid/frontal cortical coupling during tasks of moral reasoning or empathy.

Moreover, the adolescent peak of violence isn't caused by the surge in testosterone; harking back to chapter 4, testosterone no more causes violence in adolescents than it does in adult males. Moreover, testosterone levels peak during early adolescence, but violence peaks later.

The next chapter considers some of the roots of adolescent violence. For now, the important point is that an average adolescent doesn't have the self-regulation or judgment of an average adult. This can prompt us to view teenage offenders as having less responsibility than adults for criminal acts. An alternative view is that even amid poorer judgment and self-regulation, there is still enough to merit equivalent sentencing. The former view has held in two landmark Supreme Court decisions.

In the first, 2005's *Roper v. Simmons*, the Court ruled 5–4 that executing someone for crimes committed before age eighteen is unconstitutional, violating the Eighth Amendment ban on cruel and unusual punishment. Then in 2012's *Miller v. Alabama*, in another 5–4 split, the Court banned mandatory life sentences without the chance of parole for juvenile offenders, on similar grounds.[31]

The Court's reasoning was straight out of this chapter. Writing for the majority in *Roper v. Simmons*, Justice Anthony Kennedy said:

> First, [as everyone knows, a] lack of maturity and an underdeveloped sense of responsibility are found in youth more often than in adults and are more understandable among the young. These qualities often result in impetuous and ill-considered actions and decisions.[32]

I fully agree with these rulings. But, to show my hand early, I think this is just window dressing. As will be covered in the screed that constitutes chapter 16, I think the science encapsulated in this book should transform every nook and cranny of the criminal justice system.

A FINAL THOUGHT: WHY CAN'T THE FRONTAL CORTEX JUST ACT ITS AGE?

As promised, this chapter's dominant fact has been the delayed maturation of the frontal cortex. Why should the delay occur? Is it because the frontal cortex is the brain's most complicated construction project?

Probably not. The frontal cortex uses the same neurotransmitter systems as the rest of the brain and uses the same basic neurons. Neuronal density and complexity of interconnections are similar to the rest of the (fancy) cortex. It isn't markedly harder to build frontal cortex than any other cortical region.

Thus, it is not likely that if the brain "could" grow a frontal cortex as fast as the rest of the cortex, it "would." Instead I think there was evolutionary selection for delayed frontal cortex maturation.

If the frontal cortex matured as fast as the rest of the brain, there'd be none of the adolescent turbulence, none of the antsy, itchy exploration and creativity, none of the long line of pimply adolescent geniuses who dropped out of school and worked away in their garages to invent fire, cave painting, and the wheel.

Maybe. But this just-so story must accommodate behavior evolving to pass on copies of the genes of individuals, not for the good of the species (stay tuned for chapter 10). And for every individual who scored big time reproductively thanks to adolescent inventiveness, there've been far more who instead broke their necks from adolescent imprudence. I don't think delayed frontal cortical maturation evolved so that adolescents could act over the top.

Instead, I think it is delayed so that the brain gets it right. Well, duh; the brain needs to "get it right" with all its parts. But in a distinctive way in the frontal cortex. The point of the previous chapter was the brain's plasticity—new synapses form, new neurons are born, circuits rewire, brain regions expand or contract—we learn, change, adapt. This is nowhere more important than in the frontal cortex.

An oft-repeated fact about adolescents is how "emotional intelligence" and "social intelligence" predict adult success and happiness better than do IQ or SAT scores.[33] It's all about social memory, emotional perspective taking, impulse control, empathy, ability to work with others, self-regulation. There is a parallel in other primates, with their big, slowly maturing frontal cortices. For example, what makes for a "successful" male baboon in his dominance hierarchy? *Attaining* high rank is about muscle, sharp canines, well-timed aggression. But once high status is achieved, *maintaining* it is all about social smarts—knowing which coalitions to form, how to intimidate a rival, having sufficient impulse control to ignore most provocations and to keep displacement aggression to a reasonable level. Similarly, as noted in chapter 2, among male rhesus monkeys a large prefrontal cortex goes hand in hand with social dominance.

Adult life is filled with consequential forks in the road where the right thing is definitely harder. Navigating these successfully is the portfolio of the frontal cortex, and developing the ability to do this right in each context requires profound shaping by experience.

This may be the answer. As we will see in chapter 8, the brain is heavily influenced by genes. But from birth through young adulthood, the part of the human brain that most defines us is less a product of the genes with which you started life than of what life has thrown at you. Because it is the last to mature, by definition the frontal cortex is the brain region least constrained by genes and most sculpted by experience. This must be so, to be the supremely complex social species that we are. Ironically, it seems that the genetic program of human brain development has evolved to, as much as possible, free the frontal cortex from genes.

Seven

Back to the Crib, Back to the Womb

After journeying to Planet Adolescence, we resume our basic approach. Our behavior—good, bad, or ambiguous—has occurred. Why? When seeking the roots of behavior, long before neurons or hormones come to mind, we typically look first at childhood.

COMPLEXIFICATION

Childhood is obviously about increasing complexity in every realm of behavior, thought, and emotion. Crucially, such increasing complexity typically emerges in stereotypical, universal sequences of stages. Most child behavioral development research is implicitly stage oriented, concerning: (a) the sequence with which stages emerge; (b) how experience influences the speed and surety with which that sequential tape of maturation unreels; and (c) how this helps create the adult a child ultimately becomes. We start by examining the neurobiology of the "stage" nature of development.

A BRIEF TOUR OF BRAIN DEVELOPMENT

The stages of human brain development make sense. A few weeks after conception, a wave of neurons are born and migrate to their correct locations. Around twenty weeks, there is a burst of synapse formation—neurons start talking to one another. And then axons start being wrapped in myelin, the glial cell insulation (forming "white matter") that speeds up action.

Neuron formation, migration, and synaptogenesis are mostly prenatal in humans.[1] In contrast, there is little myelin at birth, particularly in evolutionarily newer brain regions; as we've seen, myelination proceeds for a quarter century. The stages of myelination and consequent functional development are stereotypical. For example, the cortical region central to language comprehension myelinates a few months earlier than that for language production—kids understand language before producing it.

Myelination is most consequential when enwrapping the longest axons, in neurons that communicate the greatest distances. Thus myelination particularly facilitates brain regions *talking to one another.* No brain region is an island, and the formation of circuits connecting far-flung brain regions is crucial—how else can the frontal cortex use its few myelinated neurons to talk to neurons in the brain's subbasement to make you toilet trained?[2]

As we saw, mammalian fetuses overproduce neurons and synapses; ineffective or unessential synapses and neurons are pruned, producing leaner, meaner, more efficient circuitry. To reiterate a theme from the last chapter, the later a particular brain region matures, the less it is shaped by genes and the more by environment.[3]

STAGES

What stages of child development help explain the good/bad/in-between adult behavior that got the ball rolling in chapter 1?

The mother of all developmental stage theories was supplied in 1923, pioneered by Jean Piaget's clever, elegant experiments revealing four stages of cognitive development:[4]

- *Sensorimotor stage* (birth to ~24 months). Thought concerns only what the child can directly sense and explore. During this stage, typically at around 8 months, children develop "object permanence," understanding that even if they can't see an object, it still exists—the infant can generate a mental image of something no longer there.*
- *Preoperational stage* (~2 to 7 years). The child can maintain ideas about how the world works without explicit examples in front of him. Thoughts are increasingly symbolic; imaginary play abounds. However, reasoning is intuitive—no logic, no cause and effect. This is when kids can't yet demonstrate "conservation of volume." Identical beakers A and B are filled with equal amounts of water. Pour the contents of beaker B into beaker C, which is taller and thinner. Ask the child, "Which has more water, A or C?" Kids in the preoperational stage use incorrect folk intuition—the water line in C is higher than that in A; it must contain more water.
- *Concrete operational stage* (7 to 12 years). Kids think logically, no longer falling for that different-shaped-beakers nonsense. However, generalizing logic from specific cases is iffy. As is abstract thinking—for example, proverbs are interpreted literally ("'Birds of a feather flock together' means that similar birds form flocks").

* How do you demonstrate object permanence in a preverbal infant? Show a child who is not yet at this stage a stuffie, which you then place in a box. For her, the stuffie no longer exists. Now take it out, and she thinks, "OMG, where'd that stuffie come from?" Her heart rate increases. Once the kid masters object permanence, pull the stuffie out of the box and (yawn) "Of course that's where you put the stuffie"—no heart-rate increase. Even better: put the stuffie in a box and then pull something different (say, a ball) out of the box. Pre-object-permanence kid isn't surprised—the stuffie stopped existing, and the ball just came into existence. Older kid with object permanence: "Wait, that stuffie turned into a ball!"—heart rate increases.

Kid playing hide-and-seek while in the "If I can't see you (or even if I can't see you as easily as usual), then you can't see me" stage.

- *Formal operational stage* (adolescence onward). Approaching adult levels of abstraction, reasoning, and metacognition.

Other aspects of cognitive development are also conceptualized in stages. An early stage occurs when toddlers form ego boundaries—"There is a 'me,' separate from everyone else." A lack of ego boundaries is shown when a toddler isn't all that solid on where he ends and Mommy starts—she's cut her finger, and he claims his finger hurts.[5]

Next comes the stage of realizing that other individuals have different information than you do. Nine-month-olds look where someone points (as can other apes and dogs), knowing the pointer has information that they don't. This is fueled by motivation: Where *is* that toy? Where's she looking? Older kids understand more broadly that other people have different thoughts, beliefs, and knowledge than they, the landmark of achieving Theory of Mind (ToM).[6]

Here's what not having ToM looks like. A two-year-old and an adult see a cookie placed in box A. The adult leaves, and the researcher switches the cookie to box B. Ask the child, "When that person comes back, where will he look for the cookie?" Box B—the child knows it's there and thus everyone knows. Around age three or four the child can reason, "They'll think it's in A, even though *I* know it's in B." Shazam: ToM.

Mastering such "false belief" tests is a major developmental landmark. ToM then progresses to fancier insightfulness—e.g., grasping irony, perspective taking, or secondary ToM (understanding person A's ToM about person B).[7]

Various cortical regions mediate ToM: parts of the medial PFC (surprise!) and some new players, including the precuneus, the superior temporal sulcus, and the temporoparietal junction (TPJ). This is shown with neuroimaging; by ToM deficits if these regions are damaged (autistic individuals, who have limited ToM, have decreased gray matter and activity in the superior temporal sulcus); and by the fact that if you temporarily inactivate the TPJ, people don't consider someone's intentions when judging them morally.[8]

Thus there are stages of gaze following, followed by primary ToM, then secondary ToM, then perspective taking, with the speed of transitions influenced by experience (e.g., kids with older siblings achieve ToM earlier than average).[9]

Naturally, there are criticisms of stage approaches to cognitive development. One is at the heart of this book: a Piagetian framework sits in a "cognition" bucket, ignoring the impact of social and emotional factors.

One example to be discussed in chapter 12 concerns preverbal infants, who sure don't grasp transitivity (if A > B, and B > C, then A > C). Show a violation of transitivity in interactions between shapes on a screen (shape A should knock over shape C, but the opposite occurs), and the kid is unbothered, doesn't look for long. But personify the shapes with eyes and a mouth, and now heart rate increases, the kid looks longer—"Whoa, *character* C is supposed to move out of *character* A's way, not the reverse." Humans understand logical operations between individuals earlier than between objects.[10]

Social and motivational state can shift cognitive stage as well. Rudiments of ToM are more demonstrable in chimps who are interacting with another chimp (versus a human) and if there is something motivating—food—involved.*[11]

* How would you test this? Two humans stand in front of a monkey, one blindfolded. A treat for the monkey is hidden somewhere. Blindfold is removed; monkey chooses which human goes to look for the treat. "Don't choose the one who was blindfolded. They don't know where the treat is," thinks the ToM Master of the Universe monkey.

Emotion and affect can alter cognitive stage in remarkably local ways. I saw a wonderful example of this when my daughter displayed both ToM *and* failure of ToM in the same breath. She had changed preschools and was visiting her old class. She told everyone about life in her new school: "Then, after lunch, we play on the swings. There are swings at my new school. And then, after that, we go inside and Carolee reads us a story. Then, after that . . ." ToM: "play on the swings"—wait, they don't know that my school has swings; I need to tell them. Failure of ToM: "Carolee reads us a story." Carolee, the teacher at her new school. The same logic should apply—tell them who Carolee is. But because Carolee was the most wonderful teacher alive, ToM failed. Afterward I asked her, "Hey, why didn't you tell everyone that Carolee is your teacher?" "Oh, everyone knows Carolee." How could everyone not?

Feeling Someone Else's Pain

ToM leads to a next step—people can have different *feelings* than me, including pained ones.[12] This realization is not sufficient for empathy. After all, sociopaths, who pathologically lack empathy, use superb ToM to stay three manipulative, remorseless steps ahead of everyone. Nor is this realization strictly necessary for empathy, as kids too young for ToM show rudiments of feeling someone else's pain—a toddler will try to comfort someone feigning crying, offering them her pacifier (and the empathy is rudimentary in that the toddler can't imagine someone being comforted by different things than she is).

Yes, very rudimentary. Maybe the toddler feels profound empathy. Or maybe she's just distressed by the crying and is self-interestedly trying to quiet the adult. The childhood capacity for empathy progresses from feeling someone's pain because you *are* them, to feeling *for* the other person, to feeling *as* them.

The neurobiology of kid empathy makes sense. As introduced in chapter 2, in adults the anterior cingulate cortex activates when they see someone hurt. Ditto for the amygdala and insula, especially in instances of intentional harm—there is anger and disgust. PFC regions including the (emotional) vmPFC are on board. Observing physical pain (e.g., a finger being poked with a needle) produces a concrete, vicarious pattern: there is activation of the periaqueductal gray (PAG), a region central to your *own* pain perception, in parts of the sensory cortex receiving sensation from your *own* fingers, and in motor neurons that command your *own* fingers to move.* You clench your fingers.

Work by Jean Decety of the University of Chicago shows that when seven-year-olds watch someone in pain, activation is greatest in the more concrete regions—the PAG and the sensory and motor cortices—with PAG activity coupled to the minimal vmPFC activation there is. In older kids the vmPFC is coupled to increasingly activated limbic structures.[13] And by adolescence the stronger vmPFC activation is coupled to ToM regions. What's happening? Empathy is shifting from the concrete world of "Her finger must *hurt*, I'm suddenly conscious of my own finger" to ToM-ish focusing on the pokee's emotions and experience.

Young kids' empathy doesn't distinguish between intentional and unintentional harm or between harm to a person and to an object. Those distinctions emerge with age, around the time when the PAG part of empathic responses lessens and there is more engagement of the vmPFC and ToM regions; moreover, intentional harm now activates the amygdala and insula—anger and disgust at the perpetrator.† This is also when kids first distinguish between self- and other-inflicted pain.

More sophistication—by around age seven, kids are expressing their empathy. By ages ten through twelve, empathy is more generalized and abstracted—empathy for "poor people," rather than one individual (downside: this is also when kids first negatively stereotype categories of people).

There are also hints of a sense of justice. Preschoolers tend to be egalitarians (e.g., it's better that the friend gets a cookie when she does). But

* This "sensory motor resonance" might bring "mirror neurons" to mind. Chapter 14 examines what mirror neurons do (often in sharp contrast to what they've been speculated to do). The involvement of the PAG also brings to mind sociopaths, with their lack of capacity for empathy; as discussed in chapter 2, such individuals have atypically blunted pain perception.

† The paper by Decety mentioned in the previous chapter had another interesting finding: For acts that harm people, the typical adult pattern is to advocate greater punishment for intentional acts. There is far less distinction made between intentional and unintentional when it comes to harm to objects. "Damn, I don't care if he meant to Krazy Glue the fan belt or not—we have to buy a new one."

before we get carried away with the generosity of youth, there is already in-group bias; if the other child is a stranger, there is less egalitarianism.[14]

There is also a growing tendency of kids to respond to an injustice, when someone has been treated unfairly.[15] But once again, before getting carried away with things, it comes with a bias. By ages four through six, kids in cultures from around the world respond negatively when *they* are the ones being shortchanged. It isn't until ages eight through ten that kids respond negatively to someone *else* being treated unfairly. Moreover, there is considerable cross-cultural variability as to whether that later stage even emerges. The sense of justice in young kids is a very self-interested one.

Soon after kids start responding negatively to someone else being treated unjustly, they begin attempting to rectify previous inequalities ("He should get more now because he got less before").[16] By preadolescence, egalitarianism gives way to acceptance of inequality because of merit or effort or for a greater good ("She should play more than him; she's better/worked harder/is more important to the team"). Some kids even manage self-sacrifice for the greater good ("She should play more than me; she's better").* By adolescence, boys tend to accept inequality more than girls do, on utilitarian grounds. And both sexes are acquiescing to inequality as social convention—"Nothing can be done; that's the way it is."

Moral Development

With ToM, perspective taking, nuanced empathy, and a sense of justice in place, a child can start wrestling with telling right from wrong.

Piaget emphasized how much kids' play is about working out rules of appropriate behavior (rules that can differ from those of adults)† and how

* "Greater good" for kids, as at any age, is in the eye of the beholder. In psychologist Robert Coles's classic *The Moral Life of Children* (New York: Atlantic Monthly Press, 1986), he describes his fieldwork in the American South during desegregation, and how older children on both sides of the divide were willing to undergo sacrifice for the good of *their* ideological group.

† I once received a lesson in kids' private world of rule making from my then-four-year-old son. We had gone to a public bathroom together; we stood side by side at two urinals, and I finished a bit earlier than he did. "I wish we had finished at the same time," he said. Why? "We get more points that way."

this involves stages of increasing complexity. This inspired a younger psychologist to investigate the topic more rigorously, with enormously influential consequences.

In the 1950s Lawrence Kohlberg, then a graduate student at the University of Chicago and later a professor at Harvard, began formulating his monumental stages of moral development.[17]

Kids would be presented with moral conundrums. For example: The only dose of the only drug that will save a poor woman from dying is prohibitively expensive. Should she steal it? Why?

Kohlberg concluded that moral judgment is a *cognitive* process, built around increasingly complex reasoning as kids mature. He proposed his famed three stages of moral development, each with two subparts.

You've been told not to eat the tempting cookie in front of you. Should you eat it? Here are the painfully simplified stages of reasoning that go into the decision:

Level 1: Should I Eat the Cookie? Preconventional Reasoning

Stage 1. It depends. How likely am I to get punished? Being punished is unpleasant. Aggression typically peaks around ages two through four, after which kids are reined in by adults' punishment ("Go sit in the corner") and peers (i.e., being ostracized).

Stage 2. It depends. If I refrain, will I get rewarded? Being rewarded is nice.

Both stages are ego-oriented—obedience and self-interest (what's in it for me?). Kohlberg found that children are typically at this level up to around ages eight through ten.

Concern arises when aggression, particularly if callous and remorseless, doesn't wane around these ages—this predicts an increased risk of adult sociopathy (aka antisocial personality).* Crucially, the behavior of future sociopaths seems impervious to negative feedback. As noted, high

* The callous aggression speaks to another childhood predictor of adult sociopathy, namely abuse of animals.

pain thresholds in sociopaths help explain their lack of empathy—it's hard to feel someone else's pain when you can't feel your own. It also helps explain the imperviousness to negative feedback—why change your behavior if punishment doesn't register?

It is also around this stage that kids first reconcile after conflicts and derive comfort from reconciliation (e.g., decreasing glucocorticoid secretion and anxiety). Those benefits certainly suggest self-interest motivating reconciliation. This is shown in another, realpolitik way—kids reconcile more readily when the relationship matters to them.

Level 2: Should I Eat the Cookie? Conventional Reasoning

Stage 3. It depends. Who will be deprived if I do? Do I like them? What would other people do? What will people think of me for eating the cookie? It's nice to think of others; it's good to be well regarded.

Stage 4. It depends. What's the law? Are laws sacrosanct? What if everyone broke this law? It's nice to have order. This is the judge who, considering predatory but legal lending practices by a bank, thinks, "I feel sorry for these victims . . . but I'm here to decide whether the bank broke a law . . . and it didn't."

Conventional moral reasoning is relational (about your interactions with others and their consequences); most adolescents and adults are at this level.

Level 3: Should I Eat the Cookie? Postconventional Reasoning

Stage 5: It depends. What circumstances placed the cookie there? Who decided that I shouldn't take it? Would I save a life by taking the cookie? It's nice when clear rules are applied flexibly. Now the judge would think: "Yes, the bank's actions were legal, but ultimately laws exist to protect the weak from the mighty, so signed contract or otherwise, that bank must be stopped."

Stage 6: It depends. Is my moral stance regarding this more vital than some law, a stance for which I'd pay the ultimate price if need be? It's nice to know there are things for which I'd repeatedly sing, "We Will Not Be Moved."

This level is egoistic in that rules and their application come from within and reflect conscience, where a transgression exacts the ultimate cost—having to live with yourself afterward. It recognizes that being good and being law-abiding aren't synonymous. As Woody Guthrie wrote in "Pretty Boy Floyd," "I love a good man outside the law, just as much as I hate a bad man inside the law."*

Stage 6 is also egotistical, implicitly built on self-righteousness that trumps conventional petty bourgeois rule makers and bean counters, The Man, those sheep who just follow, etc. To quote Emerson, as is often done when considering the postconventional stage, "Every heroic act measures itself by its contempt of some external good." Stage 6 reasoning can inspire. But it can also be insufferable, premised on "being good" and "being law abiding" as opposites. "To live outside the law, you must be honest," wrote Bob Dylan.

Kohlbergians found hardly anyone consistently at stage 5 or stage 6.

Kohlberg basically invented the scientific study of moral development in children. His stage model is so canonical that people in the business dis someone by suggesting they're stuck in the primordial soup of a primitive Kohlberg stage. As we'll see in chapter 12, there is even evidence that conservatives and liberals reason at different Kohlberg stages.

Naturally, Kohlberg's work has problems.

The usual: Don't take any stage model too seriously—there are exceptions, maturational transitions are not clean cut, and someone's stage can be context dependent.

The problem of tunnel vision and wrong emphases: Kohlberg

* I haven't a clue whether that should apply to Floyd, the Depression-era bank robber (and murderer) who nonetheless became somewhat of a folk hero to the poor, and whose Oklahoma funeral was attended by somewhere between twenty thousand and forty thousand people.

initially studied the usual unrepresentative humans, namely Americans, and as we will see in later chapters, moral judgments differ cross-culturally. Moreover, subjects were male, something challenged in the 1980s by Carol Gilligan of NYU. The two agreed on the general sequence of stages. However, Gilligan and others showed that in making moral judgments, girls and women generally value care over justice, in contrast to boys and men. As a result, females tilt toward conventional thinking and its emphasis on relationships, while males tilt toward postconventional abstractions.[18]

The cognitive emphasis: Are moral judgments more the outcome of reasoning or of intuition and emotion? Kohlbergians favor the former. But as will be seen in chapter 13, plenty of organisms with limited cognitive skills, including young kids and nonhuman primates, display rudimentary senses of fairness and justice. Such findings anchor "social intuitionist" views of moral decision making, associated with psychologists Martin Hoffman and Jonathan Haidt, both of NYU.[19] Naturally, the question becomes how moral reasoning and moral intuitionism interact. As we'll see, (a) rather than being solely about emotion, moral intuition is a different style of cognition from conscious reasoning; and (b) conversely, moral reasoning is often flagrantly illogical. Stay tuned.

The lack of predictability: Does any of this actually predict who does the harder thing when it's the right thing to do? Are gold medalists at Kohlbergian reasoning the ones willing to pay the price for whistle-blowing, subduing the shooter, sheltering refugees? Heck, forget the heroics; are they even more likely to be honest in dinky psych experiments? In other words, does moral reasoning predict moral *action*? Rarely; as we will see in chapter 13, moral heroism rarely arises from super-duper frontal cortical willpower. Instead, it happens when the right thing isn't the harder thing.

Marshmallows

The frontal cortex and its increasing connectivity with the rest of the brain anchors the neurobiology of kids' growing sophistication, most importantly in their capacity to regulate emotions and behavior. The most iconic demonstration of this revolves around an unlikely object—the marshmallow.[20]

In the 1960s Stanford psychologist Walter Mischel developed the "marshmallow test" to study gratification postponement. A child is presented with a marshmallow. The experimenter says, "I'm going out of the room for a while. You can eat the marshmallow after I leave. But if you wait and don't eat it until I get back, I'll give you another marshmallow," and leaves. And the child, observed through a two-way mirror, begins the lonely challenge of holding out for fifteen minutes until the researcher returns.

Studying hundreds of three- to six-year-olds, Mischel saw enormous variability—a few ate the marshmallow before the experimenter left the room. About a third lasted the fifteen minutes. The rest were scattered in between, averaging a delay of eleven minutes. Kids' strategies for resisting the marshmallow's siren call differed, as can be seen on contemporary versions of the test on YouTube. Some kids cover their eyes, hide the marshmallow, sing to distract themselves. Others grimace, sit on their hands. Others sniff the marshmallow, pinch off an infinitely tiny piece to eat, hold it reverentially, kiss it, pet it.

Various factors modulated kids' fortitude (shown in later studies described in Mischel's book where, for some reason, it was pretzels instead of marshmallows). Trusting the system mattered—if experimenters had previously betrayed on promises, kids wouldn't wait as long. Prompting kids to think about how crunchy and yummy pretzels are (what Mischel calls "hot ideation") nuked self-discipline; prompts to think about a "cold ideation" (e.g., the shape of pretzels) or an alternative hot ideation (e.g., ice cream) bolstered resistance.

As expected, older kids hold out longer, using more effective strategies. Younger kids describe strategies like "I kept thinking about how good that second marshmallow would taste." The problem, of course, is that this strategy is about two synapses away from thinking about the marshmallow in front of you. In contrast, older kids use strategies of distraction—thinking about toys, pets, their birthday. This progresses to reappraisal strategies ("This isn't about marshmallows. This is about the kind of person I am"). To Mischel, maturation of willpower is more about distraction and reappraisal strategies than about stoicism.

So kids improve at delayed gratification. Mischel's next step made his studies iconic—he tracked the kids afterward, seeing if marshmallow wait time predicted anything about their adulthoods.

Did it ever. Five-year-old champs at marshmallow patience averaged higher SAT scores in high school (compared with those who couldn't wait), with more social success and resilience and less aggressive* and oppositional behavior. *Forty years* postmarshmallow, they excelled at frontal function, had more PFC activation during a frontal task, and had lower BMIs.[21] A gazillion-dollar brain scanner doesn't hold more predictive power than one marshmallow. Every anxious middle-class parent obsesses over these findings, has made marshmallows fetish items.

CONSEQUENCES

We've now gotten a sense of various domains of behavioral development. Time to frame things with this book's central question. Our adult has carried out that wonderful or crummy or ambiguous behavior. What childhood events contributed to that occurring?

A first challenge is to truly incorporate biology into our thinking. A child suffers malnutrition and, as an adult, has poor cognitive skills. That's easy to frame biologically—malnutrition impairs brain development. Alternatively, a child is raised by cold, inexpressive parents and, as an adult, feels unlovable. It's harder to link those two biologically, to resist thinking that somehow this is a *less biological* phenomenon than the malnutrition/cognition link. There may be less *known* about the biological changes explaining the link between the cold parents and the adult with poor self-esteem than about the malnutrition/cognition one. It may be less *convenient* to articulate

* A recent study adds an important twist to this story. There are the kids with problems with impulse control—"I'm absolutely going to hold out for two marshmallow"—who then instantly eat that first one. That profile is a statistical predictor of adult violent crime. In contrast, there are kids with steep time-discounting curves—"Wait fifteen minutes for two marshmallows when I can have one right now? What kind of fool waits fifteen minutes?" That is a predictor of adult property crime.

the former biologically than the latter. It may be harder to *apply* a proximal biological therapy for the former than for the latter (e.g., an imaginary neural growth factor drug that improves self-esteem versus cognition). But biology mediates both links. A cloud may be less tangible than a brick, but it's constructed with the same rules about how atoms interact.

How does biology link childhood with the behaviors of adulthood? Chapter 5's neural plasticity writ large and early. The developing brain epitomizes neural plasticity, and every hiccup of experience has an effect, albeit usually a miniscule one, on that brain.

We now examine ways in which different types of childhoods produce different sorts of adults.

LET'S START AT THE VERY BEGINNING: THE IMPORTANCE OF MOTHERS

Nothing like a section heading stating the obvious. Everybody needs a mother. Even rodents; separate rat pups from Mom a few hours daily and, as adults, they have elevated glucocorticoid levels and poor cognitive skills, are anxious, and, if male, are more aggressive.[22] Mothers are crucial. Except that well into the twentieth century, most experts didn't think so. The West developed child-rearing techniques where, when compared with traditional cultures, children had less physical contact with their mothers, slept alone at earlier ages, and had longer latencies to be picked up when crying. Around 1900 the leading expert Luther Holt of Columbia University warned against the "vicious practice" of picking up a crying child or handling her too often. This was the world of children of the wealthy, raised by nannies and presented to their parents before bedtime to be briefly seen but not heard.

This period brought one of history's strangest one-night stands, namely when the Freudians and the behaviorists hooked up to explain why infants become attached to their mothers. To behaviorists, obviously, it's because mothers reinforce them, providing calories when they're hungry. For Freudians, also obviously, infants lack the "ego development" to form a relationship with anything/anyone other than Mom's breasts. When combined with children-should-be-seen-but-not-heard-ism, this suggested that once you've addressed a child's need for nutrition, proper temperature, plus other odds and ends, they're set to go. Affection, warmth, physical contact? Superfluous.

Such thinking produced at least one disaster. When a child was hospitalized for a stretch, dogma was that the mother was unnecessary—she just added emotional tumult, and everything essential was supplied by the staff. Typically, mothers could visit their children once a week for a few minutes. And when kids were hospitalized for extended periods, they wasted away with "hospitalism," dying in droves from nonspecific infections and gastrointestinal maladies unrelated to their original illness.[23] This was an era when the germ theory had mutated into the belief that hospitalized children do best when untouched, in antiseptic isolation. Remarkably, hospitalism soared in hospitals with newfangled incubators (adapted from poultry farming); the safest hospitals were poor ones that relied on the primitive act of humans actually touching and interacting with infants.

In the 1950s the British psychiatrist John Bowlby challenged the view of infants as simple organisms with few emotional needs; his "attachment theory" birthed our modern view of the mother-infant bond.*[24] In his trilogy *Attachment and Loss*, Bowlby summarized the no-brainer answers we'd give today to the question "What do children need from their mothers?": love, warmth, affection, responsiveness, stimulation, consistency, reliability. What is produced in their absence? Anxious, depressed, and/or poorly attached adults.†

Bowlby inspired one of the most iconic experiments in psychology's

* Bowlby, unlike most Freudians and behaviorists, actually had extensive experience with children, including children in the 1940s separated from their mothers—London children sent to the countryside during the Blitz, Central European Jewish children shipped to England in the Kindertransport rescues one step ahead of Hitler, and of course war orphans. BTW, what was Bowlby's childhood like? He was the son of Sir Anthony Bowlby, the king's own surgeon, and was raised by nannies.

† Naturally, by now Bowlby's offspring, the school of "attachment parenting," is so established as to have generated endless misconceptions, fads, cults, lunatic fringes, and crazy-making senses of neurotic inadequacy or self-righteous superiority among parents. To open a can of worms a smidgen, there is no scientific support for concluding that a woman has irreparably damaged her child if she does not breast-feed, breast-feeds for less than the first decade of the child's life, can't successfully breast-feed her child within seconds after birth, or ever leaves the child alone for more than two seconds, let alone works outside of the house. And nothing about the science says that the same good effects of attachment can't be provided by a man, a working single mother, two mommies, or two daddies.

history, by Harry Harlow of the University of Wisconsin; it destroyed Freudian and behaviorist dogma about mother-infant bonding.[25] Harlow would raise an infant rhesus monkey without a mother but with two "surrogates" instead. Both were made of a chicken-wire tube approximating a torso, with a monkey-ish plastic head on top. One surrogate had a bottle of milk coming from its "torso." The other had terry cloth wrapped around the torso. In other words, one gave calories, the other a poignant approximation of a mother monkey's fur. Freud and B. F. Skinner would have wrestled over access to chicken-wire mom. But infant monkeys chose the terry-cloth mom.* "Man cannot live by milk alone. Love is an emotion that does not need to be bottle- or spoon-fed," wrote Harlow.

Evidence for the most basic need provided by a mother comes from a controversial quarter. Starting in the 1990s, crime rates plummeted across the United States. Why? For liberals the answer was the thriving economy. For conservatives it was the larger budgets for policing, expanded prisons, and three-strikes sentencing laws. Meanwhile, a partial explanation was provided by legal scholar John Donohue of Stanford and economist Steven Levitt of the University of Chicago—it was the legalization of abortions. The authors' state-by-state analysis of the liberalization of abortion laws and the demographics of the crime drop showed that when abortions become readily available in an area, rates of crime by young adults decline about twenty years later. Surprise—this was highly controversial, but it makes perfect, depressing sense to me. What majorly predicts a life of crime? Being born to a mother who, if she could, would have

* The iconic nature of this study is such that I've heard psychologists sardonically reference Harlow, as in, "I had a pretty crappy childhood; my father was never around and my mother was a chicken-wire mom."

chosen that you not be. What's the most basic thing provided by a mother? Knowing that she is happy that you exist.*[26]

Harlow also helped demonstrate a cornerstone of this book, namely what mothers (and later peers) provide as children grow. To do so, he performed some of the most inflammatory research in psychology's history. This involved raising infant monkeys in isolation, absent mother or peers; they spent the first months, even years, of their lives without contact with another living being, before being placed in a social group.†

Predictably, they'd be wrecks. Some would sit alone, clutching themselves, rocking "autistically." Others would be markedly inappropriate in their hierarchical or sexual behaviors.

There was something interesting. It wasn't that these ex-isolates did behaviors wrong—they didn't aggressively display like an ostrich, make the sexually solicitive gestures of a gecko. Behaviors were normal but occurred at the wrong time and place—say, giving subordination gestures to pipsqueaks half their size, threatening alphas they should cower before. Mothers and peers don't teach the motoric features of fixed action patterns; those are hardwired. They teach when, where, and to whom—the appropriate *context* for those behaviors. They give the first lessons about when touching someone's arm or pulling a trigger can be among the best or worst of our behaviors.

I saw a striking example of this among the baboons that I study in Kenya, when both a high-ranking and a low-ranking female gave birth to daughters the same week. The former's kid hit every developmental landmark earlier than the other, the playing field already unlevel. When the infants were a few weeks old, they nearly had their first interaction. Daughter of subordinate mom spotted daughter of dominant one, toddled

* Interestingly, Bowlby's first published paper reported that thieves had an increased rate of extended maternal separation in childhood. Related to that, a 1994 study showed that individuals who suffered the combination of birth complications and maternal rejection at age one had a markedly increased likelihood of committing violent (but not nonviolent) crimes eighteen years later.

† The brutality of these studies helped spawn the animal-rights movement. I've been deeply conflicted about Harlow's work since I was first moved to tears reading about it as a teenager. He was appallingly callous, readily admitted that he felt nothing for the monkeys, and did too many of the deprivation studies. But at the same time, the work helped, among other things, lay the groundwork for understanding the biology of how early-life loss predisposes toward adult depression. Given the prevailing wisdom at the time concerning child rearing and the perceived irrelevance of features of parenting that we now view as vital, the irony is that it was Harlow's pioneering work that most clearly demonstrated the immorality of doing such research.

over to say hello. And as she got near, her low-ranking mother grabbed her by the tail and pulled her back.

This was her first lesson about her place in that world. "You see her? She's *much* higher ranking than you, so you don't just go and hang with her. If she's around, you sit still and avoid eye contact and hope she doesn't take whatever you're eating." Amazingly, in twenty years those two infants would be old ladies, sitting in the savanna, still displaying the rank asymmetries they learned that morning.

ANY KIND OF MOTHER IN A STORM

Harlow provided another important lesson, thanks to another study painful to contemplate. Infant monkeys were raised with chicken-wire surrogates with air jets in the middle of their torsos. When an infant clung, she'd receive an aversive blast of air. What would a behaviorist predict that the monkey would do when faced with such punishment? Flee. But, as in the world of abused children and battered partners, infants held harder.

Why do we often become attached to a source of negative reinforcement, seek solace when distressed from the cause of that distress? Why do we ever love the wrong person, get abused, and return for more?

Psychological insights abound. Because of poor self-esteem, believing you'll never do better. Or a codependent conviction that it's your calling to change the person. Maybe you identify with your oppressor, or have decided it's your fault and the abuser is justified, so they seem less irrational and terrifying. These are valid and can have huge explanatory and therapeutic power. But work by Regina Sullivan of NYU demonstrates bits of this phenomenon miles from the human psyche.

Sullivan would condition rat pups to associate a neutral odor with a shock.[27] If a pup that had been conditioned at ten days of age or older ("older pups") was exposed to that odor, logical things happened—amygdala activation, glucocorticoid secretion, and avoidance of the odor.

But do the same to a younger pup and none of that would occur; remarkably, the pup would be *attracted* to the odor.

Why? There is an interesting wrinkle related to stress in newborns. Rodent fetuses are perfectly capable of secreting glucocorticoids. But within hours of birth, the adrenal glands atrophy dramatically, becoming barely able to secrete glucocorticoids. This "stress hyporesponsive period" (SHRP) wanes over the coming weeks.[28]

What is the SHRP about? Glucocorticoids have so many adverse effects on brain development (stay tuned) that the SHRP represents a gamble—"I won't secrete glucocorticoids in response to stress, so that I develop optimally; if something stressful happens, Mom will handle it for me." Accordingly, deprive infant rats of their mothers, and within hours their adrenals expand and regain the ability to secrete plenty of glucocorticoids.

During the SHRP infants seem to use a further rule: "If Mom is around (and I thus don't secrete glucocorticoids), I should get attached to any strong stimulus. It couldn't be bad for me; Mom wouldn't allow that." As evidence, inject glucocorticoids into the amygdalae of young pups during the conditioning, and the amygdalae would activate and the pups would develop an aversion to the odor. Conversely, block glucocorticoid secretion in older pups during conditioning, and they'd become attracted to the odor. Or condition them with their mother present, and they wouldn't secrete glucocorticoids and would develop an attraction. In other words, in young rats even aversive things are reinforcing in Mom's presence, even if Mom is the *source* of the aversive stimuli. As Sullivan and colleagues wrote, "attachment [by such an infant] to the caretaker has evolved to ensure that the infant forms a bond to that caregiver regardless of the quality of care received." Any kind of mother in a storm.

If this applies to humans, it helps explain why individuals abused as kids are as adults prone toward relationships in which they are abused by their partner.[29] But what about the flip side? Why is it that about 33 percent of adults who were abused as children become abusers themselves?

Again, useful psychological insights abound, built around identification with the abuser and rationalizing away the terror: "I love my kids, but I smack them around when they need it. My father did that to me, so he could have loved me too." But once again something biologically deeper also occurs—infant monkeys abused by their mothers are more likely to become abusive mothers.[30]

DIFFERENT ROUTES TO THE SAME PLACE

I anticipated that, with mothers now covered, we'd next examine the adult consequences of, say, paternal deprivation, or childhood poverty, or exposure to violence or natural disasters. And there'd be the same question—what specific biological changes did each cause in children that increased the odds of specific adult behaviors?

But this plan didn't work—the similarities of effects of these varied traumas are greater than the differences. Sure, there are specific links (e.g., childhood exposure to domestic violence makes adult antisocial violence more likely than does childhood exposure to hurricanes). But they all converge sufficiently that I will group them together, as is done in the field, as examples of "childhood adversity."

Basically, childhood adversity increases the odds of an adult having (a) depression, anxiety, and/or substance abuse; (b) impaired cognitive capabilities, particularly related to frontocortical function; (c) impaired impulse control and emotion regulation; (d) antisocial behavior, including violence; and (e) relationships that replicate the adversities of childhood (e.g., staying with an abusive partner).[31] And despite that, some individuals endure miserable childhoods just fine. More on this to come.

We'll now examine the biological links between childhood adversity and increased risk of these adult outcomes.

THE BIOLOGICAL PROFILE

All these forms of adversity are obviously stressful and cause abnormalities in stress physiology. Across numerous species, major early-life stressors produce both kids and adults with elevated levels of glucocorticoids (along with CRH and ACTH, the hypothalamic and pituitary hormones that regulate glucocorticoid release) and hyperactivity of the sympathetic nervous system.[32] Basal glucocorticoid levels are elevated—the stress response is always somewhat activated—and there is delayed recovery back to baseline after a stressor. Michael Meaney of McGill University has shown how early-life stress permanently blunts the ability of the brain to rein in glucocorticoid secretion.

As covered in chapter 4, marinating the brain in excess glucocorticoids, particularly during development, adversely effects cognition, impulse control, empathy, and so on.[33] There is impaired hippocampal-dependent learning in adulthood. For example, abused children who develop PTSD have decreased volume of the hippocampus in adulthood. Stanford psychiatrist Victor Carrion has shown decreased hippocampal growth within months of the abuse. As a likely cause, glucocorticoids decrease hippocampal production of the growth factor BDNF (brain-derived neurotrophic factor).

So childhood adversity impairs learning and memory. Crucially, it also impairs maturation and function of the frontal cortex; again, glucocorticoids, via inhibiting BDNF, are likely culprits.

The connection between childhood adversity and frontocortical maturation pertains to childhood poverty. Work by Martha Farah of the University of Pennsylvania, Tom Boyce of UCSF, and others demonstrates something outrageous: By age five, the lower a child's socioeconomic status, on the average, the (a) higher the basal glucocorticoid levels and/or the more reactive the glucocorticoid stress response, (b) the thinner the frontal cortex and the lower its metabolism, and (c) the poorer the frontal function concerning working memory, emotion regulation, impulse control, and executive decision making; moreover, to achieve equivalent frontal regulation, lower-SES kids must activate more frontal cortex than do higher-SES kids. In addition, childhood poverty impairs maturation of the corpus callosum, a bundle of axonal fibers connecting the two hemispheres and integrating their function. This is *so* wrong—foolishly pick a poor family to be born into, and by kindergarten, the odds of your succeeding at life's marshmallow tests are already stacked against you.[34]

Considerable research focuses on how poverty "gets under the skin." Some mechanisms are human specific—if you're poor, you're more likely to grow up near environmental toxins,*[35] in a dangerous neighborhood with more liquor stores than markets selling produce; you're less likely to attend a good school or have parents with time to read to you. Your community is likely to have poor social capital, and you, poor self-esteem. But part of the link reflects the corrosive effects of subordination in all hierarchical species. For example, having a low-ranking mother predicts elevated glucocorticoids in adulthood in baboons.[36]

Thus, childhood adversity can atrophy and blunt the functioning of the hippocampus and frontal cortex. But it's the opposite in the amygdala—lots of adversity and the amygdala becomes larger and hyperreactive. One consequence is increased risk of anxiety disorders; when coupled with the poor frontocortical development, it explains problems with emotion and behavior regulation, especially impulse control.[37]

Childhood adversity accelerates amygdaloid maturation in a particular way. Normally, around adolescence the frontal cortex gains the ability to inhibit the amygdala, saying, "I wouldn't do this if I were you." But after childhood adversity, the amygdala develops the ability to inhibit the frontal cortex, saying, "I'm doing this and just try to stop me."

Childhood adversity also damages the dopamine system (with its role in reward, anticipation, and goal-directed behavior) in two ways.

First, early adversity produces an adult organism more vulnerable to drug and alcohol addiction. The pathway to this vulnerability is probably threefold: (a) effects on the developing dopamine system; (b) the excessive adult exposure to glucocorticoids, which increases drug craving; (c) that poorly developed frontal cortex.[38]

Childhood adversity also substantially increases the risk of adult de-

* For example, early-childhood lead exposure—a strong correlate of living in a poor neighborhood—impairs brain development and predicts poor cognitive and emotional regulatory skills and increased incidence of criminality in adulthood.

pression. Depression's defining symptom is anhedonia, the inability to feel, anticipate, or pursue pleasure. Chronic stress depletes the mesolimbic system of dopamine, generating anhedonia.* The link between childhood adversity and adult depression involves both organizational effects on the developing mesolimbic system and elevated adult glucocorticoid levels, which can deplete dopamine.[39]

Childhood adversity increases depression risk via "second hit" scenarios—lowering thresholds so that adult stressors that people typically manage instead trigger depressive episodes. This vulnerability makes sense. Depression is fundamentally a pathological sense of loss of control (explaining the classic description of depression as "learned helplessness"). If a child experiences severe, uncontrollable adversity, the most fortunate conclusion in adulthood is "Those were terrible circumstances over which I had no control." But when childhood traumas produce depression, there is cognitively distorted overgeneralization: "And life will always be uncontrollably awful."

TWO SIDE TOPICS

So varied types of childhood adversity converge in producing similar adult problems. Nonetheless, two types of adversity should be considered separately.

Observing Violence

What happens when children observe domestic violence, warfare, a gang murder, a school massacre? For weeks afterward there is impaired concentration and impulse control. Witnessing gun violence doubles a child's likelihood of serious violence within the succeeding two years.

* What's anhedonia like in a rat? Give a normal rat two water bottles to choose from, one with water, the other with water sweetened with sucrose. The rat prefers the sucrose water. But a stressed, anhedonic rat shows no preference. Same result for other pleasurable things.

And adulthood brings the usual increased risks of depression, anxiety, and aggression. Consistent with that, violent criminals are more likely than nonviolent ones to have witnessed violence as kids.*[40]

This fits our general picture of childhood adversity. A separate topic is the effects of *media* violence on kids.

Endless studies have analyzed the effects of kids witnessing violence on TV, in movies, in the news, and in music videos, and both witnessing and participating in violent video games. A summary:

Exposing children to a violent TV or film clip increases their odds of aggression soon after.[41] Interestingly, the effect is stronger in girls (amid their having lower overall levels of aggression). Effects are stronger when kids are younger or when the violence is more realistic and/or is presented as heroic. Such exposure can make kids more accepting of aggression—in one study, watching violent music videos increased adolescent girls' acceptance of dating violence. The violence is key—aggression isn't boosted by material that's merely exciting, arousing, or frustrating.

Heavy childhood exposure to media violence predicts higher levels of aggression in young adults of both sexes ("aggression" ranging from behavior in an experimental setting to violent criminality). The effect typically remains after controlling for total media-watching time, maltreatment or neglect, socioeconomic status, levels of neighborhood violence, parental education, psychiatric illness, and IQ. This is a reliable finding of large magnitude. The link between exposure to childhood media violence and increased adult aggression is stronger than the link between lead exposure and IQ, calcium intake and bone mass, or asbestos and laryngeal cancer.

Two caveats: (a) there is no evidence that catastrophically violent individuals (e.g., mass shooters) are that way because of childhood exposure to violent media; (b) exposure does not remotely guarantee increased aggression—instead, effects are strongest on kids already prone toward violence. For them, exposure desensitizes and normalizes their own aggression.†

* Remarkably, exposure to multiple incidents of violence even accelerates the aging of children's chromosomes.

† I want to thank a really excellent undergrad, Dylan Alegria, who helped me tread water in this voluminous literature.

Bullying

Being bullied is mostly another garden-variety childhood adversity, with adult consequences on par with childhood maltreatment at home.[42]

There is a complication, though. As most of us observed, exploited, or experienced as kids, bullying targets aren't selected at random. Kids with the metaphorical "kick me" signs on their backs are more likely to have personal or family psychiatric issues and poor social and emotional intelligence. These are kids already at risk for bad adult outcomes, and adding bullying to the mix just makes the child's future even bleaker.

The picture of the bullies is no surprise either, starting with their disproportionately coming from families of single moms or younger parents with poor education and employment prospects. There are generally two profiles of the kids themselves—the more typical is an anxious, isolated kid with poor social skills, who bullies out of frustration and to achieve acceptance. Such kids typically mature out of bullying. The second profile is the confident, unempathic, socially intelligent kid with an imperturbable sympathetic nervous system; this is the future sociopath.

There is an additional striking finding. You want to see a kid who's really likely to be a mess as an adult? Find someone who both bullies *and* is bullied, who terrorizes the weaker at school and returns home to be terrorized by someone stronger.[43] Of the three categories (bully, bullied, bully/bullied), they're most likely to have prior psychiatric problems, poor school performance, and poor emotional adjustment. They're more likely than pure bullies to use weapons and inflict serious damage. As adults, they're most at risk for depression, anxiety, and suicidality.

In one study kids from these three categories read scenarios of bullying.[44] Bullied victims would condemn bullying and express sympathy. Bullies would condemn bullying but rationalize the scenario (e.g., this time it was the victim's fault). And bully/bullied kids? They would say bullying is okay. No wonder they have the worst outcome. "The weak deserve to be bullied, so it's fine when I bully. But that means I deserve to be bullied at home. But I don't, and that relative bullying me is awful.

Maybe then I'm awful when I bully someone. But I'm not, because the weak deserve to be bullied. . . ." A Möbius strip from hell.*

A KEY QUESTION

We've now examined adult consequences of childhood adversity and their biological mediators. A key question persists. Yes, childhood abuse increases the odds of being an abusive adult; witnessing violence raises the risk for PTSD; loss of a parent to death means more chance of adult depression. Nevertheless, many, maybe even most victims of such adversity turn into reasonably functional adults. There is a shadow over childhood, demons lurk in corners of the mind, but overall things are okay. What explains such resilience?

As we'll see, genes and fetal environment are relevant. But most important, recall the logic of collapsing different types of trauma into a single category. What counts is the sheer number of times a child is bludgeoned by life and the number of protective factors. Be sexually abused as a child, or witness violence, and your adult prognosis is better than if you had experienced both. Experience childhood poverty, and your future prospects are better if your family is stable and loving than broken and acrimonious. Pretty straightforwardly, the more categories of adversities a child suffers, the dimmer his or her chances of a happy, functional adulthood.[45]

A SLEDGEHAMMER

What happens when *everything* goes wrong—no mother or family, minimal peer interactions, sensory and cognitive neglect, plus some malnutrition?[46]

---

* I thank another superb undergrad, Ali Maggioncalda, for help with this topic.

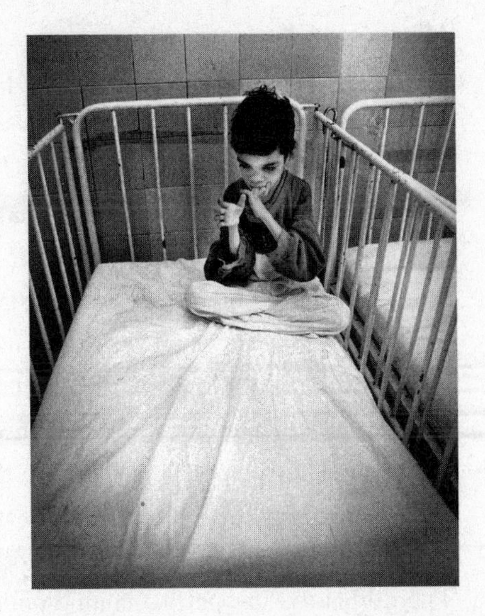

These are the Romanian institution kids, poster children for just how nightmarish childhood can be. In the 1980s, the Romanian dictator Nicolae Ceauşescu banned contraceptives and abortions for any women who had not yet given birth to five children. Soon institutions filled with thousands of infants and kids abandoned by impoverished families (many intent on reclaiming their child when finances improved).* Kids were warehoused in overwhelmed institutions, resulting in severe neglect and deprivation. The story broke after Ceauşescu's 1989 overthrow. Many kids were adopted by Westerners, and international attention led to some improvements in the institutions. Since then, children adopted in the West, those eventually returned to their families, and those who remained institutionalized have been studied, primarily by Charles Nelson of Harvard.

As adults, these kids are mostly what you'd expect. Low IQ and poor cognitive skills. Problems with forming attachments, often bordering on autistic. Anxiety and depression galore. The longer the institutionalization, the worse the prognosis.

And their brains? Decreased total brain size, gray matter, white matter, frontal cortical metabolism, connectivity between regions, sizes of individual brain regions. Except for the amygdala. Which is enlarged. That pretty much says it all.

* A shocking part of the story: Roma children were regularly abandoned at orphanages and left there until they were adolescents—and could work.

CULTURE, WITH BOTH A BIG AND A LITTLE *C*

Chapter 9 considers the effects of culture on our best and worst behaviors. We now preview that chapter, focusing on two facts—childhood is when culture is inculcated, and parents mediate that process.

There is huge cultural variability in how childhood is experienced—how long and often kids are nursed; how often they are in contact with parents and other adults; how often they're spoken to; how long they cry before someone responds; at what age they sleep alone.

Considering cross-cultural child rearing often brings out the most invidious and neurotic in parents—do other cultures do a better job at it? There must be the perfect combo out there, a mixture of the Kwakiutl baby diet, the Trobriand sleeping program, and the Ituri approach to watching Baby Mozart videos. But there is no anthropological ideal of child rearing. Cultures (starting with parents) raise children to become adults who behave in the ways valued by that culture, a point emphasized by the anthropologist Meredith Small of Cornell University.[47]

We begin with parenting style, a child's first encounter with cultural values. Interestingly, the most influential typology of parenting style, writ small, grew from thinking about cultural styles, writ large.

Amid the post–World War II ruins, scholars tried to understand where Hitler, Franco, Mussolini, Tojo, and their minions came from. What are the roots of fascism? Two particularly influential scholars were refugees from Hitler, namely Hannah Arendt (with her 1951 book *The Origins of Totalitarianism*) and Theodor Adorno (with the 1950 book *The Authoritarian Personality*, coauthored with Else Frenkel-Brunswik, Daniel Levinson, and Nevitt Sanford). Adorno in particular explored the personality traits of fascists, including extreme conformity, submission to and belief in authority, aggressiveness, and hostility toward intellectualism and introspection—traits typically rooted in childhood.[48]

This influenced the Berkeley psychologist Diana Baumrind, who in the 1960s identified three key parenting styles (in work since replicated and extended to various cultures).[49] First is *authoritative* parenting. Rules and expectations are clear, consistent, and explicable—"Because I said so" is anathema—with room for flexibility; praise and forgiveness trump punishment; parents welcome children's input; developing children's potential and autonomy is paramount. By the standards of the educated neurotics who would read (let alone write . . .) this book, this produces a good adult outcome—happy, emotionally and socially mature and fulfilled, independent and self-reliant.

Next is *authoritarian* parenting. Rules and demands are numerous, arbitrary, and rigid and need no justification; behavior is mostly shaped by punishment; children's emotional needs are low priorities. Parental motivation is often that it's a tough, unforgiving world and kids better be prepared. Authoritarian parenting tends to produce adults who may be narrowly successful, obedient, conformist (often with an undercurrent of resentment that can explode), and not particularly happy. Moreover, social skills are often poor because, instead of learning by experience, they grew up following orders.

And then there is *permissive* parenting, the aberration that supposedly let Boomers invent the 1960s. There are few demands or expectations, rules are rarely enforced, and children set the agenda. Adult outcome: self-indulgent individuals with poor impulse control, low frustration tolerance, plus poor social skills thanks to living consequence-free childhoods.

Baumrind's trio was expanded by Stanford psychologists Eleanor Maccoby and John Martin to include *neglectful* parenting.[50] This addition produces a two-by-two matrix: parenting is authoritative (high demand, high responsiveness), authoritarian (high demand, low responsiveness), permissive (low demand, high responsiveness), or neglectful (low demand, low responsiveness).

Importantly, each style usually produces adults with that same approach, with different cultures valuing different styles.

Then comes the next way cultural values are transmitted to kids, namely by peers. This was emphasized in Judith Rich Harris's *The Nurture Assumption*. Harris, a psychologist without an academic affiliation or doctorate, took the field by storm, arguing that the importance of parenting in shaping a child's adult personality is exaggerated.[51] Instead, once kids pass a surprisingly young age, peers are most influential. Elements of her argument included: (a) Parental influence is often actually mediated via peers. For example, being raised by a single mother increases the risk of adult antisocial behavior, but not because of the parenting; instead, because of typically lower income, kids more likely live in a neighborhood with tough peers. (b) Peers have impact on linguistic development (e.g., children acquire the accent of their peers, not their parents). (c) Other young primates are mostly socialized by peers, not mothers.

The book was controversial (partially because the theme begged to be distorted—"Psychologist proves that parents don't matter"), drawing criticism and acclaim.* As the dust has settled, current opinion tends to be that peer influences are underappreciated, but parents still are plenty important, including by influencing what peer groups their kids experience.

Why are peers so important? Peer interactions teach social competence—context-dependent behavior, when to be friend or foe, where you fit in hierarchies. Young organisms utilize the greatest teaching tool ever to acquire such information—play.[52]

What is social play in the young? Writ large, it's an array of behaviors that train individuals in social competence. Writ medium, it's fragments of the real thing, bits and pieces of fixed action patterns, a chance to safely try out roles and improve motor skills. Writ small and endocrine, it's a demonstration that moderate and transient stress—"stimulation"—is great. Writ small and neurobiological, it's a tool for deciding which excess synapses to prune.

The historian Johan Huizinga characterized humans as "Homo Ludens," Man the Player, with our structured, rule-bound play—i.e., games. Nevertheless, play is universal among socially complex species, ubiquitous among the young and peaking at puberty, and all play involves similar behaviors, after some ethological translating (e.g., a dominant dog signals the benevolence needed to initiate play by crouching, diminish-

* An irresistible irony: In the aftermath of the book's publication, Harris received a major award from the august American Psychological Association, an award named in honor of . . . the man who, decades before, as chairman of the psychology department at Harvard, had thrown Harris out of the PhD program for her lack of potential.

ing herself; translated into baboon, a dominant kid presents her rear to someone lower ranking).

Play is vital. In order to play, animals forgo foraging, expend calories, make themselves distracted and conspicuous to predators. Young organisms squander energy on play during famines. A child deprived of or disinterested in play rarely has a socially fulfilling adult life.

Most of all, play is intrinsically pleasurable—why else perform a smidgen of a behavioral sequence in an irrelevant setting? Dopaminergic pathways activate during play; juvenile rats, when playing, emit the same vocalizations as when rewarded with food; dogs spend half their calories wagging their tails to pheromonally announce their presence and availability for play. As emphasized by the psychiatrist Stuart Brown, founder of the National Institute for Play, the opposite of play is not work—it's depression. A challenge is to understand how the brain codes for the reinforcing properties of the variety of play. After all, play encompasses everything from mathematicians besting each other with hilarious calculus jokes to kids besting each other by making hilarious fart sounds with their armpits.

One significant type of play involves fragments of aggression, what Harlow called "rough and tumble" play—kids wrestling, adolescent impalas butting heads, puppies play-biting each other.[53] Males typically do it more than females, and as we'll see soon, it's boosted by prenatal testosterone. Is rough-and-tumble play practice for life's looming status tournament, or are you already in the arena? A mixture of both.

Expanding beyond peers, neighborhoods readily communicate culture to kids. Is there garbage everywhere? Are houses decrepit? What's ubiquitous—bars, churches, libraries, or gun shops? Are there many parks, and are they safe to enter? Do billboards, ads, and bumper stickers sell religious or material paradises, celebrate acts of martyrdom or kindness and inclusiveness?

And then we get to culture at the level of tribes, nations, and states. Here, briefly, are some of the broadest cultural differences in child-rearing practices.

Collectivist Versus Individualist Cultures

As will be seen in chapter 9, this is the most studied cultural contrast, typically comparing collectivist East Asian cultures with überindividualist America. Collectivist cultures emphasize interdependence, harmony, fitting in, the needs and responsibilities of the group; in contrast, individualist cultures value independence, competition, the needs and rights of the individual.

On average, mothers in individualist cultures, when compared with those in collectivist ones, speak louder, play music louder, have more animated expressions.[54] They view themselves as teachers rather than protectors, abhor a bored child, value high-energy affect. Their games emphasize individual competition, urge hobbies involving doing rather than observing. Kids are trained in verbal assertiveness, to be autonomous and influential. Show a cartoon of a school of fish with one out front, and she'll describe it to her child as the leader.*

Mothers in collectivist cultures, in contrast, spend more time than individualist mothers soothing their child, maintaining contact, and facilitating contact with other adults. They value low arousal affect and sleep with their child to a later age. Games are about cooperation and fitting in; if playing with her child with, say, a toy car, the point is not exploring what a car does (i.e., being automobile), but the process of sharing ("Thank you for giving me your car; now I'll give it back to you"). Kids are trained to get along, think of others, accept and adapt, rather than change situations; morality and conformity are nearly synonymous. Show the cartoon of the school of fish, and the fish out front must have done something wrong, because no one will play with him.

Logically, kids in individualist cultures acquire ToM later than collectivist-culture kids and activate pertinent circuits more to achieve the same degree of competence. For a collectivist child, social competence is all about taking someone else's perspective.[55]

Interestingly, kids in (collectivist) Japan play more violent video

* All of these differences are typical in fathers as well but have been studied far more in mothers.

games than do American kids, yet are less aggressive. Moreover, exposing Japanese kids to media violence boosts aggression less than in American kids.[56] Why the difference? Three possible contributing factors: (a) American kids play alone more often, a lone-wolf breeding ground; (b) Japanese kids rarely have a computer or TV in their bedroom, so they play near their parents; (c) Japanese video-game violence is more likely to have prosocial, collectivist themes.

More in chapter 9 on collectivist versus individualist cultures.

Cultures of Honor

These cultures emphasize rules of civility, courtesy, and hospitality. Taking retribution is expected for affronts to the honor of one's self, family, or clan; failing to do so is shameful. These are cultures filled with vendettas, revenge, and honor killings; cheeks aren't turned. A classic culture of honor is the American South, but as we'll see in chapter 9, such cultures occur worldwide and with certain ecological correlates. A particularly lethal combo is when a culture of victimization—we were wronged last week, last decade, last millennium—is coupled with a culture of honor's ethos of retribution.

Parenting in cultures of honor tends to be authoritarian.[57] Kids are aggressive, particularly following honor violations, and staunchly endorse aggressive responses to scenarios of honor violation.

Class Differences

As noted, an infant baboon learns her place in the hierarchy from her mother. A human child's lessons about status are more complex—there is implicit cuing, subtle language cues, the cognitive and emotional weight of remembering the past ("When your grandparents emigrated here they couldn't even . . .") and hoping about the future ("When you grow up, you're going to . . ."). Baboon mothers teach their young appropriate behavioral context; human parents teach their young what to bother dreaming about.

Class differences in parenting in Western countries resemble parenting differences between Western countries and those in the developing world. In the West a parent teaches and facilitates her child exploring the world. In the toughest corners of the developing world, little more is expected than the awesome task of keeping your child alive and buffered from the menacing world.*

In Western cultures, class differences in parenting sort by Baumrind's typologies. In higher-SES strata, parenting tends to be authoritative or permissive. In contrast, parenting in society's lower-SES rungs is typically authoritarian, reflecting two themes. One concerns protecting. When are higher-SES parents authoritarian? When there is danger. "Sweetie, I love that you question things, but if you run into the street and I scream 'Stop,' you stop." A lower-SES childhood is rife with threat. The other theme is preparing the child for the tough world out there—for the poor, adulthood consists of the socially dominant treating them in an authoritarian manner.

Class differences in parenting were explored in a classic study by the anthropologist Adrie Kusserow of St. Michael's College, who did fieldwork observing parents in three tribes—wealthy families on Manhattan's Upper East Side; a stable, blue-collar community; and a poor, crime-ridden one (the last two both in Queens).[58] The differences were fascinating.

Parenting in the poor neighborhood involved "hard defensive individualism." The neighborhood was rife with addiction, homelessness, incarceration, death—and parents' aim was to shelter their child from the literal and metaphorical street. Their speech was full of metaphors about not losing what was achieved—standing your ground, keeping up your pride, not letting others get under your skin. Parenting was authoritarian, toughening the goal. For example, parents teased kids far more than in the other neighborhoods.

In contrast, working-class parenting involved "hard offensive individid-

* I've experienced a manifestation of such parenting in my decades of fieldwork in Kenya, where my nearest neighbors have been highly un-Westernized members of the Maasai tribe. Sometimes I'd run into someone I hadn't seen in a while who had had a baby in the interim, and it took me years to get out of my ridiculous Western reflex—"A new baby! That's wonderful! Mazel tov! What's her name?" Awkward silence—you don't give a baby a name (or perhaps aren't willing to utter it) until she's survived her first malarial rainy season and hungry dry season.

ualism." Parents had some socioeconomic momentum, and kids were meant to maintain that precarious trajectory. Parents' speech about their hopes for their kids contained images of movement, progress, and athletics—getting ahead, testing the waters, going for the gold. With hard work and the impetus of generations of expectations, your child might pioneer landfall in the middle class.

Parenting in both neighborhoods emphasized respect for authority, particularly within the family. Moreover, kids were fungible members of a category, rather than individualized—"You kids get over here."

Then there was the "soft individualism" of upper-middle-class parenting.* Children's eventual success, by conventional standards, was a given, as were expectations of physical health. Far more vulnerable was a child's psychological health; when children could become anything, parents' responsibility was to facilitate their epic journey toward an individuated "fulfillment." Moreover, the image of fulfillment was often postconventional—"I hope my child will never work an unsatisfying job just for money." This, after all, is a tribe giddied by tales of the shark in line to become CEO chucking it to learn carpentry or oboe. Parents' speech brimmed with metaphors of potential being fulfilled—flowering, blooming, growing, blossoming. Parenting was authoritative or permissive, riddled with ambivalence about parent-child power differentials. Rather than "You kids, clean up this mess," there'd be the individuated, justifying request—"Caitlin, Zach, Dakota, could you clean things up a bit please? Malala is coming for dinner."†

We've now seen how childhood events—from the first mother-infant interaction to the effects of culture—have persistent influences, and how biology mediates such influences. When combined with the preceding chapters, we have finished our tour of environmental effects on behavior,

* Which, as noted by Kusserow, included the largest percentage of fathers who were willing to be interviewed.

† I once got a poignant reminder of just how permeating the consequences of lack of privilege are in my professional world. I was interviewing candidates for a position in my lab. In the process, I'd ask each person about how they handled interpersonal conflict, looking for people who promptly addressed a social tension, rather than letting it fester into passive-aggressiveness. I was interviewing a guy whose background was in Queens, rather than the Upper East Side. And when asked, instead of giving the Upper East Side answer I hoped for ("Yeah, I know how bad things get when you don't communicate; I'd be pretty good at just asking the person to be considerate and please return my pipette when they borrow it"), I got the correct answer from Queens: "Nope, no problems there. I know that a lab is no place for fighting; you take it outside. You've got nothing to worry about with me."

from the second before a behavior occurs to a second after birth. In effect, we've done "environment"; time for next chapter's "genes."

But this ignores something crucial: environment doesn't begin at birth.

NINE LONG MONTHS

The Cat in the Hat *in the Womb*

The existence of prenatal environmental influences caught the public's imagination with some charming studies demonstrating that near-term fetuses hear (what's going on outside the womb), taste (amniotic fluid), and remember and prefer those stimuli after birth.

This was shown experimentally—inject lemon-flavored saline into a pregnant rat's amniotic fluid, and her pups are born preferring that flavor. Moreover, some spices consumed by pregnant women get into amniotic fluid. Thus we may be born preferring foods our mothers ate during pregnancy—pretty unorthodox cultural transmission.[59]

Prenatal effects can also be auditory, as shown by inspired research by Anthony DeCasper of the University of North Carolina.[60] A pregnant woman's voice is audible in the womb, and newborns recognize and prefer the sound of their mother's voice.* DeCasper used ethology's playbook to show this: A newborn can learn to suck a pacifier in two different patterns of long and short sucks. Generate one pattern, and you hear Mom's voice; the other, another woman's voice. Newborns want Mom's voice. Elements of language are also learned in utero—the contours of a newborn's cry are similar to the contours of speech in the mother's language.

The cognitive capacities of near-term fetuses are even more remarkable. For example, fetuses can distinguish between two pairs of nonsense syllables ("biba" versus "babi"). How do you know? Get this—Mom says "Biba, biba, biba" repeatedly while fetal heart rate is monitored. "Boring

* In contrast, newborns recognize but show no preference for their father's voice.

(or perhaps lulling)," thinks the fetus, and heart rate slows. Then Mom switches to "babi." If the fetus doesn't distinguish between the two, heart rate deceleration continues. But if the difference is noted—"Whoa, what happened?"—heart rate increases. Which is what DeCasper reported.[61]

DeCasper and colleague Melanie Spence then showed (using the pacifier-sucking-pattern detection system) that newborns typically don't distinguish between the sounds of their mother reading a passage from *The Cat in the Hat* and from the rhythmically similar *The King, the Mice, and the Cheese.*[62] But newborns whose mothers had read *The Cat in the Hat* out loud for hours during the last trimester preferred Dr. Seuss. Wow.

Despite the charm of these findings, this book's concerns aren't rooted in such prenatal learning—few infants are born with a preference for passages from, say, *Mein Kampf.* However, other prenatal environmental effects are quite consequential.

BOY AND GIRL BRAINS, WHATEVER THAT MIGHT MEAN

We start with a simple version of what "environment" means for a fetal brain: the nutrients, immune messengers, and, most important, hormones carried to the brain in the fetal circulation.

Once the pertinent glands have developed in a fetus, they are perfectly capable of secreting their characteristic hormones. This is particularly consequential. When hormones first made their entrance in chapter 4, our discussion concerned their "activational" effects that lasted on the order of hours to days. In contrast, hormones in the fetus have "organizational" effects on the brain, causing lifelong changes in structure and function.

Around eight weeks postconception, human fetal gonads start secreting their steroid hormones (testosterone in males; estrogen and progesterone in females). Crucially, testosterone plus "anti-Müllerian hormone" (also from the testes) masculinize the brain.

Three complications, of increasing messiness:

- In many rodents the brain isn't quite sexually differentiated at birth, and these hormonal effects continue postnatally.
- A messier complication: Surprisingly few testosterone effects in the brain result from the hormone binding to androgen receptors. Instead, testosterone enters targets cells and, bizarrely, is converted to estrogen, then binds to intracellular estrogen receptors (while testosterone has its effects outside the brain either as itself or, after intracellular conversion to a related androgen, dihydrotestosterone). Thus testosterone has much of its masculinizing effect in the brain by becoming estrogen. The conversion of testosterone to estrogen also occurs in the fetal brain. Wait. Regardless of fetal sex, fetal circulation is full of maternal estrogen, plus female fetuses secrete estrogen. Thus female fetal brains are bathed in estrogen. Why doesn't that masculinize the female fetal brain? Most likely it's because fetuses make something called alpha-fetoprotein, which binds circulating estrogen, taking it out of action. So neither Mom's estrogen nor fetal-derived estrogen masculinizes the brain in female fetuses. And it turns out that unless there is testosterone and anti-Müllerian hormone around, fetal mammalian brains automatically feminize.[63]
- Now for the übermessy complication. What exactly is a "female" or "male" brain? This is where the arguments begin.

To start, male brains merely consistently drool reproductive hormones out of the hypothalamus, whereas female brains must master the cyclic secretion of ovulatory cycles. Thus fetal life produces a hypothalamus that is more complexly wired in females.

But how about sex differences in the behaviors that interest us? The question is, how much of male aggression is due to prenatal masculinizing of the brain?

Virtually all of it, if we're talking rodents. Work in the 1950s by Robert

Goy of the University of Wisconsin showed that in guinea pigs an organizational effect of perinatal testosterone is to make the brain responsive to testosterone in adulthood.[64] Near-term pregnant females would be treated with testosterone. This produced female offspring who, as adults, appeared normal but were behaviorally "masculinized"—they were more sensitive than control females to an injection of testosterone, with a greater increase in aggression and male-typical sexual behavior (i.e., mounting other females). Moreover, estrogen was less effective at eliciting female-typical sexual behavior (i.e., a back-arching reflex called lordosis). Thus prenatal testosterone exposure had masculinizing organizational effects, so that these females as adults responded to the activational effects of testosterone and estrogen as males would.

This challenged dogma that sexual identity is due to social, not biological, influences. This was the view of sociologists who hated high school biology . . . and of the medical establishment as well. According to this view, if an infant was born with sexually ambiguous genitalia (roughly 1 to 2 percent of births), it didn't matter which gender they were raised, as long as you decided within the first eighteen months—just do whichever reconstructive surgery was more convenient.*[65]

So here's Goy reporting that prenatal hormone environment, not social factors, determines adult sex-typical behaviors. "But these are guinea pigs" was the retort. Goy and crew then studied nonhuman primates.

A quick tour of sexually dimorphic (i.e., differing by sex) primate behavior: South American species such as marmosets and tamarins, who form pair-bonds, show few sex differences in behavior. In contrast, most Old World primates are highly dimorphic; males are more aggressive, and females spend more time at affiliative behaviors (e.g., social grooming, interacting with infants). How's this for a sex difference: in one study, adult male rhesus monkeys were far more interested in playing with "masculine" human toys (e.g., wheeled toys) than "feminine" ones (stuffed animals), while females had a slight preference for feminine.[66]

* This view held in most medical circles for many years afterward. For an example of how wrong that approach can be, see John Colapinto, *As Nature Made Him: The Boy Who Was Raised as a Girl* (New York: Harper Perennial, 2006).

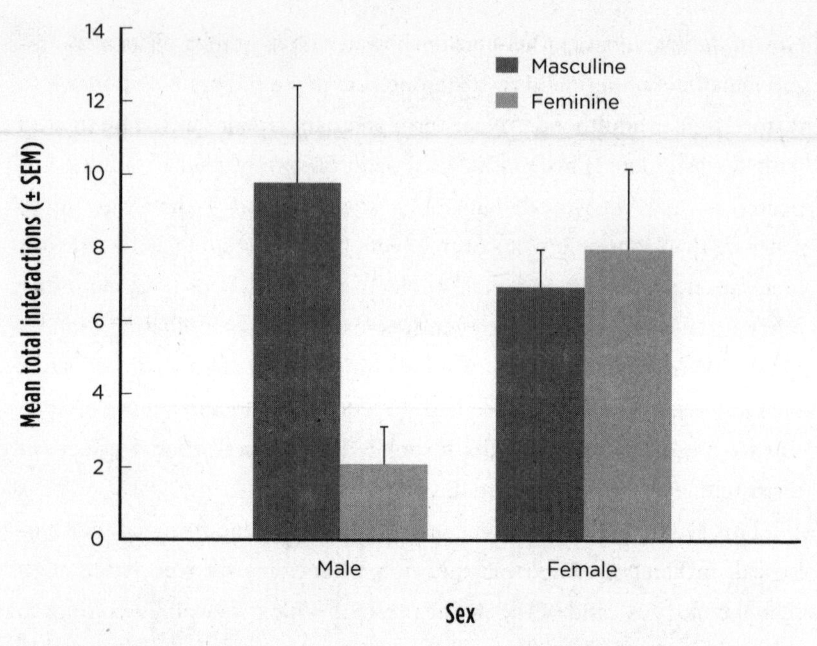

Male rhesus monkeys show a strong preference for playing with stereotypically "masculine" versus "feminine" human toys.

What next, female monkeys prefer young-adult fantasy novels with female protagonists? Why should human toys be relevant to sex differences in monkeys? The authors speculate that this reflects the higher activity levels in males, and how masculine toys facilitate more active play.

Goy studied highly sexually dimorphic rhesus monkeys. There were already hints that testosterone has organizational effects on their behavior—within weeks of birth, males are more active than females and spend more time in rough-and-tumble play. This is long before puberty and its burst of testosterone secretion. Furthermore, even if you suppress their testosterone levels at birth (low, but nevertheless still higher than those of females), males still do more roughing and tumbling. This suggested that the sex difference arose from fetal hormone differences.

Goy proved this by treating pregnant monkeys with testosterone and examining their female offspring. Testosterone exposure throughout pregnancy produced daughters who were "pseudohermaphrodites"—looked like males on the outside but had female gonads on the inside. When compared with control females, these androgenized females did more rough-and-tumble play, were more aggressive, and displayed male-typical mounting behavior and vocalizations (as much as males, by some measures). Importantly, most but not all behaviors were masculinized, and these androgenized females were as interested as control females in infants. Thus, testosterone has prenatal organizational effects on some but not all behaviors.

In further studies, many carried out by Goy's student Kim Wallen of Emory University, pregnant females received lower doses of testosterone, and only in the last trimester.[67] This produced daughters with normal genitalia but masculinized behavior. The authors noted the relevance of this to transgender individuals—the external appearance of one sex but the brain, if you will, of the other.*

And Us

Initially it seemed clear that prenatal testosterone exposure is also responsible for male aggression in humans. This was based on studies of a rare disorder, congenital adrenal hyperplasia (CAH). An enzyme in the adrenal glands has a mutation, and instead of making glucocorticoids, they make testosterone and other androgens, starting during fetal life.

The lack of glucocorticoids causes serious metabolic problems requiring replacement hormones. And what about the excessive androgens in CAH girls (who are typically born with ambiguous genitals and are infertile as adults)?

In the 1950s psychologist John Money of Johns Hopkins University reported that CAH girls had pathologically high levels of male-typical behaviors, a paucity of female-typical ones, and elevated IQ.

That sure stopped everyone in their tracks. But the research had some problems. First, the IQ finding was spurious—parents willing to enroll their CAH child in these studies averaged higher levels of education than did controls. And the gender-typical behaviors? "Normal" was judged by

* Remarkably, studies have examined brains of transgender individuals, concentrating on brain regions that, on the average, differ in size between men and women. And consistently, regardless of the desired direction of the sex change and, in fact, regardless of whether the person had undergone a sex change yet, the dimorphic brain regions in transgender individuals resembled the sex of the person they had always felt themselves to be, not their "actual" sex. In other words, it's not the case that transgender individuals think they're a different gender than they actually are. It's more like they got stuck with the bodies of a different sex from who they actually are.

1950s Ozzie and Harriet standards—CAH girls were pathologically interested in having careers and disinterested in having babies.

Oops, back to the drawing board. Careful contemporary CAH research has been conducted by Melissa Hines of the University of Cambridge.[68] When compared with non-CAH girls, CAH girls do more rough-and-tumble play, fighting, and physical aggression. Moreover, they prefer "masculine" toys over dolls. As adults they score lower on measures of tenderness and higher in aggressiveness and self-report more aggression and less interest in infants. In addition, CAH women are more likely to be gay or bisexual or have a transgender sexual identity.*

Importantly, drug treatments begun soon after birth normalize androgen levels in these girls, so that the excessive androgen exposure is solely prenatal. Thus prenatal testosterone exposure appears to cause organizational changes that increase the incidence of male-typical behaviors.

A similar conclusion is reached by an inverse of CAH, namely androgen insensitivity syndrome (AIS, historically called "testicular feminization syndrome").[69] A fetus is male—XY chromosomes, testes that secrete testosterone. But a mutation in the androgen receptor makes it insensitive to testosterone. Thus the testes can secrete testosterone till the cows come home but there won't be any masculinization. And often the individual is born with a female external phenotype and is raised as a girl. Along comes puberty, she's not getting periods, and a trip to the doctor reveals that the "girl" is actually a "boy" (with testes typically near the stomach, plus a shortened vagina that dead-ends). The individual usually continues with a female identity but is infertile as an adult. In other words, when human males don't experience the organizational prenatal effects of testosterone, you get female-typical behaviors and identification.

Between CAH and AIS, the issue seems settled—prenatal testosterone plays a major role in explaining sex differences in aggression and various affiliative prosocial behaviors in humans.

---

* Prenatal screening for CAH is now possible, and the fetal masculinization can be prevented to some extent with fetal hormone treatments. This has been framed by some clinicians as a means to increase the odds of a CAH female having a heterosexual orientation, something that has drawn the ire of bioethicists and the LGBTQ community.

Careful readers may have spotted two whopping big problems with this conclusion:[70]

- Remember that CAH girls are born with a "something's *very* different" Post-it—the ambiguous genitalia, typically requiring multiple reconstructive surgeries. CAH females are not merely prenatally androgenized. They're also raised by parents who know something is different, have slews of doctors mighty interested in their privates, and are treated with all sorts of hormones. It's impossible to attribute the behavioral profile solely to the prenatal androgens.
- Testosterone has no effects in AIS individuals because of the androgen receptor mutation. But doesn't testosterone have most of its fetal brain effects as estrogen, interacting with the estrogen receptor? That aspect of brain masculinization should have occurred despite the mutation. Complicating things, some of the masculinizing effects of prenatal testosterone in monkeys don't require conversion to estrogen. So we have genetically and gonadally male individuals with at least some brain masculinization raised successfully as females.

The picture is complicated further—AIS individuals raised female have higher-than-expected rates of being gay, and of having an other-than-female or neither-female-nor-male-sex/gender self-identification.

Argh. All we can say is that there is (imperfect) evidence that testosterone has masculinizing prenatal effects in humans, as in other primates. The question becomes how *big* these effects are.

Answering that question would be easy if you knew how much testosterone people were exposed to as fetuses. Which brings up a truly quirky finding, one likely to cause readers to start futzing awkwardly with a ruler.

Weirdly, prenatal testosterone exposure influences digit length.[71] Specifically, while the second finger is usually shorter than the fourth finger, the difference (the "2D:4D ratio") is greater in men than in women, something first noted in the 1880s. The difference is demonstrable in third-trimester fetuses, and the more fetal testosterone exposure (as assessed by amniocentesis), the more pronounced the ratio. Moreover, CAH females have a more masculine ratio, as do females who shared their fetal environment (and thus some testosterone) with a male twin, while AIS males have a more feminine ratio. The sex difference in the ratio occurs in other primates and rodents. And no one knows why this difference exists. Moreover, this oddity is not alone. A barely discernible background noise generated by the inner ear ("otoacoustic emissions") shows a sex difference that reflects prenatal testosterone exposure. Go explain that.

The 2D:4D ratio is so variable, and the sex difference so small, that you can't determine someone's sex by knowing it. But it does tell you something about the extent of fetal testosterone exposure.

So what does the extent of exposure (as assessed by the ratio) predict about adult behavior? Men with more "masculine" 2D:4D ratios tend toward higher levels of aggression and math scores; more assertive personalities; higher rates of ADHD and autism (diseases with strong male biases); and decreased risk of depression and anxiety (disorders with a female skew). The faces and handwriting of such men are judged to be more "masculine." Furthermore, some reports show a decreased likelihood of being gay.

Women having a more "feminine" ratio have less chance of autism and more of anorexia (a female-biased disease). They're less likely to be left-handed (a male-skewed trait). Moreover, they exhibit less athletic ability and more attraction to highly masculine faces. And they're more likely to be straight or, if lesbian, more likely to take stereotypical female sexual roles.[72]

This constitutes some of the strongest evidence that (a) fetal androgen exposure has organizational effects on adult behavior in humans as in other species, and (b) that *individual differences* in the extent of such exposure predict individual differences in adult behavior.*[73] Prenatal endocrine environment is destiny.

* Meanwhile, there is no consistent evidence that the extent of androgen exposure in the hours to weeks after birth predicts anything about subsequent behavior.

Well, not exactly. These effects are small and variable, producing a meaningful relationship only when considering large numbers of individuals. Do testosterone's organizational effects determine the quality and/or quantity of aggression? No. How about the organizational *plus* the activational effects? Not those either.

Expanding the Scope of "Environment"

Thus the fetal brain can be influenced by hormones secreted by the fetus. But in addition, the outside world alters a pregnant woman's physiology, which in turn affects the fetal brain.

The most obvious version of this is how food ingested by a pregnant female influences what nutrients are delivered to the fetal circulation.* At an extreme, maternal malnutrition broadly impairs fetal brain development.†[74] Moreover, pathogens acquired by the mother can be passed to the fetus—for example, the protozoan parasite *Toxoplasma gondii* can infect someone pregnant (typically after exposure to infected cat feces) and eventually reach the fetal nervous system, potentially wreaking serious havoc. And this is also the world of maternal substance abuse producing heroin and crack babies or fetal alcohol syndrome.

Importantly, maternal stress impacts fetal development. There are indirect routes—for example, stressed people consume less healthy diets and consume more substances of abuse. More directly, stress alters maternal blood pressure and immune defenses, which impact a fetus. Most important, stressed mothers secrete glucocorticoids, which enter fetal circulation and basically have the same bad consequences as in stressed infants and children.

Glucocorticoids accomplish this through organizational effects on fetal brain construction and decreasing levels of growth factors, numbers of neurons and synapses, and so on. Just as prenatal testosterone exposure generates an adult brain that is more sensitive to environmental triggers

* Why not "determines" instead of "influences"? Because the female's body can convert one nutrient into another before passing it to the fetus.

† Third-trimester malnutrition also alters aspects of physiology, so that the fetus has a lifelong increased risk of diabetes, obesity, and metabolic syndrome, something known as the "Dutch Hunger Winter effect."

of aggression, excessive prenatal glucocorticoid exposure produces an adult brain more sensitive to environmental triggers of depression and anxiety.

In addition, prenatal glucocorticoid exposure has effects that blend classical developmental biology with molecular biology. To appreciate this, here's a highly simplified version of the next chapter's focus on genes: (a) each gene specifies the production of a specific type of protein; (b) a gene has to be "activated" for the protein to be produced and "deactivated" to stop producing it—thus genes come with on/off switches; (c) every cell in our bodies contains the same library of genes; (d) during development, the pattern of which genes are activated determines which cells turn into nose, which into toes, and so on; (e) forever after, nose, toes, and other cells retain distinctive patterns of gene activation.

Chapter 4 discussed how some hormones have activational effects by altering on/off switches on particular genes (e.g., testosterone-activating genes related to increased growth in muscle cells). The field of "epigenetics" concerns how some hormonal organizational effects arise from *permanently* turning particular genes on or off in particular cells.[75] Plenty more on this in the next chapter.

This helps explain why your toes and nose work differently. More important, epigenetic changes also occur in the brain.

This domain of epigenetics was uncovered in a landmark 2004 study by Meaney and colleagues, one of the most cited papers published in the prestigious journal *Nature Neuroscience.* They had shown previously that offspring of more "attentive" rat mothers (those that frequently nurse, groom, and lick their pups) become adults with lower glucocorticoid levels, less anxiety, better learning, and delayed brain aging. The paper showed that these changes were epigenetic—that mothering style altered the on/off switch in a gene relevant to the brain's stress response.* Whoa—mothering *style* alters gene regulation in pups' brains. Remarkably, Meaney, along with Darlene Francis of the University of California, Berkeley, then showed that such rat pups, as adults, are more attentive mothers—passing

* FYI, the gene that codes for the glucocorticoid receptor.

this trait epigenetically to the next generation.* Thus, adult behavior produces persistent molecular brain changes in offspring, "programming" them to be likely to replicate that distinctive behavior in adulthood.[76]

More findings flooded in, many provided by Meaney, his collaborator Moshe Szyf, also of McGill, and Frances Champagne of Columbia University.[77] Hormonal responses to various fetal and childhood experiences have epigenetic effects on genes related to the growth factor BDNF, to the vasopressin and oxytocin system, and to estrogen sensitivity. These effects are pertinent to adult cognition, personality, emotionality, and psychiatric health. Childhood abuse, for example, causes epigenetic changes in hundreds of genes in the human hippocampus. Moreover, Stephen Suomi of the National Institutes of Health and Szyf found that mothering style in monkeys has epigenetic effects on more than a *thousand* frontocortical genes.†

This is totally revolutionary. Sort of. Which segues to a chapter summary.

CONCLUSIONS

Epigenetic environmental effects on the developing brain are hugely exciting. Nonetheless, curbing of enthusiasm is needed. Findings have been overinterpreted, and as more researchers flock to the subject, the quality of studies has declined. Moreover, there is the temptation to conclude that epigenetics explains "everything," whatever that might be; most effects of childhood experience on adult outcomes probably don't involve epigenetics and (stay tuned) most epigenetic changes are transient. Particularly strong criticisms come from molecular geneticists rather than behavioral scientists (who generally embrace the topic); some of the

* The next chapter discusses how this nongenetic but rather epigenetic transmission of traits multigenerationally resembles a long-discredited idea about acquired inheritance proposed by the eighteenth-century scientist Jean-Baptiste Lamarck.

† Note: This does not mean that every neuron in the frontal cortex had the regulation of a thousand-odd genes changed. Instead, there are glial cells in addition to neurons, and the neurons are of a variety of types. So in reality the average number of changes within any given cell was probably far less than a thousand. Note about note: which doesn't make any of this less interesting, just harder to study.

negativity from the former, I suspect, is fueled by the indignity of having to incorporate the likes of rat mothers licking their pups into their beautiful world of gene regulation.

But the excitement should be restrained on a deeper level, one relevant to the entire chapter. Stimulating environments, harsh parents, good neighborhoods, uninspiring teachers, optimal diets—all alter genes in the brain. Wow. And not that long ago the revolution was about how environment and experience change the excitability of synapses, their number, neuronal circuits, even the number of neurons. Whoa. And earlier the revolution was about how environment and experience can change the sizes of different parts of the brain. Amazing.

But none of this is truly amazing. Because things *must* work these ways. While little in childhood determines an adult behavior, virtually everything in childhood changes propensities toward some adult behavior. Freud, Bowlby, Harlow, Meaney, from their differing perspectives, all make the same fundamental and once-revolutionary point: childhood *matters.* All that the likes of growth factors, on/off switches, and rates of myelination do is provide insights into the innards of that fact.

Such insight is plenty useful. It shows the steps linking childhood point A to adult point Z. It shows how parents can produce offspring whose behaviors resemble their own. It identifies Achilles' heels that explain how childhood adversity can make for damaged and damaging adults. And it hints at how bad outcomes might be reversed and good outcomes reinforced.

There is another use. In chapter 2 I recounted how it required the demonstration of hippocampal volume loss in combat vets with PTSD to finally convince many in power that the disorder is "real." Similarly, it shouldn't require molecular genetics or neuroendocrinology factoids to prove that childhood matters and thus that it profoundly matters to provide childhoods filled with good health and safety, love and nurturance and opportunity. But insofar as it seems to require precisely that sort of scientific validation at times, more power to those factoids.

Eight

Back to When You Were Just a Fertilized Egg

I'm reminded of a cartoon where one lab-coated scientist is telling the other, "You know how you're on the phone, and the other person wants to get off but won't say it, so they say, 'Well, you probably need to get going,' like *you're* the one who wants to get off, when it's really *them*? I think I found the gene for that."

This chapter is about progress in finding "the gene for that."

Our prototypical behavior has occurred. How was it influenced by events when the egg and sperm that formed that person joined, creating their genome—the chromosomes, the sequences of DNA—destined to be duplicated in every cell in that future person's body? What role did those genes play in causing that behavior?

Genes are relevant to, say, aggression, which is why we're less alarmed if a toddler pulls at the ears of a basset hound rather than a pit bull. Genes are relevant to everything in this book. Many neurotransmitters and hormones are coded for by genes. As are molecules that construct or degrade those messengers, as are their receptors. Ditto for growth factors guiding brain plasticity. Genes typically come in different versions; we each consist of an individuated orchestration of the different versions of our approximately twenty thousand genes.

This topic carries two burdens. The first reflects many people being troubled by linking genes with behavior—in one incident from my academic youth, a federally funded conference was canceled for suggesting that genes were pertinent to violence. This suspicion of gene/behavior links exists because of the pseudoscientific genetics used to justify various "isms," prejudice, and discrimination. Such pseudoscience has fostered racism and sexism, birthed eugenics and forced sterilizations, allowed scientifically meaningless versions of words like "innate" to justify the neglect of have-nots. And monstrous distortions of genetics have fueled those who lynch, ethnically cleanse, or march children into gas chambers.*[1]

But studying the genetics of behavior also carries the opposite burden of people who are overly enthusiastic about the subject. After all, this is the genomics era, with personalized genomic medicine, people getting their genomes sequenced, and popular writing about genomics giddy with terms like "the holy grail" and "the code of codes." In a reductionist view, understanding something complex requires breaking it down into its components; understand those parts, add them together, and you'll understand the big picture. And in this reductionist world, to understand cells, organs, bodies, and behavior, the best constituent part to study is genes.

Overenthusiasm for genes can reflect a sense that people possess an immutable, distinctive essence (although essentialism predates genomics). Consider a study concerning "moral spillover" based on kinship.[2] Suppose a person harmed people two generations ago; are this person's grandchildren obliged to help his victims' grandchildren? Subjects viewed a biological grandchild as more obligated than one adopted into the family at birth; the biological relationship carried a taint. Moreover, subjects

* The strongest ideological criticisms of genetics have typically been leftist in flavor. Despite that, and to my surprise, the one study I know of that has examined the subject showed no Left/Right ideological differences in the tendency to attribute individual differences to genetics. Where they differ concerns what sorts of differences are attributed. Thus right-wing ideology is more associated with genetic interpretations for race or class differences, while left-wing ideology is more associative when it comes to sexual orientation.

were more willing to jail two long-lost identical twins for a crime committed by one of them than to jail two unrelated but perfect look-alikes—the former, raised in different environments, share a moral taint because of their identical genes. People see essentialism embedded in bloodlines—i.e., genes.*

This chapter threads between these two extremes, concluding that while genes are important to this book's concerns, they're far less so than often thought. The chapter first introduces gene function and regulation, showing the limits of genes' power. Next it examines genetic influences on behavior in general. Finally we'll examine genetic influences on our best and worst behaviors.

PART I: GENES FROM THE BOTTOM UP

We start by considering the limited power of genes. If you are shaky about topics such as the central dogma (DNA codes for RNA, which codes for protein sequence), protein structure determining function, the three-nucleotide codon code, or the basics of point, insertion, and deletion mutations, first read the primer in appendix 3.

Do Genes Know What They Are Doing? The Triumph of the Environment

So genes specify protein structure, shape, and function. And since proteins do virtually everything, this makes DNA the holy grail of life. But no—genes don't "decide" when a new protein is made.

* My own personal experience with extreme essentialism: During 1976 and 1977 the New York City area was terrorized by the string of "Son of Sam" murders (I was home from college in Brooklyn during the summer of 1977 and can attest that the psychological impact of the murder spree was enormous). In August 1977 it ended with the arrest of David Berkowitz, a twenty-three-year-old petty criminal and arsonist who claimed that he killed under the command of a neighbor's dog, said dog ostensibly being demonically possessed. A month later, back at college, the phone rang. My roommate answered and handed me the phone, looking a bit puzzled. "It's your mother; she seems kind of excited." "Hi, Mom, what's new?" And in a euphoric, relieved, triumphant tone, she shouted: "David Berkowitz! He's adopted. Adopted! HE'S NOT REALLY JEWISH!" Ironic ending department for my mother: The biological mother of Berkowitz, born Richard David Falco, was Jewish. As was his biological father, who was not Falco.

Dogma was that there'd be a stretch of DNA in a chromosome, constituting a single gene, followed by a stop codon, followed immediately by the next gene, and then the next. . . . But genes don't actually come one after another—not all DNA constitutes genes. Instead there are stretches of DNA between genes that are noncoding, that are not "transcribed."* And now a flabbergasting number—95 percent of DNA is noncoding. *Ninety-five percent.*

What is that 95 percent? Some is junk—remnants of pseudogenes inactivated by evolution.†[3] But buried in that are the keys to the kingdom, the instruction manual for *when* to transcribe particular genes, the on/off switches for gene transcription. A gene doesn't "decide" when to be photocopied into RNA, to generate its protein. Instead, before the start of the stretch of DNA coding for that gene is a short stretch called a promoter‡—the "on" switch. What turns the promoter switch on? Something called a transcription factor (TF) binds to the promoter. This causes the recruitment of enzymes that transcribe the gene into RNA. Meanwhile, other transcription factors deactivate genes.

This is huge. Saying that a gene "decides" when it is transcribed§ is like saying that a recipe decides when a cake is baked.

Thus transcription factors regulate genes. What regulates transcription factors? The answer devastates the concept of genetic determinism: the environment.

To start unexcitingly, "environment" can mean intracellular environment. Suppose a hardworking neuron is low on energy. This state activates a particular transcription factor, which binds to a specific promoter, which activates the next gene in line (the "downstream" gene). This gene codes for a glucose transporter; more glucose transporter proteins are made and inserted into the cell membrane, improving the neuron's ability to access circulating glucose.

* Terminology: A gene is "transcribed" when the RNA template of its DNA sequence is made, which is then used to generate the protein that it codes for.
† Note that "junk" DNA may be junk or, more likely, DNA whose function hasn't been discovered yet. There are reasons to go with the latter interpretation.
‡ There are related stretches of noncoding DNA that are part of on/off switches called enhancers and operators. For our purposes, we'll just use the term promoters.
§ Or to use other jargon in the field, when it is "activated" or "expressed"—I'll use these terms interchangeably.

Next consider "environment," including the neuron next door, which releases serotonin onto the neuron in question. Suppose less serotonin has been released lately. Sentinel transcription factors in dendritic spines sense this, travel to the DNA, and bind to the promoter upstream of the serotonin receptor gene. More receptor is made and placed in the dendritic spines, and they become more sensitive to the faint serotonin signal.

Sometimes "environment" can be far-flung within an organism. A male secretes testosterone, which travels through the bloodstream and binds to androgen receptors in muscle cells. This activates a transcription-factor cascade that results in more intracellular scaffolding proteins, enlarging the cell (i.e., muscle mass increases).

Finally, and most important, there is "environment," meaning the outside world. A female smells her newborn, meaning that odorant molecules that floated off the baby bind to receptors in her nose. The receptors activate and (many steps later in the hypothalamus) a transcription factor activates, leading to the production of more oxytocin. Once secreted, the oxytocin causes milk letdown. Genes are not the deterministic holy grail if they can be regulated by the smell of a baby's tushy. Genes are regulated by all the incarnations of environment.

In other words, *genes don't make sense outside the context of environment.* Promoters and transcription factor introduce if/then clauses: "If you smell your baby, then activate the oxytocin gene."

Now the plot thickens.

There are multiple types of transcription factors in a cell, each binding to a particular DNA sequence constituting a particular promoter.

Consider a genome containing one gene. In that imaginary organism there is only a single profile of transcription (i.e., the gene is transcribed), requiring only one transcription factor.

Now consider a genome consisting of genes A and B, meaning three different transcription profiles—A is transcribed, B is transcribed, A and B are transcribed—requiring three different TFs (assuming you activate only one at a time).

Three genes, seven transcription profiles: A, B, C, A + B, A + C, B + C, A + B + C. Seven different TFs.

Four genes, fifteen profiles. Five genes, thirty-one profiles.*

As the number of genes in a genome increases, the number of possible expression profiles increases exponentially. As does the number of TFs needed to produce those profiles.

Now another wrinkle that, in the lingo of an ancient generation, will blow your mind.

TFs are usually proteins, coded for by genes. Back to genes A and B. To fully exploit them, you need the TF that activates gene A, and the TF that activates gene B, and the TF that activates genes A and B. Thus there must exist three more genes, each coding for one of those TFs. Requiring TFs that activate *those* genes. And TFs for the genes coding for those TFs . . .

Whoa. Genomes aren't infinite; instead TFs regulate one another's transcription, solving that pesky infinity problem. Importantly, across the species whose genomes have been sequenced, the longer the genome (i.e., roughly the more genes there are), the greater the percentage of genes coding for TFs. In other words, *the more genomically complex the organism, the larger the percentage of the genome devoted to gene regulation by the environment.*

Back to mutations. Can there be mutations in DNA stretches constituting promoters? Yes, and more often than in genes themselves. In the 1970s Allan Wilson and Mary-Claire King at Berkeley correctly theorized that the evolution of genes is less important than the evolution of regulatory sequences upstream of genes (and thus how the environment regulates genes). Reflecting that, a disproportionate share of genetic differences between chimps and humans are in genes for TFs.

Time for more complexity. Suppose you have genes 1–10, and transcription factors A, B, and C. TF-A induces the transcription of genes 1, 3, 5, 7, and 9. TF-B induces genes 1, 2, 5, and 6. TF-C induces 1, 5, and 10. Thus, upstream of gene 1 are separate promoters responding to TFs A, B, and C—thus genes can be regulated by multiple TFs. Conversely, each

* For those who care, the number of unique transcriptional profiles for *n* number of genes is $(2^n) - 1$, not counting the state where no genes are being transcribed. So plug the approximately 20,000 human genes into the equation, and you get a gargantuan number of possible transcriptional profiles.

TF usually activates more than one gene, meaning that multiple genes are typically activated in *networks* (for example, cell injury causes a TF called NF-κB to activate a network of inflammation genes). Suppose the promoter upstream of gene 3 that responds to promoter TF-A has a mutation making it responsive to TF-B. Result? Gene 3 is now activated as part of a different network. Same networkwide outcome if there is a mutation in a gene for a TF, producing a protein that binds to a different promoter type.[4]

Consider this: the human genome codes for about 1,500 different TFs, contains 4,000,000 TF-binding sites, and the average cell uses about 200,000 such sites to generate its distinctive gene-expression profile.[5] This is boggling.

Epigenetics

The last chapter introduced the phenomenon of environmental influences freezing genetic on/off in one position. Such "epigenetic" changes* were relevant to events, particularly in childhood, causing persistent effects on the brain and behavior. For example, recall those pair-bonding prairie voles; when females and males first mate, there are epigenetic changes in regulation of oxytocin and vasopressin receptor genes in the nucleus accumbens, that target of mesolimbic dopamine projection.[6]

Let's translate the last chapter's imagery of "freezing on/off switches" into molecular biology.[7] What mechanisms underlie epigenetic changes in gene regulation? An environmental input results in a chemical being attached tightly to a promoter, or to some nearby structural proteins surrounding DNA. The result of either is that TFs can no longer access or properly bind to the promoter, thus silencing the gene.

As emphasized in the last chapter, epigenetic changes can be multigenerational.[8] Dogma was that all the epigenetic marks (i.e., changes in the DNA or surrounding proteins) were erased in eggs and sperm. But it turns out that epigenetic marks can be passed on by both (e.g., make male

* "Epigenetic" technically refers to altering the regulation of genes, rather than the sequence of genes. Therefore a transcription factor activating some gene for ten minutes counts as epigenetic as well. When neuroscientists talk of the "epigenetics revolution," however, they're almost always referring to the long-lasting mechanisms discussed here.

mice diabetic, and they pass the trait to their offspring via epigenetic changes in sperm).

Recall one of the great punching bags of science history, the eighteenth-century French biologist Jean-Baptiste Lamarck.[9] All anybody knows about the guy now is that he was wrong about heredity. Suppose a giraffe habitually stretches her neck to reach leaves high in a tree; this lengthens her neck. According to Lamarck, when she has babies, they will have longer necks because of "acquired inheritance."* Lunatic! Buffoon! Epigenetically mediated mechanisms of inheritance—now often called "neo-Lamarckian inheritance"—prove Lamarck right in this narrow domain. Centuries late, the guy's getting some acclaim.

Thus, not only does environment regulate genes, but it can do so with effects that last days to lifetimes.

The Modular Construction of Genes: Exons and Introns

Time to do in another dogma about DNA. It turns out that most genes are not coded for by a continuous stretch of DNA. Instead there might be a stretch of noncoding DNA in the middle. In that case, the two separate stretches of coding DNA are called "exons," separated by an "intron." Many genes are broken into numerous exons (with, logically, one less intron than the number of exons).

How do you produce a protein from an "exonic" gene? The RNA photocopy of the gene initially contains the exons and introns; an enzyme removes the intronic parts and splices together the exons. Clunky, but with big implications.

Back to each particular gene coding for a particular protein.[10] Introns and exons destroy this simplicity. Imagine a gene consisting of exons 1, 2, and 3, separated by introns A and B. In one part of the body a splicing enzyme exists that splices out the introns and also trashes exon 3, producing a protein coded for by exons 1 and 2. Meanwhile, elsewhere in the

* Note that Lamarck was talking about the concept of species evolution long before Darwin and Wallace. The latter two didn't invent the idea of evolution; rather, they figured out how evolution works, namely by natural selection.

body, a different splicing enzyme jettisons exon 2 along with the introns, producing a protein derived from exons 1 and 3. In another cell type a protein is made solely from exon 1. . . . Thus "alternative splicing" can generate multiple unique proteins from a single stretch of DNA; so much for "one gene specifies one protein"—this gene specifies seven (A, B, C, A-B, A-C, B-C, and A-B-C). Remarkably, 90 percent of human genes with exons are alternatively spliced. Moreover, when a gene is regulated by multiple TFs, each can direct the transcription of a different combination of exons. Oh, and splicing enzymes are proteins, meaning that each is coded for by a gene. Loops and loops.

Transposable Genetic Elements, the Stability of the Genome, and Neurogenesis

Time to unmoor another cherished idea, namely that genes inherited from your parents (i.e., what you started with as a fertilized egg) are immutable. This calls up a great chapter of science history. In the 1940s an accomplished plant geneticist named Barbara McClintock observed something impossible. She was studying the inheritance of kernel color in maize (a frequent tool of geneticists) and found patterns of mutations unexplained by any known mechanism. The only possibility, she concluded, was that stretches of DNA had been copied, with the copy then randomly inserted into another stretch of DNA.

Yeah, right.

Clearly McClintock, with her (derisively named) "jumping genes," had gone mad, and so she was ignored (not exactly true, but this detracts from the drama). She soldiered on in epic isolation. And finally, with the molecular revolution of the 1970s, she was vindicated about her (now termed) transposable genetic elements, or transposons. She was lionized, canonized, Nobel Prized (and was wonderfully inspirational, as disinterested in acclaim as in her ostracism, working until her nineties).

Transpositional events rarely produce great outcomes. Consider a hypothetical stretch of DNA coding for "The fertilized egg is implanted in the uterus."

There has been a transpositional event, where the underlined stretch of message was copied and randomly plunked down elsewhere: "The fertilized eggterus is implanted in the uterus."

Gibberish.

But sometimes "The fertilized egg is implanted in the uterus" becomes "The fertilized eggplant is implanted in the uterus."

Now, that's not an everyday occurrence.

Plants utilize transposons. Suppose there is a drought; plants can't move to wetter pastures like animals can. Plant "stress" such as drought induces transpositions in particular cells, where the plant metaphorically shuffles its DNA deck, hoping to generate some novel savior of a protein.

Mammals have fewer transposons than plants. The immune system is one transposon hot spot, in the enormous stretches of DNA coding for antibodies. A novel virus invades; shuffling the DNA increases the odds of coming up with an antibody that will target the invader.*

The main point here is that transposons occur in the brain.[11] In humans transpositional events occur in stem cells in the brain when they are becoming neurons, making the brain a mosaic of neurons with different DNA sequences. In other words, when you make neurons, that boring DNA sequence you inherited isn't good enough. Remarkably, transpositional events occur in neurons that form memories in *fruit flies.* Even flies evolved such that their neurons are freed from the strict genetic marching orders they inherit.

Chance

Finally, chance lessens genes as the Code of Codes. Chance, driven by Brownian motion—the random movement of particles in a fluid—has

* And as a brilliant counterstrategy, some parasites use transposons to shuffle the DNA coding for their surface proteins every few weeks. In other words, just as the infected host is building up stocks of antibody to recognize the surface protein, the parasite switches identities, making the host immune system start all over.

big effects on tiny things like molecules floating in cells, including molecules regulating gene transcription.[12] This influences how quickly an activated TF reaches the DNA, splicing enzymes bump into target stretches of RNA, and an enzyme synthesizing something grabs the two precursor molecules needed for the synthesis. I'll stop here; otherwise, I'll go on for hours.

Some Key Points, Completing This Part of the Chapter

a. Genes are not autonomous agents commanding biological events.

b. Instead, genes are regulated by the environment, with "environment" consisting of everything from events inside the cell to the universe.

c. Much of your DNA turns environmental influences into gene transcription, rather than coding for genes themselves; moreover, evolution is heavily about changing regulation of gene transcription, rather than genes themselves.

d. Epigenetics can allow environmental effects to be lifelong, or even multigenerational.

e. And thanks to transposons, neurons contain a mosaic of different genomes.

In other words, genes don't *determine* much. This theme continues as we focus on the effects of genes on behavior.

PART 2: GENES FROM THE TOP DOWN—BEHAVIOR GENETICS

Long before anything was known about promoters, exons, or transcription factors, it became clear that you study genetics top down, by observing traits shared by relatives. Early in the last century, this emerged as the science of "behavior genetics." As we'll see, the field has often been mired in controversy, typically because of disagreements over the magnitude of genetic effects on things like IQ or sexual orientation.

First Attempts

The field began with the primitive idea that, if everyone in a family does it, it must be genetic. This was confounded by environment running in families as well.

The next approach depended on closer relatives having more genes in common than distant ones. Thus, if a trait runs in a family and is more common among closer relatives, it's genetic. But obviously, closer relatives share more environment as well—think of a child and parent versus a child and grandparent.

Research grew subtler. Consider someone's biological aunt (i.e., the sister of a parent), and the uncle who married the aunt. The uncle shares some degree of environment with the individual, while the aunt shares the same, plus genes. Therefore, the extent to which the aunt is more similar to the individual than the uncle reflects the genetic influence. But as we'll see, this approach has problems.

More sophistication was needed.

Twins, Adoptees, and Adopted Twins

A major advance came with "twin studies." Initially, examining twins helped rule out the possibility of genetic determination of a behavior. Consider pairs of identical twins, sharing 100 percent of their genes. Suppose one of each pair has schizophrenia; does the twin as well? If there are any cases where the other twin doesn't (i.e., where the "concordance rate" is less than 100 percent), you've shown that the genome and epigenetic profile inherited at birth do not solely determine the incidence of schizophrenia (in fact the concordance rate is about 50 percent).

But then a more elegant twin approach emerged, involving the key distinction between identical (monozygotic, or MZ) twins, who share 100 percent of their genes, and fraternal, nonidentical (dizygotic, or DZ) twins, who, like all other sibling pairs, share 50 percent of their genes. Compare pairs of MZ twins with same-sex DZ twins. Each pair is the same age, was raised in the same environment, and shared a fetal environment; the only difference is the percentage of genes shared. Examine a trait occurring in one member of the twin pair; is it there in the other? The logic ran that, if a trait is shared more among MZ than among DZ twins, that increased degree of sharing reflects the genetic contribution to the trait.

Another major advance came in the 1960s. Identify individuals adopted soon after birth. All they share with their biological parents is genes; all they share with their adoptive parents is environment. Thus, if adopted individuals share a trait more with their biological than with their adoptive parents, you've uncovered a genetic influence. This replicates a classic tool in animal studies, namely "cross-fostering"—switching newborn rat pups between two mothers. The approach was pioneered in revealing a strong genetic component to schizophrenia.[13]

Then came the most wonderful, amazing, like, totally awesome thing ever in behavior genetics, started by Thomas Bouchard of the University of Minnesota. In 1979 Bouchard found a pair of identical twins who were—get this—separated at birth and adopted into different homes, with no knowledge of each other's existence until being reunited as adults.[14] Identical twins separated at birth are so spectacular and rare that behavior geneticists swoon over them, want to collect them all. Bouchard eventually studied more than a hundred such pairs.

The attraction was obvious—same genes, different environments (and the more different the better); thus, similarities in behavior probably reflect genetic influences. Here's an imaginary twin pair that would be God's gift to behavior geneticists—identical twin boys separated at birth. One, Shmuel, is raised as an Orthodox Jew in the Amazon; the other, Wolfie, is raised as a Nazi in the Sahara. Reunite them as adults and see if they do similar quirky things like, say, flushing the toilet before using it. Flabbergastingly, one twin pair came close to that. They were born in 1933 in Trinidad to a German Catholic mother and a Jewish father; when the boys were six months of age, the parents separated; the mother returned to Germany with one son, and the other remained in Trinidad with the father. The latter was raised there and in Israel as Jack Yufe, an observant Jew whose first language was Yiddish. The other, Oskar Stohr, was raised in Germany as a Hitler Youth zealot. Reunited and studied by Bouchard, they warily got to know each other, discovering numerous shared behavioral and personality traits including . . . flushing the toilet before use. (As we'll see, studies were more systematic than just documenting bathroom quirks. The flushing detail, however, always comes up in accounts of the pair.)

Behavior geneticists, wielding adoption and twin approaches, generated scads of studies, filling specialized journals like *Genes, Brain and Behavior* and *Twin Research and Human Genetics*. Collectively, the research consistently showed that genetics plays a major role in a gamut of domains of behavior, including IQ and its subcomponents (i.e., verbal ability, and spatial ability),*[15] schizophrenia, depression, bipolar disorder, autism, attention-deficit disorder, compulsive gambling, and alcoholism.

Nearly as strong genetic influences were shown for personality measures related to extroversion, agreeableness, conscientiousness, neuroticism, and openness to experience (known as the "Big Five" personality traits).[16] Likewise with genetic influences on degree of religiosity, attitude toward authority, attitude toward homosexuality,† and propensities toward cooperation and risk taking in games.

Other twin studies showed genetic influences on the likelihood of risky sexual behavior and on people's degree of attraction to secondary sexual characteristics (e.g., musculature in men, breast size in women).[17]

Meanwhile, some social scientists report genetic influences on the extent of political involvement and sophistication (independent of political orientation); there are behavior genetics papers in the *American Journal of Political Science*.[18]

* There have even been reports of heritability of intelligence in chimpanzees.

† I was pleased to see this study. Numerous studies, stretching back decades, have attempted to uncover the biological roots of sexual orientation; the earlier literature overwhelmingly came with the political agenda of trying to figure out what is biologically "wrong" with homosexuals. Thus, it was about time for people to study what's wrong with homophobes.

Genes, genes, everywhere. Large genetic contributions have even been uncovered for everything from the frequency with which teenagers text to the occurrence of dental phobias.[19]

So does this mean there is a gene "for" finding chest hair on guys to be hot, for likelihood of voting, for feelings about dentists? Vanishingly unlikely. Instead, gene and behavior are often connected by tortuous routes.[20] Consider the genetic influence on voter participation; the mediating factor between the two turns out to be sense of control and efficacy. People who vote regularly feel that their actions matter, and this central locus of control reflects some genetically influenced personality traits (e.g., high optimism, low neuroticism). Or how about the link between genes and self-confidence? Some studies show that the intervening variable is genetic effects on height; taller people are considered more attractive and treated better, boosting their self-confidence, dammit.*

In other words, genetic influences on behavior often work through very indirect routes, something rarely emphasized when news broadcasts toss out behavior genetics sound bites—"Scientists report genetic influence on strategy when playing Candyland."

The Debates About Twin and Adoption Studies

Many scientists have heavily criticized the assumptions in twin and adoption studies, showing that they generally lead to overestimates of the importance of genes.† Most behavior geneticists recognize these problems but argue that the overestimates are tiny.[21] A summary of this technical but important debate:

Criticism #1: Twin studies are premised on MZ and same-sex DZ twin pairs sharing environment equally (while sharing genes to very different extents). This "equal environment assumption" (EEA) is simply wrong; starting with parents, MZ twins are treated more similarly than

* Yeah, I'm shorter than average.

† Historically, the most hyperoxygenated criticisms of behavioral genetics as a discipline have come from nongeneticists questioning the motives and hidden sociopolitical agendas behind behavior genetics findings. It is historically justified to conclude this at many junctures; however, it's utterly inapplicable to the behavior geneticists I know. The next chapter will look at a related version of a "there's a hidden agenda" controversy.

DZ twins, creating more similar environments for them. If this isn't recognized, greater similarity between MZs will be misattributed to genes.[22]

Scientists such as Kenneth Kendler of Virginia Commonwealth University, a dean of the field, have tried to control for this by (a) quantifying just how similar childhoods were for twins (with respect to variables like sharing rooms, clothing, friends, teachers, and adversity); (b) examining cases of "mistaken zygosity," where parents were wrong about their twins' MZ/DZ status (thus, for example, raising their DZ twins as if they were MZ); and (c) comparing full-, half-, and step-siblings who were reared together for differing lengths of time. Most of these studies show that controlling for the assumption of MZs sharing more environment than do DZs doesn't significantly reduce the size of genetic influences.*[23] Hold that thought.

Criticism #2: MZ twins experience life more similarly starting as fetuses. DZ twins are "dichorionic," meaning that they have separate placentas. In contrast, 75 percent of MZ twins share one placenta (i.e., are "monochorionic").† Thus most MZ twin fetuses share maternal blood flow more than do DZ twins, and thus are exposed to more similar levels of maternal hormones and nutrients. If that isn't recognized, greater similarity in MZs will be misattributed to genes.

Various studies have determined what the chorionic status was in different MZ pairs and then examined end points related to cognition, personality, and psychiatric disease. By a small margin, most studies show that chorionic status does make a difference, leading to overestimates of genetic influence. How big of an overestimation? As stated in one review, "small but not negligible."[24]

Criticism #3: Recall that adoption studies assume that if a child is adopted soon after birth, she shares genes but no environment with her biological parents. But what about prenatal environmental effects? A newborn just spent nine months sharing the circulatory environment with Mom. Moreover, eggs and sperm can carry epigenetic changes into the

* And roughly similar conclusions can be reached concerning end points like weight, height, BMI, and various metabolic measures.

† Whether MZ twins wind up as mono- or dichorionic depends on when the new embryo divides.

next generation. If these various effects are ignored, an environmentally based similarity between mother and child would be misattributed to genes.

Epigenetic transmission via sperm seems of small significance. But prenatal and epigenetic effects from the mother can be huge—for example, the Dutch Hunger Winter phenomenon showed that third-trimester malnutrition increased the risk of some adult diseases more than *tenfold.*

This confound can be controlled for. Roughly half your genes come from each parent, but prenatal environment comes from Mom. Thus, traits shared more with biological mothers than with fathers argue against a genetic influence.* The few tests of this, concerning the genetic influence on schizophrenia demonstrated in twin studies, suggest that prenatal effects aren't big.

Criticism #4: Adoption studies assume that a child and adoptive parents share environment but not genes.[25] That might approach being true if adoption involved choosing adoptive parents randomly among everyone on earth. Instead, adoption agencies prefer to place children with families of similar racial or ethnic background as the biological parents (a policy advocated by the National Association of Black Social Workers and the Child Welfare League).† Thus, kids and adoptive parents typically share genes at a higher-than-chance level; if this isn't recognized, a similarity between them will be misattributed to environment.

Researchers admit there is selective placement but argue over whether it's consequential. This remains unsettled. Bouchard, with his twins separated at birth, controlled for cultural, material, and technological similarities between the separate homes of twin pairs, concluding that shared similarity of home environments due to selective placement was a negligible factor. A similar conclusion was reached in a larger study carried out by both Kendler and another dean of the field, Robert Plomin of King's College London.

These conclusions have been challenged. The most fire-breathing

* Not always—there are some truly weird mechanisms of gene transmission involving "imprinted genes" that violate this, but we're ignoring that.

† I thank an excellent student assistant, Katrina Hui, for help in this area.

critic has been Princeton psychologist Leon Kamin, who argues that concluding that selective placement isn't important is wrong because of misinterpretation of results, use of wimpy analytical tests, and overreliance on questionable retrospective data. He wrote: "We suggest that no scientific purpose is served by the flood of heritability estimates generated by these studies."[26]

Here's where I give up—if super smart people who think about this issue all the time can't agree, I sure don't know how seriously selective placement distorts the literature.

Criticism #5: Adoptive parents tend to be more educated, wealthier, and more psychiatrically healthy than biological parents.[27] Thus, adoptive households show "range restriction," being more homogeneous than biological ones, which decreases the ability to detect environmental effects on behavior. Predictably, attempts to control for this satisfy only some critics.

So what do we know after this slog through the criticisms and countercriticisms about adoption and twin studies?

- Everyone agrees that confounds from prenatal environment, epigenetics, selective placement, range restriction, and assumptions about equal environment are unavoidable.
- Most of these confounds inflate the perceived importance of genes.
- Efforts have been made to control for these confounds and generally have shown that they are of less magnitude than charged by many critics.
- Crucially, these studies have mostly been about psychiatric disorders, which, while plenty interesting, aren't terribly relevant to the concerns of this book. In other words, no one has studied whether these confounds matter when considering genetic influences on, say, people's tendency to endorse their culture's moral rules yet

rationalize why those rules don't apply to them today, because they're stressed and it's their birthday. Lots more work to be done.

The Fragile Nature of Heritability Estimates

Now starts a bruising, difficult, immensely important subject. I review its logic every time I teach it, because it's so unintuitive, and I'm still always just words away from getting it wrong when I open my mouth in class.

Behavior genetics studies usually produce a number called a heritability score.[28] For example, studies have reported heritability scores in the 40 to 60 percent range for traits related to prosocial behavior, resilience after psychosocial stress, social responsiveness, political attitudes, aggression, and leadership potential.

What's a heritability score? "What does a gene do?" is at least two questions. How does a gene influence average levels of a trait? How does a gene influence *variation* among people in levels of that trait?

These are crucially different. For example, how much do genes have to do with people's scores averaging 100 on this thing called an IQ test? Then how much do genes have to do with one person scoring higher than another?

Or how much do genes help in explaining why humans usually enjoy ice cream? How much in explaining why people like different flavors?

These issues utilize two terms with similar sounds but different meanings. If genes strongly influence average levels of a trait, that trait is strongly inherited. If genes strongly influence the extent of variability around that average level, that trait has high heritability.* It is a population measure, where a heritability score indicates the percentage of total variation attributable to genetics.

The difference between an inherited trait and heritability generates at least two problems that inflate the putative influence of genes. First,

* Although many purists in the field would say that we don't actually inherit a trait; we inherit the material needed to construct a trait.

people confuse the two terms (things would be easier if heritability were called something like "gene tendency"), and in a consistent direction. People often mistakenly believe that if a trait is strongly inherited, it's thus highly heritable. And it's particularly bad that confusion is typically in that direction, because people are usually more interested in variability of traits among humans than in average levels of traits. For example, it's more interesting to consider why some people are smarter than others than why humans are smarter than turnips.

The second problem is that research consistently inflates heritability measures, leading people to conclude that genes influence individual differences more than they do.

Let's slowly work through this, because it's really important.

The Difference Between a Trait Being Inherited and Having a High Degree of Heritability

You can appreciate the difference by considering cases where they dissociate.

First, an example of a trait that is highly inherited but has low heritability, offered by the philosopher Ned Block:[29] What do genes have to do with humans averaging five fingers per hand? Tons; it's an inherited trait. What do genes have to do with variation around that average? Not much—cases of other than five fingers on a hand are mostly due to accidents. While average finger number is an inherited trait, the heritability of finger number is low—genes don't explain individual differences much. Or stated differently: Say you want to guess whether some organism's limb has five fingers or a hoof. Knowing their genetic makeup will help by identifying their species. Alternatively, you're trying to guess whether a particular person is likely to have five or four fingers on his hand. Knowing whether he uses buzz saws while blindfolded is more useful than knowing the sequence of his genome.

Next consider the opposite—a trait that is not highly inherited but which has high heritability. What do genes directly have to do with humans being more likely than chimps to wear earrings? Not much. Now consider individual differences among humans—how much do genes help predict which individuals are wearing earrings at a high school dance in 1958? Tons. Basically, if you had two X chromosomes, you probably wore earrings, but if you had a Y chromosome, you wouldn't have been caught dead doing so. Thus, while genes had little to do with the prevalence of earrings among Americans in 1958 being around 50 percent, they had lots to do with determining *which* Americans wore them. Thus, in that time and place, wearing earrings, while not a strongly inherited trait, had high heritability.

The Reliability of Heritability Measures

We're now clear on the difference between inherited traits and their degree of heritability and recognize that people are usually more interested in the latter—you versus your neighbor—than the former—you versus a wildebeest. As we saw, scads of behavioral and personality traits have heritability scores of 40 to 60 percent, meaning that genetics explains about half the variability in the trait. The point of this section is that the nature of research typically inflates such scores.*[30]

Say a plant geneticist sits in the desert, studying a particular species of plant. In this imaginary scenario a single gene, gene 3127, regulates the plant's growth. Gene 3127 comes in versions, A, B, and C. Plants with version A always grow to be one inch tall; version B, two inches; C, three inches.† What single fact gives you the most power in predicting a plant's height? Obviously, whether it has version A, B, or C—that explains all the variation in height between plants, meaning 100 percent heritability.

Meanwhile, twelve thousand miles away in a rain forest, a second plant geneticist is studying a clone of that same plant. And in that environment plants with version A, B, or C are 101, 102, or 103 inches tall, respectively. This geneticist also concludes that plant height in this case shows 100 percent heritability.

---

* This next section has been heavily influenced by the writings of the geneticists Richard Lewontin of Harvard and David Moore of Pitzer College and the science writer Matt Ridley.

† Genetics savants will note that I've simplified things here by ignoring heterozygosity; it doesn't matter.

Then, as required by the plot line, the two stand side by side at a conference, one brandishing 1/2/3 inch data, the other 101/102/103. They combine data sets. Now you want to predict the height of one example of that plant, taken from anywhere on the planet. You can either know which version of gene 3127 it possesses or what environment it is growing in. Which is more useful? Knowing which environment. When you study this plant species in two environments, you discover that heritability of height is miniscule.

Neon lights! This is crucial: Study a gene in only one environment and, by definition, you've eliminated the ability to see if it works differently in other environments (in other words, if other environments regulate the gene differently). And thus you've artificially inflated the importance of the genetic contribution. The more environments in which you study a genetic trait, the more novel environmental effects will be revealed, decreasing the heritability score.

Scientists study things in controlled settings to minimize variation in extraneous factors and thus get cleaner, more interpretable results—for example, making sure that the plants all have their height measured around the same time of year. This inflates heritability scores, because you've prevented yourself from ever discovering that some extraneous environmental factor isn't actually extraneous.* Thus a heritability score tells how much variation in a trait is explained by genes *in the environment(s) in which it's been studied*. As you study the trait in more environments, the heritability score will decrease. This is recognized by Bouchard: "These conclusions [derived from a behavior genetics study] can be generalized, of course only to new populations exposed to a range of environments similar to those studied."[31]

Okay, that was slick on my part, inventing a plant that grows in both desert and rain forest, just to trash heritability scores. Real plants rarely occur in both of those environments. Instead, in one rain forest the three gene versions might produce plants of heights 1, 2, and 3 inches, while in

* Here's a cool example pointed out to me by a colleague, Bud Ruby. All those twin studies generate heritability scores, indicating the strength of genes in explaining individual variation. But those studies, by definition, have eliminated an important nongenetic source of variation—birth order.

another they are 1.1, 2.1, and 3.1, producing a heritability score that, while less than 100 percent, is still extremely high.

Genes typically still play hefty roles in explaining individual variability, given that any given species lives in a limited range of environments—capybaras stick to the tropics, polar bears to the Arctic. This business about heterogeneous environments driving down heritability scores is important only in considering some hypothetical species that, say, lives in both tundra and desert, in various population densities, in nomadic bands, sedentary farming communities, and urban apartment buildings.

Oh, that's right, humans. Of all species, heritability scores in humans plummet the most when shifting from a controlled experimental setting to considering the species' full range of habitats. Just consider how much the heritability score for wearing earrings, with its gender split, has declined since 1958.

Now to consider an extremely important complication.

Gene/Environment Interactions

Back to our plant. Imagine a growth pattern in environment A of 1, 1, and 1 for the three gene variants, while in environment B it's 10, 10, and 10. When considering the combined data from both environments, heritability is zero—variation is entirely explained by which environment the plant grew in.

Now, instead, in environment A it's 1, 2, and 3, while in environment B it's also 1, 2, and 3. Heritability is 100 percent, with all variability in height explained by genetic variation.

Now say environment A is 1, 2, and 3, and environment B is 1.5, 2.5, 3.5. Heritability is somewhere between 0 percent and 100 percent.

Now for something different: Environment A: 1, 2, 3. Environment B: *3, 2, 1.* In this case even talking about a heritability score is problematic, because different gene variants have diametrically opposite effects in different environments. We have an example of a central concept in genetics, a *gene/environment interaction*, where qualitative, rather than just quantitative, effects of a gene differ by environment. Here's a rule of thumb for recognizing gene/environment interactions, translated into English: You are studying the behavioral effects of a gene in two environments. Someone asks, "What are the effects of the gene on some behavior?" You answer, "It depends on the environment." Then they ask, "What are the effects of environment on this behavior?" And you answer, "It depends on the version of the gene." "It depends" = a gene/environment interaction.

Here are some classic examples concerning behavior:[32]

The disease phenylketonuria arises from a single gene mutation; skipping over details, the mutation disables an enzyme that converts a potentially neurotoxic dietary constituent, phenylalanine, into something safe. Thus, if you eat a normal diet, phenylalanine accumulates, damaging the brain. But eat a phenylalanine-free diet from birth, and there is no damage. What are the effects of this mutation on brain development? *It depends* on your diet. What's the effect of diet on brain development? *It depends* on whether you have this (rare) mutation.

Another gene/environment interaction pertains to depression, a disease involving serotonin abnormalities.[33] A gene called 5HTT codes for a transporter that removes serotonin from the synapse; having a particular 5HTT variant increases the risk of depression . . . but only when coupled with childhood trauma.* What's the effect of 5HTT variant on depression risk? It depends on childhood trauma exposure. What's the effect of childhood trauma exposure on depression risk? It depends on 5HTT variant (plus loads of other genes, but you get the point).

Another example concerns FADS2, a gene involved in fat metabolism.[34] One variant is associated with higher IQ, but only in breast-fed children. Same pair of "what's the effect" questions, same "it depends" answers.

One final gene/environment interaction was revealed in an important

* There has been some controversy about the replicability of this immensely important observation, and I've followed it closely. When considering only the carefully done studies with adequate sample sizes and clearly and narrowly defined end points, I believe that it's been amply replicated.

1999 *Science* paper. The study was a collaboration among three behavioral geneticists—one at Oregon Health Sciences University, one at the University of Alberta, and one at the State University of New York in Albany.[35] They studied mouse strains known to have genetic variants relevant to particular behaviors (e.g., addiction or anxiety). First they ensured that the mice from a particular strain were essentially genetically identical in all three labs. Then the scientists did cartwheels to test the animals in identical conditions in the labs.

They standardized everything. Because some mice were born in the lab but others came from breeders, homegrowns were given bouncy van rides to simulate the jostling that commercially bred mice undergo during shipping, just in case that was important. Animals were tested at the same day of age on the same date at the same local time. Animals had been weaned at the same age and lived in the same brand of cage with the same brand and thickness of sawdust bedding, changed on the same day of the week. They were handled the same number of times by humans wearing the same brand of surgical gloves. They were fed the same food and kept in the same lighting environment at the same temperature. The environments of these animals could hardly have been more similar if the three scientists had been identical triplets separated at birth.

What did they observe? Some gene variants showed massive gene/environment interactions, with variants having radically different effects in different labs.

Here's the sort of data they got: Take a strain called 129/SvEvTac and a test measuring the effects of cocaine on activity. In Oregon cocaine increased activity in these mice by 667 centimeters of movement per fifteen minutes. In Albany, an increase of 701. Those are pretty similar numbers; good. And in Alberta? More than 5,000. That's like identical triplets pole-vaulting, each in a different location; they've all had the same training, equipment, running surface, night's rest, breakfast, and brand of underwear. The first two vault 18 feet and 18 feet one inch, and the third vaults 108 feet.

Maybe the scientists didn't know what they were doing; maybe the labs were chaotic. But variability was small within each lab, showing stable environmental conditions. And crucially, a few variants didn't show a gene/environment interaction, producing similar effects in the three labs.

What does this mean? That most of the gene variants were so sensitive to environment that gene/environment interactions occurred even in these obsessively similar lab settings, where incredibly subtle (and still unidentified) environmental differences made huge differences in what the gene did.

Citing "gene/environment interactions" is a time-honored genetics cliché.[36] My students roll their eyes when I mention them. *I* roll my eyes when I mention them. Eat your vegetables, floss your teeth, remember to say, "It's difficult to quantitatively assess the relative contributions of genes and environment to a particular trait when they interact." This suggests a radical conclusion: *it's not meaningful to ask what a gene does, just what it does in a particular environment.* This is summarized wonderfully by the neurobiologist Donald Hebb: "It is no more appropriate to say things like characteristic A is more influenced by nature than nurture than . . . to say that the area of a rectangle is more influenced by its length than its width." It's appropriate to figure out if lengths or widths explain more of the variability in a population of rectangles. But not in individual ones.

As we conclude part 2 of this chapter, some key points:

a. A gene's influence on the average value of a trait (i.e., whether it is inherited) differs from its influence on variability of that trait across individuals (its heritability).

b. Even in the realm of inherited traits—say, the inheritance of five fingers as the human average—you can't really say that there is genetic determination in the classically hard-assed sense of the word. This is because the inheritance of a gene's effect requires not just passing on the gene but also the context that regulates the gene in that manner.

c. Heritability scores are relevant only to the environments in which the traits have been studied. The more environments you study a trait in, the lower the heritability is likely to be.

d. Gene/environment interactions are ubiquitous and can be dramatic. Thus, you can't really say what a gene "does," only what it does in the environments in which it's been studied.

Current research actively explores gene/environment interactions.[37] How's this for fascinating: Heritability of various aspects of cognitive development is very high (e.g., around 70 percent for IQ) in kids from high–socioeconomic status (SES) families but is only around 10 percent in low-SES kids. Thus, higher SES allows the full range of genetic influences on cognition to flourish, whereas lower-SES settings restrict them. In other words, genes are nearly irrelevant to cognitive development if you're growing up in awful poverty—poverty's adverse effects trump the genetics.* Similarly, heritability of alcohol use is lower among religious than nonreligious subjects—i.e., your genes don't matter much if you're in a religious environment that condemns drinking. Domains like these showcase the potential power of classical behavior genetics.

PART 3: SO WHAT DO GENES ACTUALLY HAVE TO DO WITH BEHAVIORS WE'RE INTERESTED IN?

The Marriage of Behavior Genetics and Molecular Genetics

Behavior genetics has gotten a huge boost by incorporating molecular approaches—after examining similarities and differences between twins

* A subtle point for which I thank Stephen Manuck of the University of Pittsburgh: This example represents an exception to the rule that heritability scores go down as you study a trait in more environments. If you started by studying only low-SES individuals, you'd generate a very low heritability score (~10 percent). Thus, if one studies both low- and high-SES subjects (the latter with a high heritability score of about 70 percent), the score will rise.

or adoptees, find the actual genes that underlie those similarities and differences. This powerful approach has identified various genes relevant to our interests. But first, our usual caveats: (a) not all of these findings consistently replicate; (b) effect sizes are typically small (in other words, some gene may be involved, but not in a major way); and (c) the most interesting findings show gene/environment interactions.

Studying Candidate Genes

Gene searches can take a "candidate" approach or a genomewide association approach (stay tuned). The former requires a list of plausible suspects—genes already known to be related to some behavior. For example, if you're interested in a behavior that involves serotonin, obvious candidate genes would include those coding for enzymes that make or degrade serotonin, pumps that remove it from the synapse, or serotonin receptors. Pick one that interests you, and study it in animals using molecular tools to generate "knockout" mice (where you've eliminated that gene) or "transgenic" mice (with an extra copy of the gene). Make manipulations like these only in certain brain regions or at certain times. Then examine what's different about behavior. Once you're convinced of an effect, ask whether variants of that gene help explain individual differences in human versions of the behavior. I start with the topic that has gotten the most attention, for better or worse, mostly "worse."

The Serotonin System

What do genes related to serotonin have to do with our best and worst behaviors? Plenty.

Chapter 2 presented a fairly clear picture of low levels of serotonin fostering impulsive antisocial behavior. There are lower-than-average levels of serotonin breakdown products in the bloodstreams of people with that profile, and of serotonin itself in the frontal cortex of such animals. Even more convincingly, drugs that decrease "serotonergic tone" (i.e., decreasing serotonin levels or sensitivity to serotonin) increase impulsive aggression; raising the tone does the opposite.

This generates some simple predictions—all of the following should be associated with impulsive aggression, as they will produce low serotonin signaling:

- **a.** Low-activity variants of the gene for tryptophan hydroxylase (TH), which makes serotonin
- **b.** High-activity variants of the gene for monoamine oxidase-A (MAO-A), which degrades serotonin
- **c.** High-activity variants of the gene for the serotonin transporter (5HTT), which removes serotonin from the synapse
- **d.** Variants of genes for serotonin receptors that are less sensitive to serotonin

An extensive literature shows that for each of those genes the results are inconsistent and generally go in the *opposite* direction from "low serotonin = aggression" dogma. Ugh.

Studies of genes for TH and serotonin receptors are inconsistent messes.[38] In contrast, the picture of 5HTT, the serotonin transporter gene, is consistently in the opposite direction from what's expected. Two variants exist, with one producing less transporter protein, meaning less serotonin removed from the synapse.* And counter to expectations, this variant, producing more serotonin in the synapse, is associated with more impulsive aggression, not less. Thus, according to these findings, "high serotonin = aggression" (recognizing this as simplified shorthand).

The clearest and most counterintuitive studies concern MAO-A. It burst on the scene in a hugely influential 1993 *Science* paper reporting a Dutch family with an MAO-A gene mutation that eliminated the protein.[39] Thus serotonin isn't broken down and accumulates in the synapse.

* Harking back to how the noncoding regulatory regions of the genome are at least as important as regions that code for genes themselves, the 5HTT variants do not differ in the DNA sequence of the gene but rather in the sequence of a promoter for the gene. As a result, the two variants differ in their sensitivity to a transcription factor, and thus in the amount of transporter protein made.

And counter to chapter 2's predictions, the family was characterized by varied antisocial and aggressive behaviors.

Mouse studies in which the MAO-A gene was "knocked out" (producing the equivalent of the Dutch family's mutation) produced the same—elevated serotonin levels in the synapse and hyperaggressive animals with enhanced fear responses.[40]

This finding, of course, concerned a *mutation* in MAO-A resulting in the complete absence of the protein. Research soon focused on low-activity MAO-A variants that produced elevated serotonin levels.*[41] People with that variant averaged higher levels of aggression and impulsivity and, when looking at angry or fearful faces, more activation of the amygdala and insula and less activation of the prefrontal cortex. This suggests a scenario of more fear reactivity and less frontal capacity to restrain such fear, a perfect storm for reactive aggression. Related studies showed decreased activation of frontal cortical regions during various attentional tasks and enhanced anterior cingulate activity in response to social rejection in such individuals.

So studies where serotonin breakdown products are measured in the body, or where serotonin levels are manipulated with drugs, say that low serotonin = aggression.[42] And the genetic studies, particularly of MAO-A, say high serotonin = aggression. What explains this discrepancy? The key probably is that a drug manipulation lasts for a few hours or days, while genetic variants have their effects on serotonin for a lifetime. Possible explanations: (a) The low-activity MAO-A variants don't produce higher synaptic levels of serotonin all that consistently because the 5HTT serotonin reuptake pump works harder at removing serotonin from the synapse, compensating, and maybe even *over*compensating. There is evidence for this, just to make life really complicated. (b) Those variants do produce chronically elevated serotonin levels in the synapse, but the postsynaptic neurons compensate or overcompensate by decreasing serotonin receptor numbers, thereby reducing sensitivity to all that serotonin; there is evidence for that too. (c) The lifelong consequences of differences in

* Again, the variation in the DNA sequence was not in the MAO-A gene but in its promoter.

serotonin signaling due to gene variants (versus transient differences due to drugs) produce structural changes in the developing brain. There is evidence there as well, and in accordance with that, while temporarily inhibiting MAO-A activity with a drug in an adult rodent decreases impulsive aggression, doing the same in fetal rodents produces adults with increased impulsive aggression.

Yikes, this is complicated. Why go through the agony of all these explanatory twists and turns? Because this obscure corner of neurogenetics has caught the public's fancy, with—I kid you not—the low-activity MAO-A variant being referred to as the "warrior gene" by both scientists and in the media.*[43] And that warrior hoo-hah is worsened by the MAO-A gene being X linked and its variants being more consequential in males than females. Amazingly, prison sentences for murderers have now been lessened in at least two cases because it was argued that the criminal, having the "warrior gene" variant of MAO-A, was inevitably fated to be uncontrollably violent. OMG.

Responsible people in the field have recoiled in horror at this sort of unfounded genetic determinism seeping into the courtroom. The effects of MAO-A variants are tiny. There is nonspecificity in the sense that MAO-A degrades not only serotonin but norepinephrine as well. Most of all, there is nonspecificity in the behavioral effects of the variants. For example, while nearly everyone seems to remember that the landmark MAO-A paper that started all the excitement was about aggression (one authoritative review referred to the Dutch family with the mutation as "notorious for the persistent and extreme reactive aggression demonstrated by some of its males"), in actuality members of the family with the mutation had borderline mental retardation. Moreover, while some individuals with the mutation were quite violent, the antisocial behavior of others consisted of arson and exhibitionism. So maybe the gene has something to do with the extreme reactive aggression of some family members. But it is just as responsible for explaining why other family

* Part of what may explain the "warrior gene" frothiness is the "aggressive" variant being found at a high rate among Maori populations and traditional Maori culture having very high rates of warfare. Nonetheless, it is far from the case that *every* Maori individual with the "warrior" variant is highly aggressive, or that every highly aggressive Maori has the warrior variant.

members, rather than being aggressive, were flashers. In other words, there is as much rationale for going on about the "drop your pants gene" as the "warrior gene."

Probably the biggest reason to reject warrior-gene determinism nonsense is something that should be utterly predictable by now: MAO-A effects on behavior show strong gene/environment interactions.

This brings us to a hugely important 2002 study, one of my favorites, by Avshalom Caspi and colleagues at Duke University.[44] The authors followed a large cohort of children from birth to age twenty-six, studying their genetics, upbringing, and adult behavior. Did MAO-A variant status predict antisocial behavior in twenty-six-year-olds (as measured by a composite of standard psychological assessments and convictions for violent crimes)? No. But MAO-A status coupled with something else powerfully did. Having the low-activity version of MAO-A tripled the likelihood . . . but only in people with a history of severe childhood abuse. And if there was no such history, the variant was not predictive of anything. This is the essence of gene/environment interaction. What does having a particular variant of the MAO-A gene have to do with antisocial behavior? It depends on the environment. "Warrior gene" my ass.

This study is important not just for its demonstration of a powerful gene/environment interaction but for what the interaction is, namely the ability of an abusive childhood environment to collaborate with a particular genetic constitution. To quote a major review on the subject, "In a healthy environment, increased threat sensitivity, poor emotion control and enhanced fear memory in MAOA-L [i.e., the "warrior" variant] men might only manifest as variation in temperament within a 'normal' or subclinical range. However, these same characteristics in an abusive childhood environment—one typified by persistent uncertainty, unpredictable threat, poor behavioral modeling and social referencing, and inconsistent reinforcement for prosocial decision making—might predispose toward frank aggression and impulsive violence in the adult." In a similar vein, the low-activity variant of the serotonin transporter gene was reported to be associated with adult aggressiveness . . . but only when coupled with childhood adversity.[45] This is straight out of the lessons of the previous chapter.

Since then, this MAO-A variant/childhood abuse interaction has been frequently replicated, and even demonstrated with respect to aggressive behavior in rhesus monkeys.[46] There have also been hints as to how this interaction works—the MAO-A gene promoter is regulated by stress and glucocorticoids.

MAO-A variants show other important gene/environment interactions. For example, in one study the low-activity MAO-A variant predicts criminality, but only if coupled with high testosterone levels (consistent with that, the MAO-A gene also has a promoter responsive to androgens). In another study low-activity MAO-A participants in an economic game were more likely than high-activity ones to retaliate aggressively when exploited by the other player—but only if that exploitation produced a large economic loss; if the loss was small, there was no difference. In another study low-activity individuals were more aggressive than others—but only in circumstances of social exclusion. Thus the effects of this genetic variant can be understood only by considering other, nongenetic factors in individuals' lives, such as childhood adversity and adult provocation.[47]

The Dopamine System

Chapter 2 introduced the role of dopamine in the anticipation of reward and in goal-directed behavior. Lots of work has examined the genes involved, most broadly showing that variants that produce lowered dopamine signaling (less dopamine in the synapse, fewer dopamine receptors, or lower responsiveness of these receptors) are associated with sensation seeking, risk taking, attentional problems, and extroversion. Such individuals have to seek experiences of greater intensity to compensate for the blunted dopamine signaling.

Much of the research has focused on one particular dopamine receptor; there are at least five kinds (found in different parts of the brain, binding dopamine with differing strengths and duration), each coded for by a gene.[48] Work has focused on the gene for the D4 dopamine receptor (the gene is called DRD4), which mostly occurs in the neurons in the cortex and nucleus accumbens. The DRD4 gene is super variable, coming in at least ten different flavors in humans. One stretch of the gene is repeated a variable number of times, and the version with seven repeats (the "7R" form) produces a receptor protein that is sparse in the cortex and relatively unresponsive to dopamine. This is the variant associated with a host of related traits—sensation and novelty seeking, extroversion, alcoholism, promiscuity, less sensitive parenting, financial risk taking, impulsivity, and, probably most consistently, ADHD (attention-deficit/hyperactivity disorder).

The implications cut both ways—the 7R could make you more likely to impulsively steal the old lady's kidney dialysis machine, or to impulsively give the deed of your house to a homeless family. In come gene/environment interactions. For example, kids with the 7R variant are less generous than average. But only if they show insecure attachment to their parents. Secure-attachment 7Rs show *more* generosity than average. Thus 7R has something to do with generosity—but its effect is entirely context dependent. In another study 7R students expressed the least interest in organizations advocating prosocial causes, unless they were given a religious prime,* in which case they were the *most* prosocial. One more—7Rs are worse at gratification-postponement tasks, but only if they grew up poor. Repeat the mantra: don't ask what a gene does; ask what it does in a particular context.[49]

Interestingly, the next chapter considers the extremely varied frequency of the 7R variant in different populations. As we'll see, it tells you a lot about the history of human migration, as well as about differences between collectivist and individualist cultures.[50]

We shift now to other parts of the dopamine system. As introduced in chapter 2, after dopamine binds to receptors, it floats off and must be removed from the synapse.[51] One route involves its being degraded by the

* Control subjects had the task of unscrambling jumbled strings of words into coherent phrases. The religion-prime group did the same with word strings that contained religious terms.

enzyme catechol-O-methyltransferase (COMT). Among the variants of the COMT gene is one associated with a more efficient enzyme. "More efficient" = better at degrading dopamine = less dopamine in the synapse = less dopamine signaling. The highly efficient COMT variant is associated with higher rates of extroversion, aggression, criminality, and conduct disorder. Moreover, in a gene/environment interaction straight out of the MAO-A playbook, that COMT variant is associated with anger traits, but only when coupled with childhood sexual abuse. Intriguingly, the variants seem pertinent to frontal regulation of behavior and cognition, especially during stress.

In addition to degradation, neurotransmitters can be removed from the synapse by being taken back up into the axon terminal for recycling.[52] Dopamine reuptake is accomplished by the dopamine transporter (DAT). Naturally, the DAT gene comes in different variants, and those that produce higher levels of synaptic dopamine (i.e., transporter variants that are less efficient) in the striatum are associated with people who are more oriented toward social signaling—they're drawn more than average to happy faces, are more repelled by angry faces, and have more positive parenting styles. How these findings merge with the findings from the DRD4 and COMT studies (i.e., fitting risk taking with a preference for happy faces) is not immediately apparent.

Cool people with certain versions of these dopamine-related genes are more likely to engage in all sorts of interesting behaviors, ranging from the healthy to the pathological. But not so fast:

- These findings are not consistent, no doubt reflecting unrecognized gene/environment interactions.
- Again, why should the COMT world be related to sensation seeking, while there are the DAT people and their happy faces? Both genes are about ending dopamine signaling. This is probably related to different parts of the brain differing as to whether DAT or COMT plays a bigger role.[53]
- The COMT literature is majorly messy, for the inconvenient reason that the enzyme also degrades norepinephrine. So COMT

variants are pertinent to two totally different neurotransmitter systems.

- These effects are tiny. For example, knowing which DRD4 variant someone has explains only 3 to 4 percent of the variation in novelty-seeking behavior.
- The final piece of confusion seems most important but is least considered in the literature (probably because it would be premature). Suppose that every study shows with whopping clarity and consistency that a DRD4 variant is highly predictive of novelty seeking. That still doesn't tell us why for some people novelty seeking means frequently switching their openings in chess games, while for others it means looking for a new locale because it's getting stale being a mercenary in the Congo. No gene or handful of genes that we are aware of will tell us much about that.

The Neuropeptides Oxytocin and Vasopressin

Time for a quick recap from chapter 4. Oxytocin and vasopressin are involved in prosociality, ranging from parent/offspring bonds to monogamous bonds to trust, empathy, generosity, and social intelligence. Recall the caveats: (a) sometimes these neuropeptides are more about sociality than prosociality (in other words, boosting social information gathering, rather than acting prosocially with that information); (b) they most consistently boost prosociality in people who already lean in that direction (e.g., making generous people more generous, while having no effect on ungenerous people); and (c) the prosocial effects are within groups, and these neuropeptides can make people crappier to outsiders—more xenophobic and preemptively aggressive.

Chapter 4 also touched on oxytocin and vasopressin genetics, showing that individuals with genetic variants that result in higher levels of either the hormones or their receptors tend toward more stable monogamous relationships, more actively engaged parenting, better skill at perspective taking, more empathy, and stronger fusiform cortex responses to faces. These are fairly consistent effects of moderate magnitude.

Meanwhile, there are studies showing that one oxytocin receptor gene variant is associated with extreme aggression in kids, as well as a callous, unemotional style that foreshadows adult psychopathy.[54] Moreover, another variant is associated with social disconnection in kids and unstable adult relationships. But unfortunately these findings are uninterpretable because no one knows if these variants produce more, less, or the usual amount of oxytocin signaling.

Of course, there are cool gene/environment interactions. For example, having a particular oxytocin receptor gene variant predicts less sensitive mothering—but only when coupled with childhood adversity. Another variant is associated with aggression—but only when people have been drinking. Yet another variant is associated with greater seeking of emotional support during times of stress—among Americans (including first generation Korean Americans) but not Koreans (stay tuned for more in the next chapter).

Genes Related to Steroid Hormones

We start with testosterone. The hormone is not a protein (none of the steroid hormones are), meaning there isn't a testosterone gene. However, there are genes for the enzymes that construct testosterone, for the enzyme that converts it to estrogen, and for the testosterone (androgen) receptor. The most work has focused on the gene for the receptor, which comes in variants that differ in their responsiveness to testosterone.*

Intriguingly, a few studies have shown that among criminals, having the more potent variant is associated with violent crimes.[55] A related finding concerns sex differences in structure of the cortex, and adolescent boys with the more potent variant show more dramatic "masculinization" of the cortex. An interaction between receptor variant and testosterone levels occurs. High basal testosterone levels do not predict elevated levels

* For aficionados: The testosterone receptor contains what is called a polyglutamine repeat—a stretch of the protein where the same amino acid, called glutamine, is repeated. Importantly, there is tremendous variability among people as to how many glutamine repeats there are; the fewer, the more potently the androgen receptor works. Recall that receptors for steroid hormones like testosterone work as transcription factors, and proteins that have polyglutamine repeats are often transcription factors.

of aggressive mood or of amygdaloid reactivity to threatening faces in males—except in those with that variant. Interestingly, the equivalent variant predicts aggressiveness in Akita dogs.

How important are these findings? A key theme in chapter 4 was how little individual differences in testosterone levels in the normal range predict individual differences in behavior. How much more predictability is there when combining knowledge of testosterone levels *and* of receptor sensitivity? Not much. How about hormone levels *and* receptor sensitivity *and* number of receptors? Still not much. But definitely an improvement in predictive power.

Similar themes concern the genetics of the estrogen receptor.[56] For example, different receptor variants are associated with higher rates of anxiety among women, but not men, and higher rates of antisocial behavior and conduct disorder in men, but not women. Meanwhile, in genetically manipulated mice, the presence or absence of the receptor gene influences aggression in females . . . depending on how many brothers there were in the litter in utero—gene/environment again. Once again, the magnitude of these genetic influences is tiny.

Finally, there is work on genes related to glucocorticoids, particularly regarding gene/environment interactions.[57] For example, there is an interaction between one variant of the gene for a type of receptor for glucocorticoids (for mavens: it's the MR receptor) and childhood abuse in producing an amygdala that is hyperreactive to threat. Then there is a protein called FKBP5, which modifies the activity of another type of receptor for glucocorticoids (the GR receptor); one FKBP5 variant is associated with aggression, hostility, PTSD, and hyperreactivity of the amygdala to threat—but only when coupled with childhood abuse.

Buoyed by these findings, some researchers have examined two candidate genes simultaneously. For example, having both "risk" variants of 5HTT and DRD4 synergistically increases the risk of disruptive behavior in kids—an effect exacerbated by low socioeconomic status.[58]

Phew; all these pages and we've only gotten to thinking about two genes and one environmental variable simultaneously. And despite this, things still aren't great:

- The usual—results aren't terribly consistent from one study to the next.
- The usual—effect sizes are small. Knowing what variant of a candidate gene someone has (or even what variants of a collection of genes) doesn't help much in predicting their behavior.
- A major reason is that, after getting a handle on 5HTT and DRD4 interactions, there are still roughly 19,998 more human genes and a gazillion more environments to study. Time to switch to the other main approach—looking at all those 20,000 genes at once.

Fishing Expeditions, Instead of Looking Where the Light Is

The small effect sizes reflect a limitation in the candidate gene approach; in scientific lingo, the problem is that one is only looking where the light is. The cliché harks back to a joke: You discover someone at night, searching the ground under a street lamp. "What's wrong?" "I dropped my ring; I'm looking for it." Trying to be helpful, you ask, "Were you standing on this side or that side of the lamp when you dropped it?" "Oh, no, I was over by those trees when I dropped it." "Then why are you searching here?" "This is where the light is." With candidate gene approaches, you look only where the light is, examine only genes that you already know are involved. And with twenty thousand or so genes, it's pretty safe to assume there are still some interesting genes that you don't know about yet. The challenge is to find them.

The most common way of trying to find them all is with genomewide association studies (GWAS).[59] Examine, say, the gene for hemoglobin and look at the eleventh nucleotide in the sequence; everyone will pretty much have the same DNA letter in that spot. However, there are little hot spots of variability, single nucleotides where, say, two different DNA letters occur, each in about 50 percent of the population (and where this typically doesn't change the amino acid being specified, because of DNA redundancy). There are more than a million of such "SNPs" (single-nucleotide polymorphisms) scattered throughout the genome—in stretches of DNA coding for genes, for promoters, for mysterious DNA junk. Collect DNA from a huge number of people, and examine whether particular SNPs associate with particular traits. If an SNP that's implicated occurs in a gene, you've just gotten a hint that the gene may be involved in that trait.*

A GWAS study might implicate scads of genes as being associated with a trait. Hopefully, some will be candidate genes already known to be related to the trait. But other identified genes may be mysterious. Now go check out what they do.

In a related approach, suppose you have two populations, one with and one without a degenerative muscle disease. Take a muscle biopsy from everyone, and see which of the ~20,000 genes are transcriptionally active in the muscle cells. With this "microarray" or "gene chip" approach, you look for genes that are transcriptionally active only in diseased or in healthy muscle, not in both. Identify them, and you have some new candidate genes to explore.†

These fishing expeditions‡ show why we're so ignorant about the genetics of behavior.[60] Consider a classic GWAS that looked for genes related to height. This was a crazy difficult study involving examining the genomes of 183,727 people. *183,727.* It must have taken an army of scientists just to label the test tubes. And reflecting that, the paper reporting the findings in *Nature* had approximately 280 authors.

And the results? *Hundreds* of genetic variants were implicated in regulating height. A handful of genes identified were known to be involved in skeletal growth, but the rest was terra incognita. The single genetic variant identified that most powerfully predicted height explained all of 0.4 percent—four tenths of one percent—of the variation in height, and all those hundreds of variants put together explained only about 10 percent of the variation.

* And, following that logic, if a trait associates with a particular version of an SNP in the promoter of a gene, you've just gotten a hint that the regulation of the gene (as opposed to the gene itself) may be involved in that trait. As an example, the gene for one type of serotonin receptor contains an SNP in the third base of the codon coding for the thirty-fourth amino acid in the protein; and one of the variants of that SNP is associated with responsiveness to a particular drug in schizophrenics.

† For lovers of details: Note that the GWAS and microarray approaches are usually telling different things. In the former you are looking for genes that have a variant that is associated with whatever disease or behavior you're studying. In microarray studies you're looking for genes whose expression profiles are associated with the disease or behavior.

‡ More scientific lingo—pass a big fishing net through a stretch of the ocean, and see what you wind up catching.

Meanwhile, an equally acclaimed study did a GWAS regarding body mass index (BMI). Similar amazingness—almost a quarter million genomes examined, even more authors than the height study. And in this case the single most explanatory genetic variant identified accounted for only 0.3 percent of the variation in BMI. Thus both height and BMI are highly "polygenic" traits. Same for age of menarche (when girls menstruate for the first time). Moreover, additional genes are being missed because their variants are too rare to be picked up by current GWAS techniques. Thus these traits are probably influenced by hundreds of genes.[61]

What about behavior? A superb 2013 GWAS study examined the genetic variants associated with educational attainment.[62] The usual over-the-top numbers—126,559 study subjects, about 180 authors. And the most predictive genetic variant accounted for 0.02 percent—two hundredths of one percent—of the variation. All the identified variants together accounted for about 2 percent of the variation. A commentary accompanying the paper contained this landmark of understatement: "In short, educational attainment looks to be a very polygenic trait."

Educational attainment—how many years of high school or college one completes—is relatively easy to measure. How about the subtler, messier behaviors that fill this book's pages? A handful of studies have tackled that, and the findings are much the same—at the end, you have a list of scores of genes implicated and can then go figure out what they do (logically, starting with the ones that showed the strongest statistical associations). Hard, hard approaches that are still in their infancy. Made worse by a GWAS missing more subtly variable spots,* meaning even more genes are likely involved.[63]

As we conclude this section, some key points:[64]

a. This review of candidate genes barely scratches even the surface of the surface. Go on PubMed (a major search engine of the biomedical

* This would be if there is a gene that has an SNP that is unbelievably powerfully associated with something, but the alternate letter only occurs in a thousand humans. That will be missed with current GWAS.

literature) and search "MAO gene/behavior"—up come more than 500 research papers. "Serotonin transporter gene/behavior"—1,250 papers. "Dopamine receptor gene/behavior"—nearly 2,000.

b. The candidate gene approaches show that the effect of a single gene on a behavior is typically tiny. In other words, having the "warrior gene" variant of MAO probably has less effect on your behavior than does believing that you have it.

c. Genomewide survey approaches show that these behaviors are influenced by huge numbers of genes, each one playing only a tiny role.

d. What this translates into is nonspecificity. For example, serotonin transporter gene variants have been linked to risk of depression, but also anxiety, obsessive-compulsive disorder, schizophrenia, bipolar disorder, Tourette's syndrome, and borderline personality disorder. In other words, that gene is part of a network of hundreds of genes pertinent to depression, but also part of another equally large and partially overlapping network relevant to anxiety, another relevant to OCD, and so on. And meanwhile, we're plugging away, trying to understand interactions of two genes at a time.

e. And, of course, gene and environment, gene and environment.

CONCLUSIONS

At long last, you (and I!) have gotten to the end of this excruciatingly but necessarily long chapter. Amid all these tiny effects and technical limitations, it's important to not throw out the genetic baby with the bathwater, as has been an agitated sociopolitical goal at times (during my intellectual youth in the 1970s, sandwiched between the geologic periods of Cranberry Bell-bottoms and of John Travolta White Suits was the Genes-Have-Nothing-to-Do-with-Behavior Ice Age).

Genes have plenty to do with behavior. Even more appropriately, all behavioral traits are affected to some degree by genetic variability.[65] They have to be, given that they specify the structure of all the proteins pertinent to every neurotransmitter, hormone, receptor, etc. that there is. And they have plenty to do with individual differences in behavior, given the large percentage of genes that are polymorphic, coming in different flavors. But their effects are supremely context dependent. Ask not what a gene does. Ask what it does in a particular environment and when expressed in a particular network of other genes (i.e., gene/gene/gene/gene . . . /environment).

Thus, for our purposes, genes aren't about inevitability. Instead they're about context-dependent tendencies, propensities, potentials, and vulnerabilities. All embedded in the fabric of the other factors, biological and otherwise, that fill these pages.

Now that this chapter's done, why don't we all take a bathroom break and then see what's in the refrigerator.

Nine

Centuries to Millennia Before

Let's start with a seeming digression. Parts of chapters 4 and 7 have debunked some supposed sex differences concerning the brain, hormones, and behavior. One difference, however, is persistent. It's far from issues that concern this book, but bear with me.

A remarkably consistent finding, starting with elementary school students, is that males are better at math than females. While the difference is minor when it comes to considering average scores, there is a huge difference when it comes to math stars at the upper extreme of the distribution. For example, in 1983, for every girl scoring in the highest percentile on the math SAT, there were eleven boys.

Why the difference? There have always been suggestions that testosterone is central. During development, testosterone fuels the growth of a brain region involved in mathematical thinking, and giving adults testosterone enhances some math skills. Oh, okay, it's biological.

But consider a paper published in *Science* in 2008.[1] The authors examined the relationship between math scores and sexual equality in forty countries (based on economic, educational, and political indices of gender equality; the worst was Turkey, the United States was middling, and, naturally, the Scandinavians were tops). Lo and behold, the more gender

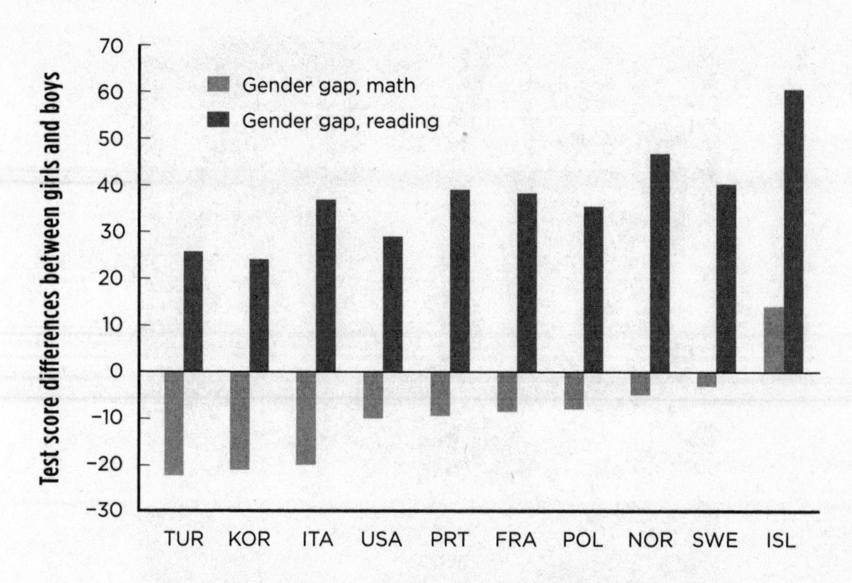

L. Guiso et al., "Culture, Gender, and Math," Sci *320 (2008): 1164.*

equal the country, the less of a discrepancy in math scores. By the time you get to the Scandinavian countries, it's statistically insignificant. And by the time you examine the most gender-equal country on earth at the time, Iceland, girls are *better* at math than boys.*

In other words, while you can never be certain, the Rajasthani girl pictured on top, on the next page, seated next to her husband, is less likely than the Swedish girl pictured below her to solve the Erdös-Hajnal conjecture in graph theory.

In other, other words, culture matters. We carry it with us wherever we go. As one example, the level of corruption—a government's lack of transparency regarding use of power and finances—in UN diplomats' home countries predicts their likelihood of racking up unpaid parking tickets in Manhattan. Culture leaves long-lasting residues—Shiites and Sunnis slaughter each other over a succession issue fourteen centuries old; across thirty-three countries population density in the year *1500* significantly predicts how authoritarian the government was in 2000; over

* Note that the other reliable sex difference in cognition, namely better reading performance by girls than by boys, doesn't disappear in more gender-equal societies. It gets bigger.

the course of millennia, earlier adoption of the hoe over the plow predicts gender equality today.[2]

And in other, other, other words, when we contemplate our iconic acts—the pulling of a trigger, the touching of an arm—and want to explain why they happened using a biological framework, culture better be on our list of explanatory factors.

Thus, the goals of this chapter:

- Look at systematic patterns of cultural variation as they pertain to the best and worst of our behaviors.

- Explore how different types of brains produce different culture and different types of culture produce different brains. In other words, how culture and biology coevolve.[3]
- See the role of ecology in shaping culture.

DEFINITIONS, SIMILARITIES, AND DIFFERENCES

"Culture," naturally, has been defined various ways. One influential definition comes from Edward Tylor, a distinguished nineteenth-century cultural anthropologist. For him culture is "that complex whole which includes knowledge, belief, art, morals, law, custom, and any other capabilities and habits acquired by man [*sic*] as a member of society."[4]

This definition, obviously, is oriented toward something that is specific to humans. Jane Goodall blew off everyone's socks in the 1960s by reporting the now-iconic fact that chimps make tools. Her study subjects modified twigs by stripping off the leaves and pushing them into termite mounds; termites would bite the twig, still holding on when it was pulled out, yielding a snack for the chimps.

This was just the start. Chimps were subsequently found to use various tools—wood or rock anvils for cracking open nuts, wads of chewed leaves to sponge up hard-to-reach water, and, in a real shocker, sharpened sticks for spearing bush babies.[5] Different populations make different tools; new techniques spread across social networks (among chimps who hang with one another); kids learn the ropes by watching their moms; techniques spread from one group to another when someone emigrates; chimp nut-cracking tools in excess of four thousand years old have been excavated. And in my favorite example, floating between tool use and accessorizing, a female in Zambia got it into her head to go around with a strawlike blade of grass in her ear. The action had no obvious function; apparently she just liked having a piece of grass sticking out of her ear. So sue her. She did it for years, and over that time the practice spread throughout her group. A fashionista.

*E. van Leeuwen et al., "A Group-Specific Arbitrary Tradition in Chimpanzees (*Pan troglodytes*)," *Animal Cog* 17 (2014): 1421.*

In the decades since Goodall's discovery, tool use has been observed in apes and monkeys, elephants, sea otters, mongoose.[6] Dolphins use sea sponges to dig up fish burrowed into the sea floor. Birds use tools for nest building or food acquisition—jays and crows, for example, use twigs to fish for insects, much as chimps do. And there's tool use in cephalopods, reptiles, and fish.

All this is mighty impressive. Nonetheless, such cultural transmission doesn't show progression—this year's chimp nut-cracking tool is pretty much the same as that of four thousand years ago. And with few exceptions (more later), nonhuman culture is solely about material culture (versus, say, social organization).

So the classical definition of culture isn't specific to humans.[7] Most cultural anthropologists weren't thrilled with Goodall's revolution—great, next the zoologists will report that Rafiki persuaded Simba to become the Lion King—and now often emphasize definitions of culture that cut chimps and other hoi polloi out of the party. There's a fondness for the thinking of Alfred Kroeber, Clyde Kluckhohn, and Clifford Geertz, three heavyweight social anthropologists who focused on how culture is about *ideas and symbols*, rather than the mere behaviors in which they instantiate, or material products like flint blades or iPhones. Contemporary anthropologists like Richard Shweder have emphasized a more affective but still human-centric view of culture as being about moral and visceral versions of right and wrong. And of course these views have been critiqued by postmodernists for reasons I can't begin to follow.

Basically, I don't want to go anywhere near these debates. For our purposes we'll rely on an intuitive definition of culture that has been emphasized by Frans de Waal: "culture" is how we do and think about things, transmitted by nongenetic means.

Working with that broad definition, is the most striking thing about the array of human cultures the similarities or the differences? Depends on your taste.

If the similarities seem most interesting, there are plenty—after all, multiple groups of humans independently invented agriculture, writing, pottery, embalming, astronomy, and coinage. At the extreme of similarities are human universals, and numerous scholars have proposed lists of them. One of the lengthiest and most cited comes from the anthropologist Donald Brown.[8] Here's a partial list of his proposed cultural universals: the existence of and concern with aesthetics, magic, males and females seen as having different natures, baby talk, gods, induction of altered states, marriage, body adornment, murder, prohibition of some type of murder, kinship terms, numbers, cooking, private sex, names, dance, play, distinctions between right and wrong, nepotism, prohibitions on certain types of sex, empathy, reciprocity, rituals, concepts of fairness, myths about afterlife, music, color terms, prohibitions, gossip, binary sex terms, in-group favoritism, language, humor, lying, symbolism, the linguistic concept of "and," tools, trade, and toilet training. And that's a partial list.

For the purposes of this chapter, the staggeringly large cultural differences in how life is experienced, in resources and privileges available, in opportunities and trajectories, are most interesting. Just to start with some breathtaking demographic statistics born of cultural differences: a girl born in Monaco has a ninety-three-year life expectancy; one in Angola, thirty-nine. Latvia has 99.9 percent literacy; Niger, 19 percent. More than 10 percent of children in Afghanistan die in their first year, about 0.2 percent in Iceland. Per-capita GDP is $137,000 in Qatar, $609 in the Central African Republic. A woman in South Sudan is roughly a thousand times more likely to die in childbirth than a woman in Estonia.[9]

The experience of violence also varies enormously by culture. Someone in Honduras is 450 times more likely to be murdered than someone in Singapore. 65 percent of women experience intimate-partner violence in Central Africa, 16 percent in East Asia. A South African woman is more than one hundred times more likely to be raped than one in Japan. Be a school kid in Romania, Bulgaria, or Ukraine, and you're about ten times more likely to be chronically bullied than a kid in Sweden, Iceland, or Denmark (stay tuned for a closer look at this).[10]

Of course, there are the well-known gender-related cultural differences. There are the Scandinavian countries approaching total gender equality and Rwanda, with 63 percent of its lower-house parliamentary seats filled by women, compared with Saudi Arabia, where women are not allowed outside the house unless accompanied by a male guardian, and Yemen, Qatar, and Tonga, with 0 percent female legislators (and with the United States running around 20 percent).[11]

Then there's the Philippines, where 93 percent of people say they feel happy and loved, versus 29 percent of Armenians. In economic games, people in Greece and Oman are more likely to spend resources to punish overly generous players than to punish those who are cheaters, whereas among Australians such "antisocial punishment" is nonexistent. And there are wildly different criteria for prosocial behavior. In a study of employees throughout the world working for the same multinational bank, what was the most important reason cited to help someone? Among Americans it was that the person had previously helped them; for Chinese it was that the person was higher ranking; in Spain, that they were a friend or acquaintance.[12]

Your life will be unrecognizably different, depending on which culture the stork deposited you into. In wading through this variability, there are some pertinent patterns, contrasts, and dichotomies.

COLLECTIVIST VERSUS INDIVIDUALIST CULTURES

As introduced in chapter 7, a large percentage of cross-cultural psychology studies compare collectivist with individualist cultures. This almost always means comparisons between subjects from collectivist East Asian cultures and Americans, coming from that mother of all individualist cultures.* As defined, collectivist cultures are about harmony, interdependence, conformity, and having the needs of the group guiding behavior, whereas individualist cultures are about autonomy, personal achievement, uniqueness, and the needs and rights of the individual. Just to be a wee bit caustic, individualist culture can be summarized by that classic American concept of "looking out for number one"; collectivist culture can be summarized by the archetypical experience of American Peace Corps teachers in such countries—pose your students a math question, and no one will volunteer the correct answer because they don't want to stand out and shame their classmates.

Individualist/collectivist contrasts are striking. In individualist cultures, people more frequently seek uniqueness and personal accomplishment, use first-person singular pronouns more often, define themselves

* In reading about Americans versus East Asians in this section, and Americans versus other cultures in later sections, you'll realize that in some ways it's Americans (and Western Europeans) versus the rest of the world in many cultural ways. They are just plain "WEIRD"—Westernized, educated, industrialized, rich, and democratic.

in terms that are personal ("I'm a contractor") rather than relational ("I'm a parent"), attribute their successes to intrinsic attributes ("I'm really good at X") rather than to situational ones ("I was in the right place at the right time"). The past is more likely to be remembered via events ("That's the summer I learned to swim") rather than social interactions ("That's the summer we became friends"). Motivation and satisfaction are gained from self- rather than group-derived effort (reflecting the extent to which American individualism is about noncooperation, rather than nonconformity). Competitive drive is about getting ahead of everyone else. When asked to draw a "sociogram"—a diagram of their social network, with circles representing themselves and their friends, connected by lines—Americans tend to place the circle representing themselves in the middle of the page and make it the largest.[13]

In contrast, those from collectivist cultures show more social comprehension; some reports suggest that they are better at Theory of Mind tasks, more accurate in understanding someone else's perspective—with "perspective" ranging from the other person's abstract thoughts to how objects appear from where she is sitting. There is more blame of the group when someone violates a norm due to peer pressure, and a greater tendency to give situational explanations for behavior. Competitive drive is about not falling behind everyone else. And when drawing sociograms, the circle representing "yourself" is far from the center, and far from the biggest.

Naturally, these cultural differences have biological correlates. For example, subjects from individualist cultures strongly activate the (emotional) mPFC when looking at a picture of themselves, compared to looking at a picture of a relative or friend; in contrast, the activation is far less for East Asian subjects.* Another example is a favorite demonstration of mine of cross-cultural differences in psychological stress—when asked in free recall, Americans are more likely than East Asians to remember

* These are tough studies to pull off, as neuroimaging is a bit of an art in addition to a science, and being able to quantitatively compare data derived from two scanners and scanning protocols on opposite sides of the globe is challenging. The alternative—having subjects from both cultures studied in the same scanner—is challenging as well; those aren't going to be representative subjects, since half of them are probably international students—connected, well off, and adventurous enough to be in an American college town, volunteering for a Psych 101 study.

times in which they influenced someone; conversely, East Asians are more likely to remember times when someone influenced them. Force Americans to talk at length about a time someone influenced them, or force East Asians to detail their influencing someone, and both secrete glucocorticoids from the stressfulness of having to recount this discomfiting event. And work by my Stanford colleagues and friends Jeanne Tsai and Brian Knutson shows that mesolimbic dopamine systems activate in European Americans when looking at excited facial expressions; in Chinese, when looking at calm expressions.

As we will see in chapter 13, these cultural differences produce different moral systems. In the most traditional of collectivist societies, conformity and morality are virtually synonymous and norm enforcement is more about shame ("What will people think if I did that?") than guilt ("How could I live with myself?"). Collectivist cultures foster more utilitarian and consequentialist moral stances (for example, a greater willingness for an innocent person to be jailed in order to prevent a riot). The tremendous collectivist emphasis on the group produces a greater degree of in-group bias than among individualist culture members. In one study, for example, Korean and European American subjects observed pictures of either in- or out-group members in pain. All subjects reported more subjective empathy and showed more activation of Theory of Mind brain regions (i.e., the temporoparietal junction) when observing in-group members, but the bias was significantly greater among Korean subjects. In addition, subjects from both individualist and collectivist cultures denigrate out-group members, but only the former inflate assessments of their own group. In other words, East Asians, unlike Americans, don't have to puff up their own group to view others as inferior.[14]

What is fascinating is the direction that some of these differences take, as shown in approaches pioneered by one of the giants in this field, Richard Nisbett of the University of Michigan. Westerners problem-solve in a more linear fashion, with more reliance on linguistic rather than spatial coding. When asked to explain the movement of a ball, East Asians are more likely to invoke relational explanations built around the interactions of the ball with its environment—friction—while Westerners focus on intrinsic properties like weight and density. Westerners are more accurate at estimating length in absolute terms ("How long is that line?") while East Asians are better with relational estimates ("How much longer is this line than that?"). Or how's this one: Consider a monkey, a bear, and a banana. Which two go together? Westerners think categorically and choose the monkey and bear—they're both animals. East Asians think relationally and link the monkey and banana—if you're thinking of a monkey, also think of food it will need.[15]

Remarkably, the cultural differences extend to sensory processing, where Westerners process information in a more focused manner, East Asians in a more holistic one.[16] Show a picture of a person standing in the middle of a complex scene; East Asians will be more accurate at remembering the scene, the context, while Westerners remember the person in the middle. Remarkably, this is even observed on the level of eye tracking—typically Westerners' eyes first look at a picture's center, while East Asians scan the overall scene. Moreover, force Westerners to focus on the holistic context of a picture, or East Asians on the central subject, and the frontal cortex works harder, activating more.

As covered in chapter 7, cultural values are first inculcated early in life. So it's no surprise that culture shapes our attitudes about success, morality, happiness, love, and so on. But what is startling to me is how these cultural differences also shape where your eyes focus on a picture or how you think about monkeys and bananas or the physics of a ball's trajectory. Culture's impact is enormous.

Naturally, there are various caveats concerning collectivist/individualist comparisons:

- The most obvious is the perpetual "on the average"—there are plenty of Westerners, for example, who are more collectivist than plenty of East Asians. In general, people who are most individualist by various personality measures are most individualist in neuroimaging studies.[17]
- Cultures change over time. For example, levels of conformity in East Asian cultures are declining (one study, for example, shows

increased rates of babies in Japan receiving unique names). Moreover, one's degree of inculcation into one's culture can be altered rapidly. For example, priming someone beforehand with individualist or collectivist cultural cues shifts how holistically he processes a picture. This is especially true for bicultural individuals.[18]

- We will soon see about some genetic differences between collectivist and individualist populations. There is nothing resembling genetic destiny about this—the best evidence for this conclusion comes from one of the control groups in many of these studies, namely East Asian Americans. In general, it takes about a generation for the descendants of East Asian immigrants to America to be as individualist as European Americans.[19]
- Obviously, "East Asians" and "Westerners" are not monolithic entities. Just ask someone from Beijing versus the Tibetan steppes. Or stick three people from Berkeley, Brooklyn, and Biloxi in a stalled elevator for a few hours and see what happens. As we will see, there is striking variation within cultures.

Why should people in one part of the globe have developed collectivist cultures, while others went individualist? The United States is the individualism poster child for at least two reasons. First there's immigration. Currently, 12 percent of Americans are immigrants, another 12 percent are (like me) children of immigrants, and everyone else except for the 0.9 percent pure Native Americans descend from people who emigrated within the last five hundred years.[20] And who were the immigrants? Those in the settled world who were cranks, malcontents, restless, heretical, black sheep, hyperactive, hypomanic, misanthropic, itchy, unconventional, yearning to be free, yearning to be rich, yearning to be out of their damn boring repressive little hamlet, yearning. Couple that with the second reason—for the majority of its colonial and independent history, America has had a moving frontier luring those whose extreme prickly optimism made merely booking passage to the New World insufficiently novel—and you've got America the individualistic.

Why has East Asia provided textbook examples of collectivism?[21] The

key is how culture is shaped by the way people traditionally made a living, which in turn is shaped by ecology. And in East Asia it's all about rice. Rice, which was domesticated there roughly ten thousand years ago, requires massive amounts of communal work. Not just backbreaking planting and harvesting, which are done in rotation because the entire village is needed to harvest each family's rice.* Collective labor is first needed to transform the ecosystem—terracing mountains and building and maintaining irrigation systems for controlled flooding of paddies. And there's the issue of dividing up water fairly—in Bali, religious authority regulates water access, symbolized by iconic water temples. How's this for amazing—the Dujiuangyan irrigation system irrigates more than five thousand square kilometers of rice farms near Chengdu, China, and it is more than *two thousand* years old. The roots of collectivism, like those of rice, run deep in East Asia.†

A fascinating 2014 *Science* paper strengthens the rice/collectivism connection by exploring an exception.[22] In parts of northern China it's difficult to grow rice, and instead people have grown wheat for millennia; this involves individual rather than collective farming. And by the standard

* The United States was not without labor-intensive agriculture historically. But rather than solving that with collectivism, it solved it with slavery.

† I have no idea if rice roots actually run deep, but the metaphor was begging to be written.

tests of individualist versus collectivist cultures (e.g., draw a sociogram, which two are most similar of a rabbit, dog, and carrot?)—they look like Westerners. The region has two other markers of individualism, namely higher rates of divorce and of inventiveness—patent filings—than in rice-growing regions. The roots of individualism, likes those of wheat, run deep in northern China.

The links between ecology, mode of production, and culture are shown in a rare collectivist/individualist study not comparing Asians and Westerners.[23] The authors studied a Turkish region on the Black Sea, where mountains hug the coastline. There, in close proximity, people live by fishing, by farming the narrow ribbon of land between the sea and the mountains, or as mountain shepherds. All three groups had the same language, religion, and gene stock.

Herding is solitary; while Turkish farmers and fishermen (and women) were no Chinese rice farmers, they at least worked their fields in groups and manned their fishing boats in crews. Herders thought less holistically than farmers or fishermen—the former were better at judging absolute length of lines, the other two at relative judgments; when shown a glove, a scarf, and a hand, herders grouped gloves and scarves categorically, while the others grouped relationally, pairing gloves and hands. In the authors' words, "social interdependence fosters holistic thinking."

This theme appears in another study, comparing Jewish boys from either observant Orthodox homes (dominated by endless shared rules about beliefs and behaviors) with ones from far more individualist secular homes. Visual processing was more holistic in the Orthodox, more focused in the secular.[24]

The East Asian/Western collectivist/individualist dichotomy has a fascinating genetic correlate.[25] Recall from the last chapter dopamine and DRD4, the gene for the D4 receptor. It's extraordinarily variable, with at least twenty-five human variants (with lesser variability in other primates). Moreover, the variation isn't random, inconsequential drift of DNA sequences; instead there has been strong positive selection for variants. Most common is the 4R variant, occurring in about half of East Asians and European Americans. There's also the 7R variant, producing

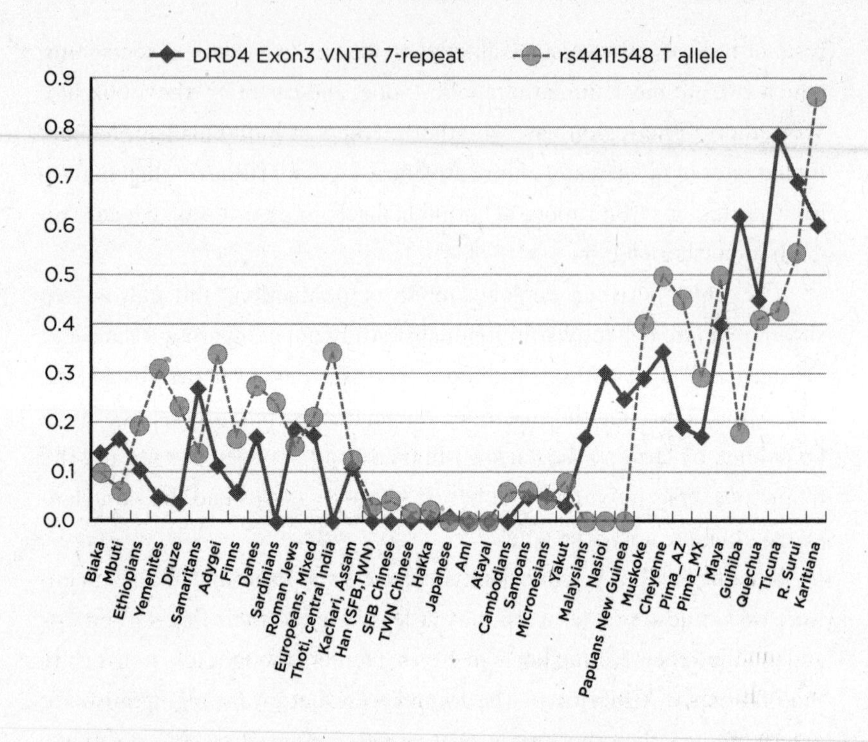

K. Kidd et al., "An Historical Perspective on 'The World-Wide Distribution of Allele Frequencies at the Human Dopamine D4 Receptor Locus,'" Human Genetics, *133 (2014): 431.*

a receptor less responsive to dopamine in the cortex, associated with novelty seeking, extroversion, and impulsivity. It predates modern humans but became dramatically more common ten to twenty thousand years ago. The 7R variant occurs in about 23 percent of Europeans and European Americans. And in East Asians? 1 percent.

So which came first, 7R frequency or cultural style? The 4R and 7R variants, along with the 2R, occur worldwide, implying they already existed when humans radiated out of Africa 60,000 to 130,000 years ago. Classic work by Kenneth Kidd of Yale, examining the distribution of 7R, shows something remarkable.

Starting at the left of the figure above, there's roughly a 10 to 25 percent incidence of 7R in various African, European, and Middle Eastern populations. Jumping to the right side of the figure, there's a slightly higher incidence among the descendants of those who started island-hopping from mainland Asia to Malaysia and New Guinea. The same for folks whose ancestors migrated to North America via the Bering land bridge about fifteen thousand years ago—the Muskoke, Cheyenne, and Pima tribes of Native Americans. Then the Maya in Central America—up to around 40 percent. Then the Guihiba and Quechua of the northern parts of South America, at around 55 percent. Finally there are the descendants of folks who made it all the way to the Amazon basin—the Ticuna, Surui, and Karitiana—with a roughly 70 percent incidence of 7R, the highest in the world. In other words, the descendants of people who, having made it to the future downtown Anchorage, decided to just keep going for another six thousand miles.* A high incidence of 7R, associated with impulsivity and novelty seeking, is the legacy of humans who made the greatest migrations in human history.

And then in the middle of the chart is the near-zero incidence of 7R in China, Cambodia, Japan, and Taiwan (among the Ami and Atayal). When East Asians domesticated rice and invented collectivist society, there was massive selection against the 7R variant; in Kidd's words, it was "nearly lost" in these populations.† Maybe the bearers of 7R broke their necks inventing hang gliding or got antsy and tried to walk to Alaska but drowned because there was no longer a Bering land bridge. Maybe they were less desirable mates. Regardless of the cause, East Asian cultural collectivism coevolved with selection against the 7R variant.‡

Thus, in this most studied of cultural contrasts, we see clustering of ecological factors, modes of production, cultural differences, and differences in endocrinology, neurobiology, and gene frequencies.§ The cultural

* Obviously, no individual actually migrated that far—the slow creep of migration southward in the Western Hemisphere took millennia.

† For genetics fans with more of a background than chapter 8, the near-zero incidence of 7R means that in these cultures there isn't even any benefit to heterozygous versions of 7R.

‡ As noted earlier, within a few generations of immigration, East Asian Americans are typically as individualist as European Americans. This raises the question of whether East Asians who chose to immigrate had a higher frequency of 7R than East Asians in general (one might also wonder whether there is a higher incidence of 7R in the wheat-growing regions of China than in the rice districts). Unfortunately, according to Kenneth Kidd, no one knows about either.

§ Another striking difference in gene variant frequencies concerns the gene coding for the serotonin transporter, which removes serotonin from the synapse and which, as we saw in the last chapter, is associated with impulsive aggression in vastly confusing ways. One variant of the gene is associated with negative emotion, an attentional bias toward negative stimuli, anxiety, and depression risk when coupled with stressful risk factors. Its incidence is less than 50 percent worldwide but 70 to 80 percent in East Asian populations.

contrasts appear in likely ways—e.g., morality, empathy, child-rearing practices, competition, cooperation, definitions of happiness—but also in unexpected ones—e.g., where, within milliseconds, your eyes look at a picture, or you're thinking about bunnies and carrots.

PASTORALISTS AND SOUTHERNERS

Another important link among ecology, mode of production, and culture is seen in dry, hardscrabble, wide-open environments too tough for agriculture. This is the world of nomadic pastoralism—people wandering the desert, steppes, or tundra with their herds.

There are Bedouins in Arabia, Tuareg in North Africa, Somalis and Maasai in East Africa, Sami of northern Scandinavia, Gujjars in India, Yörük in Turkey, Tuvans of Mongolia, Aymara in the Andes. There are herds of sheep, goats, cows, llamas, camels, yaks, horses, or reindeer, with the pastoralists living off their animals' meat, milk, and blood and trading their wool and hides.

Anthropologists have long noted similarities in pastoralist cultures born of their tough environments and the typically minimal impact of centralized government and the rule of law. In that isolated toughness stands a central fact of pastoralism: thieves can't steal the crops on someone's farm or the hundreds of edible plants eaten by hunter-gatherers, but they can steal someone's herd. This is the vulnerability of pastoralism, a world of rustlers and raiders.

This generates various correlates of pastoralism:[26]

> Militarism abounds. Pastoralists, particularly in deserts, with their far-flung members tending the herds, are a spawning ground for warrior classes. And with them typically come (a) military trophies as stepping-stones to societal status; (b) death in battle as a guarantee of a glorious afterlife; (c) high rates of economic polygamy and mistreatment of

women; and (d) authoritarian parenting. It is rare for pastoralists to be pastoral, in the sense of Beethoven's Sixth Symphony.

Worldwide, monotheism is relatively rare; to the extent that it does occur, it is disproportionately likely among desert pastoralists (while rain forest dwellers are atypically likely to be polytheistic). This makes sense. Deserts teach tough, singular things, a world reduced to simple, desiccated, furnace-blasted basics that are approached with a deep fatalism. "I am the Lord your God" and "There is but one god and his name is Allah" and "There will be no gods before me"—dictates like these proliferate. As implied in the final quote, desert monotheism does not always come with only one supernatural being—monotheistic religions are replete with angels and djinns and devils. But they sure come with a hierarchy, minor deities paling before the Omnipotent One, who tends to be highly interventionist both in the heavens and on earth. In contrast, think of tropical rain forest, teeming with life, where you can find more species of ants on a single tree than in all of Britain. Letting a hundred deities bloom in equilibrium must seem the most natural thing in the world.

Pastoralism fosters cultures of honor. As introduced in chapter 7, these are about rules of civility, courtesy, and hospitality, especially to the weary traveler because, after all, aren't all herders often weary travelers? Even more so, cultures of honor are about taking retribution after affronts to self, family, or clan, and reputational consequences for failing to do so. If they take your camel today and you do nothing, tomorrow they will take the rest of your herd, plus your wives and daughters.*

* I once got to experience what this looks like for an extended stretch, as I traveled with a group of Somalis who were driving empty gasoline tankers back from Sudan to the Indian Ocean in Kenya for refilling. At the end of each day of driving through the desert, we'd sit around a campfire between the trucks, cooking a pot of spaghetti and camel's milk. (Why that particular combo? That's a whole other story. . . .) And inevitably one of the six Somalis would do something that was perceived as insulting by someone. There would be snarling, angry words, knives drawn from boots, two guys circling and lunging at each other until everyone else roused themselves to get the two to settle down. And then, the hospitality flip side of the culture on display, everyone would hurry over to make sure I got the best of the spaghetti/milk glob. "Eat, eat. You are our brother," they'd say, including whichever two had just been slashing at each other.

Few of humanity's low or high points are due to the culturally based actions of, say, Sami wandering the north of Finland with their reindeer, or Maasai cow herders in the Serengeti. Instead the most pertinent cultures of honor are ones in more Westernized settings. "Culture of honor" has been used to describe the workings of the Mafia in Sicily, the patterns of violence in rural nineteenth-century Ireland, and the causes and consequences of retributive killings by inner-city gangs. All occur in circumstances of resource competition (including the singular resource of being the last side to do a retributive killing in a vendetta), of a power vacuum provided by the minimal presence of the rule of law, and where prestige is ruinously lost if challenges are left unanswered and where the answer is typically a violent one. Amid those, the most famous example of a Westernized culture of honor is the American South, the subject of books, academic journals, conferences, and Southern studies majors in universities. Much of this work was pioneered by Nisbett.[27]

Hospitality, chivalry toward women, and emphasis on social decorum and etiquette are long associated with the South.[28] In addition, the South traditionally emphasizes legacy, long cultural memory, and continuity of family—in rural Kentucky in the 1940s, for example, 70 percent of men had the same first name as their father, far more than in the North. When coupled with lesser mobility in the South, honor in need of defense readily extends to family, clan, and place. For example, by the time the Hatfields and McCoys famously began their nearly thirty-year feud in 1863,* they had lived in the same region of the West Virginia/Kentucky border for nearly a century. The Southern sense of honor in place is also seen in Robert E. Lee; he opposed Southern secession, even made some ambiguous statements that could be viewed as opposed to slavery. Yet when offered the command of the Union Army by Lincoln, Lee wrote, "I wish to live under no other government and there is no sacrifice I am not ready to make for the preservation of the Union save that of honor." When

* Well, whether the feud actually ended in the 1890s is open to interpretation. While the families declared a truce and stopped the killings in 1891, their descendants battled for a week in 1979 on the game show *Family Feud*. The McCoys won three of the five games, while the Hatfields won more money.

Virginia chose secession, he regretfully fulfilled his sense of honor to his home and led the Confederate Army of Northern Virginia.

In the South, defense of honor was, above all, an act of self-reliance.[29] The Southerner Andrew Jackson was advised by his dying mother to never seek redress from the law for any grievances, to instead be a man and take things into his own hands. That he certainly did, with a history of dueling (even fatally) and brawling; on his final day as president, he articulated two regrets in leaving office—that he "had been unable to shoot Henry Clay or to hang John C. Calhoun." Carrying out justice personally was viewed as a requirement in the absence of a functional legal system. At best, legal justice and individual justice were in uneasy equilibrium in the nineteenth-century South; in the words of the Southern historian Bertram Wyatt-Brown, "Common law and lynch law were ethically compatible. The first enabled the legal profession to present traditional order, and the second conferred upon ordinary men the prerogative of ensuring that community values held ultimate sovereignty."

The core of retribution for honor violations was, of course, violence. Sticks and stones might break your bones, but names will cause you to break the offender's bones. Dueling was commonplace, the point being not that you were willing to kill but that you were willing to die for your honor. Many a Confederate boy went off to war advised by his mother that better he come back in a coffin than as a coward who fled.

The result of this all is a long, still-extant history of high rates of violence in the South. But crucially, violence of a particular type. I once heard it summarized by a Southern studies scholar describing the weirdness of leaving the rural South to start grad school in a strange place, Cambridge, Massachusetts, where families would get together at Fourth of July picnics and *no one would shoot each other*. Nisbett and Dov Cohen have shown that the high rates of violence, particularly of murder, by white Southern males are not features of large cities or about attempts to gain material goods—we're not talking liquor store stickups. Instead the violence is disproportionately rural, among people who know each other, and concerns slights to honor (that sleazebag cousin thought it was okay

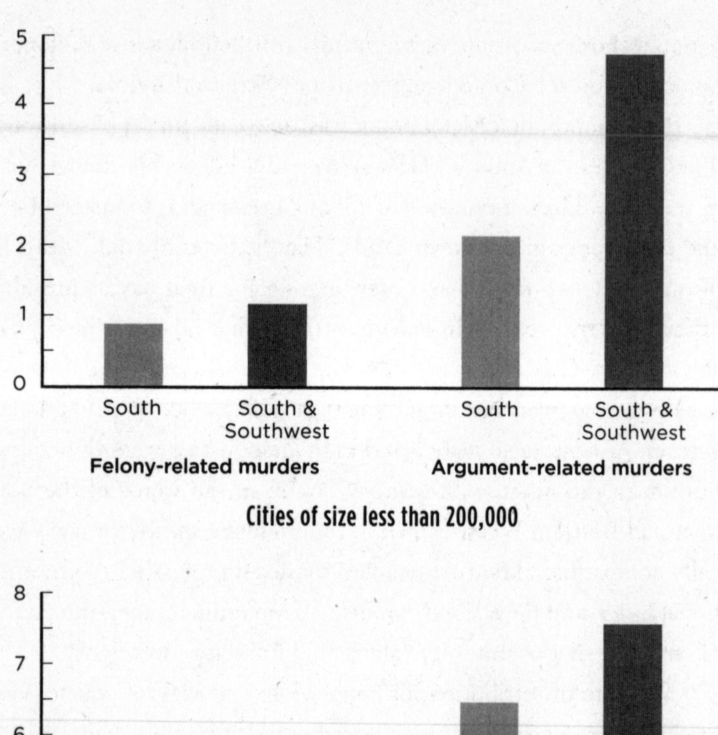

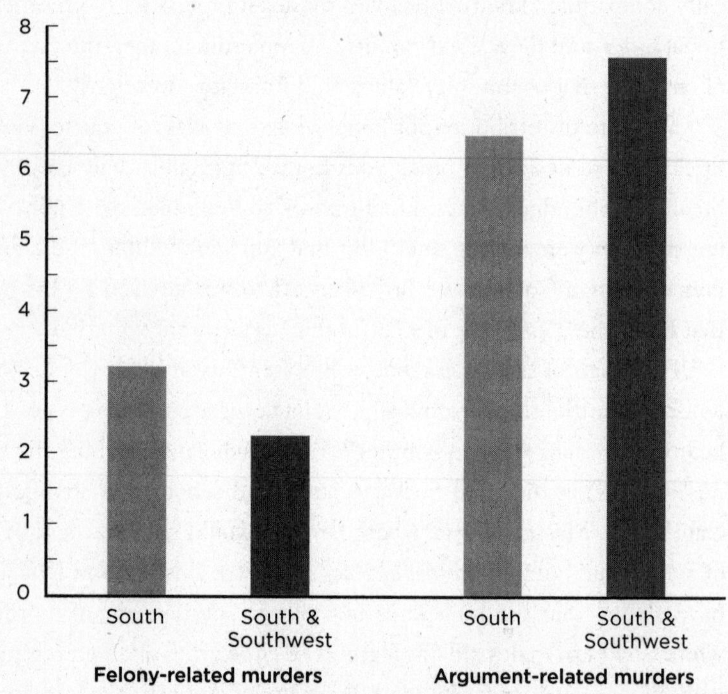

R. Nisbett and D. Cohen, Culture of Honor: The Psychology of Violence in the South *(Boulder, CO: Westview Press, 1996).*

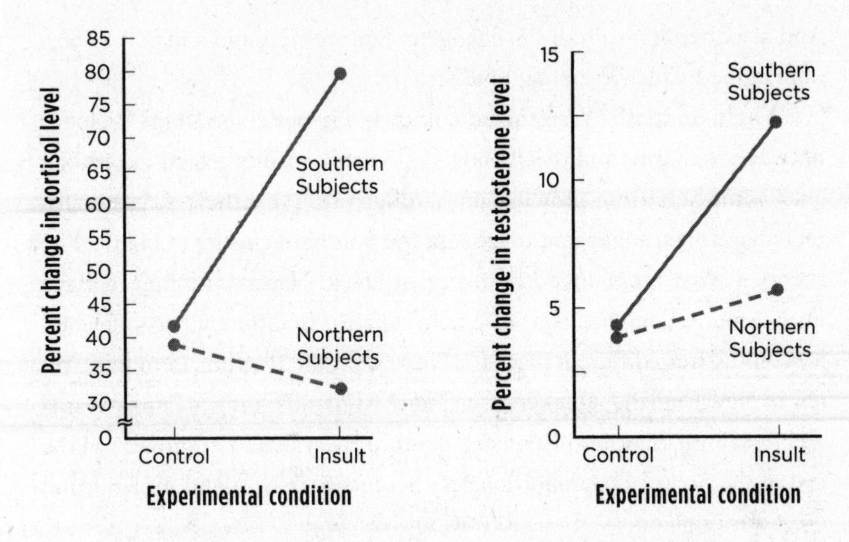

Southern, but not Northern, college studies show strong physiological responses to a social provocation.

to flirt with your wife at the family reunion, so you shot him). Moreover, Southern juries are atypically forgiving of such acts.[30]

Southern violence was explored in one of the all-time coolest psychology studies, involving the use of a word rare in science journals, conducted by Nisbett and Cohen. Undergraduate male subjects had a blood sample taken. They then filled out a questionnaire about something and were then supposed to drop it off down the hall. It was in the narrow hallway, filled with file cabinets, that the experiment happened. Half the subjects traversed the corridor uneventfully. But with half, a confederate (get it? ha-ha) of the psychologists, a big beefy guy, approached from the opposite direction. As the subject and the plant squeezed by each other, the latter would jostle the subject and, in an irritated voice, say the magic word—"asshole"—and march on. Subject would continue down the hall to drop off the questionnaire.

What was the response to this insult? It depended. Subjects from the South, but not from elsewhere, showed massive increases in levels of testosterone and glucocorticoids—anger, rage, stress. Subjects were then told a scenario where a guy observes a male acquaintance making a pass at his fiancée—what happens next in the story? In control subjects, Southerners were a bit more likely than Northerners to imagine a violent outcome.

And after being insulted? No change in Northerners and a massive boost in imagined violence among Southerners.

Where do these Westernized cultures of honor come from? Violence between the Crips and the Bloods in LA is not readily traced to combatants' mind-sets from growing up herding yak. Nonetheless, pastoralist roots have been suggested to explain the Southern culture of honor. The theory as first propounded by historian David Hackett Fischer in 1989: Early American regionalism arose from colonists in different parts of America coming from different places.[31] There were the Pilgrims from East Anglia in New England. Quakers from North Midlands going to Pennsylvania and Delaware. Southern English indentured servants to Virginia. And the rest of the South? Disproportionately herders from Scotland, Ireland, and northern England.

Naturally, the idea has some problems. Pastoralists from the British Isles mostly settled in the hill country of the South, whereas the honor culture is stronger in the Southern lowlands. Others have suggested that the Southern ethos of retributive violence was born from the white Southern nightmare scenario of slave uprisings. But most historians have found a lot of validity in Fischer's idea.

Violence Turned Inward

Culture-of-honor violence is not just about outside threat—the camel rustlers from the next tribe, the jerk at the roadhouse who came on to some guy's girlfriend. Instead it is equally defined by its role when honor is threatened from within. Chapter 11 examines when norm violations by members of your own group provoke cover-ups, excuses, or leniency, and when they provoke severe public punishment. The latter is when "you've dishonored us in front of everyone," a culture-of-honor specialty. Which raises the issue of honor killings.

What constitutes an honor killing? Someone does something considered to tarnish the reputation of the family. A family member then kills the despoiler, often publicly, thereby regaining face. Mind-boggling.

Some characteristics of honor killings:

- While they have been widespread historically, contemporary ones are mostly restricted to traditional Muslim, Hindu, and Sikh communities.
- Victims are usually young women.
- Their most common crimes? Refusing an arranged marriage. Seeking to divorce an abusive spouse and/or a spouse to whom they were forcibly married as a child. Seeking education. Resisting constraining religious orthodoxy, such as covering their head. Marrying, living with, dating, interacting with, or speaking to an unapproved male. Infidelity. Religious conversion. In other words, a woman resisting being the property of her male relatives. And also, stunningly, staggeringly, a frequent cause of honor killings is being raped.
- In the rare instances of men being subject to honor killings, the typical cause is homosexuality.

There has been debate as to whether honor killings are "just" domestic violence, and whether morbid Western fascination with them reflects anti-Muslim bias;[32] if some Baptist guy in Alabama murders his wife because she wants a divorce, no one frames it as a "Christian honor killing" reflecting deep religious barbarity. But honor killings typically differ from garden-variety domestic violence in several ways: (a) The latter is usually committed by a male partner; the former are usually committed by male blood relatives, often with the approval of and facilitation by female relatives. (b) The former is rarely an act of spontaneous passion but instead is often planned with the approval of family members. (c) Honor killings are often rationalized on religious grounds, presented without remorse, and approved by religious leaders. (d) Honor killings are carried out openly—after all, how else can "honor" be regained for the family?—and the chosen perpetrator is often an underage relative (e.g., a younger brother), to minimize the extent of sentencing for the act.

Left to right, starting top column: Shafilea Ahmed, England, killed by father and mother after resisting an arranged marriage; age 17. Anooshe Sediq Ghulam, Norway, married at 13; killed by husband after requesting a divorce; age 22. Palestina Isa, USA, killed by parents for dating someone outside the faith, listening to American music, and secretly getting a part-time job; age 16. Aqsa Parvez, Canada, killed by father and brother for refusing to wear a hajib; age 16. Ghazala Khan, Denmark, killed by nine family members for refusing an arranged marriage; age 19. Fadime Sahindal, Sweden, killed by father for refusing an arranged marriage; age 27. Hatun Surucu Kird, Germany, killed by brother after divorcing the cousin she was forced to marry at age 16; age 23. Hina Salem, Italy, killed by father for refusing an arranged marriage; age 20. Amina and Sarah Said, USA, both sisters killed by parents who perceived them as becoming too Westernized; ages 18 and 17.

By some pretty meaningful criteria, this is not "just" domestic violence. According to estimates by the UN and advocacy groups, five to twenty thousand honor killings occur annually. And they are not restricted to far-off, alien lands. Instead they occur throughout the West, where patriarchs expect their daughters to be untouched by the world they moved them to, where a daughter's successful assimilation into this world proclaims the irrelevance of that patriarch.

STRATIFIED VERSUS EGALITARIAN CULTURES

Another meaningful way to think about cross-cultural variation concerns how unequally resources (e.g., land, food, material goods, power, or prestige) are distributed.[33] Hunter-gatherer societies have typically been egalitarian, as we'll soon see, throughout hominin history. Inequality emerged when "stuff"—things to possess and accumulate—was invented following animal domestication and the development of agriculture. The more stuff, reflecting surplus, job specialization, and technological sophistication, the greater the potential inequality. Moreover, inequality expands enormously when cultures invent inheritance within families. Once invented, inequality became pervasive. Among traditional pastoralist or small-scale agricultural societies, levels of wealth inequality match or exceed those in the most unequal industrialized societies.

Why have stratified cultures dominated the planet, generally replacing more egalitarian ones? For population biologist Peter Turchin, the answer is that stratified cultures are ideally suited to being conquerors—they come with chains of command.[34] Both empirical and theoretical work suggests that in addition, in unstable environments stratified societies are "better able to survive resource shortages [than egalitarian cultures] by sequestering mortality in the lower classes." In other words, when times are tough, the unequal access to wealth becomes the unequal distribution of misery and death. Notably, though, stratification is not the only solution to such instability—this is where hunter-gatherers benefit from being able to pick up and move.

A score of millennia after the invention of inequality, Westernized societies at the extremes of the inequality continuum differ strikingly.

One difference concerns "social capital." Economic capital is the collective quantity of goods, services, and financial resources. Social capital is the collective quantity of resources such as trust, reciprocity, and cooperation. You learn a ton about a community's social capital with two simple questions. First: "Can people usually be trusted?" A community in which most people answer yes is one with fewer locks, with people watching out for one another's kids and intervening in situations where one could easily look away. The second question is how many organizations someone participates in—from the purely recreational (e.g., a bowling league) to the vital (e.g., unions, tenant groups, co-op banks). A community with high levels of such participation is one where people feel efficacious, where institutions work transparently enough that people believe they can effect change. People who feel helpless don't join organizations.

Put simply, cultures with more income inequality have less social capital.[35] Trust requires reciprocity, and reciprocity requires equality, whereas hierarchy is about domination and asymmetry. Moreover, a culture highly unequal in material resources is almost always also unequal in the ability to pull the strings of power, to have efficacy, to be visible. (For example, as income inequality grows, the percentage of people who bother voting generally declines.) Almost by definition, you can't have a society with both dramatic income inequality and plentiful social capital. Or translated from social science–ese, marked inequality makes people crummier to one another.

This can be shown in various ways, studied on the levels of Westernized countries, states, provinces, cities, and towns. The more income inequality, the less likely people are to help someone (in an experimental setting) and the less generous and cooperative they are in economic games. Early in the chapter, I discussed cross-cultural rates of bullying and of "antisocial punishment," where people in economic games punish overly generous players more than they punish cheaters.* Studies of these phenomena show that high levels of inequality and/or low levels of social

* What's antisocial punishment about? The general interpretation is that people are being punished for being generous because they make everyone else look bad and increase the expectation of generosity from everyone else.

capital in a country predict high rates of bullying and of antisocial punishment.[36]

Chapter 11 examines the psychology with which we think about people of different socioeconomic status; no surprise, in unequal societies, people on top generate justifications for their status.[37] And the more inequality, the more the powerful adhere to myths about the hidden blessings of subordination—"They may be poor, but at least they're happy/honest/loved." In the words of the authors of one paper, "Unequal societies may need ambivalence for system stability: Income inequality compensates groups with partially positive social images."

Thus unequal cultures make people less kind. Inequality also makes people less healthy. This helps explain a hugely important phenomenon in public health, namely the "socioeconomic status (SES)/health gradient"—as noted, in culture after culture, the poorer you are, the worse your health, the higher the incidence and impact of numerous diseases, and the shorter your life expectancy.[38]

Extensive research has examined the SES/health gradient. Four quick rule-outs: (a) The gradient isn't due to poor health driving down people's SES. Instead low SES, beginning in childhood, predicts subsequent poor health in adulthood. (b) It's not that the poor have lousy health and everyone else is equally healthy. Instead, for every step down the SES ladder, starting from the top, average health worsens. (c) The gradient isn't due to less health-care access for the poor; it occurs in countries with universal health care, is unrelated to utilization of health-care systems, and occurs for diseases unrelated to health-care access (e.g., juvenile diabetes, where having five checkups a day wouldn't change its incidence). (d) Only about a third of the gradient is explained by lower SES equaling more health risk factors (e.g., lead in your water, nearby toxic waste dump, more smoking and drinking) and fewer protective factors (e.g., everything from better mattresses for overworked backs to health club memberships).

What then is the principal cause of the gradient? Key work by Nancy Adler at UCSF showed that it's not so much *being* poor that predicts poor health. It's *feeling* poor—someone's subjective SES (e.g., the answer to "How do you feel you're doing financially when you compare yourself with other people?") is at least as good a predictor of health as is objective SES.

Crucial work by the social epidemiologist Richard Wilkinson of the University of Nottingham added to this picture: it's not so much that poverty predicts poor health; it's poverty amid plenty—income inequality. The surest way to make someone feel poor is to rub their nose in what they don't have.

Why should high degrees of income inequality (independent of absolute levels of poverty) make the poor unhealthy? Two overlapping pathways:

A *psychosocial* explanation has been championed by Ichiro Kawachi of Harvard. When social capital decreases (thanks to inequality), up goes psychological stress. A mammoth amount of literature explores how such stress—lack of control, predictability, outlets for frustration, and social support—chronically activates the stress response, which, as we saw in chapter 4, corrodes health in numerous ways.

A *neomaterialist* explanation has been offered by Robert Evans of the University of British Columbia and George Kaplan of the University of Michigan. If you want to improve health and quality of life for the average person in a society, you spend money on public goods—better public transit, safer streets, cleaner water, better public schools, universal health care. But the more income inequality, the greater the financial distance between the wealthy and the average and thus the less direct benefit the wealthy feel from improving public goods. Instead they benefit more from dodging taxes and spending on their private good—a chauffeur, a gated community, bottled water, private schools, private health insurance. As Evans writes, "The more unequal are incomes in a society, the more pronounced will be the disadvantages to its better-off members from public expenditure, and the more resources will those members have [available to them] to mount effective political opposition" (e.g., lobbying). Evans notes how this "secession of the wealthy" promotes "private affluence and public squalor." Meaning worse health for the have-nots.[39]

The inequality/health link paves the way for understanding how inequality also makes for more crime and violence. I could copy and paste the previous stretch of writing, replacing "poor health" with "high crime," and I'd be set. Poverty is not a predictor of crime as much as poverty amid plenty is. For example, extent of income inequality is a major predictor of rates of violent crime across American states and across industrialized nations.[40]

Why does income inequality lead to more crime? Again, there's the psychosocial angle—inequality means less social capital, less trust, cooperation, and people watching out for one another. And there's the neomaterialist angle—inequality means more secession of the wealthy from contributing to the public good. Kaplan has shown, for example, that states with more income inequality spend proportionately less money on that key crime-fighting tool, education. As with inequality and health, the psychosocial and neomaterial routes synergize.

A final depressing point about inequality and violence. As we've seen, a rat being shocked activates a stress response. But a rat being shocked who can then bite the hell out of another rat has less of a stress response. Likewise with baboons—if you are low ranking, a reliable way to reduce glucocorticoid secretion is to displace aggression onto those even lower in the pecking order. It's something similar here—despite the conservative nightmare of class warfare, of the poor rising up to slaughter the wealthy, when inequality fuels violence, it is mostly the poor preying on the poor.

This point is made with a great metaphor for the consequences of societal inequality.[41] The frequency of "air rage"—a passenger majorly, disruptively, dangerously losing it over something on a flight—has been increasing. Turns out there's a substantial predictor of it: if the plane has a first-class section, there's almost a fourfold increase in the odds of a coach passenger having air rage. Force coach passengers to walk through first class when boarding, and you more than double the chances further.

Nothing like starting a flight by being reminded of where you fit into the class hierarchy. And completing the parallel with violent crime, when air rage is boosted in coach by reminders of inequality, the result is not a crazed coach passenger sprinting into first class to shout Marxist slogans. It's the guy being awful to the old woman sitting next to him, or to the flight attendant.*

POPULATION SIZE, POPULATION DENSITY, POPULATION HETEROGENEITY

The year 2008 marked a human milestone, a transition point nine thousand years in the making: for the first time, the majority of humans lived in cities.

The human trajectory from semipermanent settlements to the megalopolis has been beneficial. In the developed world, when compared with rural populations, city dwellers are typically healthier and wealthier; larger social networks facilitate innovation; because of economies of scale, cities leave a smaller per-capita ecological footprint.[42]

Urban living makes for a different sort of brain. This was shown in a 2011 study of subjects from a range of cities, towns, and rural settings who underwent an experimental social stressor while being brain-scanned. The key finding was that the larger the population where someone lived, the more reactive their amygdala was during that stressor.†[43]

Most important for our purposes, urbanized humans do something completely unprecedented among primates—regularly encountering strangers who are never seen again, fostering the invention of the anonymous act. After all, it wasn't until nineteenth-century urbanization that crime fiction was invented, typically set in cities—in traditional settings there's no whodunit, since everyone knows what everyone dun.

* Ironic footnote: when coach passengers board through first class, the rate of air rage related to a sense of entitlement increases even more among the first-class passengers.

† The paper generated an astonishing number of articles in the lay press whose titles were variants on "Stress and the City."

Growing cultures had to invent mechanisms for norm enforcement among strangers. For example, across numerous traditional cultures, the larger the group, the greater the punishment for norm violations and the more cultural emphases on equitable treatment of strangers. Moreover, larger groups evolved "third-party punishment" (stay tuned for more in the next chapter)—rather than victims punishing norm violators, punishment is meted out by objective third parties, such as police and courts. At an extreme, a crime not only victimizes its victim but also is an affront to the collective population—hence "The People Versus Joe Blow."*[44]

Finally, life in larger populations fosters the ultimate third-party punisher. As documented by Ara Norenzayan of the University of British Columbia, it is only when societies grow large enough that people regularly encounter strangers that "Big Gods" emerge—deities who are concerned with human morality and punish our transgressions.[45] Societies with frequent anonymous interactions tend to outsource punishment to gods.† In contrast, hunter-gatherers' gods are less likely than chance to care whether we've been naughty or nice. Moreover, in further work across a range of traditional cultures, Norenzayan has shown that the more informed and punitive people consider their moralistic gods to be, the more generous they are to coreligionist strangers in a financial allocation game.

Separate from the size of a population, how about its density? One study surveying thirty-three developed countries characterized each nation's "tightness"—the extent to which the government is autocratic, dissent suppressed, behavior monitored, transgressions punished, life regulated by religious orthodoxy, citizens viewing various behaviors as inappropriate (e.g., singing in an elevator, cursing at a job interview).[46] Higher population density predicted tighter cultures—both high density in the present and, remarkably, historically, in the year 1500.

The issue of population density's effects on behavior gave rise to a well-known phenomenon, mostly well known incorrectly.

* The online world is now undergoing cultural evolution, wrestling with how to deal with the toxic behavior of some people online when they are shielded behind anonymity. Psychologists are even doing experiments, gifted with mammoth databases, to see how best to curb such behavior with top-down approaches (e.g., being banned by authorities) and interventions driven by peers (i.e., other players).

† And there are remarkable similarities among such moralizing religions.

In the 1950s John Calhoun at the National Institute of Mental Health asked what happens to rat behavior at higher population densities, research prompted by America's ever-growing cities.[47] And in papers for both scientists and the lay public, Calhoun gave a clear answer: high-density living produced "deviant" behavior and "social pathology." Rats became violent; adults killed and cannibalized one another; females were aggressive to their infants; there was indiscriminate hypersexuality among males (e.g., trying to mate with females who weren't in estrus).

The writing about the subject, starting with Calhoun, was colorful. The bloodless description of "high-density living" was replaced with "crowding." Aggressive males were described as "going berserk," aggressive females as "Amazons." Rats living in these "rat slums" became "social dropouts," "autistic," or "juvenile delinquents." One expert on rat behavior, A. S. Parkes, described Calhoun's rats as "unmaternal mothers, homosexuals and zombies" (quite the trio you'd invite to dinner in the 1950s).[48]

The work was hugely influential, taught to psychologists, architects, and urban planners; more than a million reprints were requested of Calhoun's original *Scientific American* report; sociologists, journalists, and politicians explicitly compared residents of particular housing projects and Calhoun's rats. The take-home message sent ripples through the American heartland destined for the chaotic sixties: inner cities breed violence, pathology, and social deviance.

Calhoun's rats were more complicated than this (something underemphasized in his lay writing). High-density living doesn't make rats more aggressive. Instead it makes aggressive rats more aggressive. (This echoes the findings that neither testosterone, nor alcohol, nor media violence uniformly increases violence. Instead they make violent individuals more sensitive to violence-evoking social cues.) In contrast, crowding makes unaggressive individuals more timid. In other words, it exaggerates preexisting social tendencies.

Calhoun's erroneous conclusions about rats don't even hold for humans. In some cities—Chicago, for example, circa 1970—higher population density in neighborhoods does indeed predict more violence. Nevertheless, some of the highest-density places on earth—Hong Kong,

Singapore, and Tokyo—have miniscule rates of violence. High-density living is not synonymous with aggression in rats or humans.

The preceding sections examined the effects of living with lots of people, and in close quarters. How about the effects of living with different *kinds* of people? Diversity. Heterogeneity. Admixture. Mosaicism.

Two opposite narratives come to mind:

> Mister Rogers' neighborhood: When people of differing ethnicities, races, or religions live together, they experience the similarities rather than the differences and view one another as individuals, transcending stereotypes. Trade flows, fostering fairness and mutuality. Inevitably, dichotomies dissolve with intermarriage, and soon you're happily watching your grandkid in the school play on "their" side of town. Just visualize whirled peas.

> Sharks versus the Jets: Differing sorts of people living in close proximity rub, and thus abrade, elbows regularly. One side's act of proud cultural identification feels like a hostile dig to the other side, public spaces become proving grounds for turf battles, commons become tragedies.

Surprise: both outcomes occur; the final chapter explores circumstances where intergroup contact leads to one rather than the other. Most interesting at this juncture is the importance of the spatial qualities of the heterogeneity. Consider a region filled with people from Elbonia and Kerplakistan, two hostile groups, each providing half the population. At one extreme, the land is split down the middle, each group occupying one side, producing a single boundary between the two. At the other extreme is a microcheckerboard of alternating ethnicities, where each square on the grid is one person large, producing a vast quantity of boundaries between Elbonians and Kerplakis.

Intuitively, both scenarios should bias against conflict. In the condition of maximal separation, each group has a critical mass to be locally sovereign, and the total length of border, and thus of the amount of intergroup elbow rubbing, is minimized. In the scenario of maximal mixing, no patch of ethnic homogeneity is big enough to foster a self-identity that can dominate a public space—big deal if someone raises a flag between their feet and declares their square meter to be the Elbonian Empire or a Kerplakistani Republic.

But in the real world things are always in between the two extremes, and with variation in the average size of each "ethnic patch." Does patch size, and thus amount of border, influence relations?

This was explored in a fascinating paper from the aptly named New England Complex Systems Institute, down the block from MIT.[49] The authors first constructed an Elbonian/Kerplaki mixture, with individuals randomly distributed as pixels on a grid. Pixels were given a certain degree of mobility plus a tendency to assort with other pixels of the same type. As self-assortment progresses, something emerges—islands and peninsulas of Elbonians amid a sea of Kerplakis, or the reverse, a condition that intuitively seems rife with potential intergroup violence. As self-assortment continues, the number of such isolated islands and peninsulas declines. The intermediate stage that maximizes the number of islands and peninsulas maximizes the number of people living within a surrounded enclave.*

The authors then considered a balkanized region, namely the Balkans, ex-Yugoslavia, in 1990. This was just before Serbians, Bosnians, Croatians, and Albanians commenced Europe's worst war since World War II, the war that taught us the names of places like Srebrenica and people like Slobodan Milošević. Using a similar analysis, with ethnic island size varying from roughly twenty to sixty kilometers in diameter, they identified the spots theoretically most rife for violence; remarkably, this predicted the sites of major fighting and massacres in the war.

* The authors used math straight out of chemistry for analyzing the extent of mixing between different types of solutions, plus some math from physics usually used to disentangle the contributions made by overlapping waves. I understood exactly zero of any of this and am putting faith in the vetting process of the journal, *Science*, the most selective science journal in the country.

In the words of the authors, violence can arise "due to the structure of boundaries between groups rather than as a result of inherent conflicts between the groups themselves." They then showed that the *clarity* of borders matters as well. Good, clear-cut fences—e.g., mountain ranges or rivers between groups—make for good neighbors. "Peace does not depend on integrated coexistence, but rather on well defined topographical and political boundaries separating groups, allowing for partial autonomy within a single country," the authors concluded.

Thus, not only do size, density, and heterogeneity of populations help explain intergroup violence, but patterns and clarity of fragmentation do as well. These issues will be revisited in the final chapter.

THE RESIDUES OF CULTURAL CRISES

In times of crisis—the London Blitz, New York after 9/11, San Francisco after the 1989 Loma Prieta earthquake—people pull together.* That's cool. But in contrast, chronic, pervasive, corrosive menace doesn't necessarily do the same to people or cultures.

The primal menace of hunger has left historical marks. Back to that study of differences between countries' tightness (where "tight" countries were characterized by autocracy, suppression of dissent, and omnipresence and enforcement of behavior norms).[50] What sorts of countries are tighter?† In addition to the high population-density correlates mentioned earlier, there are also more historical food shortages, lower food intake, and lower levels of protein and fat in the diet. In other words, these are cultures chronically menaced by empty stomachs.

* I was in San Francisco for the quake, and much was made of the fact that fancy downtown hotels opened their doors to house people needing shelter. It's worth noting that this generosity was for people *made* homeless by the quake, not people who were *already* homeless. For them the earthquake was just another day of scrabbling. The hotels supposedly required a credit card from people, not because they'd be charged for the room, but as evidence that this was the sort of person whose homelessness mattered. This well could have been apocryphal; it's hard to imagine that the staff at reception needed to see someone's plastic to tell the difference.

† What were the "tightest" countries? Pakistan, Malaysia, India, Singapore, and South Korea. The least tight? Ukraine, Estonia, Hungary, Israel, and the Netherlands.

Cultural tightness was also predicted by environmental degradation—less available farmland or clean water, more pollution. Similarly, habitat degradation and depletion of animal populations worsens conflict in cultures dependent on bush meat. And a major theme of Jared Diamond's magisterial *Collapse: How Societies Choose to Fail or Succeed* is how environmental degradation explains the violent collapse of many civilizations.

Then there's disease. In chapter 15 we'll touch on "behavioral immunity," the ability of numerous species to detect cues of illness in other individuals; as we'll see, implicit cues about infectious disease make people more xenophobic. Similarly, historical prevalence of infectious disease predicts a culture's openness to outsiders. Moreover, other predictors of cultural tightness include having high historical incidence of pandemics, of high infant and child mortality rates, and of higher cumulative average number of years lost to communicable disease.

Obviously, weather influences the incidence of organized violence—consider the centuries of European wars taking a hiatus during the worst of winter and the growing season.[51] Even broader is the capacity of weather and climate to shape culture. The Kenyan historian Ali Mazrui has suggested that one reason for Europe's historical success, relative to Africa, has been the weather—Western-style planning ahead arose from the annual reality of winter coming.* Larger-scale changes in weather are known to be consequential. In the tightness study, cultural tightness was also predicted by a history of floods, droughts, and cyclones. Another pertinent aspect of weather concerns the Southern Oscillation, known as El Niño, the multiyear fluctuation of average water temperatures in the equatorial Pacific Ocean. El Niños, occurring about every dozen years, involve warmer, drier weather (with the opposite during La Niña years) and are associated in many developing countries with droughts and food shortages. Over the last fifty years El Niños have roughly doubled the likelihood of civil conflict, mostly by stoking the fires of preexisting conflicts.

* As a counter to this, though, people in the tropics also have to foresee annual fluctuations in weather, and no Swede has ever had to plan ahead for the monsoon season.

The relationship between drought and violence is tricky. The civil conflict referred to in the previous paragraph concerned deaths caused by battle between governmental and nongovernmental forces (i.e., civil wars or insurgencies). Thus, rather than fighting over a watering hole or a field for grazing, this was fighting for modern perks of power. But in traditional settings drought may mean spending more time foraging or hauling water for your crops. Raiding to steal the other group's women isn't a high priority, and why rustle someone else's cows when you can't even feed your own? Conflict declines.

Interestingly, something similar occurs in baboons. Normally, baboons in rich ecosystems like the Serengeti need forage only a few hours a day. Part of what endears baboons to primatologists is that this leaves them about nine hours daily to devote to social machinations—trysting and jousting and backbiting. In 1984 there was a devastating drought in East Africa. Among baboons, while there was still sufficient food, it took every waking moment to get enough calories; aggression decreased.[52]

So ecological duress can increase or decrease aggression. This raises the key issue of what global warming will do to our best and worst behaviors. There will definitely be some upsides. Some regions will have longer growing seasons, increasing the food supply and reducing tensions. Some people will eschew conflict, being preoccupied with saving their homes from the encroaching ocean or growing pineapples in the Arctic. But amid squabbling about the details in predictive models, the consensus is that global warming won't do good things to global conflict. For starters, warmer temperatures rile people up—in cities during the summers, for every three degree increase in temperature, there was a 4 percent increase in interpersonal violence and 14 percent in group violence. But global warming's bad news is more global—desertification, loss of arable land due to rising seas, more droughts. One influential meta-analysis projected 16 percent and 50 percent increases in interpersonal and group violence, respectively, in some regions by 2050.[53]

OH, WHY NOT: RELIGION

Time for a quick hit-and-run about religion before considering it in the final chapter.

Theories abound as to why humans keep inventing religions. It's more than a human pull toward the supernatural; as stated in one review, "Mickey Mouse has supernatural powers, but no one worships or would fight—or kill—for him. Our social brains may help explain why children the world over are attracted to talking teacups, but religion is much more than that." Why does religion arise? Because it makes in-groups more cooperative and viable (stay tuned for more in the next chapter). Because humans need personification and to see agency and causality when facing the unknown. Or maybe inventing deities is an emergent by-product of the architecture of our social brains.[54]

Amid these speculations, far more boggling is the variety of the thousands of religions we've invented. They vary as to number and gender of deities; whether there's an afterlife, what it's like, and what it takes to enter; whether deities judge or interfere with humans; whether we are born sinful or pure and whether sexuality changes those states; whether the myth of a religion's founder is of sacredness from the start (so much so that, say, wise men visit the infant founder) or of a sybarite who reforms (e.g., Siddhārtha's transition from palace life to being the Buddha); whether the religion's goal is attracting new followers (say, with exciting news—e.g., an angel visited me in Manchester, New York, and gave me golden plates) or retaining members (we've got a covenant with God, so stick with us). On and on.

There are some pertinent patterns amid this variation. As noted, desert cultures are prone toward monotheistic religions; rain forest dwellers, polytheistic ones. Nomadic pastoralists' deities tend to value war and valor in battle as an entrée to a good afterlife. Agriculturalists invent gods who alter the weather. As noted, once cultures get large enough that anonymous acts are possible, they start inventing moralizing gods. Gods and religious orthodoxy dominate more in cultures with frequent threats (war, natural disasters), inequality, and high infant mortality rates.

Before turfing this subject to the final chapter, three obvious points: (a) a religion reflects the values of the culture that invented or adopted it, and very effectively transmits those values; (b) religion fosters the best and worst of our behaviors; (c) it's complicated.

We've now looked at various cultural factors—collectivism versus individualism, egalitarian versus hierarchical distribution of resources, and so on. While there are others to consider, it's time to shift to the chapter's final topic. This is one that has generated shit storms of debate as old as the weathered layers of Olduvai Gorge and as fresh as a newborn baby's tush, a topic that has scientists who study peace at one another's throats.

HOBBES OR ROUSSEAU

Yes, those guys.

To invoke some estimates, anatomically modern humans emerged about 200,000 years ago, and behaviorally modern ones about 40,000 to 50,000 years ago; animal domestication is 10,000 to 20,000 years old, agriculture around 12,000. After plant domestication, it was roughly 5,000 more years until "history" began with civilizations in Egypt, the Mideast, China, and the New World. When in this arc of history was war invented? Does material culture lessen or worsen tendencies toward war? Do successful warriors leave more copies of their genes? Has the centralization of authority by civilization actually civilized us, providing a veneer of socially contractual restraint? Have humans become more or less decent to one another over the course of history? Yes, it's short/nasty/brutish versus noble savage.

In contrast to the centuries of food fights among philosophers, contemporary Hobbes-versus-Rousseau is about actual data. Some of it is archaeological, where researchers have sought to determine the prevalence and antiquity of warfare from the archaeological record.

Predictably, half of each conference on the subject consists of definitional squabbles. Is "war" solely organized and sustained violence between groups? Does it require weapons? A standing army (even if only seasonally)? An army with hierarchy and chain of command? If fighting is mostly along lines of relatedness, is it a vendetta or clan feud instead of a war?

Fractured Bones

For most archaeologists the operational definition has been streamlined to numerous people simultaneously meeting violent deaths. In 1996 the archaeologist Lawrence Keeley of the University of Illinois synthesized the existing literature in his highly influential *War Before Civilization: The Myth of the Peaceful Savage*, ostensibly showing that the archaeological evidence for war is broad and ancient.[55]

A similar conclusion comes in the 2011 book *The Better Angels of Our Nature: Why Violence Has Declined*, by Harvard's Steven Pinker.[56] Cliché police be damned, you can't mention this book without calling it "monumental." In this monumental work Pinker argued that (a) violence and the worst horrors of inhumanity have been declining for the last half millennium, thanks to the constraining forces of civilization; and (b) the warfare and barbarity preceding that transition are as old as the human species.

Keeley and Pinker document savagery galore in prehistoric tribal societies—mass graves filled with skeletons bearing multiple fractures, caved-in skulls, "parrying" fractures (which arise from raising your arm to fend off a blow), stone projectiles embedded in bone. Some sites suggest the outcome of battle—a preponderance of young adult male skeletons. Others suggest indiscriminate massacre—butchered skeletons of both sexes and all ages. Other sites suggest cannibalism of the vanquished.

In their separate surveys of the literature, Keeley and Pinker present evidence of prestate tribal violence comes from sites in Ukraine, France, Sweden, Niger, India, and numerous precontact American locations.[57] This collection of sites includes the oldest such massacre, the 12,000- to 14,000-year-old Jebel Sahaba site along the Nile in northern Sudan, a cemetery of fifty-nine men, women, and children, nearly half of whom

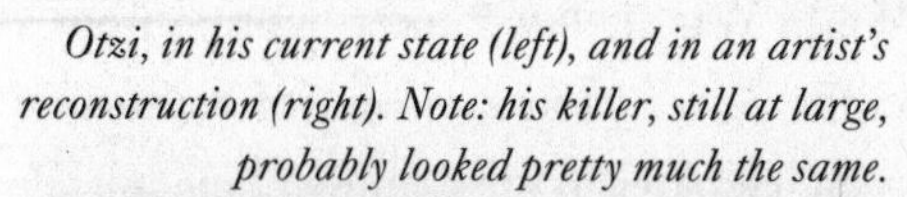

Otzi, in his current state (left), and in an artist's reconstruction (right). Note: his killer, still at large, probably looked pretty much the same.

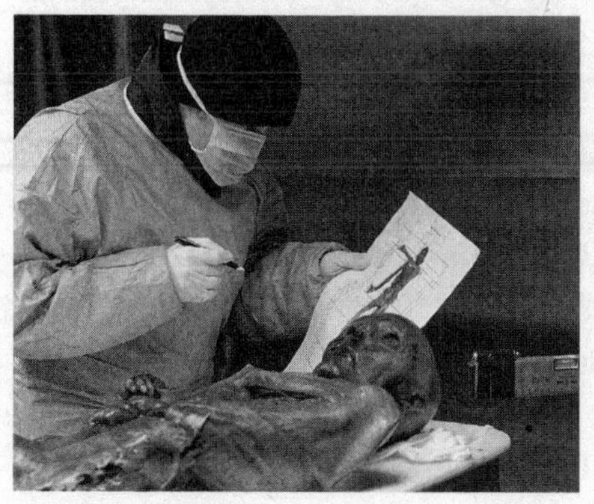

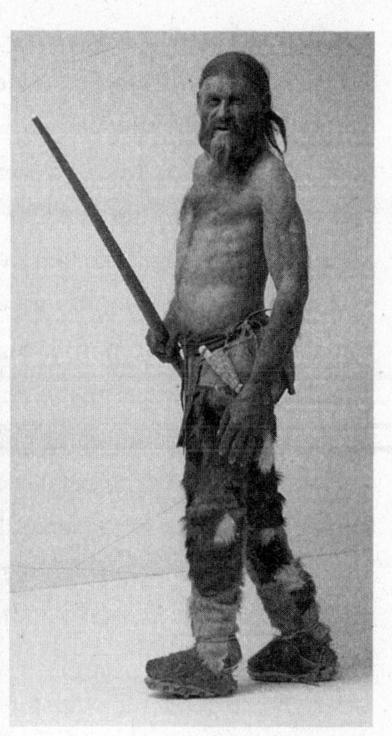

have stone projectiles embedded in their bones. And it includes the largest massacre site, the 700-year-old Crow Creek in South Dakota, a mass grave of more than four hundred skeletons, with 60 percent showing evidence of violent deaths. Across the twenty-one sites surveyed, about 15 percent of skeletons showed evidence of "death in warfare." One can, of course, be killed in war in a way that doesn't leave fractures or projectiles embedded in bone, suggesting that the percentage of deaths due to warfare was higher.

Keeley and Pinker also document how prehistoric settlements frequently protected themselves with defensive palisades and fortifications. And, of course, as the poster child for prehistoric violence, there is Otzi, the 5,300-year-old Tyrolean "iceman" found in a melting glacier in 1991 on the Italian/Austrian border. In his shoulder was a freshly embedded arrowhead.

Thus, Keeley and Pinker document mass casualties of warfare long predating civilization. Just as important, both (starting with Keeley's subtitle) suggest a hidden agenda among archaeologists to ignore that evidence. Why had there been, to use Keeley's phrase, "pacification of the past"? In chapter 7 we saw how World War II produced a generation of social scientists trying to understand the roots of fascism. In Keeley's view, the post–World War II generations of archaeologists recoiled from the trauma of the war by recoiling from the evidence that humans had been prepping a long time for World War II. For Pinker, writing from a younger generation's perspective, the current whitewashing of prehistoric violence has the flavor of today's archaeological graybeards being nostalgic about getting stoned in high school and listening to John Lennon's "Imagine."

Keeley and Pinker generated a strong backlash among many notable archaeologists, who charged them with "war-ifying the past." Most vocal has been R. Brian Ferguson of Rutgers University, with publications with titles like "Pinker's List: Exaggerating Prehistoric War Mortality." Keeley and Pinker are criticized for numerous reasons:[58]

a. Some of the sites supposedly presenting evidence for warfare actually contain only a single case of violent death, suggesting homicide, not war.

b. The criteria for inferring violent death include skeletons in close proximity to arrowheads. However, many such artifacts were actually tools used for other purposes, or simply chips and flakes. For example, Fred Wendorf, who excavated Jebel Sahaba, considered most of the projectiles associated with skeletons to have been mere debris.[59]

c. Many fractured bones were actually healed. Instead of reflecting war, they might indicate the ritualized club fighting seen in many tribal societies.

d. Proving that a human bone was gnawed on by a fellow human instead of another carnivore is tough. One tour-de-force paper demonstrated cannibalism in a Pueblo village from around the year 1100—human feces there contained the human version of the muscle-specific protein myoglobin.[60] In other words, those humans had been eating human meat. Nonetheless, even when cannibalism is clearly documented, it doesn't indicate whether there was exo- or endocannibalism (i.e., eating vanquished enemies versus deceased relatives, as is done in some tribal cultures).

e. Most important, Keeley and Pinker are accused of cherry-picking their data, discussing only sites of putative war deaths, rather than the entire literature.* When you survey the thousands of prehistoric skeletal remains from hundreds of sites worldwide, rates of violent deaths are far lower than 15 percent. Moreover, there are regions and periods lacking any evidence of warlike violence. The glee in refuting the broadest conclusions of Keeley and Pinker is unmistakable (e.g., Ferguson in the previously cited work: "For 10,000 years in the Southern Levant, *there is not one single instance where it can be said with confidence, 'war was there.'* [his emphasis] Am I wrong? Name the place."). Thus these critics conclude that wars were rare prior to human civilizations. Supporters of Keeley and Pinker retort that you can't ignore bloodbaths like Crow Creek or Jebel Sahaba and that the absence of proof (of early war in so many of these sites) is not proof of absence.

This suggests a second strategy for contemporary Hobbes-versus-Rousseau debates, namely to study contemporary humans in prestate tribal societies. How frequently do they war?

Prehistorians in the Flesh

Well, if researchers endlessly argue about who or what gnawed on a ten-thousand-year-old human bone, imagine the disagreements about actual living humans.

Keeley and Pinker, along with Samuel Bowles of the Santa Fe Institute, conclude that warfare is nearly universal in contemporary nonstate

---

* Pinker's response to the cherry-picking charge is as follows: "*Better Angels* reports all the published estimates of per capita rates of violent death in the archaeological and anthropological literature I could find." S. Pinker, "Violence: Clarified," *Sci* 338 (2012): 327. If I understand what he is saying accurately, this feels a bit facile. To be facetious, this would be like not including Quakers in one's analysis of violence because no one studying them had published something along the lines of "Estimated per-capita rates of death in Quaker communities due to gangland-style executions in nightclubs: zero; due to targeted drone missile strikes: zero; due to dirty bombs made with stolen plutonium: zero . . ."

Clockwise, top left: *New Guinea, Masai, Amazonian, Zulu*

societies. This is the world of headhunters in New Guinea and Borneo, Maasai and Zulu warriors in Africa, Amazonians on raiding parties in the rain forest. Keeley estimates that, in the absence of pacification enforced by outside forces such as a government, 90 to 95 percent of tribal societies engage in warfare, many constantly, and a much higher percentage are at war at any time than is the case for state societies. For Keeley the rare peaceful tribal societies are usually so because they have been defeated and dominated by a neighboring tribe. Keeley charges that there has been systematic underreporting of violence by contemporary anthropologists intent on pacifying living relics of the past.

Keeley also tries to debunk the view that tribal violence is mostly ritualistic—an arrow in someone's thigh, a head or two bashed with a war club, and time to call it a day. Instead violence in nonstate cultures is lethal. Keeley seems to take pride in this, documenting how various cultures use weapons designed for warfare, meant to cause festering damage. He often has an almost testy, offended tone about those pacifying anthropologists who think indigenous groups lack the organization, self-discipline, and Puritan work ethic to inflict bloodbaths. He writes about the superiority of tribal warriors to Westernized armies, e.g., describing how in the Anglo-Zulu War, Zulu spears were more accurate than nineteenth-century British guns, and how the Brits won the war not because they were superior fighters but because their logistical sophistication allowed them to fight prolonged wars.

Like Keeley, Pinker concludes that warfare is nearly ubiquitous in traditional cultures, reporting 10 to 30 percent of deaths as being war related in New Guinea tribes such as the Gebusi and Mae Enga, and a 35 to 60 percent range for Waorani and Jivaro tribes in the Amazon. Pinker estimates rates of death due to violence. Europe currently is in the range of 1 death per 100,000 people per year. During the crime waves of the 1970s and 1980s, the United States approached 10; Detroit was around 45. Germany and Russia, during their twentieth-century wars, averaged 144 and 135, respectively. In contrast, the twenty-seven nonstate societies surveyed by Pinker average 524 deaths. There are the Grand Valley Dani of New Guinea, the Piegan Blackfoot of the American Great Plains, and the Dinka of Sudan, all of whom in their prime approached 1,000 deaths, roughly equivalent to losing one acquaintance per year. Taking the gold are the Kato, a California tribe that in the 1840s crossed the finish line near 1,500 deaths per 100,000 people per year.

No tour of violence in indigenous cultures is complete without the Yanomamö, a tribe living in the Brazilian and Venezuelan Amazon. According to dogma, there is almost always raiding between villages;

30 percent of adult male deaths are due to violence; nearly 70 percent of adults have had a close relative killed by violence; 44 percent of men have murdered.[61] Fun folks.

The Yanomamö are renowned because of Napoleon Chagnon, one of the most famous and controversial anthropologists, a tough, pugnacious, no-holds-barred academic brawler who first studied them in the 1960s. He established the Yanomamös' street cred with his 1967 monograph *Yanomamo: The Fierce People*, an anthropology classic. Thanks to his publications and his ethnographic films about Yanomamö violence, both their fierceness and his are well-known tropes in anthropology.*

A central concept in the next chapter is that evolution is about passing copies of your genes into the next generation. In 1988 Chagnon published the remarkable report that Yanomamö men who were killers had more wives and offspring than average—thus passing on more copies of their genes. This suggested that if you excel at waging it, war can do wonders for your genetic legacy.

Thus, among these nonstate tribal cultures standing in for our prehistoric past, nearly all have histories of lethal warfare, some virtually nonstop, and those who excel at killing are more evolutionarily successful. Pretty grim.

And numerous anthropologists object strenuously to every aspect of that picture:[62]

- Again with the cherry-picking. In Pinker's analysis of violence among hunter-horticulturalists and other tribal groups, all but one of his examples come from the Amazon or the New Guinea

* When Chagnon was a guest lecturer in an anthro class of mine when I was an undergrad, students dressed up as Yanomamö in a salute to him (hell no, I didn't—I'm too inhibited); apparently it was standard fare for anthro students to crash his road-trip lectures that way, which was probably totally irksome after a while, as he'd have to act all surprised and then pose for pictures with them. Chagnon was at the center of a firestorm of controversy in 2000 when the journalist Patrick Tierney, in his book *Darkness in El Dorado: How Scientists and Journalists Devastated the Amazon*, accused Chagnon and a collaborator of causing a genocidal measles epidemic among the Yanomamö, along with other ethical abuses of them as research subjects. The American Anthropological Association initially condemned Chagnon, which was generally interpreted as his being convicted as much for being an abrasive, anti–old boy enfant terrible as for the factuality of the charges. Eventually both the AAA and independent investigators exonerated Chagnon entirely, showing Tierney's charges to range from the sloppy to the fraudulent. Chagnon's most recent book, a memoir, is entitled *Noble Savages: My Life Among Two Dangerous Tribes—the Yanomamö and the Anthropologists*.

highlands. Global surveys yield much lower rates of warfare and violence.

- Pinker had foreseen this criticism by playing the Keeley pacification-of-the-past card, questioning those lower rates. In particular he has leveled this charge against the anthropologists (whom he somewhat pejoratively calls "anthropologists of peace," somewhat akin to "believers in the Easter Bunny") who have reported on the remarkably nonviolent Semai people of Malaysia. This produced a testy letter to *Science* from this group that, in addition to saying that they are "peace anthropologists," not "anthropologists of peace,"* stated that they are objective scientists who studied the Semai without preconceived notions, rather than a gaggle of hippies (they even felt obliged to declare that most of them are not pacifists). Pinker's response was "It is encouraging that 'anthropologists of peace' now see their discipline as empirical rather than ideological, a welcome change from the days when many anthropologists signed manifestos on which their position on violence was 'correct,' and censured, shut down, or spread libelous rumors about colleagues who disagreed." Whoof, accusing your academic adversaries of signing manifestos is like a sharp knee to the groin.[63]
- Other anthropologists have studied the Yanomamö, and no one else reports violence like Chagnon has.[64] Moreover, his report of increased reproductive success among more murderous Yanomamö has been demolished by the anthropologist Douglas Fry of the University of Alabama at Birmingham, who showed that Chagnon's finding is an artifact of poor data analysis: Chagnon compared the number of descendants of older men who had killed people in battle with those who had not, finding significantly more kids among the former. However: (a) Chagnon did not control for age differences—the killers happened to be an average of more than a

* A distinction somehow reminiscent of Charlie the Tuna, back in those TV ads from my youth, being told that StarKist wants tuna that taste good, not tuna with good taste.

decade older than the nonkillers, meaning more than a decade more time to accumulate descendants. (b) More important, this was the wrong analysis to answer the question posed—the issue isn't the reproductive success of elderly men who had been killers in their youth. You need to consider the reproductive success of *all* killers, including the many who were themselves killed as young warriors, distinctly curtailing their reproductive success. Not doing so is like concluding that war is not lethal, based solely on studies of war veterans.

- Moreover, Chagnon's finding does not generalize—at least three studies of other cultures fail to find a violence/reproductive success link. For example, a study by Luke Glowacki and Richard Wrangham of Harvard examined a nomadic pastoralist tribe, the Nyangatom of southern Ethiopia. Like other pastoralists in their region, the Nyangatom regularly raid one another for cattle.[65] The authors found that frequent participation in large, open battle raiding did not predict increased lifelong reproductive success. Instead such success was predicted by frequent participation in "stealth raiding," where a small group furtively steals cows from the enemy at night. In other words, in this culture being a warrior on 'roids does not predict amply passing on your genes; being a low-down sneaky varmint of a cattle rustler does.
- These indigenous groups are *not* stand-ins for our prehistoric past. For one thing, many have obtained weapons more lethal than those of prehistory (a damning criticism of Chagnon is that he often traded axes, machetes, and shotguns to Yanomamö for their cooperation in his studies). For another, these groups often live in degraded habitats that increase resource competition, thanks to being increasingly hemmed in by the outside world. And outside contact can be catastrophic. Pinker cites research showing high rates of violence among the Amazonian Aché and Hiwi tribes. However, in examining the original reports, Fry found that *all* of the Aché and Hiwi deaths were due to killings by frontier ranchers intent on forcing them off their land.[66] This tells nothing about our prehistoric past.

Both sides in these debates see much at stake. Near the end of his book, Keeley airs a pretty weird worry: "The doctrines of the pacified past unequivocally imply that the only answer to the 'mighty scourge of war' is a return to tribal conditions and the destruction of all civilization." In other words, unless this tomfoolery of archaeologists pacifying the past stops, people will throw away their antibiotics and microwaves, do some scarification rituals, and switch to loincloths—and where will that leave us?

Critics on the other side of these debates have deeper worries. For one thing, the false picture of, say, Amazonian tribes as ceaselessly violent has been used to justify stealing their land. According to Stephen Corry of Survival International, a human-rights organization that advocates for indigenous tribal peoples, "Pinker is promoting a fictitious, colonialist image of a backward 'Brutal Savage', which pushes the debate back over a century and is still used to destroy tribes."[67]

Amid these roiling debates, let's keep sight of what got us to this point. A behavior has occurred that is good, bad, or ambiguous. How have cultural factors stretching back to the origins of humans contributed to that behavior? And rustling cattle on a moonless night; or setting aside tending your cassava garden to raid your Amazonian neighbors; or building fortifications; or butchering every man, woman, and child in a village is irrelevant to that question. That's because all these study subjects are pastoralists, agriculturalists, or horticulturalists, lifestyles that emerged only in the last ten thousand to fourteen thousand years, after the domestication of plants and animals. In the context of hominin history stretching back hundreds of thousands of years, being a camel herder or farmer is nearly as newfangled as being a lobbyist advocating for legal rights for robots. For most of history, humans have been hunter-gatherers, a whole different kettle of fish.

War and Hunter-Gatherers, Past and Present

Roughly 95 to 99 percent of hominin history has been spent in small, nomadic bands that foraged for edible plants and hunted cooperatively.

What is known about hunter-gatherer (for sanity's sake, henceforth HG) violence?

Given that prehistoric HGs didn't have lots of material possessions that have lasted tens of thousands of years, they haven't left much of an archaeological record. Insight into their minds and lifestyle comes from cave paintings dating back as much as forty thousand years. Though paintings from around the world show humans hunting, hardly any unambiguously depict interhuman violence.

The paleontological record is even sparser. To date, there has been discovered one site of an HG massacre, dating back ten thousand years in northern Kenya; this will be discussed later.

What to do with this void of information? One approach is comparative, inferring the nature of our distant ancestors by comparing them with extant nonhuman primates. Early versions of this approach were the writings of Konrad Lorenz and of Robert Ardrey, who argued in his 1966 best seller *The Territorial Imperative* that human origins are rooted in violent territoriality.[68] The most influential modern incarnation comes from Richard Wrangham, particularly in his 1997 book (with Dale Peterson) *Demonic Males: Apes and the Origins of Human Violence.* For Wrangham chimps provide the clearest guide to the behavior of earliest humans, and the picture is a bloody one. He essentially leapfrogs HGs entirely: "So we come back to the Yanomamo. Do they suggest to us that chimpanzee violence is linked to human war? Clearly they do." Wrangham summarizes his stance:

> The mysterious history before history, the blank slate of knowledge about ourselves before Jericho, has licensed our collective imagination and authorized the creation of primitive Edens for some, forgotten matriarchies for others. It is good to dream, but a sober, waking rationality suggests that if we start with ancestors like chimpanzees and end up with modern humans building walls and fighting platforms, the 5-million-year-long trail to our modern selves was lined, along its full stretch, by a male aggression that structured our ancestors' social lives and technology and minds.

It's Hobbes all the way down, plus Keeley-esque contempt for pacification-of-the-past dreamers.

This view has been strongly criticized: (a) We're neither chimps nor their descendants; they've been evolving at nearly the same pace as humans since our ancestral split. (b) Wrangham picks and chooses in his cross-species linkages; for example, he argues that the human evolutionary legacy of violence is rooted not only in our close relationship to chimps but also in our nearly-as-close kinship with gorillas, who practice competitive infanticide. The problem is that, overall, gorillas display minimal aggression, something Wrangham ignores in linking human violence to gorillas. (c) As the most significant species cherry-picking, Wrangham pretty much ignores bonobos, with their far lower levels of violence than chimps, female social dominance, and absence of hostile territoriality. Crucially, humans share as much of their genes with bonobos as with chimps, something unknown when *Demonic Males* was published (and, notably, Wrangham has since softened his views).

For most in the field, most insights into the behavior of our HG ancestors come from study of contemporary HGs.

Once, the world of humans consisted of nothing but HGs; today the remnants of that world are in the few remaining pockets of peoples who live pure HG lives. These include the Hadza of northern Tanzania, Mbuti "Pygmies" in the Congo, Batwa in Rwanda, Gunwinggu of the Australian outback, Andaman Islanders in India, Batak in the Philippines, Semang in Malaysia, and various Inuit cultures in northern Canada.

To start, it was once assumed that among HGs, women do the gathering while men supply most of the calories by hunting. In actuality, the majority of calories come from foraging; men spend lots of time talking about how awesome they were in the last hunt and how much awesomer they'll be in the next—among some Hadza, maternal grandmothers supply more calories to families than do the Man the Hunter men.[69]

The arc of human history is readily equated with an arc of progress, and key to the latter is the view that agriculture was the best thing humans ever invented; I'll rant about that later. A cornerstone of the

Clockwise from top: *Hadza; Mbuti, Andaman, Semang*

agriculture lobby is the idea that primordial HGs were half starved. In reality, HGs typically work fewer hours for their daily bread than do traditional farmers and are longer-lived and healthier. In the words of anthropologist Marshall Sahlins, HGs were the original affluent society.

There are some demographic themes shared among contemporary HGs.[70] Dogma used to be that HG bands had fairly stable group membership, producing considerable in-group relatedness. More recent work suggests less relatedness than thought, reflecting fluid fusion/fission groupings in nomadic HGs. The Hadza show one consequence of such fluidity, namely that particularly cooperative hunters find one another and work together. More on this in the next chapter.

What about our best and worst behaviors in contemporary HGs? Up into the 1970s, the clear answer was that HGs are peaceful, cooperative, and egalitarian. Interband fluidity serves as a safety valve preventing individual violence (i.e., when people are at each other's throats, someone moves to another group), and nomadicism as a safety valve preventing intergroup violence (i.e., rather than warring with the neighboring band, just hunt in a different valley from them).

The standard-bearers for HG grooviness were the Kalahari !Kung.*[71] The title of an early monograph about them—Elizabeth Marshall Thomas's 1959 *The Harmless People*—says it all.† !Kung are to the Yanomamö as Joan Baez is to Sid Vicious and the Sex Pistols.

Naturally, this picture of the !Kung in particular and HGs in general was ripe for revisionism. This occurred when field studies were sufficiently long term to document HGs killing one another, as summarized in an influential 1978 publication by Carol Ember of Yale.[72] Basically, if you're observing a band of thirty people, it will take a long time to see that, on a per-capita basis, they have murder rates approximating Detroit's (the standard comparison always made). Admitting that HGs were violent was seen as a purging of sixties anthropological romanticism, a bracing slap in the face for anthropologists who had jettisoned objectivity in order to dance with wolves.

By the time of Pinker's synthesis, HG violence was established, and the percentage of their deaths attributed to warfare averaged around 15 percent, way more than in modern Western societies. Contemporary HG

* !Kung speak a click language, with the exclamation mark in their name the notation for the click sound. Informally known as "Bushmen," they are part of the larger cultural group of Khoisan San found in Botswana, Namibia, Angola, and South Africa. As orientation, the movie *The Gods Must Be Crazy* featured the !Kung. Of note, while "!Kung" is the most familiar term most often used for these people, both they and most contemporary anthropologists use the term "Ju/'hoansi" instead.

† I was raised in an anthropology department that was a major stronghold of !Kung fandom and generalized this to a huge fondness for all things African HG (probably in part because they're all short). A tiny remnant HG tribe alternately called the Ndorobo or Okiek lives in forests north of the Serengeti in Kenya. They have an oddly symbiotic relationship with the neighboring Maasai, emerging from the forest to trade things or to serve a shamanistic role in some Maasai ceremony. They are short and silent, dressed in animal skins, and I've taken great pleasure in seeing how they unnerve tall, spear-toting Maasai. My Maasai friends would make fun of me for how obsessed I was with the Ndorobo.

Kalahari !Kung hunter-gatherers

violence constitutes a big vote for the Hobbesian view of warfare and violence permeating all of human history.

Time for the criticisms:[73]

- Mislabeling—some HGs cited by Pinker, Keeley, and Bowles are, in fact, hunter-horticulturalists.
- Many instances of supposed HG warfare, on closer inspection, were actually singular homicides.
- Some violent Great Plains HG cultures were untraditional in the sense of using something crucial that didn't exist in the Pleistocene—domesticated horses ridden into battle.
- Like non-Western agriculturalists or pastoralists, contemporary HG are not equivalent to our ancestors. Weapons invented in the last ten thousand years have been introduced through trade; most HG cultures have spent millennia being displaced by agriculturalists and pastoralists, pushed into ever tougher, resource-sparse ecosystems.
- Once again, the cherry-picking issue, i.e., failure to cite cases of peaceful HGs.
- Most crucially, there's more than one type of HG. Nomadic HGs are the original brand, stretching back hundreds of thousands of years.[74] But in addition to HG 2.0 equestrians, there are "complex HGs," who are different—violent, not particularly egalitarian, and sedentary, typically because they're sitting on a rich food source that they defend from outsiders. In other words, a transitional form from pure HGs. And many of the cultures cited by Ember, Keeley, and Pinker are complex HGs. This difference is relevant to Nataruk, that northern Kenyan site of a ten-thousand-year-old massacre—skeletons of twenty-seven unburied people, killed by clubbing, stabbing, or stone projectiles. The victims were sedentary HGs, living alongside a shallow bay on Lake Turkana, prime beachfront property with easy fishing and plentiful game animals coming to the water to drink. Just the sort of real estate that someone else would try to muscle in on.

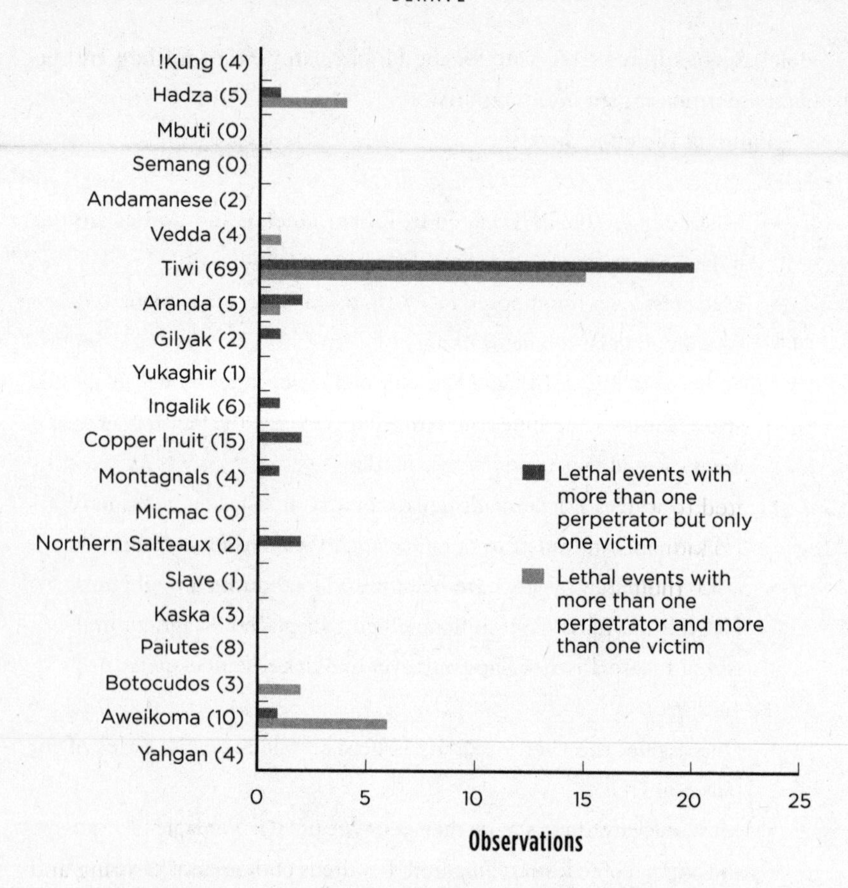

D. P. Fry and P. Söderberg, "Lethal Aggression in Mobile Forager Bands and Implications for the Origins of War," Sci *341 (2013): 270.*

The most thoughtful and insightful analyses of HG violence come from Fry and from Christopher Boehm of the University of Southern California. They paint a complex picture.

Fry has provided what I consider the cleanest assessment of warfare in such cultures. In a notable 2013 *Science* paper, he and Finnish anthropologist Patrik Söderberg reviewed all cases of lethal violence in the ethnographic literature in "pure" nomadic HGs (i.e., well studied before extensive contact with outsiders and living in a stable ecosystem). The sample consisted of twenty-one such groups from around the world. Fry and Söderberg observed what might be called warfare (defined by the fairly unstringent criterion of conflict producing multiple casualties) in only a minority of the cultures. Not exactly widespread. This is probably the best approximation we'll ever get about warfare in our HG ancestors. Nonetheless, these pure HGs are no tie-dyed pacifists; 86 percent of the cultures experienced lethal violence. What are their causes?

In his 2012 book *Moral Origins: The Evolution of Virtue, Altruism, and Shame*, Boehm also surveys the literature, using slightly less stringent criteria than Fry uses, producing a list of about fifty relatively "pure" nomadic HG cultures (heavily skewed toward Inuit groups from the Arctic).[75] As expected, violence is mostly committed by men. Most common is killing related to women—two men fighting over a particular woman, or attempts to kidnap a woman from a neighboring group. Naturally, there are men killing their wives, usually over accusations of adultery. There's female infanticide and killing arising from accusations of witchcraft. There are occasional killings over garden-variety stealing of food or refusals to share food. And lots of revenge killings by relatives of someone killed.

Both Fry and Boehm report killings akin to capital punishment for severe norm violations. What norms do nomadic HGs value most? Fairness, indirect reciprocity, and avoidance of despotism.

Fairness. As noted, HGs pioneered human cooperative hunting and sharing among nonrelatives.[76] This is most striking with meat. It's typically shared by successful hunters with unsuccessful ones (and their families); individuals playing dominant roles in hunts don't necessarily get much more meat than everyone else; crucially, the most successful hunter rarely decides how the meat is divided—instead this is typically done by a third party. There are fascinating hints about the antiquity of this. Big-game hunting by hominins 400,000 years ago has been documented; bones from animals butchered then show cut marks that are chaotic, overlapping at different angles, suggesting a free-for-all. But by 200,000 years ago the contemporary HG pattern is there—cut marks are evenly spaced and parallel, suggesting that single individuals butchered and dispensed the meat.

This does not mean, though, that sharing is effortless for pure HGs. Boehm notes how, for example, the !Kung perpetually kvetch about being shortchanged on meat. It's the background hum of social regulation.

Indirect reciprocity. The next chapter discusses reciprocal altruism between pairs of individuals. Boehm emphasizes how nomadic HGs specialize, instead, in indirect reciprocity. Person A is altruistic to B; B's social obligation now isn't necessarily as much being altruistic to A as paying the altruism forward to C. C pays it forward to D, etc. . . . This stabilizing cooperation is ideal for big-game hunters, where two rules hold: (a) your hunts are usually unsuccessful; and (b) when they are successful, you typically have more meat than your family can consume, so you might as well share it around. As has been said, an HG's best investment against future hunger is to put meat in other people's stomachs now.

Avoidance of despotism. As also covered in the next chapter, there's considerable evolutionary pressure for detecting cheating (when someone reneges on their half of a reciprocal relationship). For nomadic HGs, policing covert cheating is less of a concern than overt evidence of intimidation and powermongering. HGs are constantly on guard against bullies throwing their weight around.

HG societies expend lots of *collective* effort on enforcing fairness, indirect reciprocity, and avoidance of despotism. This is accomplished with that terrific norm-enforcement mechanism, gossip. HGs gossip endlessly, and as studied by Polly Wiessner of the University of Utah, it's mostly about the usual: norm violation by high-status individuals.[77] *People* magazine around the campfire.* Gossiping serves numerous purposes. It helps for reality testing ("Is it just me, or was he being a total jerk?"), passing news ("Two guesses who just happened to get a foot cramp during the hairiest part of the hunt today"), and building consensus ("Something needs to be done about this guy"). Gossip is the weapon of norm enforcement.

* Boehm emphasizes that anthropologists never really know what truly is going on among their research subjects until they're privy to gossip. In doing my baboon research, I spent many seasons sharing camp with Maasai guys whom I knew relatively well and would hear about big goings-on in their community. Eventually, my soon-to-be wife started joining me in the field, and it was only then that we started to hear about the good stuff, via her becoming friends with some of the women—the usual of who was or wasn't sleeping with whom.

HG cultures take similar actions—collectively subjecting miscreants to criticism, shaming and mockery, ostracizing and shunning, refusing to share meat, nonlethal physical punishment, expulsion from the group, or, as a last resort, killing the person (done either by the whole group or by a designated executioner).

Boehm documents such judicial killings in nearly half the pure HG cultures. What transgressions merit them? Murder, attempts at grabbing power, use of malicious sorcery, stealing, refusal to share, betrayal of the group to outsiders, and of course breaking of sexual taboos. All typically punished this way after other interventions have failed repeatedly.

So, Hobbes or Rousseau? Well, a mixture of the two, I say unhelpfully. This lengthy section makes clear that you have to make some careful distinctions: (a) HGs versus other traditional ways of making a living; (b) nomadic HGs versus sedentary ones; (c) data sets that canvass an entire literature versus those that concentrate on extreme examples; (d) members of traditional societies killing one another versus members being killed by gun-toting, land-grabbing outsiders; (e) chimps as our cousins versus chimps erroneously viewed as our ancestors; (f) chimps as our closest relatives versus chimps and bonobos as our closest relatives; (g) warfare versus homicide, where lots of the former can decrease the latter in the name of in-group cooperation; (h) contemporary HGs living in stable, resource-filled habitats with minimal interactions with the outside world versus contemporary HGs pushed into marginal habitats and interacting with non-HGs. Once you've done that, I think a pretty clear answer emerges. The HGs who peopled earth for hundreds of thousands of years were probably no angels, being perfectly capable of murder. However, "war"—both in the sense that haunts our modern world and in the stripped-down sense that haunted our ancestors—seems to have been rare until most humans abandoned the nomadic HG lifestyle. Our history as a species has not been soaked in escalated conflict. And ironically Keeley tacitly concludes the same—he estimates that 90 to 95 percent of societies engage in war. And whom does he note as the exceptions? Nomadic HGs.

Which brings us to agriculture. I won't pull any punches—I think that its invention was one of the all-time human blunders, up there with, say, the New Coke debacle and the Edsel. Agriculture makes people dependent on a few domesticated crops and animals instead of hundreds of wild food sources, creating vulnerability to droughts and blights and zoonotic diseases. Agriculture makes for sedentary living, leading humans to do something that no primate with a concern for hygiene and public health would ever do, namely living in close proximity to their feces. Agriculture makes for surplus and thus almost inevitably the unequal distribution of surplus, generating socioeconomic status differences that dwarf anything that other primates cook up with their hierarchies. And from there it's just a hop, skip, and a jump until we've got Mr. McGregor persecuting Peter Rabbit and people incessantly singing "Oklahoma."

Maybe this is a bit over the top. Nonetheless, I do think it is reasonably clear that it wasn't until humans began the massive transformation of life that came from domesticating teosinte and wild tubers, aurochs and einkorn, and of course wolves, that it became possible to let loose the dogs of war.

SOME CONCLUSIONS

The first half of the chapter explored where we are; the second, how we most likely got here.

"Where we are" is awash in cultural variation. From our biological perspective, the most fascinating point is how brains shape cultures, which shape brains, which shape . . . That's why it's called *co*evolution. We've seen some evidence of coevolution in the technical sense—where there are significant differences between different cultures in the distribution of gene variants pertinent to behavior. But those influences are pretty small. Instead what is most consequential is childhood, the time when cultures inculcate individuals into further propagating their culture. In that regard, probably the most important fact about genetics and culture is the delayed maturation of the frontal cortex—the genetic programming for the young frontal cortex to be freer from genes than other brain regions, to be sculpted instead by environment, to sop up cultural norms. To hark back to a theme from the first pages of this book, it doesn't take a particularly fancy brain to learn how to motorically, say, throw a punch. But it takes a fancy, environmentally malleable frontal cortex to learn culture-specific rules about when it's okay to throw punches.

In another theme from the first half, cultural differences manifest themselves in monumentally important, expected ways—say, whom it is okay to kill (an enemy soldier, a cheating spouse, a newborn of the "wrong" sex, an elderly parent too old to hunt, a teenage daughter who is absorbing the culture around her rather than the culture her parents departed). But the manifestations can occur in unlikely places—e.g., where your eyes look within milliseconds of seeing a picture, or whether thinking of a rabbit prompts you to think of other animals or of what rabbits eat.

Another key theme is the paradoxical influence of ecology. Ecosystems majorly shape culture—but then that culture can be exported and persist in radically different places for millennia. Stated most straightforwardly, most of earth's humans have inherited their beliefs about the nature of birth and death and everything in between and thereafter from preliterate Middle Eastern pastoralists.

The second half of the chapter, just concluded, addresses the key issue of how we got here—has it been hundreds of thousands of years of Hobbes or of Rousseau? Your answer to that question greatly shapes what you'll make of something we'll consider in the final chapter, namely that over the last half millennium people have arguably gotten a lot less awful to one another.

Ten

The Evolution of Behavior

At last we reach the foundations. Genes and promoters evolve. As do transcription factors, transposases, and splicing enzymes. As has every trait touched by genetic influences (i.e., everything). In the words of the geneticist Theodosius Dobzhansky, "Nothing in biology makes sense except in the light of evolution." Including this book.[1]

EVOLUTION 101

Evolution rests on three steps: (a) certain biological traits are inherited by genetic means; (b) mutations and gene recombination produce variation in those traits; (c) some of those variants confer more "fitness" than others. Given those conditions, over time the frequency of more "fit" gene variants increases in a population.

We start by trashing some common misconceptions.

First, that evolution favors *survival* of the fittest. Instead evolution is about reproduction, passing on copies of genes. An organism living centuries but not reproducing is evolutionarily invisible.* The difference

* We'll soon see an exception to that, involving the nonreproducing individual helping relatives to reproduce.

between survival and reproduction is shown with "antagonistic pleiotropy," referring to traits that increase reproductive fitness early in life yet decrease life span. For example, primates' prostates have high metabolic rates, enhancing sperm motility. Upside: enhanced fertility; downside: increased risk of prostate cancer. Antagonistic pleiotropy occurs dramatically in salmon, who epically journey to their spawning grounds to reproduce and then die. If evolution were about survival rather than passing on copies of genes, there'd be no antagonistic pleiotropy.[2]

Another misconception is that evolution can select for preadaptations—neutral traits that prove useful in the future. This doesn't happen; selection is for traits pertinent to the present. Related to this is the misconception that living species are somehow better adapted than extinct species. Instead, the latter were just as well adapted, until environmental conditions changed sufficiently to do them in; the same awaits us. Finally, there's the misconception that evolution directionally selects for greater complexity. Yes, if once there were only single-celled organisms and there are multicellular ones now, average complexity has increased. Nonetheless, evolution doesn't necessarily select for greater complexity—just consider bacteria decimating humans with some plague.

The final misconception is that evolution is "just a theory." I will boldly assume that readers who have gotten this far believe in evolution. Opponents inevitably bring up that irritating canard that evolution is unproven, because (following an unuseful convention in the field) it is a "theory" (like, say, germ theory). Evidence for the reality of evolution includes:

- Numerous examples where changing selective pressures have changed gene frequencies in populations within generations (e.g., bacteria evolving antibiotic resistance). Moreover, there are also examples (mostly insects, given their short generation times) of a species in the process of splitting into two.
- Voluminous fossil evidence of intermediate forms in numerous taxonomic lineages.
- Molecular evidence. We share ~98 percent of our genes with the other apes, ~96 percent with monkeys, ~75 percent with dogs, ~20

percent with fruit flies. This indicates that our last common ancestor with other apes lived more recently than our last common ancestor with monkeys, and so on.

- Geographic evidence. To use Richard Dawkins's suggestion for dealing with a fundamentalist insisting that all species emerged in their current forms from Noah's ark—how come all thirty-seven species of lemurs that made landfall on Mt. Ararat in the Armenian highlands hiked over to Madagascar, none dying and leaving fossils in transit?
- Unintelligent design—oddities explained only by evolution. Why do whales and dolphins have vestigial leg bones? Because they descend from a four-legged terrestrial mammal. Why should we have arrector pili muscles in our skin that produce thoroughly useless gooseflesh? Because of our recent speciation from other apes whose arrector pili muscles were attached to hair, and whose hair stands up during emotional arousal.

Enough. Don't get me started.

Evolution sculpts the traits of an organism in two broad ways. "Sexual selection" selects for traits that attract members of the opposite sex, "natural selection" for traits that enhance the passing on of copies of genes through any other route—e.g., good health, foraging skills, predator avoidance.

The two processes can work in opposition.[3] For example, among wild sheep one gene influences the size of horns in males. One variant produces large horns, improving social dominance, a plus for sexual selection. The other produces small horns, which are metabolically cheaper, allowing males to live and mate (albeit at low rates) longer. Which wins—transient but major reproductive success, or persistent but minor success? An intermediate form.* Or consider male peacocks paying a price, in

* That is, the heterozygotic state. I made the tough decision to bypass homozygosity and heterozygosity in the main text, in the interest of simplicity for the newcomer, and to instead exile the subject to footnotes. A brief primer: A point that I blithely ignored in the genetics chapter is that most species, including humans, are "diploid," which means that there are actually two sets of chromosomes in each cell, with the same variety of genes. Eggs and sperm are

terms of natural selection, for their garish plumage—it costs a fortune metabolically to grow, restricts mobility, and is conspicuous to predators. But it sure boosts fitness via sexual selection.

Importantly, neither type of selection necessarily selects for "the" most adaptive version of a trait, which replaces all others. There can be frequency-dependent selection, where the rarer version of two traits is preferable, or balanced selection, where multiple versions of traits are maintained in equilibrium.

BEHAVIOR CAN BE SHAPED BY EVOLUTION

Organisms are amazingly well adapted. A desert rodent has kidneys that excel at retaining water; a giraffe's huge heart can pump blood up to its brain; elephants' leg bones are strong enough to support an elephant. Well, yes—it *has* to work that way: desert rodents whose kidneys weren't great at retaining water didn't pass on copies of their genes. Thus there is a logic to evolution, where natural selection sculpts traits into adaptiveness.

Importantly, natural selection works not only on anatomy and physiology but on behavior as well—in other words, behavior evolves, can be optimized by selection into being adaptive.

Various branches of biology focus on the evolution of behavior. Probably best known is sociobiology, premised on social behavior being sculpted by evolution to be optimized, just as biomechanical optimization sculpts the size of a giraffe's heart.[4] Sociobiology emerged in the 1970s, eventually generating the offshoot evolutionary psychology—the study of the

specialized cells, being haploid (i.e., containing only a single copy of a chromosome). Put them together, and the egg that is destined to make you you is now fertilized (i.e., diploid). Thus you actually have two copies of each gene, one from each parent. (Footnote to a footnote: the exception is a specialized collection of genes in mitochondria, which come almost entirely from the mother). If both copies of the gene have sequences that code for identical copies of a protein, the gene is "homozygotic." If there are two different versions, the gene is "heterozygotic." What sort of trait is specified by the heterozygotic mixing of a gene? Some of the time, the result is a trait that is intermediate between the two possible forms of homozygosity. More often the heterozygotic form produces a trait that is identical to one of the two homozygotic forms. In other words, one of the versions "wins" out over the other and is called a "dominant" version of the gene. In contrast, versions of genes that produce a trait only when in the homozygotic form are "recessive." If this is vastly confusing, I promise that you'll be okay reading the rest of the book, nonetheless.

evolutionary optimization of psychological traits; as we'll see, both have been plenty controversial. As a simplifying convenience, I'll refer to people who study the evolution of social behavior as "sociobiologists."

THE DEMISE OF GROUP SELECTION

We start by grappling with an entrenched misconception about the evolution of behavior. This is because Americans were taught about the subject in the 1960s by Marlin Perkins on the TV program *Mutual of Omaha's Wild Kingdom.*

It was great. Perkins would host. Jim, his sidekick, did dangerous things with snakes. And there were always seamless segues from the program to ads from Mutual of Omaha—"Just as lions mate for hours, you'll want fire insurance for your home."

Unfortunately, Perkins espoused wildly wrong evolutionary thinking. Here's how it looked on the program: It's dawn on the savanna; there's a herd of wildebeest on a river's edge. The grass is greener on the other side, and everyone wants some, but the river teems with predatory crocodiles. The wildebeest are hemming and hawing in agitation when suddenly an elderly wildebeest pushes to the front, says, "I sacrifice myself for you, my children," and leaps in. And while the crocs are busy with him, the other wildebeest cross the river.

Why would the old wildebeest do that? Marlin Perkins would answer with patrician authority: because animals behave For the Good of the Species.

Yes, behavior evolves by "group selection" for the good of the species. This idea was championed in the early 1960s by V. C. Wynne-Edwards, whose wrongness made him modern evolutionary biology's Lamarck.*[5]

Animals don't behave for the good of the species. But what about that

* Poor Wynne-Edwards was actually a major figure in evolution and behavior but, thanks to shallow and superficial people, he is remembered only for having blown it with group selection. I, for example, haven't a clue what else the guy ever did. His full name was Vero Copner Wynne-Edwards, which probably explains why he's always called "V. C. Wynne-Edwards," no doubt even as an infant.

wildebeest? Look closely and you'll see what really happens. Why did he wind up saving the day? Because he was old and weak. "Good of the species" my keister. They pushed the old guy in.

Group selection was done in by theoretical and empirical studies showing patterns of behavior incompatible with it. Key work was done by two gods of evolutionary biology, George Williams of SUNY Stony Brook and Oxford's Bill ("W.D.") Hamilton.[6] Consider "eusocial insects," where most individuals are nonreproductive workers. Why forgo reproduction to aid the queen? Group selection, obviously. Hamilton showed that eusocial insects' unique genetic system makes a colony of ants, bees, or termites a single superorganism; asking why worker ants forgo reproduction is like asking why your nose cells forgo reproduction. In other words, eusocial insects constitute a unique type of "group." Williams then elaborated on how the more standard genetic system, in species from noneusocial insects to us, was incompatible with group selection. Animals don't behave for the good of the species. They behave to maximize the number of copies of their genes passed into the next generation.*

This is the cornerstone of sociobiology and was summarized in Dawkins's famed sound bite that evolution is about "selfish genes." Time to see its building blocks.

INDIVIDUAL SELECTION

Passing on lots of copies of one's genes is accomplished most directly by maximizing reproduction. This is summarized by the aphorism "A chicken is an egg's way of making another egg"—behavior is just an epiphenomenon, a means of getting copies of genes into the next generation.

Individual selection fares better than group selection in explaining basic behaviors. A hyena bears down on some zebras. What would the

* The unique feature of the eusocial insect genetic system is that a sterile worker passes on more copies of her genes by helping the queen to reproduce than by reproducing herself. Meanwhile, the eusocial insect world has been shaken up by the fact that in some species (e.g., termites) there is a more conventional genetic system in place. People are still sorting that one out.

nearest one do if she's a group selectionist? Stand there, sacrificing herself for the group. In contrast, an individual selectionist zebra would run like hell. Zebras run like hell. Or consider hyenas that have just killed a zebra. Group selection mind-set—everyone calmly takes turns eating. Individual selection—frenzied free-for-all. Which is what occurs.

But wait, says the group selectionist, wouldn't the zebra species benefit if it is the fastest animals who survive and pass on those fast-running genes? Ditto for the group benefits of the fiercest hyena getting the most food.

As more nuances of behavior are observed, clinging to group selection requires increasingly tortuous arguments. But one single observation devastates group selection.

In 1977 the Harvard primatologist Sarah Blaffer Hrdy documented something remarkable—langur monkeys in the Mount Abu region of India kill one another.[7] People already knew that some male primates kill one another, fighting for dominance—okay, makes sense, boys will be boys. But that's not what Hrdy reported; male langurs were killing infants.

Once people believed her careful documentation, there was an easy answer—since babies are cute and inhibit aggression, something pathological must be happening.[8] Maybe the Abu langur population density was too high and everyone was starving, or male aggression was overflowing, or infanticidal males were zombies. Something certifiably abnormal.

Hrdy eliminated these explanations and showed a telling pattern to the infanticide. Female langurs live in groups with a single resident breeding male. Elsewhere are all-male groups that intermittently drive out the resident male; after infighting, one male then drives out the rest. Here's his new domain, consisting of females with the babies of the previous male. And crucially, the average tenure of a breeding male (about twenty-seven months) is shorter than the average interbirth interval. No females are ovulating, because they're nursing infants; thus this new stud will be booted out himself before any females wean their kids and resume ovulating. All for nothing, none of his genes passed on.

What, logically, should he do? Kill the infants. This decreases the reproductive success of the previous male and, thanks to the females ceasing to nurse, they start ovulating.*

That's the male perspective. What about the females? They're also into maximizing copies of genes passed on. They fight the new male, protecting their infants. Females have also evolved the strategy of going into "pseudoestrus"—falsely appearing to be in heat. They mate with the male. And since males know squat about female langur biology, they fall for it—"Hey, I mated with her this morning and now she's got an infant; I am one major stud." They'll often cease their infanticidal attacks.

Despite initial skepticism, competitive infanticide has been documented in similar circumstances in 119 species, including lions, hippos, and chimps.[9]

A variant occurs in hamsters; because males are nomadic, any infant a male encounters is unlikely to be his, and thus he attempts to kill it (remember that rule about never putting a pet male hamster in a cage with babies?). Another version occurs among wild horses and gelada baboons; a new male harasses pregnant females into miscarrying. Or suppose you're a pregnant mouse and a new, infanticidal male has arrived. Once you give birth, your infants will be killed, wasting all the energy of pregnancy. Logical response? Cut your losses with the "Bruce effect," where pregnant females miscarry if they smell a new male.[10]

Thus competitive infanticide occurs in numerous species (including among female chimps, who sometimes kill infants of unrelated females).[11] None of this makes sense outside of gene-based individual selection.

Individual selection is shown with heartbreaking clarity by mountain gorillas, my favorite primate.[12] They're highly endangered, hanging on in pockets of high-altitude rain forest on the borders of Uganda, Rwanda, and the Democratic Republic of the Congo. There are only about a thousand gorillas left, because of habitat degradation, disease caught from nearby humans, poaching, and spasms of warfare rolling across those

* Note: No one is claiming that a langur monkey is thinking this through, any more than would be some brine shrimp who has evolved some sort of optimal behavioral reproductive strategy. An animal has the "goal" of "wanting" to pass on copies of their genes and thus "decides" to do X. This is just shorthand for saying something like "Over the course of millennia, individuals who do X have passed on copies of their genes at a higher rate, and this has become a common behavioral feature of this species." Animals don't know about evolutionary biology, just as prototypes of airplane wings in a wind tunnel don't know about aerodynamics.

borders. And also because mountain gorillas practice competitive infanticide. Logical for an individual intent on maximizing the copies of his genes in the next generation, but simultaneously pushing these wondrous animals toward extinction. This isn't behaving for the good of the species.

KIN SELECTION

To understand the next foundational concept, reflect on what it means to be related to someone and to pass on copies of "your" genes.

Suppose you have an identical twin, with the same genome as you. As a startling, irrefutable fact, in terms of the genes being passed on to the next generation, it doesn't matter if you reproduce or sacrifice yourself so that your twin reproduces.

What about a full sibling who isn't an identical twin? Recall from chapter 8 that you'd share 50 percent of your genes with him.* Thus reproducing once and dying so that he reproduces twice are evolutionarily identical. Half sibling, 25 percent of genes in common, calculate accordingly. . . .

The geneticist J. B. S. Haldane, who, when asked if he'd sacrifice his life for a brother, is credited to have quipped, "I'll gladly lay down my life for two brothers or eight cousins." You can leave copies of your genes in the next generation by reproducing, but also by helping relatives reproduce, especially closer relatives. Hamilton formalized this with an equation factoring in the costs and benefits of helping someone, weighted by their degree of relatedness to you. This is the essence of kin selection.† This explains the crucial fact that in countless species, whom you cooperate with, compete with, or mate with depends on their degree of relatedness to you.

Mammals first encounter kin selection soon after birth, reflecting something monumentally obvious: females rarely nurse someone else's infants. Next, among numerous primates the mother of a newborn and an

* Or, more correctly, for each gene there'd be a 50 percent chance you'd share the same variant.

† Also known as "inclusive fitness," because a gene-based focus *includes* not only direct reproductive success (Darwinian fitness) but also payoffs derived from the success of other relatives, weighted by their degree of relatedness.

adolescent female may commence a relationship fraught with pluses and minuses—the mother occasionally lets the adolescent care for her offspring. For the mother the plus is getting time to forage without baby on board; the minus is that the babysitter may be incompetent. For the adolescent the plus is getting mothering experience; the minus, the effort of child care. Lynn Fairbanks of UCLA has quantified the pluses and minuses of such "allomothering" (including that adolescents who practiced mothering have a better survival rate for their own kids). And who is a frequent "allomother"? The female's kid sister.[13]

An extension of allomothering is the cooperative breeding of New World monkeys like marmosets. In their social groups only one female breeds, while the others—typically younger relatives—help with child care.[14]

The extent to which a male primate cares for infants reflects his certainty of paternity.[15] Among marmosets, who form stable pair-bonds, males do most of the child care. In contrast, among baboons, where a female mates with multiple males during her estrus cycle, it's only the likely fathers (i.e., males who mated on the female's most fertile day, when she had her most conspicuous estrus swelling) who invest in the well-being of the child, aiding him in a fight.*

Among many primates, how often you groom someone depends on how closely related they are to you. Among baboons, females spend their whole life in their natal troop (whereas males migrate to a new troop at puberty); as a result, adult females have complex cooperative kinship relations and inherit their dominance rank from their mother. Among chimps it's the opposite; females leave home at puberty, and kin-based adult cooperation occurs only among males (for example, where groups of related males attack solitary males from neighboring groups). And among langurs, when a female defends her infant against a new male, she most often is helped by elderly female relatives.

Moreover, primates understand kinship. Dorothy Cheney and Robert Seyfarth of the University of Pennsylvania, studying wild vervet monkeys,

* Note the term used—"invest"—reflecting a quasi-economic orientation to some of the analyses in this field.

have shown that if animal A is crummy to animal B, afterward, B is more likely to be crummy to A's *relatives*. And if A is lousy to B, B's *relatives* are more likely to be crummy to A. Furthermore, if A is lousy to B, B's relatives are more likely to be crummy to A's *relatives*.[16]

In beautiful "playback" experiments, Cheney and Seyfarth first recorded vocalizations from each vervet in a group. They'd place a speaker in some bushes, and when everyone was sitting around, they'd play a recording of some kid giving a distress call. And the females would all look at the kid's mother—"Hey, that's Madge's kid. What's she going to do?" (Note that this also shows that monkeys recognize voices.)

In a study of wild baboons, Cheney and Seyfarth would wait for two unrelated females to sit near the bush with the speaker and then play one of three vocalizations: (a) sounds of the two females' relatives fighting with each other; (b) a relative of one fighting with a third party; (c) two other random females fighting.[17] If a female's relative was involved in the fighting, she'd look toward the speaker longer than if there were no relatives involved. And if it was relatives of the two females fighting each other, the higher-ranking one would remind the subordinate of her place by supplanting her from her spot.

Another playback study created some baboon virtual reality.[18] Baboon A dominates baboon B. Thanks to cutting and splicing of recordings of vocalizations, baboon A is heard making a dominance vocalization, B making a subordination one. When this happens, no baboons looked at the bushes—A > B, boring status quo. But if baboon A is heard making a *subordination* vocalization after B makes a *dominance* one—a rank reversal—everyone orients to the bushes ("Did you hear what I just heard?"). Then a third scenario—a dominance reversal between two members of the same family. And no one looks, because it's uninteresting. ("Families, they're crazy. You should see mine—we have these huge dominance reversals and are hugging an hour later.") Baboons "classify others simultaneously according to both individual rank and kinship."

Thus other primates contemplate kinship with remarkable sophistication, with kinship determining patterns of cooperation and competition.

Nonprimates are also into kin selection. Consider this—sperm in a female's vaginal tract can aggregate, allowing them to swim faster. Among a deer mouse species where females mate with multiple males, sperm aggregate only with sperm from the same individual or a close relative.[19]

As behavioral examples, squirrels and prairie dogs give alarm vocalizations when spotting a predator. It's risky, calling attention to the caller, and such altruism is more common when in the proximity of relatives. Social groups built around female relatives occur in numerous species (e.g., lion prides, where related females nurse one another's cubs). Moreover, while prides typically contain a single breeding male, on those occasions when it's two males, better than chance that they're brothers. There is a striking similarity in humans. Most cultures have historically allowed polygyny, with monogamy as the rarer beast. Even rarer is polyandry—multiple men married to one woman. This occurs in northern India, Tibet, and Nepal, where the polyandry is "adelphic" (aka "fraternal")—a woman marries all the brothers of one family, from the strapping young man to his infant brother.*[20]

A challenging implication of kin selection arises.

Those hot cousins. If one accrues fitness benefits by helping relatives pass on copies of their genes, why not help them do that by mating with them? Yech; inbreeding produces decreased fertility and those genetic unpleasantnesses in European royalty.†[21] So the dangers of inbreeding counter the kin-selection advantages. Theoretical models suggest that the optimal balance is third-cousin matings. And indeed, numerous species prefer to mate with between a first and a third cousin.[22]

This occurs in insects, lizards, and fish, where, on top of that, cousin-mating pairs invest more in the rearing of their offspring than do unrelated parents. A preference for cousin matings occurs in quail, frigate

* Such fraternal polyandry occurs in resource-poor regions, basically acting as a means to decrease population growth and prevent family plots from winding up being below subsistance level when subdivided and inherited among all the sons in a family. Instead, all the brothers are married to the one woman, who has equal sexual access to all of them; the brothers "believe" that all of them, down to their infant brother, are equally biologically responsible for the children.

† There is good evidence that inbreeding was responsible for the demise of the Spanish branch of the Habsburg dynasty. G. Alvarez et al., "The Role of Inbreeding in the Extinction of a European Royal Dynasty," *PLoS ONE* 4 (2009): e5174.

birds, and zebra finches, while among pair-bonded barn swallows and ground tits, females sneak out on their partner to mate with cousins. Similar preferences occur in some rodents (including the Malagasy giant jumping rat, a species that sounds disturbing even without cousins shacking up with each other).[23]

And what about humans? Something similar. Women prefer the smell of moderately related over unrelated men. And in a study of 160 years of data concerning every couple in Iceland (which is a mecca for human geneticists, given its genetic and socioeconomic homogeneity), the highest reproductive success arose from third- and fourth-cousin marriages.[24]

Recognizing Relatives?

These findings concerning kin selection require animals to recognize degrees of relatedness. How do they do this?

Some species have innate recognition. For example, place a mouse in an arena; at one end is an unrelated female, at the other, a full sister from a different litter, never encountered before. The mouse spends more time with the sister, suggesting genetically based kin recognition.

How does this work? Rodents produce pheromonal odors with individual signatures, derived from genes called the major histocompatibility complex (MHC). This is a super variable gene cluster that produces unique proteins that form a signature for an individual. This was first studied by immunologists. What does the immune system do? It differentiates between you and invaders—"self" and "nonself"—and attacks the latter. All your cells carry your unique MHC-derived protein, and surveillance immune cells attack any cell lacking this protein password. And MHC-derived proteins also wind up in pheromones, producing a distinctive olfactory signature.

This system can indicate that this mouse is John Smith. How does it also tell that he's your never-before-encountered brother? The closer the relative, the more similar their cluster of MHC genes and the more similar their olfactory signature. Olfactory neurons in a mouse contain receptors that respond most strongly to the mouse's own MHC protein. Thus, if the receptor is maximally stimulated, it means the mouse is sniffing its armpit. If near maximally stimulated, it's a close relative. Moderately, a distant relative. Not at all (though the MHC protein is being detected by other olfactory receptors), it's a hippo's armpit.*

Olfactory recognition of kin accounts for a fascinating phenomenon. Recall from chapter 5 how the adult brain makes new neurons. In rats, pregnancy triggers neurogenesis in the olfactory system. Why there? So that olfactory recognition is in top form when it's time to recognize your newborn; if the neurogenesis doesn't occur, maternal behavior is impaired.[25]

Then there is kin recognition based on imprinted sensory cues. How do I know which newborn to nurse? The one who smells like my vaginal fluid. Which kid do I hang out near? The one who smells like Mom's milk. Many ungulates use such rules. So do birds. Which bird do I know is Mom? The bird whose distinctive song I learned before hatching.

And there are species that figure out relatedness by reasoning; my guess is that male baboons make statistical inferences when identifying their likely offspring: "How much of this mom's peak estrus swelling was spent with me? All. Okay, this is my kid; act accordingly." Which brings us to the most cognitively strategic species, namely us. How do we do kin recognition? In ways that are far from accurate, with interesting consequences.

We start with a long theorized type of pseudo–kin recognition. What if you operate with the rule that you cooperate with (i.e., act related to) individuals who share conspicuous traits with you? This facilitates passing on copies of genes if you possess a gene (or genes) with three properties: (a) it generates that conspicuous signal; (b) recognizes it in others; and (c) makes you cooperate with others who have that signal. It's a kind of primitive, stripped-down kin selection.

Hamilton speculated about the existence of such a "green-beard

* Note: not all olfactory kin recognition is based on MHC proteins; there are numerous other sources of an individual olfactory signature. Note also how this can explain the kin-selection phenomenon mentioned earlier, where sperm form cooperative swimming aggregates only with sperm from the same individual or a close relative. How to pull this off? Use the MHC proteins on the surface of the sperm as Velcro—if two sperm have identical proteins (i.e., they're from the same person), they aggregate very tightly; close relative, not as tightly but still pretty tightly; more distant relative, less tightly, etc.

effect"; if an organism has a gene that codes for both growing a green beard and cooperating with other green bearders, green bearders will flourish when mixed with non–green bearders.[26] Thus, "the crucial requirement for altruism is genetic relatedness at the altruism locus [i.e., merely a multifaceted green-beard gene] and not genealogical relationship over the whole genome."[27]

Green-beard genes exist. Among yeast, cells form cooperative aggregates that need not be identical or even closely related. Instead, it could be any yeast that expresses a gene coding for a cell-surface adhesion protein that sticks to copies of the same molecule on other cells.[28]

Humans show green-beard effects. Crucially, we differ as to what counts as a green-beard trait. Define it narrowly, and we call it parochialism. Include enmity toward those without that green-beard trait and it's xenophobia. Define the green-beard trait as being a member of your species, and you've described a deep sense of humanity.

RECIPROCAL ALTRUISM

So sometimes a chicken is an egg's way of making another egg, genes can be selfish, and sometimes we gladly lay down our lives for two brothers or eight cousins. Does everything have to be about competition, about individuals or groups of relatives leaving *more* copies of their genes than the others, being *more* fit, having *more* reproductive success?* Is the driving force of behavioral evolution always that someone be vanquished?

Not at all. One exception is elegant, if specialized. Remember rock/

* Antisocial behavior in the name of kin selection reaches its apogee in the animal kingdom, as far as I'm concerned, with a phenomenon reported in a 2008 article in the *Wall Street Journal*. What restaurant/fast-food chain has the highest rate of fights among clientele, nationwide? Yup, you guessed it—Chuck E. Cheese's, where the fighting is among parents on edge about anything that would detract from the perfection of their child's birthday party. A particularly common scenario might be where a parent takes exception to some kid hogging a video game and forcefully intervenes to allow their own child to play, leading to an altercation between the parents—Cheney and Seyfarth's monkeys would have no trouble following that one. As reported in another journalistic exposé, such incidents can also involve attacks on the Chuck E. Cheese's mascot, including a case of a father accusing Chuck of having pinned his boy against a wall, while the mouse said he was just trying to squeeze by a crowd of overexcited kids: "The man ripped the mouse's head off and yelled at him in front of said rowdy children, who probably were forever traumatized by the sight of the frightened 19-year-old kid's head sticking out of the giant mouse's neck."

paper/scissors? Paper envelops rock; rock breaks scissors; scissors cut paper. Would rocks want to bash every scissors into extinction? No way. Because then all those papers would enwrap the rocks into extinction. Each participant has an incentive for restraint, producing an equilibrium.

Remarkably, such equilibriums occur in living systems, as shown in a study of the bacteria *Escherichia coli*.[29] The authors generated three colonies of *E. coli*, each with a strength and a weakness. To simplify: Strain 1 secretes a toxin. Strength: it can kill competitor cells. Weakness: making the toxin is energetically costly. Strain 2 is vulnerable to the toxin, in that it has a membrane transporter that absorbs nutrients, and the toxin slips in via that transporter. Strength: it's good at getting food. Weakness: vulnerability to the toxin. Strain 3 doesn't have the transporter and thus isn't vulnerable to the toxin, and it doesn't make the toxin. Strength: it doesn't bear the cost of making the toxin and is insensitive to it. Weakness: it doesn't absorb as much nutrients. Thus, destruction of strain 2 by strain 1 causes the demise of strain 1 thanks to strain 3. The study showed that the strains could exist in equilibrium, each limiting its growth.

Cool. But it doesn't quite fit our intuitions about cooperation. Rock/paper/scissors is to cooperation as peace due to nuclear weapons–based mutually assured destruction is to the Garden of Eden.

Which raises a third fundamental, alongside individual selection and kin selection: reciprocal altruism. "I'll scratch your back if you scratch mine. I'd rather not actually scratch yours if I can get away with it. And I'm watching you in case you try the same."

Despite what you might expect from kin selection, unrelated animals frequently cooperate. Fish swarm in a school, birds fly in formation. Meerkats take risks by giving alarm calls that aid everyone, vampire bats who maintain communal colonies feed one another's babies.*[30] Depending on the species, unrelated primates groom one another, mob predators, and share meat.

Why should nonrelatives cooperate? Because many hands lighten the

* This one is a bit controversial, in that the bat colonies are often made up of somewhat related females, making way for a kin-selection argument.

load. School with other fish, and you're less likely to be eaten (competition for the safest spot—the center—produces what Hamilton termed the "geometry of the selfish herd"). Birds flying in a *V* formation save energy by catching the updraft of the bird in front (raising the question of who gets stuck there).[31] If chimps groom one another, there are fewer parasites.

In a key 1971 paper biologist Robert Trivers laid out the evolutionary logic and parameters by which unrelated organisms engage in "reciprocal altruism"—incurring a fitness cost to enhance a nonrelative's fitness, with the expectation of reciprocation.[32]

It doesn't require consciousness to evolve reciprocal altruism; back to the metaphor of the airplane wing in the wind tunnel. But there are some requirements for its occurrence. Obviously, the species must be social. Furthermore, social interactions have to be frequent enough that the altruist and the indebted are likely to encounter each other again. And individuals must be able to recognize each other.

Amid reciprocal altruism occurring in numerous species, individuals often attempt to cheat (i.e., to not reciprocate) and monitor attempts by others to do the same to them. This raises the realpolitik world of cheating and counterstrategies, the two coevolving in escalating arms races. This is called a "Red Queen" scenario, for the Red Queen in *Through the Looking-Glass*, who must run faster and faster to stay in place.[33]

This raises two key interrelated questions:

- Amid the cold calculations of evolutionary fitness, when is it optimal to cooperate, when to cheat?
- In a world of noncooperators it's disadvantageous to be the first altruist. How do systems of cooperation ever start?*

* To rein in the length of this chapter, I've had to force myself to relegate to this footnote a description of a system of reciprical altruism found in single-cell amoeba called *Dictyostelium discoideum* (aka slime mold). In order to reproduce, individual cells join in a structured colony in which about 80 percent of the cells reproduce and the rest play nonreproductive supporting roles. When the colony consists of two different genetic lines of amoebas, there is cooperation in that each line contributes about 20 percent of its cells to the unfun supporting role. Except that lines evolve to try to cheat by sneaking all their cells into the reproductive pool, and other lines evolve to detect cheaters and refuse to interact with them. For example, the amoebas express a cell-surface protein "adhesion molecule" that lets cells adhere to one another, forming the colony; an anticheater mechanism is to express an adhesion molecule that doesn't recognize (i.e., attach to) the adhesion protein of a cheater line.

Gigantic Question #1: What Strategy for Cooperating Is Optimal?

While biologists were formulating these questions, other scientists were already starting to answer them. In the 1940s "game theory" was founded by the polymath John von Neumann, one of the fathers of computer science. Game theory is the study of strategic decision making. Framed slightly differently, it's the mathematical study of when to cooperate and when to cheat. The topic was already being explored with respect to economics, diplomacy, and warfare. What was needed was for game theorists and biologists to start talking. This occurred around 1980 concerning the Prisoner's Dilemma (PD), introduced in chapter 3. Time to see its parameters in detail.

Two members of a gang, A and B, are arrested. Prosecutors lack evidence to convict them of a major crime but can get them on a lesser charge, for which they'll serve a year in prison. A and B can't communicate with each other. Prosecutors offer each a deal—inform on the other and your sentence is reduced. There are four possible outcomes:

- Both A and B refuse to inform on each other: each serves one year.
- Both A and B inform on each other: each serves two years.
- A informs on B, who remains silent: A walks free and B serves three years.
- B informs on A, who remains silent: B walks and A serves three years.

Thus, each prisoner's dilemma is whether to be loyal to your partner ("cooperate") or betray him ("defect"). The thinking might go, "Best to cooperate. This is my partner; he'll also cooperate, and we'll each serve only a year. But what if I cooperate and he stabs me in the back? He walks, and I'm in for three years. Better defect. But what if we both defect—that's two years. But maybe defect, in case he cooperates . . ." Round and round.*

---

* Some years ago a game show called *Golden Balls* ran in Britain. As the final step in a series of competitions, two competitors would face each other and play a modified version of the PD. There'd be a pot of money (potentially tens of thousands of pounds); each player would have to independently choose either "Split" or "Steal." If both chose Split, they split the money. If one chose Split and the other Steal, the sucker got zero and the defector got everything. If they both chose Steal, they got nothing. YouTube is full of clips from various episodes, and they're embarrassingly

If you play PD once, there is a rational solution. If you, prisoner A, defect, your sentence averages out to one year (zero years if B cooperates, two years if B defects); if you cooperate, the average is two years (one year if B cooperates, three years if B defects). Thus you should defect. In single-round versions of PD, it's always optimal to defect. Not very encouraging for the state of the world.

Suppose there are two rounds of PD. The optimal strategy for the second round is just like in a single-round version—always defect. Given that, the first-round defaults into being like a single-round game—and thus, defect during it also.

What about a three-round game? Defect in the third, meaning that things default into a two-round game. In which case, defect in the second, meaning defect in the first.

It's always optimal to defect in round Z, the final round. And thus it's always optimal to defect in round Z – 1, and thus round Z – 2. . . . In other words, when two individuals play for a *known* number of rounds, the optimal strategy precludes cooperation.

But what if the number of rounds is unknown (an "iterated" PD)? Things get interesting. Which is when the game theorists and biologists met.

The catalyst was political scientist Robert Axelrod of the University of Michigan. He explained to his colleagues how PD works and asked them what strategy they'd use in a game with an unknown number of rounds. The strategies offered varied enormously, with some being hair-raisingly complicated. Axelrod then programmed the various strategies and pitted them against each other in a simulated massive round-robin tournament. Which strategy won, was most optimal?

It was provided by a mathematician at the University of Toronto, Anatol Rapoport; as the mythic path-of-the-hero story goes, it was the simplest strategy. Cooperate in the first round. After that, you do whatever the other player did in the previous round. It was called Tit for Tat. More details:

---

addictive. Also, see this *Radiolab* analysis of the show: www.youtube.com/watch?annotation_id=annotation_1155372699&feature=iv&src_vid=S0qjK3TWZE8&v=zUdBd7BDNu8.

You cooperate (C) in the first round, and if the other player always cooperates (C), you both happily cooperate into the sunset:

Example 1:
You: C C C C C C C C C C. . . .
Her: C C C C C C C C C C. . . .

Suppose the other player starts cooperating but then, tempted by Satan, defects (D) in round 10. You cooperated, and thus you take a hit:

Example 2:
You: C C C C C C C C C C
Her: C C C C C C C C C D

Thus, you Tit for Tat her, punishing her in the next round:

Example 3:
You: C C C C C C C C C C D
Her: C C C C C C C C C D ?

If by then she's resumed cooperating, you do as well; peace returns:

Example 4:
You: C C C C C C C C C C D C C C. . . .
Her: C C C C C C C C C D C C C C. . . .

If she continues defecting, you do as well:

Example 5:
You: C C C C C C C C C C D D D D D. . . .
Her: C C C C C C C C C D D D D D D. . . .

Suppose you play against someone who always defects. Things look like this:

Example 6:
You: C D D D D D D D D D. . . .
Her: D D D D D D D D D D. . . .

This is the Tit for Tat strategy. Note that it can never win. Best case is a draw, if playing against another person using Tit for Tat or someone using an "always cooperate" strategy. Otherwise it loses by a small margin. Every other strategy would always beat Tit for Tat by a small margin. However, other strategies playing against each other can produce catastrophic losses. And when everything is summed, Tit for Tat wins. It lost nearly every battle but won the war. Or rather, the peace. In other words, Tit for Tat drives other strategies to extinction.

Tit for Tat has four things going for it: Its proclivity is to cooperate (i.e., that's its starting state). But it isn't a sucker and punishes defectors. It's forgiving—if the defector resumes cooperating, so will Tit for Tat. And the strategy is simple.

Axelrod's tournament launched a zillion papers about Tit for Tat in PD and related games (more later). Then something crucial occurred—Axelrod and Hamilton hooked up. Biologists studying the evolution of behavior longed to be as quantitative as those studying the evolution of kidneys in desert rats. And here was this world of social scientists studying this very topic, even if they didn't know it. PD provided a framework for thinking about the strategic evolution of cooperation and competition, as Axelrod and Hamilton explored in a 1981 paper (famous enough that it's a buzz phrase—e.g., "How'd your lecture go today?" "Terrible, way behind schedule; I didn't even get to Axelrod and Hamilton").[34]

As the evolutionary biologists started hanging with the political scientists, they inserted real-world possibilities into game scenarios. One addressed a flaw in Tit for Tat.

Let's introduce signal errors—a message is misunderstood, someone forgets to tell someone something, or there's a hiccup of noise in the system. Like in the real world.

There has been a signal error in round 5, with two individuals using a Tit for Tat strategy. This is what everyone means:

Example 7:
You: C C C C C
Her: C C C C C

But thanks to a signal error, this is what you think happened:

Example 8:
You: C C C C C
Her: C C C C D

You think, "What a creep, defecting like that." You defect in the next round. Thus, what you think has happened:

Example 9:
You: C C C C C D
Her: C C C C D C

What she thinks is happening, being unaware of the signal error:

Example 10:
You: C C C C C D
Her: C C C C C C

She thinks, "What a creep, defecting like that." Thus she defects the next round. "Oh, so you want more? I'll give you more," you think, and defect. "Oh, so you want more? I'll give you more," she thinks:

Example 11:
You: C C C C C D C D C D C D C D C D. . . .
Her: C C C C D C D C D C D C D C D C. . . .

When signal errors are possible, a pair of Tit for Tat players are vulnerable to being locked forever in this seesawing of defection.*

* The 1962 geopolitical thriller *Fail-Safe*, by Eugene Burdick and Harvey Wheeler, was premised on a Tit for Tat solution to a signal error. An electronic glitch causes an air force squadron of bombers with nuclear weapons to

The discovery of this vulnerability prompted evolutionary biologists Martin Nowak of Harvard, Karl Sigmund of the University of Vienna, and Robert Boyd of UCLA to provide two solutions.[35] "Contrite Tit for Tat" retaliates only if the other side has defected twice in a row. "Forgiving Tit for Tat" automatically forgives one third of defections. Both avoid doomsday signal-error scenarios but are vulnerable to exploitation.*

A solution to this vulnerability is to shift the frequency of forgiveness in accordance with the likelihood of signal error ("Sorry I'm late again; the train was delayed" being assessed as more plausible and forgivable than "Sorry I'm late again; a meteorite hit my driveway *again*").

Another solution to Tit for Tat's signal-error vulnerability is to use a shifting strategy. At the beginning, in an ocean of heterogeneous strategies, many heavily biased toward defection, start with Tit for Tat. Once they've become extinct, switch to Forgiving Tit for Tat, which outcompetes Tit for Tat when signal errors occur. What is this transition from hard-assed, punitive Tit for Tat to incorporating forgiveness? Establishing trust.

Other elaborations simulate living systems. The computer scientist John Holland of the University of Michigan introduced "genetic algorithms"—strategies that mutate over time.

Another real-world elaboration was to factor in the "cost" of certain strategies—for example, with Tit for Tat, the costs of monitoring for and then punishing cheating—costly alarm systems, police salaries, and jail construction. These are superfluous in a world of no signal errors and nothing but Tit for Tat–ers, and Tit for Tat can be replaced by the cheaper Always Cooperate.

Thus, when there are signal errors, differing costs to different strategies, and the existence of mutations, a cycle emerges: a heterogeneous

---

believe that the United States is under nuclear attack by the USSR; they are to destroy Moscow. The Americans and Soviets see what is happening, and the U.S. military unsuccessfully tries to get the planes to turn back; the Soviets assume the American "Oops, sorry" is a ruse and prepare an all-out counterattack. The American president (modeled on JFK) tries to show his sincerity and stop the attack by sending up fighters to help Soviet jets shoot down the bomber squadron. A few are shot down, but a few get through, and most of the Soviet brass are still convinced it is a ruse. Finally, as the only means to prevent an all-out nuclear exchange, the president does a Tit for Tat, ordering a bomber to drop an equivalent bomb on New York City. Bummer of a signal error. This book scared the willies out of me when I was a kid. I'd regularly scan the skies over my hometown, New York, waiting to see the inevitable bomber.

* I.e., "Oops, sorry, our bad taking out St. Petersburg. We thought we'd sorted out that bug after that Moscow snafu."

population of strategies, including exploitative, noncooperative ones, are replaced by Tit for Tat, then replaced by Forgiving Tit for Tat, then by Always Cooperate—until a mutation reintroduces an exploitative strategy that spreads like wildfire, a wolf among Always Cooperate sheep, starting the cycle all over again. . . .*[36] More and more modifications made the models closer to the real world. Soon the computerized game strategies were having sex with each other, which must have been the most exciting thing ever for the mathematicians involved.

The evolutionary biologists were delighted to generate increasingly sophisticated models with the theoretical economists and theoretical diplomats and theoretical war strategists. The real question was whether animal behavior actually fits any of these models.

One bizarre animal system suggests Tit for Tat enforcement of cooperation involving the black hamlet fish, which form stable pair-bonds.[37] Nothing strange there. The fish can change sex (something that occurs in some fish species). As per usual, reproduction is more metabolically costly for the female than the male. So the fish in a pair take turns being the more expensive female. Say fish A and fish B have been doing their sex-change tango, and most recently A was the expensive female and B the cheap male. Suppose B cheats by staying male, forcing A to continue as female; A switches to male and stays that way until B regains his social conscience and becomes female.

Another widely cited study suggested a Tit for Tat strategy among stickleback fish.[38] The fish is in a tank, and on the other side of a glass partition is something scary—a bigger cichlid fish. The stickleback tentatively darts forward and back, investigating. Now put a mirror in its tank, perpendicular to the axis of the two fish. In other words, thanks to the mirror, there appears to be a *second* cichlid next to the first. Terrifying, except from out of nowhere there's this mysterious second stickleback

---

* A particularly clever exploitative strategy is called Pavlov. If you're playing PD, the most advantageous outcomes for you, in rank order, are: (a) you defect while the other person is a sucker who cooperates; (b) you both cooperate; (c) you both defect; (d) you're the sucker who cooperates while the other defects. Pavlov's basic temperament is to cooperate, but every so often, randomly, it defects, and the rule is that, independent of those occasional random actions, if your play resulted in one of the two better outcomes, you do the same thing again next time; if the result was one of the two worse outcomes, you switch your behavior the next time. What that means is that if you are playing against Always Cooperate or a highly forgiving version of Forgiving Tit for Tat, your occasional random defections are either never or rarely punished, allowing you to exploit the other player at length.

who checks out the second cichlid every time our hero checks out the first—"I have no idea who this guy is, but we're an amazing, coordinated team."

Now convince the stickleback his partner is defecting. Angle the mirror so that the stickleback's reflection is deflected backward. Now when the fish darts forward, his reflection does as well, but—that *jerk!*—it looks like he's hanging back safely (lagging back even half a body length decreases the likelihood of a fish being predated). When the fish believes his partner is defecting, he stops darting forward.

Greater complexity in Tit for Tat–ing is suggested by some animals having multiple roles in their social groups.[39] Back to the playback technique with lions, where the roar of a strange male emanated from a speaker in the bushes (or from a life-sized model of a lion). Lions tentatively came forward to investigate, a risky proposition. Consistently, certain lions hung back. The toleration of these habitual scaredy-cats seemed to violate the demands of reciprocity, until it was recognized that such animals took the lead in other domains (e.g., in hunts). A similar punch line emerges concerning the Damaraland mole rat. The social groups of it and its relative, the naked mole rat, resemble those of social insects, with nonreproductive workers and a single breeding queen.* Researchers noted some workers who never worked and were considerably fatter than the rest. It turns out that they have two specialized jobs—during the rains, they dig through flooded, collapsed tunnels of the burrows, and when necessary, they disperse with the risky task of starting a new colony.

I'm not convinced that a Tit for Tat reciprocity has been clearly demonstrated in other species. But evidence of its strict use would be hard for Martian zoologists to document in humans—after all, there are frequently pairs where one human does all the labor, the other doing

* This doesn't begin to scratch the surface of the weirdness of the naked mole rat. They live underground, have giant incisors and no body hair so that they look like saber-toothed sausages, have remarkably little need for oxygen, have next to no pain receptors in their skin, live about ten times longer than other rodents (up to around thirty years), and are remarkably resistant to cancer. For this reason the prestigious science journal *Nature* named the naked mole rat its Vertebrate of the Year a few years ago, which is much cooler and more impressive than making *People* magazine's list of the Fifty Most Beautiful People in the World.

nothing other than intermittently handing him some green pieces of paper. The point is that animals have systems of reciprocity with sensitivity to cheating.

Gigantic Question #2: How Can Cooperation Ever Start?

So a handful of Tit for Tat–ers can outcompete a mix of other strategies, including highly exploitative, uncooperative ones, losing the battles but winning the war. But what if there's only one Tit for Tat–er in a population of ninety-nine Always Defect–ers? Tit for Tat doesn't stand a chance. Always Defect–ers playing each other produces the second-worst outcome for each. But a Tit for Tat–er playing an Always Defect–er does worse, getting the sucker payoff that first round before becoming a de facto Always Defect–er. This raises the second great challenge for reciprocal altruism: forget which strategy is best at fostering cooperation—how do you ever start *any* type? Amid a sea of Always Defect–ers, the first black hamlet fish, mole rat, or *Dictyostelium* amoeba who, after reading Gandhi, Mandela, Axelrod, and Hamilton, takes the first altruistic step is screwed, lagging behind everyone else forever. One can practically hear the Always Defect amoebas chortling derisively.

Let's make it slightly easier for Tit for Tat to gain a foothold. Consider two Tit for Tat–ers amid ninety-eight Always Defect–ers. Both will crash and burn . . . unless they find each other and form a stable cooperative core, where the Always Defect–ers either must switch to Tit for Tat or go extinct. A nidus of cooperation crystallizes outward through the population.

This is where green-beard effects help, conspicuous features of cooperators that help them recognize one another. Another mechanism is spatial, where the cooperative trait itself facilitates cooperators finding one another.

Another route has been suggested for jump-starting reciprocal altruism. Occasionally a geographic event occurs (say, a land bridge disappears), isolating a subset of a population for generations. What happens in such a "founder population"? Inbreeding, fostering cooperation via kin selection. Eventually the land bridge reappears, the inbred cooperative founder population rejoins the main group, and cooperation propagates outward.*

We return to the issue of starting cooperation in the final chapter.

STANDING ON THREE LEGS

We've now seen the three foundations of thinking about the evolution of behavior—individual selection, kin selection, and reciprocal altruism. Moreover, we've seen how these three concepts can explain otherwise puzzling behaviors. Some concern individual selection, with competitive infanticide as the canonical example. Other behaviors are most explicable with kin selection—why there's male-male aggression between groups in only some primate species; why many species have hereditary ranking systems; why cousin matings are more frequent than one might expect. And some behaviors are all about reciprocal altruism. Why else would a vampire bat, aware of the vanquishing power of group selection, regurgitate blood for someone else's kid?

Let's consider a few more examples.

Pair-Bonding Versus Tournament Species

Suppose you've discovered two new species of primates. Despite watching both for years, here's all you know: In species A, male and females have similar body sizes, coloration, and musculature; in species B, males are far bigger and more muscular than females and have flashy,

* The importance of founder populations was something championed by one of the giants of evolutionary biology, Ernst Mayr of Harvard; in his view, small founder populations were the driving force for new species forming; it is an extension of his thinking to view transient founder populations as a means to establish cooperation in a larger population. Remarkably, Mayr published four well-received books when he was over age ninety, the last one (*What Makes Biology Unique?*) in 2004 at age one hundred, shortly before his death. Inspirational guy, for a bunch of reasons.

Male-female pairs of tamarins (top) and mandrills (bottom)

conspicuous facial coloration (jargon: species B is highly "sexually dimorphic"). We'll now see how these two facts allow you to accurately predict a ton of things about these species.

First off, which species has dramatic, aggressive conflict among males for high dominance rank? Species B, where males have been selected evolutionarily for fighting skills and display. Species A males, in contrast, are minimally aggressive—that's why males haven't been selected for muscle.

What about variability in male reproductive success? In one species 5 percent of the males do nearly all the mating; in the other, all males reproduce a few times. The former describes species B—that's what all the rank competition is about—the latter, species A.

Next, in one species, if a male mates with a female and she conceives, he'll do a ton of child care. In contrast, no such male "parental investment" is seen in the other species. No-brainer: the former describes species A; the few species B males who father most of the kids sure aren't doing child care.

One species has a tendency to twin, the other not. Easy—the twinning is in species A, with two sets of parental hands available.

How picky are males about whom they mate with? In species B, males mate with anyone, anywhere, anytime—it only costs the price of some sperm. In contrast, males of species A, with its rule of "You get her pregnant, you do child care," are more selective. Related to that, which species forms stable pair-bonds? Species A, of course.

After correcting for body size, which species' males have bigger testes and higher sperm count? It's species B, ever prepared for mating, should the opportunity arise.

What do females look for in a potential mate? Species B females get nothing from a male except genes, so they should be good ones. This helps explain the flamboyant secondary sexual characteristics of males—"If I can afford to waste all this energy on muscle plus these ridiculous neon antlers, I must be in great shape, with the sorts of genes you'd want in your kids." In contrast, species A females look for stable, affiliative behavior and good parenting skills in males. This is seen in bird species with this pattern, where males display parenting expertise during courtship—symbolically feeding the female with worms, proof that he'd be a competent worm winner. Related to that, among bird versions of species A and B, in which is a female more likely to abandon her offspring, passing on more copies of her genes by breeding with another male? Species A, where you see "cuckoldry"—because the male is going to stick there, caring for the kids.

Related to that, in species A, females compete aggressively to pair-bond with a particularly desirable (i.e., paternal) male. In contrast, species B females don't need to compete, since all they get from males is sperm, and there's enough to go around from desirable males.

Remarkably, what we've described here is a broad and reliable dichotomy between two social systems, where A is a "pair-bonding" species, B a "tournament" species.*

	Pair-Bonded	Tournament
Male parental behavior	Extensive	Minimal
Male mating pickiness	High	Low
Variability in male reproductive success	Low	High
Testes size, sperm count	Small/low	Large/high
Levels of male-male aggression	Low	High
Degree of sexual dimorphism in body weight, physiology, coloration, and life span	Low	High
Females select for	Parenting skill	Good genes
Rates of cuckoldry	High	Low

Primates that pair-bond include South American monkeys like marmosets, tamarins, and owl monkeys, and among the apes, gibbons (with nonprimate examples including swans, jackals, beavers, and, of course, chapter 4's prairie voles). Classic tournament species include baboons, mandrills, rhesus monkeys, vervets, and chimps (with nonprimate examples including gazelles, lions, sheep, peacocks, and elephant seals). Not all species fit perfectly into either extreme (stay tuned). Nonetheless, the

* Two technical notes. Social monogamy of pair-bonding species doesn't always translate into sexual monogamy. "Tournament" is used by some solely to describe species in which the male-male competition literally takes the form of all the males gathering for competitive displays (as in sage grouse or some ungulate species) but is also used by many, as here, to more broadly describe multimale, multifemale promiscuous mating systems.

point is the internal logic with which the traits of each of these types of species cluster, based on these evolutionary principles.

Parent-Offspring Conflict

Another feature of behavior turns kin selection on its head. The emphasis until now has been on the fact that relatives share many genes and evolutionary goals. Nonetheless, except for identical twins, just as pertinent is relatives not sharing *all* their genes or goals. Which can cause conflict.

There's *parent-offspring conflict.* One classic example is whether a female should give her child great nutrition, guaranteeing his survival, but at the cost of nutrition for her other children (either current or future). This is weaning conflict.[40]

This causes endless primate tantrums.[41] Some female baboon looks frazzled and cranky. Three steps behind is her toddler, making the most pitiful whimpering and whining sounds imaginable. Every few minutes the kid tries to nurse; Mom irritably pushes him away, even slaps him. More wailing. It's parent-offspring weaning conflict; as long as Mom nurses, she's unlikely to ovulate, curtailing her future reproductive potential. Baboon moms evolved to wean their kids at the age where they can feed themselves, and baboon kids evolved to try to delay that day. Interestingly, as females age, with decreasing likelihood of a future child, they become less forceful in weaning.*

There's also mother-fetus conflict. You're a fetus with an evolutionary agenda. What do you want? Maximal nutrition from Mom, and who cares if that impacts her future reproductive potential? Meanwhile, Mom wants to balance current and future reproductive prospects. Remarkably, fetus and Mom have a metabolic struggle involving insulin, the pancreatic hormone secreted when blood glucose levels rise, which triggers glucose en-

* Goodall, in her chimp fieldwork, reported the case of Flint, the youngest child of the very aged Flo; she never fully weaned him, and he remained highly dependent on her, even into adolescence. When she died of old age, Flint underwent what can only be described as a reactive depression, failing to forage or socially interact; he died a month later.

try into target cells. The fetus releases a hormone that makes Mom's cells unresponsive to insulin (i.e., "insulin resistant"), as well as an enzyme that degrades Mom's insulin. Thus Mom absorbs less glucose from her bloodstream, leaving more for the fetus.*

Intersexual Genetic Conflict

In some species the fetus has an ally during maternal/fetal conflict—the father. Consider a species where males are migratory, mating with females and then moving on, never to be seen again. What's a male's opinion about maternal/fetal conflict? Make sure the fetus, i.e., his child, grabs as much nutrition as possible, even if that lessens Mom's future reproductive potential—who cares, that won't be his kid down the line. He's more than just rooting for his fetus.

This helps explain a mysterious, quirky feature of genetics. Normally a gene works the same way, regardless of which parent it comes from. But certain rare genes are "imprinted," working differently, or only being activated, depending on the parent of origin. Their purpose was discovered in a creative synthesis by evolutionary biologist David Haig of Harvard. Paternal imprinted genes bias toward more fetal growth, while maternal imprinted genes counter this. For example, some paternal genes code for potent versions of growth factors, while the maternal genes code for growth factor receptors that are relatively unresponsive. A paternally derived gene expressed in the brain makes newborns more avid nursers; the maternally derived version counters this. It's an arms race, with Dad genetically egging on his offspring toward more growth at the cost of the female's future reproductive plans, and Mom genetically countering this with a more balanced reproductive strategy.†

* What is an extreme version of such insulin resistance called by doctors? Gestational diabetes. In other words, we're back to disciplinary buckets—if you're an OB/GYN, we're talking about a disease. If you're an evolutionary biologist, we're talking about a particularly tumultuous struggle between mom and fetus.

† This arms race is revealed in two classes of diseases. Normal development represents a balancing of paternally derived progrowth genes and maternally derived ones doing the opposite. What if there is a mutation in a paternally derived imprinted gene, removing it from the equation? The counteracting maternal genes, unopposed, greatly inhibit fetal growth, and the fetus doesn't implant. And what if the opposite occurs, with a mutation incapacitating the female gene, leaving the progrowth paternal genes working unopposed? Out-of-control growth of the placenta, resulting in an aggressive cancer, choriocarcinoma.

Tournament species, where males have minimal investment in a female's future reproductive success, have numerous imprinted genes, while pair-bonders don't.[42] What about humans? Stay tuned.

MULTILEVEL SELECTION

So we've got individual selection, kin selection, and reciprocal altruism. And then what happened in recent years? Group selection reappeared, sneaking in the back door.

"Neo–group selection" crashed a long-standing debate as to the "unit of selection."

Genotype Versus Phenotype, and the Most Meaningful Level of Selection

To appreciate this, let's contrast *genotype* and *phenotype.* Genotype = someone's genetic makeup. Phenotype = the traits observable to the outside world produced by that genotype.*

Suppose there's a gene that influences whether your eyebrows come in two separate halves or form a continuous unibrow. You've noted that unibrow prevalence is decreasing in a population. Which is the more important level for understanding why—the gene variant or the eyebrow phenotype? We know after chapter 8 that genotype and phenotype are not synonymous, because of gene/environment interactions. Maybe some prenatal environmental effect silences one version of the gene but not the other. Maybe a subset of the population belongs to a religion where you must cover your eyebrows when around the opposite sex, and thus eyebrow phenotype is untouched by sexual selection.

* Neuroscientists often use the term "endophenotype," which basically means "a trait that we used to be unable to detect at the phenotypic level but now can, thanks to some invention, so we're going to call it an *endo*phenotype, meaning a newly observable trait that is kind of inside you." Your blood type is an endophenotype, detectable with an assay on a blood sample; the size of your amygdala is an endophenotype, detectable with a brain scanner.

You're a grad student researching unibrow decline, and you must choose whether to study things at the genotypic or phenotypic level. Genotypic: sequencing eyebrow gene variants, trying to understand their regulation. Phenotypic: examining, say, eyebrow appearance and mate choice, or whether unibrows absorb more heat from sunlight, thereby damaging the frontal cortex, producing inappropriate social behavior and decreased reproductive success.

This was the debate—is evolution best understood by focusing on genotype or phenotype?

The most visible proponent of the gene-centered view has long been Dawkins, with his iconic "selfish gene" meme—it is the gene that is passed to the next generation, the thing whose variants spread or decline over time. Moreover, a gene is a clear and distinctive sequence of letters, reductive and irrefutable, while phenotypic traits are much fuzzier and less distinct.

This is the core of the concept of "a chicken is just an egg's way of making another egg"—the organism is just a vehicle for the genome to be replicated in the next generation, and behavior is just this wispy epiphenomenon that facilitates the replication.

This gene-centered view can be divided in two. One is that the genome (i.e., the collection of all the genes, regulatory elements, and so on) is the best level to think about things. The more radical view, held by Dawkins, is that the most appropriate level is that of individual genes—i.e., selfish genes, rather than selfish genomes.

Amid some evidence for single-gene selection (an obscure phenomenon called intragenomic conflict, which we won't go into), most people who vote for the importance of gene(s) over phenotype view single-gene selfishness as a bit of a sideshow and vote for the genome level of selection being most important.

Meanwhile, there's the view that phenotype trumps genotype, something championed by Ernst Mayr, Stephen Jay Gould, and others. The core of their argument is that it's phenotypes rather than genotypes that are selected for. As Gould wrote, "No matter how much power Dawkins wishes to assign to genes, there is one thing he cannot give them—direct visibility to natural selection." In that view, genes and the frequencies of their variants are merely the record of what arose from phenotypic selection.[43]

Dawkins introduced a great metaphor: a cake recipe is a genotype, and how the cake tastes is the phenotype.* Genotype chauvinists emphasize that the recipe is what is passed on, the sequence of words that make for a stable replicator. But people select for taste, not recipe, say the phenotypists, and taste reflects more than just the recipe—after all, there are recipe/environment interactions where bakers differ in their skill levels, cakes bake differently at various altitudes, etc. The recipe-versus-taste question can be framed practically: Your cake company isn't selling enough cakes. Do you change the recipe or the baker?

Can't we all get along? There's the obvious bleeding-heart answer, namely that there's room for a range of views and mechanisms in our rainbow-colored tent of evolutionary diversity. Different circumstances bring different levels of selection to the forefront. Sometimes the most informative level is the single gene, sometimes the genome, sometimes a single phenotypic trait, sometimes the collection of all the organism's phenotypic traits.[44] We've just arrived at the reasonable idea of multilevel selection.

---

* By now it should be clear how often thinking about evolution is helped by metaphors and analogies. This prompted a great meta-analogy that everyone attributes to the biologist Steve Jones of University College London: "Evolution is to analogy as statues are to birdshit."

The Resurrection of Group Selection

Hooray, progress. Sometimes it makes the most sense to pay attention to the recipe, sometimes to the baking process; the recipe is what is replicated, the taste what is chosen.

But there's another level. Sometimes cake sales can be changed most consequentially by altering something other than recipe or taste—advertisements, packaging, or the perception of whether the cake is a staple or a luxury. Sometimes sales are changed by tying the product to a particular audience—think of products that advertise fair-trade practices, the Nation of Islam's Your Black Muslim Bakery, or the Christian fundamentalist ideology of Chick-fil-A restaurants. And in those cases recipe and taste can both be trumped by ideology in purchasing decisions.

This is where neo–group selection fits into multilevel selection—the idea that some heritable traits may be maladaptive for the individual but adaptive for a group. This has cooperation and prosociality written all over it, straight out of the analysis of Tit for Tat–ers finding one another in a sea of Always Defect–ers. Stated more formally, it's when A dominates B but a *group* of Bs dominates a group of As.

Here's a great example of neo–group selectionism: As a poultry farmer, you want your groups of chickens to lay as many eggs as possible. Take the most prolific egg layer in each group, forming them into a group of superstar chickens who, presumably, will be hugely productive. Instead, egg production is miniscule.[45]

Why was each superstar the egg queen in her original group? Because she would aggressively peck subordinates enough to stress them into reduced fertility. Put all these mean ones together, and a group of subordinated chickens will outproduce them.

This is a world away from "animals behave for the good of the species." Instead, this is the circumstance of a genetically influenced trait that, while adaptive on an individual level, emerges as maladaptive when shared by a group and where there is competition between groups (e.g., for an ecological niche).

There's been considerable resistance to neo–group selectionism. Part of it is visceral, often pronounced among the old guard—"Great, we've finally confiscated all the *Wild Kingdom* videos, and now we're back to playing Whac-A-Mole with group selection sentimentality?" But the more fundamental resistance is from people who distinguish bad old group selection from neo–group selection, accept that the latter can occur, but think it's very rare.

Maybe so, across the animal kingdom. But neo–group selection plays out with great frequency and consequence in humans. Groups compete for hunting grounds, pastures, water sources. Cultures magnify the intensity of between-group selection and lessen within-group selection with ethnocentrism, religious intolerance, race-based politics, and so on. The economist Samuel Bowles, of the Santa Fe Institute, emphasizes how intergroup conflict like war is the driving force for intragroup cooperation ("parochial altruism"); he refers to intergroup conflict as "altruism's midwife."[46]

Most in the field now both accept multilevel selection and see room for instances of neo–group selection, especially in humans. Much of this reemergence is the work of two scientists. The first is David Sloan Wilson of the State University of New York at Binghamton, who spent decades pushing for neo–group selection (although he sees it not really as "neo" but rather as old-style group selection finally getting some scientific rigor), generally being dismissed, and arguing his case with research of his own, studies ranging from fish sociality to the evolution of religion. He slowly convinced some people, most importantly the second scientist, Edward O. Wilson of Harvard (no relation). E. O. Wilson is arguably the most important naturalist of the last half of the twentieth century, an architect of the sociobiology synthesis along with a number of other fields, a biology god. E. O. Wilson had long dismissed David Sloan Wilson's ideas. And then a few years back, the octogenarian E. O. Wilson did something extraordinary—he decided he was wrong. And then he published a key paper with the other Wilson—"Rethinking the Theoretical Foundation of Sociobiology." My respect for these two, both as people and as scientists, is enormous.[47]

Thus something resembling détente has occurred among the advocates for the importance of differing levels of selection. Our three-legged chair of individual selection, kin selection, and reciprocal altruism seems more stable with four legs.

AND US

Where do humans fit into all this? Our behavior closely matches the predictions of these evolutionary models. Until you look more closely.[48]

Let's start by clearing up some misconceptions. First, we are not descended from chimps. Or from any extant animal. We and chimps share a common ancestor from roughly five million years ago (and genomics show that chimps have been as busy evolving since then as we have).[49]

And there are misconceptions as to which ape is our "closest relative." In my experience, someone who is fond of duck hunting and country music usually votes chimp, but if you eat organic food and know about oxytocin, it's bonobo. The reality is that we're equally related to both, sharing roughly 98 to 99 percent of our DNA with each. Svante Pääbo of the Max Planck Institutes in Germany has shown that 1.6 percent of the human genome is more related to bonobos than to chimps; 1.7 percent more to chimps than to bonobos.*[50] Despite the combination of some of our most fervent wishes and excuses, we're neither bonobos nor chimps.

On to how the conceptual building blocks of behavioral evolution apply to humans.

Promiscuous Tournament or Monogamous Pair-Bonded?

I can't resist starting with an irresistible question—so, are we a pair-bonded or tournament species?[51]

Western civilization doesn't give a clear answer. We praise stable,

* Pääbo, who is a stunningly good scientist, pioneered the sequencing of ancient DNA, being the first to sequence the genomes of mammoths and Neanderthals.

devoted relationships yet are titillated, tempted, and succumb to alternatives at a high rate. Once divorces are legalized, a large percentage of marriages end in them, yet a smaller percentage of married people get divorced—i.e., the high divorce rate arises from serial divorcers.

Anthropology doesn't help either. Most cultures have allowed polygyny. But within such cultures most people are (socially) monogamous. But most of those men would presumably be polygamous if they could buy more wives.

What about human sexual dimorphism? Men are roughly 10 percent taller and 20 percent heavier than women, need 20 percent more calories, and have life spans 6 percent shorter—more dimorphic than monogamous species, less than polygamous ones. Likewise with subtle secondary sexual characteristics like canine length, where men average slightly longer canines than women. Moreover, compared with, say, monogamous gibbons, human males have proportionately bigger testes and higher sperm counts . . . but pale in comparison with polygamous chimps. And back to imprinted genes, reflecting intersexual genetic competition, which are numerous in tournament species and nonexistent in pair-bonders. What about humans? Some such genes, but not many.

Measure after measure, it's the same. We aren't classically monogamous or polygamous. As everyone from poets to divorce attorneys can attest, we are by nature profoundly confused—mildly polygynous, floating somewhere in between.*

Individual Selection

At first pass we seem like a great example of a species where the driving force on behavior is maximizing reproductive success, where a person can be an egg's way of making another egg, where selfish genes triumph. Just consider the traditional perk of powerful men: being polygamous. Pharaoh Ramses II, incongruously now associated with a brand of condoms, had 160 children and probably couldn't tell any of them from

* A great analysis of this can be found in *The Myth of Monogamy* (New York: Henry Holt, 2002), by University of Washington psychologist David Barash and psychiatrist Judith Lipton.

Moses. Within half a century of his death in 1953, Ibn Saud, the founder of the Saudi dynasty, had more than three thousand descendants. Genetic studies suggest that around sixteen million people today are descended from Genghis Khan. And in recent decades more than one hundred children each were fathered by King Sobhuza II of Swaziland, Ibn Saud's son King Saud, the dictator Jean-Bédel Bokassa of the Central African Republic, plus various fundamentalist Mormon leaders.[52]

The human male drive to maximize reproductive success is shown by a key fact—the most common cause of individual human violence is male-male competition for direct or indirect reproductive access to females. And then there is the dizzyingly common male violence against females for coercive sex or as a response to rejection.

So plenty of human behaviors would make sense to a baboon or elephant seal. But that's only half the story. Despite Ramses, Ibn Saud, and Bokassa, numerous people forgo reproducing, often because of theology or ideology. And an entire sect—the United Society of Believers in Christ's Second Appearing, aka the Shakers, will soon be extinct because of its adherents' celibacy. And finally, the supposed selfishness of human genes driving individual selection must accommodate individuals sacrificing themselves for strangers.

Earlier in the chapter I presented competitive infanticide as stark evidence of the importance of individual selection. Does anything like that occur in humans? The psychologists Martin Daly and (the late) Margo Wilson of McMaster University in Canada looked at patterns of child abuse and made a striking observation—a child is far more likely to be abused or killed by a stepparent than by a parent. This is readily framed as parallel to competitive infanticide.[53]

This finding, termed the "Cinderella effect," while embraced by human sociobiologists, has also been robustly criticized. Some charge that socioeconomic status was not sufficiently controlled for (homes with a stepparent, rather than two biological parents, generally have less income and more economic stress, known causes of displacement aggression). Others think there's a detection bias—the same degree of abuse is more likely to be identified by authorities when it's committed by a stepparent.

And the finding has been independently replicated in some but not all studies. I think the jury is still out on the subject.

Kin Selection

Where do humans fit in terms of kin selection? We've already seen examples that fit well—e.g., the fraternal polyandry in Tibet, the weirdness of women liking the smell of their male cousins, the universality of nepotism.

Moreover, humans are obsessed with kin relations in culture after culture, with elaborate systems of kinship terms (just go into a store and look at the Hallmark cards organized by kinship category—for a sister, a brother, an uncle, and so on). And in contrast to other primates who leave their natal group around adolescence, when humans in traditional society marry someone from another group and go live with them, they maintain contact with their family of origin.[54]

Moreover, from New Guinea highlanders to the Hatfields and McCoys, feuds and vendettas occur along clan lines of relatedness. We typically bequeath our money and land among our descendants rather than among strangers. From ancient Egypt to North Korea and on to the Kennedys and Bushes, we have dynastic rule. How's this for a display of human kin selection: Subjects were given a scenario of a bus hurtling toward a human and a nondescript dog, and they could only save one. Whom would they pick? It depended on degree of relatedness, as one progressed from sibling (1 percent chose the dog over the sibling) to grandparent (2 percent) to distant cousin (16 percent) to foreigner (26 percent).[55]

As another measure of the importance of kinship in human interactions, people can't be compelled to testify in court against a first-degree relative in many countries and American states. And when humans have damage to the (emotional) vmPFC, they become so unemotionally utilitarian that they would choose to harm family members in order to save strangers.[56]

There's a fascinating historical example of how wrong it feels when someone chooses strangers over kin. This is the story of Pavlik Morozov, a boy in Stalin's Soviet Union.[57] Young Pavlik, according to the official story, was a model citizen, an ardent flag-waving patriot. In 1932 he chose the state over his kin, denouncing his father (for supposed black marketeering), who was promptly arrested and executed. Soon afterward the boy was killed, allegedly by relatives who felt more strongly about kin selection than he did.

The regime's propagandists embraced the story. Statues of the young martyr to the revolution were erected. Poems and songs were written; schools were named for him. An opera was composed, a hagiographic movie made.

As the story emerged, Stalin was told about the boy. And what was the response of the man most benefiting from such fealty to the state? Was it "If only all my citizens were that righteous; this lad gives me hope for our future"? No. According to historian Vejas Liulevicius of the University of Tennessee, when told about Pavlik, Stalin snorted derisively and said, "What a little pig, to have done such a thing to his own family." And then he turned the propagandists loose.

Thus even Stalin was of the same opinion as most mammals: something's wrong with that kid. Human social interactions are profoundly organized around kin selection; with the rare exception of a Pavlik Morozov, blood is thicker than water.

Naturally, until you look more closely.

For starters, yes, across cultures we are obsessed with kinship terms, but the terms often don't overlap with actual biological relatedness.

We certainly have clan vendettas, but we also have wars where combatants on opposing sides have higher degrees of relatedness than do fighters on the same side. Brothers fought on opposing sides in the Battle of Gettysburg.[58]

Relatives and their armies battle over royal succession; the cousins George V of England, Nicholas II of Russia, and Wilhelm II of Germany happily oversaw and sponsored World War I. And intrafamily individual violence occurs (although at extremely low rates when corrected for amount of time spent together). There's patricide, often an act of revenge for a long history of abuse, and fratricide. Rarely due to conflicts over issues of economic or reproductive importance—stolen birthrights of biblical proportion, someone sleeping with their sibling's spouse—fratricide is most often about long-standing irritants and disagreements that just happen to boil over into lethality (in early May 2016, for example, a Florida man was charged with second-degree murder in the killing of his brother—during a dispute over a cheeseburger). And then there is the hideous commonness of honor killings in parts of the world, as we've seen.[59]

The most puzzling cases of intrafamily violence, in terms of kin selection, are of parents killing children, a phenomenon most commonly arising from combined homicide/suicide, profound mental illness, or abuse that unintentionally proves fatal.*[60] And then there are cases where a mother kills an unwanted child who is viewed as a hindrance—parent/offspring conflict flecked with the spittle of madness.[61]

While we bequeath money to our descendants, we also give charitably to strangers on the other side of the planet (thank you, Bill and Melinda Gates) and adopt orphans from other continents. (Sure, as we'll see in a later chapter, being charitable is tinged with self-interest, and most people who adopt kids do so because they cannot have biological offspring—but the occurrence of either act violates strict kin selection.) And in the primogeniture system of land inheritance, birth order trumps degree of relatedness.

Thus we have textbook examples of kin selection, but also dramatic exceptions.

Why do humans have such marked deviations from kin selection? I think this often reflects how humans go about recognizing relatives. We don't do it with certainty, by innate recognition of MHC-derived pheromones, the way rodents do (despite our being able to distinguish degrees of relatedness to some extent by smell). Nor do we do it by imprinting on

* I recently read in the *Kenya Daily Nation* about a case that takes one's breath away, in its challenge not just to kin-selection thinking but to our notion of what boundaries of inhumanity would never be crossed. In parts of Tanzania there is the widespread belief that the organs of albinos have magical healing powers, and a shocking number of albino individuals are murdered there for that purpose. The story reports on a five-year-old albino girl in neighboring Kenya and the plot to smuggle her into Tanzania to sell her to a shaman to be sacrificed for her organs. The plotters? The girl's stepfather and father.

sensory cues, deciding, "This person is my mother because I remember that her voice was the loudest when I was a fetus."

Instead we do kin recognition cognitively, by thinking about it. But crucially, not always rationally—as a general rule, we treat people like relatives when they *feel* like relatives.

One fascinating example is the Westermarck effect, demonstrated by marriage patterns among people raised in the Israeli kibbutz system.[62] Communal child rearing is central to the ethos of the traditional socialist agricultural kibbutz approach. Children know who their parents are and interact with them a few hours a day. But otherwise they live, learn, play, eat, and sleep with the cohort of kids their age in communal quarters staffed by nurses and teachers.

In the 1970s anthropologist Joseph Shepher examined records of all the marriages that had ever occurred between people from the same kibbutz. And out of the nearly three thousand occurrences, there was no instance of two individuals marrying who had been in the same age group during their first six years of life. Oh, people from the same peer group typically had loving, close, lifelong relationships. But no sexual attraction. "I love him/her to pieces, but am I attracted? Yech—he/she feels like my sibling." Who feels like a relative (and thus not like a potential mate)? Someone with whom you took a lot of baths when you both were kids.

How's this for irrationality? Back to people deciding whether to save the person or the dog. The decision depended not only on who the person was (sibling, cousin, stranger) but also on who the dog was—a strange dog or your own. Remarkably, *46 percent* of women would save their dog over a foreign tourist. What would any rational baboon, pika, or lion conclude? That those women believe they are more related to a neotenized wolf than to another human. Why else act that way? "I'll gladly lay down my life for eight cousins or my awesome labradoodle, Sadie."

Human irrationality in distinguishing kin from nonkin takes us to the heart of our best and worst behaviors. This is because of something crucial—we can be *manipulated* into feeling more or less related to someone than we actually are. When it is the former, wonderful things happen—we adopt, donate, advocate for, empathize with. We look at someone very different from us and see similarities. It is called pseudokinship. And the converse? One of the tools of the propagandist and ideologue drumming up hatred of the out-group—blacks, Jews, Muslims, Tutsis, Armenians, Roma—is to characterize them as animals, vermin, cockroaches, pathogens. So different that they hardly count as human. It's called pseudospeciation, and as will be seen in chapter 15, it underpins many of our worst moments.

Reciprocal Altruism and Neo–Group Selectionism

There's not much to say here other than that this is the most interesting stuff in the chapter. When Axelrod got his round-robin tournament all fired up, he didn't canvass, say, fish for their Prisoner's Dilemma strategies. He asked humans.

We're the species with unprecedented cooperation among unrelated individuals, even total strangers; *Dictyostelium* colonies are green with envy at the human ability to do a wave in a football stadium. We work collectively as hunter-gatherers or as IT execs. Likewise when we go to war or help disaster victims a world away. We work as teams to hijack planes and fly them into buildings, or to award a Nobel Peace Prize.

Rules, laws, treaties, penalties, social conscience, an inner voice, morals, ethics, divine retribution, kindergarten songs about sharing—all driven by the third leg of the evolution of behavior, namely that it is evolutionarily advantageous for nonrelatives to cooperate. Sometimes.

One manifestation of this strong human tendency has been appreciated recently by anthropologists. The standard view of hunter-gatherers was that their cooperative, egalitarian nature reflected high degrees of relatedness within groups—i.e., kin selection. The man-the-hunter version of hunter-gatherers viewed this as arising from patrilocality (i.e., where a woman, when marrying, moves to live with the group of her new husband), while the groovy-hunter-gatherers version tied it to matrilocality (i.e., the opposite). However, a study of more than five thousand people from thirty-two hunter-gatherer societies from around the world* showed that only around 40 percent of people within bands are blood relatives.[63] In other words, hunter-gatherer cooperativeness, the social building block of 99 percent of hominin history, rests at least as much on reciprocal altruism among nonrelatives as on kin selection (with chapter 9's caveat that this assumes that contemporary hunter-gatherers are good stand-ins for ancestral ones).

So humans excel at cooperation among nonrelatives. We've already considered circumstances that favor reciprocal altruism; this will be returned to in the final chapter. Moreover, it's not just groups of nice chickens outcompeting groups of mean ones that has revivified group selectionism. It is at the core of cooperation and competition among human groups and cultures.

Thus humans deviate from the strict predictions concerning the evolution of behavior. And this is pertinent when considering three major criticisms of sociobiology.

THE USUAL: WHERE ARE THE GENES?

I pointed out earlier a requirement for neo–group selection, namely that genes be involved in a trait that differs more between than within groups. This applies to everything in this chapter. The first requirement for a trait to evolve is that it be heritable. But this is often forgotten along the way, as evolutionary models tacitly assume genetic influences. Chapter 8 showed how tenuous is the idea that there is "the gene," or even genes, "for" aggression, intelligence, empathy, and so on. Given that, even

* For example, !Kung bushmen from the Kalahari in Botswana, aboriginal Australian groups, Mbuti Pygmies from the Congo, northern Canadian Inuits, and Amazonian populations.

more tenuous would be the idea of a gene(s) for maximizing your reproductive success by, say, "mating indiscriminately with every available female," or by "abandoning your kids and finding a new mate, because the father will raise them."

So critics will often demand, "Show me the gene that you assume is there." And sociobiologists will respond, "Show me a more parsimonious explanation than this assumption."

THE NEXT CHALLENGE: IS EVOLUTIONARY CHANGE CONTINUOUS AND GRADUAL?

The term "evolution" carries context-dependent baggage. If you're in the Bible Belt, evolution is leftist besmirching of God, morality, and human exceptionalism. But to extreme leftists, "evolution" is a reactionary term, the slow change that impedes real change—"All reform undermines revolution." This next challenge addresses whether evolution is actually more about rapid revolution than about slow reform.

A basic sociobiological premise is that evolutionary change is gradual, incremental. As a selective pressure gradually changes, a useful gene variant grows more common in a population's gene pool. As enough changes accrue, the population may even constitute a new species ("phyletic gradualism"). Over millions of years, dinosaurs gradually turn into chickens, organisms emerge that qualify as mammals as glandular secretions slowly evolve into milk, thumbs increasingly oppose in proto-primates. Evolution is gradual, continuous.

In 1972 Stephen Jay Gould and paleontologist Niles Eldredge of the American Museum of Natural History proposed an idea that simmered and then caught fire in the 1980s. They argued that evolution isn't gradual; instead, most of the time nothing happens, and evolution occurs in intermittent rapid, dramatic lurches.[64]

Punctuated Equilibrium

Their idea, which they called punctuated equilibrium, was anchored in paleontology. Fossil records, we all know, show gradualism—human ancestors show progressively larger skulls, more upright posture, and so on. And if two fossils in chronological progression differ a lot, a jump in the gradualism, there must be an intermediate form that is the "missing link" from a time between those two fossils. With enough fossils in a lineage, things will look gradualist.

Eldredge and Gould focused on there being plenty of fossil records that were complete chronologically (for example, trilobites and snails, Eldredge's and Gould's specialties, respectively) and didn't show gradualism. Instead there were long periods of stasis, of unchanged fossils, and then, in a paleontological blink of an eye, there'd be a rapid transition to a very different form. Maybe evolution is mostly like this, they argued. What triggers punctuated events of sudden change? A sudden, massive selective factor that kills most of a species, the only survivors being ones with some obscure genetic trait that turned out to be vital—an "evolutionary bottleneck."

Why does punctuated equilibrium challenge sociobiological thinking? Sociobiological gradualism implies that every smidgen of difference in fitness counts, that every slight advantage of one individual over another at leaving copies of genes in future generations translates into evolutionary change. At every juncture, optimizing competition, cooperation, aggression, parental investment, all of it, is evolutionarily consequential. And if instead there is mostly evolutionary stasis, much of this chapter becomes mostly irrelevant.*

The sociobiologists were not amused. They called the punctuated equilibrium people "jerks" (while the punctuated equilibrium people

* Related to this was the notion that most of the evolution of behavior was not about dealing with the social complexities of fellow species members but about dealing with abiotic (i.e., nonbiological) pressures. In other words, that behavior evolved mostly for dealing with the environment, rather than for competing with other individuals. Again, the main implication of that for our purposes is that it would be another way in which the gradualist importance of interindividual competition was less than the sociobiologists thought. This emphasis on the importance of abiotic selective pressures was common among Soviet evolutionary biologists, probably reflecting not only the Marxist ideology but also the awful winters.

called them "creeps"—get it? PE = evolution in a series of jerks; sociobiology = evolution as a gradual, creeping process).* Gradualist sociobiologists responded with strong rebuttals, taking a number of forms:

They're just snail shells. First, there are some very complete fossil lineages that are gradualist. And don't forget, said the gradualists, these punctuated equilibrium guys are talking about trilobite and snail fossils. The fossil record we're most interested in—primates, hominins—is too spotty to tell if it is gradualist or punctuated.

How fast do their eyes blink? Next, said the gradualists, remember, these punctuated equilibrium fans are paleontologists. They see long periods of stasis and then extremely rapid blink-of-the-eye changes in the fossil record. But with fossils the blink of an eye, a stretch of time unresolvably short in the fossil record, could be 50,000 to 100,000 years. That's plenty of time for evolution bloody in tooth and claw to happen. This is only a partial refutation, since if a paleontological blink of the eye is so long, paleontological stasis is humongously long.

They're missing the important stuff. A key rebuttal is to remind everyone that paleontologists study things that are fossilized. Bones, shells, bugs in amber. Not organs—brains, pituitaries, ovaries. Not cells—neurons, endocrine cells, eggs, sperm. Not molecules—neurotransmitters, hormones, enzymes. In other words, none of the interesting stuff. Those punctuated equilibrium nudniks spend their careers measuring zillions of snail shells and, based on that, say we're wrong about the evolution of behavior?

This opens the way for some compromise. Maybe the hominin pelvis did indeed evolve in a punctuated manner, with long periods of stasis and bursts of rapid change. And maybe the pituitary's evolution was punctuated as well, but with punctuations at different times. And maybe steroid hormone receptors and the organization of frontocortical neurons and the inventions of oxytocin and vasopressin all evolved that way also, but each undergoing punctuated change at a different time. Overlap and average these punctuated patterns, and it will be gradualist. This only gets you so

* Who says a scientist can't be the life of the party?

far, though, since it assumes the occurrence of numerous evolutionary bottlenecks.

Where's the molecular biology? One of the strongest gradualist retorts was a molecular one. Micromutation, consisting of point, insertion, and deletion mutations that subtly change the function of preexisting proteins, is all about gradualism. But what mechanisms of molecular evolution explain rapid, dramatic change and long periods of stasis?

As we saw in chapter 8, recent decades have provided many possible molecular mechanisms for rapid change. This is the world of macromutations: (a) traditional point, insertion, and deletion mutations in genes whose proteins have amplifying network effects (transcription factors, splicing enzymes, transposes) in an exon expressed in multiple proteins in genes for enzymes involved in epigenetics; (b) traditional mutations in promoters, transforming the when/where/how-much of gene expression (remember that promoter change that makes polygamous voles monogamous); (c) untraditional mutations such as the duplication or deletion of entire genes. All means for big, rapid changes.

But what about a molecular mechanism for the stasis? Plunk a random mutation into a transcription-factor gene, thereby creating a new cluster of genes never before expressed simultaneously. What are the odds that it won't be a disaster? Randomly mutate a gene for an enzyme that mediates epigenetic changes, thereby producing randomly different patterns of gene silencing. Right, that's bound to work out swell. Parachute a transposable genetic element into the middle of some gene, change a splicing enzyme so that it mixes and matches different exons in multiple proteins. Both asking for major trouble. Implicit in this is stasis, a conservatism about evolutionary change—it takes very unique macro changes during times of very unique challenge to luck out.

Show us some actual rapid change. A final rebuttal from gradualists was to demand real-time evidence of rapid evolutionary change in species. And plenty exist. One example was wonderful research by the Russian geneticist Dmitry Belyaev, who in the 1950s domesticated Siberian silver foxes.[65] He bred captive ones for their willingness to be in proximity to

humans, and within thirty-five generations he'd generated tame foxes who'd cuddle in your arms. Pretty punctuated, I'd say. The problem here is that this is artificial rather than natural selection.

Interestingly, the opposite has occurred in Moscow, which has a population of thirty thousand feral dogs dating back to the nineteenth century (and where some contemporary dogs have famously mastered riding the Moscow subway system).[66] Most Moscow dogs are now descendants of generations of feral dogs, and over that time they have evolved to have a unique pack structure, avoid humans, and no longer wag their tails. In other words, they're evolving into something wolflike. Most likely, the first generations of these feral populations were subject to fierce selection for these traits, and it's their descendants who comprise the current population.*[67]

* Something fascinating about the foxes and Moscow dogs: Both were selected primarily or exclusively for behavioral traits. But along with those traits came changes in appearance: The foxes are *cute*—shorter snouts, rounder ears and foreheads, curly tails, more varied coloration than standard foxes. And the Moscow dogs, exactly the opposite. If you want to domesticate a species, breed it for arrested development—a dog is basically a baby wolf, interacting with humans as if they're all Mommy, and with the cute baby features. Same with the foxes, and just the opposite with the Moscow dogs. There is evidence that domestication mostly works on genes disproportionately related to brain development.

Rapid change in the human gene pool has occurred as well with the spread of lactase persistence—a change in the gene for the enzyme lactase, which digests lactose, such that it persists into adulthood, allowing adults to consume dairy.[68] The new variant is common in populations that subsist on dairy—pastoralists like Mongolian nomads or East African Maasai—and is virtually nonexistent in populations that don't use dairy after weaning—Chinese and Southeast Asians. Lactase persistence evolved and spread in a fraction of a geologic blink of an eye—in the last ten thousand years or so, coevolving with domestication of dairy animals.

Feral Moscow dogs

Other genes have spread in humans even faster. For example, a variant of a gene called ASPM, which is involved in cell division during brain development, has emerged and spread to about 20 percent of humans in the last 5,800 years.[69] And genes that confer resistance to malaria (at the cost of other diseases, such as sickle-cell disease or the thalassemias) are even younger.

Still, thousands of years counts as fast only for snail shell obsessives. However, evolution has been observed in real time. A classic example is the work of the Princeton evolutionary biologists Peter and Rosemary Grant, who, over the course of decades of work in the Galapagos, demonstrated substantial evolutionary change in Darwin's finches. Evolutionary change in humans has occurred in genes related to metabolism, when populations transition from traditional to Westernized diets (e.g., Pacific Islanders from Nauru, Native Americans of the Pima tribe in Arizona). The first generations with Westernized diets develop catastrophically high rates of obesity, hypertension, adult-onset diabetes, and death at early ages, thanks to "thrifty" genotypes that are great at storing nutrients, honed by millennia of sparser diets. But within a few generations diabetes rates begin to subside, as there is an increased prevalence in the population of "sloppier" metabolic genotypes.[70]

Thus, there are examples of rapid changes in gene frequencies in real time. Are there examples of gradualism? That's hard to show because gradual change is, er, gradual. A great example, however, comes from decades of work by Richard Lenski of Michigan State University. He has cultured *E. coli* bacteria colonies under constant conditions for 58,000 generations, roughly equivalent to a million years of human evolution. Over that time, different colonies have *gradually* evolved in distinctive ways, becoming more adapted.[71]

Thus, both gradualism and punctuated change occur in evolution, probably depending upon the genes involved—for example, there has been faster evolution of genes expressed in some brain regions than others. And no matter how rapid the changes, there's always some degree of gradualism—no female has ever given birth to a member of a new species.[72]

A FINAL CHALLENGE LACED WITH POLITICS: IS EVERYTHING ADAPTIVE?

As we've seen, variants of genes that make organisms more adapted to their environment increase in frequency over time. But what about the reverse—if a trait is prevalent in a population, must it mean that it evolved in the past *because* it was adaptive?[73]

"Adaptationism" assumes this is typically the case; an adaptationist approach is to determine whether a trait is indeed adaptive and, if so, what the selective forces were that brought it about. Much of sociobiological thinking is adaptationist in flavor.

This was subject to scathing criticism by the likes of Stephen Jay

Gould and Harvard geneticist Richard Lewontin, who mocked the approach as "just so" stories, after Kipling's absurdist fantasies about how certain traits came to be: how the elephant got its trunk (because of a tug-of-war with a crocodile), how the zebra got its stripes, how the giraffe got a long neck. So why not, supposedly ask the sociobiologists in this critique, how the baboon male got big cojones while the gorilla male got little ones? Observe a behavior, generate a just-so story that assumes adaptation, and the person with the best just-so story wins. How the evolutionary biologist got his tenure. In their view, sociobiological standards lack rigor. As one critic, Andrew Brown, stated, "The problem was that sociobiology explained too much and predicted too little."[74]

According to Gould, traits often evolve for one reason and are later co-opted for another use (fancy term: "exaptation"); for example, feathers predate the evolution of bird flight and originally evolved for insulation.[75] Only later did their aerodynamic uses become relevant. Similarly, the duplication of a gene for a steroid hormone receptor (as mentioned many chapters ago) allowed one copy to randomly drift in its DNA sequence, producing an "orphan" receptor with no use—until a novel steroid hormone was synthesized that happened to bind to it. This haphazard, jury-rigged quality evokes the aphorism "Evolution is a tinkerer, not an inventor." It works with whatever's available as selective pressures change, producing a result that may not be the most adaptive but is good enough, given the starting materials. Squid are not great swimmers compared with sailfish (maximum speed: sixty-eight miles per hour). But they're damn good for something whose great-great-grandparents were mollusks.

Meanwhile, ran the criticism, some traits exist not because they're adaptive, or were adapted for something else but got co-opted, but because they're baggage carried along with other traits that were selected for. It was here that Gould and Lewontin famously introduced "spandrels" in their 1979 paper "The Spandrels of San Marco and the Panglossian Paradigm: A Critique of the Adaptationist Programme." A spandrel is an architectural term for the space between two arches, and Gould and

Lewontin considered the artwork on the spandrels of the Basilica San Marco in Venice.*

Gould and Lewontin's stereotypical adaptationist would look at these spandrels and conclude that they were built to provide spaces for the artwork. In other words, that these spandrels evolved for their adaptive value in providing space for art. In reality they didn't evolve for a purpose—if you're going to have a series of arches (which most definitely exist for the adaptive purpose of holding up a dome), a space between each pair is an inevitable by-product. No adaptation. And as long as these spaces were carried along as evolutionary baggage as a result of selection for the adaptive arches, might as well paint on them. In that view, male nipples are spandrels—they serve an adaptive role in females and came along for the ride as baggage in males because there's been no particular selection *against* males having them.† Gould and Lewontin argued that numerous traits that prompted just-so stories from adaptationists are merely spandrels.

Sociobiologists responded to spandrelism by noting that the rigor in pronouncing something a spandrel was not intrinsically greater than that in pronouncing it adaptive.[76] In other words, the former provide just-*not*-so stories. Psychologist David Barash and psychiatrist Judith Lipton compared spandrelites to the character Topsy in *Uncle Tom's Cabin*, who states that she "just growed"—when faced with evidence of adaptation in traits,

* Apparently there has been a brouhaha over the fact that San Marco's arches don't quite fit the technical architectural definition of a spandrel. Whatever.

† Considerable debate and speculation have gone into the question of whether female orgasm is a spandrel, carried along as baggage by the selection that gave rise to it in males. Enough said; fools rush in. . . .

they'd conclude that those traits are mere baggage, without adaptive purpose, providing explanations that explained nothing—"just growed stories."

Furthermore, sociobiologists argued, adaptationist approaches were more rigorous than Gouldian caricature; rather than explaining everything and predicting nothing, sociobiological approaches predict plenty. Is, say, competitive infanticide a just-so story? Not when you can predict with some accuracy whether it will occur in a species based on its social structure. Nor is the pair-bond/tournament comparison, when you can predict a vast amount of information about the behavior, physiology, and genetics of species ranging across the animal kingdom simply by knowing their degree of sexual dimorphism. Furthermore, evolution leaves an echo of selection for adaptive traits when there is evidence of "special design"—complex, beneficial functions where a number of traits converge on the same function.

All this would be your basic, fun academic squabble, except that underlying the criticisms of adaptationism, gradualism, and sociobiology is a political issue. This is embedded in the title of the spandrel paper: the "Panglossian paradigm." This refers to Voltaire's Dr. Pangloss and his absurd belief, despite life's miseries, that this is the "best of all possible worlds." In this criticism, adaptationism stinks of the naturalistic fallacy, the view that if nature has produced something, it must be a good thing. That furthermore, "good" in the sense of, say, solving the selective problem of water retention in deserts, is in some indefinable way also morally "good." That if ant species make slaves, if male orangutans frequently rape females, and if for hundreds of thousands of years hominin males drink milk directly out of the container, it is because it is somehow "meant" to be that way.

When aired as a criticism in this context, the naturalistic fallacy had an edge to it. In its early years human sociobiology was wildly controversial, with conferences picketed and talks disrupted, with zoologists guarded by police at lectures, all sorts of outlandish things. On one storied occasion,

E. O. Wilson was physically attacked while giving a talk.* Anthropology departments split in two, collegial relationships were destroyed. This was particularly so at Harvard, where many of the principals could be found—Wilson, Gould, Lewontin, Trivers, Hrdy, the primatologist Irven DeVore, the geneticist Jonathan Beckwith.

Things were so febrile because sociobiology was accused of using biology to justify the status quo—conservative social Darwinism that implied that if societies are filled with violence, unequal distribution of resources, capitalistic stratification, male dominance, xenophobia, and so on, these things are in our nature and probably evolved for good reasons. The critics used the "is versus ought" contrast, saying, "Sociobiologists imply that when an unfair feature of life *is* the case, it is because it *ought* to be." And the sociobiologists responded by flipping is/ought around: "We agree that life *ought* to be fair, but nonetheless, this *is* reality. Saying that we advocate something just because we report it is like saying oncologists advocate cancer."

The conflict had a personal tinge. This was because by chance (or not, depending on your viewpoint), that first generation of American sociobiologists were all white Southerners—Wilson, Trivers,† DeVore, Hrdy; in contrast, the first generation of its loudest critics were all Northeastern, urban, Jewish leftists—Harvard's Gould, Lewontin, Beckwith, Ruth Hubbard, Princeton's Leon Kamin, and MIT's Noam Chomsky. You can see how the "there's a hidden agenda here" charge arose from both sides.‡

It's easy to see how punctuated equilibrium generated similar ideolog-

* Well, maybe not all that dramatically—someone poured a pitcher of water on his head. But still.

† This easy picture was complicated by Trivers being a friend and coauthor of Black Panther Party founder Huey Newton.

‡ By great fortune, I arrived at Harvard as a freshman bio/anthropology major the season when Wilson published *Sociobiology* and all hell broke loose. And while it was fantastic, giddying fun for me, watching the fireworks, the personal nature of the attacks was clearly devastating to some of the principals—for example, protestors at Wilson's talks regularly and absurdly chanted about his being a genocidal racist. Those years afforded me the chance to observe some of the players up close, to even get to know a few of them slightly, and both camps had roughly equal distributions of terrific, admirable role models and arrogant, insufferable egotists. My favorite story from this period: Many sociobiologists favored a macho, hard-edged persona. One day I rushed into the office of one of them, Professor X, holding a new paper I had just read. This guy was famed for a sociobiological model about some behavior, and this paper, by his adversary Professor Z, ripped apart the model with page after page of statistical analyses. "Wow, did you see this paper? Whaddya think?" I stupidly asked. Professor X flipped through the paper *backward*, glancing at the equations now and then. Finally he dismissively tossed the paper on his desk and delivered the ultimate sociobiological putdown: "Professor Z has a slide rule instead of a penis."

ical battles, given its premise that evolution is mostly about long periods of stasis pierced by revolutionary upheaval. In their original publication, Gould and Eldredge asserted that the law of nature "holds that a new quality emerges in a leap as the slow accumulation of quantitative changes, long resisted by a stable system, finally forces it rapidly from one state into another." This was a bold assertion that the heuristic of dialectical materialism not only extends beyond the economic world into the naturalistic one, but is ontologically rooted in the essential sameness of both worlds' dynamic of resolution of irresolvable contradictions.* It is Marx and Engels as trilobite and snail.†

Eventually the paroxysms about adaptationism versus spandrels, gradualism versus punctuated change, and the very notion of a science of human sociobiology subsided. The political posturing lost steam, the demographic contrasts between the two camps softened, the general quality of research improved considerably, and everybody got some gray hair and a bit more calm.

This has paved the way for a sensible, middle-of-the-road middle age for the field. There's clear empirical evidence for both gradualism and punctuated change, and for molecular mechanisms underlying both. There's less adaptation than extreme adaptationists claim, but fewer spandrels than touted by spandrelites. While sociobiology may explain too much and predict too little, it does predict many broad features of behavior and social systems across species. Moreover, even though the notion of selection happening at the level of groups has been resurrected from the graves of self-sacrificial elderly wildebeest, it is probably a rare occurrence; nonetheless, it is most likely to occur in the species that is the focus of this book. Finally, all of this is anchored in evolution being a fact, albeit a wildly complex one.

* I have no idea what it is that I just wrote. . . .
† Ditto.

Remarkably, we've finished this first part of the book. A behavior has occurred; what happened in everything from a second to a million years earlier that helps explain why it happened? Some themes have come up repeatedly:

- The context and meaning of a behavior are usually more interesting and complex than the mechanics of the behavior.
- To understand things, you must incorporate neurons *and* hormones *and* early development *and* genes, etc., etc.
- These aren't separate categories—there are few clear-cut causal agents, so don't count on there being *the* brain region, *the* neurotransmitter, *the* gene, *the* cultural influence, or *the* single anything that explains a behavior.
- Instead of causes, biology is repeatedly about propensities, potentials, vulnerabilities, predispositions, proclivities, interactions, modulations, contingencies, if/then clauses, context dependencies, exacerbation or diminution of preexisting tendencies. Circles and loops and spirals and Möbius strips.
- No one said this was easy. But the subject matters.

And thus we transition to the second part, synthesizing this material in order to look at realms of behavior where this matters the most.

Eleven

Us Versus Them

As a kid, I saw the original 1968 version of *Planet of the Apes*. As a future primatologist, I was mesmerized, saw it repeatedly, and loved the cheesy ape costumes.

Years later I discovered a great anecdote about the filming of the movie, related by both Charlton Heston and Kim Hunter, its stars: at lunchtime, the people playing chimps and those playing gorillas ate in separate groups.[1]

As it's been said (most often attributed to Robert Benchley), "There are two kinds of people in the world: those who divide the world into two kinds of people and those who don't." There are more of the first. And it is vastly consequential when people are divided into Us and Them, in-group and out-group, "the people" (i.e., our kind) and the Others.

This chapter explores our tendency to form Us/Them dichotomies and to favor the former. Is this mind-set universal? How malleable are "Us" and "Them" categories? Is there hope that human clannishness and xenophobia can be vanquished so that Hollywood-extra chimps and gorillas break bread together?

THE STRENGTH OF US/THEM

Our brains form Us/Them dichotomies (henceforth, "Us/Theming," for brevity) with stunning speed.[2] As discussed in chapter 3, fifty-millisecond exposure to the face of someone of another race activates the amygdala, while failing to activate the fusiform face area as much as same-race faces do—all within a few hundred milliseconds. Similarly, the brain groups faces by gender or social status at roughly the same speed.

Rapid, automatic biases against a Them can be demonstrated with the fiendishly clever Implicit Association Test (IAT).[3]

Suppose you are unconsciously prejudiced against trolls. To simplify the IAT enormously: A computer screen flashes either pictures of humans or trolls or words with positive connotations (e.g., "honest") or negative ones ("deceitful"). Sometimes the rule is "If you see a human or a positive term, press the red button; if it's a troll or a negative term, press the blue button." And sometimes it's "Human or negative term, press red; troll or positive term, press blue." Because of your antitroll bias, pairing a troll with a positive term, or a human with a negative, is discordant and slightly distracting. Thus you pause for a few milliseconds before pressing a button.

It's automatic—you're not fuming about clannish troll business practices or troll brutality in the Battle of Somewhere in 1523. You're processing words and pictures, and unconsciously you pause, stopped by the dissonance linking troll and "lovely," or human and "malodorous." Run enough rounds and that pattern of delay emerges, revealing your bias.

The brain's fault lines dividing Us from Them were shown in chapter 4's discussion of oxytocin. Recall how the hormone prompts trust, generosity, and cooperation toward Us but crappier behavior toward Them—more preemptive aggression in economic play, more advocacy of sacrificing Them (but not Us) for the greater good. Oxytocin exaggerates Us/Them-ing.

This is hugely interesting. If you like broccoli but spurn cauliflower, no hormone amplifies both preferences. Ditto for liking chess and disdaining backgammon. Oxytocin's opposing effects on Us and Them demonstrate the salience of such dichotomizing.

Our depth of Us/Them-ing is supported further by something remarkable—other species do it as well. Initially this doesn't seem profound. After all, chimps kill males from other groups, baboon troops bristle when encountering each other, animals of all stripes tense at strangers.

This simply reflects not taking kindly to someone new, a Them. But some other species have a broader concept of Us and Them.[4] For example, chimp groups that have swollen in number might divide; murderous animosities soon emerge between ex-groupmates. Remarkably, you can show automatic Us/Them-ing in other primates with a monkey equivalent of the IAT. In one study animals were shown pictures of either members of their own or the neighboring group, interspersed with positive things (e.g., fruit) or negative (e.g., spiders). And monkeys looked longer at discordant pairings (e.g., group members with spiders). These monkeys don't just fight neighbors over resources. They have negative associations with them—"Those guys are like yucky spiders, but us, *us*, we're like luscious tropical fruit."*

Numerous experiments confirm that the brain differentially processes images within milliseconds based on minimal cues about race or gender.[5] Similarly, consider "minimal group" paradigms, pioneered in the 1970s by Henri Tajfel of the University of Bristol. He showed that even if groupings

* Two important points: This intergroup bias effect was demonstrable in males but not females and was most pronounced when males were looking at pictures of other males. Second, shortly after publication the paper was retracted; apparently a data-coding error called into question some of the findings; however, those described here were unaltered by this error, and I think they are perfectly valid. Out of commendable cautiousness, the authors, all top researchers, retracted the paper.

are based on flimsy differences (e.g., whether someone over- or underestimated the number of dots in a picture), in-group biases, such as higher levels of cooperation, still soon develop. Such prosociality is about group identification—people preferentially allocate resources to anonymous in-group individuals.

Merely grouping people activates parochial biases, no matter how tenuous the basis of the grouping. In general, minimal group paradigms enhance our opinion of Us rather than lessening our opinion of Them. I guess that's meager good news—at least we resist thinking that people who came up heads on the coin toss (in contrast to our admirable tails) eat their dead.

The power of minimal, arbitrary groupings to elicit Us/Them-ing recalls "green-beard effects" from chapter 10. Recall how these hover between prosociality due to kin selection and due to reciprocal altruism—they require an arbitrary, conspicuous, genetically based trait (e.g., a green beard) that indicates a tendency to act altruistically toward other green-bearders—under those conditions, green-bearders flourish.

Us/Them-ing based on minimal shared traits is like psychological rather than genetic green-beard effects. We feel positive associations with people who share the most meaningless traits with us.

As a great example, in one study subjects conversed with a researcher who, unbeknownst to them, did or didn't mimic their movements (for example, leg crossing).[6] Not only is mimicry pleasing, activating mesolimbic dopamine, but it also made subjects more likely to help the researcher, picking up their dropped pen. An unconscious Us-ness born from someone slouching in a chair like you do.

Thus an invisible strategy becomes yoked to an arbitrary green-beard marker. What helps define a particular culture? Values, beliefs, attributions, ideologies. All invisible, until they are yoked with arbitrary markers such as dress, ornamentation, or regional accent. Consider two value-laden approaches to what to do to a cow: (A) eat it; (B) worship it. Two As or two Bs would be more peaceful when sorting out cow options than an A and B together. What might reliably mark someone who uses approach A? Maybe a Stetson and cowboy boots. And a B person? Perhaps a sari or a Nehru jacket. Those markers were initially arbitrary—nothing about the object called a sari intrinsically suggests a belief that cows are sacred because a god tends them. And there's no inevitable link between carnivory and a Stetson's shape—it keeps the sun out of your eyes and off your neck, useful whether you tend cows because you love steak or because Lord Krishna tended cows. Minimal group studies show our propensity for generating biased Us/Thems from arbitrary differences. What we then do is link arbitrary markers to meaningful differences in values and beliefs.

And then something happens with those arbitrary markers. We (e.g., primates, rats, Pavlov's dogs) can be conditioned to associate something arbitrary, like a bell, with a reward.[7] As the association solidifies, is the ringing bell still "just" a marker symbolizing impending pleasure, or does it become pleasurable itself? Elegant work related to the mesolimbic dopamine system shows that in a substantial subset of rats, the arbitrary signal itself becomes rewarding. Similarly, an arbitrary symbol of an Us core value gradually takes on a life and power of its own, becoming the signified instead of the signifier. Thus, for example, the scattering of colors and patterns on cloth that constitutes a nation's flag becomes something that people will kill and die for.*

The strength of Us/Them-ing is shown by its emergence in kids. By age three to four, kids already group people by race and gender, have more negative views of such Thems, and perceive other-race faces as being angrier than same-race faces.[8]

And even earlier. Infants learn same-race faces better than other-race. (How can you tell? Show an infant a picture of someone repeatedly; she looks at it less each time. Now show a different face—if she can't tell

* A powerful example of this is seen in the first war of Indian independence, also known as the Sepoy Mutiny, of 1857. Indian soldiers—sepoys—serving in the British East India Company's army rebelled when it became known that the bullets they were issued were greased in either tallow, derived from cows, or lard, from pigs—major offenses to the Hindu and Muslim soldiers, respectively. Mind you, this was not the British colonial overlords doing something offensive to the core cultural values of either group—for example, declaring Allah a false prophet or banning polytheistic worship. Virtually every culture on earth has food prohibitions, often pretty arbitrary ones meant to merely signal core values (kosher laws for Orthodox Jews, for example, revolve around zoological arcana about whether a species has a cloven hoof) but that eventually gain a huge power. Before it was over, the Sepoy Mutiny killed more than 100,000 Indians.

the two apart, she barely glances at it. But if it's recognized as being new, there's excitement, and longer looking).[9]

Four important thoughts about kids dichotomizing:

- Are children learning these prejudices from their parents? Not necessarily. Kids grow in environments whose nonrandom stimuli tacitly pave the way for dichotomizing. If an infant sees faces of only one skin color, the salient thing about the first face with a different skin color will be the skin color.
- Racial dichotomies are formed during a crucial developmental period. As evidence, children adopted before age eight by someone of a different race develop the expertise at face recognition of the adoptive parent's race.[10]
- Kids learn dichotomies in the absence of any ill intent. When a kindergarten teacher says, "Good morning, boys and girls," the kids are being taught that dividing the world that way is more meaningful than saying, "Good morning, those of you who have lost a tooth and those of you who haven't yet." It's everywhere, from "she" and "he" meaning different things to those languages so taken with gender dichotomizing that inanimate objects are given honorary gonads.*[11]
- Racial Us/Them-ing can seem indelibly entrenched in kids because the parents most intent on preventing it are often lousy at it. As shown in studies, liberals are typically uncomfortable discussing race with their children. Instead they counter the lure of Us/Them-ing with abstractions that mean squat to kids—"It's wonderful that everyone can be friends" or "Barney is purple, and we love Barney."

Thus, the strength of Us/Them-ing is shown by: (a) the speed and minimal sensory stimuli required for the brain to process group differences; (b) the unconscious automaticity of such processes; (c) its presence in other

* Animate too, in ways that no doubt make historical sense, but still. For example, in French the kidney is masculine but the bladder feminine; the trachea is feminine, the esophagus masculine.

primates and very young humans; and (d) the tendency to group according to arbitrary differences, and to then imbue those markers with power.

US

Us/Them-ing typically involves inflating the merits of Us concerning core values—we are more correct, wise, moral, and worthy when it comes to knowing what the gods want/running the economy/raising kids/fighting this war. Us-ness also involves inflating the merits of our arbitrary markers, and that can take some work—rationalizing why our food is tastier, our music more moving, our language more logical or poetic.

Perhaps even more than superiority, feelings about Us center on shared obligations, on willingness and expectation of mutuality.[12] The essence of an Us mind-set is nonrandom clustering producing higher-than-expected frequencies of positive interactions. As we saw in chapter 10, the logical strategy in one-round Prisoner's Dilemma is to defect. Cooperation flourishes when games have an uncertain number of rounds, and with the capacity for our reputations to precede us. Groups, by definition, have multiple-round games and the means to spread news of someone being a jerk.

This sense of obligation and reciprocity among Us is shown in economic games, where players are more trusting, generous, and cooperative with in-group than with out-group members (even with minimal group paradigms, where players know that groupings are arbitrary).[13] Chimps even show this trust element where they have to choose between (a) being guaranteed to receive some unexciting food and (b) getting some fabulous food if another chimp will share it with them. Chimps opt for the second scenario, requiring trust, when the other chimp is a grooming partner.

Moreover, priming people to think of a victim of violence as an Us, rather than a Them, increases the odds of their intervening. And recall from chapter 3 how fans at a soccer match are more likely to aid an injured spectator if he's wearing home-team insignias.[14]

Enhanced prosociality for in-group members does not even require face-to-face interactions. In one study subjects from an ethnically polarized neighborhood encountered an open, stamped questionnaire on the sidewalk near a mailbox. Subjects were more likely to mail it if the questionnaire indicated support for a value of the subject's ethnic group.[15]

In-group obligation is shown by people feeling more need to make amends for transgressions against an Us than against a Them. For the former, people usually make amends to the wronged individual and act more prosocially to the group overall. But people often make in-group amends by being more antisocial to another group. Moreover, in such scenarios, the guiltier the person feels about her in-group violation, the worse she is to Thems.[16]

Thus, sometimes you help Us by directly helping Us, sometimes by hurting Them. This raises a broad issue about in-group parochialism: is the goal that your group do well, or simply better than Them? If the former, maximizing absolute levels of in-group well-being is the goal, and the levels of rewards to Them is irrelevant; if the latter, the goal is maximizing the gap between Us and Them.

Both occur. Doing better rather than doing well makes sense in zero-sum games where, say, only one team can win, and where winning with scores of 1–0, 10–0, and 10–9 are equivalent. Moreover, for sectarian sports fans, there is similar mesolimbic dopamine activation when the home team wins or when a hated rival loses to a third party.*[17] This is schadenfreude, gloating, where their pain is your gain.

It's problematic when non–zero sum games are viewed as zero-sum (winner-take-all).[18] It's not a great mind-set to think you've won World War III if afterward Us have two mud huts and three fire sticks and They have only one of each.† A horrific version of this thinking occurred late

* This study, of avid Yankees and Red Sox fans, also showed that this neuroimaging pattern was strongest among individuals who self-reported the highest likelihood of feeling aggressive toward a fan from the other side (after controlling for the person's general level of aggression).

† I heard a brutally cynical joke years ago built around the zero-sum notion that *anything* that is bad for Them is automatically good for Us: So God appears to all the leaders on earth and announces that he is destroying the world because of human sinfulness. The American president assembles his cabinet and says, "I have good news and bad news. God exists, but he is going to destroy earth." The premier of the Soviet Union (this was told during the atheistic days of the USSR) pulls together his advisers and says, "I have bad news and worse news. God exists, and he's going to destroy earth." And the prime minister of Israel tells his cabinet, "I have good news and great news. God exists, and he's going to destroy the Palestinians for us."

during World War I, when the Allies knew they had more resources (i.e., soldiers) than Germany. Therefore, the British commander, Douglas Haig, declared a strategy of "ceaseless attrition," where Britain went on the offensive, no matter how many of his men were killed—as long as the Germans lost at least as many.

So in-group parochialism is often more concerned about Us beating Them than with Us simply doing well. This is the essence of tolerating inequality in the name of loyalty. Consistent with that, priming loyalty strengthens in-group favoritism and identification, while priming equality does the opposite.[19]

Intertwined with in-group loyalty and favoritism is an enhanced capacity for empathy. For example, the amygdala activates when viewing fearful faces, but only of group members; when it's an out-group member, Them showing fear might even be good news—if it scares Them, bring it on. Moreover, recall from chapter 3 the "isomorphic sensorimotor" reflex of tensing your hand when watching a hand being poked with a needle; the reflex is stronger if it is a same-race hand.[20]

As we saw, people are more likely to make amends for transgressions against Us than against Them. What about responses to other in-group members violating a norm?

Most common is forgiving Us more readily than Them. As we will see, this is often rationalized—we screw up because of special circumstances; They screw up because that's how They are.

Something interesting can happen when someone's transgression constitutes airing the group's dirty laundry that affirms a negative stereotype. The resulting in-group shame can provoke high levels of punishment as a signal to outsiders.[21]

The United States, with its rationalizations and ambivalences about ethnicity, provides many such examples. Consider Rudy Giuliani, who grew up in Brooklyn in an Italian American enclave dominated by organized crime (Giuliani's father served time for armed robbery and then worked for his brother-in-law, a Mob loan shark). Giuliani rose to national prominence in 1985 as the attorney prosecuting the "Five Families" in the Mafia Commission Trial, effectively destroying them. He was strongly motivated to counter the stereotype of "Italian American" as synonymous with organized crime. When referring to his accomplishment, he said, "And if that's not enough to remove the Mafia prejudice, then there probably could not be anything you could do to remove it." If you want someone to prosecute mafiosi with tireless intensity, get a proud Italian American outraged by the stereotypes generated by the Mob.[22]

Similar motivations were widely ascribed to Chris Darden, the African American attorney who was cocounsel for the prosecution in the O. J. Simpson trial. Ditto for the trial of Julius and Ethel Rosenberg and Morton Sobell, all Jewish, accused of spying for the Soviet Union. The very public prosecution was by two Jews, Roy Cohn and Irving Saypol, and presided over by a Jewish judge, Irving Kaufman, all eager to counter the stereotype of Jews as disloyal "internationalists." After death sentences were doled out, Kaufman was honored by the American Jewish Committee, the Anti-Defamation League, and the Jewish War Veterans.*[23] Giuliani, Darden, Cohn, Saypol, and Kaufman show that being in a group means that someone else's behaviors can make you look bad.†[24]

This raises a larger issue, namely our sense of obligation and loyalty to Us as a whole. At one extreme it can be contractual. This can be literal, with professional athletes in team sports. It is expected that when jocks sign contracts, they will play their hardest, putting the team's fortune above showboating. But the obligations are finite—they're not expected to sacrifice their lives for their team. And when athletes are traded, they don't serve as a fifth column, throwing games in their new uniform to benefit their old team. The core of such a contractual relationship is the fungibility of both employer and employee.

At the other extreme, of course, are Us memberships that are not fungible and transcend negotiation. People aren't traded from the Shiites to

* These scenarios of members of ethnic, religious, or racial groups eager to publicly punish a shameful in-group member can cut both ways—which behavior constitutes acting shamefully? During the 1969 Chicago Seven trial, presided over by a Jewish judge, Julius Hoffman, the chief provocateur of the defendants, the Jewish Abbie Hoffman (no relation), would insult and taunt the judge by yelling, "You are a shanda fur die goyim [Yiddish for "disgrace in front of the gentiles"]. You would have served Hitler better."

† This plays out currently with the deep resentment of many in the Muslim American community that they are especially obligated to condemn Islamic fundamentalist terrorism and will be under a cloud of suspicion if they do not. "I refuse to condemn, not because I don't condemn, but . . . because doing so would mean that I agree that I deserve to be asked," states the Arab American writer Amer Zahr.

the Sunnis, or from the Iraqi Kurds to the Sami herders in Finland. It would be a rare Kurd who would want to be Sami, and his ancestors would likely turn over in their graves when he nuzzled his first reindeer. Converts are often subject to ferocious retribution by those they left—consider Meriam Ibrahim, sentenced to death in Sudan in 2014 for converting to Christianity—and suspicion from those they've joined. With the sense of one's lot being permanent come distinctive elements of Us-ness. You don't sign a faith-based baseball contract with vague promises of a salary. But Us-ness based on sacred values, with wholes bigger than sums of parts, where unenforceable obligations stretch across generations, millennia, even into afterlives, where it's Us, right or wrong, is the essence of faith-based relationships.

Naturally, things are more complicated. Sometimes an athlete choosing to switch teams is viewed as betraying a sacred trust. Consider the perceived treachery when LeBron James chose to leave the Cavaliers of his hometown, Cleveland, and the perception of his choice to return as akin to the Second Coming. At the other extreme of group membership, people do convert, emigrate, assimilate, and, especially in the United States, wind up a pretty atypical Us—consider ex-Governor Bobby Jindal of Louisiana, with his rich Southern accent and Christian faith, born Piyush Jindal to Hindu immigrant parents from India. And consider the complexities in, to use a horrible phrase, the unidirectionality of fungibility—Muslim fundamentalists who would execute Meriam Ibrahim while advocating forced conversions *to* Islam at the point of a sword.

The nature of group membership can be bloodily contentious concerning people's relationship to the state. Is it contractual? The people pay taxes, obey laws, serve in the army; the government provides social services, builds roads, and helps after hurricanes. Or is it one of sacred values? The people give absolute obedience and the state provides the myths of the Fatherland. Few such citizens can conceive that if the stork had arbitrarily deposited them elsewhere, they'd fervently feel the innate rightness of a different brand of exceptionalism, goose-stepping to different martial music.

THOSE THEMS

Just as we view Us in standardized ways, there are patterns in how we view Them. A consistent one is viewing Them as threatening, angry, and untrustworthy. Take space aliens in movies, as an interesting example. In an analysis of nearly a hundred pertinent movies, starting with Georges Méliès's pioneering 1902 *A Trip to the Moon*, nearly 80 percent present aliens as malevolent, with the remainder either benevolent or neutral.* In economic games people implicitly treat members of other races as less trustworthy or reciprocating. Whites judge African American faces as angrier than white faces, and racially ambiguous faces with angry expressions are more likely to be categorized as the other race. White subjects become more likely to support juvenile criminals being tried as adults when primed to think about black (versus white) offenders. And the unconscious sense of Them as menacing can be remarkably abstract—baseball fans tend to underestimate the distance to a rival team's stadium, while Americans hostile to Mexican immigrants underestimate the distance to Mexico City.

But Thems do not solely evoke a sense of menace; sometimes it's disgust. Back to the insular cortex, which in most animals is about gustatory disgust—biting into rotten food—but whose human portfolio includes moral and aesthetic disgust. Pictures of drug addicts or the homeless typically activate the insula, not the amygdala.[25]

Being disgusted by another group's abstract beliefs isn't naturally the role of the insula, which evolved to care about disgusting tastes and smells. Us/Them markers provide a stepping-stone. Feeling disgusted by Them because they eat repulsive, sacred, or adorable things, slather themselves with rancid scents, dress in scandalous ways—these are things the insula

* Examples of "good alien" movies include *The Day the Earth Stood Still* (1951), *Close Encounters of the Third Kind* (1977), *Cocoon* (1985), *Avatar* (2009), and, of course, *E.T. the Extra-Terrestrial* (1982). Meanwhile, the numerous "bad alien" movies include *The Blob* (1958), *Liquid Sky* (1982), *Devil Girl from Mars* (1954), and, naturally, *Alien* (1979). The bad/good alien ratio is consistent over the decades (in other words, it's not the case that the 1950s were disproportionately filled with scary alien movies so that the directors weren't called by the House Un-American Activities Committee, and the 1960s filled with good-alien efforts of stoned directors just back from Kathmandu). I thank Katrina Hui, a student research assistant, who did this analysis.

can sink its teeth into. In the words of the psychologist Paul Rozin of the University of Pennsylvania, "Disgust serves as an ethnic or out-group marker." Establishing that They eat disgusting things provides momentum for deciding that They also have disgusting ideas about, say, deontological ethics.[26]

The role of disgust in Them-ing explains some individual differences in its magnitude. Specifically, people with the strongest negative attitudes toward immigrants, foreigners, and socially deviant groups tend to have low thresholds for interpersonal disgust (e.g., are resistant to wearing a stranger's clothes or sitting in a warm seat just vacated).[27] We will return to this finding in chapter 15.

Some Thems are ridiculous, i.e., subject to ridicule and mockery, humor as hostility.[28] Out-groups mocking the in-group is a weapon of the weak, damaging the mighty and lessening the sting of subordination. When an in-group mocks an out-group, it's to solidify negative stereotypes and reify the hierarchy. In line with this, individuals with a high "social dominance orientation" (acceptance of hierarchy and group inequality) are most likely to enjoy jokes about out-groups.

Thems are also frequently viewed as simpler and more homogeneous than Us, with simpler emotions and less sensitivity to pain. David Berreby, in his superb book *Us and Them: The Science of Identity*, gives a striking example, namely that whether it was ancient Rome, medieval England, imperial China, or the antebellum South, the elite had the system-justifying stereotype of slaves as simple, childlike, and incapable of independence.[29]

Essentialism is all about viewing Them as homogeneous and interchangeable, the idea that while we are individuals, they have a monolithic, immutable, icky essence. A long history of bad relations with Thems fuels essentialist thinking—"They've always been like this and always will be." As does having few personal interactions with Thems—after all, the more interactions with Thems, the more exceptions accumulate that challenge essentialist stereotyping. But infrequency of interactions is not required, as evidenced by essentialist thinking about the opposite sex.[30]

Thus, Thems come in different flavors—threatening and angry, disgusting and repellent, primitive and undifferentiated.

Thoughts Versus Feelings About Them

How much are our thoughts about Them post-hoc rationalizations for our feelings about Them? Back to interactions between cognition and affect.

Us/Them-ing is readily framed cognitively. John Jost of NYU has explored one domain of this, namely the cognitive cartwheels of those on top to justify the existing system's unequal status quo. Cognitive gymnastics also occur when our negative, homogeneous view of a type of Them must accommodate the appealing celebrity Them, the Them neighbor, the Them who has saved our ass—"Ah, *this* Them is different" (no doubt followed by a self-congratulatory sense of open-mindedness).[31]

Cognitive subtlety can be needed in viewing Thems as threats.[32] Being afraid that the Them approaching you will rob you is rife with affect and particularism. But fearing that those Thems will take our jobs, manipulate the banks, dilute our bloodlines, make our children gay, etc., requires future-oriented cognition about economics, sociology, political science, and pseudoscience.

Thus Us/Them-ing can arise from cognitive capacities to generalize, imagine the future, infer hidden motivations, and use language to align these cognitions with other Us-es. As we saw, other primates not only kill individuals because they are Thems but have negative associations about them as well. Nonetheless, no other primate kills over ideology, theology, or aesthetics.

Despite the importance of thought in Us/Them-ing, its core is emotional and automatic.[33] In the words of Berreby in his book, "Stereotyping isn't a case of lazy, short-cutting cognition. It isn't conscious cognition at all." Such automaticity generates statements like "I can't put my finger on why, but it's just wrong when They do that." Work by Jonathan Haidt of NYU shows that in such circumstances, cognitions are post-hoc justifications for feelings and intuitions, to convince yourself that you have indeed rationally put your finger on why.

The automaticity of Us/Them-ing is shown by the speed of the amygdala and insula in making such dichotomies—the brain weighing in affectively precedes conscious awareness, or there never is conscious awareness, as with subliminal stimuli. Another measure of the affective core of Them-ing is when no one even knows the basis of a prejudice. Consider the Cagots, a minority in France whose persecution began in the eleventh century and continued well into the last one.[34] Cagots were required to live outside villages, dress distinctively, sit separately in church, and do menial jobs. Yet they didn't differ in appearance, religion, accent, or names, and no one knows why they were pariahs. They may have descended from Moorish soldiers in the Islamic invasion of Spain and thus were discriminated against by Christians. Or they might have been early Christians, and discrimination against them was started by *non*-Christians. No one knew the sins of ancestral Cagots or how to recognize Cagots beyond community knowledge. During the French Revolution, Cagots burned birth certificates in government offices to destroy proof of their status.

The automaticity is seen in another way. Consider an individual with an impassioned hatred for an array of out-groups.[35] There are two ways to explain this. Option 1: He has carefully concluded that group A's trade policies hurt the economy *and* just happens to also believe that group B's ancestors were blasphemous, *and* thinks that group C members don't express sufficient contrition for a war started by their grandparents, *and* perceives group D members as pushy, *and* thinks that group E undermines family values. That's a lot of cognitive just-happens-to's. Option 2: The guy's authoritarian temperament is unsettled by novelty and ambiguity about hierarchies; this isn't a set of coherent cognitions. As we saw in chapter 7, Theodor Adorno, in trying to understand the roots of fascism, formalized this authoritarian temperament. Individuals prejudiced against one type of out-group tend toward being prejudiced against other ones, and for affective reasons.*[36] More on this in the next chapter.

* Interestingly, research has shown a similar pattern with conspiracy theorists. People who believe that aliens landed in New Mexico way back when have a higher-than-chance likelihood of also believing that Princess Di was murdered at the behest of the other royals. And just to show how irrational this all is, as long as you don't ask them about both

The strongest evidence that abrasive Them-ing originates in emotions and automatic processes is that supposed rational cognitions about Thems can be unconsciously manipulated. In an example cited earlier, subjects unconsciously primed about "loyalty" sit closer to Us-es and farther from Thems, while those primed about "equality" do the opposite.* In another study, subjects watched a slide show of basic, unexciting information about a country they knew nothing about ("There's a country called 'Moldova'?"). For half the subjects, faces with positive expressions were flashed at subliminal speeds between slides; for the other half, it was negative expressions. The former developed more positive views of the country than the latter.[37]

Conscious judgments about Thems are unconsciously manipulated in the real world. In an important experiment discussed in chapter 3, morning commuters at train stations in predominantly white suburbs filled out questionnaires about political views. Then, at half the stations, a pair of young Mexicans, conservatively dressed, appeared each morning for two weeks, chatting quietly in Spanish before boarding the train. Then commuters filled out second questionnaires.

Remarkably, the presence of such pairs made people more supportive of decreasing *legal* immigration from Mexico and of making English the official language, and more opposed to amnesty for illegal immigrants. The manipulation was selective, not changing attitudes about Asian Americans, African Americans, or Middle Easterners.

How's this for a fascinating influence on Us/Them-ing, way below the level of awareness: Chapter 4 noted that when women are ovulating, their fusiform face areas respond more to faces, with the ("emotional") vmPFCs responding more to men's faces in particular. Carlos Navarrete at Michigan State University has shown that white women, when ovulating, have more negative attitudes toward African American men.†[38] Thus the

---

scenarios too close in time to each other, people who believe that Di was murdered . . . also . . . believe at a higher-than-chance level that she faked her own death and is, say, living under an assumed name in Wisconsin.

* How is such priming done? The subject is given a series of scrambled sentences and has to unscramble them. In one group most of the sentences discuss things that tap into the concept of loyalty ("teammates helps Jane her"), while the other group's sentences are about equality ("fairness advocates Chris for").

† In a follow-up study, one that, incongruously, I was involved in, similar issues were examined concerning one target individual—Barack Obama—during the 2008 election season. Subjects were presented with swatches of different shades of brown and were asked which most accurately matched Obama's skin color. Women who viewed him as

intensity of Us/Them-ing is being modulated by hormones. Our feelings about Thems can be shaped by subterranean forces we haven't a clue about.

Automatic features of Us/Them-ing can extend to magical contagion, a belief that the essentialism of people can transfer to objects or other organisms.[39] This can be a plus or a minus—one study showed that washing a sweater worn by JFK would decrease its value at auction, whereas sterilizing one worn by Bernie Madoff would increase its value. This is sheer irrationality—it's not like an unwashed JFK sweater still contains his magical armpit essence, while an unwashed Madoff sweater swarms with moral-taint cooties. And magical contagion has occurred elsewhere—Nazis killed supposedly contaminated "Jewish dogs" along with their owners.*[40]

The heart of cognition catching up with affect is, of course, rationalization. A great example of this occurred in 2000, when everyone learned the phrase "hanging chads" following the election of Al Gore and the Supreme Court's selection of George W. Bush.† For those who missed that fun, a chad is the piece of paper knocked out of a punch-card ballot when someone votes, and a hanging chad is one that doesn't completely detach; does this justify disqualifying the vote, even though it is clear who the person voted for? And obviously, if one millisecond before chads reared their hanging heads, you had asked pundits what would be the hanging-chad stances of the party of Reagan and trickle-down economics, and the party of FDR and the Great Society, they wouldn't have had a clue. And yet there we were, one millisecond postchads, with each party passionately explaining why the view of the opposing Thems threatened Mom, apple pie, and the legacy of the Alamo.

The "confirmation biases" used to rationalize and justify automatic Them-ing are numerous—remembering supportive better than opposing evidence; testing things in ways that can support but not negate your

more white were more likely to vote for him at their time of ovulation; women who viewed him as more black showed the opposite. Of note, these are small effects. Electability is in the eye, and the hormone status, of the beholder.

* As a historical oddity, Nazi Germany had the strictest laws in the world concerning the humane treatment and euthanizing of animals. The dogs went with far less suffering than their owners.

† Just to subtly hint at what I thought of the debacle.

hypothesis; skeptically probing outcomes you don't like more than ones you do.

Moreover, manipulating implicit Them-ing alters justification processes. In one study Scottish students read about a game where Scottish participants either did or didn't treat English participants unfairly. Students who read about Scots being prejudicial became more positive in their stereotypes about Scots and more negative about Brits—justifying the bias by the Scottish participants.[41]

Our cognitions run to catch up with our affective selves, searching for the minute factoid or plausible fabrication that explains why we hate Them.[42]

Individual Intergroup Interactions Versus Group Intergroup Interactions

Thus, we tend to think of Us as noble, loyal, and composed of distinctive individuals whose failings are due to circumstance. Thems, in contrast, seem disgusting, ridiculous, simple, homogeneous, undifferentiated, and interchangeable. All frequently backed up by rationalizations for our intuitions.

That is a picture of an individual navigating Us/Them in his mind. Interactions between *groups* tend to be more competitive and aggressive than are interactions between individual Us-es and Thems. In the words of Reinhold Niebuhr, writing during World War II, "The group is more arrogant, hypocritical, self-centered and more ruthless in the pursuit of its ends than the individual."[43]

There is often an inverse relationship between levels of intragroup and intergroup aggression. In other words, groups with highly hostile interactions with neighbors tend to have minimal internal conflict. Or, to spin this another way, groups with high levels of internal conflict are too distracted to focus hostility on the Others.[44]

Crucially, is that inverse relationship causal? Must a society be internally peaceful to muster the large-scale cooperation needed for major intergroup hostilities? Must a society suppress homicide to accomplish genocide? Or to reverse the causality, do threats from Thems make societies more internally cooperative? This is a view advanced by the economist Samuel Bowles of the Santa Fe Institute who has framed this as "Conflict: Altruism's Midwife."[45] Stay tuned.

UNIQUE REALMS OF HUMAN US/THEM-ING

Despite other primates displaying rudimentary abstractions of Us/Them-ing, humans are in a stratosphere of uniqueness. In this section I consider how:

- we all belong to multiple categories of Us, and their relative importance can rapidly change;
- all Thems are not the same, and we have complex taxonomies about different types of Thems and the responses they evoke;
- we can feel badly about Us/Them-ing and try to conceal it;
- cultural mechanisms can sharpen or soften the edges of our dichotomizing.

Multiple Us-es

I am a vertebrate, mammal, primate, ape, human, male, scientist, lefty, sun sneezer, *Breaking Bad* obsessive, and Green Bay Packers fan.* All

* This one puzzles me intensely. When I was a kid, I decided that bullies would be nicer to me if I knew vast amounts about football. This was during the glory days of the Packers in the Vince Lombardi era; thus I decided that they were the team I favored. I memorized and irritatingly spouted every pointless bit of trivia I could find on them, watched my first (and pretty much only) football game, which turned out to be the Packers legendarily defeating the Cowboys for the 1967 championship with a fourth-down touchdown from the one-yard line, with sixteen seconds remaining, played in minus-fifteen-degree weather. And that's it. My football obsession faded when I decided that knowing baseball factoids would be more advantageous (this was fortuitous, living in Brooklyn—soon after came the miraculous 1969 championship season for the hapless Mets). I've never been to a professional football game, can't tell you anything about the Packers since (I don't even know if Bart Starr is still their quarterback, but it wouldn't surprise me one bit if he's retired by now), basically ignore football. Yet almost fifty years later, if I happen to hear once every few years that the Packers are having a good or bad season, my mood is briefly influenced by the news; if I see a photograph of people playing football and they include the Packers, I'm sure I preferentially look at them versus the other team, am made fleetingly happy by it being them; I felt excited the one time I met someone from Green Bay and, after thirty seconds of chatting pointlessly with them about the Packers of the sixties, felt a near-spiritual connection with them. It's just plain weird. And sure demonstrates the unlikely power that "belonging" can have.

grounds for concocting an Us/Them. Crucially, which Us is most important to me constantly shifts—if some octopus moved in next door, I would feel hostile superiority because I have a spine and it didn't, but that animosity might melt into a sense of kinship when I discovered that the octopus, like me, loved playing Twister as a kid.

We all belong to multiple Us/Them dichotomies. Sometimes one can be a surrogate for another—for example, the dichotomy of people who are/aren't knowledgeable about caviar is a good stand-in for a dichotomy about socioeconomic status.

As noted, the most important thing about our membership in multiple Us/Thems is the ease with which their prioritizing shifts. A famed example, discussed in chapter 3, concerned math performance in Asian American women, built around the stereotypes of Asians being good at math, and women not. Half the subjects were primed to think of themselves as Asian before a math test; their scores improved. Half were primed about gender; scores declined. Moreover, levels of activity in cortical regions involved in math skills changed in parallel.*[46]

We also recognize that other individuals belong to multiple categories, and shift which we consider most relevant. Not surprisingly, lots of that literature concerns race, with the core question being whether it is an Us/Them that trumps all others.

The primacy of race has a lot of folk-intuition appeal. First, race is a

* I once got pulled into this silly, fun venture. There is a diner called Buck's near Stanford that is famed as a place where venture capitalists come to make deals over power breakfasts; apparently, legendary Silicon Valley companies have been born at its tables. A Silicon Valley newspaper persuaded me, as a primatologist, to tag along with a reporter and do ethological observations of venture capitalist dominance interactions in their natural habitat at Buck's. We monitored one table with two opposing pairs of business guys negotiating something. Each side had a tanned, fit alpha male, presumably the boss; each side had a subaltern toady, weighed down with folders and spreadsheets. The toadies interacted with each other constantly, pushing papers at each other, jabbing fingers in the air, grimacing. The two bosses floated above it, their chairs angled to conspicuously ignore each other, their cell phones miraculously ringing each time the other side attempted to talk to them—they'd wave an imperious, dismissive hand at the opponent and take the call. Occasionally, the toadie would ask his boss something privately and, with a display of Mandarin minimalism, the boss would briefly nod his head and change the course of history. Negotiations concluded, seemingly to everyone's satisfaction, hands were shaken, breakfast was left ritualistically untouched, and everyone departed. The reporter and I scrambled over to the window to observe them in the parking lot. Adversarial interactions over with, the Us/Them-ing shifted—the subalterns scurried off to their sensible little Priuses while the two Masters of the Universe remained chatting, each retrieving a tennis racket from his SUV, amiably comparing them, each trying out a swing or two with the other's. At that moment the face of each one's faithful toady probably wouldn't have even activated the fusiform face area in his boss; instead, the most important Us concerned the enjoyable presence of someone else who could commiserate about the hassles of alimony for a third ex-wife.

biological attribute, a conspicuous fixed identity that readily prompts essentialist thinking.[47] This also fuels intuitions about evolution—humans evolved under conditions where different skin color is the clearest signal that someone is a distant Them. And the salience of race is seen cross-culturally—an astonishing percentage of cultures have historically made status distinctions by skin color, including in traditional cultures before Western contact, where with few exceptions (e.g., the low-status Ainu ethnic minority in Japan) lighter skin tone confers higher status both within and between groups.

But these intuitions are flimsy. First, while there are obvious biological contributions to racial differences, "race" is a biological continuum rather than a discrete category—for example, unless you cherry-pick the data, genetic variation within race is generally as great as between races. And this really is no surprise when looking at the range of variation with a racial rubric—compare Sicilians with Swedes or a Senegalese farmer with an Ethiopian herder.*

The evolutionary argument doesn't hold up either. Racial differences, which have only relatively recently emerged, are of little Us/Them significance. For the hunter-gatherers of our hominin history, the most different person you'd ever encounter in your life came from perhaps a couple of dozen miles away, while the nearest person of a different race lived thousands of miles away—there is no evolutionary legacy of humans encountering people of markedly different skin color.

Furthermore, the notion of race as a fixed, biologically based classification system doesn't work either. At various times in the history of the U.S. census, "Mexican" and "Armenian" were classified as distinctive races; southern Italians were of a different race from northern Europeans; someone with one black great-grandparent and seven white ones was classified as white in Oregon but not Florida. This is race as a cultural rather than biological construct.[48]

* Such heterogeneity is hard to appreciate in the United States, where most African Americans descend from a few West African tribes that constitute 1 to 2 percent of the total tribal variability in Africa. One of the consequences of this, the fact that drugs are now marketed that preferentially target hypertension among African Americans, seemingly reifies the biological race concept but actually tells you more about the biology of descendants of a small subset of West Africans than about race as a whole.

Given facts like these, it is not surprising that racial Us/Them dichotomies are frequently trumped by other classifications. The most frequent is gender. Recall the finding that it is more difficult to "extinguish" a conditioned fear association with an other- than a same-race face. Navarrete has shown that this occurs only when the conditioned faces are male; gender outweighs race as an automatic classification in this case.* Age as a classification readily trumps race as well. Even occupation can—for example, in one study white subjects showed an automatic preference for white politicians over black athletes when they were primed to think of race, but the opposite when primed to think of occupation.[49]

Race as a salient Us/Them category can be shoved aside by subtle reclassification. In one study subjects saw pictures of individuals, each black or white, each associated with a statement, and then had to recall which face went with which statement.[50] There was automatic racial categorization—if subjects misattributed a quote, the face picked and the one actually associated with the statement were likely to be of the same race. Next, half the black and half the white individuals pictured wore the same distinctive yellow shirt; the other half wore gray. Now subjects most often confused faces by shirt color.

Wonderful research by Mary Wheeler and Susan Fiske of Princeton showed how categorization is shifted, studying the phenomenon of amygdala activation by pictures of other-race faces.[51] In one group subjects tried to find a distinctive dot in each picture. An other-race face didn't activate the amygdala; face-ness wasn't being processed. In a second group subjects judged whether each face looked older than some age. Amygdaloid responses to other-race faces enlarged—thinking categorically about age strengthened thinking categorically about race. In a third group a vegetable was displayed before each face; subjects judged whether the person liked that vegetable. The amygdala didn't respond to other-race faces.

At least two interpretations come to mind to explain this last result:

* This is not always the case, though. Much analysis went into the acquittal of O. J. Simpson by a jury that included eight African American women. Would their most pertinent group identification be one of gender—and thus responsive to Simpson's history of domestic violence—or of race—yet another African American man potentially being framed by the criminal justice system? The rest, as they say, is history.

a. Distraction. Subjects were too busy thinking about, say, carrots to do automatic categorization by race. This would resemble the effect of searching for the dot.

b. Recategorization. You look at a Them face, thinking about what food they'd like. You picture the person shopping, ordering a meal in a restaurant, sitting down to dinner at home and enjoying a particular food. . . . In other words, you think of the person as an individual. This is the readily accepted interpretation.

But recategorization can occur in the real world under the most brutal and unlikely circumstances. Here are examples that I find to be intensely poignant:

In the Battle of Gettysburg, Confederate general Lewis Armistead was mortally wounded while leading a charge. As he lay on the battlefield, he gave a secret Masonic sign, in hopes of its being recognized by a fellow Mason. It was, by a Union officer, Hiram Bingham, who protected him, got him to a Union field hospital, and guarded his personal effects. In an instant the Us/Them of Union/Confederate became less important than that of Mason/non-Mason.*[52]

Another shifting of Thems also occurred during the Civil War. Both armies were filled with Irish immigrant soldiers; Irish typically had picked sides haphazardly, joining what they thought would be a short conflict to gain some military training—useful for returning home to fight for Irish independence. Before battle, Irish soldiers put identifying sprigs of green in their hats, so that, should they lie dead or dying, they'd shed the arbitrary Us/Them of this American war and revert to the Us that mattered—to be recognized and aided by their fellow Irish.[53] A green sprig as a green beard.

* The story has a double layer of poignancies. Prior to the war, one of Armistead's closest friends was Winfield Scott Hancock, commanding a brigade at the battle . . . on the Union side. The dying Armistead asked after Hancock's well-being and requested that Bingham send his warm greetings to his old friend.

Rapid shifting of Us/Them dichotomies is seen during World War II, when British commandos kidnapped German general Heinrich Kreipe in Crete, followed by a dangerous eighteen-day march to the coast to rendezvous with a British ship. One day the party saw the snows of Crete's highest peak. Kreipe mumbled to himself the first line (in Latin) of an ode by Horace about a snowcapped mountain. At which point the British commander, Patrick Leigh Fermor, continued the recitation. The two men realized that they had, in Leigh Fermor's words, "drunk at the same fountains." A recategorization. Leigh Fermor had Kreipe's wounds treated and personally ensured his safety through the remainder of the march. The two stayed in touch after the war and were reunited decades later on Greek television. "No hard feelings," said Kreipe, praising Leigh Fermor's "daring operation."[54]

And finally there is the World War I Christmas truce, something I will consider at length in the final chapter. This is the famed event where soldiers on both sides spent the day singing, praying, and partying together, playing soccer, and exchanging gifts, and soldiers up and down the lines struggled to extend the truce. It took all of one day for British-versus-German to be subordinated to something more important—*all* of us in the trenches versus the officers in the rear who want us to go back to killing each other.

Thus Us/Them dichotomies can wither away into being historical trivia questions like the Cagots and can have their boundaries shifted at the whims of a census. Most important, we have multiple dichotomies in our heads, and ones that seem inevitable and crucial can, under the right circumstances, have their importance evaporate in an instant.

Cold and/or Incompetent

That both a gibbering schizophrenic homeless man and a successful businessman from a resented ethnic group can be a Them demonstrates something crucial—different types of Thems evoke different feelings in us, anchored in differences in the neurobiologies of fear and disgust.[55] As but one example, fear-evoking faces cause us to watch vigilantly and activate the visual cortex; disgust-evoking faces do the opposite.

We carry various taxonomies in our heads as to our relationships with different types of Others. Thinking about some Thems is simple. Consider someone who pushes all our judgmental buttons—say, a homeless junkie whose wife threw him out of the house because of his abusiveness, and who now mugs elderly people. Throw 'im under a trolley—people are most likely to agree to sacrifice one to save five when the five are in-group members and the one is this extreme of an out-group.*[56]

But what about Thems who evoke more complex feelings? Tremendously influential work has been done by Fiske, with her "stereotype content model."[57] This entire section concerns that work.

We tend to categorize Thems along two axes: "warmth" (is the individual or group a friend or foe, benevolent or malevolent?) and "competence" (how effectively can the individual or group carry out their intentions?).

The axes are independent. Ask subjects to assess someone about whom they have only minimal information. Priming them with cues about the person's status alters ratings of competence but not of warmth. Prime them about the person's competitiveness and you do the opposite. These two axes produce a matrix with four corners. There are groups that we rate as being high in both warmth and competence—Us, naturally. And Americans typically view this group as containing good Christians, African American professionals, and the middle class.

And there's the other extreme, low in both warmth and competence—our homeless, addicted mugger. Subjects typically hand out low-warmth/low-competence assessments for the homeless, people on welfare, and poor people of any race.

Then there's the high-warmth/low-competence categorization—the

* The punch line here is how such individuals barely register with us as people—as we'll see, neuroimaging supports this. A recent finding highlights the opposite concerning the weird American legal notion of "corporate personhood"—when people contemplate the morality of corporate actions, they activate Theory of Mind networks, just as when contemplating the morality of actions of fellow humans.

mentally disabled, people with handicaps, the elderly.* And the categorization of low warmth/high competence. It's how people in the developing world tend to view the European culture that used to rule them,† and how many minority Americans view whites. It's the hostile stereotype of Asian Americans by white America, of Jews in Europe, of Indo-Pakistanis in East Africa, of Lebanese in West Africa, and of ethnic Chinese in Indonesia (and, to a lesser extent, of rich people by poorer people everywhere). And it's the same derogation—they're cold, greedy, cleverly devious, clannish, don't assimilate,‡ have loyalties elsewhere—but, dang, they sure know how to make money, and you probably should go to one who is a doctor if you have something serious.

People tend toward consistent feelings evoked by each of the extremes. For high warmth, high competence (i.e., Us), there's pride. Low warmth, high competence—envy. High warmth, low competence—pity. Low warmth, low competence—disgust. Stick someone in a brain scanner, show them pictures of low-warmth/low-competence people, and there's activation of the amygdala and insula but not of the fusiform face area or the (emotional) vmPFC—a profile evoked by viewing disgusting objects (although, once again, this pattern shifts if you get subjects to individuate, asking them to think about what food this homeless person likes, rather than "anything they can find in garbage cans").§ In contrast, viewing low-warmth/high-competence or high-warmth/low-competence individuals activates the vmPFC.

The places between the extremes evoke their own characteristic responses. Individuals who evoke a reaction between pity and pride evoke a desire to help them. Floating between pity and disgust is a desire to

* With the reminder that "competence" is used not in the everyday sense where "low competence" would seem pejorative but simply as a measure of agency.

† With "competence" here not being skill at being rocket scientists but rather the efficacy those people had when they got it into their heads to, say, steal your ancestral lands.

‡ In my experience in East Africa, the charge by African men that the "Hindis" (i.e., Indo-Pakistanis, most of whose families have lived in East Africa for generations) are not "real Africans" is often code for "They won't sleep with us."

§ Here's an example of how things, naturally, are more complicated than this simple matrix. Insofar as we view low-warmth/low-competence individuals as dehumanized objects, we objectify them. But "objectify" more frequently denotes sexualization of women. In one study men with high degrees of hostile sexism showed less activation of the medial PFC (along with other brain regions associated with Theory of Mind and perspective taking) when looking at pictures of women. But only if the pictures were particularly sexualized. And there was a world of difference between how a hostilely sexist male viewed a sexually provocative picture of a woman versus a picture of a homeless person. In the words of the authors, the study shows that "diminished mental state attribution is not unique to targets that people prefer to avoid."

exclude and demean. Between pride and envy is a desire to associate, to derive benefits from. And between envy and disgust are our most hostile urges to attack.

What fascinates me is when someone's categorization changes. The most straightforward ones concern shifts from high-warmth/high-competence (HH) status:

HH to HL: This is watching a parent decline into dementia, a situation evoking extremes of poignant protectiveness.

HH to LH: This is the business partner who turns out to have been embezzling for decades. Betrayal.

And the rare transition from HH to LL—a buddy who made partner in your law firm, but then "something happened" and now he's homeless. Disgust mingled with bafflement—what went wrong?

Equally interesting are shifts from other categorizations. There's when you shift your perception of someone from HL to LL—the janitor whom you condescendingly greet each day turns out to think you're a jerk. Ingrate.

There's the shift from LL to LH. When I was a kid in the sixties, the parochial American view of Japan was LL—the shadow of World War II generating dislike and contempt—"Made in Japan" was about cheap plastic gewgaws. And then, suddenly, "Made in Japan" meant outcompeting American car and steel manufacturers. Whoa. A sense of alarm, of being caught napping at your post.

Then there's the shift from LL to HL. This is when a homeless guy finds someone's wallet and does cartwheels to return it—and you realize that he's more decent than half your friends.

Most interesting to me is the transition from LH status to LL, which invokes gloating, glee, schadenfreude. I remember a great example of this in the 1970s, when Nigeria nationalized its oil industry and there was the (delusionally misplaced, it turns out) belief that this would usher in wealth and stability. I recall a Nigerian commentator crowing that within the decade, Nigeria would be sending foreign aid to its ex-colonial overlord, Great Britain (i.e., Brits would be shifting from LH to LL).

The glee explains a feature of persecution of LH out-groups, namely to first degrade and humiliate them to LL. During China's Cultural Revolution, resented elites were first paraded in dunce caps before being shipped to labor camps. Nazis eliminated the mentally ill, already LL, by unceremoniously murdering them; in contrast, premurder treatment of the LH Jews involved forcing them to wear degrading yellow armbands, to cut one another's beards, to scrub sidewalks with toothbrushes before jeering crowds. When Idi Amin expelled tens of thousands of LH Indo-Pakistani citizens from Uganda, he first invited his army to rob, beat, and rape them. Turning LH Thems into LL Thems accounts for some of the worst human savagery.

These variations are sure more complicated than chimps associating rivals with spiders.

One strange human domain is the phenomenon of developing a grudging respect, even a sense of camaraderie, with an enemy. This is the world of the probably-apocryphal mutual respect between opposing World War I flying aces: "Ah, monsieur, if it were another time, I would delight in discussing aeronautics with you over some good wine." "Baron, it is an honor that it is you who shoots me out of the sky."

This one's easy to understand—these were knights, dueling to the gallant death, their Us-ness being their shared mastery of the new art of aerial combat, soaring above the little people below.

But surprisingly, the same is shown by combatants who, instead of soaring, were cannon fodder, faceless cogs in their nation's war machine. In the words of a British infantry grunt serving in the bloodbath of trench warfare in World War I, "At home one abuses the enemy, and draws insulting caricatures. How tired I am of grotesque Kaisers. Out here, one can respect a brave, skillful, and resourceful enemy. They have people they love at home, they too have to endure mud, rain and steel." Whispers of Us-ness with people trying to kill you.[58]

And then there is the even stranger world of differing feelings about the economic versus the cultural enemy, the relatively new enemy versus the ancient one, or the distant alien enemy versus the neighboring enemy whose miniscule differences have been inflated. These are the differing subjugations that the British Empire inflicted on the Irish next door versus on Australian aborigines. Or Ho Chi Minh, rejecting the offer of Chinese troops on the ground during the Vietnam War, with a statement to the effect of "The Americans will leave in a year or a decade, but the Chinese will stay for a thousand years if we let them in." And what's most pertinent about byzantine Iranian geopolitics—the millennia-old Persian antipathy toward the Mesopotamians next door, the centuries-old Shiite conflicts with Sunnis, or the decades-old Islamic hatred of the Great Satan, the West?*

No discussion of the oddities of human Us/Them-ing is complete without the phenomenon of the self-hating ________ (take your pick of the out-group), where out-group members buy into the negative stereotypes and develop favoritism for the in-group.[59] This was shown by psychologists Kenneth and Mamie Clark in their famed "doll studies," begun in the 1940s. They demonstrated with shocking clarity that African American children, along with white children, preferred to play with white dolls over black ones, ascribing more positive attributes (e.g., nice, pretty) to them. That this effect was most pronounced in black kids in segregated schools was cited in *Brown v. Board of Education*.† Roughly 40 to 50 percent of African Americans, gays and lesbians, and women show automatic IAT biases in favor of whites, heterosexuals, and men, respectively.

Some of My Best Friends

The "honorable enemy" phenomenon raises another domain of human peculiarity. Even if he could, no chimp would ever deny that the neighboring chimps remind him of spiders. None would feel bad about it, urge others to overcome that tendency, teach their children to never call one of those neighboring chimps a "spider." None would proclaim that he

* As I write, the Shiite/Sunni dichotomy dominates, producing the profound incongruity of both Iranian and American forces battling ISIS fighters in Iraq. The enemy of my enemy is my friend.

† To see how little has changed, see the 2005 documentary *A Girl Like Me*, by the then seventeen-year-old filmmaker Kiri Davis: www.youtube.com/watch?v=z0BxFRu_Sow.

can't distinguish between Us chimps and Them chimps. And these are all commonplace in progressive Western cultures.

Young humans are like chimps—six-year-olds not only prefer to be with kids like themselves (by whatever criteria) but readily say so. It isn't until around age ten that kids learn that some feelings and thoughts about Thems are expressed only at home, that communication about Us/Them is charged and contextual.[60]

Thus there can be striking discrepancies in Us/Them relations between what people claim they believe and how they act—consider differences between election poll results and election results. This is shown experimentally as well—in one depressing study, subjects claimed they'd be very likely to proactively confront someone expressing racist views; yet actual rates were far lower when they were unknowingly put in that position (note—this is not to say that this reflected racist sentiments; instead it likely reflected social-norm inhibitions being stronger than the subjects' principles).[61]

Attempts to control and repress Us/Them antipathies have the frontal cortex written all over them. As we saw, a subliminal fifty-millisecond exposure to the face of another can activate the amygdala, and if exposure is long enough for conscious detection (about five hundred milliseconds or more), the initial amygdala activation is followed by PFC activation and amygdala damping; the more PFC activation, particularly of the "cognitive" dlPFC, the more amygdala silencing. It's the PFC regulating discomfiting emotions.[62]

Behavioral data also implicate the frontal cortex. For example, for the same degree of implicit racial prejudice (as shown with the IAT), the bias is more likely to be expressed behaviorally in individuals with poor frontal executive control (as shown with an abstract cognitive task).[63]

Chapter 2 introduced the concept of "cognitive load," where a taxing frontal executive task diminishes performance on a subsequent frontal task. This occurs with Us/Them-ing. White subjects do better on certain behavior tests when the tester is white as opposed to black; subjects whose performance declines most dramatically in the latter scenario show the greatest dlPFC activation when viewing other-race faces.[64]

The cognitive load generated by frontal executive control during interracial interactions can be modulated. If white subjects are told, "Most people are more prejudiced than they think they are," before taking the test with a black tester, performance plummets more than if they are told, "Most people perform worse [on a frontal cortical cognitive test] than they think they did." Moreover, if white subjects are primed with a command reeking of frontal regulation ("avoid prejudice" during an interracial interaction), performance declines more than when they are told to "have a positive intercultural exchange."[65]

A different type of executive control can occur in minority Thems when dealing with individuals of the dominant culture—be certain to interact with them in a positive manner, to counter their assumed prejudice against you. In one startling study African American subjects were primed to think about either racial or age prejudice, followed by an interaction with someone white.[66] When the prime was racial, subjects became more talkative, solicited the other person's opinion more, smiled more, and leaned forward more; the same didn't occur when subjects interacted with another African American. Think about chapter 3's African American grad student intentionally whistling Vivaldi each evening on the way home.

Two points are worth making about these studies about executive control and interactions with Thems:

> Frontal cortical activation during an interracial interaction could reflect: (a) being prejudiced and trying to hide it; (b) being prejudiced and feeling bad about it; (c) feeling no prejudice and working to communicate that; (d) who knows what else. Activation merely implies that the interracial nature of the interaction is weighing on the subject (implicitly or otherwise) and prompting executive control.

> As per usual, subjects in these studies were mostly university students fulfilling some Psych 101 requirement. In other words, individuals of an age associated with openness to novelty, residing in a privileged place where Us/Them cultural and economic differences are less than

in society at large, and where there is not only institutionalized celebration of diversity but also some actual diversity (beyond the university's home page with the obligatory picture of smiling, conventionally good-looking students of all races and ethnicities peering in microscopes plus, for good measure, a cheerleader type fawning over a nerdy guy in a wheelchair). That even this population demonstrates more implicit antipathy to Thems than they like to admit is pretty depressing.

MANIPULATING THE EXTENT OF US/THEM-ING

What situations lessen or exacerbate Us/Them-ing? (I define "lessen" as decreased antipathy toward Thems and/or decreased perception of the size or importance of contrasts between Us and Them.) Here are some brief summaries, a warm-up for the final two chapters.

The Subterranean Forces of Cuing and Priming

Subliminally flash a picture of a hostile and/or aggressive face, and people are subsequently more likely to perceive a Them as the same (an effect that does not occur in-group).[67] Prime subjects subliminally with negative stereotypes of Thems, and you exacerbate Them-ing. As noted in chapter 3, amygdala activation* in white subjects seeing black faces is increased if rap music is playing in the background and lessened if music associated with negative white stereotypes—heavy metal—is played instead. Moreover, implicit racial bias is lessened after subliminal exposure to counterstereotypes—faces of popular celebrities of that race.

Such priming can work within seconds to minutes and can persist; for example, the counterstereotype effect lasted at least twenty-four hours.[68] Priming can also be extraordinarily abstract and subtle. An example

* Where activation is, I think, an appropriate marker for negative Them-ing.

concerned differences in electroencephalographic (EEG) responses in the brain when looking at same- versus different-race faces. In the study the other-race response lessened if subjects unconsciously felt they were drawing the person toward them—if they were pulling a joystick toward themselves (versus pushing it away) at the time.

Finally, priming is not equally effective at altering all domains of Them-ing; it is easier to subliminally manipulate warmth than competence ratings.

These can be powerful effects. And to be more than merely semantic, the malleability of automatic responses (e.g., of the amygdala) shows that "automatic" does not equal "inevitable."

The Conscious, Cognitive Level

Various overt strategies have been found to decrease implicit biases. A classic one is perspective taking, which enhances identification with Them. For example, in a study concerning age bias, having subjects take the perspective of older individuals more effectively reduced bias than merely instructing subjects to inhibit stereotypical thoughts. Another is to consciously focus on counterstereotypes. In one such study automatic sexual biases were lessened more when men were instructed to imagine a strong woman with positive attributes, than when they were instructed to attempt stereotype suppression. Another strategy is to make implicit biases explicit—show people evidence of their automatic biases. More to come concerning these strategies.[69]

Changing the Rank Ordering of Us/Them Categories

This refers to the multiple Us/Them dichotomies we carry and the ease of shifting their priority—shifting automatic categorizing by race to categorizing by shirt color or manipulating math performance by emphasizing either gender or ethnicity. Shifting which categorization is at the forefront isn't necessarily a great thing and may just constitute six of one, half a dozen of the other—for example, among European American men, a photograph of an Asian woman applying makeup makes gender automaticity stronger than ethnic automaticity, while a picture of her using chopsticks does the opposite. More effective than getting people to shift a Them of one category into merely another type of Them, of course, is to shift the Them to being perceived as an Us—emphasizing attributes in common.[70] Which brings us to . . .

Contact

In the 1950s the psychologist Gordon Allport proposed "contact theory."[71] Inaccurate version: if you bring Us-es and Thems together (say, teenagers from two hostile nations brought together in a summer camp), animosities disappear, similarities become more important than differences, and everyone becomes an Us. More accurate version: put Us-es and Thems together under very narrow circumstances and something sort of resembling that happens, but you can also blow it and worsen things.

Some of those effective narrower circumstances: there are roughly equal numbers from each side; everyone's treated equally and unambiguously; contact is lengthy and on neutral, benevolent territory; there are "superordinate" goals where everyone works together on a task they care about (say, the summer campers turning an overgrown meadow into a soccer field).[72]

Essentialism Versus Individuation

This harks back to two important earlier points. First, that Thems tend to be viewed as homogeneous, simple, and having an unchangeable (and negative) essence. Second, that being forced to think of a Them as an individual can make them seem more like an Us. Decreasing essentialist thinking via individuation is a powerful tool.

One elegant study showed this. White subjects were given a questionnaire assessing the extent of their acceptance of racial inequalities, after being given one of two primes.[73] The first bolstered essentialist thinking about race as invariant and homogeneous—"Scientists pinpoint the genetic underpinnings of race." The other prime was antiessentialist—"Scientists reveal that race has no genetic basis." Being primed toward essentialism made subjects more accepting of racial inequalities.

Hierarchy

Predictably, making hierarchies steeper, more consequential, or more overt worsens Them-ing; the need for justification fuels those on top to pour the stereotypes of, at best, high warmth/low competence or, worse, low warmth/low competence onto the heads of those struggling at the bottom, and those on the bottom reciprocate with the simmering time bomb that is the perception of the ruling class as low warmth/high competence.[74] Fiske has explored how those on top perceiving the underclass as high warmth/low competence can stabilize the status quo; the powerful feel self-congratulatory about their presumed benevolence, while the subordinated are placated by the sops of respect. Supporting this, across thirty-seven countries, higher levels of income inequality correlate with more of the condescension of HL perceptions trickling down. Jost has explored this in a related way, examining how myths of "No one has it all" can reinforce the status quo. For example, the cultural trope of "poor but happy"—the poor are more carefree, more in touch with and able to enjoy the simple things in life—and the myth of the rich as unhappy, stressed, and burdened with responsibility (think of miserable, miserly Scrooge and those warm, loving Cratchits) are great ways to keep things from changing. The trope of "poor but honest," by throwing a sop of prestige to Thems, is another great means of rationalizing the system.*

Individual differences in how people feel about hierarchy help explain

* An extensive health psychology literature shows that "poor but happy" is mostly nonsense—poverty gives rise to higher rates of major depressive and anxiety disorders, suicide, and stress-related disease. As we'll see in a later chapter, "poor but honest" has more truth to it.

variation in the extent of Them-ing. This is shown in studies examining social-dominance orientation (SDO: how much someone values prestige and power) and right-wing authoritarianism (RWA: how much someone values centralized authority, the rule of law, and convention).[75] High-SDO individuals show the greatest increases in automatic prejudices when feeling threatened; more acceptance of bias against low-status out-groups; if male, more tolerance of sexism. And as discussed, people high in SDO (and/or in RWA) are less bothered by hostile humor about out-groups.

Related to our all being part of multiple Us/Them dichotomies is our simultaneous membership in multiple hierarchies.[76] No surprise, people emphasize the importance of the hierarchy in which they rank highest—being captain of the company's weekend softball team takes on more significance than the lousy, lowly nine-to-five job during the week. Particularly interesting are hierarchies that tend to map onto Us/Them categories (for example, when race and ethnicity overlap heavily with socioeconomic status). In those cases, those on top tend to emphasize the convergence of the hierarchies and the importance of assimilating the values of the core hierarchy ("Why can't they all just call themselves 'Americans' instead of 'Ethnicity-Americans'?"). Interestingly, this is a local phenomenon—whites tend to favor assimilationist, unitary adherence to national values while African Americans favor more pluralism; however, the opposite occurs concerning campus life and policies among white and African American students at traditionally black universities. We can keep two contradictory things in our heads at the same time if that works to our benefit.

Thus, in order to lessen the adverse effects of Us/Them-ing, a shopping list would include emphasizing individuation and shared attributes, perspective taking, more benign dichotomies, lessening hierarchical differences, and bringing people together on equal terms with shared goals. All to be revisited.

CONCLUSIONS

An analogy concerning health: Stress can be bad for you. We no longer die of smallpox or the plague and instead die of stress-related diseases of lifestyle, like heart disease or diabetes, where damage slowly accumulates over time. It is understood how stress can cause or worsen disease or make you more vulnerable to other risk factors. Much of this is even understood on the molecular level. Stress can even cause your immune system to abnormally target hair follicles, causing your hair to turn gray.

All true. Yet stress researchers do not aim to eliminate, to "cure," us of stress. It can't be done, and even if it could, we wouldn't want that—we love stress when it's the right kind; we call it "stimulation."

The analogy is obvious. From massive, breathtaking barbarity to countless pinpricks of microaggression, Us versus Them has produced oceans of pain. Yet our generic goal is not to "cure" us of Us/Them dichotomizing. It can't be done, unless your amygdala is destroyed, in which case everyone seems like an Us. But even if we could, we wouldn't want to eliminate Us/Them-ing.

I'm a fairly solitary person—after all, I've spent a significant amount of my life studying a different species from my own, living alone in a tent in Africa. Yet some of the most exquisitely happy moments of my life have come from feeling like an Us, feeling accepted and not alone, safe and understood, feeling part of something enveloping and larger than myself, filled with a sense of being on the right side and doing both well and good. There are even Us/Thems that I—eggheady, meek, and amorphously pacifistic—would be willing to kill or die for.[77]

If we accept that there will always be sides, it's a nontrivial to-do list item to always be on the side of angels. Distrust essentialism. Keep in mind that what seems like rationality is often just rationalization, playing catch-up with subterranean forces that we never suspect. Focus on the larger, shared goals. Practice perspective taking. Individuate, individuate,

individuate. Recall the historical lessons of how often the truly malignant Thems keep themselves hidden and make third parties the fall guy.

And in the meantime, give the right-of-way to people driving cars with the "Mean people suck" bumper sticker, and remind everyone that we're all in it together against Lord Voldemort and the House Slytherin.

Twelve

Hierarchy, Obedience, and Resistance

At first glance, this chapter simply complements the previous one. Us/Them-ing is about relations between groups and our automatic tendency to favor in-groups over out-groups. Similarly, hierarchies are about a type of relations within groups, our automatic tendency to favor people close in rank to us over those who are distant. Other themes repeat as well—the appearance of these tendencies early in life and in other species, and the intertwined cognitive and affective underpinnings.

Moreover, Us/Them categorization and hierarchical position interact. In one study subjects gave racial designations to pictures of racially ambiguous individuals; those dressed in low-status attire were more likely to be categorized as black, high-status attire, white.[1] Thus, among these American subjects, Us/Them dichotomizing by race and the hierarchy of socioeconomic status overlap.

But as we'll see, hierarchy heads in different directions from Us/Them-ing, and in uniquely human ways: Like other hierarchical species, we have alpha individuals, but unlike most others, we occasionally get to choose them. Moreover, they often are not merely highest ranking but also "lead," attempting to maximize this thing called the common good.

Furthermore, individuals vie for leadership with differing visions of how best to attain that common good—political ideologies. And finally, we express obedience both to an authority and to the idea of Authority.

THE NATURE AND VARIETIES OF HIERARCHIES

For starters, a hierarchy is a ranking system that formalizes unequal access to limited resources, ranging from meat to that nebulous thing called "prestige." We begin by examining hierarchies in other species (with the proviso that not all social species have hierarchies).

The textbook 1960s picture of hierarchies in other species was straightforward. A group forms a stable, linear hierarchy where the alpha individual dominates everyone, the beta individual dominates everyone except the alpha, gamma everyone except alpha and beta, and so on.

Hierarchies establish a status quo by ritualizing inequalities. Two baboons encounter something good—say, a spot shaded from the sun. Without stable dominance relations, there's a potentially injurious fight. Likewise over the figs in a fruiting tree an hour later, and for the chance to be groomed by someone after that, etc. Instead, fights rarely occur, and if a subordinate forgets his status, a "threat yawn"—a ritualistic display of canines—from the dominant male usually suffices.*,†[2]

Why have ranking systems? The answer, circa 1960s, was Marlin Perkins group selection, where a species benefits from a stable social system in which everyone knows their place. This view was fostered by the primatological belief that in a hierarchy the alpha individual (i.e., the one who gets first dibs on anything good) was in some manner a "leader" who does something useful for the group. This was emphasized by the

* My apologies for how baboon-centric the examples in the coming pages will be; it reflects my thirty-plus years hanging around them.

† Nice evidence that we're not always just like other animals: Those antihierarchical Buddhists have a text, the Vinaya Pitaka, that instructs monks to defecate not in the order of seniority but in the order of arrival at the toilet. There is hope for this planet.

Harvard primatologist Irven DeVore, who reported that among savanna baboons, the alpha male led the troop in each day's direction of foraging, led communal hunts, defended everyone against lions, disciplined the kids, changed the lightbulbs, etc. This turned out to be nonsense. Alpha males don't know which direction to go (given that they transfer into troops as adolescents). No one follows them anyway; instead everyone follows the old females, who do know. Hunts are disorganized free-for-alls. And an alpha male might face down a lion to protect a kid—if the kid is probably his own. Otherwise, he'd grab the safest spot.

Male baboon giving a (hopefully) intimidating threat yawn.

When viewed without Perkins-colored glasses, the benefits of hierarchy are individualistic. Interactions that proclaim the status quo obviously help the upper crust. Meanwhile, for subordinates, better to not get a shady spot than to not get it after receiving a canine slash. This is logical in a static, hereditary ranking system. In systems where ranks change, this caution must be balanced with occasionally challenging things—because the alpha male may be past his prime and getting by on bluff.

This is a classic "pecking order" (a term derived from the hierarchical system of hens). Variations begin. A first concern is whether there's actually a hierarchy, in the sense of gradations of rank. Instead, in some species (e.g., South American marmoset monkeys) there's the alpha and there's everyone else, with fairly equal relations.

In species with gradations, there's the issue of what a "rank" actually means. If your rank is number six in a hierarchy, in your mind are numbers one through five interchangeable guys you'd better kowtow to, while seven through infinity are undifferentiated peons? If so, it would be irrelevant to you if numbers two and three, or numbers nine and ten, were having tensions; rank gradations would be in the eyes of the primatologist, not the primate.

In reality, such primates think about gradations of rank. For example, a baboon typically interacts differently with, say, the guy one step above him in rank than the one five steps above. Furthermore, primates note gradations that don't directly concern them. Recall from chapter 10 how researchers recorded vocalizations of individuals in a troop, splicing them to invent social scenarios. Play a recording of number ten giving a dominance call and number one responding with a subordination call, and everyone pays attention: whoa, Bill Gates just panhandled a homeless guy.

This can be abstracted further, as shown with ravens, which are outrageously smart birds. As with baboons, vocalizations implying dominance reversals command more attention than does the status quo. Remarkably, this even occurs for reversals between birds in a *neighboring* flock. Ravens discern dominance relations just by listening and are interested in hierarchical gossip about a different group.

Next is the issue of variation within and among species as to what life is *like* having a particular rank. Does being high ranking merely mean that everyone keeps tabs on your mood or, at the other extreme, that no one else is getting enough calories to ovulate, lactate, or survive? How often do subordinates challenge dominant individuals? How readily do dominant individuals vent their frustrations on subordinates? How much do such subordinates have coping outlets (e.g., someone to groom with)?

Then there is the issue of how high rank is attained. In many cases (e.g., female baboons, as noted) rank is inherited, a system with kin selection written all over it. In contrast, in other species/sexes (male baboons, for example) ranks shift over time, changing as a function of fights, showdowns, and Shakespearean melodrama, where rising in the hierarchy is about brawn, sharp canines, and winning the right fight.*

Hurrah for clawing your way to the top, for sweaty, zero-sum, muscular capitalism. But what about the more interesting issue of how high rank, once attained, is maintained? As we'll see, this has less to do with muscle than with social skills.

This ushers in a key point—social competence is challenging, and this is reflected in the brain. The British anthropologist Robin Dunbar has shown that across various taxa (e.g., "birds," "ungulates" or "primates"), the bigger the average size of the social group in the species, (a) the larger the brain, relative to total body size, and (b) the larger the neocortex, relative to total brain size. Dunbar's influential "social brain hypothesis" posits that increases in social complexity and the evolutionary expansion of the neocortex have been linked. This link also occurs within species. Among some primates, group size can vary tenfold (depending on the richness of the ecosystem). This was modeled in a fascinating neuroimaging study, in which captive macaque monkeys were housed in different-sized groups; the bigger the group, the more thickening of the prefrontal cortex and the superior temporal gyrus, a cortical region involved in Theory of Mind, and the tighter the activity coupling between the two.[†3]

Thus primate social complexity and big brains go together. This is shown further by examining fission-fusion species, where the size of the social group regularly changes dramatically. Baboons, for example, start and end the day in a large, coherent troop, whereas midday foraging occurs in small groups. As other examples, hyenas hunt in groups but scavenge individually, and wolves often do the opposite.

Sociality is more complex in fission-fusion species. You must remember if someone's rank differs when in a subgroup versus the entire group.

---

* Implicit in this is the fact these males and females have separate hierarchies. In general, females of the highest-ranking family can push around the lowest-quartile rankings of males, though males otherwise dominate females.

† Note: The correlation between neocortex size and group size across primate species probably reflects each trait influencing the other, i.e., coevolution of the two traits. The neuroimaging study shows that a bigger social group can *cause* interesting parts of the brain to expand (in ways having far more to do with the neural plasticity of chapter 5 than with genes and evolution).

Being away from someone all day makes it tempting to see if dominance relations have changed since breakfast.

One study compared fission-fusion primates (chimps, bonobos, orangutans, spider monkeys) and non-fission-fusion (gorillas, capuchins, long-tailed macaques).[4] Among these captive animals, fission-fusion species were better at frontocortical tasks and had larger neocortices relative to total brain size. Studies of corvids (crows, ravens, magpies, jackdaws) showed the same thing.

Thus "rank" and "hierarchy" in other animals is anything but straightforward and varies considerably depending on the species, gender, and social group.

RANK AND HIERARCHY IN HUMANS

Human hierarchies resemble those of other species in many ways. For example, there's the distinction between stable and unstable hierarchies—centuries of czarist rule versus the first inning of the Russian Revolution. As we'll see below, those situations evoke different patterns of brain activation.

Group size also matters—primate species with bigger social groups have larger cortices relative to the rest of the brain (with humans topping off both measures).[5] If you graph the size of the neocortex against the average size of the social group across primate species, you get "Dunbar's number," the predicted average group size in traditional human cultures. It's 150 people, and there's much evidence supporting that prediction.

This also plays out in the Western world, where the larger the size of someone's social network (often calculated by the number of e-mail/texting relationships), the larger the vmPFC, orbital PFC, and amygdala, and the better the person's Theory of Mind–related skills.[6]

Do these brain regions expand when someone has a larger social network, or do larger sizes of these regions predispose people toward forming larger networks? Naturally, some of both.

As with other species, human quality of life also varies with the consequences of rank inequalities—there's a big difference between the powerful getting seated at a restaurant before you and the powerful getting to behead you if the fancy strikes them. Recall the study of thirty-seven countries showing that the more income inequality, the more preadolescent bullying in schools. In other words, countries with more brutal socioeconomic hierarchies produce children who enforce their own hierarchies more brutally.[7]

Amid these cross-species similarities, there are unique things about humans, including the following.

Membership in Multiple Hierarchies

We belong to multiple hierarchies and can have very different ranks in them.* Naturally, this invites rationalization and system justification—deciding why hierarchies where we flounder are crap and the one where we reign really counts.

Implicit in being part of multiple hierarchies is their potential overlap. Consider socioeconomic status, which encompasses both local and global hierarchies. I'm doing great socioeconomically—my car's fancier than yours. I'm doing terribly—I'm not richer than Bill Gates.

The Specialization of Some Ranking Systems

A high-ranking chimp is generally good at related things. But humans can dwell in incredibly specialized hierarchies. Example: There's a guy named Joey Chestnut who's a god in one subculture—he's the most successful competitive hot dog eater in history. However, whether Chestnut's gift generalizes to other domains is unclear.

---

* An example of this that I found to be excruciatingly uncomfortable: I used to play in a regular pickup soccer game at Stanford. I was terrible, which was widely and tolerantly recognized by all. One of the best, most respected players was a Guatemalan guy who happened to be a janitor in my building. At soccer he'd call me Robert (on the rare occasion when anything I did was relevant to play). And when he came to empty the garbage from my office and lab, no matter how much I tried to get him to stop, it would be "Dr. Sapolsky."

Internal Standards

This is the circumstance of having internal standards independent of the outside world. As an example, winning or losing at a team sport generally increases or decreases, respectively, testosterone levels in men. But things are subtler than that—testosterone more closely tracks winning through skill (rather than luck), and also more closely tracks individual (rather than team) performance.[8]

Thus, as usual, we are just like other animals but totally different. We now consider the biology of individual ranks.

THE VIEW FROM THE TOP, THE VIEW FROM THE BOTTOM

Detecting Rank

Much as with our ability to detect Thems, we're intensely interested in and adept at spotting rank differences. For example, forty milliseconds is all we need to reliably distinguish between a dominant face (with direct gaze) and a subordinate one (with averted gaze and lowered eyebrows). Status is also signaled in the body, albeit to a less accurate extent—dominance with an exposed torso with arms wide open, subordination with arms sheltering a bent torso, intent on invisibility. Again, we recognize those cues at automatic speeds.[9]

Human infants also recognize status differences, as shown in a truly clever study. Show an infant a computer screen displaying a big square and little square; each has eyes and a mouth.[10] The two squares are at opposite ends of the screen and repeatedly move to the other side, passing each other in the process. Then show a version where the two bump into each other—conflict. The squares bump repeatedly until one of them "gives in" by lying down, letting the other one pass. Toddlers look at the interaction longer when it's the big square that gives in, rather than the little one. The first scenario is more interesting because it violates expectations—"Hey, I thought big squares dominated little squares." Just like monkeys and corvids.

But wait, this may just reflect folk physics, not attunement to hierarchy—big things knock over little things, not the other way around. This confound was eliminated. First, the adversarial squares were not touching when one gave in. Second, the subordinating one would fall in the opposite direction from that predicted by physics—rather than being knocked backward, it prostrates itself before the alpha square.

Along with this expertise comes intense interest in hierarchy—as emphasized in chapter 9, gossip is mostly about the status of status: Are there any fallen mighty? Have the meek inherited anything lately? Regardless of which square wins, infants look longer at the conflict situation than when the squares peacefully glide past each other.

This is logical self-interest. Knowing the hierarchical lay of the land helps you navigate it better. But there's more to that than just self-interest. Those monkeys and corvids not only pay attention when there are rank reversals in their group; they do the same when eavesdropping on the neighbors. Same with us.[11]

What's happening in our brains when we contemplate rank?[12] Naturally, the prefrontal cortex weighs in. Frontal damage impairs skill at recognizing dominance relations (along with recognizing kinship, deception, or intimacy in faces). The vlPFC and dlPFC activate and become coupled when we figure out dominance relations or look at a dominant face, reflecting the combined affective and cognitive components to the process. These responses are most pronounced when considering someone of the opposite sex (which may reflect mating goals more than mere academic interest about hierarchy).

Seeing a dominant face also activates the superior temporal gyrus (STG, with its role in Theory of Mind) and increases its coupling to the PFC—we're more interested in what dominant individuals are thinking.[13] Moreover, individual "social status" neurons occur in the monkeys. And as noted in chapter 2, contemplating an unstable hierarchy does all of the above, plus activates the amygdala, reflecting the unsettling effects of instability. Of course, though, none of this tells us *what* we are contemplating at these times.

Your Brain and Your Own Status

Your own rank does logical things to your brain. In macaque monkeys an increase in rank increases mesolimbic dopamine signaling. And back to that rhesus monkey study showing that being in a larger social group causes expansion and functional coupling of the STG and PFC. The study also showed that the higher the rank attained within each group, the greater the expansion and coupling. Consistent with that, a study of mice showed that higher-ranking animals had stronger excitatory inputs into the mouse equivalent of the (cognitive) dlPFC.[14]

I love these findings. As I said, in lots of social species, attaining high rank is about sharp teeth and good fighting skills. But *maintaining* the high rank is about social intelligence and impulse control: knowing which provocations to ignore and which coalitions to form, understanding other individuals' actions.

Does the monkey make history, or does history make the monkey? Once groups were formed, did individuals who became dominant respond with the biggest expansions of those brain regions? Or, prior to group formation, were the individuals destined to become dominant already endowed in those regions?

Unfortunately, animals weren't imaged before and after group formation in the study. However, subsequent work showed that the larger the size of the group, the larger the association between dominance and those brain changes, suggesting that attaining high rank drives the enlargement.* In contrast, the mouse study showed that when synaptic excitability was increased or decreased in the dlPFC, rank rose or declined, respectively, suggesting that enlargement drives attainment of high rank. The brain can shape behavior can shape the brain can shape . . .[15]

* Given how unlikely it is that those soon-to-be-dominant individuals with the largest-of-the-large PFC/STCs just happened to be placed in the largest groups.

Your Body and Your Own Status

What about biological differences outside the brain as a function of rank? For example, do high- and low-ranking males differ in their testosterone profiles and, if there are differences, are they causes, consequences, or mere correlates of the rank differences?

Folk endocrinology has always held that high rank (in any species) and elevated testosterone levels go hand in hand, with the latter powering the former. But as covered at length in chapter 4, neither is the case in primates. As a reminder:

- In stable hierarchies high-ranking males typically don't have the highest testosterone concentrations. Instead it's usually low-ranking adolescent males, starting fights they can't finish. When there is an association between high rank and high testosterone, it generally reflects the higher rates of sexual behavior among dominant individuals driving secretion.
- An exception to the above is during unstable times. For example, among a number of primate species, high-ranking males have the highest testosterone levels for the first months but not years after group formation. During unstable times, the high-testosterone/high-rank relationship is more a consequence of the high rates of fighting among the high-ranking cohort than of rank itself.[16]
- Reiterating the "challenge hypothesis," the elevation in testosterone levels caused by fighting is not so much about aggression as about challenge. If status is maintained aggressively, testosterone fosters aggression; if status were maintained by writing beautiful, delicate haikus, testosterone would foster that.

Next we consider the relationship between rank and stress. Are different ranks associated with different levels of stress hormones, styles of coping, and incidences of stress-related disease? Is it more stressful to be dominant or subordinate?

An extensive literature shows that a sense of control and predictability reduces stress. Yet monkey research conducted by Joseph Brady in 1958 produced a different view. Half the animals could press a bar to delay shocks ("executive" monkeys); the passive other half received a shock whenever the executive did. And the executive monkeys, with their control and predictability, were more likely to ulcerate. This birthed the "executive stress syndrome"—those on top are burdened with the stressors of control, leadership, and responsibility.[17]

Executive stress syndrome became a meme. But a huge problem was that monkeys were not randomly assigned to be "executives" and "non-executives." Instead, those that pressed the bar soonest in pilot studies were made executives.* Such monkeys were subsequently shown to be more emotionally reactive, so Brady had inadvertently stacked the executive side with the ulcer-prone neurotics.

So much for ulcerating executives; contemporary studies show that the worst stress-related health typically occurs in middle management, with its killer combo of high work demands but little autonomy—responsibility without control.

By the 1970s dogma held that subordinate organisms are the most stressed and unhealthy. This was first shown with lab rodents, where subordinate animals typically had elevated *resting* levels of glucocorticoids. In other words, even in the absence of stress, they showed signs of chronically activating the stress response. The same is observed in primates ranging from rhesus monkeys to lemurs. Same for hamsters, guinea pigs, wolves, rabbits, and pigs. Even fish. Even sugar gliders, whatever they are. In a pair of unintentional studies of captive monkeys in which subordinate individuals were basically subordinated to death, such animals had extensive damage to the hippocampus, a brain region very sensitive to the damaging effects of glucocorticoid excess.[18]

My own work with baboons in Africa showed the same (being the first such studies of wild primates). In general, low-ranking male baboons had

* Probably to speed things along, using animals that would learn the shock/bar-pressing relationship the fastest.

elevated basal glucocorticoid levels. When something stressful did occur, their glucocorticoid stress response was relatively sluggish. When the stressor was over, their levels returned to that elevated baseline more slowly. In other words, too much of the stuff in the bloodstream when you don't need it, and too little when you do. Remarkably, at the nuts-and-bolts level of brain, pituitary, and adrenals, the elevated basal glucocorticoid levels of a subordinate occurred for the same reasons as the elevated levels in humans with major depression. For a baboon, social subordination resembles the learned helplessness of depression.

Excessive glucocorticoids get you into trouble in various ways, helping explain why chronic stress makes you sick. Subordinate baboons paid a price in other realms as well. They had (a) elevated blood pressure and a sluggish cardiovascular response to a stressor; (b) lower levels of "good" HDL cholesterol; (c) subtle immune impairments, a higher frequency of getting sick and slower wound healing; (d) a testicular system more easily disrupted by stress than that of dominant males; and (e) lower circulating levels of a key growth factor. Try not to be a subordinate baboon.

Chickens and eggs reappear—does a particular physiological attribute contribute to rank, or the reverse? This is impossible to determine in wild animals, but in captive primate populations the distinctive physiological features of a rank generally follow, rather than precede, the establishment of the rank.[19]

At that point I'd happily proclaim that these findings reflected the nature of Hierarchy, with a capital *H*, and of the stressfulness of social subordination. Which turned out to be totally wrong.

A first wrinkle was provided by Jeanne Altmann of Princeton and Susan Alberts of Duke, studying wild baboons with stable hierarchies. They found the familiar picture, namely of subordination associated with elevated basal glucocorticoid levels. However, unexpectedly, levels in alphas were elevated into the range seen in lowest-ranking males. Why is life more stressful for alpha than beta males? The two ranks had similar rates of being challenged by lower-ranking males (a source of stress) and being groomed by females (a source of coping). However, alpha males fight

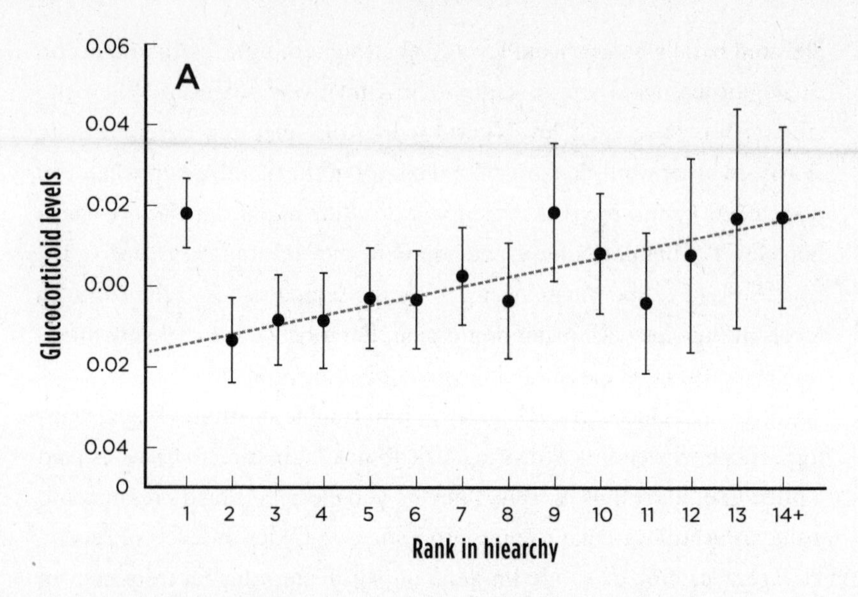

Modified from R. Sapolsky, "Sympathy for the CEO," Sci *333 (2011): 293.*

more often and spend more time in sexual consortships with females (which is majorly stressful, as the male has to defend his consortship from harassing males). Ironically, a chief benefit of alphadom—sexual consortships—can be a major stressor. Be careful what you wish for.[20]

Okay, so except for the curse of being alpha, social subordination is generally stressful. But this is also wrong. It's not just the rank that matters but what the rank *means*.

Consider the primate species in which a relationship has been found between rank and glucocorticoid levels. Across these species, basal glucocorticoid levels are relatively elevated in subordinate animals if: (a) dominant individuals in bad moods frequently displace aggression onto subordinates; (b) subordinates lack coping outlets (such as a grooming partner); and/or (c) the social structure is such that subordinate animals have no relatives present. And when the profile was the opposite, it was dominant animals with the highest glucocorticoid levels.[21]

The "meaning" of rank and its physiological correlates also vary between different groups of the same species. For example, while health of subordinate baboons fared particularly badly in a troop with high rates of dominant males displacing aggression, the health of dominant males fared badly in a troop during a period of instability centered around the top of the hierarchy.

And superimposed on all this, personality shapes the perception of the reality of rank. Using the word "personality" about other species used to cost you tenure, but it's now a hot topic in primatology. Individuals in other species have stable differences in temperament—how likely someone is to displace aggression when frustrated, how socially affiliative they are, how rattled they get by novelty, and so on. Primates differ as to whether they see watering holes as half empty or full; in the context of hierarchy, some individuals who are number two care only that they're not number one, and some individuals who are number nine gain comfort from at least not being number ten.

Not surprisingly, personality influences the rank/health relationship. For the same high rank, an individual is likely to be less healthy if he (a) is particularly reactive to novelty; (b) sees threats in benign circumstances (e.g., his rival showing up and merely taking a nap nearby); (c) doesn't take advantage of social control (e.g., letting a rival determine the start of an obvious showdown); (d) doesn't differentiate between good and bad news (e.g., distinguishing behaviorally between winning and losing a fight); and/or (e) doesn't have social outlets when frustrated. You could make a living giving baboons "how to succeed in business" seminars built around these factors.[22]

Meanwhile, on the flip side, for the same low rank, an individual tends to be healthier if (a) he has lots of grooming relationships; and/or (b) there's someone even lower ranking than him to serve as a target for displaced aggression.

Thus, in other species, how does rank affect the body? It depends on what it's like to have a certain rank in that species and particular social group, and the personality traits that filter the perception of those variables. What about humans?

And Us

A smidgen of neurobiology research has examined differences in how people feel about hierarchy. Back to a concept from the last chapter, social-dominance orientation (SDO), the measure of how much people value power and prestige. In one study subjects viewed someone in emotional pain. As reviewed in chapter 2, this activates the anterior cingulate cortex and insular cortex—empathy, and disgust at the circumstance that evoked the pain. The higher someone's SDO score, the less activation of those two regions. Those with the most interest in prestige and power seem least likely to feel for those less fortunate.[23]

What about the biological correlates of a human having a particular rank? In some ways we're more subtle than other primates; in others, far less so.

Two studies examined high-status individuals in government or the military (in the latter case, officers up to the level of colonel). As compared with low-status controls, these folks had lower basal glucocorticoid levels, less self-reported anxiety, and an enhanced sense of control (this telling us nothing, however, as to which came first—the rank or the unstressed profile).[24]

Baboons redux. But something subtler was happening. The authors deconstructed high rank with three questions: (a) How many people ranked lower than the subject in his organization? (b) How much autonomy did he have (e.g., to hire and fire)? (c) How many people did he directly supervise? And high rank came with low glucocorticoids and anxiety only insofar as the position was about the first two variables—lots of subordinates, lots of authority. In contrast, having to directly supervise lots of subordinates did not predict those good outcomes.

This gives credence to executives' bellyaching about how they aren't supervising eleventy people; instead they have eleventy bosses. To accrue the full physiological benefits of high status, don't supervise people; instead, glide through the workplace like a master of the universe while minions whom you never interact with smile obsequiously. It's not just rank; it's what rank means and entails.

In what sense is the status/health relationship in humans less subtle than in other primates?[25] In that it reflects the most permeating form of status any primate has invented, namely socioeconomic status (SES). Numerous studies examine the "health/SES" gradient, the fact that life expectancy and the incidence and morbidity of numerous diseases are worse in poor people.

To summarize this sprawling topic that was reviewed in chapter 9:

- Which comes first—poverty or poor health? Overwhelmingly the former. Recall that developing in a low-SES womb makes poorer health as an adult more likely.
- It's not that the poor have poor health and everyone else is equally healthy. For every step down the SES ladder, health is worse.
- The problem isn't that poor people have less health-care access. The gradient occurs in countries with socialized medicine and universal health care and for diseases whose incidence is independent of health-care access.
- Only about a third of the variability is explained by the poor being exposed to more health risk factors (e.g., pollution) and fewer protective factors (e.g., health club memberships).
- The gradient seems to be about the psychological baggage of SES. (a) Subjective SES predicts health at least as accurately as objective SES, meaning that it's not about being poor. It's about *feeling* poor. (b) Independent of absolute levels of income, the more income inequality in a community—meaning the more frequently the poor have their noses rubbed in their low status—the steeper the health gradient. (c) Lots of inequality in a community makes for low social capital (trust and a sense of efficacy), and that's the most direct cause of the poor health. Collectively these studies show that the psychological stress of low SES is what decreases health. Consistent with that, it is diseases that are most sensitive to stress (cardiovascular, gastrointestinal, and psychiatric disorders) that show the steepest SES/health gradients.

The SES/health gradient is ubiquitous. Regardless of gender, age, or race. With or without universal health care. In societies that are ethnically homogeneous and those rife with ethnic tensions. In societies in which the central mythology is a capitalist credo of "Living well is the best revenge" and those in which it is a socialist anthem of "From each according to his ability, to each according to his need." When humans invented material inequality, they came up with a way of subjugating the low ranking like nothing ever before seen in the primate world.

A Really Odd Thing That We Do Now and Then

Amid the unique features of human hierarchies, one of the most distinctive and recent is this business of having leaders and choosing who they are.

As discussed, outdated primatology confused high rank with "leadership" in silly ways. An alpha male baboon is not a leader; he just gets the best stuff. And while everyone follows a knowledgeable old female when she chooses her foraging route in the morning, there is every indication that she is "going" rather than "leading."

But humans have leaders, anchored in the unique notion of the common good. What counts as the common good, and the leader's role in fostering it, obviously varies, ranging from leading the horde in the siege of a castle to leading a bird-watching walk.

Even more newfangled is humans choosing their leaders, whether selecting a clan head by acclamation around the campfire, or a three-year-long presidential campaign season topped with the bizarreness of the Electoral College. How do we choose leaders?

A frequent conscious component of decision making is to vote for experience or competence rather than for stances on specific issues. This is so common that in one study faces judged to look more competent won elections 68 percent of the time.[26] People also make conscious voting choices based on single, potentially irrelevant issues (e.g., voting for assistant county dogcatcher based on the candidates' stances on drone warfare in Pakistan). And then there's the facet of American decision making that baffles citizens of other democracies, namely voting for "likability." Just consider Bush v. Kerry in 2004, where Republican pundits suggested that people's choice for the most powerful position on earth should reflect which guy you'd rather have a beer with.

At least as interesting are the automatic and unconscious elements of decision making. As probably the strongest factor, of candidates with identical political stances, people are more likely to vote for the better-looking one. Given the preponderance of male candidates and officeholders, this mostly translates into voting for masculine traits—tall, healthy-looking, symmetrical features, high forehead, prominent brow ridges, jutting jaw.[27]

As first raised in chapter 3, this fits into the larger phenomenon of attractive people typically being rated as having better personalities and higher moral standards and as being kinder, more honest, more friendly, and more trustworthy. And they are treated better—for the same résumé, being more likely to be hired; for the same job, getting a higher salary; for the same crime, being less likely to be convicted. This is the beauty-is-good stereotype, summarized in an 1882 quote by Friedrich Schiller: "Physical beauty is the sign of an interior beauty, a spiritual and moral beauty."[28] This is the flip side of the view that disfigurement, illness, and injury are karmic payback for one's sins. And as we saw in chapter 3, we use the same circuitry in the orbitofrontal PFC when we evaluate the moral goodness of an act and the beauty of a face.

Other implicit factors come into play. One study examined the campaign speeches of candidates in every prime minister election in Australian history.[29] In 80 percent of elections the winner was the one to use more collective pronouns (i.e., "we" and "us"), suggesting an attraction to candidates who speak on everyone's behalf.

There are also contingent automatic preferences. For example, in scenarios concerning war, both Western and East Asian subjects prefer candidates with older, more masculine faces; during peacetime, it's younger, more feminine faces. Furthermore, in scenarios involving fostering cooperation between groups, intelligent-looking faces are preferred;

at other times more intelligent faces are viewed as less masculine or desirable.[30]

These automatic biases fall into place early in life. One study showed kids, ages five to thirteen, pairs of faces of candidates from obscure elections and asked them whom they'd prefer as captain on a hypothetical boat trip. And kids picked the winner *71 percent* of the time.[31]

Scientists doing these studies often speculate as to why such preferences have evolved; frankly, much of this feels like just-so stories. For example, in analyzing the preference for leaders with more masculine faces during war, the authors noted that high testosterone levels produce both more masculine facial features (generally true) and more aggressive behavior (not true, back to chapter 4), and that aggressiveness is what you want in a leader during times of war (personally, not so sure about that one). Thus, preferring candidates with more masculine faces increases the odds of landing the aggressive leader you need to triumph in war. And everyone then passes on more copies of their genes. Voilà.

Regardless of causes, the main point is the power of these forces—five-year-olds with 71 percent accuracy demonstrate that these are some very generalized, deeply entrenched biases. And then our conscious cognitions play catch-up to make our decision seems careful and wise.

OH, WHY NOT TAKE THIS ONE ON? POLITICS AND POLITICAL ORIENTATIONS

So humans keep getting weirder—multiple hierarchies *and* having leaders *and* occasionally choosing them *and* doing so with some silly, implicit criteria. Now let's plunge into politics.

Frans de Waal introduced the term "politics" into primatology with his classic book *Chimpanzee Politics*, using it in the sense of "Machiavellian intelligence"—nonhuman primates struggling in socially complex manners to control access to resources. The book documents chimpanzee genius for such maneuvering.

This is "politics" in the traditional human sense as well. But I will use a more restricted, starry-eyed sense, which is politics being the struggle among the powerful with differing visions of the common good. Forget liberals accusing conservatives of waging war on the poor. Ditto conservatives accusing those depraved liberals of destroying family values. Cutting through this posturing, we'll assume that everyone equally desires that people do as well as possible, but differs as to how best to accomplish this. In this section we'll focus on three issues:

a. Do political orientations tend to be internally consistent (e.g., do people's opinions about their town's littering policy and about military actions in Somewhere-istan come as an ideological package)? Quick answer: usually.

b. Do such consistent orientations arise from deep, implicit factors with remarkably little to do with specific political issues? Yup.

c. Can one begin to detect the bits of biology underlying these factors? Of course.

The Internal Consistency of Political Orientation

The previous chapter examined the remarkable consistency in Us/Them orientations—people who dislike a particular out-group on economic grounds are likelier than chance to dislike another group on historical grounds, another on cultural, and so on.[32] Much the same is true here—social, economic, environmental, and international political orientations tend to come in a package. This consistency explains the humor behind a *New Yorker* cartoon (pointed out by the political psychologist John Jost) showing a woman modeling a dress for her husband and asking, "Does this dress make me look Republican?" Another example concerns the bioethicist Leon Kass, who not only has had influential conservative stances on human cloning, finding the possibility "repugnant," but also finds it repugnant when people display the "catlike activity" of licking ice cream cones in public. More to come on his issues, including with licking ice cream cones. What this internal consistency suggests is that political ideology is merely one manifestation of broader, underlying ideology—as we'll see, this helps explain conservatives being more likely than liberals to have cleaning supplies in their bedrooms.

Naturally, strict consistency in political ideology isn't always the rule. Libertarians are a mixture of social liberalism and economic conservatism; conversely, black Baptist churches are traditionally economically liberal but socially conservative (for example, rejecting both gay rights and the idea that gay rights are a form of civil rights). Moreover, neither extreme of political ideology is monolithic (and ignoring that, I'll be simplifying throughout by using "liberal" and "left-wing" interchangeably, as well as "conservative" and "right-wing").

Nonetheless, the building blocks of political orientation tend to be stable and internally consistent. It's usually possible to dress like a Republican or lick ice cream like a Democrat.

Implicit Factors Underlying Political Orientation

If political ideology is but one manifestation of larger internal forces pertinent to everything from cleaning supplies in the bedroom to ice cream consumption, are there psychological, affective, cognitive, and visceral ways in which leftists and rightists tend to differ? This question has produced deeply fascinating findings; I'll try to corral them into some categories.

INTELLIGENCE

Oh, what the hell? Let's begin with something inflammatory. Starting with Theodor Adorno in the 1950s, people have suggested that lower intelligence predicts adherence to conservative ideology.[33] Some but not all studies since then have supported this conclusion. More consistent has been a link between lower intelligence and a subtype of conservatism, namely right-wing authoritarianism (RWA, a fondness for hierarchy). One particularly thorough demonstration of this involved more than fifteen thousand subjects in the UK and United States; importantly, the links among low IQ, RWA, and intergroup prejudice were there after controlling for education and socioeconomic status. The standard, convincing explanation for the link is that RWA provides simple answers, ideal for people with poor abstract reasoning skills.

INTELLECTUAL STYLE

This literature has two broad themes. One is that rightists are relatively uncomfortable intellectually with ambiguity; this is covered below. The other is that leftists, well, think *harder*, have a greater capacity for what the political scientist Philip Tetlock of the University of Pennsylvania calls "integrative complexity."

In one study conservatives and liberals, when asked about the causes of poverty, both tended toward personal attributions ("They're poor because they're lazy"). But only if they had to make snap judgments. Give people more time, and liberals shifted toward situational explanations ("Wait, things are stacked against the poor"). In other words, conservatives start gut and stay gut; liberals go from gut to head.[34]

This differing attributional style extends way beyond politics. Tell liberals or conservatives about a guy who trips over someone's feet while learning a dance, ask for a snap assessment, and everyone does personal attribution—the guy's clumsy. It's only with time that the liberals go situational—maybe the dance is really difficult.

Obviously this dichotomy isn't perfect. Rightists did personal attribution for Lewinsky-gate (Bill Clinton's rotten) while leftists did situational (it's a vast right-wing conspiracy), and things ran the opposite with Nixon and Watergate. However, they are pretty reliable.

Why the difference? Liberals and conservatives are equally capable of thinking past gut personal attributions to subtler situational ones—when asked to do so, both are equally adept at dispassionately presenting the viewpoints of the opposite camp. It's that liberals are more motivated to push toward situational explanations.

Why? Some have suggested it's a greater respect for thinking, which readily becomes an unhelpful tautology. Linda Skitka of the University of

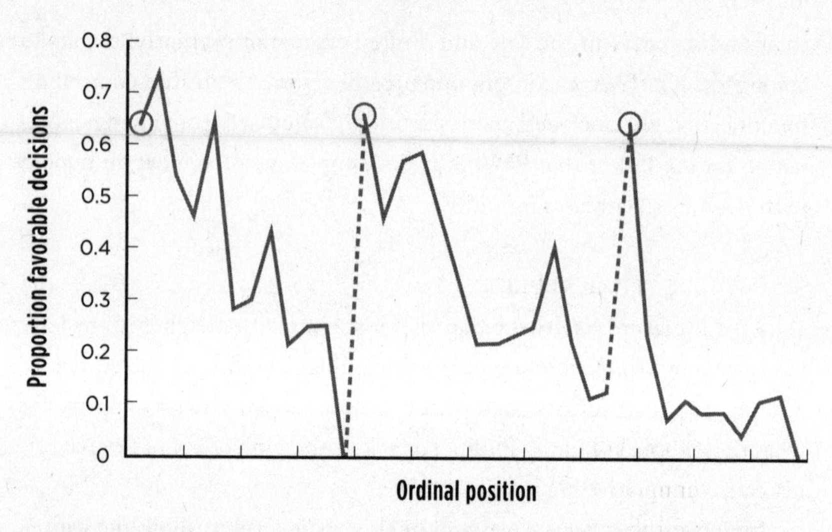

Proportion of rulings in favor of the prisoners by ordinal position. Circled points indicate the first decision in each of the three decision sessions; tick marks on x axis denote every third case; dotted line denotes food break. Because unequal session lengths resulted in a low number of cases for some of the later ordinal positions, the graph is based on the first 95% of the data from each session.

Illinois emphasizes how the personal attributions of snap judgments readily feel dissonant to liberals, at odds with their principles; thus they are motivated to think their way to a more consonant view. In contrast, even with more time, conservatives don't become more situational, because there's no dissonance.

While logical, this just turfs us to asking where the liberal ideology causing the dissonance comes from in the first place. As we'll see, it comes from factors having little to do with cognitive style.

These findings suggest that it's easier to make a liberal think like a conservative than the other way around.[35] Or, stated in a familiar way, increasing cognitive load* should make people more conservative. This is precisely the case. The time pressure of snap judgments is a version of increased cognitive load. Likewise, people become more conservative when tired, in pain or distracted with a cognitive task, or when blood alcohol levels rise.

* This applies to more global periods of duress as well; it turns out that despite the image of such periods as increasing polarization, it is the rare leftist who becomes more implicitly leftist at such times (stay tuned).

Recall from chapter 3 that willpower takes metabolic power, thanks to the glucose demands of the frontal cortex. This was the finding that when people are hungry, they become less generous in economic games. A real-world example of this is startling (see graph on previous page)—in a study of more than 1,100 judicial rulings, prisoners were granted parole at about a 60 percent rate when judges had recently eaten, and at essentially a 0 percent rate just before judges ate (note also the overall decline over the course of a tiring day). Justice may be blind, but she's sure sensitive to her stomach gurgling.[36]

MORAL COGNITION

Another minefield. Surprise, people at both ends of the political spectrum accuse the other side of impoverished moral thought.[37] One direction of this is seemingly bolstered by chapter 7's Kohlberg stages of moral development. Liberals, steeped in civil disobedience, tend to be at a "higher" Kohlberg stage than are conservatives, with their fondness for law and order. Are rightists less intellectually *capable* of reasoning at a more advanced Kohlberg stage, or are they less *motivated* to do so? Seemingly the latter—rightists and leftists are equally capable of presenting the other's perspective.

Jonathan Haidt of NYU provides a very different view.[38] He identifies six foundations of morality—care versus harm; fairness versus cheating; liberty versus oppression; loyalty versus betrayal; authority versus subversion; sanctity versus degradation. Both experimental and real-world data show that liberals preferentially value the first three goals, namely care, fairness, and liberty (and, showing an overlap with Kohlbergian formulations, undervaluing loyalty, authority, and sanctity is in many ways synonymous with postconventional thinking). In contrast, conservatives heavily value loyalty, authority, and sanctity. Obviously, this is a big difference. Is it okay to criticize your group to outsiders? Rightists: no, that's disloyal. Leftists: yes, if justified. Should you ever disobey a law? Rightists: no, that undermines authority. Leftists: of course, if it's a bad law. Is it okay to burn the flag? Rightists: never, it's sacred. Leftists: come on, it's a piece of cloth.

These differing emphases explain a lot—for example, the classical liberal view is that everyone has equal rights to happiness; rightists instead discount fairness in favor of expedient authority, generating the classical conservative view that some socioeconomic inequality is a tolerable price for things running smoothly.

What does it mean that, in Haidt's view, conservatives count up six (moral foundations) on their toes and liberals only three? Here is where internecine sniping starts. Conservatives embrace Haidt's characterization of liberals as being morally impoverished, with half their moral foundations atrophied.* The opposite interpretation, espoused by Jost, and Joshua Greene of Harvard, is that liberals have more refined moral foundations, having jettisoned the less important, more historically damaging ones that conservatives perseverate on—in effect, liberals count from one to three, while conservatives really only count four to six.

Why are conservatives more concerned with "binding foundations" like loyalty, authority, and sanctity, often stepping-stones to right-wing authoritarianism and social-dominance orientation? This segues to the next section.

AFFECTIVE PSYCHOLOGICAL DIFFERENCES

Research consistently shows that leftists and rightists differ in overlapping categories of emotional makeup. To summarize: on the average, rightists are made more anxious by ambiguity and have a stronger need for closure, dislike novelty, are more comforted by structure and hierarchy, more readily perceive circumstances as threatening, and are more parochial in their empathy.

The conservative dislike of ambiguity has been demonstrated in numerous apolitical contexts (e.g., responses to visual illusions, taste in entertainment) and is closely related to the differing feelings about novelty, which by definition evokes ambiguity and uncertainty.[39] The differing views of novelty certainly explain the liberal view that with correct reforms, our best days are ahead of us in a novel future, whereas conservatives view our best days as behind us, in familiar circumstances that

* Interestingly, Haidt does not characterize himself as conservative, although recent interviews suggest that this is shifting.

should be returned to, to make things great again. Once again, these differences in psychological makeup play out in apolitical realms as well—liberals are more likely to own travel books than are conservatives.

The conservative need for predictability and structure obviously fuels the emphases on loyalty, obedience, and law and order.[40] It also gives insights into a puzzling feature of the political landscape: how is it that over the last fifty years, Republicans have persuaded impoverished white Americans to so often vote against their own economic self-interest? Do they actually believe that they're going to win the lottery and then get to enjoy the privileged side of American inequality? Nah. The psychological issues of needing structured familiarity show that for poor whites, voting Republican constitutes an implicit act of system justification and risk aversion. Better to resist change and deal with the devil that you know. Harking back to the last chapter, gay conservatives show more implicit *anti*gay biases than do gay liberals. Better to hate who you are, if that bolsters a system whose stability and predictability are sources of comfort.

Intertwined with these variables is the Left/Right difference in tendency to see things as threatening, particularly when conservatism is anchored in authoritarianism. Life is filled with ambiguity, most of all with the novel future, and if those make you anxious, lots of things will seem threatening. Now, a "threat" can be abstract, such as to your self-esteem; there are few political differences in the perception of such threats. The differences concern concrete threats to your keister.

This helps explain political stances—"I have a list here of two hundred communist spies working in the State Department" is a pretty good demonstration of imagined threat.* The difference in threat perception can be apolitical. In one study subjects had to rapidly do a task when a word flashed up on a screen. Authoritarian conservatives, but not liberals, responded more rapidly to threatening words like "cancer," "snake," or "mugger" than to nonthreatening words (e.g., "telescope," "tree," "canteen"). Moreover, as compared with liberals, such conservatives are more likely to associate "arms" with "weapons" (rather than with "legs"), more

* While it is debatable whether McCarthy actually felt threatened (or even believed a word of what he spewed), he certainly knew how to exploit others with that tendency.

likely to interpret ambiguous faces as threatening, and more easily conditioned to associate negative (but not positive) stimuli with neutral stimuli. Republicans report three times as many nightmares as do Democrats, particularly ones involving loss of personal power. As the saying goes, a conservative is a liberal who has been mugged.

Related to this is "terror-management theory," which suggests that conservatism is psychologically rooted in a pronounced fear of death; supporting this is the finding that priming people to think about their mortality makes them more conservative.[41]

These differences in threat perception help explain the differing views as to role of government—providing for people (the leftist view; social services, education, etc.) or protecting people (the rightist view; law and order, the military, etc.).*

Fear, anxiety, the terror of mortality—it must be a drag being right-wing. But despite that, in a multinational study, rightists were happier than leftists.[42] Why? Perhaps it's having simpler answers, unburdened by motivated correction. Or, as favored by the authors, because system justification allows conservatives to rationalize and be less discomfited by inequality. And as economic inequality rises, the happiness gap between the Right and the Left increases.

As emphasized, political ideology is just one manifestation of intellectual and emotional style. As a great example, a *four*-year-old's openness to a new toy predicts how open she'll be as an adult to, say, the United States forging new relations with Iran or Cuba.[43]

And of Course, Some Underlying Biology

We've now seen that political orientation is usually stable and internally consistent across a range of disparate issues, and that it is typically merely one manifestation of a package of cognitive and affective style.

* Importantly, while conservatives may be more sensitive to feeling threatened, they're not necessarily more empathic to threat to someone else—conservatives are more likely to be skeptical about the validity of someone else's physical pain, more likely to frame it as malingering and dependent manipulation..

Stepping deeper, what are the biological correlates of differing political orientations?

Back to the insular cortex and its role in mediating gustatory and olfactory disgust in mammals and in mediating moral disgust in humans. Recall from the last chapter how you can reliably stoke hatred of Thems by making them seem viscerally disgusting. When people's insulae activate at the thought of Thems, you can check one thing off your genocide to-do list.

This recalls a remarkable finding—stick subjects in a room with a smelly garbage can, and they become more socially conservative.[44] If your insula is gagging from the smell of dead fish, you're more likely to decide that a social practice of an Other that is merely different is, instead, just plain wrong.

This leads to a thoroughly fascinating finding—social conservatives tend toward lower thresholds for disgust than liberals. In one study subjects were exposed to either positively or negatively charged emotional images,* and galvanic skin resistance (GSR, an indirect measure of sympathetic nervous system arousal) was measured. The biggest autonomic responses to negative (but not positive) emotional images were in conservatives opposed to gay marriage or premarital sex (while GSR response was unrelated to nonsocial issues like free trade or gun control). Concerns about hygiene and purity sure predict valuing of sanctity.[45]

Related to that, when confronted with something viscerally disturbing, conservatives are less likely to use reappraisal strategies (e.g., when seeing something gory, thinking, "This isn't real; it's staged"). Moreover, when conservatives, but not liberals, are instructed to use reappraisal techniques (e.g., "Try to view the images in a detached, unemotional way"), they express less conservative political sentiments. In contrast, a suppression strategy ("Don't let your feelings show when you're looking at this image") doesn't work. As we saw, make a liberal tired, hungry, rushed, distracted, or disgusted, and they become more conservative. Make a

* Negative images included someone eating worms, excrement floating in a toilet, a bloody wound, and an open sore teeming with maggots. Fun.

conservative more detached about something viscerally disturbing, and they become more liberal.[46]

Thus political orientation about social issues reflects sensitivity to visceral disgust and strategies for coping with such disgust. In addition, conservatives are more likely to think that disgust is a good metric for deciding whether something is moral. Which recalls Leon Kass, the bioethicist with the ice cream–licking issues. He headed George W. Bush's bioethics panel, one that, thanks to Kass's antiabortion ideology, greatly restricted embryonic stem cell research. Kass has argued for what he calls "the wisdom of repugnance," where disgust at something like human cloning can be "the emotional expression of deep wisdom, beyond wisdom's power completely to articulate it." The visceral level, with or without post-hoc rationalization, is all you need in order to know what's right. If it makes you puke, then you must rebuke.[47]

The monumental flaw is obvious. Different things disgust different people; whose gag reflex wins? Moreover, things once viewed as disgusting are viewed differently now (e.g., the idea of slaves having the same rights as whites would probably have struck most white Americans circa 1800 as not just economically unworkable but disgusting as well). It's disgusting, the things people weren't disgusted by in the past. Disgust is a moving target.

Thus issues anchored in the insula help explain differences in political orientation; this point will be returned to in chapter 17.[48] Additional neurobiological differences have been demonstrated. Liberalism has been associated with larger amounts of gray matter in the cingulate cortex (with its involvement in empathy), whereas conservatism has been associated with an enlarged amygdala (with, of course, its starring role in threat perception). Moreover, there's more amygdala activation in conservatives than in liberals when viewing a disgusting image or doing a risky task.

But not all the findings fit easily. For example, when looking at disgusting images, conservatives also show relatively greater activation of a hodgepodge of other brain regions—the basal ganglia, thalamus, periaqueductal gray, (cognitive) dlPFC, middle/superior temporal gyrus, pre-supplementary motor, fusiform, and inferior frontal gyrus. How all those fit together isn't clear.

Naturally, one must ask: have behavior geneticists reported genetic influences on political orientation? Twin studies report heritability of about 50 percent for political orientation. Genomewide survey approaches have identified genes whose polymorphic variants were associated with political orientation. Most of the genes had no known functions, or were previously thought to be unrelated to the brain; those whose brain-related functions were known (for example, one coded for a receptor for the neurotransmitter glutamate) don't teach much about political orientation. As an interesting gene/environment interaction, the "risk-taking" version of the D4 dopamine receptor gene is associated with liberals—but only in people with lots of friends. Moreover, some studies show a genetic association with people's likelihood of voting, independent of political orientation.[49]

Interesting. However, the approach comes with all of chapter 8's caveats—most findings haven't been replicated, reported effects are small, and these are published in political science journals rather than genetics journals. Finally, to the extent that genes are related to political orientation, links are likely to be via intervening factors, such as the tendency toward anxiety.

OBEDIENCE AND CONFORMITY, DISOBEDIENCE AND NONCONFORMITY

So humans have multiple simultaneous hierarchies and hierarchies built around abstractions, and occasionally choose leaders who labor for the common good.[50] Add to that obedience to leaders. This is utterly different from a schlub of a baboon obediently surrendering his spot in the shade to the looming alpha male. Instead humans show obedience to authority that transcends any given occupant of a throne (the king is dead; long live the king), to the very notion of authority. Its elements range from loyalty, admiration, and emulation to brownnosing, sycophancy, and instrumental self-interest, and can range from mere compliance (i.e., the public conformity of going along, without actually agreeing) to drinking the Kool-Aid (i.e., identifying with the authority and internalizing and extending its beliefs).

Obedience is closely intertwined with conformity, a concept central to the previous chapter but considered here. Both consist of going along; the former with authority, the latter with the group. And for us the commonalities are what matter. Moreover, the opposites—disobedience and nonconformity—are also intertwined and range from the independence of marching to a different drummer to the intentionality and mirrored determinism of anticonformity.

Importantly, these are value-free terms. Conformity can be great—it's helpful if everyone in a culture agrees on whether shaking your head up and down means yes or no. Conforming is necessary for the benefits of the wisdom of the crowd. And it can be truly comforting. But obviously conformity can be horrendous—joining in on bullying, oppressing, shunning, expelling, killing, just because everyone else is on board.

Obedience can be swell too, ranging from everyone stopping at stop signs to (to the embarrassment of my pseudoanarchist adolescence) my kids listening when my wife and I say it's bedtime. And malign obedience obviously underlies "just following orders"—from goose-stepping to Jonestown's wretched obeying the command to kill their children.

Roots

Conformity and obedience have deep roots, as evidenced by their presence in other species and in very young humans.

Animal conformity is a type of social learning—a subordinate primate does not have to be thrashed by some bruiser in order to express subordination to him; everyone else's doing so can be sufficient.*[51] The confor-

* This has even been shown to involve formal transitive logic. Animal A loses a dominance interaction with animal B. Animal A then observes animal B losing a dominance interaction to animal C. Animal A then, on his first encounter with animal C, gives a subordination signal. This has been shown in various primate species, rats, birds, and even fish.

mity has a familiar human tinge to it. For example, a chimp is more likely to copy an action if he sees three other individuals do it once each than if one other individual does it three times.* Moreover, learning can include "cultural transmission"—in chimps, for example, this includes learning types of tool construction. Conformity relates to social and emotional contagion where, say, a primate aggressively targets an individual just because someone else is already doing so. Such contagion even works between groups. For example, among marmosets aggression in a group becomes more likely if aggressive vocalizations are heard from the neighboring group. Other primates are even subject to the social contagion of yawning.†[52]

My favorite example of nonhuman conformity is so familiar that it could come right out of high school. A male grouse courts a female who, alas, doesn't feel magic in the air and rebuffs him. The researchers then make him seem like the hottest stud on the prairie—by surrounding him with some rapt, stuffed female grouse. Soon the reluctant maiden is all over him, pushing her statuesque rivals aside.[53]

An even clearer demonstration of animal conformity was shown in a beautiful study of chimpanzees by Frans de Waal. In each of two groups the alpha female was separated from the rest and shown how to open a puzzle box containing food. Crucially, the two were shown different, equally difficult ways of doing it. Once the females had mastered their approaches, the chimps in each group got to watch their alpha female strut her stuff repeatedly with the puzzle box. Finally everyone got access to the puzzle box and promptly copied their alpha's technique.[54]

Thus this is a cool demonstration of the spread of cultural information. But something even more interesting happened. A chimp in the group would occasionally stumble onto the alternative method—and would then abandon it, going back to doing it the "normal" way. Just because

* The study also showed, with nice ethological logic, that the same conforming did not occur in orangutans, who are solitary primates.

† Chimp yawning is most readily evoked by watching yawning by another familiar chimp, next-most readily by watching a familiar human yawning, followed by an unfamiliar human yawning; however, contagious yawning is not evoked by an unfamiliar chimp or an unfamiliar primate species (a baboon).

everyone else was doing so.* The same phenomenon was subsequently shown in capuchin monkeys and wild birds.

Thus animals will perform one version of a behavior not because it is better but simply because everyone else does. Even more striking, animal conformity can be detrimental. In a 2013 study Andrew Whiten of the University of St Andrews presented wild vervet monkeys with two bins of maize, dyed either pink or blue.[55] One color tasted fine; the other had a bitter additive. The monkeys learned to avoid the latter and months later still ate only the "safe"-colored maize—even after the additive was omitted.

As infants were born or adults who had grown up elsewhere migrated into the group, they conformed to this food choice, learning to eat only the same color food as everyone else. In other words, forgoing half the potential food just because of the need to fit in—monkeys joining the herd, acting like sheep, going over the cliff like lemmings. One example starkly displays the same in humans: in life-threatening emergencies (e.g., a restaurant fire), people frequently attempt to escape by following the crowd in what they know to be the wrong direction.

The deeply ingrained nature of human conformity and obedience in humans is shown by the ages when they are apparent. As detailed in chapter 7, zillions of pages have been written about conformity and peer pressure in kids. One study nicely demonstrates the continuity of conformity between us and other species. This was the report that a chimp was more likely to conform to the behaviors of three individuals doing a behavior once each than to one individual doing the behavior three times. The study showed the same in two-year-old humans.

The depths of human conformity and obedience are shown by the speed with which they occur—it takes less than 200 milliseconds for your brain to register that the group has picked a different answer from yours, and less than 380 milliseconds for a profile of activation that predicts

* I'd love to know what is going on in the heads of chimps when they abandon their alternative method. Are they activating the amygdala, initiating a stress response? What is a chimp's equivalent of worrying about seeming like a dork?

changing your opinion. Our brains are biased to get along by going along in less than a second.[56]

Neural Bases

This last study raises the question of what occurs in the brain under these circumstances. Our usual cast of brain regions pops up in informative ways.

The influential "social identity theory" posits that our concept of who we are is heavily shaped by social context—by the groups we do or don't identify with.*[57] In that view, conformity and obedience, while certainly about avoiding punishment, are at least as much about the positives of fitting in. When we imitate someone's actions, our mesolimbic dopamine system activates.† When we choose incorrectly in a task, the dopaminergic decline is less if we made the decision as part of a group than if we did so as an individual. Belonging is safety.

In numerous studies a subject in a group answers some question, finds out after that—oh no!—everyone else disagrees, and can then change their answer.[58] No surprise, the discovery that you are out of step activates the amygdala and insular cortex; the more activation, the greater your likelihood of changing your mind, and the more persistent the change (as opposed to the transient change of compliant public conformity). This is a profoundly social phenomenon—people are more likely to change their answer if you show them a picture of the person(s) who disagrees with them.

When you get the news that everyone else disagrees with you, there is also activation of the (emotional) vmPFC, the anterior cingulate cortex, and the nucleus accumbens. This is a network mobilized during reinforcement learning, where you learn to modify your behavior when there

* Social identity theory is most associated with the Polish/French/British psychologist Henri Tajfel. As will be seen, Tajfel, pondering why normal people join the herd in doing awful things, was but one of the many scientists in this field whose lives had been personally scorched by the Holocaust.

† What, if anything, such mimicry has to do with "mirror neurons" will await the discussion in chapter 14 as to what, if anything, mirror neurons have to do with empathy.

is a mismatch between what you expected to happen and what actually did. Find out that everyone disagrees with you and this network activates. What is it basically telling you? Not just that you're *different* from everyone else. That you're *wrong*. Being different = being wrong. The greater the activation of this circuit, the greater the likelihood of changing answers to conform.[59]

Like most of the neuroimaging literature, these studies are merely correlative. Thus, particularly important is a 2011 study that used transcranial magnetic stimulation techniques to temporarily inactivate the vmPFC; subjects became less likely to change their answer to conform.[60]

Back to the contrast between conforming taking the form of "You know what, if everyone says they saw B, I guess I did too; whatever" and its taking the form of "Now that I think about it, I didn't actually see A; I think I saw B; in fact I'm certain of it." The latter is associated with activation of the hippocampus, the brain region central to learning and memory—the revisionism involves you literally revising your memory. Remarkably, in another study this process of conforming was also associated with activation of the occipital cortex, the brain region that does the primary processing of vision—you can almost hear the frontal and limbic parts of the brain trying to convince the occipital cortex that it saw something different from what it actually saw. As has been said, winners (in this case, in the court of public opinion) get to write the history books, and everyone else better revise theirs accordingly. War is peace. Freedom is slavery. That dot you saw was actually blue, not red.[61]

Thus the neurobiology of conforming consists of a first wave of anxiety where we equate differentness with wrongness, followed by the cognitive work needed to change our opinion. These findings obviously come from an artificial world of psych experiments. Thus they're only a faint whisper of what occurs when it's you against the rest of the jury, when it's you being urged to join the lynch mob, when it's you choosing between conforming and being deeply lonely.

What is the neurobiology of obedience to authority, when you're being ordered to do something wrong? A similar mixture as with conformity, with the vmPFC and the dlPFC mud-wrestling, with indices of anxiety and glucocorticoid stress hormones showing up to bias you toward subordination. Which leads us to consider classic studies of "just following orders."

Asch, Milgram, and Zimbardo

The neurobiology of conformity and obedience won't soon be revealing much about the core question in this field: if the circumstances are right, is every human capable of doing something appalling simply because they've been ordered to, because everyone else is doing it?

It is virtually required by law to discuss three of the most influential, daring, disturbing, and controversial studies in the history of psychology, namely the conformity experiments of Solomon Asch, the shock/obedience studies of Stanley Milgram, and the Stanford Prison Experiment of Philip Zimbardo.

The grandparent of the trio was Asch, working in the early 1950s at Swarthmore College.[62] The format of his studies was simple. A volunteer, thinking that this was a study of perception, would be given a pair of cards. One card would have a line on it, the other a trio of different-length lines, one of which matched the length of the singleton line. Which line of the trio is the same length as the singleton? Easy; volunteers sitting alone in a room had about a 1 percent error rate over a series of cases.

Meanwhile, the volunteers in the experimental group take the test in a room with seven others, each saying his choice out loud. Unbeknownst to the volunteer, the other seven worked on the project. The volunteer would "just happen" to go last, and the first seven would unanimously pick a glaringly wrong answer. Stunningly, volunteers would now agree with that incorrect answer about a third of the time, something replicated frequently in the cottage industry of research spawned by Asch. Whether due to the person's actually changing their mind or their merely deciding to go along, this was a startling demonstration of conformity.

On to the Milgram obedience experiment, whose first incarnations appeared in the early 1960s at Yale.[63] A pair of volunteers would show up for a psychology "study of memory"; one would arbitrarily be designated the "teacher," the other the "learner." Learner and teacher would be in separate rooms, hearing but not seeing each other. In the room with the teacher would be the lab-coated scientist supervising the study.

The teacher would recite pairs of words (from a list given by the scientist); the learner was supposed to remember their pairing. After a series of these, the teacher would test the learner's memory of the pairings. Each time the learner made a mistake, the teacher was supposed to shock them; with each mistake, shock intensity increased, up to a life-threatening 450 volts, ending the session.

Teachers thought the shocks were real—at the start they'd been given a real shock, supposedly of the intensity of the first punitive one. It hurt. In reality no punitive shocks were given—the "learner" worked on the project. As the intensity of the supposed shocks increased, the teacher would hear the learner responding in pain, crying out, begging for the teacher to stop.* (In one variant the "volunteer" who became the learner mentioned in passing that he had a heart condition. As shock intensity increased, this learner would scream about chest pains and then go silent, seemingly having passed out.)

Amid the screams of pain, teachers would typically become hesitant, at which point they'd be urged on by the scientist with commands of increasing intensity: "Please continue." "The experiment requires that you continue." "It is absolutely essential that you continue." "You have no other choice. You must go on." And, the scientist assured them, they weren't responsible; the learner had been informed of the risks.

And the famed result was that most volunteers complied, shocking the learner repeatedly. Teachers would typically try to stop, argue with the scientist, would even weep in distress—but would obey. In the original study, horrifically, 65 percent of them administered the maximum shock of 450 volts.

* As a slick part of the design, it wasn't the actor in the next room doing this emoting. Instead, pressing the shock button of each particular intensity triggered the playing of a particular recording of sounds commensurate with that shock intensity. This would standardize the supposed agony of the learner from one subject to the next.

And then there's the Stanford Prison Experiment (SPE), carried out by Zimbardo in 1971.[64] Twenty-four young male volunteers, mostly college students, were randomly split into a group of twelve "prisoners" and twelve "guards." The prisoners were to spend seven to fourteen days jailed in a pseudoprison in the basement of Stanford's psychology department. The guards were to keep order.

Tremendous effort went into making the SPE realistic. The future prisoners thought they were scheduled to show up at the building at a particular time to start the study. Instead, Palo Alto police helped Zimbardo by showing up earlier at each prisoner's home, arresting him, and taking him to the police station for booking—fingerprinting, mug shots, the works. Prisoners were then deposited in the "prison," strip-searched, given prison garb, along with stocking hats to simulate their heads being shaved, and dumped as trios in cells.

The guards, in surplus military khakis, batons, and reflective sunglasses, ruled. They had been informed that while there was no violence allowed, they could make the prisoners feel bored, afraid, helpless, humiliated, and without a sense of privacy or individuality.

And the result was just as famously horrific as that of the Milgram experiment. The guards put prisoners through pointless, humiliating rituals of obedience, forced painful exercise, deprived them of sleep and food, forced them to relieve themselves in unemptied buckets in the cells (rather than escorting them to the bathroom), put people in solitary, set prisoners against each other, addressed them by number, rather than by name. The prisoners, meanwhile, had a range of responses. One cell revolted on the second day, refusing to obey the guards and barricading the entrance to their cell; guards subdued them with fire extinguishers. Other prisoners resisted more individualistically; most eventually sank into passivity and despair.

The experiment ended famously. Six days into it, as the brutality and degradation worsened, Zimbardo was persuaded to halt the study by a graduate student, Christina Maslach. They later married.

Situational Forces and What Lurks in All of Us

These studies are famed, have inspired movies and novels, have entered the common culture (with predictably horrendous misrepresentations).*[65] They brought renown and notoriety to Asch, Milgram, and Zimbardo.† And they were vastly influential in scientific circles—according to Google Scholar, Asch's work is cited more than 4,000 times in the literature, Milgram's more than 27,000 times, the SPE more than 58,000.‡[66] The number of times your average science paper is cited can be counted on one hand, with most of the citations by the scientist's mother. The trio is a cornerstone of social psychology. In the words of Harvard psychologist Mahzarin Banaji, "The primary simple lesson the SPE [and, by extension, Asch and Milgram] teaches is that *situations matter*" (her emphasis).

What did they show? Thanks to Asch, that average people will go along with absurdly incorrect assertions in the name of conformity. And thanks to the other two studies, that average people will do stunningly bad things in the name of obedience and conformity.

The larger implications of this are enormous. Asch and Milgram (the former a Jewish Eastern European immigrant, the latter the son of Jewish Eastern European immigrants) worked in the era of the intellectual challenge of making sense of Germans "just following orders." Milgram's study was prompted by the start, a few months earlier, of the war-crimes trial of Adolf Eichmann, the man who famously epitomized the "banality of evil" because of his seeming normalcy. Zimbardo's work burst forth during the Vietnam War era with the likes of the My Lai Massacre, and the SPE became bitingly relevant thirty years later with the abuse and torture of Iraqis at Abu Ghraib Prison by perfectly normal American soldiers.§[67]

---

* E.g., "So, the scientists found that 65 percent of the subjects were willing to shock the learner to death and then eat his heart. And in the prison study, get this, 65 percent of the guards also became cannibals. It's, like, freaky that they got the same number."

† Cool real-life coincidence that isn't coincidental—Milgram and Zimbardo knew each other as classmates in their high school in the Bronx.

‡ One study inspired by Milgram was the Hofling hospital experiment, in which nurses, unaware that they were in an experiment, would be ordered by an unknown doctor to give a dangerously high dose of a medication to a patient. Despite their knowing of the danger, twenty-one out of twenty-two nurses were willing to comply.

§ Ironic Beginning Department: the SPE was funded by the U.S. military, which was interested in making military prisons run better.

Zimbardo took a particularly extreme stance as to what these findings mean, namely his "bad barrel" theory—the issue isn't how a few bad apples can ruin the whole barrel; it's how a bad barrel can turn any apple bad. In another apt metaphor, rather than concentrating on one evil person at a time, what Zimbardo calls a "medical" approach, one must understand how some environments cause epidemics of evil, a "public health" approach. As he states: "Any deed, for good or evil, that any human being has ever done, you and I could also do—given the same situational forces." Anyone could potentially be an abusive Milgram teacher, Zimbardo guard, or goose-stepping Nazi. In a similar vein, Milgram stated, "If a system of death camps were set up in the US of the sorts we had seen in Nazi Germany one would be able to find sufficient personnel for those camps in any medium-sized American town." And as stated by Aleksandr Solzhenitsyn in *The Gulag Archipelago*, in a quote perpetually cited in this literature, "The line dividing good and evil cuts through the heart of every human being. And who is willing to destroy a piece of his own heart?"[68]

Some Different Takes

Big surprise—the studies and their conclusions, especially those of Milgram and Zimbardo, have been controversial. Those two attracted firestorms of controversy because of the unethical nature of the work; some teachers and guards were psychological wrecks afterward, seeing what they had proven capable of;* it changed the course of a number of their lives.† No human-subjects committee would approve the Milgram study these days; in contemporary versions subjects are ordered to, for example, say increasingly insulting things to the learner or administer virtual shocks, evoking virtual pain, in avatars (stay tuned).[69]

The controversies about the science itself in the Milgram and

* Remember, these were predominantly psychologically sound college students. In the SPE nearly all of them had indicated at the beginning that they would rather be a prisoner than a guard, and a number indicated that they had volunteered in order to learn what prison would be like, expecting to be jailed at some point for civil-rights or antiwar activity. And as is often underemphasized in accounts of the SPE, many of the prisoners, as well as the guards, were deeply distressed afterward, having seen how readily they were broken into passivity.

† One teacher, for example, became a conscientious objector during Vietnam, prompted by his horror at his behavior in the study.

Zimbardo studies are more pertinent. The Milgram edifice has been questioned in three ways, most piercingly by the psychologist Gina Perry:

- Milgram seems to have fudged some of his work. Perry has analyzed Milgram's unpublished papers and recordings of sessions, finding that teachers refused to shock much more frequently than reported. However, despite the seemingly inflated results, the finding of roughly 60 percent compliance rates has been replicated.[70]
- Few of the replicating studies were traditional academic ones published in peer-reviewed journals. Instead most have been recreations for films and television programs.
- Perhaps most important, as analyzed by Perry, far more teachers than Milgram indicated realized that the learner was an actor and that there were no actual shocks. This problem probably extends to the replications as well.

The SPE has arguably attracted the most controversy.

- The biggest lightning rod was the role of Zimbardo himself. Rather than being a detached observer, he served as the prison's "superintendent." He set the ground rules (e.g., telling guards that they could make the prisoners feel afraid and helpless) and met regularly with the guards throughout. He was clearly excited as hell to see what was happening in the study. Zimbardo is a larger-than-life force of nature, someone whom you'd very much wish to please. Thus guards were subject to pressure not only to conform with their cohort but also to obey and please Zimbardo; his role, consciously or otherwise, almost certainly prompted the guards to more extreme behavior. Zimbardo, a humane, decent man who is a friend and colleague, has written at length about this distortive impact that he had on the study.
- At the beginning of the study, volunteers were randomly assigned to be guards or prisoners, and the resulting two groups did not

differ on various personality measures. While that's great, what was not appreciated was the possibility that the volunteers as a whole were distinctive. This was tested in a 2007 study in which volunteers were recruited through one of two newspaper ads. The first described "a psychological study of prison life"—the words used in the advertisement for the SPE—while in the other the word "prison" was omitted. The two groups of volunteers then underwent personality testing. Importantly, volunteers for the "prison" study scored higher than the others on measures of aggressiveness, authoritarianism, and social dominance and lower for empathy and altruism. Insofar as both guards and prisoners in the SPE might have had this makeup, it's not clear why that would have biased toward the famously brutal outcome.[71]

- Finally, there's science's gold standard, independent replication. If you redid the SPE, down to matching the brand of the guards' socks, would you get the same result? Any study this big, idiosyncratic, and expensive would be difficult to match perfectly in the replication attempt. Moreover, Zimbardo actually published remarkably little of the data about the SPE in professional journals; instead he mostly wrote for the lay public (hard to resist, given the attention the study garnered). Thus there's only really been one attempted replication.

The 2001 "BBC Prison Study" was run by two respected British psychologists, Stephen Reicher of the University of St Andrews and Alex Haslam of the University of Exeter.[72] As the name implies, it was carried out (i.e., among other things, paid for) by the BBC, which filmed it for a documentary. Its design replicated the broad features of the SPE.

As is so often the case, there was a completely different outcome. To summarize a book's worth of complex events:

- Prisoners organized to resist any abuse by the guards.
- Prisoner morale soared while guards became demoralized and divided.

- This led to a collapse of the guard/prisoner power differential and ushered in a cooperative, power-sharing commune.
- Which lasted only briefly before three ex-prisoners and one ex-guard overthrew the utopians and instituted a draconian regime; fascinatingly, those four had scored highest on scales of authoritarianism before the study began. As the new regime settled into repressive power, the study was terminated.

Thus, rather than a replication of the SPE, this wound up being more like a replication of the FRE and the RRE (i.e., the French Revolution and the Russian Revolution): a hierarchical regime is overthrown by wet-nosed idealists who know all the songs from *Les Mis*, who are then devoured by Bolsheviks or Reign of Terror–ists. Importantly, the ruling junta at the end having entered the study with the strongest predispositions toward authoritarianism certainly suggests bad apples rather than bad barrels.

Even bigger surprise—stop the presses—Zimbardo criticized the study, arguing that its structure invalidated it as a chance to replicate the SPE; that guard/prisoner assignments could not have really been random; and that filming made this a TV spectacle rather than science; and asking, how can this be a model for anything when the prisoners take over the prison?[73]

Naturally, Reicher and Haslam disagreed with his disagreement, pointing out that prisoners have de facto taken over some prisons, such as the Maze in Northern Ireland, which the Brits filled with IRA political prisoners, and the Robben Island prison, in which Nelson Mandela spent his endless years.

Zimbardo called Reicher and Haslam "scientifically irresponsible" and "fraudulent." They pulled out all the stops by quoting Foucault: "Where there is [coercive] power there is resistance."

Let's calm down. Amid the controversies over Milgram and the SPE, two deeply vital things are indisputable:

- When pressured to conform and obey, a far higher percentage of perfectly normal people than most would predict succumb and do awful things. Contemporary work using a variant on the Milgram paradigm shows "just following orders" in action, where the pattern of neurobiological activation differs when the same act is carried out volitionally versus obediently.[74]

- Nonetheless, there are always those who resist.

This second finding is no surprise, given Hutus who died shielding Tutsi neighbors from Hutu death squads, Germans with every opportunity to look the other way who risked everything to save people from the Nazis, the informant who exposed Abu Ghraib. Some apples, even in the worst of barrels, do not go bad.*

Thus what becomes vital is to understand the circumstances that push us toward actions we thought we were far better than or that reveal strength we never suspected we had.

Modulators of the Pressures to Conform and Obey

The end of the previous chapter examined factors that lessen Us/Them dichotomizing. These included becoming aware of implicit, automatic biases; becoming aware of our sensitivity to disgust, resentment, and envy; recognizing the multiplicity of Us/Them dichotomies that we harbor and emphasizing ones in which a Them becomes an Us; contact with a Them under the right circumstances; resisting essentialism; perspective taking; and, most of all, individuating Thems.

Similar factors decrease the likelihood of people doing appalling things in the name of conformity or obedience. These include:

* And reflecting this, Zimbardo's recent work examines defiance of unjust authority.

THE NATURE OF THE AUTHORITY OR GROUP PRESSING FOR CONFORMITY

Does the authority(s) evoke veneration, identification, pants-wetting terror? Is the authority in close proximity? Milgram follow-ups showed that when the authority (i.e., the scientist) was in a different room, compliance decreased. Does the authority come cloaked in prestige? When the experiment was conducted in some nondescript warehouse in New Haven, instead of on the Yale campus, compliance declined. And, as emphasized by Tajfel in his writing, is the authority perceived as legitimate and stable? I'd more likely comply with, say, lifestyle advice issued by the Dalai Lama than by the head of Boko Haram.

Similar issues of prestige, proximity, legitimacy, and stability influence whether people conform to a group. Obviously, groups of Us-es evoke more conformity than do groups of Thems. Consider the invoking of Us in Konrad Lorenz's attempt to justify becoming a Nazi: "Practically all my friends and teachers did so, including my own father who certainly was a kindly and humane man."[75]

With groups, issues of numbers come into play—how many other voices are urging you to join the cool kids? Recall how among chimps or two-year-old humans, one individual doing something three times does not evoke the conformity of three individuals doing the same act once each. Echoing this, follow-up studies by Asch showed that conformity first kicks in when there are at least three people unanimously contradicting what the subject thinks, with maximum conformity first evoked by around half a dozen contradictors. But this is the artificial world of lab subjects judging the length of a line—in the real world the conforming power of a lynch mob of six doesn't approach that of a mob of a thousand.[76]

WHAT IS BEING REQUIRED AND IN WHAT CONTEXT

Two issues stand out. The first is the persuasive power of the incremental. "You were okay shocking the guy with 225 volts, but not with 226? That's illogical." "Come on, we're all boycotting their businesses. Let's shut them down; it's not like anyone patronizes them. Come on, we've shut down their businesses, let's loot them; it's not like the stores are doing them any good." We rarely have a rational explanation for an intuitive sense that a line has been crossed on a continuum. What incrementalism does is put the potential resister on the defensive, making the savagery seem like an issue of rationality rather than of morality. This represents an ironic inversion of our tendency to think in categories, to irrationally inflate the importance of an arbitrary boundary. The descent into savagery can be so incremental as to come with nothing but arbitrary boundaries, and our descent becomes like the proverbial frog cooked alive without noticing. When your conscience finally rebels and draws a line in the sand, we know that it is likely to be an arbitrary one, fueled by implicit subterranean forces—despite your best attempts at pseudospeciation, this victim's face reminds you of a loved one's; a smell just wafted by that took you back to childhood and reminds you of how life once felt innocent; your anterior cingulate neurons just had breakfast. At such times, a line having finally been drawn must be more important than its arbitrariness.

The second issue concerns responsibility. When debriefed afterward, compliant teachers typically cited how persuasive they found the information that the learner had been informed of the risks and had given consent. "Don't worry, you won't be held responsible." The Milgram phenomenon also showed the coercive power of misdirecting responsibility, when researchers would seek compliance by emphasizing that the teacher's responsibility was to the project, not the learner—"I thought you said you were here to help." "You're a team member." "You're ruining things." "You signed a form." It's hard enough to respond with "This isn't the job I signed up for." It's that much harder when the fine print reveals that this *is* what you signed up for.

Compliance increases when guilt is diffused—even if I hadn't done it, it still would have happened.[77] Statistical guilt. This is why, historically, people were not executed with five shots fired from one gun. Instead there were five guns fired simultaneously—a firing squad. Firing squads traditionally took the diffusion of responsibility a step further, where one member was randomly given a blank instead of a real bullet. That way, a shooter could shift from the comforting irrationality that "I only one fifth

killed him" to the even better "I may not even have shot him." This tradition was translated into modern execution technology. Lethal injection machines used in prison executions come with a dual control system—two syringes, each filled with a lethal dose, two separate delivery systems, two buttons pressed simultaneously by two different people—at which point a random binary generator would secretly determine which syringe was emptied into a bucket and which into a human. And then the record would be erased, allowing each person to think, "Hey, I may not even have given him any drug."

Finally, responsibility is diffused by anonymity.[78] This comes de facto if the group is large enough, and large groups also facilitate individual efforts at anonymity—during the Chicago riots of 1968, many police notoriously covered their name tags before setting on the unarmed antiwar demonstrators. Groups also facilitate conformity by institutionalizing anonymity; examples range from the KKK to *Star Wars*' Imperial Storm Troopers to the finding that in traditional human societies, warriors who transform and standardize their appearance before battle are more likely to torture and mutilate their enemies than warriors from cultures that don't transform themselves. All use means to deindividuate, where the goal may not be to ensure that a victimized Them won't be able to recognize you afterward as much as to facilitate moral disengagement so that *you* won't be able to recognize you afterward.

THE NATURE OF THE VICTIM

No surprise, compliance becomes easier when the victim is an abstraction—say, the future generations who will inherit this planet. In Milgram follow-ups, compliance declined if the learner was in the same room as the teacher and would plummet if the two had shaken hands. Ditto if psychological distance was shortened by perspective taking—what would it feel like if you were in their shoes?

Predictably, compliance is also decreased when the victim is individuated.[79] However, don't let the authority individuate victims for you. In one classic Milgram-esque study, the scientists would "accidentally" allow a teacher to overhear their opinion of the learner. "Seems like a nice guy" versus "This guy seems like an animal." Guess who'd get more shocks?

Authorities rarely ask us to administer shocks to those whom they label as nice guys. It's always to the animals. Implicit in the latter categorization's evoking more compliance is our having ceded power to the authorities or to the group to create the narrative. One of the greatest wellsprings of resistance is to seize back the narrative. From "children of exceptionalities" to the Paralympics, from gay-pride marches to "never again," from Hispanic Heritage Month to James Brown singing, "Say It Out Loud, I'm Black and I'm Proud," a major step toward victims' resistance is to gain the power to define themselves.

THINGS BROUGHT TO THE TABLE BY THE PERSON BEING PRESSURED

Some personality traits predict resistance to the pressure to comply: not valuing being conscientious or agreeable; being low in neuroticism; scoring low on right-wing authoritarianism (any particular authority is more likely to be questioned if you already question the very concept of authority); social intelligence, which may be mediated by an enhanced ability to understand things like scapegoating or ulterior motives. And where these individual differences come from is, of course, the end product of most of the preceding chapters.[80]

What about gender? Milgram-like studies have shown that women average higher rates than men of voicing resistance to the demands to obey . . . but higher rates, nonetheless, of ultimately complying. Other studies show that women have higher rates than men of public conformity and lower rates of private conformity. Overall, though, gender is not much of a predictor. Interestingly, rates of conformity in Asch-like studies increase in mixed-sex groups. When in the presence of the opposite sex, perhaps there's less desire to seem like a rugged individualist than fear of seeming foolish.[81]

Finally, of course, we are the products of our culture. In broad cross-cultural surveys, Milgram and others showed more compliance in subjects from collectivist than from individualist cultures.[82]

STRESS

Exactly as with Us/Them-ing, people are more likely to conform and obey at times of stress, ranging from time pressure to a real or imagined outside threat to a novel context. In stressful settings rules gain power.

ALTERNATIVES

Finally, there is the key issue of whether you perceive alternatives to the actions demanded of you. It can be a solitary task to reframe and reappraise a situation, to make the implicit explicit, to engage in perspective taking, to question. To imagine that resistance is *not* futile.

A huge help in doing that is evidence that you are not alone. From Asch and Milgram on, it's clear that the presence of anyone else pushing back against the pressure can be galvanizing. Ten against two in a jury room is a world of difference from eleven against one. One lone voice crying out in the wilderness is a crank. Two voices joined together form a nidus of resistance, offer the start of an oppositional social identity.

It certainly helps to know that you are not alone, that there are others who are willing to resist, that there are those who have done so in the past. But often something still holds us back. Eichmann's seeming normalcy supplied us, thanks to Hannah Arendt, with the notion of the banality of evil. Zimbardo, in his recent writing, emphasizes the "banality of heroism." As discussed in various chapters, people who heroically refuse to look the other way, who do the right thing even when it carries the ultimate cost—tend to be surprisingly normal. The stars didn't align at their births; doves of peace did not envelop them where they strode. They put their pants on one leg at a time. This should be a huge source of strength for us.

SUMMARY AND CONCLUSIONS

- We're just like numerous other social species in terms of having marked status differences among individuals and hierarchies that emerge from those differences. Like many of these other species, we're fantastically attuned to status differences, are sufficiently fascinated by them that we monitor status relations in individuals who are irrelevant to us, and can perceive status differences in a blink of an eye. And we find it deeply unsettling, with the amygdala leading front and center, when status relations are ambiguous and shifting.
- As in so many other species, our brains, particularly the neocortex and most particularly the frontal cortex, have coevolved with the social complexity of status differences. It takes a lot of brainpower to make sense of the subtleties of dominance relations. This is no surprise, given that "knowing your place" can be so contextual. Navigating status differences is most challenging when it comes to attaining and maintaining high rank; this requires cognitive mastery of Theory of Mind and perspective taking; of manipulation, intimidation, and deceit; and of impulse control and emotion regulation. As with so many other primates, the biographies of our most

hierarchically successful members are built around what provocations are ignored during occasions where the frontal cortex kept a level head.

- Our bodies and brains, like those of other social species, bear the imprint of social status, and having the "wrong" rank can be corrosively pathogenic. Moreover, the physiology is not so much about rank per se as about its social meaning in your species and particular group, the behavioral advantages and disadvantages, and the psychological baggage of a particular rank.
- And then we're unlike any other species on earth in that we belong to multiple hierarchies, are psychologically adept at overvaluing those in which we excel, and maintain internal standards that can trump objective rank in their impact.
- Humans committed themselves to a unique trajectory when we invented socioeconomic status. In terms of its caustic, scarring impact on minds and bodies, nothing in the history of animals being crappy to one another about status differences comes within light-years of our invention of poverty.
- We're really out there as a species in that sometimes our high-status individuals don't merely plunder and instead actually lead, actually attempt to facilitate the common good. We've even developed bottom-up mechanisms for collectively choosing such leaders on occasion. A magnificent achievement. Which we then soil by having our choosing of leaders be shaped by implicit, automatic factors more suitable to five-year-olds deciding who should captain their boat on a voyage with the Teletubbies to Candyland.
- Stripped to their idealistic core, our political differences concern differing visions of how best to bring about the common good. We tend to come as internally consistent packages of political stances ranging from the small and local to the mammoth and global. And with remarkable regularity our stances reflect our implicit, affective makeup, with cognition playing post-hoc catch up. If you really want to understand someone's politics, understand their cognitive load, how prone they are to snap judgments, their approaches to

reappraisal and resolving cognitive dissonance. Even more important, understand how they *feel* about novelty, ambiguity, empathy, hygiene, disease and dis-ease, and whether things used to be better and the future is a scary place.

- Like so many other animals, we have an often-frantic need to conform, belong, and obey. Such conformity can be markedly maladaptive, as we forgo better solutions in the name of the foolishness of the crowd. When we discover we are out of step with everyone else, our amygdalae spasm with anxiety, our memories are revised, and our sensory-processing regions are even pressured to experience what is not true. All to fit in.
- Finally, the pull of conformity and obedience can lead us to some of our darkest, most appalling places, and far more of us can be led there than we'd like to think. But despite that, even the worst of barrels doesn't turn all apples bad, and "Resistance" and "Heroism" are often more accessible and less rarefied and capitalized than assumed. We're rarely alone in thinking this is wrong, wrong, wrong. And we are usually no less special or unique than those before us who have fought back.

Thirteen

Morality and Doing the Right Thing, Once You've Figured Out What That Is

The two previous chapters examined the thoroughly unique contexts for some human behaviors that are on a continuum with behaviors in other species. Like some other species, we make automatic Us/Them dichotomies and favor the former—though only humans rationalize that tendency with ideology. Like many other species, we are implicitly hierarchical—though only humans view the gap between haves and have-nots as a divine plan.

This chapter considers another domain rife with human uniqueness, namely morality. For us, morality is not only belief in norms of appropriate behavior but also the belief that they should be shared and transmitted culturally.

Work in the field is dominated by a familiar sort of question. When we make a decision regarding morality, is it mostly the outcome of moral reasoning or of moral intuition? Do we mostly think or feel our way to deciding what is right?

This raises a related question. Is human morality as new as the cultural institutions we've hatched in recent millennia, or are its rudiments a far older primate legacy?

This raises more questions. What's more impressive, consistencies and universalities of human moral behavior or variability and its correlation with cultural and ecological factors?

Finally, there will be unapologetically prescriptive questions. When it comes to moral decision making, when is it "better" to rely on intuition, when on reasoning? And when we resist temptation, is it mostly an act of will or of grace?

People have confronted these issues since students attended intro philosophy in togas. Naturally, these questions are informed by science.

THE PRIMACY OF REASONING IN MORAL DECISION MAKING

One single fact perfectly demonstrates moral decision making being based on cognition and reasoning. Have you ever picked up a law textbook? They're humongous.

Every society has rules about moral and ethical behavior that are reasoned and call upon logical operations. Applying the rules requires reconstructing scenarios, understanding proximal and distal causes of events, and assessing magnitudes and probabilities of consequences of actions. Assessing individual behavior requires perspective taking, Theory of Mind, and distinguishing between outcome and intent. Moreover, in many cultures rule implementation is typically entrusted to people (e.g., lawyers, clergy) who have undergone long training.

Harking back to chapter 7, the primacy of reasoning in moral decision making is anchored in child development. The Kohlbergian emergence of increasingly complex stages of moral development is built on the Piagetian emergence of increasingly complex logical operations. They are similar, neurobiologically. Logical and moral reasoning about the correctness of an economic or ethical decision, respectively, both activate the (cognitive) dlPFC. People with obsessive-compulsive disorder get mired in both everyday decision making and moral decision making, and their dlPFCs go wild with activity for both.[1]

Similarly, there's activation of the temporoparietal junction (TPJ) during Theory of Mind tasks, whether they are perceptual (e.g., visualizing a complex scene from another viewer's perspective), amoral (e.g., keeping straight who's in love with whom in *A Midsummer Night's Dream*), or moral/social (e.g., inferring the ethical motivation behind a person's act). Moreover, the more the TPJ activation, the more people take intent into account when making moral judgments, particularly when there was intent to harm but no actual harm done. Most important, inhibit the TPJ with transcranial magnetic stimulation, and subjects become less concerned about intent.[2]

The cognitive processes we bring to moral reasoning aren't perfect, in that there are fault lines of vulnerability, imbalances, and asymmetries.[3] For example, doing harm is worse than allowing it—for equivalent outcomes we typically judge commission more harshly than omission and must activate the dlPFC more to judge them as equal. This makes sense—when we do one thing, there are innumerable other things we didn't do; no wonder the former is psychologically weightier. As another cognitive skew, as discussed in chapter 10, we're better at detecting violations of social contracts that have malevolent rather than benevolent consequences (e.g., giving less versus more than promised). We also search harder for causality (and come up with more false attributions) for malevolent than for benevolent events.

This was shown in one study. First scenario: A worker proposes a plan to the boss, saying, "If we do this, there'll be big profits, and we'll harm the environment in the process." The boss answers: "I don't care about the environment. Just do it." Second scenario: Same setup, but this time there'll be big profits and *benefits* to the environment. Boss: "I don't care about the environment. Just do it." In the first scenario 85 percent of subjects stated that the boss harmed the environment *in order to* increase profits; however, in the second scenario only 23 percent said that the boss helped the environment *in order to* increase profits.[4]

Okay, we're not perfect reasoning machines. But that's our goal, and numerous moral philosophers emphasize the preeminence of reasoning, where emotion and intuition, if they happen to show up, just soil the carpet. Such philosophers range from Kant, with his search for a mathematics of morality, to Princeton philosopher Peter Singer, who kvetches that if things like sex and bodily functions are pertinent to philosophizing, time to hang up his spurs: "It would be best to forget all about our particular moral judgments." Morality is anchored in reason.[5]

YEAH, SURE IT IS: SOCIAL INTUITIONISM

Except there's a problem with this conclusion—people often haven't a clue *why* they've made some judgment, yet they fervently believe it's correct.

This is straight out of chapter 11's rapid implicit assessments of Us versus Them and our post-hoc rational justifications for visceral prejudice. Scientists studying moral philosophy increasingly emphasize moral decision making as implicit, intuitive, and anchored in emotion.

The king of this "social intuitionist" school is Jonathan Haidt, whom we've encountered previously.[6] Haidt views moral decisions as primarily based on intuition and believes reasoning is what we then use to convince everyone, including ourselves, that we're making sense. In an apt phrase of Haidt's, "moral thinking is for social doing," and sociality always has an emotional component.

The evidence for the social intuitionist school is plentiful:

When contemplating moral decisions, we don't just activate the egg-heady dlPFC.[7] There's also activation of the usual emotional cast—amygdala, vmPFC and the related orbitofrontal cortex, insular cortex, anterior cingulate. Different types of moral transgressions preferentially activate different subsets of these regions. For example, moral quandaries eliciting pity preferentially activate the insula; those eliciting indignation activate the orbitofrontal cortex. Quandaries generating intense conflict preferentially activate the anterior cingulate. Finally, for acts assessed as equally morally wrong, those involving nonsexual transgression (e.g., stealing from a sibling) activate the amygdala, whereas those involving sexual transgressions (e.g., sex with a sibling) also activate the insula.*

Moreover, when such activation is strong enough, we also activate the sympathetic nervous system and feel arousal—and we know how those peripheral effects feedback and influence behavior. When we confront a moral choice, the dlPFC doesn't adjudicate in contemplative silence. The waters roil below.

The pattern of activation in these regions predicts moral decisions better than does the dlPFC's profile. And this matches behavior—people punish to the extent that they *feel* angered by someone acting unethically.[8]

People tend toward instantaneous moral reactions; moreover, when subjects shift from judging nonmoral elements of acts to moral ones, they make assessments faster, the antithesis of moral decision making being about grinding cognition. Most strikingly, when facing a moral quandary, activation in the amygdala, vmPFC, and insula typically *precedes* dlPFC activation.[9]

Damage to these intuitionist brain regions makes moral judgments more pragmatic, even coldhearted. Recall from chapter 10 how people with damage to the (emotional) vmPFC readily advocate sacrificing one relative to save five strangers, something control subjects never do.

* The authors of the study also included a category of repellent acts that were nevertheless not moral transgressions, once again matched for a sibling involvement—drinking your sibling's urine, eating your sibling's scab.

Most telling is when we have strong moral opinions but can't tell why, something Haidt calls "moral dumbfounding"—followed by clunky post-hoc rationalizing.[10] Moreover, such moral decisions can differ markedly in different affective or visceral circumstances, generating very different rationalizations. Recall from the last chapter how people become more conservative in their social judgments when they're smelling a foul odor or sitting at a dirty desk. And then there's that doozy of a finding—knowing a judge's opinions about Plato, Nietzsche, Rawls, and any other philosopher whose name I just looked up gives you less predictive power about her judicial decisions than knowing if she's hungry.

The social intuitionist roots of morality are bolstered further by evidence of moral judgment in two classes of individuals with limited capacities for moral reasoning.

AGAIN WITH BABIES AND ANIMALS

Much as infants demonstrate the rudiments of hierarchical and Us/Them thinking, they possess building blocks of moral reasoning as well. For starters, infants have the bias concerning commission versus omission. In one clever study, six-month-olds watched a scene containing two of the same objects, one blue and one red; repeatedly, the scene would show a person picking the blue object. Then, one time, the red one is picked. The kid becomes interested, looks more, breathes faster, showing that this seems discrepant. Now, the scene shows two of the same objects, one blue, one a different color. In each repetition of the scene, a person picks the one that is not blue (its color changes with each repetition). Suddenly, the blue one is picked. The kid isn't particularly interested. "He always picks the blue one" is easier to comprehend than "He never picks the blue one." Commission is weightier.[11]

Infants and toddlers also have hints of a sense of justice, as shown by

Kiley Hamlin of the University of British Columbia, and Paul Bloom and Karen Wynn of Yale. Six- to twelve-month-olds watch a circle moving up a hill. A nice triangle helps to push it. A mean square blocks it. Afterward the infants can reach for a triangle or a square. They choose the triangle.* Do infants prefer nice beings, or shun mean ones? Both. Nice triangles were preferred over neutral shapes, which were preferred over mean squares.

Such infants advocate punishing bad acts. A kid watches puppets, one good, one bad (sharing versus not). The child is then presented with the puppets, each sitting on a pile of sweets. Who should lose a sweet? The bad puppet. Who should gain one? The good puppet.

Remarkably, toddlers even assess secondary punishment. The good and bad puppets then interact with two additional puppets, who can be nice or bad. And whom did kids prefer of those second-layer puppets? Those who were nice to nice puppets and those who punished mean ones.

Other primates also show the beginnings of moral judgments. Things started with a superb 2003 paper by Frans de Waal and Sarah Brosnan.[12] Capuchin monkeys were trained in a task: A human gives them a mildly interesting small object—a pebble. The human then extends her hand palm up, a capuchin begging gesture. If the monkey puts the pebble in her hand, there's a food reward. In other words, the animals learned how to buy food.

Now there are two capuchins, side by side. Each gets a pebble. Each gives it to the human. Each gets a grape, very rewarding.

Now change things. Both monkeys pay their pebble. Monkey 1 gets a grape. But monkey 2 gets some cucumber, which blows compared with grapes—capuchins prefer grapes to cucumber 90 percent of the time. Monkey 2 was shortchanged.

And monkey 2 would then typically fling the cucumber at the human or bash around in frustration. Most consistently, they wouldn't give the

* And to demonstrate how much this is tapping into the social brains of these kids, this works only if the shapes are personified with eyes.

pebble the next time. As the *Nature* paper was entitled, "Monkeys reject unequal pay."

This response has since been demonstrated in various macaque monkey species, crows, ravens, and dogs (where the dog's "work" would be shaking her paw).*[13]

Subsequent work by Brosnan, de Waal, and others fleshed out this phenomenon further:[14]

- One criticism of the original study was that maybe capuchins refused to work for cucumbers because grapes were visible, regardless of whether the other guy was getting paid in grapes. But no—the phenomenon required unfair payment.
- Both animals are getting grapes, then one gets switched to cucumber. What's key—that the other guy is still getting grapes, or that I no longer am? The former—if doing the study with a single monkey, switching from grapes to cucumbers would not evoke refusal. Nor would it if both monkeys got cucumbers.
- Across the various species, males were more likely than females to reject "lower pay"; dominant animals were more likely than subordinates to reject.
- It's about the work—give one monkey a free grape, the other free cucumber, and the latter doesn't get pissed.
- The closer in proximity the two animals are, the more likely the one getting cucumber is to go on strike.
- Finally, rejection of unfair pay isn't seen in species that are solitary (e.g., orangutans) or have minimal social cooperation (e.g., owl monkeys).

Okay, very impressive—other social species show hints of a sense of justice, reacting negatively to unequal reward. But this is worlds away

* Dogs differed from primates in two interesting ways that make sense, both dogwise and primatewise. Primates would get pissed and stop working if there was a difference in the *quality* of the reward (grape versus cucumber); in contrast, dogs didn't distinguish quality (bread versus sausage), only whether one was rewarded and the other not. Second, while many monkeys refused to accept an eventual reward and would never cooperate again, dogs all eventually came around, after enough entreaties to "shake" by the human.

from juries awarding money to plaintiffs harmed by employers. Instead it's self-interest—"This isn't fair; I'm getting screwed."

How about evidence of a sense of fairness in the treatment of another individual? Two studies have examined this in a chimp version of the Ultimatum Game. Recall the human version—in repeated rounds, player 1 in a pair decides how money is divided between the two of them. Player 2 is powerless in the decision making but, if unhappy with the split, can refuse, and no one gets any money. In other words, player 2 can forgo immediate reward to punish selfish player 1. As we saw in chapter 10, Player 2s tend to accept 60:40 splits.

In the chimp version, chimp 1, the proposer, has two tokens. One indicates that each chimp gets two grapes. The other indicates that the proposer gets three grapes, the partner only one. The proposer chooses a token and passes it to chimp 2, who then decides whether to pass the token to the human grape dispenser. In other words, if chimp 2 thinks chimp 1 is being unfair, no one gets grapes.

In one such study, Michael Tomasello (a frequent critic of de Waal—stay tuned) at the Max Planck Institutes in Germany, found no evidence of chimp fairness—the proposer always chose, and the partner always accepted unfair splits.[15] De Waal and Brosnan did the study in more ethologically valid conditions and reported something different: proposer chimps tended toward equitable splits, but if they could give the token directly to the human (robbing chimp 2 of veto power), they'd favor unfair splits. So chimps will opt for fairer splits—but only when there is a downside to being unfair.

Sometimes other primates are fair when it's at no cost to themselves. Back to capuchin monkeys. Monkey 1 chooses whether both he and the other guy get marshmallows or it's a marshmallow for him and yucky celery for the other guy. Monkeys tended to choose marshmallows for the other guy.* Similar "other-regarding preference" was shown with marmoset monkeys, where the first individual got nothing and merely chose

* But what if the monkey chooses the two-marshmallow option over the marshmallow/celery one because, well, having any sort of situation with two marshmallows on the scene was just so much more exciting? The authors did a nice control—when there was no monkey in the adjacent space, the choice would be random as to what food was deposited in the second space.

whether the other guy got a cricket to eat (of note, a number of studies have failed to find other-regarding preference in chimps).[16]

Really interesting evidence for a nonhuman sense of justice comes in a small side study in a Brosnan/de Waal paper. Back to the two monkeys getting cucumbers for work. Suddenly one guy gets shifted to grapes. As we saw, the one still getting the cucumber refuses to work. Fascinatingly, the grape mogul often refuses as well.

What is this? Solidarity? "I'm no strike-breaking scab"? Self-interest, but with an atypically long view about the possible consequences of the cucumber victim's resentment? Scratch an altruistic capuchin and a hypocritical one bleeds? In other words, all the questions raised by human altruism.

Given the relatively limited reasoning capacities of monkeys, these findings support the importance of social intuitionism. De Waal perceives even deeper implications—the roots of human morality are older than our cultural institutions, than our laws and sermons. Rather than human morality being spiritually transcendent (enter deities, stage right), it transcends our species boundaries.[17]

MR. SPOCK AND JOSEPH STALIN

Many moral philosophers believe not only that moral judgment is built on reasoning but also that it *should* be. This is obvious to fans of Mr. Spock, since the emotional component of moral intuitionism just introduces sentimentality, self-interest, and parochial biases. But one remarkable finding counters this.

Relatives are special. Chapter 10 attests to that. Any social organism would tell you so. Joseph Stalin thought so concerning Pavlik Morozov ratting out his father. As do most American courts, where there is either de facto or de jure resistance to making someone testify against their own parent or child. Relatives are special. But not to people lacking social intuitionism. As noted, people with vmPFC damage make extraordinarily practical, unemotional moral decisions. And in the process they do something that everyone, from clonal yeast to Uncle Joe to the Texas Rules of Criminal Evidence considers morally suspect: they advocate harming kin as readily as strangers in an "Is it okay to sacrifice one person to save five?" scenario.[18]

Emotion and social intuition are not some primordial ooze that gums up that human specialty of moral reasoning. Instead, they anchor some of the few moral judgments that most humans agree upon.

CONTEXT

So social intuitions can have large, useful roles in moral decision making. Should we now debate whether reasoning or intuition is more important? This is silly, not least of all because there is considerable overlap between the two. Consider, for example, protesters shutting down a capital to highlight income inequity. This could be framed as the Kohlbergian reasoning of people in a postconventional stage. But it could also be framed à la Haidt in a social intuitionist way—these are people who resonate more with moral intuitions about fairness than with respect for authority.

More interesting than squabbling about the relative importance of reasoning and intuition are two related questions: What circumstances bias toward emphasizing one over the other? Can the differing emphases produce different decisions?

As we've seen, then–graduate student Josh Greene and colleagues helped jump-start "neuroethics" by exploring these questions using the poster child of "Do the ends justify the means?" philosophizing, namely the runaway trolley problem. A trolley's brake has failed, and it is hurtling down the tracks and will hit and kill five people. Is it okay to do something that saves the five but kills someone else in the process?

People have pondered this since Aristotle took his first trolley ride;* Greene et al. added neuroscience. Subjects were neuroimaged while pondering trolley ethics. Crucially, they considered two scenarios. Scenario 1:

* Actually, the trolley problem was invented by the British philosopher Philippa Foot in 1967.

Here comes the trolley; five people are goners. Would you pull a lever that diverts the trolley onto a different track, where it will hit and kill someone (the original scenario)? Scenario 2: Same circumstance. Would you *push* the person onto the tracks to stop the trolley?[19]

By now I bet readers can predict which brain region(s) activates in each circumstance. Contemplate pulling the lever, and dlPFC activity predominates, the detached, cerebral profile of moral reasoning. Contemplate consigning the person to death by pushing them, and it's vmPFC (and amygdala), the visceral profile of moral intuition.

Would you pull the lever? Consistently, 60 to 70 percent of people, with their dlPFCs churning away, say yes to this utilitarian solution—kill one to save five. Would you push the person with your own hands? Only 30 percent are willing; the more the vmPFC and/or amygdaloid activation, the more likely they are to refuse.* This is hugely important—a relatively minor variable determines whether people emphasize moral reasoning or intuition, and they engage different brain circuits in the process, producing radically different decisions. Greene has explored this further.

Are people resistant to the utilitarian trade-off of killing one to save five in the pushing scenario because of the visceral reality of actually touching the person whom they have consigned to death? Greene's work suggests not—if instead of pushing with your hands, you push with a pole, people are still resistant. There's something about the personal force involved that fuels the resistance.

Are people willing in the lever scenario because the victim is at a distance, rather than right in front of them? Probably not—people are just as willing if the lever is right next to the person who will die.

Greene suggests that intuitions about intentionality are key. In the lever scenario, the five people are saved because the trolley has been diverted to another track; the killing of the individual is a side effect and the five would still have been saved if that person hadn't been standing on the tracks. In contrast, in the pushing scenario the five are saved *because*

* And as alluded to earlier, people with vmPFC damage are strongly and equally willing to pull the lever or push the person. You see the same if you give people a benzodiazepine (a tranquilizer like Valium). The vmPFC and amygdala are calmed down (both by direct actions of the drug and secondarily via damping of the sympathetic nervous system), and people are more willing to push.

the person is killed, and the intentionality feels intuitively wrong. As evidence, Greene would give subjects another scenario: Here comes the trolley, and you are rushing to throw a switch that will halt it. Is it okay to do this if you know that in the process of lunging for the switch, you must push a person out of the way, who falls to the ground and dies? About 80 percent of people say yes. Same pushing the person, same proximity, but done unintentionally, as a side effect. The person wasn't killed as a *means* to save the five. Which seems much more okay.

Now a complication. In the "loop" scenario, you pull a lever that diverts the trolley to another track. But—oh no!—it's just a loop; it merges back on to the original track. The trolley will still kill the five people—except that there's a person on the side loop who will be killed, stopping the trolley. This is as intentional a scenario as is pushing with your hands—diverting to another track isn't enough; the person has to be killed. By all logic only about 30 percent of people should sign on, but instead it's in the 60 to 70 percent range.

Greene concludes (from this and additional scenarios resembling the loop) that the intuitionist universe is very local. Killing someone intentionally as a means to save five feels intuitively wrong, but the intuition is strongest when the killing would occur right here, right now; doing it in more complicated sequences of intentionality doesn't feel as bad. This is not because of a cognitive limit—it's not that subjects don't realize the necessity of killing the person in the loop scenario. It just doesn't *feel* the same. In other words, intuitions discount heavily over space and time. Exactly the myopia about cause and effect you'd expect from a brain system that operates rapidly and automatically. This is the same sort of myopia that makes sins of commission feel worse than those of omission.

Thus these studies suggest that when a sacrifice of one requires active, intentional, and local actions, more intuitive brain circuitry is engaged, and ends don't justify means. And in circumstances where either the harm is unintentional or the intentionality plays out at a psychological distance, different neural circuitry predominates, producing an opposite conclusion about the morality of ends and means.

These trolleyology studies raise a larger point, which is that moral decision making can be wildly context dependent.[20] Often the key thing that a change in context does is alter the locality of one's intuitionist morals, as summarized by Dan Ariely of Duke University in his wonderful book *Predictably Irrational.* Leave money around a common work area and no one takes it; it's not okay to steal money. Leave some cans of Coke and they're all taken; the one-step distance from the money involved blunts the intuitions about the wrongness of stealing, making it easier to start rationalizing (e.g., someone must have left them out for the taking).

The effects of proximity on moral intuitionism are shown in a thought experiment by Peter Singer.[21] You're walking by a river in your hometown. You see that a child has fallen in. Most people feel morally obliged to jump in and save the child, even if the water destroys their $500 suit. Alternatively, a friend in Somalia calls and tells you about a poor child there who will die without $500 worth of medical care. Can you send money? Typically not. The locality and moral discounting over distance is obvious—the child in danger in your hometown is far more of an Us than is this dying child far away. And this is an intuitive rather than cognitive core—if you were walking along in Somalia and saw a child fall into a river, you'd be more likely to jump in and sacrifice the suit than to send $500 to that friend making the phone call. Someone being right there, in the flesh, in front of our eyes is a strong implicit prime that they are an Us.

Moral context dependency can also revolve around language, as noted in chapter 3.[22] Recall, for example, people using different rules about the morality of cooperation if you call the same economic game the "Wall Street game" or the "community game." Framing an experimental drug as having a "5 percent mortality rate" versus a "95 percent survival rate" produces different decisions about the ethics of using it.

Framing also taps into the themes of people having multiple identities, belonging to multiple Us groups and hierarchies. This was shown in a hugely interesting 2014 *Nature* paper by Alain Cohn and colleagues at the University of Zurich.[23] Subjects, who worked for an (unnamed) international bank, played a coin-toss game with financial rewards for guessing outcomes correctly. Crucially, the game's design made it possible for subjects to cheat at various points (and for the investigators to detect the cheating).

In one version subjects first completed a questionnaire filled with mundane questions about their everyday lives (e.g., "How many hours of television do you watch each week?"). This produced a low, baseline level of cheating.

Then, in the experimental version, the questionnaire was about their bank job. Questions like these primed the subjects to implicitly think more about banking (e.g., they became more likely in a word task to complete "__oker" with "broker" than with "smoker").

So subjects were thinking about their banking identity. And when they did, rates of cheating rose 20 percent. Priming people in other professions (e.g., manufacturing) to think about their jobs, or about the banking world, didn't increase cheating. These bankers carried in their heads two different sets of ethical rules concerning cheating (banking and nonbanking), and unconscious cuing brought one or the other to the forefront.* Know thyself. Especially in differing contexts.

"But This Circumstance Is Different"

The context dependency of morality is crucial in an additional realm.

It is a nightmare of a person who, with remorseless sociopathy, believes it is okay to steal, kill, rape, and plunder. But far more of humanity's worst behaviors are due to a different kind of person, namely most of the rest of us, who will say that of course it is wrong to do X . . . but here is why these special circumstances make me an exception right now.

We use different brain circuits when contemplating our own moral failings (heavy activation of the vmPFC) versus those of others (more of the insula and dlPFC).[24] And we consistently make different judgments, being more likely to exempt ourselves than others from moral condemnation. Why? Part of it is simply self-serving; sometimes a hypocrite bleeds

* I kind of wish the authors had indicated the name of the bank, just in case I'm ever considering depositing money in a Swiss bank and want to be able to immediately check one candidate off my list of possible banks.

because you've scratched a hypocrite. The difference may also reflect different emotions being involved when we analyze our own actions versus those of others. Considering the moral failings of the latter may evoke anger and indignation, while their moral triumphs prompt emulation and inspiration. In contrast, considering our own moral failings calls forth shame and guilt, while our triumphs elicit pride.

The affective aspects of going easy on ourselves are shown when stress makes us more this way.[25] When experimentally stressed, subjects make more egoistic, rationalizing judgments regarding emotional moral dilemmas and are less likely to make utilitarian judgments—but only when the latter involve a personal moral issue. Moreover, the bigger the glucocorticoid response to the stressor, the more this is the case.

Going easy on ourselves also reflects a key cognitive fact: we judge ourselves by our internal motives and everyone else by their external actions.[26] And thus, in considering our own misdeeds, we have more access to mitigating, situational information. This is straight out of Us/Them—when Thems do something wrong, it's because they're simply rotten; when Us-es do it, it's because of an extenuating circumstance, and "Me" is the most focal Us there is, coming with the most insight into internal state. Thus, on this cognitive level there is no inconsistency or hypocrisy, and we might readily perceive a wrong to be mitigated by internal motives in the case of anyone's misdeeds. It's just easier to know those motives when we are the perpetrator.

The adverse consequences of this are wide and deep. Moreover, the pull toward judging yourself less harshly than others easily resists the rationality of deterrence. As Ariely writes in his book, "Overall cheating is not limited by risk; it is limited by our ability to rationalize the cheating to ourselves."

Cultural Context

So people make different moral judgments about the same circumstance depending on whether it's about them or someone else, which of their identities has been primed, the language used, how many steps the intentionality is removed, and even the levels of their stress hormones, the fullness of their stomach, or the smelliness of their environment. After chapter 9 it is no surprise that moral decision making can also vary dramatically by culture. One culture's sacred cow is another's meal, and the discrepancy can be agonizing.

When thinking about cross-cultural differences in morality, key issues are what universals of moral judgment exist and whether the universals or the differences are more interesting and important.

Chapter 9 noted some moral stances that are virtually universal, whether de facto or de jure. These include condemnation of at least some forms of murder and of theft. Oh, and of some form of sexual practice.

More broadly, there is the near universal of the Golden Rule (with cultures differing as to whether it is framed as "Do only things you'd want done to you" or "Don't do things you wouldn't want done to you"). Amid the power of its simplicity, the Golden Rule does not incorporate people differing as to what they would/wouldn't want done to them; we have entered complicated terrain when we can make sense of an interchange where a masochist says, "Beat me," and the sadist sadistically answers, "No."

This criticism is overcome with the use of a more generalized, common currency of reciprocity, where we are enjoined to give concern and legitimacy to the needs and desires of people in circumstances where we would want the same done for us.

Cross-cultural universals of morality arise from shared categories of rules of moral behavior. The anthropologist Richard Shweder has proposed that all cultures recognize rules of morality pertinent to autonomy, community, and divinity. As we saw in the last chapter, Jonathan Haidt breaks this continuum into his foundations of morality that humans have strong intuitions about. These are issues related to harm, fairness and reciprocity (both of which Shweder would call autonomy), in-group loyalty and respect for authority (both of which Shweder would call community), and issues of purity and sanctity (i.e., Shweder's realm of divinity).*[27]

* And recall from the last chapter how Haidt has shown that liberals place more of an emphasis than conservatives do on harm and fairness issues, while conservatives disproportionately value loyalty, respect, and purity. Haidt drolly refers to these studies as his "cross-cultural" research, conjuring images of him with pith helmet and mosquito net, trekking through the likes of Berkeley and Provo.

The existence of universals of morality raises the issue of whether that means that they should trump more local, provincial moral rules. Between the moral absolutists on one side and the relativists on the other, people like the historian of science Michael Shermer argue reasonably for provisional morality—if a moral stance is widespread across cultures, start off by giving the benefit of the doubt to its importance, but watch your wallet.[28]

It's certainly interesting that, for example, all cultures designate certain things as sacred; but it is far more so to look at the variability in what is considered sacred, how worked up people get when such sanctity is violated,* and what is done to keep such violations from reoccurring. I'll touch on this huge topic with three subjects—cross-cultural differences concerning the morality of cooperation and competition, affronts to honor, and the reliance on shame versus guilt.

COOPERATION AND COMPETITION

Some of the most dramatic cross-cultural variability in moral judgment concerns cooperation and competition. This was shown to an extraordinary extent in a 2008 *Science* paper by a team of British and Swiss economists.

Subjects played a "public good" economic game where players begin with a certain number of tokens and then decide, in each of a series of rounds, how many to contribute to a shared pool; the pool is then multiplied and shared evenly among all the players. The alternative to contributing is for subjects to keep the tokens for themselves. Thus, the worst payoff for an individual player would be if they contributed all their tokens to the pool, while no other player contributed any; the best would be if the individual contributed no tokens and everyone else contributed everything. As a feature of the design, subjects could "pay" to punish other players for the size of their contribution. Subjects were from around the world.

* Just as an off-the-cuff example, if I were to find myself in the middle of a religious service and suddenly suffer from hideously loud, malodorous flatulence, I'd sure hope I was hanging out with Quakers rather than, say, a bunch of the boys from the Taliban at Friday prayer.

First finding: Across all cultures, people were more prosocial than sheer economic rationality would predict. If everyone played in the most brutally asocial, realpolitik manner, no one would contribute to the pool. Instead subjects from all cultures consistently contributed. Perhaps as an explanation, subject from all cultures punished people who made lowball contributions, and to roughly equal extents.

Where the startling difference came was with a behavior that I'd never even seen before in the behavioral economics literature, something called "antisocial punishment." Free-riding punishment is when you punish another player for contributing less than you (i.e., being selfish). Antisocial punishment is when you punish another player for contributing *more* than you (i.e., being generous).

What is that about? Interpretation: This hostility toward someone being overly generous is because they're going to up the ante, and soon everyone (i.e., me) will be expected to be generous. Kill 'em, spoiling things for everyone. It's a phenomenon where you punish someone for being nice, because what if that sort of crazy deviance becomes the norm and you feel pressure to be nice back?

At one extreme were subjects from countries (the United States and Australia) where this weird antisocial punishment was nearly nonexistent. And at the mind-boggling other extreme were subjects from Oman and Greece, who were willing to spend *more* to punish generosity than to punish selfishness. And this was not a comparison of, say, theologians in Boston with Omani pirates. Subjects were all urban university students.

So what's different among these cities? The authors found a key correlation—the lower the social capital in a country, the higher the rates of antisocial punishment. In other words, when do people's moral systems include the idea that being generous deserves punishment? When they live in a society where people don't trust one another and feel as if they have no efficacy.

Fascinating work has also been done specifically on people in non-Western cultures, as reported in a pair of studies by Joseph Henrich, of the University of British Columbia, and colleagues.[29] Subjects were in

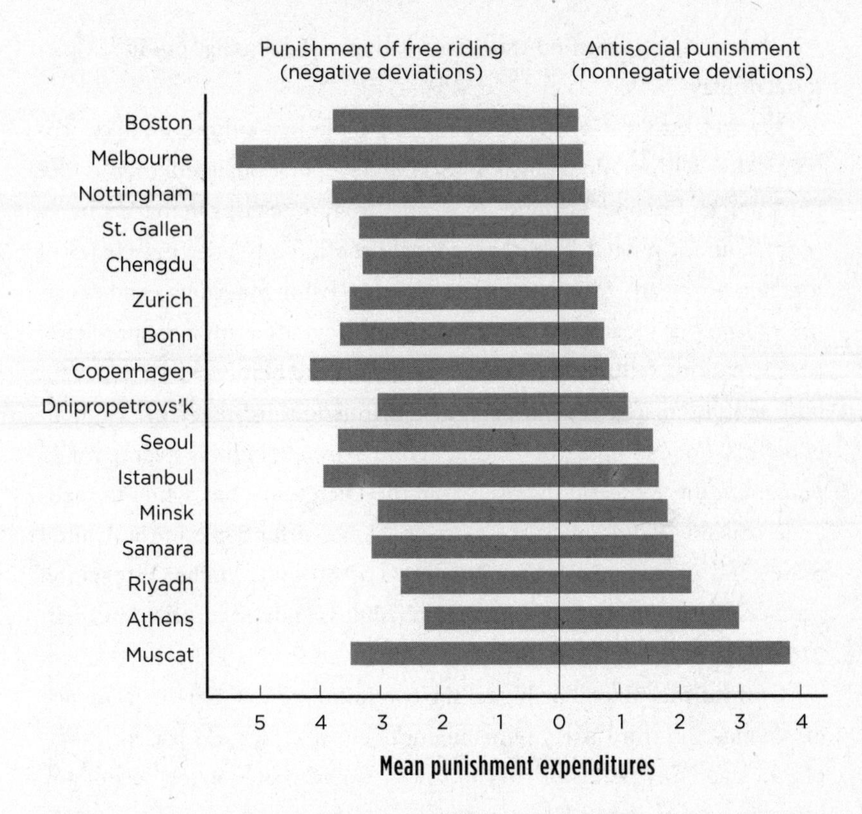

B. Herrmann et al., "Antisocial Punishment Across Societies,"
Sci *319 (2008): 1362.*

the thousands and came from twenty-five different "small-scale" cultures from around the world—they were nomadic pastoralists, hunter-gatherers, sedentary forager/horticulturalists, and subsistence farmers/wage earners. There were two control groups, namely urbanites from Missouri and Accra, Ghana. As a particularly thorough feature of the study, subjects played three economic games: (a) The Dictator Game, where the subject simply decides how money is split between them and another player. This measures a pure sense of fairness, independent of consequence. (b) The Ultimatum Game, where you can pay to punish someone treating you unfairly (i.e., self-interested second-party punishment). (c) A third-party punishment scenario, where you can pay to punish someone treating a third party unfairly (i.e., altruistic punishment).

The authors identified three fascinating variables that predicted patterns of play:

Market integration: How much do people in a culture interact economically, with trade items? The authors operationalized this as the percentage of people's calories that came from purchases in market interactions, and it ranged from 0 percent for the hunter-gathering Hadza of Tanzania to nearly 90 percent for sedentary fishing cultures. And across the cultures a greater degree of market integration strongly predicted people making fairer offers in all three games and being willing to pay for both self-interested second-party and altruistic third-party punishment of creeps. For example, the Hadza, at one extreme, kept an average of 73 percent of the spoils for themselves in the Dictator Game, while the sedentary fishing Sanquianga of Colombia, along with people in the United States and Accra, approached dictating a 50:50 split. Market integration predicts more willingness to punish selfishness and, no surprise, less selfishness.

Community size: The bigger the community, the more the incidence of second- and third-party punishment of cheapskates. Hadza, for example, in their tiny bands of fifty or fewer, would pretty much accept any offer above zero in the Ultimatum Game—there was no punishment. At the other extreme, in communities of five thousand or more (sedentary agriculturalists and aquaculturalists, plus the Ghanaian and American urbanites), offers that weren't in the ballpark of 50:50 were typically rejected and/or punished.

Religion: What percentage of the population belonged to a worldwide religion (i.e., Christianity or Islam)? This ranged from none of the Hadza to 60 to 100 percent for all the other groups. The greater the incidence of belonging to a Western religion, the more third-party punishment (i.e., willingness to pay to punish person A for being unfair to person B).

What to make of these findings?

First the religion angle. This was a finding not about religiosity generally but about religiosity within a worldwide religion, and not about generosity or fairness but about altruistic third-party punishment. What is it about worldwide religions? As we saw in chapter 9, it is only when groups get large enough that people regularly interact with strangers that cultures invent moralizing gods. These are not gods who sit around the banquet table laughing with detachment at the foibles of humans down below, or gods who punish humans for lousy sacrificial offerings. These are gods who punish humans for being rotten to other humans—in other words, the large religions invent gods who do third-party punishment. No wonder this predicts these religions' adherents being third-party punishers themselves.

Next the twin findings that more market integration and bigger community size were associated with fairer offers (for the former) and more willingness to punish unfair players (for both). I find this to be a particularly challenging pair of findings, especially when framed as the authors thoughtfully did.

The authors ask where the uniquely extreme sense of fairness comes from in humans, particularly in the context of large-scale societies with strangers frequently interacting. And they offer two traditional types of explanations that are closely related to our dichotomies of intuition versus reasoning and animal roots versus cultural inventions:

- Our moral anchoring in fairness in large-scale societies is a residue and extension of our hunter-gatherer and nonhuman primate past. This was life in small bands, where fairness was mostly driven by kin selection and easy scenarios of reciprocal altruism. As our community size has expanded and we now mostly have one-shot interactions with unrelated strangers, our prosociality just represents an expansion of our small-band mind-set, as we use various green-beard marker shibboleths as proxies for relatedness. I'd gladly lay down my life for two brothers, eight cousins, or a guy who is a fellow Packers fan.
- The moral underpinnings of a sense of fairness lie in cultural institutions and mind-sets that we invented as our groups became larger and more sophisticated (as reflected in the emergence of markets, cash economies, and the like).

This many pages in, it's obvious that I think the former scenario is pretty powerful—look, we see the roots of a sense of fairness and justice in the egalitarian nature of nomadic hunter-gatherers, in other primates, in infants, in the preeminent limbic rather than cortical involvement. But, inconveniently for that viewpoint, that's totally counter to what emerges from these studies—across the twenty-five cultures it's the hunter-gatherers, the ones most like our ancestors, living in the smallest groups, with the highest degrees of relatedness and with the least reliance on market interactions, who show the least tendency toward making fair offers and are least likely to punish unfairness, whether to themselves or to the other guy. None of that prosociality is there, a picture counter to what we saw in chapter 9.

I think an explanation is that these economic games tap into a very specific and artificial type of prosociality. We tend to think of market interactions as being the epitome of complexity—finding a literal common currency for the array of human needs and desires in the form of this abstraction called money. But at their core, market interactions represent an impoverishment of human reciprocity. In its natural form, human reciprocity is a triumph of comfortably and intuitively doing long-term math with apples and oranges—this guy over here is a superstar hunter; that other guy isn't in his league but has your back if there's a lion around; meanwhile, she's amazing at finding the best mongongo nuts, that older woman knows all about medicinal herbs, and that geeky guy remembers the best stories. We know where one another live, the debit columns even out over time, and if someone is really abusing the system, we'll get around to collectively dealing with them.

In contrast, at its core, a cash-economy market interaction strips it all down to "I give you this now, so you give me that now"—myopic present-tense interactions whose obligations of reciprocity must be balanced immediately. People in small-scale societies are relatively new to functioning this way. It's not the case that small-scale cultures that are growing big and market reliant are newly schooled in how to be fair. Instead they're newly schooled in how to be fair in the artificial circumstances modeled by something like the Ultimatum Game.

HONOR AND REVENGE

Another realm of cross-cultural differences in moral systems concerns what constitutes appropriate response to personal affronts. This harks back to chapter 9's cultures of honor, from Maasai tribesmen to traditional American Southerners. As we saw, such cultures have historical links to monotheism, warrior age groups, and pastoralism.

To recap, such cultures typically see an unanswered challenge to honor as the start of a disastrous slippery slope, rooted in the intrinsic vulnerability of pastoralism—while no one can raid farmers and steal all their crops, someone can rustle a herd overnight—and if this sum'a bitch gets away with insulting my family, he'll be coming for my cattle next. These are cultures that place a high moral emphasis on revenge, and revenge at least in kind—after all, an eye for an eye was probably the invention of Judaic pastoralists. The result is a world of Hatfields and McCoys, with their escalating vendettas. This helps explain why the elevated murder rates in the American South are not due to urban violence or things like robberies but are instead about affronts to honor between people who know each other. And it helps explain why Southern prosecutors and juries are typically more forgiving of such crimes of affronted honor. And it also helps explain the command apparently given by many Southern matriarchs to their sons marching off to join the Confederate fight: come back a winner or come back in a coffin. The shame of surrender is not an option.

SHAMED COLLECTIVISTS AND GUILTY INDIVIDUALISTS

We return to our contrast between collectivist and individualistic cultures (in the studies, as a reminder, "collectivist" has mostly meant East Asian societies, while "individualistic" equals Western Europeans and North Americans). Implicit in the very nature of the contrast are markedly different approaches to the morality of ends and means. By definition, collectivist cultures are more comfortable than individualistic ones with people being used as a means to a utilitarian end. Moreover, moral imperatives in collectivist cultures tend to be about social roles and duties to the group, whereas those in individualistic cultures are typically about individual rights.

Collectivist and individualistic cultures also differ in how moral behavior is enforced. As first emphasized by the anthropologist Ruth Benedict in 1946, collectivist cultures enforce with shame, while individualistic cultures use guilt. This is a doozy of a contrast, as explored in two excellent books, Stanford psychiatrist Herant Katchadourian's *Guilt: The Bite of Conscience* and NYU environmental scientist Jennifer Jacquet's *Is Shame Necessary?*[30]

In the sense used by most in the field, including these authors, shame is external judgment by the group, while guilt is internal judgment of yourself. Shame requires an audience, is about honor. Guilt is for cultures that treasure privacy and is about conscience. Shame is a negative assessment of the entire individual, guilt that of an act, making it possible to hate the sin but love the sinner. Effective shaming requires a conformist, homogeneous population; effective guilt requires respect for law. Feeling shame is about wanting to hide; feeling guilt is about wanting to make amends. Shame is when everyone says, "You can no longer live with us"; guilt is when you say, "How am I going to live with myself?"*

From the time that Benedict first articulated this contrast, there has been a self-congratulatory view in the West that shame is somehow more primitive than guilt, as the West has left behind dunce caps, public flogging, and scarlet letters. Shame is the mob; guilt is internalizing rules, laws, edicts, decrees, and statutes. Yet, Jacquet convincingly argues for the continued usefulness of shaming in the West, calling for its rebirth in a postmodernist form. For her, shaming is particularly useful when the powerful show no evidence of feeling guilt and evade punishment. We have no shortage of examples of such evasion in the American legal system, where one can benefit from the best defense that money or power can buy; shaming can often step into that vacuum. Consider a 1999 scandal at UCLA, when more than a dozen healthy, strapping football players

* Just to bring another term into the mix, most in the field seem to categorize embarrassment as a transient, low-rent version of shame. Its regulatory power is shown by the Semai people of the Malay Peninsula, who say, "There is no authority here but embarrassment."

were discovered to have used connections, made-up disabilities, and forged doctors' signatures to get handicapped parking permits. Their privileged positions resulted in what was generally seen as slaps on the wrist by both the courts and UCLA. However, the element of shaming may well have made up for it—as they left the courthouse in front of the press, they walked past a phalanx of disabled, wheelchair-bound individuals jeering them.[31]

Anthropologists, studying everyone from hunter-gatherers to urbanites, have found that about two thirds of everyday conversation is gossip, with the vast majority of it being negative. As has been said, gossip (with the goal of shaming) is a weapon of the weak against the powerful. It has always been fast and cheap and is infinitely more so now in the era of the Scarlet Internet.

Shaming is also effective when dealing with outrages by corporations.[32] Bizarrely, the American legal system considers a corporation to be an individual in many ways, one that is psychopathic in the sense of having no conscience and being solely interested in profits. The people running a corporation are occasionally criminally responsible when the corporation has done something illegal; however, they are not when the corporation does something legal yet immoral—it is outside the realm of guilt. Jacquet emphasizes the potential power of shaming campaigns, such as those that forced Nike to change its policies about the horrific working conditions in its overseas sweatshops, or paper giant Kimberly-Clark to address the cutting of old-growth forests.

Amid the potential good that can come from such shaming, Jacquet also emphasizes the dangers of contemporary shaming, which is the savagery with which people can be attacked online and the distance such venom can travel—in a world where getting to anonymously hate the sinner seems more important than anything about the sin itself.

FOOLS RUSH IN: APPLYING THE FINDINGS OF THE SCIENCE OF MORALITY

How can the insights we already have in hand be used to foster the best of our behaviors and lessen the worst?

Which Dead White Male Was Right?

Let's start with a question that has kept folks busy for millennia, namely, what is the optimal moral philosophy?

People pondering this question have grouped the different approaches into three broad categories. Say there's money sitting there, and it's not yours but no one is looking; why not grab it?

Virtue ethics, with its emphasis on the actor, would answer: because you are a better person than that, because you'll have to live with yourself afterward, etc.

Deontology, with its emphasis on the act: because it's not okay to steal.

Consequentialism, with its emphasis on the outcome: what if everyone started acting that way, think about the impact on the person whose money you've stolen, etc.

Virtue ethics has generally taken a backseat to the other two in recent years, having acquired a quaint veneer of antiquarian fretting over how an improper act tarnishes one's soul. As we'll see, I think that virtue ethics returns through the back door with considerable relevance.

By focusing on deontology versus consequentialism, we are back on the familiar ground of whether ends justify means. For deontologists the answer is "No, people can never be pawns." For the consequentialist the answer is "Yes, for the right outcome." Consequentialism comes in a number of stripes, taken seriously to varying degrees, depending on its features—for example, yes, the end justifies the means if the end is to maximize my pleasure (hedonism), to maximize overall levels of wealth,* to strengthen the powers that be (state consequentialism). For most, though, consequentialism is about classical utilitarianism—it is okay to use people as a means to the end of maximizing overall levels of happiness.

When deontologism and consequentialism contemplate trolleys, the former is about moral intuitions rooted in the vmPFC, amygdala, and insula, while the latter is the domain of the dlPFC and moral reasoning. Why is it that our automatic, intuitive moral judgments tend to be nonutilitarian? Because, as Greene states in his book, "Our moral brains evolved to help us spread our genes, not to maximize our collective happiness."

The trolley studies show people's moral heterogeneity. In them approximately 30 percent of subjects were consistently deontologists, unwilling to either pull a lever or push a person, even at the cost of those five lives. Another 30 percent were always utilitarian, willing to pull or push. And for everyone else, moral philosophies were context dependent. The fact that a plurality of people fall into this category prompts Greene's "dual process" model, stating that we are usually a mixture of valuing means and ends. What's your moral philosophy? If harm to the person who is the means is unintentional or if the intentionality is really convoluted and indirect, I'm a utilitarian consequentialist, and if the intentionality is right in front of my nose, I'm a deontologist.

The different trolley scenarios reveal what circumstances push us toward intuitive deontology, which toward utilitarian reasoning. Which outcome is better?

For the sort of person reading this book (i.e., who reads and thinks, things to be justifiably self-congratulatory about), when considering this issue at a calm distance, utilitarianism seems like the place to start—maximizing collective happiness. There is the emphasis on equity—not

* Which, to emphasize something that we all truly know but have trouble remembering, is not synonymous with happiness. Extensive research on happiness, ranging from longitudinal studies of the same individuals over time to huge cross-cultural studies of tens of thousands of subjects in dozens of countries, all show the same thing: when people rise out of abject poverty, they most definitely tend to become happier. But above the level of struggling to eke out an existence, there is remarkably little relationship between income and happiness.

equal treatment but taking everyone's well-being into equal consideration. And there is the paramount emphasis on impartiality: if someone thinks the situation being proposed is morally equitable, they should be willing to flip a coin to determine which role they play.

Utilitarianism can be critiqued on practical grounds—it's hard to find a common currency of people's differing versions of happiness, the emphasis on ends over means requires that you be good at predicting what the actual ends will be, and true impartiality is damn hard with our Us/Them minds. But in theory, at least, there is a solid, logical appeal to utilitarianism.

Except that there's a problem—unless someone is missing their vmPFC, the appeal of utilitarianism inevitably comes to a screeching halt at some point. For most people it's pushing the person in front of the trolley. Or smothering a crying baby to save a group of people hiding from Nazis. Or killing a healthy person to harvest his organs and save five lives. As Greene emphasizes, virtually everyone immediately grasps the logic and appeal of utilitarianism yet eventually hits a point where it is clear that it's not a good guide for everyday moral decision making.

Greene and, independently, the neuroscientist John Allman of Caltech and historian of science James Woodward of the University of Pittsburgh have explored the neurobiological underpinnings of a key point—the utilitarianism being considered here is unidimensional and artificial; it hobbles the sophistication of both our moral intuitions and our moral reasoning. A pretty convincing case can be made for utilitarian consequentialism. As long as you consider the immediate consequences. And the longer-term consequences. And the long-long-term consequences. And then go and consider them all over again a few times.

When people hit a wall with utilitarianism, it's because what is on paper a palatable trade-off in the short run ("Intentionally kill one to save five—that obviously increases collective happiness") turns out not to be so in the long run. "Sure, that healthy person's involuntary organ donation just saved five lives, but who else is going to get dissected that way? What if they come for me? I kinda like my liver. What else might they start doing?" Slippery slopes, desensitization, unintended consequences, intended consequences. When shortsighted utilitarianism (what Woodward and Allman call "parametric" consequentialism) is replaced with a longer-viewed version (what they call "strategic" consequentialism and what Greene calls "pragmatic utilitarianism"), you get better outcomes.

Our overview of moral intuition versus moral reasoning has generated a dichotomy, something akin to how guys can't have lots of blood flow to their crotch and their brain at the same time; they have to choose. Similarly, you have to choose whether your moral decision making will be about the amygdala or the dlPFC. But this is a false dichotomy, because we reach our best long-term, strategic, consequentialist decisions when we engage both our reasoning and our intuition. "Sure, being willing to do X in order to accomplish Y seems like a good trade-off in the short run. But in the long run, if we do that often enough, doing Z is going to start to seem okay also, and I'd feel awful if Z were done to me, and there's also a good chance that W would happen, and that's going to generate really bad feelings in people, which will result in . . ." And the "feel" part of that process is not the way Mr. Spock would do it, logically and dispassionately remembering that those humans are irrational, flighty creatures and incorporating that into his rational thinking about them. Instead, this is feeling what the feelings would feel like. This is straight out of chapter 2's overview of Damasio's somatic marker hypothesis: when we are making decisions, we are running not only thought experiments but somatic feeling experiments as well—how is it going to *feel* if this happens?—and this combination is the goal in moral decision making.

Thus, "No way I'd push someone onto the trolley tracks; it's just wrong" is about the amygdala, insula, and vmPFC. "Sacrifice one life to save five, sure" is the dlPFC. But do long-term strategic consequentialism, and all those regions are engaged. And this yields something more powerful than the cocksureness of knee-jerk intuitionism, the "I can't tell you why, but this is simply wrong." When you've engaged all those brain systems, when you've done the thought experiments and feeling experiments of how things might play out in the long run, and when you've prioritized the inputs—gut reactions are taken seriously, but they're sure not given veto power—you'll know exactly why something seems right or wrong.

The synergistic advantages of combining reasoning with intuition raise an important point. If you're a fan of moral intuitions, you'd frame them as being foundational and primordial. If you don't like them, you'd present them as simplistic, reflexive, and primitive. But as emphasized by Woodward and Allman, our moral intuitions are neither primordial nor reflexively primitive. They are the end products of learning; they are cognitive conclusions to which we have been exposed so often that they have become automatic, as implicit as riding a bicycle or reciting the days of the week forward rather than backward. In the West we nearly all have strong moral intuitions about the wrongness of slavery, child labor, or animal cruelty. But that sure didn't used to be the case. Their wrongness has become an implicit moral intuition, a gut instinct concerning moral truth, only because of the fierce moral reasoning (and activism) of those who came before us, when the average person's moral intuitions were unrecognizably different. Our guts learn their intuitions.

Slow and Fast: The Separate Problems of "Me Versus Us" and "Us Versus Them"

The contrast between rapid, automatic moral intuitionism and conscious, deliberative moral reasoning plays out in another crucial realm and is the subject of Greene's superb 2014 book *Moral Tribes: Emotion, Reason, and the Gap Between Us and Them.*[33]

Greene starts with the classic tragedy of the commons. Shepherds bring their flocks to a common grazing field. There are so many sheep that there is the danger of destroying the commons, unless people decrease the size of their herds. And the tragedy is that if it is truly a commons, there is no incentive to ever cooperate—you'd range from being a fool if no one else was cooperating to being a successful free rider if everyone else was.

This issue, namely how to jump-start and then maintain cooperation in a sea of noncooperators, ran through all of chapter 10 and, as shown in the widespread existence of social species that cooperate, this is solvable (stay tuned for more in the final chapter). When framed in the context of morality, averting the tragedy of the commons requires getting people in groups to not be selfish; it is an issue of Me versus Us.

But Greene outlines a second type of tragedy. Now there are two different *groups* of shepherds, and the challenge is that each group has a different approach to grazing. One, for example, treats the pasture as a classic commons, while the other believes that the pasture should be divided up into parcels of land belonging to individual shepherds, with high, strong fences in between. In other words, mutually contradictory views about using the pasture.

The thing that fuels the danger and tragedy of this situation is that each group has such a tightly reasoned structure in their heads as to why their way is correct that it can acquire moral weight, be seen as a "right." Greene dissects that word brilliantly. For each side, perceiving themselves as having a "right" to do things their way mostly means that they have slathered enough post-hoc, Haidtian rationalizations on a shapeless, self-serving, parochial moral intuition; have lined up enough of their gray-bearded philosopher-king shepherds to proclaim the moral force of their stance; feel in the most sincere, pained way that the very essence of what they value and who they are is at stake, that the very moral rightness of the universe is wobbling; all of that so strongly that they can't recognize the "right" for what it is, namely "I can't tell you why, but this is how things should be done." To cite a quote attributed to Oscar Wilde, "Morality is simply the attitude we adopt towards people whom we personally dislike."

It's Us versus Them framed morally, and the importance of what Greene calls "the Tragedy of Commonsense Morality" is shown by the fact that most intergroup conflicts on our planet ultimately are cultural disagreements about whose "right" is righter.

This is an intellectualized, bloodless way of framing the issue. Here's a different way.

Say I decide that it would be a good thing to have pictures here

demonstrating cultural relativism, displaying an act that is commonsensical in one culture but deeply distressing in another. "I know," I think, "I'll get some pictures of a Southeast Asian dog-meat market; like me, most readers will likely resonate with dogs." Good plan. On to Google Images, and the result is that I spend hours transfixed, unable to stop, torturing myself with picture after picture of dogs being carted off to market, dogs being butchered, cooked, and sold, pictures of humans going about their day's work in a market, indifferent to a crate stuffed to the top with suffering dogs.

I imagine the fear those dogs feel, how they are hot, thirsty, in pain. I think, "What if these dogs had come to trust humans?" I think of their fear and confusion. I think, "What if one of the dogs whom I've loved had to experience that? What if this happened to a dog my children loved?" And with my heart racing, I realize that I hate these people, *hate* every last one of them and despise their culture.

And it takes a locomotive's worth of effort for me to admit that I can't justify that hatred and contempt, that mine is a mere moral intuition, that there are things that I do that would evoke the same response in some distant person whose humanity and morality are certainly no less than mine, and that but for the randomness of where I happen to have been born, I could have readily had their views instead.

The thing that makes the tragedy of commonsense morality so tragic is the intensity with which you just know that They are deeply wrong.

In general, our morally tinged cultural institutions—religion, nationalism, ethnic pride, team spirit—bias us toward our best behaviors when we are single shepherds facing a potential tragedy of the commons. They make us less selfish in Me versus Us situations. But they send us hurtling toward our worst behaviors when confronting Thems and their different moralities.

The dual process nature of moral decision making gives some insights into how to avert these two very different types of tragedies.

In the context of Me versus Us, our moral intuitions are shared, and emphasizing them hums with the prosociality of our Us-ness. This was shown in a study by Greene, David Rand of Yale, and colleagues, where subjects played a one-shot public-goods game that modeled the tragedy of the commons.[34] Subjects were given differing lengths of time to decide how much money they would contribute to a common pot (versus keeping it for themselves, to everyone else's detriment). And the faster the decision required, the more cooperative people were. Ditto if you had primed subjects to value intuition (by having them relate a time when intuition led them to a good decision or where careful reasoning did the opposite)—more cooperation. Conversely, instruct subjects to "carefully consider" their decision, or prime them to value reflection over intuition, and they'd be more selfish. The more time to think, the more time to do a version of "Yes, we all agree that cooperation is a good thing . . . but here is why I should be exempt this time"—what the authors called "calculated greed."

What would happen if subjects played the game with someone screamingly different, as different a human as you could find, by whatever the subject's standards of comfort and familiarity? While the study hasn't been done (and would obviously be hard to do), you'd predict that fast, intuitive decisions would overwhelmingly be in the direction of easy, unconflicted selfishness, with "Them! Them!" xenophobia alarms ringing and automatic beliefs of "Don't trust Them!" instantly triggered.

When facing Me-versus-Us moral dilemmas of resisting selfishness, our rapid intuitions are good, honed by evolutionary selection for cooperation in a sea of green-beard markers.[35] And in such settings, regulating and formalizing the prosociality (i.e., moving it from the realm of intuition to that of cogitation) can even be counterproductive, a point emphasized by Samuel Bowles.*

In contrast, when doing moral decision making during Us-versus-Them scenarios, keep intuitions as far away as possible. Instead, think, reason, and question; be deeply pragmatic and strategically utilitarian; take their perspective, try to think what they think, try to feel what they feel. Take a deep breath, and then do it all again.†

Veracity and Mendacity

The question rang out, clear and insistent, a question that could not be ignored or evaded. Chris swallowed once, tried for a voice that was calm and steady, and answered, "No, absolutely not." It was a bald-faced lie.

Is this a good thing or bad thing? Well, it depends on what the question was: (a) "When the CEO gave you the summary, were you aware that the numbers had been manipulated to hide the third-quarter losses?" asked the prosecutor. (b) "Is this a toy you already have?" asked Grandma tentatively. (c) "What did the doctor say? Is it fatal?" (d) "Does this outfit make me look ____ ?" (e) "Did you eat the brownies that were for tonight?" (f) "Harrison, are you harboring the runaway slave named Jack?" (g) "Something's not adding up. Are you lying about being at work late last night?" (h) "OMG, did you just cut one?"

Nothing better typifies the extent to which the meanings of our behaviors are context dependent. Same untruth, same concentration on controlling your facial expression, same attempt to make just the right

* Bowles cites a great example of this, where sanctions decrease in-group prosociality: Some parents are habitually late picking up their kids at their preschool. "Please don't do it," the school e-mails all parents. "It delays our wonderful staff from being able to leave work." This helps, but still there are some parents who are habitually late. So the school institutes a sanctions program—each time you're late, we add a charge to your bill. And the rate of parental tardiness *worsens*. Why? Because the transgression has been moved from the realm of in-group social intuition ("Hey, I shouldn't be selfish toward members of *our* school community") to a more calculated realm ("Okay, I'm willing to incur an increased cost for my convenience"). This might also be a way to frame the explanation for why, in that cross-cultural study of small-scale societies discussed above, those with the most market integration had the most prosocial game play—what markets and cash economies do is shift a world of reciprocal altruism from the realm of social intuition to social calculation.

† These themes bear a strong resemblance to those of the economics Nobel laureate Daniel Kahneman, in his best seller, *Thinking, Fast and Slow*—rather than framing things in a moral arena, his analysis is of the differing strengths and weaknesses of fast intuitive thinking and slow analytical thinking in the realm of economics.

amount of eye contact. And depending on the circumstance, this could be us at our best or worst. On the converse side of context dependency, sometimes being honest is the harder thing—telling an unpleasant truth about another person activates the medial PFC (along with the insula).*[36]

Given these complexities, it is no surprise that the biology of honesty and duplicity is very muddy.

As we saw in chapter 10, the very nature of competitive evolutionary games selects for both deception and vigilance against it. We even saw protoversions of both in social yeast. Dogs attempt to deceive one another, with marginal success—when a dog is terrified, fear pheromones emanate from his anal scent glands, and it's not great if the guy you're facing off against knows you're scared. A dog can't consciously choose to be deceptive by not synthesizing and secreting those pheromones. But he can try to squelch their dissemination by putting a lid on those glands, by putting his tail between his legs—"I'm not scared, no siree," squeaked Sparky.

No surprise, nonhuman primate duplicity takes things to a whole other level.[37] If there is a good piece of food and a higher-ranking animal nearby, capuchins will give predator alarm calls to distract the other individual; if it is a lower-ranking animal, no need; just take the food. Similarly, if a low-ranking capuchin knows where food has been hidden and there is a dominant animal around, he will move away from the hiding place; if it's a subordinate animal, no problem. The same is seen in spider monkeys and macaques. And other primates don't just carry out "tactical concealment" about food. When a male gelada baboon mates with a female, he typically gives a "copulation call." Unless he is with a female who has snuck away from her nearby consortship male. In which case he doesn't make a sound. And, of course, all of these examples pale in comparison with what politico chimps can be up to. Reflecting deception as a task requiring lots of social expertise, across primate species, a larger neocortex predicts higher rates of deception, independent of group size.†

* Although the neuroscientist Sam Harris, in his book *Lying*, argues that all lying—even white lies, lies to spare someone's feelings, lies accomplishing the proverbial heroics of, say, hiding a runaway slave—are wrong.

† Just to reiterate, starting with those social yeast, deception is not limited to primates. Deception along similar lines to that observed in capuchins has been reported in those brilliant corvid birds; moreover, behaviors such as plover

That's impressive. But it is highly unlikely that there is conscious strategizing on the part of these primates. Or that they feel bad or even morally soiled about being deceptive. Or that they actually believe their lies. For those things we need humans.

The human capacity for deception is enormous. We have the most complex innervation of facial muscles and use massive numbers of motor neurons to control them—no other species can be poker-faced. And we have language, that extraordinary means of manipulating the distance between a message and its meaning.

Humans also excel at lying because our cognitive skills allow us to do something beyond the means of any perfidious gelada baboon—we can finesse the truth.

A cool study shows our propensity for this. To simplify: A subject would roll a die, with different results yielding different monetary rewards. The rolls were made in private, with the subject reporting the outcome—an opportunity to cheat.

Given chance and enough rolls, if everyone was honest, each number would be reported about one sixth of the time. If everyone always lied for maximal gain, all rolls would supposedly have produced the highest-paying number.

There was lots of lying. Subjects were over 2,500 college students from twenty-three countries, and higher rates of corruption, tax evasion, and political fraud in a subject's country predicted higher rates of lying. This is no surprise, after chapter 9's demonstration that high rates of rule violations in a community decrease social capital, which then fuels individual antisocial behavior.

What was most interesting was that across all the cultures, lying was of a particular type. Subjects actually rolled a die twice, and only the first roll counted (the second, they were told, tested whether the die was "working properly"). The lying showed a pattern that, based on prior work, could be explained by only one thing—people rarely made up a

---

birds feigning injury to lure a predator away from its nest have been interpreted as tactical deception. ("Don't eat the babies. Look, come after me! I've got more meat and I can't get away because I'm injured.") Similar deception has also been reported in other birds, some ungulates, and cuttlefish.

high-paying number. Instead they simply reported the higher roll of the two.

You can practically hear the rationalizing. "Darn, my first roll was a 1 [a bad outcome], my second a 4 [better]. Hey, rolls are random; it could just as readily have been 4 as a 1, so . . . let's just say I rolled a 4. That's not really cheating."

In other words, lying most often included rationalizing that made it feel less dishonest—not going whole hog for that filthy lucre, so that your actions feel like only slightly malodorous untruthiness.

When we are lying, naturally, regions involved in Theory of Mind are involved, particularly with circumstances of strategic social deception. Moreover, the dlPFC and related frontal regions are central to a neural circuit of deception. And then insight grinds to a halt.[38]

Back to the theme introduced in chapter 2 of the frontal cortex, and the dlPFC in particular, getting you to do the harder thing when it's the right thing to do. And in our value-free sense of "right," you'd expect the dlPFC to activate when you're struggling to do (a) the morally right thing, which is to avoid the temptation to lie, as well as (b) the strategically right thing, namely, once having decided to lie, doing it effectively. It can be *hard* to deceive effectively, having to think strategically, carefully remember what lie you're actually saying, and create a false affect ("Your Majesty, I bring terrible, sad news about your son, the heir to the throne [yeah, we ambushed him—high fives!]").* Thus activation of the dlPFC will reflect both the struggle to resist temptation and the executive effort to wallow effectively in the temptation, once you've lost that struggle. "Don't do it" + "if you're going to do it, do it right."

This confusion arises in neuroimaging studies of compulsive liars.†[39] What might one expect? These are people who habitually fail to resist the temptation of lying; I bet they have atrophy of something frontocortical.

* Prompting two great quotes, one generally attributed to the politician Sam Rayburn ("Son, always tell the truth. Then you'll never have to remember what you said the last time.") and the other from eighteenth-century Swiss philosopher Johann Lavater ("He who is passionate and hasty is generally honest; it is your cool dissembling hypocrite of whom you should be beware.").

† As assessed by scores on a subpart of a psychopathy questionnaire or by a history of successfully conning people. Importantly, the studies included not only a control group of normal individuals but also a control group of psychopaths who just happened not to be compulsive liars.

These are people who habitually lie and are good at it (and typically have high verbal IQs); I bet they have expansion of something frontocortical. And the studies bear out both predictions—compulsive liars have increased amounts of white matter (i.e., the axonal cables connecting neurons) in the frontal cortex, but lesser amounts of gray matter (i.e., the cell bodies of the neurons). It's not possible to know if there's causality in these neuroimaging/behavior correlates. All one can conclude is that frontocortical regions like the dlPFC show multiple and varied versions of "doing the harder thing."

You can dissociate the frontal task of resisting temptation from the frontal task of lying effectively by taking morality out of the equation.[40] This is done in studies where people are *told* to lie. (For example, subjects are given a series of pictures; later they are shown an array of pictures, some of which are identical to ones in their possession, and asked, "Is this a picture you have?" A signal from the computer indicates whether the subject should answer honestly or lie.) In this sort of scenario, lying is most consistently associated with activation of the dlPFC (along with the nearby and related ventrolateral PFC). This is a picture of the dlPFC going about the difficult task of lying effectively, minus worrying about the fate of its neuronal soul.

The studies tend to show activation of the anterior cingulate cortex (ACC) as well. As introduced in chapter 2, the ACC responds to circumstances of conflicting choices. This occurs for conflict in an emotional sense, as well as in a cognitive sense (e.g., having to choose between two answers when both seem to work). In the lying studies the ACC isn't activating because of moral conflict about lying, since subjects were instructed to lie. Instead, it's monitoring the conflict between reality and what you've been instructed to report, and this gums up the works slightly; people show minutely longer response times during lying trials than during honest ones.

This delay is useful in polygraph tests (i.e., lie detectors). In the classic form, the test detected arousal of the sympathetic nervous system, indicating that someone was lying and anxious about not getting caught. The trouble is that you'd get the same anxious arousal if you're telling the truth but your life's over if that fallible machine says otherwise. Moreover, sociopaths are undetectable, since they don't get anxiously aroused when lying. Plus subjects can take countermeasures to manipulate their sympathetic nervous system. As a result, this use of polygraphs is no longer admissible in courts. Contemporary polygraph techniques instead home in on that slight delay, on the physiological indices of that ACC conflict—not the moral one, since some miscreant may have no moral misgivings, but the cognitive conflict—"Yeah, I robbed the store, but no, wait, I have to say that I didn't." Unless you thoroughly believe your lie, there's likely to be that slight delay, reflecting the ACC-ish cognitive conflict between reality and your claim.

Thus, activation of the ACC, dlPFC, and nearby frontal regions is associated with lying on command.[41] At this point we have our usual issue of causality—is activation of, say, the dlPFC a cause, a consequence, or a mere correlate of lying? To answer this, transcranial direct-current stimulation has been used to inactivate the dlPFC in people during instructed-lying tasks. Result? Subjects were slower and less successful in lying—implying a causal role for the dlPFC. And to remind us of how complicated this issue is, people with damage to the dlPFC are less likely to take honesty into account when honesty and self-interest are pitted against each other in an economic game. So this most eggheady, cognitive part of the PFC is central to both resisting lying and, once having decided to lie, doing it well.

This book's focus is not really how good a liar someone is. It's whether we lie, whether we do the harder thing and resist the temptation to deceive. For more understanding of that, we turn to a pair of thoroughly cool neuroimaging studies where subjects who lied did so not because they were instructed to but because they were dirty rotten cheaters.

The first was carried out by the Swiss scientists Thomas Baumgartner, Ernst Fehr (whose work has been noted previously), and colleagues.[42] Subjects played an economic trust game where, in each round, you could be cooperative or selfish. Beforehand a subject would tell the other player what their strategy would be (always/sometimes/never cooperate). In other words, they made a promise.

Some subjects who promised to always cooperate broke their promise at least once. At such times there was activation of the dlPFC, the ACC, and, of course, the amygdala.*[43]

A pattern of brain activation before each round's decision *predicted* breaking of a promise. Fascinatingly, along with predictable activation of the ACC, there'd be activation of the insula. Does the scoundrel think, "I'm disgusted with myself, but I'm going to break my promise"? Or is it "I don't like this guy because of X; in fact, he's kind of disgusting; I owe him nothing; I'm breaking my promise"? While it's impossible to tell, given our tendency to rationalize our own transgressions, I'd bet it's the latter.

The second study comes from Greene and colleague Joseph Paxton.[44] Subjects in a scanner would predict the outcome of coin tosses, earning money for correct guesses. The study's design contained an extra layer of distracting nonsense. Subjects were told the study was about paranormal mental abilities, and for some of the coin tosses, for this concocted reason, rather than state their prediction beforehand, subjects would just think about their choice and then tell afterward if they were right. In other words, amid a financial incentive to guess correctly, there were intermittent opportunities to cheat. Crucially, this was detectable—during the periods of forced honesty, subjects averaged a 50 percent success rate. And if accuracy jumped a lot higher during opportunities to cheat, subjects were probably cheating.

The results were pretty depressing. Using this form of statistical detection, about a third of the subjects appeared to be big-time cheaters, with another sixth on the statistical border. When cheaters cheated, there was activation of the dlPFC, as we'd expect. Were they struggling with the combination of moral and cognitive conflict? Not particularly—there wasn't activation of the ACC, nor was there the slight lag time in response. Cheaters typically didn't cheat at every opportunity; what did things look like when they resisted? Here's where you saw the struggling—even

* The amygdaloid involvement is probably pertinent to a case report from some French neurologists concerning a man who had a seizure every time he lied during business negotiations. He was found to have a tumor pressing on his amygdala; once it was removed, the seizures went away (there was no mention of whether the guy was still lying at work). The authors called this "Pinocchio syndrome."

greater activation of the dlPFC (along with the vlPFC), the ACC roaring into action, and a significant delay in response time. In other words, for people capable of cheating, the occasional resistance seems to be the outcome of major neurobiological Sturm und Drang.

And now for probably the most important finding in this chapter. What about subjects who never cheated? There are two very different scenarios, as framed by Greene and Paxton: Is resisting temptation at every turn an outcome of "will," of having a stoked dlPFC putting Satan into a hammerlock of submission? Or is it an act of "grace," where there's no struggle, because it's simple; you don't cheat?

It was grace. In those who were always honest, the dlPFC, vlPFC, and ACC were in veritable comas when the chance to cheat arose. There's no conflict. There's no working hard to do the right thing. You simply don't cheat.

Resisting temptation is as implicit as walking up stairs, or thinking "Wednesday" after hearing "Monday, Tuesday," or as that first piece of regulation we mastered way back when, being potty trained. As we saw in chapter 7, it's not a function of what Kohlbergian stage you're at; it's what moral imperatives have been hammered into you with such urgency and consistency that doing the right thing has virtually become a spinal reflex.

This is not to suggest that honesty, even impeccable honesty that resists all temptation, can only be the outcome of implicit automaticity.[45] We can think and struggle and employ cognitive control to produce similar stainless records, as shown in some subsequent work. But in circumstances like the Greene and Paxton study, with repeated opportunities to cheat in rapid succession, it's not going to be a case of successfully arm wrestling the devil over and over. Instead, automaticity is required.

We've seen something equivalent with the brave act, the person who, amid the paralyzed crowd, runs into the burning building to save the child. "What were you thinking when you decided to go into the house"? (Were you thinking about the evolution of cooperation, of reciprocal altruism, of game theory and reputation?) And the answer is always "I wasn't thinking anything. Before I knew it, I had run in." Interviews of Carnegie Medal recipients about that moment shows precisely that—a first, intuitive thought of needing to help, resulting in the risking of life without a second thought. "Heroism feels and never reasons," to quote Emerson.[46]

It's the same thing here: "Why did you never cheat? Is it because of your ability to see the long-term consequences of cheating becoming normalized, or your respect for the Golden Rule, or . . . ?" The answer is "I don't know [shrug]. I just don't cheat." This isn't a deontological or a consequentialist moment. It's virtue ethics sneaking in the back door in that moment—"I don't cheat; that's not who I am." Doing the right thing *is* the easier thing.

Fourteen

Feeling Someone's Pain, Understanding Someone's Pain, Alleviating Someone's Pain

A person is in pain, frightened, or crushed with a malignant sadness. And another human, knowing that, is likely to experience something absolutely remarkable—an aversive state that is approximated by the word "empathy." As we'll see in this chapter, it is a state on a continuum with what occurs in a baby or in another species. The state takes varied forms, with varied underlying biology, reflecting its sensorimotor, emotional, and cognitive building blocks. Various logical influences sharpen or dull the state. All leading to this chapter's two key questions: When does empathy lead us to actually do something helpful? When we do act, whose benefit is it for?

"FOR" VERSUS "AS IF" AND OTHER DISTINCTIONS

Empathy, sympathy, compassion, mimicry, emotional contagion, sensorimotor contagion, perspective taking, concern, pity. Let the terminology and squabbles begin over definitions of ways in which we resonate with someone else's adversity (along with the question of whether the opposite of such resonance is gloating pleasure or indifference).

We start with, for want of a better word, primitive versions of resonating with someone's pain. There's sensorimotor contagion—you see a hand poked with a needle, and the part of your sensory cortex that maps onto your hand activates, sensitizing you to the imagined sensation. Perhaps your motor cortex will also activate, causing you to compress your own hand. Or you watch a tightrope walker and involuntarily put your arms out for balance. Or someone has a coughing fit, and your throat constricts.

Even more explicitly motoric is the act of matching movements with simple mimicry. Or there's emotional contagion, the automatic transfer of strong emotive states—such as one baby crying because another is, or someone catching the fever of a mob plunging into a riot.

Your resonance with someone's plight can carry an implicit power differential. You can pity someone in pain—recalling Fiske's categories of Thems in chapter 11, this belittling pity means you view the person as high in warmth and low in competence and agency. And we all know the everyday meaning of "sympathy" ("Look, I sympathize with your situation, but . . ."); you have the power to alleviate their distress but choose not to.

Then there are terms reflecting how much your resonance is about emotion versus cognition. In that sense "sympathy" means you *feel* sorry for someone else's pain without understanding it. In contrast, "empathy" contains the cognitive component of understanding the cause of someone's pain, taking his perspective, walking in his shoes.

And then there are distinctions introduced in chapter 6, describing how much you and your own feelings play into resonating with someone else's distress. There's the emotionally distanced sense of sympathy, of feeling *for* someone. There's the rawer, vicarious state of feeling their pain *as if* it were happening to you. And then there is the more cognitively distanced state of perspective taking, of imagining what this must be like for *her*, not you. As we'll see, an as-if state carries the danger that you experience her pain so intensely that your primary concern becomes alleviating your own distress.

Which raises a different word—"compassion," where your resonance with someone's distress leads you to actually help.[1]

Perhaps most important, these words are generally about inwardly motivated states—you can't force someone to truly feel empathy, can't induce it in them with guilt or a sense of obligation. You can generate ersatz versions of it those ways, but not the real thing. Consistent with that, some recent work shows that when you help someone out of empathy, there is a very different profile of brain activation from when you do so out of an obliged sense of reciprocity.[2]

As usual, we gain insights into the nature and biology of these states by looking at their rudiments in other species, their development in children, and their pathological manifestations.

EMOTIONALLY CONTAGIOUS, COMPASSIONATE ANIMALS

Lots of animals display building blocks of empathic states (I use "empathic state" throughout the chapter when referring to the collectivity of sympathy, empathy, compassion, etc.). There's mimicry, a cornerstone of social learning in many species—think young chimps watching Mom to learn to use tools. Ironically, humans' strong proclivity for imitation can have a downside. In one study chimps and children observed an adult human repeatedly accessing a treat inside a puzzle box; crucially, the person added various extraneous movements. When exploring the box themselves afterward, chimps imitated only the steps needed to open it, whereas kids "overimitated," copying the superfluous gestures as well.*[3]

Social animals are also constantly buffeted with emotional contagion—shared states of arousal in a pack of dogs or male chimps going on a border patrol. These are not terribly precise states, often spilling over into other behaviors. For example, say some baboons flush out something good to eat—say, a young gazelle. The gazelle is running like hell, with these baboons in pursuit. And then the male in front seems to think something along the lines of "Well, here I am running fast and—WHAT? There's my hated rival running right behind me! Why's that jerk chasing me?" He spins around for a head-on collision and fight with the baboon behind him, gazelle forgotten.

Mimicry and emotional contagion are baby steps. Do other animals feel one another's pain? Sort of. Mice can learn a specific fear association vicariously by observing another mouse experiencing the fear conditioning. Moreover, this is a social process—learning is enhanced if the mice are related or have mated.[4]

In another study a mouse would be exposed to an aggressive intruder placed in its cage.[5] As shown previously, this produces persistent adverse consequences—a month later, such mice still had elevated glucocorticoid levels and were more anxious and more vulnerable to a mouse model of depression.† Importantly, the same persistent effects would be induced in a mouse merely observing another mouse experiencing that stressful intruder paradigm.

An even more striking demonstration of "your pain is my pain" in another species came in a 2006 *Science* paper from Jeff Mogil of McGill University.[6] A mouse would observe another mouse (separated from it by Plexiglas) in pain, and, as a result, its own pain sensitivity increased.‡ In another part of the study, an irritant would be injected in a mouse's paw;

* Or stated another way, the chimps were less susceptible to superstitious behavior than were the humans.

† They gave up more readily in a difficult task and experienced less pleasure—showing less of a preference for sucrose-flavored water.

‡ This is determined with a "hot-plate test." A mouse is placed on a room-temperature hot plate; the temperature of the plate is gradually raised. You can tell the instant when the heat first becomes uncomfortable—the mouse lifts a paw (at which point the mouse is removed). What was the plate's temperature at that point? That's the mouse's pain threshold.

mice typically lick their paw at that point, with the amount of licking indicating the amount of discomfort. Thus, X amount of the irritant would produce Z amount of licking. However, if the mouse was simultaneously observing a mouse who had been exposed to more than X amount of irritant and who thus was licking more than Z amount, the subject mouse would lick more than usual. Conversely, if the subject observed a mouse licking less (having been exposed to less than X amount of irritant), it would also lick less. Thus the amount of pain a mouse was feeling was modulated by the amount of pain a nearby mouse was. Importantly, this was a social phenomenon—this shared pain only occurred between mice that were cagemates.*

Obviously we can't know the internal state of these animals. Were they feeling bad for the other mouse in pain, feeling "for" or "as if," taking the other mouse's perspective? Pretty unlikely, making the use of the word "empathy" in this literature controversial.[7]

However, we can observe overt behavior. Do other species proactively lessen the distress of another individual? Yes.

As we will see in the final chapter, numerous species show "reconciliative" behavior, where two individuals, soon after a negative interaction, show higher-than-chance levels of affiliative behaviors (grooming, sitting in contact) between them, and this decreases the odds of subsequent tensions between them. As shown by de Waal and colleagues, chimps also show third-party "consolation" behavior. This is not when, after two individuals fight, some bleeding-heart chimp indiscriminately nices both of them. Instead the consoler is preferentially affiliative to the victim over the initiator of the fight. This reflects both a cognitive component of tracking who started a tension and an affective desire to comfort. Similar consolation, focused on fight victims, also occurs in wolves, dogs, elephants, and corvids (who preen the feathers of victims). Ditto for bonobos—with some bonoboesque sex thrown in for victims along with all that platonic grooming. In contrast, such consolation doesn't occur in monkeys.[8]

Consolation is also shown among those heartwarming pair-bonding

* Reading about these animals experiencing this sure induces an empathic state.

prairie voles, as shown in a 2016 *Nature* paper from Larry Young of Emory University, a pioneer of the vole/monogamy/vasopressin story, along with de Waal.[9] Members of a vole pair would be placed in separate rooms. One of the pair would be either stressed (with a mild shock) or left undisturbed; pairs were then reunited. As compared with unstressed individuals, stressed ones would be licked and groomed more by their partner. Partners would also match the anxiety behaviors and glucocorticoid levels of their stressed pairmate. This didn't occur for a stressed stranger, nor among polygamous meadow voles. As we'll see, the neurobiology of this effect is all about oxytocin and the anterior cingulate cortex.

Animals will intervene even more proactively. In one study rats worked more (pressing a lever) to lower a distressed rat, dangling in the air in a harness, than a suspended block. In another study rats proactively worked to release a cagemate from a stressful restrainer. Subjects were as motivated to do this as to get chocolate (nirvana for rats). Moreover, when a rat could both release the cagemate and get chocolate, they'd share it more than half the time.[10]

This prosociality had an Us/Them component. The authors subsequently showed that rats would work to release even a strange rat—so long as it was of the same strain and thus nearly genetically identical.[11] Is this automatic Us/Them-ing built on the genetics of shared pheromone signatures (back to chapter 10)? No—if a rat is housed with a cagemate of another strain, it will help individuals of that other strain. And if a rat is switched at birth and raised by a female of another strain, it helps members of its adopted but not its biological strain. "Us" is malleable by experience, even among rodents.

Why do all these animals labor away consoling another individual in distress, or even helping them? It's probably not conscious application of the Golden Rule, and it's not necessarily for the social benefits—rats were just as likely to release cagemates from restrainers even if they didn't get to interact afterward. Maybe it's something resembling compassion. On the other hand, maybe it's just self-interest—"That dangling rat's incessant alarm calls are getting on my nerves. I'm going to work to lower him so he'll shut up." Scratch an altruistic rat and a hypocrite bleeds.

EMOTIONALLY CONTAGIOUS, COMPASSIONATE CHILDREN

A recap of material covered in chapters 6 and 7:

As we saw, a developmental landmark is attaining Theory of Mind, something necessary but not sufficient for empathy, which paves the way for increasing abstraction. The capacity for simple sensorimotor contagion matures into empathic states for someone's physical pain and, later, for someone's emotional pain. There's the progression from feeling sorry for an individual (e.g., someone homeless) to feeling sorry for a category (e.g., "homeless people"). There is increasing cognitive sophistication, as kids first distinguish between harming an object and harming a person. Likewise for distinguishing between intentional and unintentional harm, along with a capacity for moral indignation that is more readily evoked by the former. Along with this comes a capacity to express empathy and a sense of responsibility to act upon it, to be proactively compassionate. Perspective taking matures as well, as the child transitions from solely being capable of feeling "for" to also feeling "as if."

As we saw, the neurobiology of this developmental arc makes sense. At the age where an empathic state is evoked only by someone's physical pain, brain activation centers on the periaqueductal gray (PAG), a fairly low-level way station in the brain's pain circuitry. Once emotional pain can evoke an empathic state, the profile is mostly about coupled activation between the (emotional) vmPFC and limbic structures. As the capacity for moral indignation matures, coupling among the vmPFC, the insula, and amygdala emerges. And as perspective taking comes into play, the vmPFC is increasingly coupled to regions associated with Theory of Mind (like the temporoparietal junction).

This was our picture of empathic states in kids being built upon the cognitive foundation of Theory of Mind and perspective taking. But as we also saw, there are empathic states earlier on—infants showing emotional contagion, a toddler trying to comfort a crying adult by offering her stuffie, long before textbook Theory of Mind occurs. And just as with empathic states in other animals, one must ask whether compassion in kids is mostly about ending the sufferer's distress or ending their own.

AFFECT AND/OR COGNITION?

This again. We can predict the major punch lines, thanks to the previous three chapters: both cognitive and affective components contribute to healthy empathic states; it's silly to debate which is more important; what's interesting is seeing when one predominates over the other. Even more interesting is to look at the neurobiology of how those components interact.

The Affective Side of Things

When it comes to empathy, all neurobiological roads pass through the anterior cingulate cortex (ACC). As introduced in chapter 2, this frontal cortical structure has starred in empathy neuroscience ever since people felt someone else's pain while inside a brain scanner.[12]

Given its more traditional roles in mammals, the ACC's empathy connection is unexpected. Broadly, those roles are:

- *Processing interoceptive information.* As introduced in chapter 3, our brains monitor sensory information not only from outside us but from our internal world as well—interoceptive information about achy muscles, dry mouths, bowels in an uproar. If you unconsciously sense that your heart is racing and that makes you experience some emotion more intensely, thank the ACC. The ACC funnels literal gut feelings into intuitions and metaphorical gut feelings influencing frontal function. Pain is a key type of interoceptive information that catches the ACC's attention.[13]
- *Conflict monitoring.* The ACC responds to "conflict" in the sense of a discrepancy from what is expected. If you associate doing some behavior with a particular outcome, when that outcome doesn't

> occur, the ACC takes notice. This monitoring of discrepancy from expectation is asymmetrical—do some task that pays two brownie points and today, unexpectedly, you get three instead, and the ACC perks up and takes notice; do the task and instead of two brownie points you only get one, and the ACC activates like mad. In the words of Kevin Ochsner of Columbia University and colleagues, the ACC is an "all-purpose alarm that signals when ongoing behavior has hit a snag."[14]

Unexpected pain is at the intersection of those two roles of the ACC, a sure sign that things are amiss with your schema about the world. Even with anticipated pain, you monitor whether it turns out to be of the quality and quantity expected. As noted, the ACC doesn't concern itself with pedestrian concerns about pain (is it my finger or my toe that hurts?); that's the purview of less refined, more ancient brain circuitry. What the ACC cares about is the *meaning* of the pain. Good news or bad, and of what nature? Thus the ACC's perception of pain can be manipulated. Poke your finger with a pin and the ACC activates, along with those brain regions telling you which finger and what parameters of pain. Make someone believe that the inert cream you just smeared on his finger is a powerful painkiller, and when you poke his finger, the "it's my finger, not my toe" circuitry still activates. But the ACC falls for the placebo effect and stays silent.

Obviously the ACC receives inputs from interoceptive and exteroceptive outposts. Equally logically, it sends lots of projections into the sensorimotor cortex, making you very aware of and focused on the body part that hurts.

But the sophistication of the ACC, the reason it sits up there in the frontal cortex, is apparent when considering another type of pain. Back to chapter 6 and the Cyberball game where subjects in brain scanners play catch with a virtual ball on a computer screen, tossing it back and forth, and the other two players stop throwing the ball to you. You're being left out, and the ACC activates. Insofar as the ACC cares about the *meaning* of pain, it's just as concerned with the abstractions of social and emotional pain—social exclusion, anxiety, disgust, embarrassment—as with physical pain. Intriguingly, major depression is associated with various abnormalities in the ACC.* And the ACC is also involved during positive resonance—when their pleasure is your pleasure.[15]

All this makes the ACC sound pretty self-oriented, mighty concerned with your well-being. Which makes its empathy role initially surprising. Nonetheless, numerous studies consistently show that if someone else's pain—a poked finger, a sad face, a tale of misfortune—is evoking an empathic state in you, the ACC is involved.[16] Moreover, the more painful the other person's situation seems to be, the more ACC activation. The ACC is also central to doing something to alleviate someone else's distress.

The neuropeptide/hormone oxytocin gets into the mix. Recall from chapter 4 how it promotes bonding and affiliative behaviors, trust, and generosity.† Recall the study in which prairie voles are observed consoling their stressed partner. And we'd expect, the effect depends on the actions of oxytocin. Remarkably, the oxytocin works in the ACC—selectively block oxytocin effects in the ACC, and voles don't console.

So how do we go from the ACC as this outpost of self-interest, monitoring your pain and whether you are getting what you think you deserve, to the ACC allowing you to feel the pain of the wretched of the earth? I think the link is a key issue of this chapter—how much is an empathic state actually about yourself?[17] "Ouch, that hurt" is a good way to learn not to repeat whatever you just did. But often, even better is to monitor someone else's misfortune—"That sure seems to have hurt her; I'm staying away from doing that." Crucially, the ACC is essential for learning fear and conditioned avoidance by observation alone. Going from "She seems to be having a miserable time" to "Thus I should avoid that" requires an intervening step of shared representation of self: "Like her, I wouldn't enjoy feeling that way." *Feeling* someone else's pain can be more effective for learning than just *knowing* that they're in pain. At its core the ACC

* "Is associated with"—that's pretty uninformative. For simplicity, I've ignored there being all sorts of subparts to the ACC; depression is associated with increased activation in some and decreased in others. Overall, it fits a picture of ACC dysfunction being centrally involved in the suffocating, permeating sadness of depression.

† With that truly important proviso that this applies only to within-group interactions. When dealing with a Them, as we saw, oxytocin makes people more hostile and xenophobic.

is about self-interest, with *caring* about that other person in pain as an add-on.

Other brain regions are pertinent as well. As we saw, maturation of the circuitry of empathy involves bringing into the mix not only the ACC but the insula as well.[18] By adulthood the insula (and to a lesser degree the amygdala) is nearly as intertwined with experiencing empathy as is the ACC. The three regions are highly interconnected, and a big chunk of the amygdala texting the frontal cortex is funneled through the ACC. Numerous circumstances that evoke a sense of empathy, particularly physical pain, activate the insula along with the ACC, with the magnitude of the response correlating with the subject's basic proclivity toward empathy, or the subjective sense of empathy they are feeling in the situation.

This makes sense, given the workings of the insula and amygdala. As we saw, their involvement in empathic states emerges developmentally as kids first embed empathy in context and causality—*why* is this person in pain, and whose *fault* is it? This is obvious when pain is rooted in injustice, when disgust, indignation, and anger sweep in because we know that this pain could have been prevented, that someone profited from it. Even when it is unclear that a cause of pain lies in injustice, we seek attribution—the intertwining of the ACC with the insula and amygdala is our world of scapegoating. And that pattern is so often there even when pain is random, without human agency or villainy—literal or metaphorical tectonic plates shift, the earth opens up and swallows someone innocent, and we rail against the people who deprived that victim of a happier life before the tragedy struck, against the God behind this act of God, against the mechanistic indifference of the universe. And as we will see, the more the purity of empathy is clouded with the anger, disgust, and indignation of blame, the harder it is to actually help.

The Cognitive Side of Things

When do the more cognitive components of an empathic state—the PFC, the dlPFC in particular, along with Theory of Mind networks such as the temporoparietal juncture (TPJ) and superior central sulcus—come more to the forefront? Obviously, and uninterestingly, when it's challenging to even figure out what's going on—"Wait, who won the game?" "Do I want my pieces to surround or be surrounded by the other person's?"

More interesting is when more cognitive brain circuitry is recruited by issues of causation and intentionality: "Wait, does he have a horrible headache because he's a migrant farm worker who was sprayed with pesticide, or because he's been binge drinking with his frat bros?" "Did this AIDS patient get his HIV from a blood transfusion or drug use? (People show more activation of the ACC for the former.) This is precisely what chimps have thought through when comforting an innocent victim of aggression but not an instigator. As we saw in chapter 7, the more cognitive profile of activation emerges when kids start distinguishing between self- and other-inflicted pain. In the words of Jean Decety, who did such research, this demonstrates that "empathic arousal [was] moderated early in information processing by a priori attitudes toward other people."[19] In other words, cognitive processes serve as a gatekeeper, deciding whether a particular misfortune is worthy of empathy.

It is also a cognitive task to resonate with pain that is less overt—for example, there is more engagement of the dmPFC when observing someone in emotional pain than physical pain. Likewise when the pain is presented more abstractly—a signal on a screen indicating that someone's hand has been stuck with a needle versus the act itself being shown. Resonating with someone else's pain is also a cognitive task when it is a type of pain that you haven't experienced. "Well, I suppose I can understand the disappointment of this militia leader when he was passed over for the chance to carry out the ethnic cleansing—kinda like when I lost the election in kindergarten to be president of the random-act-of-kindness club." Now, that takes cognitive work. In one study subjects considered people suffering from a neurological disorder involving a novel type of pain sensitivity; empathizing for that novel pain involved more frontal cortical activation than for more conventional pain.[20]

As we saw, the rudimentary "empathy" of rodents is contingent, depending on whether the other individual is a cagemate or a stranger.[21] It is an enormous cognitive task for humans to overcome that, to reach an empathic state for someone who is different, unappealing. A hospital chaplain once described to me how he has to actively make sure that he is not preferentially visiting patients who were "YAVIS"—young, attractive, verbal, intelligent, or social. This is straight out of Us versus Them—recall Susan Fiske's work showing how extreme out-group members, such as the homeless or addicts, are processed differently in the frontal cortex than other people. And it is also straight out of Josh Greene's tragedy of the commons versus tragedy of the commonsense morality, where acting morally toward an Us is automatic, while doing so for a Them takes work.

The ease of empathizing with people like us starts at the level of autonomic building blocks of empathy—in one study of ritual fire walkers in Spain, heart rate changes in the walkers synchronized with spectators—but only those who were relatives. In line with that distinction, taking the perspective of a loved one in pain activates the ACC; doing the same for a stranger activates the TPJ, that region central to Theory of Mind.[22]

This extends to broader versions of Us versus Them. As introduced in chapter 3, we have a stronger sensorimotor response in our hands when the hand we see being poked with a needle is of our race; the stronger one's implicit in-group bias, the stronger this effect. Meanwhile, other studies show that the stronger the discrepancy in patterns of neural activation when observing an in-group versus an out-group person in pain, the lower the chances of helping the latter.[23] Thus it's no surprise that feeling the same degree of empathy or achieving the same level of perspective taking for a Them as for an Us requires greater frontocortical activation. This is the domain where you must suppress the automatic and implicit urges to be indifferent, if not repulsed, and do the creative, motivated work of finding the affective commonalities.*[24]

Categorical boundaries to the extension of empathy also run along socioeconomic lines, but in an asymmetrical manner. What does that mean? That when it comes to empathy and compassion, rich people tend to suck. This has been explored at length in a series of studies by Dacher

* It can be an informative political litmus test to consider whose pain you readily feel (e.g., a fetus versus a homeless person). "What it means to be liberal or conservative became ideologically solidified around the problem of [empathy for only certain types of] pain," writes one political scientist.

Keltner of UC Berkeley. Across the socioeconomic spectrum, on the average, the wealthier people are, the less empathy they report for people in distress and the less compassionately they act. Moreover, wealthier people are less adept at recognizing other people's emotions and in experimental settings are greedier and more likely to cheat or steal. Two of the findings were picked up by the media as irresistible: (a) wealthier people (as assessed by the cost of the car they were driving) are less likely than poor people to stop for pedestrians at crosswalks; (b) suppose there's a bowl of candy in the lab; invite test subjects, after they finish doing some task, to grab some candy on the way out, telling them that whatever's left over will be given to some kids—the wealthier take more candy.[25]

So do miserable, greedy, unempathic people become wealthy, or does being wealthy increase the odds of a person's becoming that way? As a cool manipulation, Keltner primed subjects to focus either on their socioeconomic success (by asking them to compare themselves with people less well off than them) or on the opposite. Make people feel wealthy, and they take more candy from children.

What explains this pattern? A number of interrelated factors, built around the system justification described in chapter 12—wealthier people are more likely to endorse greed as being good, to view the class system as fair and meritocratic, and to view their success as an act of independence—all great ways to decide that someone else's distress is beneath your notice or concern.

It is a particularly uphill battle when we are asked to empathize with the pain of people we dislike, whom we morally disapprove of—remember how their misfortune doesn't simply fail to activate the ACC but instead it activates mesolimbic dopamine reward pathways. Thus the process of taking their perspective and feeling their pain (as other than grounds for gloating) is a dramatic cognitive challenge rather than something remotely automatic.[26]

The cognitive "costs" of empathizing with someone distant are shown by increasing people's cognitive load (i.e., making their frontal cortex work harder by forcing it to override a habitual behavior)—they become less helpful to strangers but not to family members. "Empathy fatigue" can thus be viewed as the state when the cognitive load of repeated exposure to the pain of Thems whose perspective is challenging to take has exhausted the frontal cortex. The notions of cognitive work and load also help explain why people are more charitable when contemplating one person in need than a group. To quote Mother Teresa, "If I look at the mass, I will never act. If I look at the one, I will." Or to cite a quote attributed to someone who never seems to have achieved enough empathy to be vulnerable to empathy fatigue, Joseph Stalin: "The death of one man is a tragedy; the death of millions is a statistic."[27]

And probably most reliably, those mentalizing pathways are activated when we switch from focusing on what it would feel like if this were happening to us to focusing on what it must feel like for them. Thus when subjects are *instructed* to switch from first- to third-person perspective, there's not just activation of the TPJ but also frontal activation with the top-down regulatory task "Stop thinking about yourself."[28]

Thus we have themes that closely resemble those from the last few chapters. When it comes to empathic states, "emotion" and "cognition" are totally false dichotomies; you need both, but with the balance between the two shifting on a continuum, and the cognition end of it has to do the heavy lifting when the differences between you and the person in pain initially swamp the similarities.

Time now for one of the great sideshows in empathy science.

A MYTHIC LEAP FORWARD

In the early 1990s scientists at the University of Parma in Italy, led by Giacomo Rizzolatti and Vittorio Gallese, reported something that, depending on your tastes, ranged from really interesting to revolutionary. They had been studying an area of the brain called the premotor cortex (PMC) in rhesus monkeys, examining what sorts of stimuli would cause individual neurons there to activate. Back to the PMC from chapter 2. "Executive" neurons in the PFC decide something, passing the news to the rest of the frontal cortex just behind it. Which sends projections to the

PMC just behind it. Which sends projections one step further back, to the motor cortex, which then sends commands to muscles. Thus the PMC straddles the divide between thinking about and carrying out a movement.[29]

The group had discovered some mighty quirky PMC neurons. Suppose a monkey carried out a behavior—grasping some food and bringing it to her mouth. Naturally, some neurons in the PMC would have activated. If she did a different movement—grasping an object and placing it in a container—a different (partially overlapping) array of PMC neurons were involved. What the group reported was that some of the bring-food-to-mouth neurons would also activate if the monkey *observed* someone else (monkey or human) making that movement. Same for some of the place-object-in-container neurons. Same for subtler movements like facial expressions. Consistently, about 10 percent of the PMC neurons devoted to doing movement X also activated when observing someone else doing movement X—very odd for neurons a few steps away from commanding muscles to move. The neurons were concerned with the mirroring of movements. And thus were "mirror neurons" announced to the world.

Naturally, everyone looked for mirror neurons in humans, and their existence in roughly the same part of the brain*[30] was soon inferred with brain imaging studies ("inferred" because that approach tells you about the activity of large numbers of neurons at a time, rather than single ones). Then individual neurons were shown to be mirroresque in humans (in patients undergoing neurosurgery to control a rare type of epilepsy).[31]

The mirroring can be quite abstract. It can be cross-modal—see someone doing movement A, and some mirror neuron activates; hear the *sound* of someone doing movement A, and the same occurs. And the neurons can gestalt a scene, firing even if only part of the observed movement is obscured.[32]

Most interesting, mirror neurons didn't simply track movement. Find

* For those who care, it's the premotor cortex, along with the supplementary motor area and primary somatosensory cortex.

a mirror neuron that responds to the sight of someone picking up a cup of tea to drink. The sight of someone picking up the tea to clear the table doesn't activate it. In other words, mirror neurons can incorporate *intentionality* into their response.

Thus mirror neuron activity correlates with circumstances of imitation, either conscious or otherwise, including imitating the idea of an action, as well as the intent behind it. Nevertheless, no one has actually shown a causal relationship, that automatic or conscious mimicry requires mirror neuron activation. Moreover, the mirror neuron/imitation link is complicated by the cells having been first identified in rhesus monkeys—a species that does not show imitation of behavior.

But assuming that mirror neurons are indeed involved, the question becomes what purpose mimicry serves. Various possibilities have been raised and debated.

Probably the least controversial and most plausible is that mirror neurons mediate motor learning by observation.[33] Downsides of this theory, though, are that (a) mirror neurons do their thing in species with minimal learning by mimicry; (b) the amount of mirror neuron activity is unrelated to the efficacy with which observational learning of movements occurs; (c) to the extent that mirror neurons are needed for types of observational learning, it's a pretty low-level contribution in humans—after all, while we do learn how to carry out certain motoric acts by observation, far more interesting is our learning of context by observation—*when* to carry out that behavior (for example, observational learning may teach a subordinate primate the motoric features of kowtowing, but far more demanding and important is learning *whom* to kowtow to).

Related to that is the idea of mirror neurons aiding learning from another person's experience.[34] If you observe someone biting into food and they grimace at its taste, having mirror neurons at the intersection of observing that expression and experiencing it yourself will certainly make more vivid your understanding that you should probably avoid that food.

This is an idea advocated by Gregory Hickok of the University of California at Irvine, who, as we'll see, is a hard-nosed critic of mirror neuron flights of fancy.

This harks back to chapter 2 and Antonio Damasio's influential somatic marker hypothesis, the idea that when we are choosing among difficult options, the frontal cortex runs as-if experiments, canvassing your mind's and body's responses to doing X or Y—a thought experiment combined with a (gut) feeling experiment. Mirror neurons, with their putative attunement to how things worked out for observed individuals, would certainly weigh into this process.

Thus mirror neurons might be useful for learning the meaning of a movement, how to carry it out more effectively, and the consequences for someone else who did it. Nonetheless, such neuronal activity is neither necessary nor sufficient for observational learning, especially of the most interesting, abstract human kinds.

Then there's the next, more controversial realm, namely the idea that mirror neurons help you understand what someone else is thinking. This can range from understanding what behavior they are doing to understanding why they are doing it to grasping their larger motivations, all the way to peering into their souls with your mirror neurons. You can see why this has spawned debates.

In this view mirror neurons aid Theory of Mind, mind reading, and perspective taking, suggesting that part of how we understand someone else's world is by simulating (in our minds, in our PMC, in our mirror neurons) their actions.[35] This orients a mirror neuron's world in a very different way from the previous section, where mirroring is to improve your own motor performance and the most pertinent neuroanatomy about mirror neurons in the PMC is their talking to motor neurons that command muscles. In contrast, mirror neurons being concerned with understanding someone else's actions should be talking to Theory of Mind–related brain regions, for which there is evidence.

There was also the suggestion that mirror neuron–mediated perspective taking is particularly concerned with social interactions. Rizzolatti, for example, showed that mirror neuron activity was greater when the observed individual was closer.[36] But importantly, this isn't just literal distance but something resembling "social" distance; as evidence, mirror neuron activity would decrease if there was a transparent barrier between the observer and observed. In Gallese's words, "this shows the relevance of mirror neurons when mapping the potentialities for competition or cooperation between agent and observer."

The notion that mirror neurons aid us in understanding someone else's actions, leading to our understanding someone else, period, has been heavily criticized on two grounds, most notably by Hickok. First is the issue of causality—while some studies show that mirror neuron activity *correlates* with attempts at understanding someone else's perspective, there is minimal evidence that such activity *causes* the understanding. The second criticism concerns something obvious: we can understand the intent behind someone else's actions even if we can't remotely perform them ourselves. This would apply to actions of the observed individual ranging from pole-vaulting eighteen feet to explaining special relativity.

Supporters of this role for mirror neurons admit this but argue that they provide an extra level of understanding. Gallese writes, "I submit that it is only through the activation of Mirror Neurons that we can grasp the meaning of other's behavior *from within*"[37] (my emphasis). This is not my area of research, and I'm not trying to be snarky, but it seems like he's saying that there's understanding and then there's super-duper understanding, and the latter requires mirror neurons.

These mirror neuron speculations have been extended to focus on autism, a disorder in which there are profound impairments in understanding other people's actions and intentions.[38] According to the "broken mirror" hypothesis of mirror neuron pioneer Marco Iacoboni of UCLA, mirror neuron dysfunction underlies those aspects of autism. This has been examined by scads of researchers, with findings varying depending on the paradigm; most meta-analyses conclude that there is nothing flagrantly wrong with the formal features of mirror neuron function in autistic individuals.

Thus, while mirror neurons' activity correlates with attempts to understand other people's actions, their involvement seems neither necessary nor sufficient and is most pertinent to low-level, concrete aspects of such understanding. As for mirror neurons being the portal for peering into someone's soul and attaining super-duper understanding from within, I think things are best summarized by the title of Hickok's well-received 2014 book *The Myth of Mirror Neurons*.[39]

Which leads to the Wild West of mirror neuron–ology, with speculations that mirror neurons are essential to language, aesthetics, consciousness.[40] Most of all, within two seconds of people first hearing about mirror neurons, they started writing reviews where the last paragraph would say something like "Wow, mirror neurons! How cool is that? This opens up all sorts of interesting avenues. Maybe they even explain . . . EMPATHY!"

Sure, why not? Feeling someone's pain is like mirroring their experience, feeling as if you are them. Tailor made, an irresistible idea. And in the decades since mirror neurons' discovery, the "maybe they even explain empathy" reviews have continued. Gallese, for example, nearly twenty years into the mirror neuron era, speculates: "I proposed the mirroring could be a basic functional principle of our brain and that our capacity to empathize with others might be mediated by embodied stimulation mechanisms [i.e., mirroring]." Iacoboni, at the same time, writes, "Mirror neurons are likely cellular candidates for the core layer of empathy." There have been some supportive hints—for example, people who self-report being particularly empathic show stronger mirror neuron–esque responses to matching movements. But for skeptics everything else is mere speculation.[41]

That's disappointing. But worse is people skipping over the "maybe" and concluding that mirror neurons have been *proven* to mediate empathy. Iacoboni, for example, mistakes correlation for causality: "Other studies, however, show that [PMC] activity correlates with empathy even when subjects watch grasping actions without overt emotional content.

Thus, the mirror neuron activity is a *prerequisite* for experiencing empathy (my emphasis)."[42]

A flagrant example of this is the neuroscientist Vilayanur Ramachandran of UC San Diego, one of the most flamboyantly creative people in the business, doing fascinating research on phantom limbs, synesthesia, and out-of-body experiences. He's brilliant but has gotten a bit giddy with mirror neurons. A sampling: "We know that my mirror neurons can literally feel your pain." He's called them "the driving force behind the great leap forward" into human behavioral modernity sixty thousand years ago and famously said, "Mirror neurons will do for psychology what DNA did for biology." I'm not trying to harp on Ramachandran, but how can you resist someone brilliant handing out sound bites like calling mirror neurons "Gandhi neurons"? And this wasn't just in the first heady days of mirror neurons in the early 1990s. Two decades later he stated, "I don't think [the importance of mirror neurons for empathy is] being exaggerated. I think they're being played down, actually."[43]

Ramachandran is certainly not alone. British philosopher Anthony Grayling has gone for the empathy link big time, writing, "We have a great gift for empathy. This is a biologically evolved capacity, as shown by the function of 'mirror neurons.'" In a 2007 *New York Times* article about one man's heroic actions to save another, those cells featured again: "People have 'mirror neurons,' which *make them* feel what someone else is experiencing" (emphasis added). And of course there was my daughter's six-year-old classmate who, upon the class being complimented by their teacher for caring about the planet and cleaning up after their Earth Day cupcake celebration, shouted out, "It's because our neurons have mirrors."[44]

I'd like to think that I'm being a maverick here, ahead of the crowd in terms of crucial thinking, but in recent years most in the field have charged overhype. Psychologist Gary Marcus of NYU calls mirror neurons "the most oversold idea in psychology," philosopher and neuroscientist Patricia Churchland of UCSD calls them the "darling of the don't-look-too-closely crew," and Harvard's Stephen Pinker concludes, "Mirror neurons do not, in fact, explain language, empathy, society, and world peace."[45]

They simply haven't been shown to have much to do with this chapter's concerns.

THE CORE ISSUE: ACTUALLY DOING SOMETHING

The previous chapter considered the world of difference between highfalutin moral reasoning and whether, at a crucial juncture, someone actually does the right thing. As we saw, there is something consistent about that latter type of person: "What were you thinking when you leaped into that river to save the child?" "I wasn't; before I knew it, I'd jumped in." An act of implicit automaticity, the product of a childhood in which doing the right thing was ingrained as an automatic, moral imperative, light-years away from the frontal cortex calculating costs and benefits.

We face a similar situation here, one that is the core of this chapter. Sympathy versus empathy, "for" versus "as if," affect versus cognition, what we do versus what other species do—does any of this actually predict who *does* something compassionate to lessen someone's pains? Similarly, does any of this predict whether the person acting compassionately acts *effectively*, and how much it's an act of *self*-interest? As we'll see, there is a yawning gap between being in an empathic state and acting effectively in a way that is truly selfless.

Doing Something

It is far from guaranteed that an empathic state leads to a compassionate act. One reason for this is captured superbly by the essayist Leslie Jamison:

> *[Empathy] can also offer a dangerous sense of completion: that something has been done because something has been felt. It is tempting to think that feeling someone's pain is necessarily virtuous in its own right. The peril of empathy isn't simply that it can make us feel bad, but that it can make us feel good,*

which can in turn encourage us to think of empathy as an end in itself rather than part of a process, a catalyst.[46]

In such a situation, saying "I feel your pain," becomes a New Age equivalent of the unhelpful bureaucrat saying, "Look, I sympathize with your situation, but . . ." The former is so detached from action that it doesn't even require the "but" as a bridge to the "there's nothing I can/will do." Having your pain validated is swell; having it alleviated is better.

And there's a broader reason why an empathic state may not produce action, first raised in chapter 6 when considering those strange creatures, adolescents. In that discussion I emphasized a wonderful feature of so many adolescents, namely the frenzied feeling of the world's pains, but noted how that intensity often leads to little more than frenzied self-absorption. If instead of imagining how someone else is feeling (an other-oriented perspective), you are imagining how it would feel if this were happening to you (a self-oriented perspective), "you" has just come to the forefront and the main point is that feeling someone's pain feels painful.

The biological substrates of this are clear. Look at someone in pain with the instruction to take a self-oriented perspective, and the amygdala, ACC, and insular cortex activate, along with reports of distress and anxiety. Do the same with an other-oriented perspective, and all are less likely. And the more extreme the former state, the more likely that someone's focus will be to lessen their own distress, to metaphorically look the other way.[47]

This can be predicted with remarkable ease. Expose subjects to evidence of someone else in pain. If their heart rate increases a lot (a peripheral indicator of anxious, amygdaloid arousal), they are unlikely to act prosocially in the situation. The prosocial ones are those whose heart rates decrease; they can hear the sound of someone else's need instead of the distressed pounding in their own chests.*[48]

Thus, if feeling your pain makes me feel awful, I'm likely to just look out for number one, rather than helping you. Likewise if you've got your

* Back to Keltner's work—when comparing the wealthy and the poor, guess whose hearts speed up when they're forced to pay attention to someone else's suffering?

own issues. We saw this earlier, with the demonstration that if you increase people's cognitive load, they become less prosocial toward strangers. Similarly, when people are hungry, they are less charitable—hey, quit bellyaching about your problems; my belly is aching. Make people feel socially excluded and they become less generous and empathic. Stress has the same effect, working via glucocorticoids; Mogil's group (with my involvement) recently showed that if you use a drug to block glucocorticoid secretion, both mice and humans become more empathic toward strangers. Thus, if you feel highly distressed, whether due to resonating with someone else's problems or because of your own, tending to your own needs readily becomes the priority.[49]

In other words, empathic states are most likely to produce compassionate acts when we manage a detached distance. This brings to mind the anecdote from many chapters ago about the Buddhist monk I encountered who said that, yes, sometimes he cuts short his cross-legged meditation because of his knees, but not because he feels them hurting—"I do it as an act of kindness to my knees." And this is certainly in line with the Buddhist approach to compassion, which views it as a simple, detached, self-evident imperative rather than as requiring vicarious froth. You act compassionately toward one individual because of a globalized sense of wishing good things for the world.*

A handful of fascinating studies of Buddhist monks have been carried out, both by Richard Davidson of the University of Wisconsin and Tania Singer of the Max Planck Institutes in Germany. Remarkably, given the science-versus-religion culture wars, such work was given its, er, blessing and facilitated by the Dalai Lama, who is famously intrigued by neuroscience and who has said that if his Dalai Lama gig hadn't come up, he would have wanted to be a scientist or engineer. The most publicized work revolves around the neuroimaging of Matthieu Ricard, a French-born Buddhist monk (who is the Dalai Lama's French translator and who just happens to have a PhD in molecular biology from the Pasteur Institute—this is one interesting guy).[50]

* I am on astonishingly thin ice writing about Buddhist thought, which is why we're now going to quickly transition to the terra firma of considering what neuroscientists have found out about Buddhists.

When confronted with examples of human suffering and instructed to empathically feel the pain of those people, Ricard showed activation of the same circuitry as you'd see in most everyone else. And it was extremely aversive—"The empathic sharing very quickly became intolerable to me and I felt emotionally exhausted," he explained. When instead he did his Buddhist thing, focusing on thoughts of compassion, a totally different picture of activation emerged—the amygdala was silent, and instead there was heavy activation of the mesolimbic dopamine system. He described it as "a warm positive state associated with a strong prosocial motivation."

In other studies volunteers underwent either empathy training (focusing on feeling the pain of someone in distress) or compassion training (focusing on a feeling of warmth and care toward that distressed person).[51] The former would generate the typical neuroimaging profiles, including heavy amygdala activation, and a negative, anxious state. Those with compassion training did not, showing heavy activation instead in the (cognitive) dlPFC, coupling of activation between the dlPFC and dopaminergic regions, more positive emotions, and a greater tendency toward prosociality.

Okay, caveats. This is a tiny literature (i.e., not much larger than the study of Ricard). And all-star Buddhist monks apparently meditate eight hours a day, not a trivial path to take. The point is merely to emphasize this scenario of detachment. Which brings us to the next issue, which is whether compassionate acts fostered by empathy are necessarily useful.

Doing Something Effectively

In a provocatively titled 2014 article, "Against Empathy," Paul Bloom explored the ways in which empathy can lead to compassionate acts that are far from ideal.

There is the realm of what has been termed "pathological altruism," the type associated with codependency.[52] This is the scenario of someone so consumed with the vicarious pain of a loved one that they endure and facilitate his dysfunction rather than administering tough love. Then there's the danger that the empathic pain is so intense that you can only come up with solutions that would work for you, rather than ones that might help the sufferer. And there is the problem of empathy impeding your doing what's necessary—it's not great if a parent is so vicariously distressed by their child being distressed that they forgo vaccinations. A large piece of the training of health-care professionals is teaching them to keep empathy at bay.* For example, the various behavioral and neurobiological responses to seeing someone poked with a needle do not occur in acupuncturists. As Jamison describes, when anxiously seeing a doctor about something worrisome, "I needed to look at him and see the opposite of my fear, not its echo."

Bloom also emphasizes how highly aroused empathy pushes us toward psychologically easy acts that generate the least cognitive load. In those times suffering that is local, that concerns an identified appealing individual, and that is of a type with which you're familiar readily counts for more than suffering that is distant, involves a group, and is an alien form of pain.† Aroused empathy produces tunnel-visioned compassion that can wind up misplaced. As the philosopher Jesse Prinz emphasizes, the point is not whose pain pains us the most but who most needs our help.

Are There Ever Any Bloody Altruists?

Stop the presses; science has proven that it can feel good to do good, complete with activation of the mesolimbic dopamine system. This doesn't even require a brain scanner. In a 2008 study in *Science*, subjects were given either five dollars or twenty dollars; half were instructed to spend it that day on themselves, half on someone else (ranging from a friend to a charity). And comparisons of self-assessments of happiness at the beginning and end of the day showed that neither the larger amount

* With the hope that the distancing thoughts are along the lines of "This is how I do good," rather than, say, "I think I'll have a chicken salad sandwich for lunch."

† A colleague of mine used to sardonically talk about his hope that the spouse of some senator would come down with the neurological disease that he studied—then, finally, someone powerful would empathize with sufferers of that disease and steer more grant money in the direction of research.

of money nor the opportunity to spend it on oneself increased happiness; only spending it on someone else did. And particularly interesting is that other subjects, told about the design, predicted the opposite—that happiness would be raised most by spending on oneself, and that twenty dollars would buy more happiness than five.[53]

The question, of course, is why doing good can feel good, which raises the classic question: is there ever a selfless act that contains no element of self-interest? Does doing good feel good because there's something in it for you? I'm sure not going to tackle this from a philosophical perspective. For biologists the most frequent stance is anchored in chapter 10's evolutionary view of cooperation and altruism, one that always contains some element of self-interest.

Is this surprising? Pure selflessness is clearly going to be an uphill battle if the very part of the brain most central to an empathic state—the ACC—evolved to observe and learn from others' pain for your own benefit.[54] The self-oriented rewards of acting compassionately are endless. There's the interpersonal—leaving the beneficiary in your debt, thus turfing this from altruism to reciprocal altruism. There are the public benefits of reputation and acclaim—the celebrity swooping into a refugee camp for a photo op with starving kids made joyful by her incandescent presence. There's that strange version of reputation that comes in the rare cultures that have invented a moralizing god, one who monitors human behavior and rewards or punishes accordingly; as we saw in chapter 9, it is only when cultures get large enough that there are anonymous interactions among strangers that they tend to invent moralizing gods. A recent study shows that across a worldwide range of religions, the more people perceive their god(s) to monitor and punish, the more prosocial they are in an anonymous interaction. Thus there is the self-interested benefit of tipping the cosmic scale in your favor. And probably most inaccessibly, there is the purely internal reward of altruism—the warm glow of having done good, the lessened sting of guilt, the increased sense of connection to others, the solidifying sense of being able to include goodness in your self-definition.

Science has been able to catch the self-interest component of empathy in the act.[55] As noted, some of the self-interest reflects concerns about self-definition—personality profiles show that the more charitable people are, the more they tend to define themselves by their charitability. Which comes first? It's impossible to tell, but highly charitable people tend to have been brought up by parents who were charitable and who emphasized charitable acts as a moral imperative (particularly in a religious context).

How about the self-interested reputational rewards of being altruistic, the cachet of conspicuous largesse rather than conspicuous consumption? As emphasized in chapter 10, people become more prosocial when reputation rides on it, and personality profiles also show that highly charitable people tend to be particularly dependent on external approval. Two of the studies just cited that showed dopaminergic activation when people were being charitable came with a catch. Subjects were given money and, while in a brain scanner, decided whether to keep the money or donate. Being charitable activated dopamine "reward" systems—when there was an observer present. When no one was present, dopamine tended to flow most when subjects kept the money for themselves.

As emphasized by the twelfth-century philosopher Moses Maimonides, the purest form of charity, the most stripped of self-interest, is when both the giver and the recipient are anonymous.* And, as shown in those brain scanners, this is perhaps the rarest form as well.

Intuitively, if good acts must be motivated by self-interest, the reputational motive, the desire to be the biggest spender at the charity auction, seems most worthy of irony. In contrast, the motivation to think of yourself as a good person seems a pretty benign one. After all, we're all

---

* I once benefited in a Maimonides-esque scenario when I, sitting on a toilet in a Starbucks, discovered much too late that there was no toilet paper. Soon someone else entered; hearing him rummaging around one of the urinals, I tentatively begged for a charitable act—"Uh, hey, when you're done, could you tell the people at the counter that there is no toilet paper here?" "Sure," answered the anonymous voice, and soon a barista's hand appeared underneath the toilet stall door offering, if not alms for the poor, TP for the stranded. The trick now is how to re-create this scenario with subjects in brain scanners. This may not, in fact, have been the perfect anonymous interaction. While the Good Samaritan who carried my message and I were anonymous to each other, he wasn't to the baristas. And for all I know, they promptly gave him a free latte or praised him in song or offered to mate with him. So now we need to know whether the guy expected any/all of these things to happen when he agreed to help me. More research is needed.

searching for a sense of self, and better that particular sense than to assure yourself that you're tough, scary, and not to be messed with.

Is the element of self-interest ever truly absent? One 2007 study in *Science* examined this.[56] Subjects (in brain scanners, of course) were unexpectedly given varying amounts of money. Then, some of the time they were "taxed" (i.e., told that a certain percentage of that money would be forcibly given to a food bank), some of the time given the opportunity to donate that amount voluntarily. In other words, the exact same amount of public "good" was accomplished in each case, but the former constituted enforced civic duty while the latter was a purely charitable act. Thus, if someone's altruism is purely other-oriented, without a smidgen of self-interest, the two circumstances are psychologically identical—those in need are being helped, and that's all that matters. And the more different the scenarios feel, the more self-interest is coming into play.

The results were complex and interesting:

a. The more people's dopaminergic reward systems activated when they unexpectedly received money, the less activation there was when they were either taxed or asked to donate. In other words, the greater the love of money, the more painfully it is parted with. No surprise there.

b. The more dopaminergic activation there was when someone was taxed, the more voluntarily charitable they were. Being taxed could not have been welcome to the most self-interested—money was being taken from them. For subjects who instead showed heavy activation of dopaminergic systems in that circumstance, any self-interest of losing money was more than compensated for by the knowledge that people in need were being helped. This taps into the last chapter's exploration of inequity aversion and is consistent with findings that in some circumstances, when a pair of strangers are openly given unequal amounts of reward, there is typically dopaminergic activation in the one with the good luck when some of the reward is transferred afterward to make things more even. Thus it's little surprise in the present study that subjects made happy by reducing inequity, even at a cost to themselves, were also the most charitable. The authors appropriately interpret this as reflecting a compassionate act with elements independent of self-interest.[57]

c. There was more dopaminergic activation (and more self-reports of satisfaction) when people gave voluntarily than when they were taxed. In other words, a component of the charitability was about self-interest—it was more pleasing when those in need were helped by voluntary efforts than when giving was forced.

What does this show? That we're reinforced by varying things and to varying extents—getting money, knowing that the needy are being cared for, feeling the warm glow of doing a good thing. And that it is rare to be able to get the second type of pleasure with no dependence on the third—it appears to truly be rare to scratch an altruist and see an altruist bleed.

CONCLUSIONS

All things considered, it is a pretty remarkable thing that when an individual is in pain, we (i.e., we humans, primates, mammals) often are induced to be in a state of pain also. There have been some mighty interesting twists and turns for that one to have evolved.

But at the end of the day, the crucial issue is whether an empathic state actually produces a compassionate act, to avoid the trap of empathy being an end unto itself. The gap between the state and the act can be enormous, especially when the goal is for the act to be not only effective but also pristine in its motives.

For someone reading this book, a first challenge in bridging that gap is that much of the world's suffering is felt by distant masses experiencing things that we haven't an inkling of—diseases that don't touch us; poverty that precludes clean water, a place to live, the certainty of a next meal; oppression at the hands of political systems that we've been spared;

strictures due to repressive cultural norms that might as well be from another planet. And everything about us makes those the hardest scenarios for us to actually act—everything about our hominin past has honed us to be responsive to one face at a time, to a face that is local and familiar, to a source of pain that we ourselves have suffered. Yes, best that our compassion be driven by the most need rather than by the most readily shared pain. Nevertheless, there's no reason why we should expect ourselves to have particularly good intuitions when aiming to heal this far-flung, heterogeneous world. We probably need to be a bit easier on ourselves in this regard.

Likewise, we should perhaps ease up a bit on the scratching-an-altruist problem. It has always struck me as a bit mean-spirited to conclude that it is a hypocrite who bleeds. Scratch an altruist and, most of the time, the individual with unpure motives who bleeds is merely the product of "altruism" and "reciprocity" being evolutionarily inseparable. Better that our good acts be self-serving and self-aggrandizing than that they don't occur at all; better that the myths we construct and propagate about ourselves are that we are gentle and giving, rather than that we prefer to be feared than loved, and that we aim to live well as the best revenge.

Finally, there is the challenge of a compassionate act being left by the wayside when the empathic state is sufficiently real and vivid and awful. I'm not advocating that people become Buddhists in order to make the world a better place. (Nor am I advocating that people *don't* become Buddhists; what is the sound of one atheist waffling?) Most of us typically require moments of piercing, frothing shared pain to even notice those around us in need. Our intuitions run counter to doing it any other way—after all, just as one of the most frightening versions of humans at their worst is "cold-blooded" killing, one of the most puzzling and even off-putting of us at our best is "cold-blooded" kindness. Yet, as we've seen, a fair degree of detachment is just what is needed to actually act. Better that than our hearts racing in pained synchrony with the heart of someone suffering, if that cardiovascular activation mostly primes us to flee when it all becomes just too much to bear.

Which brings us to a final point. Yes, you don't act because someone else's pain is so painful—that's a scenario that begs you to flee instead. But the detachment that should be aimed for doesn't represent choosing a "cognitive" approach to doing good over an "affective" one. The detachment isn't slowly, laboriously thinking your way to acting compassionately as an ideal utilitarian solution—the danger here is the ease with which you can instead think your way to conveniently concluding this isn't your problem to worry about. The key is neither a good (limbic) heart nor a frontal cortex that can reason you to the point of action. Instead it's the case of things that have long since become implicit and automatic—being potty trained; riding a bike; telling the truth; helping someone in need.

Fifteen

Metaphors We Kill By

EXAMPLE 1

Stretching back at least to that faux pas about the golden calf at Mt. Sinai, various branches of Abrahamic religions have had a thing about graven images. Which has given us aniconism, the banning of icons, and iconoclasts, who destroy offensive images on religious grounds. Orthodox Judaism has been into that at times; ditto for Calvinists, especially when it came to those idolatrous Catholics. Currently it's branches of Sunni Islam that deploy literal graven-image police and consider the height of offense to be images of Allah and Muhammad.

In September 2005 the Danish newspaper *Jyllands-Posten* published cartoon images of Muhammad on its editorial page. It was a protest against Danish censorship and self-censorship about the subject, against Islam being a sacred cow in a Western democracy where other religions are readily criticized satirically. None of the cartoons suggested reverence or respect. Many explicitly linked Muhammad with terrorism (e.g., him wearing a bomb as a turban). Many were ironic about the ban—Muhammad as a stick figure with a turban, Muhammad (armed with a sword) with a blackened rectangle over his eyes, Muhammad in a police lineup alongside other bearded men with turbans.

And as a direct result of the cartoons, Western embassies and consulates were attacked, even burned, in Lebanon, Syria, Iraq, and Libya. Churches were burned in northern Nigeria. Protesters were killed in Afghanistan, Egypt, Gaza, Iran, Iraq, Lebanon, Libya, Nigeria, Pakistan, Somalia, and Turkey (typically either by mob stampedes or by police containing rioters). And non-Muslims were killed in Nigeria, Italy, Turkey, and Egypt as revenge for the cartoons.

In July 2007 drawings by a Swedish artist of Muhammad's head with a dog's body provoked much the same. In addition to deadly protests, the Islamic State of Iraq offered $100,000 for the artist's killing, Al-Qaeda targeted the artist for death (along with staffers from *Jyllands-Posten*), assassination plots were stopped by Western authorities, and one attempt killed two bystanders.

In May 2015 two gunmen attacked an antianiconist event in Texas where a $10,000 prize was offered for the "best" depiction of Muhammad. One person was injured before the gunmen were killed by police.

And, of course, on January 7, 2015, two brothers, French-born sons of Algerian immigrants, massacred the staff of *Charlie Hebdo*, killing twelve.

EXAMPLE 2

In the Battle of Gettysburg fierce fighting occurred between the Union First Minnesota Volunteer Infantry and Confederate Twenty-eighth Virginia Volunteer Infantry Regiment.[1] At one point Confederate soldier John Eakin, carrying the regimental flag of the Twenty-eighth Virginia, was shot three times (a typical fate of soldiers carrying the colors, who were preferential targets). Mortally wounded, Eakin handed the flag to a comrade, who was promptly killed. The flag was then taken up and displayed by Colonel Robert Allen, who was soon killed, then by Lieutenant John Lee, who was soon injured. A Union soldier, attempting to seize the colors, was killed by Confederates. Finally Private Marshall Sherman of the First Minnesota captured the flag, along with Lee.

EXAMPLES 3, 4, AND 5

In mid-2015 Tavin Price, a mentally challenged nineteen-year-old, was killed by gangbangers in Los Angeles for wearing red shoes, a rival gang's color. His dying words, in front of his mother, were "Mommy, please. I don't want to die. Mommy, please."[2]

In October 1980 Irish Republican prisoners at the Maze Prison in Northern Ireland began a hunger strike protesting, among other things, their being denied political-prisoner status by having to wear prison garb. The British government acceded to their demands as a first prisoner slipped into a coma fifty-three days later. In a similar strike a year later in the Maze, ten Irish political prisoners starved themselves to death over forty-six to seventy-three days.

By 2010 karaoke clubs throughout the Philippines had removed the Frank Sinatra song "My Way" from their playlists because of violent responses to the singing of it, including a dozen killings. Some of the "'My Way' killings" were due to poor renditions (which apparently often results in killings), but most were thought linked to the macho lyrics. "'I did it my way'—it's so arrogant. The lyrics evoke feelings of pride and arrogance in the singer, as if you're somebody when you're really nobody. It covers up your failures. That's why it leads to fights," explained the owner of a singing school in Manila to the *New York Times*.

In other words, people are willing to kill or be killed over a cartoon, a flag, a piece of clothing, a song. We have some explaining to do.

Throughout this book we've repeatedly gained insights into humans by examining other species. Some of the time the similarities have been most pertinent—dopamine is dopamine in a human or a mouse. Sometimes the interesting thing is our unique use of the identical substrate—dopamine facilitates a mouse's pressing of a lever in the hopes of getting some food and a human's praying in the hopes of entering heaven.

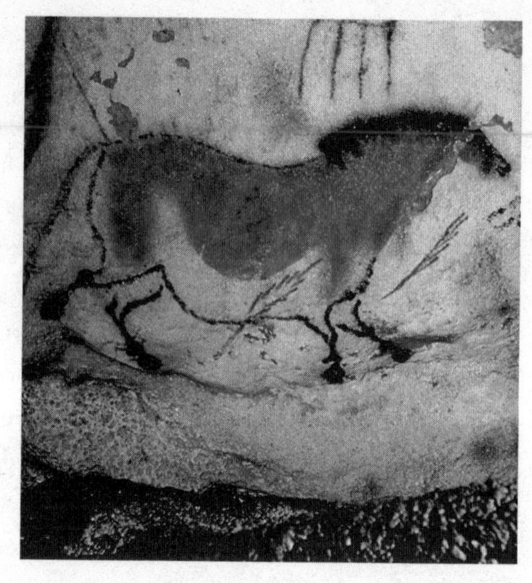

But some human behaviors stand alone, without precedent in another species. One of the most important realms of human uniqueness comes down to one simple fact, namely that this is not a horse:

Anatomically modern humans emerged around 200,000 years ago. But behavioral modernity had to wait more than another 150,000 years, as evidenced by the appearance in the archaeological record of composite tools, ornamentation, ritualistic burial, and that stunning act of putting pigment on the wall of a cave.*[3] This is not a horse. It's a great *picture* of a horse.

When René Magritte placed the words "*Ceci n'est pas une pipe*" ("This is not a pipe") beneath a picture of a pipe, in his 1928 painting *The Treachery of Images*, he was highlighting the shaky nature of imagery. The art historian Robert Hughes writes that this painting is a "visual

* Lest we get carried away with ourselves, there is good evidence that some of the most impressive cave paintings were done by Neanderthals rather than humans. But by now, who cares about those silly species designations, what with all the human/Neanderthal mating now shown to have been going on?

booby-trap" set off by thought, and that "this sense of slippage between image and object is one of the sources of modernist disquiet."[4]

Magritte's goal was to magnify and play with the distance between an object and its representation; these are coping mechanisms against that modernist disquiet. But for that human putting pigment to wall in Lascaux Cave more than seventeen thousand years ago, the point was the opposite: to minimize the distance between the two, to be as close as possible to possessing the real horse. As we say, to *capture* its likeness. To gain its power, as imbued in a symbol.

The clearest human mastery of symbolism comes with our use of language. Suppose you are being menaced by something and thus scream your head off. Someone listening can't tell if the blood-curdling "Aiiiii!" is in response to an approaching comet, suicide bomber, or Komodo dragon. It just means that things are majorly not right; the message is the meaning. Most animal communication is about such present-tense emotionality.

Symbolic language brought huge evolutionary advantages. This can be seen even in the starts of symbolism of other species. When vervet monkeys, for instance, spot a predator, they don't generically scream. They use distinct vocalizations, different "protowords," where one means "Predator on the ground, run up the tree!" and another means "Predator in the air, run down the tree!" Evolving the cognitive capacity to make that distinction is mighty useful, as it prompts you to run away from, rather than toward, something intent on eating you.

Language pries apart a message from its meaning, and as our ancestors improved at this separation, advantages accrued.[5] We became capable of representing past and future emotions, as well as messages unrelated to emotion. We evolved great expertise at separating message from reality, which, as we've seen, requires the frontal cortex to regulate the nuances of face, body, and voice: lying. This capacity creates complexities that no one else—from slime mold to chimp—deals with in life's Prisoner's Dilemmas.

The height of the symbolic features of language is our use of metaphor. And this is not just flourish metaphors, when we declare that life is a bowl of cherries. Metaphors are everywhere in language—we may literally and physically be "in" a room, but we are only metaphorically inside something when we are "in" a good mood, "in" cahoots with someone, "in" luck, a funk, a groove,* or love. We are only metaphorically standing under something when we "understand" it.†[6] The renowned cognitive linguist George Lakoff of UC Berkeley has explored the ubiquity of metaphor in language in books such as *Metaphors We Live By* (with philosopher Mark Johnson), and *Moral Politics: How Liberals and Conservatives Think* (where he demonstrates how political power involves controlling metaphors—do you favor "choice" or "life"? are you "tough on" crime, or does your "heart bleed"? are you loyal to a "fatherland" or a "motherland"? and have you captured the flag of "family values" from your opponent?). For Lakoff language is always a metaphor, transferring information from one individual to another by putting thought *into* words, as if words were shopping bags.[7]

Symbols, metaphors, analogies, parables, synecdoche, figures of speech. We understand that a captain wants more than just hands when ordering all of them on deck, that Kafka's *Metamorphosis* isn't really about a cockroach, and that June doesn't really bust out all over. If we are of a certain theological ilk, we see bread and wine intertwined with body and blood. We learn that the orchestral sounds constituting the *1812 Overture* represent Napoleon getting his ass kicked when retreating from Moscow. And that "Napoleon getting his ass kicked" represents thousands of soldiers dying cold and hungry, far from home.

This chapter explores the neurobiology of some of the most interesting outposts of symbolic and metaphorical thinking. It makes a key point: these capacities evolved so recently that our brains are, if you will, winging it and improvising on the fly when dealing with metaphor. As a result, we are actually pretty lousy at distinguishing between the

* Let alone being so much inside said groove as to be grooved yourself, i.e., groovy.

† Just consider what is inherent in the fact that numerous languages worldwide have grammatical genders, with some nouns designated as masculine, others as feminine. The cognitive scientist Lera Boroditsky has shown how grammatical gender can influence thought. In one study she showed that German speakers tend to associate the word "bridge" (which is feminine in German) with attributes such as "beautiful," "elegant," or "slender," while Spanish speakers (for whom "bridge" is masculine) tend toward associations with "big," "strong," "towering," and "sturdy."

metaphorical and literal, at remembering that "it's only a figure of speech"—with enormous consequences for our best and worst behaviors.

We start with examples of odd ways our brains handle metaphor, and the behavioral manifestations of those oddities; some have been introduced previously.

FEELING SOMEONE ELSE'S PAIN

Consider the following: You stub your toe. Pain receptors there send messages to the spine and on up to the brain, where various regions kick into action. Some of these areas tell you about the location, intensity, and quality of the pain. Is it your left toe or right ear that hurts? Was your toe stubbed or crushed by a tractor-trailer? These various pain-ometers, the meat and potatoes of pain processing, are found in every mammal.

As we first learned in chapter 2, the frontal cortical anterior cingulate cortex (ACC) also plays a role, assessing the meaning of the pain.[8] Maybe it's bad news: your painful toe signals the start of some unlikely disease. Or maybe it's good news: you're going to get your fire-walker diploma because the hot coals only made your toes throb. As we saw in the last chapter, the ACC is heavily involved in "error detection," noting discrepancies between what is anticipated and what occurs. And pain from out of nowhere surely represents a discrepancy between the pain-free setting that you anticipate versus a painful reality.

But the ACC does more than just tell you the meaning of a painful toe. As we saw in chapter 6, put a subject in a brain scanner, make them think they're tossing a Cyberball back and forth with two other players, and then make them feel excluded—the other two stop throwing the ball to them. "Hey, how come they don't want to play with me?" And the ACC activates.

In other words, rejection hurts. "Well, yeah," you might say. "But that's not like stubbing your toe." But as far as those neurons in the ACC are concerned, social and literal pain are the same. And as proof of the rooting of the former in sociality, there isn't ACC activation if the subject believes the ball isn't being thrown to them because of a glitch connecting them to the other two subjects' computers.

And the ACC can take things a step further, as we saw in chapter 14. Receive a mild shock, and there's activation of your ACC (along with activation of the more mundane pain-ometer regions). Now instead watch your beloved get shocked in the same way. Pain-ometer brain regions are silent, but the ACC activates. For those neurons, feeling someone else's pain isn't just a figure of speech.

Moreover, the brain intermixes literal and psychic pain.[9] The neurotransmitter substance P plays a central role in communicating painful signals from pain receptors in skin, muscles, and joints up into the brain. It's got pain-ometer written all over it. And remarkably, its levels are elevated in clinical depression, and drugs that block the actions of substance P can have marked antidepressant properties. Stubbed toe, stubbed psyche. Moreover, there is activation of the cortical parts of pain networks when we feel dread—anticipating an impending shock.

Furthermore, the brain becomes literal when we do the flip side of empathy.[10] It's painful watching a hated competitor succeed, and we activate the ACC at that time. Conversely, if he fails, we gloat, feel schadenfreude, get pleasure from his pain, and activate dopaminergic reward pathways. Forget "Your pain is my pain." Your pain is my gain.

DISGUST AND PURITY

This is our familiar domain of the insular cortex. If you bite into rancid food, the insula activates, just as in every other mammal. You wrinkle your nose, raise your upper lip, narrow your eyes, all to protect mouth, eyes, and nasal cavities. Your heart slows. You reflexively spit out the food, gag, perhaps even vomit. All to protect yourself from toxins and infectious pathogens.[11]

As humans we do some fancier things: Think about rancid food, and the insula activates. Look at faces showing disgust, or subjectively unattractive faces, and the same occurs. And most important, if you think about a truly reprehensible act, the same occurs. The insula mediates visceral responses to norm violations, and the more activation, the more condemnation. And this is visceral, not just metaphorically visceral—for example, when I heard about the Sandy Hook Elementary School massacre, "feeling sick to my stomach" wasn't a mere figure of speech. When I imagined the reality of the murder of twenty first-graders and the six adults protecting them, I *felt* nauseous. The insula not only prompts the stomach to purge itself of toxic food; it prompts the stomach to purge the reality of a nightmarish event. The distance between the symbolic message and the meaning disappears.[12]

The linking of visceral and moral disgust is bidirectional. As shown in a number of studies, contemplating a morally disgusting act leaves more than a metaphorical bad taste in your mouth—people eat less immediately afterward, and a neutral-tasting beverage drunk afterward is rated as having a more negative taste (and, conversely, hearing about virtuous moral acts made the drink taste better).[13]

In chapters 12 and 13 we saw the political implications of our brains intermixing visceral and moral disgust—social conservatives have a lower threshold for visceral disgust than do social progressives; the "wisdom of repugnance" school posits that being viscerally disgusted by something is a pretty good indicator that it is morally wrong; implicitly evoking a sense of visceral disgust (e.g., by sitting in close proximity to a foul odor) makes us more socially conservative.[14] This is not merely because visceral disgust is an aversive state—inducing a sense of sadness, rather than disgust, doesn't have the same effect; moreover, moralizing about purity, while predicted by people's propensity toward feeling disgust, is not predicted by propensities toward fear or anger.*

The physiological core of gustatory disgust is to protect yourself against pathogens. The core of the intermixing of visceral and moral disgust is

* Interestingly, harking back to earlier chapters on hierarchy and status, the authors also found that being of lower socioeconomic status predicted a greater degree of moralizing purity, but not moralizing justice or harm avoidance.

a sense of threat as well. A socially conservative stance about, say, gay marriage is not just that it is simply wrong in an abstract sense, or even "disgusting," but that it constitutes a threat—to the sanctity of marriage and family values. This element of threat is shown in a great study in which subjects either did or didn't read an article about the health risks of airborne bacteria.[15] All then read a history article that used imagery of America as a living organism, with statements like "Following the Civil War, the United States underwent a growth spurt." Those who read about scary bacteria before thinking about the United States as an organism were then more likely to express negative views about immigration (without changing attitudes about an economic issue). My guess is that people with a stereotypically conservative exclusionary stance about immigration rarely have the sense that they feel disgusted that people elsewhere in the world would want to come to the United States for better lives. Instead there is threat by the rabble, the unwashed masses, to the nebulous entity that is the American way of life.

How cerebral is this intertwining of moral and visceral disgust? Does the insula get involved in moral disgust only if it's of a particularly visceral nature—blood and guts, coprophagia, body parts? Paul Bloom suggests this is the case. In contrast, Jonathan Haidt feels that even the most cognitive forms of moral disgust ("He's a chess grand master and he shows off by beating that eight-year-old in three moves and reducing her to tears—that's disgusting") are heavily intertwined.[16] In support of that, something as unvisceral as getting a lousy offer in an economic game activates the insula (a lousy offer from another human, rather than a computer, that is); the more insula activation, the greater the likelihood of the offer being rejected. Amid this debate, it is clear that the intertwining of visceral and moral disgust is, at the least, greatest when the latter taps into core disgust. To repeat a neat quote from Paul Rozin, introduced in chapter 11, "Disgust serves as an ethnic or out-group marker." First you're disgusted by how Others smell, a gateway to then being disgusted by how Others think.

Of course, insofar as metaphorically being dirty and disorderly = bad, metaphorically being clean and orderly = good.*[17] Just consider the use of the word "neat" in the previous paragraph. Similarly, in Swahili the word *safi*, meaning "clean" (from *kusafisha*, "to clean"), is used in the same slangy metaphorical sense of "neat" in English. Once while in Kenya, I was hitching a ride to Nairobi from somewhere out in the boondocks and got to chatting with a local teenager who was curious about me. "Where are you going?" he asked. Nairobi. "Nairobi *ni* [is] *safi*," he said wistfully about the far-off metropolis. How are you going to keep them down on the farm once they've seen the neatness of Nairobi?

Literal cleanliness and orderliness can release us from abstract cognitive and affective distress—just consider how, during moments where life seems to be spiraling out of control, it can be calming to organize your clothes, clean the living room, get the car washed.[18] And consider how the displaced need to impose cleanliness and order runs and ruins the lives of people suffering from the archetypal anxiety disorder, obsessive-compulsive disorder. The ability of literal cleanliness to alter cognition was shown in one study. Subjects examined an array of music CDs, picked ten that they liked, and ranked them in order of liking; they were then offered a free copy of one of their midrange choices (number five or six). Subjects were then distracted with some other task and then asked to rerank the ten CDs. And they showed a common psychological phenomenon, which was to now overvalue the CD they'd been given, ranking it higher on the list than before. Unless they had just washed their hands (ostensibly to try a new brand of soap), in which case no reranking occurred. Clean hands, clean slate.

But beginning much further back than the "social hygiene" movement of the turn of the twentieth century, being metaphorically neat, pure, and hygienic could be a moral state as well—cleanliness was not just a good way to avoid uncontrolled diarrhea, dehydration, and serious electrolyte imbalance, but was also ideal for cozying up to a god.

---

* Which harks back to our confusing goodness with beauty (thus giving lesser jail sentences to those people with symmetrical faces, etc.). As first introduced in chapter 3, we use similar brain circuits, activating the medial orbitofrontal cortex, when contemplating whether an act is moral as when contemplating whether a face is beautiful.

One study was built around the phenomenon of visceral disgust making people harsher in their moral judgments. The authors first replicated this effect, showing that watching a short film clip of something physically disgusting made subjects more morally judgmental—unless they had washed their hands after watching the film. Another study suggests that the washing decreases emotional arousal, as it decreased the diameter of subjects' pupils.[19]

We intertwine physical and moral purity when it comes to our own actions. In one of my all-time favorite psychology studies, Chen-Bo Zhong of the University of Toronto and Katie Liljenquist of Northwestern University demonstrated that the brain has trouble distinguishing between being a dirty scoundrel and being in need of a bath. Subjects were asked to recount either a moral or an immoral act in their past. Afterward, as a token of appreciation, the researchers offered the volunteers a choice between the gift of a pencil and a package of antiseptic wipes. And the folks who had just wallowed in their ethical failures were more likely to go for the wipes. Another study, showing the same effect when people were instructed to lie, demonstrated that the more adversely consequential the lie was presented as being, the more washing subjects did. Lady Macbeth and Pontius Pilate weren't the only ones to at least try to absolve their sins by washing their hands, and this phenomenon of embodied cognition is referred to as the "Macbeth effect."[20]

This effect is remarkably concrete. In another study subjects were instructed to lie about something—with either their mouths (i.e., to tell a lie) or their hands (i.e., to write down a lie).[21] Afterward, remarkably, liars were more likely to pick complementary cleansing products than control subjects who communicated something truthful: the immoral mouth-ers were more likely to pick a mouthwash sample; the immoral scribes, hand soap. Furthermore, as shown with neuroimaging, when contemplating mouthwash versus soap, those who had just spoken a lie activated parts of the sensorimotor cortex related to the mouth (i.e., the subjects were more aware of their mouths at the time); those who had written the lie activated the cortical regions mapping onto their hand. Embodied cognition can be specific to parts of the body.

Another fascinating study showed the influence of culture in the Macbeth effect. The studies just cited were carried out with European or American subjects. When the same is done with East Asian subjects, the urge afterward is to wash the face, rather than the hands. If you are going to save face, it should be a clean one.[22]

Finally, most important, this intermixing of moral and physical hygiene affects the way we actually *behave.* That original study on contemplating one's moral failings and the subsequent desire to wash hands included a second experiment. As before, subjects were told to recall an immoral act of theirs. Afterward subjects either did or didn't have the opportunity to clean their hands. Those who were able to wash were less likely to respond to a subsequent (experimentally staged) request for help. In another study merely watching someone else wash their hands in this situation (versus watching them type) also decreased helpfulness afterward (although to a lesser extent than the subject washing).[23]

Many of our moments of prosociality, of altruism and Good Samaritanism, are acts of restitution, attempts to counter our antisocial moments. What these studies show is that if those metaphorically dirtied hands have been unmetaphorically washed in the interim, they're less likely to reach out to try to balance the scales.

REAL VERSUS METAPHORICAL SENSATION

Then there are ways in which we confuse literal with metaphorical sensation.

A brilliant study by John Bargh of Yale concerned haptic sensations (I had to look the word up—haptic: related to the sense of touch). Volunteers evaluated the résumés of supposed job applicants; crucially, the résumé was attached to a clipboard of one of two weights. When subjects held the heavier clipboard, they tended to judge candidates as more "serious" (while clipboard weight had no effect on other perceived traits). When you next apply for a job, hope that your résumé will be attached to a heavy clipboard. How else would the evaluator figure out that you can appreciate the gravity of a situation and deal with weighty matters, rather than being a lightweight?[24]

In the next study subjects assembled a puzzle with pieces that were either smooth or rough as sandpaper, then observed a socially ambiguous interaction. Handle the rough puzzle pieces and the interactions were rated as less coordinated, smooth, or successful (it's not clear, however, if those subjects were more likely, at home that evening, to use coarse language in describing their rough day).

Next, subjects sat in either a hard or a soft chair (to quote the authors, "We primed subjects by the seat of their pants"). Sit in the former and they were more likely to perceive individuals as stable and unemotional, to be less flexible in economic game play. This is remarkable—haptic sensations in your *butt* influencing whether you think someone is a hard-ass. Or hard-hearted instead of a softie.

Similar intermixing of the real and the metaphorical occurs with temperature sensation. In another study from Bargh's group, the researcher, hands full with something, would ask a subject to briefly hold a cup of coffee for them. Half the subjects held warm coffee, half iced coffee. Subjects then read about some individual and answered questions about them. Subjects who held the warm cup rated the individual as having a warmer personality (without altering ratings about other characteristics). In the next part of the study, the temperature of a held object altered subjects' generosity and levels of trust—cold hands, cold heart. And a more activated insula, as shown in a follow-up study.[25]

Our brains also confuse metaphorical and literal interoceptive information. Recall that remarkable study showing that in a real-world situation, a major predictor of whether a prisoner would be granted parole was how recently the judge had eaten. Empty stomach, harsher judgment. Other work has shown that when people are hungry, they become less generous with money and show more future discounting (i.e., are more likely to want reward X now, rather than wait for reward 2X). Hungering for fame and fortune are just metaphors—yet our brain pulls circuits related to real hunger into the mix. Moreover, we use more abstract levels of cognition when thinking about distant events. Ask people to make a list of the items they'd bring on a camping trip taking place either tomorrow or in a month; if the former, the list contains more specific subcategories. In another study subjects were shown a graph of the average amount of paper used by an office over time. There is a steady increase until the most recent time period:[26]

Subjects were then asked to predict what would happen in the next time period. Half the subjects were told that the office was nearby. Result: those subjects did a microanalysis, preferentially paying attention to that final X trending downward, perceiving it to be meaningful, the start of a pattern:

Down the hall

But subjects told that the office was on the other side of the planet tended to view the data points at a macro level of analysis, paying attention to the overall pattern and seeing that downturn as a mere aberration:

Far away

What's going on in these studies? Metaphors about weight, density, texture, temperature, interoceptive sensations, time, and distance are just figures of speech. Yet the brain confusedly processes them with some of the same circuits that deal with the physical properties of objects.

DUCT TAPE

The essence of a symbol is its ability to serve as a stand-in for the real thing and, remarkably, we're not the only species where a signifier, independent of what it signifies, can gain a power of its own. As discussed in chapter 2, if you condition a rat to associate a bell with a reward, about half of rats eventually come to find the bell itself rewarding.

So we've now examined cold drinks and cold personalities; lying through your teeth and then yearning for mouthwash; our hearts aching for someone else's pain. Our metaphorical symbols can gain a power all their own. But insofar as metaphors are the apogee of our capacity for symbolic thought, it's thoroughly weird that our top-of-the-line brains can't quite keep things straight and remember that those metaphors aren't literal. Why?

The answer harks back to a concept first introduced in chapter 10—evolution is a tinkerer, an improviser. So humans are evolving capacities for abstractions like morality and deep violations of it, for experiencing empathy of unprecedented intensity, and for conscious assessment of the affiliative nature of someone's temperament—moral disgust, feeling someone's pain, warm and cold personalities. Given how short a time behaviorally modern humans have existed, this has occurred in a blink of an eye. There hasn't been enough time to evolve completely new brain regions and circuits for handling these novelties. Instead, tinkering occurred—"Hmm, extreme negative affect elicited by violations of shared behavioral norms. Let's see . . . Who has any pertinent experience? I know, the insula! It does extreme negative sensory stimuli—that's, like, all that it does—so let's expand its portfolio to include this moral disgust business. That'll work. Hand me a shoehorn and some duct tape."

The key to evolution as an improviser rather than inventor is chapter 10's concept of exaptation—some trait evolves for some purpose and is co-opted when it turns out to be useful for something else. And soon feathers are aiding flight, in addition to regulating body temperature, and the insula helps get us into heaven, in addition to purging our guts of toxins. The latter is a case of what has been called "neural reuse."[27]

This isn't to say it's been an easy process, that magically one day neurons that help make you puke are suddenly involved in running the president's bioethics panel. It is insanely interesting to me that the most unique neurons in our brains, the recently evolved and slow-developing von Economo neurons, are predominantly housed in the anterior cingulate and insula. And that the neurodegenerative disease frontotemporal dementia, destined to eventually destroy the entire fancy neocortex, takes out von Economo neurons first—there's something extra fancy (and thus expensive and vulnerable) about those cells. The tinkering and improvising was inspired.

What's most interesting is that we see the beginnings of the "I know, let's persuade the ACC and insula to volunteer for these new jobs" in other species. As we saw in chapter 14, the emotional contagion and protoempathy that a rodent can feel for another one in pain is centered in the anterior cingulate. And full-blown von Economo neurons are also found in those same brain regions in the other apes, elephants, and cetaceans—evolution's Mensa club—and exist in rudimentary forms in monkeys. It's unclear if, say, a blue whale wants to wash its flippers after a social-norm violation, but a handful of other species seem to have taken the first steps into this strange new territory along with us.

THE METAPHORICAL DARK SIDE

Our brains' confusion of the metaphorical with the literal literally matters. Back to chapter 10 and the evolutionary emphasis on kin selection. We saw the array of mechanisms used by various species for recognizing kin and degree of relatedness—e.g., genetically shaped pheromonal signatures and imprinting on the female whose birdsong you heard a lot while you were still inside an egg. And we saw that among other primates there are cognitive components as well (recall male baboons' degree of paternalism being predicted by their likelihood of being the father). By the time we get to humans, the process is mostly cognitive—we can think our way to deciding who is a relative, who is an Us. And thus, as we saw, we can be manipulated into thinking that some individuals are more related to us, and others less so, than they actually are—pseudokinship and pseudospeciation. There are numerous ways to get someone to think that an Other is so different that they barely count as human. But as propagandists and ideologues have long known, if you want to get someone to *feel* that an Other hardly counts as human, there's only one way to do it—engage the insula. And the surest way to do that is with metaphor.

In 1994 many Westerners became aware of the existence of the nation of Rwanda for the first time. The mountainous Central African country is tiny, with one of the highest population densities in the world. Way back when, it had been filled with hunter-gatherers who, as per usual, had been displaced over the last millennium by agriculturalists and pastoralists, who came to form the Hutu and Tutsi tribes, respectively. It remains debated whether they arrived around the same century and whether they were actually ethnically distinct groups, but the Hutu and Tutsi Us/Them-ed with a vengeance. The minority Tutsi traditionally dominated the Hutu, reflecting the common herdsman/farmer power dynamics of

Africa; German and Belgian colonials, in the classic divide and conquer, exploited and inflamed the tribal animosities further.

With independence in 1962 came a turning of tables and Hutu domination of the government. Discrimination and violence against Tutsis drove many out of the country; over the subsequent years, many Tutsi refugee populations in neighboring countries gave rise to rebel groups seeking to invade Rwanda and establish safe havens there for Tutsis. Predictably, this increased anti-Tutsi militancy among Hutus and resulted in further discrimination and massacres. One of the ironies of what was to come, reflecting the uncertainty as to whether the Hutu and Tutsi were historically even separate people, was that it wasn't always possible to distinguish the two—identity cards were required to indicate ethnicity.

By 1994 the Rwandan president, the dictator Juvénal Habyarimana, a Hutu military man who had seized power in 1973, was under sufficient pressure from Tutsi rebel groups that he signed a power-sharing peace accord with the rebels. This was viewed as a sellout by the growing "Hutu Power" extremist bloc. On April 6, 1994, Habyarimana's plane was shot down by a missile as it approached the capital, Kigali, killing all on board. It is still unclear whether the assassination was carried out by Tutsi rebels or Hutu Power elements in the military who were intent on both eliminating Habyarimana and laying the blame on Tutsis. In any case, within a day Hutu militants had killed essentially all moderate Hutus in the government, seized power, officially laid blame for the assassination on Tutsis, and urged all Hutus to take revenge. And most Hutus complied. Thus began what is now known as the Rwandan genocide.*

The killing ran for approximately one hundred days (until it was finally halted by Tutsi rebels gaining control). During that time, there was not only a Final Solution–style attempt to kill every Tutsi in Rwanda but also killing of Hutus who were married to Tutsis, who attempted to protect Tutsis, or who refused to participate in killings. By the time it was done, approximately 75 percent of Tutsis—between 800,000 and 1,000,000 people—and around 100,000 Hutu had been killed. Roughly one out of

* Habyarimana's plane also carried Cyprien Ntaryamira, the Hutu president of neighboring Burundi, an equally small, impoverished nation with the same history of Hutu/Tutsi conflict. It soon had its own ethnic civil war.

every seven Rwandans. This translated into five times the rate of killing during the Nazi Holocaust. It was mostly ignored by the West.[28]

Five times the rate. For those of us schooled in the modern Western world's atrocities, some translation is needed. The Rwandan genocide did not involve tanks, airplanes dropping bombs, or shelling of civilians. There were no concentration camps, no transport trains, no Zyklon B. There was no bureaucratic banality of evil. There were hardly even many guns. Instead Hutu—from peasant farmers to urban professionals—bludgeoned their Tutsi neighbors, friends, spouses, business partners, patients, teachers, students. Tutsis were beaten with sticks until they were dead, killed with machetes after being gang-raped and sexually mutilated, trapped in sanctuaries that were then burned to the ground. An average of roughly ten thousand people per day. As perhaps the genocide's single most shocking atrocity, in the town of Nyange, the local Catholic priest, a Hutu named Athanase Seromba, gave sanctuary to between 1,500 and 2,000 Tutsi, many of them his parishioners, and then led the Hutu militia that ultimately killed every person inside his church. Rivers ran red, not just metaphorically.*

How could this have happened? There are many components to the answer. The populace had a long tradition of unquestioning obedience to authority, a helpful trait to develop in a brutally dictatorial nation. Hutu militants had for months before been distributing machetes to the Hutu populace. The government-controlled radio station (the main form of mass media in this marginally literate country) proclaimed that the intent of the invading Tutsi rebels was to kill every Hutu, and that one's Tutsi neighbors were a fifth column preparing to join in. And there was another meaningful factor. The anti-Tutsi propaganda was ceaselessly dehumanizing, with the infamous pseudospeciation of Tutsis being referred to only as "cockroaches." *Stamp out the cockroaches. The cockroaches are planning to*

* And continue to. In the aftermath of the predominantly Tutsi Rwandan Patriotic Front rebel army's victory, approximately two million Rwandan Hutus fled the country, fearing reprisals (of which there have been remarkably few under the government of rebel leader Paul Kagame). The massive refugee camps formed in the eastern Congo by the fleeing Hutus were soon under the control of the defeated Hutu militias and became a breeding ground for attacks on Rwanda and the two subsequent Congo wars, which killed millions.

The aftermath

kill your children. The cockroaches [the supposedly devious and seductive Tutsi women] will steal your husbands. The cockroaches [Tutsi men] will rape your wives and daughters. Stamp out the cockroaches, save yourselves, kill the cockroaches. And with insular cortices ablaze, machetes in one hand and transistor radios in the other, most Hutus did.*

Dehumanization, pseudospeciation. The tools of the propagandists of hate. Thems as disgusting. Thems as rodents, as a cancer, as a transitional species, Thems as reekingly malodorous, as living in hives of chaos that

* I've been mighty interested in the history of the Rwandan genocide. I spent time in Rwanda a few years before it occurred, looking at mountain gorillas on its border with the Congo. Predictably, pathetically, stupidly, poignantly, something-ly, I came away thinking of the people there as kind and generous. I assume that most everyone I encountered wound up dead, killers, and/or refugees. At moments when I wonder why anyone should bother writing a book like this, I taunt myself by thinking, "Golly gee, if only I'd teamed up with the Tooth Fairy and the Easter Bunny to give some lectures in Rwanda about the biology of pseudospeciation, all of this could have been prevented."

no normal human would. Thems as shit. Get the insulae of your followers to confuse the literal and metaphorical, and you're 99 percent of the way there.

A GLIMMER

A goal might be to use the good side of a double-edged sword to cut loose the silver linings of clouds and save them for rainy days. Or something metaphorically like that. The tool of the propagandist is to effectively exploit symbols of revulsion in the service of hate. But the odd literal metaphoring of our brains can also provide the peacemaker with a highly effective tool.

In a moving, important 2007 paper in *Science*, the American/French anthropologist Scott Atran, along with Robert Axelrod (of chapter 10's Prisoner's Dilemma fame) and Richard Davis, a conflict expert at Arizona

State University, considered the power of what they called "sacred values" in conflict resolution.[29] These are straight out of Greene's world of two different cultures of shepherds fighting over a commons, each with a different moral vision as to what is correct, each passionately focused on "rights" whose meaning and power are incomprehensible to the other side. Sacred values are defended far out of proportion to their material or instrumental importance or likelihood of success, because to any group such values define "who we are." And therefore, not only are attempts to reach compromises on such issues by using material incentives unlikely to be productive, but they can be insultingly counterproductive. You can't buy us off into dishonoring that which we hold sacred.

Atran and colleagues have studied the roles played by sacred values in the context of Middle East conflict. In a world of sheer rationality where the brain didn't confuse reality with symbols, bringing peace to Israel and Palestine would revolve solely around the concrete, practical, and specific—placement of borders, reparations for Palestinian land lost in 1948, water rights, the extent of militarization allowed to Palestinian police, and so on. Solving those nuts-and-bolts issues may be a way of *ending* war, but peace is not the mere absence of war, and *making* true peace requires acknowledging and respecting the sacred values of Them. Atran and colleagues found that, from the person in the street to the highest offices of power, sacred values loomed large. They interviewed senior Hamas leader Ghazi Hamad, asking what he sees as a requirement for true peace. This included, of course, reparations to Palestinians for the homes and lands they lost almost seventy years ago. Necessary but not sufficient. "Let Israel apologize for our tragedy in 1948," he added. And current Israeli prime minister Benjamin Netanyahu, in discussing with them what was needed for true peace, cited not only instrumental issues of security but also how the Palestinians must "change their textbooks and anti-Semitic characterizations." As the authors state, "In rational-choice models of decision-making, something as intangible as an apology [or getting the likes of the *Protocols of the Elders of Zion* out of schoolbooks] could not stand in the way of peace." Yet they do, because in recognizing the enemy's sacred symbols, you are de facto recognizing their

humanity, their capacity for pride, unity, and connection to their past and, probably most of all, their capacity for experiencing pain.*

"Symbolic concessions of no apparent material benefit may be key in helping to solve seemingly intractable conflicts," write the authors. In 1994 the Kingdom of Jordan became the second Arab country to sign a peace treaty with Israel. It ended war, bringing to an end decades of hostilities. And it created a successful road map for the two nations to coexist, built around addressing material and instrumental issues—water rights (e.g., Israel would give Jordan fifty million cubic meters of water annually), joint efforts to combat terrorism, joint efforts to facilitate tourism between the countries. But it wasn't until a year later that one saw evidence that something resembling a true peace was forming. It followed the creation of yet another martyr for peace, the assassination of Israeli prime minister Yitzhak Rabin, one of the architects of the Oslo Peace Accord, by a right-wing Israeli extremist. Extraordinarily, King Hussein

* As a sacred-values issue that either does or doesn't seem ironic, depending on your politics, the authors cite how the newborn State of Israel in 1948, in a terrible economic state, nonetheless refused compensation money from Germany for property of Jews murdered by the Nazis—until Germany publicly expressed contrition.

came to Rabin's funeral and eulogized him, addressing his widow in the front row:

> My sister, Mrs. Leah Rabin, my friends, I had never thought that the moment would come like this when I would grieve the loss of a brother, a colleague and a friend.

Hussein's presence and words were obviously irrelevant to any of the rational stumbling blocks to peace. And were immeasurably important.[30]

A similar arc can be seen in Northern Ireland, where an IRA ceasefire in 1994 facilitated an end to the violence of the Troubles and the 1998 Good Friday Agreement laid the groundwork for Republicans and Unionists to coexist, for ex-Unionist demagogues and ex-IRA gunmen to serve in a government together. Much of the agreement was material or instrumental, but there were elements of sacred values addressed—for example, the establishment of a Parades Commission to ensure that neither group had inflammatory, symbol-laden parades in the other's neighborhoods in Belfast. But in many ways the most palpable sign of a lasting peace came from an unexpected corner. The unity government formed after the agreement was led by Peter Robinson as first minister and Martin McGuinness as deputy first minister. The former had been a Unionist firebrand, the latter a leader of the political wing of the IRA; they were two men who epitomized the hatreds of the Troubles. They had a functional working relationship but nothing more than that and had famously refused to ever actually shake hands (something that even Rabin and Yasir Arafat had managed). What finally broke the ice? In 2010 Robinson was upended in a major scandal involving his politician wife, who had committed some major financial improprieties in the name of another type of impropriety—funneling money to her nineteen-year-old lover. And history was then made when McGuinness offered, and Robinson accepted, a commiserative handshake. A guy-code sacred-value moment.*[31]

* The coming of peace to Northern Ireland was laden with other instances of sacred values and symbolism. For example, around the time that the Reverend Ian Paisley, as bloody-handed a Unionist as there was, became first minister in Northern Ireland, the Catholic Prime Minister (Taoiseach) of the Republic of Ireland, Bertie Ahern, sent Paisley and his wife a fiftieth wedding-anniversary gift—a wooden bowl. This was rife with meaning, as it was crafted from a

Something similar happened in South Africa, much of it promulgated by Nelson Mandela, a genius at appreciating sacred values.[32] Mandela, while at Robben Island, had taught himself the Afrikaans language and studied Afrikaans culture—not just to literally understand what his captors were saying among themselves at the prison but to understand the people and their mind-set. At one point just before the birth of a free South Africa, Mandela entered into secret negotiations with the Afrikaans leader General Constand Viljoen. The latter, chief of the apartheid-era South African Defence Force and founder of the Afrikaner Volksfront group opposed to the dismantling of apartheid, commanded an Afrikaans militia of fifty to sixty thousand men. He was therefore in a position to doom South Africa's impending first free election and probably trigger a civil war that would kill thousands.

They met in Mandela's house, with the general apparently anticipating tense negotiations across a conference table. Instead the smiling, cordial Mandela led him to the warm, homey living room, sat beside him on a comfy couch designed to soften the hardest of asses, and spoke to the man in Afrikaans, including small talk about sports, leaping up now and then to get the two of them tea and snacks. While the general did not quite wind up as Mandela's soul mate, and it is impossible to assess the importance of any single thing that Mandela said or did, Viljoen was stunned by Mandela's use of Afrikaans and warm, chatty familiarity with Afrikaans culture. An act of true respect for sacred values. "Mandela wins over all who meet him," he later said. And over the course of the conversation, Mandela persuaded Viljoen to call off the armed insurrection and to instead run in the upcoming election as an opposition leader. When Mandela retired from his presidency in 1999, Viljoen gave a short, halting speech in Parliament praising Mandela . . . in the latter's native language, Xhosa.*

---

tree at the Irish site of the Battle of the Boyne, where in 1690 the Protestant William of Orange defeated Catholic James II. That victory was critical to the subsequent centuries of Protestant domination in Ireland, an endless point of pain for Catholics and pride for Protestants (who would commemorate the victory every July 12 with provocative marches through Catholic neighborhoods that usually ended in violence). For Ahern to acknowledge the sacred historical significance of the site for Unionists was enormous. Paisley soon reciprocated by visiting the site with Ahern, bringing him a 1685 musket as a gift and talking about the significance of the site for *all* the Irish people.

* How did Viljoen and Mandela come to meet secretly on that couch? It was catalyzed by a leading antiapartheid theologian . . . Viljoen's twin brother Abraham. The two had been long estranged, although the general intervened

The successful birth of the new South Africa was rife with acts of respect for sacred values. Perhaps the most famous was Mandela's public embrace of rugby, a sport highly symbolic of Afrikaans culture and historically disdained by black South Africans. And famously, as depicted in book and film, among the consequences was the tectonically symbolic act of the heavily Afrikaans national rugby team singing the ANC anthem, the hymn "Nkosi Sikelel' iAfrika," followed by a black choir singing the Afrikaans anthem, "Die Stem van Suid-Afrika," a craggy song with references to the country's craggy mountains.* This came before the South African host team's mythic underdog winning of the World Cup in 1995 in Johannesburg.

I could watch that YouTube clip of the anthems being sung at the World Cup all day long, especially after having to write the section on Rwanda. What do Hussein, McGuinness, Robinson, Viljoen, and Mandela show? That our confusion of the literal and the metaphorical, our granting of life-threatening sanctity to the symbolic, can be used to bring about the best of our behaviors. Which prepares us for the final chapter, soon to come.

---

on more than one occasion to prevent his brother from being assassinated by a right-wing death squad. The twins are teaching tools for chapter 8—same genes, radically opposite politics and worldviews. Same genes, and both charismatic leaders who devoted and risked their lives for what they viewed as a sacred cause.

* See the actual event at www.youtube.com/watch?v=Ncwee9IAu8I. South Africa's national anthem is now a hybrid of the two songs, with some Zulu, Sesotho, and English thrown in for good measure. While its existence is intensely moving, it must be hell on wheels to sing right, modulating all over the place.

Sixteen

Biology, the Criminal Justice System, and (Oh, Why Not?) Free Will*

DON'T FORGET TO CHECK THEIR TEAR DUCTS

Some years back a foundation sent a letter to various people, soliciting Big Ideas for a funding initiative of theirs. The letter said something along the lines of "Send us a provocative idea, something you'd never propose to another foundation because they'd label you crazy."

That sounded fun. So I sent them a proposal titled "Should the Criminal Justice System Be Abolished?" I argued that the answer was yes, that neuroscience shows the system makes no sense and they should fund an initiative to accomplish that.

"Ha-ha," they said. "Well, we asked for it. That certainly caught our

* I'm hugely grateful to Josh Greene and Owen Jones for closely vetting this chapter.

attention. That's a great idea to focus on interactions between neuroscience and the law. Let's do a conference."

So I went to a conference with some neuroscientists and some legal types—law professors, judges, and criminologists. We learned one another's terminology, for example seeing how we neuroscientists and the legal people use "possible," "probable," and "certainty" differently. We discovered that most of the neuroscientists, including me, knew nothing about the workings of the legal world, and that most of the legal folks had avoided science since being traumatized by ninth-grade biology. Despite the two-culture problem, all sorts of collaborations got started there, which eventually grew into a network of people studying "neurolaw."

Fun, stimulating, interdisciplinary hybrid vigor. And frustrating to me, because I kind of meant the title of the proposal that I had written. The current criminal justice system needs to be abolished and replaced with something that, while having some broad features in common with the current system,* would have utterly different underpinnings. Which I'm going to try to convince you of. And that's just the first part of this chapter.

You can't be less controversial than stating that the criminal justice system needs reform and that this should involve more science and less pseudoscience in the courtroom. If nothing else, consider this: according to the Innocence Project, nearly 350 people, a mind-boggling 20 of them on death row, imprisoned an average of fourteen years, have been exonerated by DNA fingerprinting.[1]

Despite that, I'm going to mostly ignore criminal justice reform by science. Here are some hot-button topics in that realm that I'm going to bypass entirely:

- What to do about the power and ubiquity of automatic, implicit biases (leading to, for example, juries meting out harsher decisions

* Namely keeping dangerous people far away from everyone else—just to get this one out of the way early in the chapter.

to African American defendants with darker skin). Should Implicit Association Tests be used in jury selection to eliminate people with strong, pertinent biases?

- Whether neuroimaging information regarding a defendant's brain should be admissible in a courtroom.[2] This has grown less contentious as neuroimaging has transitioned from revolutionary to a standard approach in science's tool kit. But there remains the issue of whether juries should be shown actual neuroimages—the worry is that nonexperts are readily overly impressed with exciting, color-enhanced Pictures of the Brain (it's turning out to be less of an issue than feared).
- Whether neuroimaging data regarding someone's veracity should have a place in the courtroom (or in the workplace regarding security clearances). Basically, I know of no expert who thinks the technique is sufficiently accurate. Nonetheless, there are entrepreneurs selling the approach (including, I kid you not, a company called No Lie MRI). This issue extends to lower-tech but equally unreliable versions of is-that-brain-lying? This includes electroencephalograms (EEGs), which are admissible in Indian courtrooms.[3]
- What should be the IQ cutoff for someone to be smart enough to be executed? The standard is an IQ of 70 or higher, and debate concerns whether it should be an *average* of 70 across multiple IQ tests, or if achieving that magic number even once qualifies you for being executed. This issue pertains to about 20 percent of people on death row.[4]
- What to do with the fact that scientific findings can generate new types of cognitive biases in jurors. For example, the belief that schizophrenia is a biological disorder makes jurors less likely to convict schizophrenics for their actions but more likely to view them as more incurably dangerous.[5]
- The legal system distinguishes between thoughts and actions; what to do as neuroscience increasingly reveals the former. Are we approaching precrime detection, predicting who *will* commit a

crime? In the words of one expert, "We're going to have to make a decision about the skull as a privacy domain."[6]

- And of course there's that problem of judges judging more harshly when their stomachs are gurgling.*[7]

All of these are important issues, and I think reforms are needed at the intersection of progressive politics, civil liberties, and tough standards about new science. In other words, a standard liberal agenda. Most of the time I'm a clichéd card-carrying liberal; I even know the theme songs to many of NPR's programs. Nonetheless, this chapter won't take anything resembling a liberal approach to reforming criminal justice. The reason why is summarized in the following example of a classically liberal approach to a legal issue.

It's the middle of the 1500s. Perhaps because of lax societal standards and people being morally deprived and/or depraved, Europe is overrun with witches. It's a huge problem—people fear going out at night; polls show that peasants-in-the-street list "witches" as more of a threat than "the plague" or "the Ottomans"; would-be despots gain supporters by vowing to be tough on witches.

Fortunately, there are three legal standards for deciding if someone is guilty of witchcraft:[8]

- The flotation test. Since witches reject the sacrament of baptism, water will reject their body. Take the accused, bound, and toss them into some water. If they float, they're a witch. If they sink, they're innocent. Quickly now, retrieve innocent person.

* And one thing that I'm not going anywhere near is this New Age–y notion: "Of course we have free will. You can't say that our behaviors are determined by a mechanistic universe, because the universe is indeterminate, because of quantum mechanics." Argh. What anyone sensible who has thought about this will point out is that (a) the consequences of the subatomic indeterminacy of quantum mechanics (about which I understand zero) don't ripple upward enough to influence behavior, and (b) if they did, the result wouldn't be the freedom to will your behavior. It would be the utter randomization of behavior. In the words of philosopher/neuroscientist Sam Harris, a free will trasher, if quantum mechanics actually played a role in any of this, "Every thought and action would seem to merit the statement 'I don't know what came over me.'" Except you wouldn't actually be able to make that statement, since you'd just be making gargly sounds because the muscles in your tongue would be doing all sorts of random things.

- The devil's-spot test. The devil enters someone's body to infect them with witch-ness, and that point of entry is left insensitive to pain. Systematically do something painful to every spot on the accused's body. If some spot is much less sensitive to pain than the rest, you've found a devil's spot and identified a witch.
- The tear test. Tell the accused the story of the crucifixion of Our Lord. Anyone not moved to tears is a witch.

These well-established criteria allow authorities fighting this witch wave to identify and suitably punish thousands of witches.

In 1563 a Dutch physician named Johann Weyer published a book, *De Praestigiis Daemonum*, advocating reform of the witch justice system. He, of course, acknowledged the malign existence of witches, the need to punish them sternly, and the general appropriateness of witch-fighting techniques like those three tests.

However, Weyer aired an important caveat pertinent to older female witches. Sometimes, he noted, elderly people, especially women, have had atrophy of their lachrymal glands, making it impossible to cry tears. Uh-oh—this raises the specter of false convictions of people as witches. The concerned, empathic Weyer counseled, "Make sure you're not torching some poor elderly person simply because her tear ducts don't work anymore."

Now *that's* a liberal reform of the witch justice system, imposing some sound thinking in one tiny corner of an irrational edifice. Much like what scientifically based reform of our current system does, which is why something more extreme is needed.*

THREE PERSPECTIVES

Let's get down to cases. There are three ways of viewing the place of biology in making sense of our behaviors, criminal or otherwise:

* And just to show what a bleeding heart everyone thought Weyer was, his book was banned by both the Catholic Church *and* leading Reformation clergy.

1. We have complete free will in our behavior.
2. We have none.
3. Somewhere in between.

If people are forced to carefully follow the logical extensions of their views, probably less than a thousandth of a percent would support the first proposition. Suppose someone convulsing with a grand mal epileptic seizure, flinging their arms around, strikes someone. If you truly believe we freely control our behavior, you must convict them of assault.

Virtually everyone considers that absurd. Yet that legal outcome would have occurred half a millennium ago in much of Europe.[9] That seems ludicrous because in the last few centuries the West has crossed a line and left it so far behind that a world on the other side is unimaginable. We embrace a concept that defines our progress—"It's not him. It's his disease." In other words, at times biology can overwhelm anything resembling free will. This woman didn't bump into you maliciously; she's blind. This soldier standing in formation didn't pass out because he doesn't have what it takes; he's diabetic and needs his insulin. This woman isn't heartless because she didn't help the elderly person who had fallen; she's paralyzed from a spinal cord injury. Similar shifts in the perception of criminal responsibility have occurred in other realms. For example, from two to seven centuries ago, prosecution of animals, objects, and corpses thought to have intentionally harmed someone was commonplace. Some of these trials had a weirdly modern tint to them—in a 1457 trial of a pig and her piglets for eating a child, the pig was convicted and executed, whereas the piglets were found to be too young to have been responsible for their acts. Whether the judge cited the maturational state of their frontal cortices is unknown.

Thus hardly anyone believes that we have complete conscious control over our behavior, that biology never constrains us. We'll ignore this stance forever after.

DRAWING LINES IN THE SAND

Nearly everyone believes in the third proposition, that we are somewhere between complete and no free will, that this notion of free will is compatible with the deterministic laws of the universe as embodied in biology. Only a subset of versions of this view fit the fairly narrow philosophical stance called "compatibilism." Instead this broader view is that we have something resembling a spirit, a soul, an essence that embodies our free will, from which emanates behavioral intent; and that this spirit coexists with biology that can sometimes constrain it. It's a kind of libertarian dualism ("libertarian" in the philosophical rather than political sense), what Greene calls "mitigated free will." It's encapsulated in the idea that well-intentioned spirit, while willing, can be thwarted by flesh that is sufficiently weak.

Let's start with the definitive legal framing of mitigated free will.

In 1842 a Scotsman named Daniel M'Naghten tried to assassinate British prime minister Robert Peel.[10] He mistook Peel's private secretary, Edward Drummond, for the prime minister and shot him at close range, killing him. At his arraignment M'Naghten stated, "The Tories in my native city have compelled me to do this. They follow and persecute me wherever I go, and have entirely destroyed my peace of mind. They followed me to France, into Scotland . . . wherever I go. I cannot get no rest from them night or day. I cannot sleep at night. . . . I believe they have driven me into a consumption. I am sure I shall never be the man I formerly was. . . . They wish to murder me. It can be proved by evidence. . . . I was driven to desperation by persecution."

In today's terminology M'Naghten had some form of paranoid psychosis. It may not have been schizophrenia—his delusional symptoms started many years later than the typical age of onset of the disease. Regardless of the diagnosis, M'Naghten had abandoned his business and spent the previous two years wandering Europe, hearing voices, convinced that he was being spied upon and persecuted by powerful people, with Peel his most diabolical tormentor. In the words of a doctor who testified as to his insanity, "The delusion was so strong that nothing but a physical impediment could have prevented him from committing the act [i.e., murder]." M'Naghten was so clearly impaired that the prosecution withdrew criminal charges, agreeing with the defense that he was insane. The jury agreed, and M'Naghten spent the rest of his life in insane asylums, reasonably well treated by the standards of the time.

There was bellowing protest after the jury's decision, ranging from the man in the street to Queen Victoria—M'Naghten had gotten away with murder. The presiding judge was grilled by Parliament and stood by the decision. The equivalent of the Supreme Court was tasked by Parliament with assessing the case and supported him. And out of the decision came the formalization of what is now the common criterion for finding someone innocent by reason of insanity, namely the "M'Naghten rule": if, at the time of the crime, the person is so "laboring under such a defect of reason from disease of the mind," that he cannot distinguish right from wrong.*

The M'Naghten rule was at the core of John Hinckley Jr. being found not guilty for reasons of insanity in his attempted assassination of Reagan in 1981, being hospitalized rather than jailed. There was considerable "He's getting away with it" outrage afterward; a number of states banned the M'Naghten criterion, and Congress essentially banned it for federal cases with the 1984 Insanity Defense Reform Act.† Nonetheless, the reasoning behind M'Naghten has generally withstood the test of time.

This is the essence of a stance of mitigated free will—people need to be held responsible for their actions, but being floridly psychotic can be a mitigating circumstance. It is the idea that there can be "diminished" responsibility for our actions, that something can be semivoluntary.

Here's how I've always pictured mitigated free will:

There's the brain—neurons, synapses, neurotransmitters, receptors,

---

* I thank an excellent undergrad, Tom McFadden (now a superb biology teacher at my kids' school!), for background research on M'Naghten.

† I just love the use of the word "reform" in this context.

brain-specific transcription factors, epigenetic effects, gene transpositions during neurogenesis. Aspects of brain function can be influenced by someone's prenatal environment, genes, and hormones, whether their parents were authoritative or their culture egalitarian, whether they witnessed violence in childhood, when they had breakfast. It's the whole shebang, all of this book.

And then, separate from that, in a concrete bunker tucked away in the brain, sits a little man (or woman, or agendered individual), a homunculus at a control panel. The homunculus is made of a mixture of nanochips, old vacuum tubes, crinkly ancient parchment, stalactites of your mother's admonishing voice, streaks of brimstone, rivets made out of gumption. In other words, not squishy biological brain yuck.

And the homunculus sits there controlling behavior. There are some things outside its purview—seizures blow the homunculus's fuses, requiring it to reboot the system and check for damaged files. Same with alcohol, Alzheimer's disease, a severed spinal cord, hypoglycemic shock.

There are domains where the homunculus and that brain biology stuff have worked out a détente—for example, biology is usually automatically regulating your respiration, unless you must take a deep breath before singing an aria, in which case the homunculus briefly overrides the automatic pilot.

But other than that, the homunculus makes decisions. Sure, it takes careful note of all the inputs and information from the brain, checks your hormone levels, skims the neurobiology journals, takes it all under advisement, and then, after reflecting and deliberating, decides what you do. A homunculus in your brain, but not of it, operating independently of the material rules of the universe that constitute modern science.

That's what mitigated free will is about. I see incredibly smart people recoil from this and attempt to argue against the extremity of this picture rather than accept its basic validity: "You're setting up a straw homunculus, suggesting that I think that other than the likes of seizures or brain injuries, we are making all our decisions freely. No, no, my free will is much softer and lurks around the edges of biology, like when I freely decide which socks to wear." But the frequency or significance with which free will exerts itself doesn't matter. Even if 99.99 percent of your actions are biologically determined (in the broadest sense of this book), and it is only once a decade that you claim to have chosen out of "free will" to floss your teeth from left to right instead of the reverse, you've tacitly invoked a homunculus operating outside the rules of science.

This is how most people accommodate the supposed coexistence of free will and biological influences on behavior.* For them, nearly all discussions come down to figuring what our putative homunculus should and shouldn't be expected to be capable of. To get a feel for that, let's look at some of these debates.

Age, Maturity of Groups, Maturity of Individuals

In 2005's *Roper v. Simmons* decision, the Supreme Court ruled that you can't execute someone for a crime committed before the age of eighteen. The appropriate reasoning was straight out of chapters 6 and 7: the brain, especially the frontal cortex, is not yet at adult levels of emotional regulation and impulse control. In other words, adolescents, with their adolescent brains, aren't as culpable as adults. The reasoning was the same as why the pig was executed but not her piglets.

In the years since, there have been related rulings. In 2010's *Graham v. Florida* and 2012's *Miller v. Alabama*, the Court emphasized that juvenile offenders have the highest potential for reform (because of their still-developing brains) and thus banned life sentences without parole for them.

These decisions have prompted a number of debates:

- Just because adolescents are, *on the average*, less neurobiologically and behaviorally mature than adults doesn't rule out the possibility of some individual adolescents being as mature, thus being appropriately held to adult standards of culpability. Related to that is the obvious absurdity of implying that something neurobiologically

* And I mean truly thinking that way, rather than backing into it because the alternative view would demand overwhelming changes in how society works.

magical happens on the morning of someone's eighteenth birthday, endowing them with adult levels of self-control. The usual response to these points is that, yes, these are true, but the law often relies on group-level attributes with arbitrary age boundaries (e.g., the age at which someone can vote, drink, or drive). There is this willingness because you can't test every teenager each year, month, hour, to determine whether they are mature enough yet to, say, vote. But it's worth doing so when it comes to a teenage murderer.

- In another contrarian view the issue isn't whether a seventeen-year-old is as mature as an adult but whether they are mature *enough.* Sandra Day O'Connor, in dissenting from the *Roper* decision, wrote, "The fact that juveniles are generally *less culpable* for their misconduct than adults does not necessarily mean that a 17-year-old murderer cannot be *sufficiently* culpable to merit the death penalty" (her emphasis). Another dissenter, the late Antonin Scalia, wrote that it is "absurd to think that one must be mature enough to drive carefully, to drink responsibly, or to vote intelligently, in order to be mature enough to understand that murdering another human being is profoundly wrong."[11]

Amid these differing opinions everyone, including O'Connor and Scalia, agrees that there exist age-related boundaries on free will—everyone's homunculus was once too young to have its adult powers.[12] Maybe it wasn't tall enough yet to reach all the control dials; maybe it was distracted from its job by fretting about that gross pimple on its forehead. And that needs to be considered during legal judgments. Just as with piglets and pigs, it's just an issue of when a homunculus is old enough.

The Nature and Magnitude of Brain Damage

Essentially everyone working with a model of mitigated free will accepts that if there is enough brain damage, responsibility for a criminal act goes out the window. Even Stephen Morse of the University of Pennsylvania, a strident critic of neuroscience in the courtroom (much more later), concedes, "Suppose we could show that the higher deliberative centers in the brain seem to be disabled in these cases. If these are people who cannot control episodes of gross irrationality, we've learned something that might be relevant to the legal ascription of responsibility."[13] In this view, mitigating biological factors are relevant if the capacity for reasoning has been grossly impaired.

Thus, if someone had their entire frontal cortex destroyed, you probably shouldn't hold them responsible for their actions, because their rationality is grossly impaired when deciding their own courses of action.[14] But the issue then becomes where a line is drawn on a continuum—what if 99 percent of the frontal cortex is destroyed? What if 98 percent? This is of great practical importance, since a large percentage of death row inmates have a history of damage to the frontal cortex, particularly of the most disabling type, namely early in life.

In other words, amid differing opinions about where a line should be drawn, believers in mitigated free will agree that massive amounts of brain damage overwhelm a homunculus, while it should be expected to handle at least some damage.

Responsibility at the Level of the Brain and at the Social Level

The renowned neuroscientist Michael Gazzaniga, one of the leading lights and elders of the field, has taken an extremely odd path in writing, "Free will is an illusion, but you're still responsible for your actions." This is expounded at length in a challenging book of his, *Who's in Charge? Free Will and the Science of the Brain.* Gazzaniga fully accepts the entirely material nature of the brain but nonetheless sees room for responsibility. "Responsibility exists at a different level of organization: the social level, not in our determined brains." I think either he is actually saying, "Free will is an illusion, but for practical reasons, we are still going to hold you responsible for your actions," or he is hypothesizing some manner of homunculus that exists only at a social level. In response to the latter idea, the pages of this book show how our social world is ultimately as much a product of our determined, materialist brains as are our simple motor movements.*[15]

The Time Course of Decision Making

Another well-established fault line in a stance of mitigated free will is that our capacity for free will moves to the forefront with decisions that are slow and deliberative, whereas biological factors may push free will aside in split-second-decision situations. In other words, the homunculus is not always sitting right at the helm in the bunker; instead it occasionally wanders off to grab a snack, and if something exciting suddenly arises, those neurons may fire off commands to muscles and produce a behavior before the homunculus can rush back and hit that big red button on the control panel.

Issues of getting to the red button in time intersect with issues of the adolescent brain. A number of critics of *Roper v. Simmons*, starting with O'Connor in her dissenting opinion, noted a seeming contradiction. The American Psychological Association (APA) had filed an amicus curiae brief in the case, emphasizing that adolescents (i.e., their brains) are so *immature* that they can't be held to adult criminal standards with sentencing. Turns out that the same APA had filed a brief some years earlier in a different case, emphasizing that adolescents are sufficiently *mature* that they should be able to choose whether to have an abortion, even without parental consent.

Well, that's a bit awkward, and it sure makes the APA and its ilk appear to be flip-flopping along ideological grounds, O'Connor charged. Laurence Steinberg, whose research on adolescent brain development was covered heavily in chapter 7 (and whose work was influential in the *Roper v. Simmons* decision), offers a logical resolution.[16] Deciding whether to have an abortion involves logical reasoning about moral, social, and interpersonal issues, stretching out over days to weeks. In contrast, deciding whether to, say, shoot someone can involve issues of impulse control over the course of

* I'm obviously confused by Gazzaniga's stance, and I suspect his conclusions reflect his attempts to reconcile his worldview as a neuroscientist with his being a religious man, something he discusses in his memoir, *Tales from Both Sides of the Brain: A Life in Neuroscience* (New York: Ecco, 2016).

seconds. The frontal immaturity of the adolescent brain is more pertinent to split-second issues of impulse control than to slow, deliberative reasoning processes. Or in a mitigated-free-will framework, rapid-fire, impulsive behaviors can occur while the homunculus has gone to the bathroom.

Causation and Compulsion

Some proponents of mitigated free will distinguish between the concepts of "causation" and "compulsion."[17] In a way that feels a bit nebulous, the former involves every behavior having been caused by something, of course, but the latter reflects only a subset of behaviors being *really, really* caused by something, something that compromises rational, deliberative processes. In this view some behaviors are more deterministically biological than others.

This has been relevant to schizophrenic delusions. Suppose someone suffering from schizophrenia has auditory hallucinations, including a voice telling him to commit a crime; he does so.

Some courts have viewed this as not mitigating. If your friend suggests that you mug someone, the law expects you to resist, even if it's an imaginary friend in your head.

But others see distinctions depending on qualities of the auditory hallucinations. In that view, if a schizophrenic individual commits a crime because a voice in his head demanded it, yes, his act was *caused* by that voice, but that doesn't excuse the crime. In contrast, consider a schizophrenic individual committing a crime because thundering choruses of taunting, threatening, cajoling voices in his head, complete with baying hellhounds and choirs of trombones playing loud atonal music, command him every waking moment to do the crime. When he succumbs and does so, it is deemed more excusable, because those voices constituted a *compulsion* to act.*

Thus in this view even a sensible homunculus can lose it and agree to virtually anything, just to get the hellhounds and trombones to stop.

* Many chapters ago I made reference to the "Son of Sam" murder spree in 1976 and the arrest of David Berkowitz. In his defense Berkowitz claimed that he was demonically possessed and had been commanded to murder—not by Satan, Hitler, Al Capone, or Genghis Khan but, instead, by . . . his neighbor's dog. He was convicted and given six consecutive life sentences.

Starting a Behavior Versus Halting It

It is virtually ordained that any discussion of volition and biology eventually considers the "Libet experiment."[18] In the 1980s neuroscientist Benjamin Libet of UCSF reported something fascinating. A subject is hooked up to an EEG machine, which monitors patterns of electrical excitation in the brain. She sits quietly, looking at a clock. She has been instructed to flick her wrist whenever she feels like it and to note the time, down to the second, when she decided to do so.

Libet would identify something in the EEG data called a "readiness potential"—a signal from the motor cortex and supplementary premotor areas that a movement would soon be initiated. And consistently, readiness potentials appeared about half a second before the reported time of conscious intent to move. Interpretation: your brain "decided" to move before you were even aware of it. Thus, how can you claim to have chosen when to move, evidence of free will, if the cascade of neural signaling culminating in movement started before you consciously chose? Free will is an illusion.

Naturally, this finding generated speculation, controversy, replications, elaborations, refutations, and nuances that are beyond me. One criticism concerned a necessary limitation of the approach. In this view, there's free will, you freely decide when to move your wrist, and that readiness potential is a consequence of your decision. What's the five-hundred-millisecond delay, in that case? That's the lag time between the instant when the decision to move first occurs and when (a) attention is then focused on the clock and (b) the position of the second hand is interpreted. In other words, the supposed half-second lag is an artifact of the experimental design, not a real thing. Other criticisms concerned the ambiguity of feeling that you intend to move. Other criticisms are more arcane than I can follow.

A very different interpretation of the finding was offered, interestingly, by Libet. Yes, maybe your brain prepares to initiate a behavior before there is conscious awareness of the decision, meaning that your belief that you consciously chose to move is wrong. But in that lag time is the potential to consciously choose to *veto* that action. In the pithy words of

V. S. Ramachandran (of mirror neuron speculation in chapter 14), we may not have free will, but we have "free won't."[19]

Predictably, this intriguing counterinterpretation has fueled more discussions, experiments, and counter-counterinterpretations. For us, surveying different disputes concerning mitigated free will, this entire debate is about the nature of a homunculus's control panel. How many of its buttons and switches and dials that go up to eleven are involved in initiating a behavior versus halting it?

Thus a view of mitigated free will makes room for both biological causation of behavior and free will, and all the discussions merely concern where lines in the sand are drawn and how inviolate they are. This prepares us to consider what I think is the most important line-drawing debate.

"You Must Be So Smart" Versus "You Must Have Worked So Hard"

Stanford psychologist Carol Dweck has done groundbreaking work on the psychology of motivation. In the late 1990s she reported something important. Kids do a task, take a test, something, where they do it well. You then praise them in one of two ways—"What a great score; you must be so smart" or "What a great score, you must have worked so hard." When you praise kids for working hard, they tend to work harder the next time, show more resilience, enjoy the process more, and become more likely to value the accomplishment for its own sake (rather than for the grade). Praise kids for being smart, and precisely the opposite occurs. When it becomes all about being smart, effort begins to seem suspect, beneath you—after all, if you're really so smart, you shouldn't have to work hard; you glide, you don't sweat and grunt.[20]

Beautiful work that has achieved cult status among many thoughtful parents of gifted kids, who want to understand when their child's smarts shouldn't come into the picture.

Why do "You're so smart" and "You work so hard" have such different effects? Because they fall on either side of one of the deepest lines drawn by believers in mitigated free will. It is the belief that one assigns aptitude and impulse to biology and effort and resisting impulse to free will.

It's cool to see natural ability in action. The great all-around athlete who has never seen pole-vaulting before, watches it once, tries it once, and soars like a pro. Or the singer whose voice has always had a natural timbre that evokes emotions you never knew existed. Or that student in your class who obviously just gets it, two seconds into your explaining something really abstruse.

That's impressive. But then there's inspiring. When I was a kid, I repeatedly read a book about Wilma Rudolph. She was the fastest female runner in the world in 1960, an Olympian who became a civil rights pioneer. Definitely impressive. But consider that she was born prematurely, underweight, one of twenty-two kids in a poor Tennessee family, and—get this—at age four got polio, resulting in a leg brace and a twisted foot. Polio, she was *crippled* by polio. And she defied every expert's expectations, worked and worked and worked through the pain, and became the fastest there was. That's inspiring.

In many domains we can sort of grasp the materialist building blocks of natural ability. Someone has the optimal ratio of slow-twitch to fast-twitch muscle fibers, producing a natural pole-vaulter. Or has vocal cords with the perfect degree of velvety peach fuzz (I'm winging it here) to produce an extraordinary voice. Or the ideal combination of neurotransmitters, receptors, transcription factors, and so on, producing a brain that rapidly intuits abstractions. And we can also perceive the building blocks in someone who is merely okay, or lousy, at any of these.

But Rudolph-esque accomplishments seem different. You're exhausted, demoralized, and it hurts like hell but you push on; you want to take an evening off, see a movie with a friend, but resume studying; there's that temptation, no one's looking, everyone else does it, but you know it's wrong. It seems so hard, so improbable to think of those same neurotransmitters, receptors, or transcription factors when considering feats of willpower. There seems a much easier answer—you're seeing the Calvinist work ethic of a homunculus sprinkled with the right kind of fairy dust.

Here's a great example of this dualism. Recall Jerry Sandusky, the Penn State football coach who was a horrific serial child molester. After his conviction came an opinion piece on CNN. Writing under the provocative heading of "Do pedophiles deserve sympathy?" James Cantor of the University of Toronto reviewed the neurobiology of pedophilia. For example, it runs in families in ways suggesting genes play a role. Pedophiles have atypically high rates of brain injuries during childhood. There's evidence of endocrine abnormalities during fetal life. Does this raise the possibility that a neurobiological die is cast, that some people are destined to be this way? Precisely. Cantor concludes, "One cannot choose to not be a pedophile."[21]

Brave and correct. And then Cantor does a stunning mitigated-free-will long jump. Does any of that biology lessen the condemnation and punishment that Sandusky deserved? No. "One cannot choose to not be a pedophile, but one can choose to not be a child molester."

This establishes a dichotomy of what things are supposedly made of:

Biological stuff	Homuncular grit
Destructive sexual urges	Resisting acting upon them
Delusionally hearing voices	Resisting their destructive commands
Proclivity toward alcoholism	Not drinking
Having epileptic seizures	Not driving if you didn't take your meds
Not all that bright	Getting going when the going gets tough
Not the loveliest of faces	Resisting getting that huge, hideous nose ring

Here are just a few of the things we've seen in this book that can influence the column on the right: blood glucose levels; the socioeconomic status of your family of birth; a concussive head injury; sleep quality and quantity; prenatal environment; stress and glucocorticoid levels; whether you're in pain; if you have Parkinson's disease and which medication you've been prescribed; perinatal hypoxia; your dopamine D4 receptor gene variant; if you have had a stroke in your frontal cortex; if you suffered childhood abuse; how much of a cognitive load you've borne in the last few minutes; your MAO-A gene variant; if you're infected with a particular parasite; if you have the gene for Huntington's disease; lead levels in your tap water when you were a kid; if you live in an individualist or a collectivist culture; if you're a heterosexual male and there's an attractive woman around; if you've been smelling the sweat of someone who is frightened. On and on. Of all the stances of mitigated free will, the one that assigns aptitude to biology and effort to free will, or impulse to biology and resisting it to free will, is the most permeating and destructive. "You must have worked so hard" is as much a property of the physical universe and the biology that emerged from it as is "You must be so smart." And yes, being a child molester is as much a product of biology as is being a pedophile. To think otherwise is little more than folk psychology.

BUT DOES ANYTHING USEFUL ACTUALLY COME OF THIS?

As I noted, the most formidable skeptic of the relevance of neuroscience to the legal system is Stephen Morse, who has written extensively and effectively about the subject.[22] He is the definitive advocate of free will being compatible with a deterministic world. He's fine with M'Naghten and recognizes that there can be sufficient brain damage to compromise the notion of responsibility—"Various causes can produce genuine excusing conditions, such as lack of rational or control capacity." But beyond those rare instances, he believes, neuroscience offers little that should challenge the notion of responsibility. As he has quipped, "Brains don't kill people. People kill people."

Morse epitomizes the skepticism about bringing neuroscience into the courtroom. For one thing, he viscerally cringes at how much of a fad "neurolaw" and "neurocriminology" have become. A wonderfully sardonic writer,*

* And also a very nice guy. Along with a Stanford colleague, law professor and bioethicist Hank Greely, I once got to debate against Morse and a colleague at a law school. It was both really fun, because Morse is insanely smart, and really terrifying, because he's insanely smart.

he has announced the discovery of the disorder "brain overclaim syndrome," whose sufferers have gotten carried away with the importance of neuroscience because they've been "infected and inflamed by stunning advances in our understanding of the brain," causing them to "make moral and legal claims that the new neuroscience does not entail and cannot sustain."

One absolutely valid criticism of his is a narrow, practical one. This is the worry, aired earlier, that juries will give undue weight to neuroimaging data just because of how impressive the images are. Apropos of that, Morse has called neuroscience "determinism du jour, grabbing the attention previously given to psychological or genetic determinism. . . . The only thing different about neuroscience is that we have prettier pictures and it appears more scientific."

Another valid criticism concerns findings in neuroscience usually merely being descriptive (e.g., "Brain region A projects to brain region Q") or correlative (e.g., "Elevated levels of neurotransmitter X and of behavior Z tend to go together"). Data such as those don't disprove free will. In the words of philosopher Hilary Bok, "The claim that a person chose her action does not conflict with the claim that some neural processes or states caused it; it simply redescribes it."[23]

This is a point I've made throughout the book, namely that description and correlation are nice, but actual causal data are the gold standard (e.g., "When you raise the levels of neurotransmitter X, behavior Z happens more often"). That is the source of some of our most powerful demonstrations of the material bases of our more complex behaviors—for example, transcranial magnetic stimulation techniques that transiently activate or inactivate a part of the cortex can change someone's moral decision making, decisions about punishment, or levels of generosity and empathy. That's causality.

It is when we get to the issue of causality that Morse distinguishes between causation and compulsion. He writes, "Causation is neither an excuse per se nor the equivalent of compulsion, which is an excusing condition." Morse describes himself as a "thoroughgoing materialist" and states, "We live in a causal universe, which includes human action." But try as I might, I cannot see any way of making this distinction that does not tacitly require a homunculus that is outside the causal universe, a homunculus that can be overwhelmed by "compulsion" but that can and should handle "causation." In the words of philosopher Shaun Nichols, "It seems like something has to give, either our commitment to free will or our commitment to the idea that every event is completely caused by the preceding events."[24]

Despite these criticisms of his criticisms, my stance has a major problem, one that causes Morse to conclude that the contributions of neuroscience to the legal system "are modest at best and neuroscience poses no genuine, radical challenges to concepts of personhood, responsibility, and competence."[25] The problem can be summarized in a hypothetical exchange:

Prosecutor: So, professor, you've told us about the extensive damage that the defendant sustained to his frontal cortex when he was a child. Has every person who has sustained such damage become a multiple murderer, like the defendant?

Neuroscientist testifying for the defense: No.

Prosecutor: Has every such person at least engaged in some sort of serious criminal behavior?

Neuroscientist: No.

Prosecutor: Can brain science explain why the same amount of damage produced murderous behavior in the defendant?

Neuroscientist: No.

The problem is that, even amid all these biological insights that allow us to be snitty about those silly homunculi, we still can't predict much about behavior. Perhaps at the statistical level of groups, but not when it comes to individuals.

Explaining Lots and Predicting Little

If a person's leg is fractured, how predictable is it that they will have trouble walking? I think it would be safe to predict something close to 100 percent. If they have serious inflammatory lung disease, how predictable is it that their breathing will be labored at times and that they will tire easily? Again, around 100 percent. Same for the effects of significant blockage of blood flow to the legs or extensive cirrhosis of the liver.

Let's switch to the brain and neurological dysfunction. What if someone has had a brain injury, and the neurons around the resulting scar tissue rewire so that they stimulate both themselves and one another—how predictable is it that the person will have a seizure? How about if they have congenital weaknesses in the walls of the blood vessels throughout the brain—how likely is a cerebral aneurysm at some point? How about if they have a mutation in the gene that causes Huntington's disease—how likely are they to have a neuromuscular disorder by age sixty? Really high in all cases; probably approaching 100 percent.

Let's incorporate behavior. If someone has extensive frontocortical damage, how predictable is it that you'd note something odd about them, behaviorally, after a five-minute conversation? Something like 75 percent.

Now let's consider a broader range of behaviors. How predictable is it that this person with the frontal damage will do something horrifically violent at some point? Or that someone who was abused repeatedly as a child will become an abusive adult? That a soldier who went through a battle that killed his buddies will develop PTSD? That a person with the "montane vole" polygamous version of the vasopressin receptor gene promoter will have numerous failed marriages? That a person with a particular array of glutamate receptor subtypes throughout their cortex and hippocampus will have an IQ above 140? That someone raised with extensive childhood adversity and loss will have a major depressive disorder? All under 50 percent, often way under.

So how do a fractured leg inevitably impairing locomotion and the noninevitabilities of the previous paragraph differ? Do the latter somehow involve "less" biology? Is the point that the brain contains a nonbiological homunculus but that leg bones do not?

Hopefully, after this many pages, the start of an answer is apparent. It's not that there's "less" biology in those circumstances related to social behavior. It's that it's qualitatively different biology.

When a bone shatters, there's a relatively straight line of steps leading to inflammation and pain that will impair the person's gait (should he try to walk an hour later). That straight line of biology won't be altered by conventional variation in his genome, his prenatal hormone exposure, the culture he was raised in, or when he ate lunch. But as we've seen, all of those variables can influence social behaviors that shape our best and worst moments.

The biology of the behaviors that interest us is, in all cases, *multifactorial*—that is the thesis of this book.

Let's see what "multifactorial" means in a practical sense. Consider someone with frequent depression who is visiting a friend today, pouring her heart out about her problems. How much could you have predicted the global depression and today's behavior by knowing about her biology?

Suppose "knowing about her biology" consisted only of knowing what version of the serotonin transporter gene she has. How much predictive power does that give you? As we saw in chapter 8, not much—say, 10 percent. What if "knowing about her biology" consists of knowing the status of that gene plus knowing if one of her parents died when she was a child? More, maybe 25 percent. How knowing her serotonin transporter gene status + childhood adversity status + whether she is living alone in poverty? Maybe up to 40 percent. Add knowledge of the average level of glucocorticoids in her bloodstream today. Maybe a bit more. Toss in knowing if she's living in an individualist or a collectivist culture. Some more predictability.* Know if she is menstruating (which typically exac-

* Because cross-cultural psychiatry research shows that in individualist cultures, when depressed people talk to a friend for relief, they're likely to talk about their problems, whereas in collectivist cultures they're likely to ask about the friend's problems.

erbates symptoms in seriously depressed women, making it more likely that they'll be socially withdrawn rather than reaching out to someone). Some more predictability. Maybe even above the 50 percent mark by now. Add enough factors, many of which, possibly most of which, have not yet been discovered, and eventually your multifactorial biological knowledge will give you the same predictive power as in the fractured-bone scenario. Not different *amounts* of biological causation; different *types* of causation.

The artificial intelligence pioneer Marvin Minsky once defined free will as "internal forces I do not understand."[26] People intuitively believe in free will, not just because we have this terrible human need for agency but also because most people know next to nothing about those internal forces. And even the neuroscientist on the witness stand can't accurately predict which individual with extensive frontal damage will become the serial murderer, because science as a whole still knows about only a handful of those internal forces. Shattered bone → inflammation → constricted movement is easy. Neurotransmitters + hormones + childhood + ____ + ____ + isn't.*[27]

Another factor comes into play. When I go to the Web of Science, a search engine for scanning databases of papers published in science and medical journals. Under search terms I put in "oxytocin" and "trust"—just to pick an example of the umpteen links between biology and social behavior that we've covered. And up comes the news that 193 papers have been published on the subject. Consider the following figure, showing that most of those papers have been published in the last few years.

Same with the next figure, a search for "oxytocin" and "social behavior" or, after that, "transcranial magnetic stimulation" and "decision-making," and then "brain" and "aggression."

---

* Just to give a sense of how few baby steps we've taken, the maximal number of contributing variables identified in predicting depression is serotonin transporter status + childhood adversity status + adult social support status. That's it, that's about how far the literature has gotten. For frontal damage and antisocial violence, it's neurological status of frontal cortex + D4 dopamine receptor subtype + ADHD status.

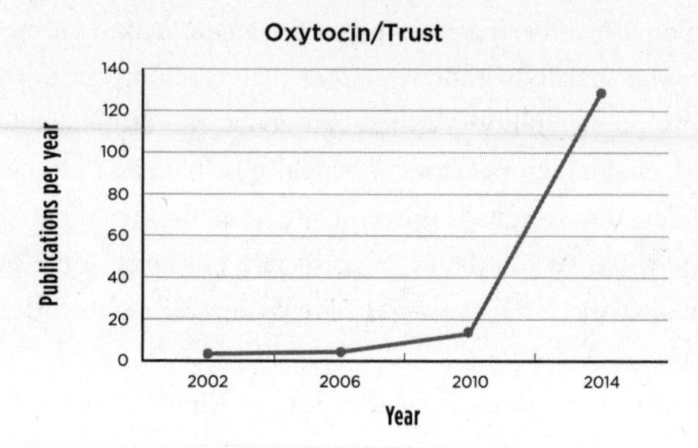
Oxytocin/Trust
Publications per year
140
120
100
80
60
40
20
0
2002
2006
2010
2014
Year

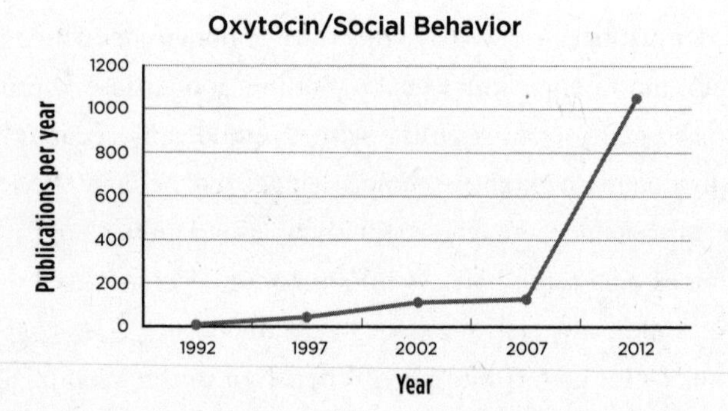
Oxytocin/Social Behavior
Publications per year
1200
1000
800
600
400
200
0
1992
1997
2002
2007
2012
Year

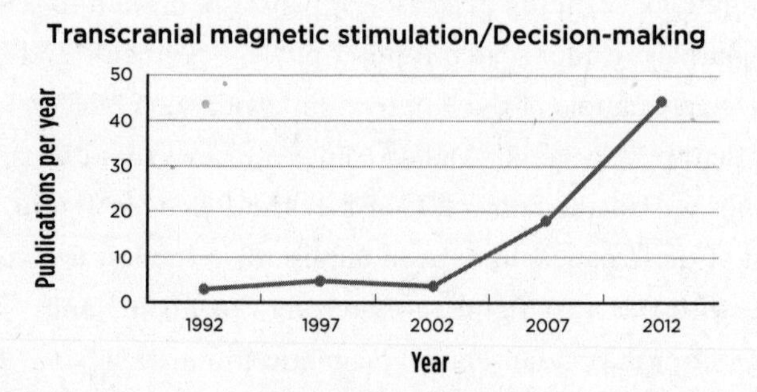
Transcranial magnetic stimulation/Decision-making
Publications per year
50
40
30
20
10
0
1992
1997
2002
2007
2012
Year

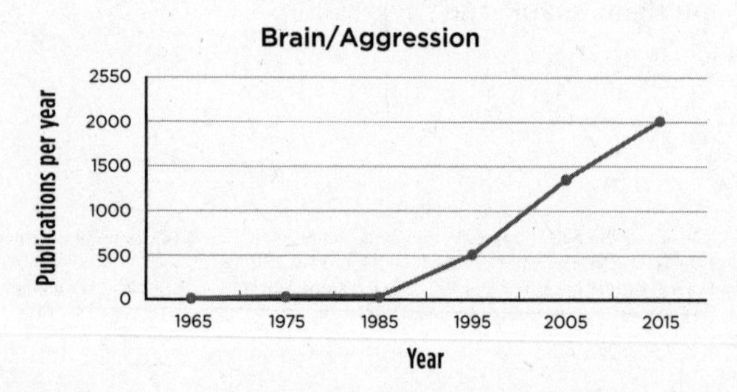
Brain/Aggression
Publications per year
2550
2000
1500
1000
500
0
1965
1975
1985
1995
2005
2015
Year

And just to give a sense of some more of these:

	Search terms				
	genes/ behavior	testoster-one/ag-gression	amygdala/ aggression	MAO/ aggression	epi-genetics/ behavior
1920–30	1	0	0	0	0
1930–40	3	0	0	0	0
1940–50	3	0	0	0	0
1950–60	10	2	0	0	0
1960–70	22	3	2	0	0
1970–80	39	24	4	1	0
1980–90	128	53	5	2	0
1990–2000	9,288	401	97	40	9
2000–2010	27,754	757	321	119	197
2010–20	52,487	1,070	560	184	1,012

(Note: 2010–20 data are prorated from 2010–2015.)

Our behaviors are constantly shaped by an array of subterranean forces. What these figures and the table show is that most of these forces involve biology that, not that long ago, we didn't know existed.

So what do we do with Minsky's definition of free will needing to be amended to "internal forces I do not understand *yet*"?

HOW THEY WILL VIEW US

If you still think there is mitigated free will, there are three possible routes to take at this juncture.

To appreciate the first, let's briefly consider epilepsy. Scientists understand a lot about the neurological bases of seizures and how they involve firing with abnormally high frequency and synchrony. But not that long ago, say, a century ago, epilepsy was viewed as a type of mental illness. And before that it was thought by many to be a communicable infectious disease. And at other times and places, it was thought to be caused by menstruation, or excessive sex, or excessive masturbation. But in 1487 two German scholars uncovered a cause of epilepsy that really seemed to hit the nail on the head.

The two Dominican friars, Heinrich Kramer and Jakob Sprenger, published *Malleus Maleficarum* (Latin for "Hammer of the Witches"), the definitive treatise about why someone becomes a witch, how to identify them, and what to do with them. What was one of the surest ways to identify a witch? If they are seized by Satan, if they convulse from the malign power of the devil within them.

Their guideline was the Gospel according to Mark, 9:14–29. A man brings his son to Jesus, saying there is something wrong with him and asking Jesus to cure him—a spirit comes and seizes him, making him mute, and then that spirit throws him to the ground, where he foams at the mouth and grinds his teeth and becomes rigid. The man presents his son, who is promptly seized by that spirit and falls to the ground, convulsing and foaming. Jesus perceives that the boy is infested with an unclean spirit and commands that vile spirit to come out and be gone. The seizing ceases.

Thus seizures were a sign of demonic possession, a certain marker of a witch. *Malleus Maleficarum* arrived in time to take advantage of mass production through the recently invented printing press. In the words of historian Jeffrey Russell, "The swift propagation of the witch hysteria by the press was the first evidence that Gutenberg had not liberated man from original sin." The book was widely read and went through more than thirty editions over the subsequent century. Estimates are that from 100,000 to a million people were persecuted, tortured, or killed as witches in the aftermath.*[28]

I don't think much of Kramer and Sprenger. My assumption is that they were sadistic monsters, but that could reflect my being influenced too much by the likes of *The Name of the Rose* or *The Da Vinci Code*. Maybe

* I thank an excellent student, Katrina Hui, for bringing my attention to *Malleus Maleficarum*.

they were opportunists who reasoned that the book would make their careers. Maybe they were utterly sincere.

Instead I imagine a scenario of an evening during the late fifteenth century. A church inquisitor comes home from work weary, burdened. His wife coaxes him to talk—"It was a usual day of condemning witches, but this one case bothered me. Everyone testified about this woman who falls and gnashes and convulses—a witch, without a doubt. I don't feel sorry for her—no one told her to open wide for Satan. But she had these two beautiful kids—you should have seen them, just so confused as to why their mother was being taken away. Distraught husband also. So that part was hard, seeing them suffer. But it is what it is—we burned her, of course." Burnings and killings and centuries were to pass until we in the West would have learned enough to say, "It's not her; it's her disease."*

We're only a first few baby steps into understanding any of this, so few that it leaves huge, unexplained gaps that perfectly smart people fill in with a homunculus. Nevertheless, even the staunchest believers in free will must admit that it is hemmed into tighter spaces than in the past. It's less than two centuries since science first taught us that the frontal cortex has something to do with appropriate behavior. Less than seventy years since we learned that schizophrenia is a biochemical disorder. Perhaps fifty years since we learned that reading problems of a type that we now call dyslexia aren't due to laziness but instead involve microscopic cortical malformations. Twenty-five since we learned that epigenetics alters behavior. The influential philosopher Daniel Dennett has written about the free will that is "worth wanting." If there really is free will, it's getting consigned to domains too mundane to be worth the effort to want—do I want briefs or boxer shorts today?[29]

Recall those charts and table showing the recentness of these scientific discoveries. If you believe that starting tonight, at midnight, something will happen and science will stop, that there will be no new publications, findings, or knowledge relevant to this book, that we now know everything there is, then it is clear what one's stance should be—there are some

* I specify "in the West" because this is by no means a universal interpretation even today.

rare domains where extremes of biological dysfunction cause involuntary changes in behavior, and we're not great at predicting who undergoes such changes. In other words, the homunculus is alive and well.

But if you believe that there will be the accrual of any more knowledge, you've just committed to either the view that any evidence for free will ultimately will be eliminated or the view that, at the very least, the homunculus will be jammed into ever tinier places. And with either of those views, you've also agreed that something else is virtually guaranteed: that people in the future will look back at us as we do at purveyors of leeches and bloodletting and trepanation, as we look back at the fifteenth-century experts who spent their days condemning witches, that those people in the future will consider us and think, "My God, the things they didn't know then. The harm that they did."

Archaeologists do something impressive, reflecting disciplinary humility. When archaeologists excavate a site, they recognize that future archaeologists will be horrified at their primitive techniques, at the destructiveness of their excavating. Thus they often leave most of a site untouched to await their more skillful disciplinary descendants. For example, astonishingly, more than forty years after excavations began, less than 1 percent of the famed Qin dynasty terra-cotta army in China has been uncovered.

Those adjudicating trials don't have the luxury of adjourning for a century until we really understand the biology of behavior. But at the very least the system needs the humility of archaeology, a sense that, above all else, we shouldn't act irrevocably.

But what do we actually do in the meantime? Simple (which is easy for me to say, looking at the legal world from the soothing distance of my laboratory): probably just three things. One is easy, one is very challenging to implement, and the third is nearly impossible.

First the easy one. If you reject free will and the discussion turns to the legal system, the crazy-making, inane challenge that always surfaces is that you'd do nothing about criminals, that they'd be free to walk the streets, wreaking havoc. Let's trash this one instantly—no rational person who rejects free will actually believes this, would argue that we should do nothing because, after all, the person has frontal damage, or because, after all, evolution has selected for the damaging trait to traditionally be adaptive, or because, after all . . . People must be protected from individuals who are dangerous. The latter can no more be allowed to walk the streets than you can allow a car whose brakes are faulty to be driven. Rehabilitate such people if you can, send them to the Island of Misfit Toys forever if you can't and they are destined to remain dangerous. Josh Greene and Jonathan Cohen of Princeton wrote an extremely clearheaded piece on this, "For the Law, Neuroscience Changes Nothing and Everything." Where neuroscience and the rest of biology change nothing is in the continued need to protect the endangered from the dangerous.[30]

Now for the nearly impossible issue, the one that "changes everything"—the issue of punishment. Maybe, just maybe, a criminal must suffer punishment at junctures in a behaviorist framework, as part of rehabilitation, part of making recidivism unlikely by fostering expanded frontal capacity. It is implicit in the very process of denying a dangerous individual their freedom by removing them from society. But precluding free will precludes punishment being an end in and of itself, punishment being imagined to "balance" the scales of justice.

It is the punisher's mind-set where everything must be changed. The difficulty of this is explored in the superb book *The Punisher's Brain: The Evolution of Judge and Jury* (2014) by Morris Hoffman, a practicing judge and legal scholar.[31] He reviews the reasons for punishment: As we see from game theory studies, because punishment fosters cooperation. Because it is in the fabric of the evolution of sociality. And most important, because it can feel good to punish, to be part of a righteous and self-righteous crowd at a public hanging, knowing that justice is being served.

This is a deep, atavistic pleasure. Put people in brain scanners, give them scenarios of norm violations. Decision making about culpability for the violation correlates with activity in the cognitive dlPFC. But decision making about appropriate punishment activates the emotional vmPFC, along with the amygdala and insula; the more activation, the more punishment.[32] The decision to punish, the passionate motivation to do so, is a frothy limbic state. As are the consequences of punishing—when subjects punish someone for making a lousy offer in an economic game, there's activation of dopaminergic reward systems. Punishment that just feels good.

It makes sense that we've evolved such that it is limbic froth that is at the center of punishing, and that a pleasurable dopaminergic surge rewards doing so. Punishment is effortful and costly, ranging from forgoing a reward when rejecting a lowball offer in the Ultimatum Game to our tax dollars paying for the dental plan of the prison guard who operates the lethal injection machine. That rush of self-righteous pleasure is what drives us to shoulder the costs. This was shown in one neuroimaging study of economic game play. Subjects alternated between being able to punish lousy offers at no cost and having to spend points they had earned to do so. And the more dopaminergic activation during no-cost punishment, the more someone would pay to punish in the other condition.[33]

Thus the nearly impossible task is to overcome that. Sure, as I said, punishment would still be used in an instrumental fashion, to acutely shape behavior. But there is simply no place for the idea that punishment is a virtue. Our dopaminergic pathways will have to find their stimulation elsewhere. I sure don't know how best to achieve that mind-set. But crucially, I sure *do* know we can do it—because we have before: Once people with epilepsy were virtuously punished for their intimacy with Lucifer. Now we mandate that if their seizures aren't under control, they can't drive. And the key point is that no one views such a driving ban as virtuous, pleasurable punishment, believing that a person with treatment-resistant seizures "deserves" to be banned from driving. Crowds of goitrous yahoos don't excitedly mass to watch the epileptic's driver's license be publicly burned. We've successfully banished the notion of punishment in that realm. It may take centuries, but we can do the same in all our current arenas of punishment.

Which brings us to the huge practical challenge. The traditional rationales behind imprisonment are to protect the public, to rehabilitate, to punish, and finally to use the threat of punishment to deter others. That last one is the practical challenge, because such threats of punishment can indeed deter. How can that be done? The broadest type of solution is incompatible with an open society—making the public *believe* that imprisonment involves horrific punishments when, in reality, it doesn't. Perhaps the loss of freedom that occurs when a dangerous person is removed from society must be deterrence enough. Perhaps some conventional punishment will still be needed if it is sufficiently deterring. But what must be abolished are the views that punishment can be deserved and that punishing can be virtuous.

None of this will be easy. When contemplating the challenge to do so, it is important to remember that some, many, maybe even most of the people who were prosecuting epileptics in the fifteenth century were no different from us—sincere, cautious, and ethical, concerned about the serious problems threatening their society, hoping to bequeath their children a safer world. Just operating with an unrecognizably different mindset. The psychological distance from them to us is vast, separated by the yawning chasm that was the discovery of "It's not her, it's her disease." Having crossed that divide, the distance we now need to go is far shorter—it merely consists of taking that same insight and being willing to see its valid extension in whatever directions science takes us.

The hope is that when it comes to dealing with humans whose behaviors are among our worst and most damaging, words like "evil" and "soul" will be as irrelevant as when considering a car with faulty brakes, that they will be as rarely spoken in a courtroom as in an auto repair shop. And crucially, the analogy holds in a key way, extending to instances of dangerous people without anything obviously wrong with their frontal cortex, genes, and so on. When a car is being dysfunctional and dangerous and we take it to a mechanic, this is not a dualistic situation where (a) if the mechanic discovers some broken widget causing the problem, we have a mechanistic explanation, but (b) if the mechanic can't find anything wrong, we're dealing with an evil car; sure, the mechanic can speculate on the source of the problem—maybe it's the blueprint from which the car was built, maybe it was the building process, maybe the environment contains some unknown pollutant that somehow impairs function, maybe someday we'll have sufficiently powerful techniques in the auto shop to spot some key molecule in the engine that is out of whack—but in the meantime we'll consider this car to be evil. Car free will also equals "internal forces we do not understand yet."*[34]

Many who are viscerally opposed to this view charge that it is dehumanizing to frame damaged humans as broken machines. But as a final, crucial point, doing that is a hell of a lot more humane than demonizing and sermonizing them as sinners.

POSTSCRIPT: NOW FOR THE HARD PART

Well, so much for the criminal justice system. Now on to the really difficult part, which is what to do when someone compliments your zygomatic arches.

If we deny free will when it comes to the worst of our behaviors, the same must also apply to the best. To our talents, displays of willpower and focus, moments of bursting creativity, decency, and compassion. Logically it should seem as ludicrous to take credit for those traits as to respond to a compliment on the beauty of your cheekbones by thanking the person for implicitly having praised your free will, instead of explaining how mechanical forces acted upon the zygomatic arches of your skull.

It will be so difficult to act that way. I am willing to admit that I have acted egregiously in this regard. My wife and I have brunch with a friend, who serves fruit salad. We proclaim, "Wow, the pineapple is delicious." "They're out of season," our host smugly responds, "but I lucked out and found a decent one." My wife and I express awestruck worship—"You really know how to pick fruit. You are a better person than we are." We

* Cars may soon be entering discussions of moral decision making—when having to do one or the other, should a self-driving car smash itself into a wall, killing the passenger, in order to save five pedestrians? Most people think that is how such cars should be programmed but, predictably, would prefer that one that they used make the opposite choice. Perhaps more expensive models will work that way, while the hoi polloi have more utilitarian cars. Or maybe the car will decide, based on how frequently you clean it and change its oil.

are praising the host for this supposed display of free will, for the choice made at the fork in life's road that is pineapple choosing. But we're wrong. In reality, genes had something to do with the olfactory receptors our host has that help detect ripeness. Maybe our host comes from a people whose deep and ancient cultural values include learning how to feel up a pineapple to tell if it's good. The sheer luck of the socioeconomic trajectory of our host's life has provided the resources to prowl an overpriced organic market playing Peruvian folk Muzak. Yet we praise our host.

I can't really imagine how to live your life as if there is no free will. It may never be possible to view ourselves as the sum of our biology. Perhaps we'll have to settle for making sure our homuncular myths are benign, and save the heavy lifting of truly thinking rationally for where it matters—when we judge others harshly.

Seventeen

War and Peace

Let's review some facts. The amygdala typically activates when seeing a face of another race. If you're poor, by the time you're five, your frontal cortical development probably lags behind average. Oxytocin makes us crappy to strangers. Empathy doesn't particularly translate into compassionate acts, nor does refined moral development translate into doing the harder, right thing. There are gene variants that, in particular settings, make you prone toward antisocial acts. And bonobos aren't perfectly peaceful—they wouldn't be masters of reconciliation if they didn't have conflicts to reconcile.

All this makes one mighty pessimistic. Yet the rationale for this book is that, nonetheless, there's ground for optimism.

Thus this final chapter's goals are (a) to evidence that things have improved, that many of our worst behaviors are in retreat, our best ones ascendant; (b) to examine ways to improve this further; (c) to derive emotional support for this venture, to see that our best behaviors can occur in the most unlikely circumstances; (d) and finally, to see if I can actually get away with calling this chapter "War and Peace."

SOMEWHAT BETTER ANGELS

When it comes to our best and worst behaviors, the world is astonishingly different from that of the not-so-distant past. At the dawn of the nineteenth century, slavery occurred worldwide, including in the colonies of a Europe basking in the Enlightenment. Child labor was universal and would soon reach its exploitative golden age with the Industrial Revolution. And there wasn't a country that punished mistreatment of animals. Now every nation has outlawed slavery, and most attempt to enforce that; most have child labor laws, rates of child labor have declined, and it increasingly consists of children working alongside their parents in their homes; most countries regulate the treatment of animals in some manner.

The world is also safer. Fifteenth-century Europe averaged 41 homicides per 100,000 people per year. Currently only El Salvador, Venezuela, and Honduras, at 62, 64, and 85, respectively, are worse; the world averages 6.9, Europe averages 1.4, and there are Iceland, Japan, and Singapore at 0.3.

Here are things that are rarer in recent centuries: Forced marriages, child brides, genital mutilation, wife beating, polygamy, widow burning. Persecution of homosexuals, epileptics, albinos. Beating of schoolchildren, beating of beasts of burden. Rule of a land by an occupying army, by a colonial overlord, by an unelected dictator. Illiteracy, death in infancy, death in childbirth, death from preventable disease. Capital punishment.

Here are things invented in the last century: Bans on the use of certain types of weapons. The World Court and the concept of crimes against humanity. The UN and the dispatching of multinational peacekeeping forces. International agreements to hinder trafficking of blood diamonds, elephant tusks, rhino horns, leopard skins, and humans. Agencies that collect money to aid disaster victims anywhere on the planet, that facilitate intercontinental adoption of orphans, that battle global pandemics and send medical personnel to any place of conflict.

Yes, I know, I'm an utter naïf if I think laws are universally enforced. For example, in 1981 Mauritania became the last country to ban slavery;

nevertheless, today roughly 20 percent of its people are slaves, and the government has prosecuted a total of one slave owner.[1] I recognize that little has changed in many places; I have spent decades in Africa living around people who believe that epileptics are possessed and that the organs of murdered albinos have healing powers, where beating of wives, children, and animals is the norm, five-year-olds herd cattle and haul firewood, pubescent girls are clitoridectomized and given to old men as third wives. Nonetheless, worldwide, things have improved.

The definitive account of this is Pinker's monumental *The Better Angels of Our Nature: Why Violence Has Declined.*[2] It's a scholarly work that's gut-wrenchingly effective in documenting just how bad things once were. Pinker graphically describes the appalling historical inhumanity of humans. Roughly half a million people died in the Roman Colosseum to supply audiences of tens of thousands the pleasure of watching captives raped, dismembered, tortured, eaten by animals. Throughout the Middle Ages, armies swept across Eurasia, destroying villages, killing every man, consigning every woman and child to slavery. Aristocracy accounted for a disproportionate share of violence, savaging peasantry with impunity. Religious and governmental authorities, ranging from Europeans to Persians, Chinese, Hindus, Polynesians, Aztecs, Africans, and Native Americans, invented means of torture. For a bored sixteenth-century Parisian, entertainment might consist of a cat burning, execution of a "criminal" animal, or bearbaiting, where a bear, chained to a post, would be torn apart by dogs. It is a sickeningly different world; Pinker quotes the writer L. P. Hartley: "The past is a foreign country: they do things differently there."

Better Angels has provoked three controversies:

Why Were People So Awful Then?

For Pinker the answer is clear. Because people had always been so awful. This is chapter 9's debate—when was war invented, was ancestral hunter-gatherer life about Hobbes or Rousseau? As we saw, Pinker is in the camp holding that organized human violence predates civilization, stretching back to our last common ancestor with chimps. And as reviewed, most experts convincingly disagree, suggesting that data have been cherry-picked, hunter-horticulturalists mislabeled as hunter-gatherers, and newfangled sedentary hunter-gatherers inappropriately grouped with traditional nomadic ones.

Why Have People Gotten Less Awful?

Pinker's answer reflects two factors. He draws on the sociologist Norbert Elias, whose notion of the "civilizing process" centered on the fact that violence declines when states monopolize force. That is coupled with spread of commerce and trade, fostering realpolitik self-restraint—recognizing that it's better to have this other person alive and trading with you. Their well-being begins to matter, prompting what Pinker calls an "escalator of reasoning"—an enlarged capacity for empathy and Us-ness. This underlies the "rights revolution"—civil rights, women's rights, children's rights, gay rights, animal rights. This view is a triumph of cognition. Pinker yokes this to the "Flynn effect," the well-documented increase in average IQ over the last century; he invokes a moral Flynn effect, as increasing intelligence and respect for reasoning fuel better Theory of Mind and perspective taking and an increased ability to appreciate the long-term advantages of peace. In the words of one reviewer, Pinker is "not too fainthearted to call his own culture civilized."[3]

Predictably, this has drawn fire from all sides. The Left charges that this giddy overvaluing of the dead-white-male Enlightenment fuels Western neoimperialism.[4] My personal political instincts run in this direction. Nonetheless, one must admit that the countries with minimal violence, extensive social safety nets, few child brides, numerous female legislators, and sacrosanct civil liberties are usually direct cultural descendants of the Enlightenment.

Meanwhile, the Right claims that Pinker ignores religion, pretending that decency was invented in the Enlightenment.[5] He is eloquently unapologetic about this—for him much of what has gone right reflects people's "shifting from valuing souls to valuing lives." For others the criticism is that this escalator of reasoning fetishizes cognition over affect—after all, sociopaths have great Theory of Mind, a (damage-induced) purely rational mind makes abhorrent moral judgments, and a sense of justice is fueled by the amygdala and insula, not the dlPFC. Obviously, this many pages into the book, I feel that the interaction of reasoning and feeling is key.

Have People Really Gotten Less Awful?

This has been very contentious. Pinker offers the sound bite "We may be living in the most peaceful era in our species' existence." The fact most driving this optimism is that, except for the Balkan wars, Europe has been at peace since 1945, the longest stretch in history. For Pinker, this "Long Peace" represents the West coming to its senses after the ruin of World War II, seeing how the advantages of being a common market outweigh those of being a perpetually warring continent, plus some expanding empathy thrown in on the side.

Critics characterize this as Eurocentrism. Western countries may kumbaya one another, but they've sure made war elsewhere—France in Indochina and Algeria, Britain in Malaya and Kenya, Portugal in Angola and Mozambique, the USSR in Afghanistan, the United States in Vietnam, Korea, and Latin America. Moreover, parts of the developing world have been continuously at war for decades—consider the eastern Congo. Most important, such wars have been made bloodier because the West invented the idea of having client states fight proxy wars for them. After all, the late twentieth century saw the United States and USSR arm the warring Somalia and Ethiopia, only to switch to arming the *other* side within a few years. The Long Peace has been for Westerners.

The claim of violence declining steadily over the last millennium also must accommodate the entire bloody twentieth century. World War II killed 55 million people, more than any conflict in history. Throw in World War I, Stalin, Mao, and the Russian and Chinese civil wars, and you're up to 130 million.

Pinker does something sensible that reflects his being a scientist. He corrects for total population size. Thus, while the eighth century's An

Lushan Rebellion and civil war in Tang dynasty China killed "only" 36 million, that represented one sixth of the world's population—the equivalent of 429 million in the midtwentieth century. When deaths are expressed as a percentage of total population, World War II is the only twentieth-century event cracking the top ten, behind An Lushan, the Mongol conquests, the Mideast slave trade, the fall of the Ming dynasty, the fall of Rome, the deaths caused by Tamerlane, the annihilation of Native Americans by Europeans, and the Atlantic slave trade.

Critics have questioned this—"Hey, stop using fudge factors to somehow make World War II's 55 million dead less than the fall of Rome's 8 million." After all, 9/11's murders would not have evoked only half as much terror if America had 600 million instead of 300 million citizens. But Pinker's analysis is appropriate, and analyzing *rates* of events is how you discover that today's London is much safer than was Dickens's or that some hunter-gatherer groups have homicide rates that match Detroit's.

But Pinker failed to take things one logical step further—also correcting for differing durations of events. Thus he compares the half dozen years of World War II with, for example, twelve *centuries* of the Mideast slave trade and four centuries of Native American genocide. When corrected for duration as well as total world population, the top ten now include World War II (number one), World War I (number three), the Russian Civil War (number eight), Mao (number ten), and an event that didn't even make Pinker's original list, the Rwandan genocide (number seven), where 700,000 people were killed in a hundred days.*

This suggests both good and bad news. Compared with the past, we are extraordinarily different in terms of whom we extend rights to and feel empathy for and what global ills we counter. And things are better in terms of fewer people acting violently and societies attempting to contain them. But the bad news is that the reach of the violent few is ever greater. They don't just rage about events on another continent—they travel there

* The full list (figures in deaths per year, approximate): (1) World War II, 11 million; (2) An Lushan Rebellion, 4.5 million; (3) World War I, 3 million; (4) and (5) Taiping Rebellion and Tamerlane, each 2.8 million; (6) fall of the Ming dynasty, 2.5 million; (7) and (8) Mongol conquests and Rwandan genocide, each 2.4 million; (9) Russian Civil War, 1.8 million; (10) Russia's sixteenth/seventeenth-century Time of Troubles, 1.5 million; (11) Mao-induced Chinese famines, 1.4 million.

and wreak havoc. The charismatically violent inspire thousands in chat rooms instead of a mob in their village. Like-minded lone wolves more readily meet and metastasize. And the chaos once let loose with a cudgel or machete occurs now with an automatic weapon or bomb, with far more horrific consequences. Things have improved. But that doesn't mean they're good.

Thus we now consider insights provided by this book that might help.

SOME TRADITIONAL ROUTES

First there's the strategy for reducing violence that stretches back tens of thousands of years—moving. If two individuals in a hunter-gatherer band are having tensions, one frequently shifts to a neighboring band, sometimes voluntarily, sometimes not. Similarly, interband tensions are reduced when one shifts to a different location, an advantage of nomadicism. A recent study of the hunter-gathering Hadza of Tanzania showed an additional benefit to this fluidity straight out of chapter 10. Specifically, it facilitates highly cooperative individuals associating with one another.[6]

Then there are the beneficial effects of trade, as emphasized by anthropologists, as well as Pinker. From trading at a village market to signing international trade agreements, it is often true that where goods do not pass frontiers, armies will. It's a version of Thomas Friedman's somewhat tongue-in-cheek Golden Arches Theory of peace—countries with McDonald's don't fight one another. While there are exceptions (e.g., the U.S. invasion of Panama, the Israeli invasion of Lebanon), Friedman's broad point holds—countries that are sufficiently stable that they are integrated into global markets with the likes of McDonald's and prosperous enough that their people keep those establishments in business likely conclude that the trade advantages of peace outweigh the imagined spoils of war.*†[7]

This isn't surefire—for example, despite being major trading partners,

* This has always been the classic interpretation of Friedman's idea. It's quite possible, however, that people don't go to war in those circumstances because they're too busy going to the doctor for their adult-onset diabetes.

† An exception is Lawrence Keeley of chapter 9, who argued that the net result of trade, with its inevitable disagreements, is more, rather than less, intergroup tension.

Germany and the UK fought World War I—and there's no shortage of people willing to go to war, even at the cost of disrupted trade and scarce commodities. Moreover, "trade" is double-edged. It's certifiably groovy when occurring between indigenous rain forest hunters; it's certifiably vile if you're protesting the WTO. But as long as countries can wage war on distant nations, long-distance trade that makes them interdependent is a good deterrent.

Cultural diffusion in general (which includes trade) can also facilitate peace. This can have a modern tint—across 189 countries, digital access predicts increased civil liberties and media freedom. Moreover, the more civil liberties in a neighboring country, the stronger this effect, as ideas flow with goods.[8]

Religion

Well, I'd love to skip this section, but I can't. That's because religion is arguably our most defining cultural invention, an incredibly powerful catalyst for both our best and worst behaviors.

When introducing the pituitary in chapter 4, I didn't feel obliged to first disclose my feelings about the gland. But the equivalent feels appropriate here. Thus: I was raised highly observant and Orthodox, felt intensely religious. But then, around age thirteen, the whole edifice collapsed; ever since, I've been incapable of any religiosity or spirituality and more readily focus on religion's destructive than its beneficial aspects. But I like being around religious people and am moved by them—while baffled by how they can believe that stuff. And I fervently wish that I could. The end.

As emphasized in chapter 9, we've created a staggering variety of religions. In considering solely religions with worldwide reaches, there are some important commonalities:

a. They all involve facets of religiosity that are intensely personal, solitary, and individualized, as well as facets that are about community; as we'll see, these are very different realms when it comes to fostering our best and worst behaviors.

b. All involve personal and communal ritualized behaviors that comfort in times of anxiety; however, many of those anxieties were created by the religion itself.

The anxiety-reducing effects of belief are logical, given that psychological stress is about lack of control, predictability, outlets, and social support. Depending on the religion, belief brings an explanation for why things happen, a conviction that there is a purpose, and the sense of a creator who is interested in us, who is benevolent, who responds to human entreaties, who preferentially responds to entreaties from people like you. No wonder religiosity has health benefits (independent of the community support that it brings and the decreased rates of substance abuse).

Recall the role of the anterior cingulate cortex (ACC) in sounding an alarm when there is a discrepancy between how you thought things worked and how they actually do. After controlling for personality and cognitive abilities, more religious people show less ACC activation when getting news of a negative discrepancy. Other studies show the anxiety-reducing effects of repetitive religious rituals.[9]

c. Finally, all the world religions distinguish between Us and Them, though they differ as to what is required to be an Us and whether the pertinent attributes are immutable.

Enough is known about the neurobiology of religiosity that there's even a journal called *Religion, Brain and Behavior.* Reciting a familiar prayer activates mesolimbic dopaminergic systems. Improvising one activates regions associated with Theory of Mind, as you try to understand a deity's perspective ("God wants me to be humble in addition to grateful; better make sure I mention that"). Moreover, more activation of this Theory of Mind network correlates with a more personified image of a deity. Believing that someone is faith healing deactivates the (cognitive) dlPFC, suspending disbelief. And performing a familiar ritual activates cortical regions associated with habit and reflexive evaluation.[10]

So are religious people nicer than nonreligious ones? It depends on whether they're interacting with in- or out-group members. Okay, are religious people nicer to in-group members? Numerous studies say yes—more volunteering (with or without a religious context), charitable giving, and spontaneous prosociality, more generosity, trust, honesty, and forgiveness in economic games. However, numerous studies show no differences.[11]

Why the discrepancy? For starters, it matters whether data are self-reported—religious people tend to inflate reports of their prosociality more than do nonreligious people. Another factor is whether the prosociality is public—conspicuous display is particularly important to those religious people who strongly need social approval. As more context dependency, in one study religious people were more charitable than nonreligious ones—but only on their Sabbath.[12]

Another important issue: what kind of religion? As introduced in chapter 9, Ara Norenzayan, Azim Shariff, and Joseph Henrich of the University of British Columbia have identified links between features of various religions and aspects of prosociality.[13] As we saw, small-band cultures (such as hunter-gatherers) rarely invent moralizing deities. It is not until cultures are large enough that people regularly interact anonymously with strangers that it becomes commonplace to invent a judgmental god—the Judeo-Christian/Muslim deity.

In such cultures overt and subliminal religious cues boost prosociality. In one study religious subjects unscrambled sentences that did or didn't contain religious terms (e.g., spirit, divine, sacred); doing the former prompted generosity afterward. This is reminiscent of chapter 3's finding that merely seeing a pair of eyes posted on a wall makes people more prosocial. And showing that this is about being monitored, unscrambling sentences with secular terms such as "jury," "police," or "contract" had the same effect.[14]

Thus reminders of a judgmental god(s) boosts prosociality. It also matters what that deity does about transgressions. Within and among cultures, the more punitive the god, the more generosity to an anonymous coreligionist. Do punitive gods make for more punitive people (at least in an economic game)? In one study, no—save your cash, God's got it covered. In another, yes—a punitive god would want me to be punitive as well. The UBC group has shown something ironic. Priming people to think of God as punitive decreases cheating; thinking of God as forgiving *increases* it. The researchers then studied subjects from sixty-seven countries, considering the prevalence in each of belief in the existence of a heaven and hell. The greater the skew toward belief in hell, rather than heaven, the lower the national crime rate. When it comes to Eternity, sticks apparently work better than carrots.

And what about religion facilitating the worst in us, with respect to Thems? Well, one piece of evidence for this is, uh, like, human history. Every major religion has historical blood on its hands—Buddhist monks led the persecution of Rohingya Muslims in Burma, and a Quaker in the White House oversaw the carpet bombing of North Vietnam for Christmas.*[15] This ranges from religious wars, which are, to cite a quote generally attributed to Napoleon, "people killing each other over who has the better imaginary friend," to secular ones where, nevertheless, omniscient support is requested and proclaimed. Religion is a particularly tenacious catalyst of violence. Catholics and Protestants have been killing each other in Europe for nearly 500 years, Shiites and Sunnis for 1,300. Violent disagreements about differing economic or governmental models never last as long—this would be like people still killing each other today over, say, Eastern Roman Emperor Heraclius's 610 decision to switch the official language from Latin to Greek. As shown in a study of six hundred terrorist groups spanning forty years, religiously based terrorism persists the longest and is least likely to subside due to fighters joining the political process.

Religious primes foster out-group hostility. In a "field study" where people were surveyed in different locations in a cosmopolitan European city, merely walking past a church made Christians express more conservative, negative attitudes toward non-Christians. Another study examined the priming effects of a violent god. Subjects read a Bible passage in which a woman is murdered by a mob from another tribe. Her husband consults

* To be fair, Richard Nixon was raised as an Evangelical Quaker; they are not pacifists.

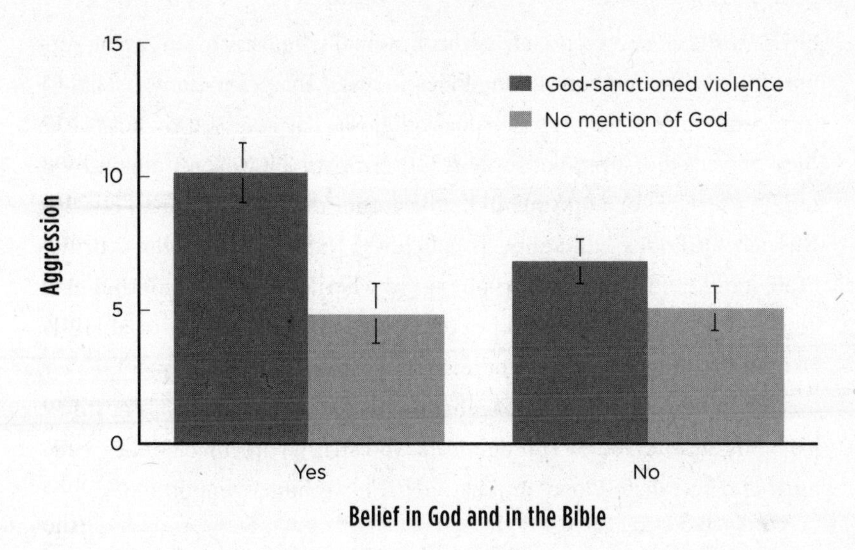

with his tribesmen and forms an army that takes revenge by attacking the other tribe (in biblical fashion, destroying their cities and killing every human and animal). Half the subjects were told this story. In the story told to the other half, while contemplating revenge, the army asks for advice from God, who sanctions them to majorly chasten the other tribe.[16]

Participants then played a competitive game in which each round's loser was blasted with a loud noise at a volume chosen by the other player. Reading the scene where God sanctions their desire for violence increased the volume with which opponents were chastened.

No surprise: the effect was bigger in males than in females. Big surprise: subjects were either devout Mormons at Brigham Young University or students of typically liberal religions at a Dutch university, and the effect was equally strong in both groups. Biggest surprise: even among subjects who did not subscribe to the Bible (a surprisingly high 1 percent of the Brigham Young students and 73 percent of the Dutch), godly sanction increased their aggressiveness (though to a lesser extent). Thus, divine sanction of violence can increase aggression even in people whose religiosity probably doesn't include a vengeful god, as well as among those who don't even believe there's divine anything.

Of course, this is not a uniform effect of religion; Norenzayan distinguishes between private and communal religiosity in surveying support for suicide bombers among Palestinians.[17] In a refutation of "Islam = terrorism" idiocy, people's personal religiosity (as assessed by how often they prayed) didn't predict support for terrorism. However, frequently attending services at a mosque did. The author then polled Indian Hindus, Russian Orthodox adherents, Israeli Jews, Indonesian Muslims, British Protestants, and Mexican Catholics as to whether they'd die for their religion and whether people of other religions caused the world's troubles. In all cases frequent attendance of religious services, but not frequent prayer, predicted those views. It's not religiosity that stokes intergroup hostility; it's being surrounded by coreligionists who affirm parochial identity, commitment, and shared loves and hatreds. This is hugely important.

What should one make of these various findings? Religiosity isn't going anywhere.* Given that, it seems that boosting in-group sociality is best done with a moralizing, punitive god. The standard, wearisome critique of atheism is that lack of a god(s) produces nihilistic amorality; the standard response is that it's pretty unimpressive if you are kind only because you fear damnation. Unimpressive or not, it appears to be useful. The big challenge is when communal aspects of religiosity fuel out-group hostility. It's useless to call for religions to broaden the extent of their Us-ness. Religions are quirky as to who is an Us, ranging from "only those who look, act, talk, and pray like people in our sect" to "all of life." It will be discouragingly tough to shift religions from the former to the latter.

Contact

As introduced in chapter 11, many have speculated that inter-group tensions are reduced by contact—when people get to know one another,

* Although it is fascinating that over the last century, while Scandinavian countries developed their enlightened and far-reaching system of governmental support of people's social needs, religiosity there plummeted dramatically; today only a small minority of Scandinavians are devoutly religious. So religiosity may not be as robust in the future as one would think, and as we saw in chapter 9, as secular institutions become better at caring for people's needs, religiosity declines. Probably more important, this is a good demonstration that religion sure isn't the only route for highly inclusive in-group prosociality.

everyone gets along. But despite that salutary possibility, intergroup contact readily elevates hostilities.[18]

As seen in chapter 9, intergroup contact worsens things when the two groups are treated unequally or are unequal in number; where the smaller group is surrounded; where intergroup boundaries are ambiguous; when the groups vie to display symbols of their sacred values (e.g., Northern Irish Protestants marching with Orangemen flags through Catholic neighborhoods). Elbows rubbed raw.

Obviously, the opposite is needed to minimize threat and anxiety—groups encountering each other in equal numbers and treatment, in a neutral setting free of agitprop and where there is institutional oversight of the venture. Most important, interactions work best when there is a shared goal, especially when it is successful. This revisits chapter 11—a shared goal reprioritizes Us/Them dichotomies, bringing this novel combined Us to the forefront.

Under those conditions, sustained intergroup contact generally decreases prejudices, often to a large extent and in a generalized, persistent manner. This was the conclusion of a 2006 meta-analysis of some five hundred studies comprising over 250,000 subjects from thirty-eight countries; beneficial effects were roughly equal for group differences in race, religion, ethnicity, or sexual orientation. As examples, a 1957 study concerning desegregation of the Merchant Marines showed that the more trips white seamen took with African Americans, the more positive their racial attitudes. Same for white cops as a function of time spent with African American partners.[19]

A more recent meta-analysis provides additional insights: (a) The beneficial effects typically involve both more knowledge about and more empathy for the Thems. (b) The workplace is a particularly effective place for contact to do its salutary thing. Decreased prejudice about the Thems at work often generalizes to Thems at large, and even sometimes to other types of Thems. (c) Contact between a traditionally dominant group and a subordinate minority usually decreases prejudice more in the former; the latter have higher thresholds. (d) Novel routes of interacting—such as sustained online relationships—can work a bit as well.[20]

All good news. Contact theory has prompted an experimental approach where people, most typically adolescents or young adults, from groups in conflict are brought together for anything from one-hour discussions to summer camps. They've most frequently involved Palestinians and Israelis, Northern Irish Catholics and Protestants, or opposing groups from the Balkans, Rwanda, or Sri Lanka, with the idea that participants will return home and spread their attitudinal shifts. This notion of germination prompted the name of one such program, Seeds of Peace.

Group pictures show Muslims and Jews, Catholics and Protestants, Tutsis and Hutu, Croats and Bosnians arm in arm; this is better than puppies. Do the programs work? Depends on what counts as "working." According to one expert, Stephen Worchel of the University of Hawaii, effects are generally positive—less fear and more positive views of Thems, more of a perception of Thems as heterogeneous, more recognition of faults of the Us, and more of a perception of oneself as an atypical Us.

This is the immediate aftermath. Disappointingly, these effects are usually transient. Individuals from across lines rarely stay in touch; in one survey of Palestinian and Israeli teenagers, 91 percent were not. Persistent reductions in prejudice usually involve exceptionalism—"Yes, *most* Thems are awful, but I hung out with a Them once who was okay." When there is major transformation, the peace-mongering convert loses street cred back home when they broadcast this. For example, no prominent peace activist has emerged from the thousands of participants in the Middle Eastern Seeds of Peace.*

Here's a way to think about contact: instead of hating a Them for what his ancestors did, you await the day that you're irritated with him for, say, eating the last s'more, or setting the office thermostat too low, or never returning to its proper place in the barn that plowshare that used to be a sword. Now, that's progress. The core of that thought is Susan Fiske's demonstration that automatic other-race-face amygdala responses can be

* Another limitation of the approach is that by definition there is self-selection for participants willing to entertain the possibility of détente with Thems. Moreover, participants often come from privileged socioeconomic backgrounds, limiting their ability to go and transform the masses afterward.

undone when subjects think of that face as belonging to a *person*, not a Them. The ability to individuate even monolithic and deindividuated monsters can be remarkable.

A moving example of this is told by Pumla Gobodo-Madikizela in her book *A Human Being Died That Night: A South African Story of Forgiveness* (Cape Town: David Philip, 2003). Gobodo-Madikizela, raised in a black township of apartheid South Africa, managed to forge an educational path all the way to a PhD in clinical psychology. As a free South Africa dawned, she worked on the Truth and Reconciliation Commission, where she had a task to give anyone pause. It concerned Eugene de Kock, the man with the most literal apartheid-era blood on his hands. De Kock had commanded the elite counterinsurgency unit of the South African Police and personally overseen kidnappings, torture, and murders of black activists. He had been tried, convicted, and given a life sentence. Gobodo-Madikizela was to interview him about his death squad; clinical psychologist that she is, over the course of over forty hours talking with him, her main focus became to understand this man.

He was a predictably multifaceted, contradictory, real human, rather than an archetype. He was remorseful in some ways, unrepentant in others; indifferent to some of his appalling brutality while proud of his patchwork of principles about whom he wouldn't kill; he pointed fingers at his bosses (who mostly escaped justice by depicting him as a rogue vigilante rather than the civil servant of apartheid that he was) while emphasizing his command of his killers. He shattered her by tentatively asking if he had killed any of her loved ones (he had not).

And Gobodo-Madikizela found herself deeply troubled by her growing empathy for de Kock.

A defining moment came one day when de Kock was recounting something that made him markedly distressed. Gobodo-Madikizela reflexively reached out and—a taboo act—touched his finger between the jail bars. The next morning her arm felt leaden, as if paralyzed by the touch. She struggled with whether her granting him this contact was a sign of her power or his (with him somehow manipulating her into the act). When she next saw him, he compounded her storm of feelings by thanking her and confessing that it was his trigger hand that she had touched. No, this was not the start of an unlikely friendship, as violins play in the background. But the automaticity, the empathy implicit in her reaching out to him, shows that somehow, remarkably, the tenuous elements of Us-ness she now shared with de Kock had dominated at that moment.

Burning and Unburning Bridges

A phenomenon in many settings of conflict is burning cultural bridges as a way to forge a new, powerful Us category. Consider the Mau Mau rebellion in Kenya in the 1950s. The brunt of British colonialism in Kenya had focused on one tribe, the Kikuyu, who had the bad luck of living on precisely the rich farm land that the colonials appropriated; Kikuyu suffering finally boiled over into the Mau Mau insurrection.*

The agricultural Kikuyu were not particularly bellicose (unlike, say, the nearby pastoralist Maasai, who had been terrorizing the Kikuyu forever), and inculcating new Mau Mau fighters required powerful symbolic effort. Oath making had great cultural significance to Kikuyus, and Mau Mau oath making notoriously involved horrendous violations of Kikuyu norms and taboos, acts guaranteeing shunning at home. The message was clear: "You have burned a bridge; your only Us is us."

This strategy is often used in a horrifying realm of modern violence, namely rebel groups transforming kidnapped children into soldiers.[21] Sometimes this involves new recruits having to burn symbolic cultural bridges. But also, perhaps reflecting recognition of kids' limited abstract cognition, something more concrete is employed—the forced killing of family members by such children. *We* are your family now.

When child soldiers are liberated, their chances of growing into healthy,

* The British eventually crushed the rebellion, at the cost of approximately 150 British lives and 10,000 to 20,000 Kikuyu lives, and then handed over power to handpicked, über-Westernized Kenyans rather than Mau Mau guerrilla fighters; as a measure of how successful the Anglicized handoff was, more than fifty years later, black Kenyan judges still wear powdered wigs when presiding.

functioning adults soars if a relative is found who will accept them. If a bridge is unburned.[22]

As I write, there's news of the rescue of a few of the two-hundred-plus Nigerian schoolgirls kidnapped in 2014 by the terrorist group Boko Haram. What these girls experienced is unimaginable—terror, pain, forced labor, endless rapes, pregnancies, AIDS. And as these few are returned home, many are shunned—for their AIDS, for the belief that they've been brainwashed into being sleeper terrorists, for the rape-born children they carry. This does not auger well for their being anything other than broken forever.

Chapter 11 emphasized pseudospeciation, when Thems are made to seem so different that they hardly count as human. Chapter 15 considered the skill of demagogues at this, framing hated Thems as insects, rodents, bacteria, malignancies, and feces. That provides a clear punch line: be wary of rabble-rousers who frame Thems as things to step on, spray with toxins, or flush down toilets. Simple.

But pseudospeciating propaganda can be subtler. In the fall of 1990 Iraq invaded Kuwait, and in the run-up to the Gulf War, Americans were sickened by a story that emerged. On October 10, 1990, a fifteen-year-old refugee from Kuwait appeared before a congressional Human Rights Caucus.[23]

The girl—she would give only her first name, Nayirah—had volunteered in a hospital in Kuwait City. She tearfully testified that Iraqi soldiers had stolen incubators to ship home as plunder, leaving over three hundred premature infants to die.

Our collective breath was taken away—"These people leave babies to die on the cold floor; they are hardly human." The testimony was seen on the news by approximately 45 million Americans, was cited by seven senators when justifying their support of war (a resolution that passed by *five* votes), and was cited more than ten times by George H. W. Bush in arguing for U.S. military involvement. And we went to war with a 92 percent approval rating of the president's decision. In the words of Representative John Porter (R-Illinois), who chaired the committee, after Nayirah's testimony, "we have never heard, in all this time, in all circumstances, a record of inhumanity, and brutality, and sadism, as the ones that [Nayirah had] given us today."

Much later it emerged that the incubator story was a pseudospeciating lie. The refugee was no refugee. She was Nayirah al-Sabah, the fifteen-year-old daughter of the Kuwaiti ambassador to the United States. The incubator story was fabricated by the public relations firm Hill + Knowlton, hired by the Kuwaiti government with the help of Porter and cochair Representative Tom Lantos (D-California). Research by the firm indicated that people would be particularly responsive to stories about atrocities against babies (ya think?), so the incubator tale was concocted, the witness coached. The story was disavowed by human rights groups (Amnesty International, Human Rights Watch) and the media, and the testimony was withdrawn from the Congressional Record—long after the war.

Be careful when our enemies are made to remind us of maggots and cancer and shit. But also beware when it is our empathic intuitions, rather than our hateful ones, that are manipulated by those who use us for their own goals.

Cooperation

As explored in chapter 10, understanding the evolution of cooperation poses two challenges.

The first is the fundamental problem of how cooperation ever starts; the dispiriting logic of the Prisoner's Dilemma shows that whoever takes the first cooperative step becomes one step behind.

As we saw, one plausible solution concerns founder populations—when a subset of a population becomes isolated and its average degree of relatedness rises, fueling cooperation through kin selection.[24] Should that founder population rejoin the general population, their cooperative tendencies will outcompete everyone else, thus propagating cooperation. Another solution involves green-beard effects, that poor man's version of kin selection, where a genetic trait generates a conspicuous marker and a cooperative bent toward bearers of that marker. In that setting the green beard–less will be outcompeted unless they also evolve cooperation. As we saw, green-beard effects occur in various species.

This raises the second challenge, namely understanding why humans are so extraordinarily cooperative with nonrelatives. We hold elevator doors open for strangers, take turns at four-way stop signs, get off buses in an orderly manner. We build cultures involving millions of people sharing conventions. This requires more than founder effects and green beards;

in the years since Hamilton and Axelrod made "tit for tat" trendy, tons of work has explored human-specific mechanisms for fostering cooperation. There are many.

Open-ended play. Two individuals play the Prisoner's Dilemma, knowing that after a single round, they'll never meet again. Rationality decrees that you defect; there'll never be a chance to catch up if you fall behind in that first round. What about two rounds? Well, the second round requires noncooperation for the same reasons the single-round game does. In other words, it never makes sense to cooperate in the final round. Thus, round 2 behavior determined, the game defaults to a single-round game—where the rational strategy is to defect. Three rounds? The same. In other words, playing for a known number of rounds biases against cooperation, and the more rational the players, the more they foresee this. It's open-ended play that fosters cooperation—an unknown number of rounds, producing the shadow of the future, where retribution is possible and the advantages of sustained mutual cooperation accumulate with increasing numbers of interactions.[25]

Multiple games. Two individuals play two games against each other simultaneously (alternating rounds between the two) where one game has a much lower threshold for establishing cooperation than the other. Once cooperation is established in that less cutthroat game, there is psychological spillover of cooperation into the other. This is why managers of tense, competitive offices bring in soothing outsiders to lead trust games, hoping that the low-threshold demands for trust there will spill over into work life.

Open-book play. This is where the other player can see if you've been a jerk to people in the past. Reputation is a powerful facilitator of cooperation. That's what a moralizing god is about—the book whose play is eternally open. As we saw in chapter 9, everyone from hunter-gatherers to urbanites gossip, doing so to open reputation books wider.[26]

Open-book play mediates a uniquely sophisticated type of human cooperation, namely "indirect reciprocity." Person A helps person B, who helps C, who helps D. . . . The reciprocity between two individuals in a closed interaction is like barter. But indirect, pay-it-forward reciprocity is like money, where the common currency is reputation.[27]

Punishment

Other animals don't have reputations or ponder whether their interactions are open-ended. However, punishment to promote cooperation occurs in numerous species—this is shown when a male baboon who is being an aggressive brute to a female is chased out of the troop for a while by the victim and her relatives. Punishment can strongly facilitate cooperation, but its implementation is potentially double-edged in humans.

All cultures show some degree of willingness to pay a cost to punish norm violators, and high degrees of willingness correlate with high levels of prosociality. One study examined rural Ethiopians who subsisted on selling charcoal made from wood from local forests—a classic tragedy of the commons scenario: no one is likely to spontaneously limit logging to keep the forest healthy. The study showed that villages with high average levels of willingness to administer costly punishment in an economic game were the ones with the most patrols to prevent overcutting of trees and the healthiest forests. And as seen in chapter 9, cultures with gods who punish norm violations are atypically prosocial.[28]

A complication in costly punishment is the cost—the danger that the costs of monitoring for and punishing violations may outweigh the benefits of the cooperation induced. A solution is to reduce surveillance after long stretches of cooperation—in other words, to trust. For example, probably very few Amish purchase costly retinal-scanner home security systems.[29]

Another complication concerns who does the punishing. In other species it is usually the victim, the second party. By definition, punishment in two-person games in humans (e.g., the Ultimatum Game) is always by the second party. In that setting the punisher forgoes the measly share offered, (a) in the hopes of deriving visceral satisfaction from depriving the first party of their larger share (and, as seen in the last chapter, that is a major motivator of punishment, fueled by the amygdala and insula); (b) in an effort to shape the first party into making fairer offers to the second party in the future; or (c) as an altruistic act, hoping to shape the first party into being more decent to whomever they play next. This is complex for second parties, balancing costs and benefits, heart and mind, birds in the hand and in the bush. It might also result in the first party being offended by the rejection and becoming even less cooperative thereafter—an outcome in some game scenarios.[30]

Humans uniquely and very effectively boost cooperation through third-party punishment meted out by objective outsiders. However, such punishing can be costly to the third party, meaning that there's the evolutionary challenge not just of jump-starting cooperation but also of jump-starting altruistic third-party punishment.[31]

The answer, as repeatedly derived by humans, is to add layers. Develop secondary punishment, punishing someone who fails to do third-party punishment—the world of honor codes, where you're punished if you don't report a violation. An alternative is to reward third-party punishers—humans make livings as cops and judges. Moreover, recent theoretical and empirical work shows that being a conspicuous third-party punisher makes people trust you. But who monitors third-party punishers? Here is where you get people to share and lower the cost by taking sociality to the max—costs are shouldered by everyone, and free riders are punished (e.g., we pay taxes and punish tax evaders). When the moving parts are balanced, you generate extraordinary levels of cooperation.[32]

The moving parts were examined in a fascinating 2010 *Science* paper. The authors studied 113,000 online participants, who each purchased an item (a souvenir photo) under one of the following conditions:[33]

- **a.** Could buy for a set price. (This was the control condition.)
- **b.** Could pay whatever they wanted; sales soared but people tended to pay tiny amounts, putting the "store" in the red.
- **c.** Were charged the original price, knowing that the company gave X percent of earnings to charity; sales increased, but less than X percent, and the store lost money.

d. Could pay whatever they wanted, with half of that going to a charity. This boosted both sales and the price voluntarily paid, yielding profits for the store and a large charitable contribution.

In other words, while evidence of corporate social responsibility (scenario C) boosts sales a bit, it's far more effective when the individual and the business *share* social responsibility and the individual determines the amount of money donated.

Choosing Your Partner

As we've seen, cooperators outcompete more numerous noncooperators to the extent that the former can find one another. This is the logic behind green beards facilitating finding a kindred soul (if not kin). Thus, when that element is introduced into a game (with the ability to refuse to play with someone), cooperation soars, and more cheaply than by punishing defectors.[34]

These findings reveal numerous theoretical routes for fostering cooperation, and with real-life equivalents; moreover, we've learned a lot about which work best when. This is how we've evolved to collectively raise barns for neighbors, plant and harvest the whole village's rice crop, or coordinate marching-band members to form a picture of their school's mascot.

And, oh yeah, to reiterate an idea aired previously, "cooperation" is a value-free term. Sometimes it takes a village to ransack a neighboring village.

Reconciliation, and Things That Are Not Synonymous with It

"So I'd caught a colobus monkey and was eating, getting to the good part, when this guy comes by, starts really begging for some. This got on my nerves and I snarled at him. Instead of taking a hint, he lunges, grabs the monkey's arm, starts yanking—so I bit his shoulder. He cleared out fast and sat at the other end of the clearing, his back to me.

"Once I calmed down, I thought a bit. To be honest, I probably should have shared some food with him. And while he definitely crossed a line when he grabbed, I probably should have nipped him instead of a real bite. So I'm feeling kind of bad. And besides, we work well together on patrols—it's probably good if we sort things out.

"So I take the monkey, sit near him. We're all awkward—he's not looking at me, I pretend there's a nettle between my toes. But eventually I give him some of the meat, he grooms me a bit. The whole thing was stupid, we should have done that in the first place."

If you're a chimp, reconciliation is easy once your heart rate returns to normal. Sometimes for us too—touch a friend's shoulder, give a self-effacing grimace, say, "Hey, look, just now I was being a—" and they cut you off, saying, "No, no, it was me. I shouldn't . . ." and things are okay.

Easy. How about when everyone's trying to patch things up after your people have slaughtered three quarters of theirs, or after they came as colonials, stole your land, and forced you to live in slum "homelands" for decades? Trickier.

We're the only species that institutionalizes reconciliation and that grapples with "truth," "apology," "forgiveness," "reparations," "amnesty," and "forgetting."

The apogee of institutionalized complexity is the truth and reconciliation commission (TRC). The first came in the 1980s, and they've been depressingly useful ever since, occurring, for example, in Bolivia, Canada, Australia, Nepal, Rwanda, and Poland. Some TRCs have been in stable countries (Canada and Australia) facing up to their long history of abuse of indigenous peoples. Most, however, have come after a nation emerged from a bloody, divisive transition—a dictator overthrown, a civil war settled, a genocide halted. The popular perception is that their purpose is for perpetrators of abuse to confess, express remorse, and beg for forgiveness from victims, who then grant it, resulting in tearful embraces between the two.

But instead TRCs are typically exercises in pragmatism, where perpetrators basically say, "This is what I did, and I vow to never harm your people again," and the victims basically say, "Okay, we vow to not seek extrajudicial retribution." An often towering achievement, if less heartwarming.

Probably the best-studied TRC was South Africa's after the defeat of apartheid. It came with enormous moral legitimacy, being overseen by Desmond Tutu, and gained further legitimacy by, though overwhelmingly focusing on the acts of whites, also examining atrocities by African liberation fighters. Hearings were public and included victims getting to tell their stories. More than six thousand of the perpetrators testified and applied for amnesty; this was granted to 13 percent.

What happened to the tearful forgiveness scenarios? What about perpetrators at least showing remorse for their actions? It was not required, and few did. The goal was not to transform those individuals; it was to increase the odds that the shattered nation would function. In follow-up studies by the South African Centre for the Study of Violence and Reconciliation, victim participants commonly felt "that the TRC had been more successful at the national than the local level." Many were outraged that there were no apologies, no reparations, that many perpetrators remained in their jobs. Interestingly, echoing chapter 15, many were equally angry about symbolic changes that had not occurred—not only is this killer still a cop, but there's still a holiday/monument/street name celebrating apartheid. A wide majority of black (but not white) South Africans saw the TRC as fair and successful, and it accompanied South Africa's miraculously transitioning to freedom, rather than descending into civil war. Thus TRCs show the differences between reconciliation and the likes of remorse and forgiveness.*[35]

As every parent knows, a transparently insincere apology accomplishes little and can even worsen things. But deep remorse is different. The *New Yorker* recounts the story of Lu Lobello, an American Iraq War veteran who accidentally killed three members of a family, collateral

* I thank a really excellent undergrad, Dawn Maxey, for her research assistance regarding TRCs and for the bulk of these insights.

damage during a firefight; haunted by it, he spent nine years tracking down the survivors to apologize. Or consider Hazel Bryan Massery, the snarling white teenager at the center of the iconic 1957 civil-rights-movement photograph of Elizabeth Eckford attempting to integrate Little Rock Central High School. A few years later Massery contacted Eckford to apologize.[36]

Do apologies "work"? It depends. One issue is what the person is apologizing for, ranging from the concrete ("I'm sorry I broke your toy") to the global and essentialist ("I'm sorry I've viewed your people as not fully human"). Another is what the apologizer aims to do about their remorse. And there's the makeup of the recipient of the apology. Studies show that (a) victims who are oriented toward the workings of a collective system respond most to apologies that emphasize failure of that system ("I'm sorry, we police are supposed to protect, not break laws"); (b) victims most oriented to relationships respond most to apologies that are empathic ("I'm sorry for the pain that I caused you, for taking your son"); and (c) victims who are most autonomous and independent respond most to apologies accompanied by offers of compensation. There is also the issue of who is apologizing. What does it mean that in 1993 Bill Clinton apologized to Japanese Americans for their World War II internment? While the apology was laudable, and accompanied by reparative money, could Clinton speak for FDR?[37]

The issue of reparations is immensely complicated. At one extreme, reparations can be the ultimate proof of sincerity. This is at the heart of the slavery reparations movement—so much of America's growth into economic privilege was built on slavery, and so many of the subsequent benefits of the successful economy have been systematically denied to African Americans, that there should be reparations to the descendants of slaves. At the other extreme, reparations meant to purchase forgiveness offend—this was the reasoning behind the newly born state of Israel's refusal of reparations from Germany, unless it was accompanied by adequate remorse.

At the end of these steps might arise one of the strangest things humans do—we forgive.[38] For starters, forgiving is not forgetting. If nothing else, that's neurobiologically unlikely. A rat learns to associate a bell with a shock and freezes when it hears it. When the next day the bell repeatedly sounds without being accompanied by a shock, causing the freezing behavior to "extinguish," the memory trace of that learning does not evaporate. Instead it is overlaid with newer learning—"Today the bell is not bad news." As proof, suppose that the day after that, the bell again signals shock. If the initial learning of "bell = shock" had been erased, it would take as long this day to learn the association as it did the first. Instead there is rapid reacquisition: "bell = shock *again*." Forgiving someone doesn't mean you've forgotten what he did.

There is a subset of victims who claim to have forgiven the perpetrator, to have relinquished their anger and desire for punishment. I include the word "claim" not to imply skepticism but to indicate that forgiveness is a self-reported state that can be claimed but not proven.

Forgiveness can occur as a religious imperative. In the June 2015 Charleston church massacre, white supremacist Dylann Roof killed nine parishioners at the Emanuel African Methodist Episcopal Church. Two days later, at Roof's arraignment, stunningly, family members of the dead were there to forgive him and pray for his soul.[39]

Forgiveness can take extraordinary cognitive reappraisal. Consider the case of Jennifer Thompson-Cannino and Ronald Cotton.[40] In 1984 Thompson-Cannino was raped by a stranger. In a police lineup she identified Cotton with great certainty; despite claiming innocence, he was convicted and sentenced to life in prison. In the years after, friends tentatively wondered if she could now put the nightmare behind her. "Like hell I'm able to" would be her response. She was consumed with her hatred for Cotton, with her desire to harm him. And then, more than ten years into his prison sentence, DNA evidence exonerated Cotton. Another man had done it; he was incarcerated in Cotton's prison for other rapes and bragged about getting away with this one. Thompson-Cannino had identified the wrong man and convinced a jury. Issues of hatred or forgiveness were now on the other foot.

When they finally met, after Cotton's release and pardon, Thompson-Cannino said, "If I spent every minute of every hour of every day for the rest of my life telling you that I'm sorry, can you ever forgive me?" And Cotton said, "Jennifer, I forgave you years ago." His ability to do so involved profound reappraisal: "Forgiving Jennifer for picking me out of that lineup as her rapist took less time than people think. I knew she was a victim and was hurting real bad. . . . We were the victims of the same injustice by the same man, and this gave us a common ground to stand on." A complete reappraisal that made them Us in their victimhood. The two now lecture together about the need for judicial reform.

Ultimately, forgiveness is usually about one thing—"This is for me, not for you." Hatred is exhausting; forgiveness, or even just indifference, is freeing. To quote Booker T. Washington, "I shall allow no man to belittle my soul by making me hate him." Belittle and distort and consume. Forgiveness seems to be at least somewhat good for your health—victims who show spontaneous forgiveness, or who have gone through forgiveness therapy (as opposed to "anger validation therapy") show improvements in general health, cardiovascular function, and symptoms of depression, anxiety, and PTSD. Chapter 14 explored how compassion readily, perhaps inevitably, contains elements of self-interest. The compassionate granting of forgiveness epitomizes this.[41]

We've now focused on forgiveness, apology, reparation, reconciliation, and the extent to which TRCs were about reconciliation rather than forgiveness. What about the "truth" part? It facilitates the healing process enormously. In the TRCs, perpetrators spilling truth—detailed, exhaustive, unflinching, and public—was the highest priority for victims. It's the need to know what happened; it's getting the villain to say the words; it's to show the world, "Look what they did to us."

Recognizing Our Irrationalities

Despite the claims of some economists, we are not rational optimization machines. We are more generous in games than logic predicts; we decide if someone is guilty based on reasoning but then decide their punishment based on emotion; roughly half of us make different decisions about sacrificing one to save five, depending if it involves pushing a person versus pulling a lever; we effortlessly resist cheating in circumstances where no one would know; we make strong moral decisions without being able to explain why. Thus it's a good idea to recognize the systematic features to our irrationality.

Sometimes we aim to eliminate these irrationalities. Perhaps the most fundamental one is the common visceral resistance to a simple fact—you don't make treaties with friends; it's to be expected that you passionately hate those whose hands you are about to shake, and that can't be an impediment to doing so. Another domain concerns discrepancies between our conscious opinions and what our implicit biases lead us to do. As we saw, Us/Them edges can be softened when implicit biases are made explicit. Doing so need not eliminate that bias—after all, you can't readily reason yourself out of a belief that you weren't originally reasoned into. Instead, revealing implicit biases indicates where to focus your monitoring to lessen their impact. This notion can be applied to all the realms of our behaviors being shaped by something implicit, subliminal, interoceptive, unconscious, subterranean—and where we then post-hoc rationalize our stance. For example, every judge should learn that judicial decisions are sensitive to how long it's been since they ate.

Another example to watch out for is the human potential for irrational optimism. For example, while people might accurately assess the risk of a behavior, they tend toward distortive optimism when assessing risk to themselves—"Nah, that couldn't happen to me." Irrational optimism can be great; it's why only about 15 percent instead of 99 percent of humans get clinically depressed. But as emphasized by the Nobel Prize–winning psychologist Daniel Kahneman, irrational optimism in warfare is disastrous. This can range from the theologically optimistic conviction that God is on your side to the tendency of military strategists to overestimate their side's capabilities and underestimate those of the opposition—"piece of cake, full steam ahead" becomes the logical conclusion.[42]

A final domain of irrationality that must be recognized concerns chapter 15's "sacred values," where purely symbolic acts can count for more than hard-nosed material concessions. Rationality may be key to establishing peace, but the irrational importance of sacred values is key to establishing lasting peace.

Our Incompetence at and Aversion to Killing

Video cameras are sufficiently ubiquitous these days to make "privacy" a threatened phenomenon. One consequence of such ubiquity is that scientists can be voyeuristic in new ways. Which has produced an interesting finding.

It concerns riots in soccer stadiums—"football hooliganism," battles between ethnic or nationalist groups, partisans of each team, or often right-wing skinheads going at it. Footage of such events shows that few people actually fight. Most are on the sidelines watching or running around like agitated, headless chickens. Of those who fight, most throw an ineffectual punch or two before discovering that punching makes your hand hurt. The actual fighters are a tiny subset. As stated by one researcher, "humans are bad at [close-range, hand-to-hand] violence, even if civilization makes us a bit better at it."[43]

Even more interesting is the evidence of our strong inhibitions against doing grievous harm to someone up close.

The definitive exploration of this is the 1995 book *On Killing: The Psychological Cost of Learning to Kill in War and Society*, by David Grossman, a professor of military science and retired U.S. Army colonel.[44]

He frames the book around something noted after the Battle of Gettysburg. Of the almost 27,000 single-load muskets recovered from the field, almost 24,000 of them were loaded and unfired; 12,000 were loaded multiple times, 6,000 loaded three to ten times. Lots of soldiers were standing there thinking, "I'm going to shoot soon, yes I am, hmm, maybe I should reload my rifle first." These weapons were recovered from the thick of the battlefield, from men whose lives were at risk while they were reloading. In Gettysburg most deaths were caused by artillery, not the infantry on the ground. In the heat of crazed battle, most men would load, tend to the wounded, shout orders, run away, or wander in a daze.

Similarly, in World War II only 15 to 20 percent of riflemen ever fired their guns. The rest? Running messages, helping people load ammunition, tending to buddies—but not aiming a rifle at someone nearby and pulling a trigger.

Psychologists of warfare emphasize how, in the heat of battle, people don't shoot another human out of hatred or obedience, or even from knowing that this enemy is trying to kill *them*. Instead it's the pseudokinship of bands of brothers—to protect your buddies, to not let the guys next to you down. But outside those motivations, humans show a strong natural aversion to killing at close range. The most resistance is against hand-to-hand combat with a knife or bayonet. Next comes short-range firing with a pistol, then long-range firing, all the way to the easiest, which is bombs and artillery.

The resistance can be psychologically modified. It's easier when you aren't targeting an identified individual—throwing a grenade into a group rather than shooting at one person. Killing as an individual is harder than in a group—while only that small subset of World War II riflemen fired their weapons, nearly all weapons operated by a team (e.g., machine guns) were fired. Responsibility is diluted, much as when a firing squad would know that one of them had received a blank, allowing every shooter to know that they might not have actually killed someone.

Grossman's premise is supported by something new and startling. Since it morphed from "battle fatigue" or "shell shock" into a formal psychiatric illness, combat PTSD has been framed as a result of the sheer terror of being under attack, of someone trying to kill you and those around you. As we've seen, it is an illness where fear conditioning is overgeneralized and pathological, an amygdala grown large, hyperreactive, and convinced that you are never safe. But consider drone pilots—soldiers who sit in control rooms in the United States, directing drones on the other side of the planet. They are not in danger. Yet their rates of PTSD are *just as high* as those of soldiers actually "in" war.

Why? Drone pilots do something horrifying and fascinating, a type of close-range, intimate killing like nothing in history, using imaging technology of extraordinary quality. A target is identified, and a drone might be positioned invisibly high in the sky over the person's house for weeks, the drone operators always watching, waiting, say, for a gathering of targets in the house. You watch the target coming and going, eating dinner, taking a nap on his deck, playing with his kids. And then comes the command to fire, to release your Hellfire missile at supersonic speed.

Here's one drone pilot, describing his first "kill"—three Afghanis targeted from his air force base in Nevada. The missile has hit, and he watches through an infrared camera, which transmits heat signatures:

> The smoke clears, and there's pieces of the two guys around the crater. And there's this guy over here, and he's missing his right leg above his knee. He's holding it, and he's rolling around, and the blood is squirting out of his leg, and it's hitting the ground, and it's hot. His blood is hot. But when it hits the ground, it starts to cool off; the pool cools fast. It took him a long time to die. I just watched him. I watched him become the same color as the ground he was lying on.[45]

But there would be more. Pilots wait to see who retrieves the bodies, who comes to the funeral, ready to perhaps release another strike. Or in other circumstances the pilot might watch as an American convoy approaches a roadside IED booby trap, unable to warn them, or watch insurgents execute a shrieking civilian begging for mercy.

The pilot above was twenty-one when he made that first kill; he would eventually accumulate 1,626 drone-mediated kills.* No personal danger, an omnipresent eye in the sky. He could finish his shift and get a doughnut on the way home. Yet he and many of his fellow drone pilots succumb to devastating PTSD.

After reading Grossman, the explanation is simple. The deepest trauma is not the fear of being killed. It's doing the close-up, individuated killing, watching someone for weeks and then turning him the color of the ground. Grossman cites that during World War II there were low rates of psychiatric breakdowns among sailors and medics—people who were

* And, it should be noted, there's enormous controversy as to what percentage of kills are accidents, collateral damage to innocent bystanders; estimates range from 2 to 20 percent.

just as endangered as infantrymen but killed either impersonally or not at all.

Militaries train soldiers to override their inhibitions against killing, and Grossman notes that the training has become more effective—trainees no longer fire at bull's-eyes; instead it's rapid-fire situations of mobile virtual-reality figures coming at you, where shooting becomes reflexive. In the Korean War, 55 percent of American riflemen fired their weapons; in the Vietnam War, over 90 percent. And this was before the rise of violent, desensitizing video games.

Maybe there will soon be completely different types of wars. Perhaps drones themselves will decide when to fire. Maybe wars will consist of autonomous weapons fighting each other, or each side racing to win with the most effective cyberattack on the other's computers. But as long as we still see the faces of those we kill, this seemingly natural inhibition will be vital.

THE POSSIBILITIES

It's remarkable the things humans can spend their lives studying. You can be a coniologist or a caliologist, studying dust or birds' nests, respectively. There are batologists and brontologists, pondering brambles and thunder, and vexillologists and zygologists, with their dazzling knowledge of flags and of methods for fastening things. On and on—odontology and odonatology, phenology and phonology, parapsychology and parasitology. A rhinologist and a nosologist fall in love and have a child who becomes a rhinological nosologist, studying the classification of nose diseases.

The preceding pages suggest the possibility of "peaceology," the scientific study of the effects of trade, demographics, religion, intergroup contact, reconciliation, and so on, on the ability of humans to live in peace. An intellectual venture with great potential to help the world.

But with each new example of us at our worst, from the pinpricks of petty meanness to massive carnage, this intellectual venture can feel like rolling a boulder uphill. And thus, to falsely separate cognition and affect, we conclude these many pages by fueling the emotional rather than intellectual certainty that there is hope, that things can change, that we can be changed, that we personally can cause change.

Rousseau with a Tail

For more than thirty years I spent my summers studying savanna baboons in the Serengeti ecosystem in East Africa. I love baboons, but I must admit that they're often violent and abusive, so that the weak suffer at the canines of the strong. Okay, some detachment—they're a highly sexually dimorphic tournament species with extensive escalated aggression and a strong propensity toward frustration displacement—i.e., they can be intensely shitty to one another.

The remainders of one of my males, the morning after being attacked by a coalition of rivals

In the mid-1980s the baboon troop adjacent to my study group hit the jackpot. Their territory included a tourist lodge; as at tourist places anywhere in the wilds, it had always been a challenge keeping wildlife from feeding on food garbage. Hidden in a grove of trees far from the lodge was a deep garbage pit, surrounded by a fence. But baboons climb fences, fences get knocked down, gates are left open—and that neighboring troop had taken to foraging daily in the dump. Like another widely dispersed primate, humans, baboons eat almost anything—fruit, plants, tubers, insects, eggs, prey they've killed, dead things they've scavenged.

This transformed the "Garbage Dump" troop. Baboons normally

Breakfast time, as garbage is dumped from a cart

descend from their sleeping trees at dawn and walk ten miles a day foraging. Garbage Dumpers slept in trees above the dump, waddled down at eight o'clock to meet the garbage tractor from the lodge, spend ten minutes in frenzied competition for discarded roast beef and drumsticks and plum pudding, and then waddle out for a nap. I'd even darted Garbage Dump animals and studied them with colleagues—they put on weight, thickened with subcutaneous fat, had elevated circulating levels of insulin and triglycerides, had the start of metabolic syndrome.[46]

Somehow baboons in "my" troop got word of the feasting over the hill, and soon half a dozen would head over each morning to join. It wasn't random who did this, who would try to compete for food against fifty or sixty Thems. The ones who tried were male, big, and aggressive. And morning is when baboons do much of their socializing—sitting in contact, grooming, playing—so going for garbage meant forgoing the socializing. The males who went each morning were the most aggressive, least affiliative members of the troop.

Not long after, there was a tuberculosis outbreak among the Garbage Dump baboons. In humans tuberculosis is a chronic disease, slowly consuming you with "consumption." In nonhuman primates, TB is wildfire, spreading rapidly, killing within weeks. Kenyan wildlife vet colleagues and I identified the cause of the outbreak—the meat inspector at the lodge was being bribed to approve tubercular cows for slaughter; animals were killed, unsightly lesionish organs discarded and then consumed by baboons. Most of the Garbage Dump troop died, as did all my males who raided the dump.[47]

This was kinda upsetting to me; I habituated a new troop at the other end of the park and wouldn't go anywhere near the remnants of my troop for half a dozen years. Finally, my soon-to-be wife visiting Kenya for the first time, I worked up the nerve to return to the troop, to show her the baboons of my youth.

They were unlike any baboon troop documented, exactly like what you'd expect if you eliminated half the adult males, producing a 2:1 female-to-male ratio instead of the typical 1:1, and if the males remaining were particularly unaggressive and affiliative.[48]

They stayed close together, sat in contact, and groomed more than average. Levels of aggression were lower, and in an informative way. Males still had a dominance hierarchy; number three would still fight with numbers four and two, defending his status and seeking a promotion. But there was minimal displacement aggression onto innocent bystanders—when number three lost a fight, he'd rarely terrorize number ten or a female. Stress hormone levels were low; the neurochemistry of anxiety and benzodiazepines worked differently in these individuals.

Here's a measure of it, a picture that, if you're a baboon-ologist, is more surprising than one showing baboons inventing the wheel—two adult males grooming. That hardly ever happens. Except in this troop.

And now the most important part. Female baboons remain in their birth troop, whereas males get itchy around puberty and leave, trying their luck anywhere from the troop next door to one thirty miles away. By the time I returned to this troop, most of the males who had avoided the TB had died; the troop was filled with males who had transferred in after the TB. In other words, adolescent males had grown up in typical baboon troops and then joined this one and adopted the style of low aggression and high affiliation. The troop's social culture was being transmitted.

How? Adolescents who joined the troop were no less aggressive or displacing than those joining other troops—there wasn't self-selection. There was no evidence of social instruction occurring. Instead the most likely explanation involved resident females. These were probably the least stressed female baboons on earth, not being subject to typical male displacement aggression. In this more relaxed state, they were more willing to risk affiliative overtures to new individuals—in a typical baboon troop it is more than two months before females first groom or sexually solicit new transfer males; in this troop it was a matter of days to weeks. Coupled with the lack of displacement aggression from resident males, this caused the new-transfer males to gradually change, assimilating into the troop culture in about six months. Thus, when treated in a less aggressive, more affiliative manner, adolescent baboons start doing the same.

In 1965 a rising star of primatology, Irven DeVore of Harvard, published the first overview of the subject.[49] Discussing his own specialty, savanna baboons, he wrote that they "have acquired an aggressive temperament as a defense against predators, and aggressiveness cannot be turned on and off like a faucet. It is an integral part of the monkeys' personalities, so deeply rooted that it makes them potential aggressors in every situation." Thus savanna baboons became, literally, textbook examples of an aggressive, stratified, male-dominated primate. Yet as we see here, this picture is not universal or inevitable.

Humans have formed both small nomadic bands and megastates and have demonstrated a flexibility whereby uprooted descendants of the former function in the latter. Human mating patterns are atypically flexible, and our societies feature monogamy, polygyny, or polyandry. We have fashioned religions where certain types of violence earn you paradise and others where the same violence consigns you to hell. Basically, if baboons unexpectedly show this much social plasticity, so can we. Anyone who says that our worst behaviors are inevitable knows too little about primates, including us.

One Person

Somewhere between neurons, hormones, and genes on one hand and culture, ecological influences, and evolution on the other, sits the individual. And with more than seven billion of us, it's easy to feel that no single individual can make much of a difference.

But we know that's not true. There's the obligatory list of those who changed everything—Mandela, Gandhi, MLK, Rosa Parks, Lincoln, Aung San Suu Kyi. Yes, they often had scads of advisers. But they were the catalysts, the ones who paid with their freedom or their lives. And there are whistle-blowers who took great risks to trigger change—Daniel Ellsberg, Karen Silkwood, W. Mark Felt (Watergate's Deep Throat), Samuel Provance (the U.S. soldier who revealed the abuses at Abu Ghraib Prison), Edward Snowden.*

But there are also lesser-known people, acting alone or in small numbers, with extraordinary impact. Take Mohamed Bouazizi, a twenty-six-year-old fruit seller in Tunisia, then in its twenty-third year of corrupt and repressive rule by a dictator. At the market the police hassled Bouazizi about an imaginary permit, expecting a bribe. He refused, not out of principle—he'd often bribed—but because he lacked the money. He was kicked and spat upon, his fruit cart overturned. His complaint at the government office was ignored. And within an hour of being preyed upon by the police, on December 10, 2010, Bouazizi stood in front of that office, doused himself with gasoline, shouted, "How do you expect me to make a living?" and set himself on fire.

Bouazizi's immolation and death triggered protests in Tunisia against

* Yes, yes, I know this isn't necessarily everyone's list, but the point is the singularity, rather than the specifics of their acts.

Antigovernment protestors displaying a picture of Bouazizi

the leader, Zine El Abidine Ben Ali, against his ruling party, against the police. The protests grew, and within a month the government and Ben Ali were overthrown. Bouazizi's act led to protests in Egypt, toppling Hosni Mubarak's thirty-year dictatorship. Likewise in Yemen, ending Ali Abdullah Saleh's thirty-four-year rule. And in Libya, leading to the overthrow and killing of Muammar Gaddafi after forty-three years in power. And in Syria, where protests morphed into civil war. And in Jordan, Oman, and Kuwait, leading to the resignations of their prime ministers. And in Algeria, Iraq, Bahrain, Morocco, and Saudi Arabia, producing semblances of governmental reform. The Arab Spring. Bouazizi wasn't thinking about political reform in the Muslim world when he lit the match; instead there was rage with nowhere to go but inward. Make what you will of the Arab Spring's brief hopefulness, followed by new strongmen, violence, refugees, and the catastrophe of Syria and ISIS. And perhaps history makes the self-immolator as much as the self-immolator makes history—regional discontent had long been brewing. Regardless, Bouazizi's singular act catalyzed millions in twenty countries to decide that they could cause change.

There have been other singular acts. In the mid-1980s a commemoration was being held at the Pearl Harbor Memorial on the anniversary of the attack. A group of survivors who had gathered were approached by an elderly man. This was his third trip to the memorial, trying to work up his nerve. He approached the survivors and, in halting English, apologized.[50]

The man, Zenji Abe, was a fighter pilot in the 1937 Japanese invasion of China and throughout the Pacific during World War II—including helping to lead the attack on Pearl Harbor.

Little in his earlier life predicted Abe's apologizing as an old man. His inculcation into war started early, when he joined a military academy as a seventh grader. His experience of war was detached—he never killed an American soldier at close quarters. The Pearl Harbor attack felt like a training exercise. His sense of responsibility could readily have been blunted, as the bomb he dropped hadn't detonated. And his country had been defeated.

Some things favored Abe's gesture. He had been captured and spent a year as a POW, treated decently by Americans. And he felt shame about the attack—pilots had been told that war against America had been declared that morning, that American defenses would be ready. He soon learned that instead it was a sneak attack.

Some larger factors favored his gesture as well. Japanese/American relations had transformed. And Americans were not traditional enemies. The racial, cultural, and geographic distance might have facilitated pseudospeciation of Americans, but it was nouveau pseudospeciation, contrasting with centuries of hatred of a nearby enemy—Abe never went to China to apologize for the Rape of Nanking. As we know, Thems come in different categories.

So these likelihoods and unlikelihoods converged, and Abe stood there, along with nine other pilots who had flown that day, apologizing. Some survivors refused the overture. Most accepted it. Abe and other pilots made subsequent trips to Pearl Harbor and had multiday meetings with American survivors; reconciliative handshakes were broadcast on the *Today* show for the fiftieth anniversary. Survivors generally considered the pilots to have just been following orders and found their current actions brave and admirable. Abe became close with one survivor, Richard Fiske, a docent at the memorial. Fiske had been on one of the ships during the attack, lost many friends among the 2,390 Americans killed,

Left: *Zenji Abe, December 6th, 1941;* right: *Abe and Richard Fiske, December 6th, 1991*

fought at Iwo Jima, described himself as so hating the Japanese that he developed a bleeding ulcer. For reasons he never fully understood, Fiske was the first to accept Abe's gesture. Other Japanese and Americans became close as well, visiting the homes and, eventually, the grave sites of their ex-enemies.

The process was rich in symbols, starting with an apology that, as we've seen, changes nothing and everything. Abe gave Fiske money so that, for the rest of his life, Fiske placed flowers monthly at the memorial. Fiske, a bugler, took to playing not only taps at the memorial, but the Japanese equivalent as well. Some semblance of Us-ness had emerged that included everyone there that infamous day.

Perhaps most important, Abe's singular act isn't singular. There are now travel agencies specializing in serving American Vietnam War vets returning to Vietnam for reconciliation ceremonies with ex–Viet Cong. Veterans have spearheaded organizations such as Friends of Danang, doing service projects in Vietnam, building schools, clinics, and literal bridges.[51]

This picture segues into another extraordinary act. Arguably the single most shocking event of the Vietnam War, an atrocity that finally shook America's self-perception as a force of good, was the My Lai Massacre.

On March 16, 1968, a company of American soldiers, under the command of Lieutenant William Calley Jr., attacked the unarmed civilians of the village of My Lai.[52] The company had been in Vietnam all of three months and had had no direct enemy contact. They had, however, suffered twenty-eight deaths or injuries due to booby traps and mines, reducing the company's number to around one hundred. The common interpretation, one that we readily recognize by now, is that they had a fierce, vengeful desire to connect faces to this faceless enemy. The official rationale was that the village harbored Viet Cong fighters and civilian sympathizers. Some of the participants reported being instructed to kill only Viet Cong fighters; others that they should kill everyone, burn houses, kill livestock, and destroy wells.

Regardless of these conflicting reports, the rest, as they say, is agonizing history. Between 350 and 500 unarmed civilians, including infants and elderly people, were killed. Bodies were mutilated and dumped down wells, huts and fields set ablaze, numerous women gang-raped before being killed. Calley was described to have personally shot children under their mothers who had died sheltering them. The Americans encountered no enemy fire, found no military-aged men. It was destruction of biblical proportions, or Roman proportions, or Crusader, or Viking, or . . . This destruction was photographed. The horror is worsened because My Lai was not a solitary atrocity, and the government labored to conceal events and slapped Calley on the wrist, sentencing him to three years of house arrest.

There was by no means universal participation by Americans (ultimately twenty-six soldiers were criminally charged, with Calley the only one convicted; "just following orders" was the order of the day).*[53] Individual thresholds varied. One soldier killed a mother and child and then refused to do more. Another helped herd civilians together but refused to fire. Some refused orders outright, even in the face of threats of court-martial or being shot. One, PFC Michael Bernhardt, refused and threat-

* Two who participated in the killings eventually committed suicide. One, Lieutenant Stephen Brooks, did so for unknown reasons while in Vietnam. The other, PFC Varnado Simpson, did so years later, after, among other things, seeing his ten-year-old son killed by a stray bullet fired by neighborhood teens. He said, "He died in my arms. And when I looked at him, his face was like the same face of the child that I had killed. And I said: This is the punishment for killing the people that I killed." He suffered from severe PTSD, sequestered himself in his home with windows shuttered for years, and succeeded on his third suicide attempt.

Iconic photos of the nightmare. Left: *civilians seconds before being killed; the woman in the back holding her child had just been raped.* Right: *dead villagers*

ened to report events to superiors; officers subsequently placed him on more dangerous patrols, perhaps hoping he'd be killed.

And three men halted the killings. Predictably, they were outsiders. The catalyst was Warrant Officer Hugh Thompson Jr., age twenty-five, who was flying a helicopter, along with two crew members, Glenn Andreotta and Lawrence Colburn. Perhaps pertinent to what occurred was the fact that Thompson descended from Native American survivors of the Trail of Tears death march; his religious parents raised him, in the 1950s in rural Georgia, to oppose segregation. Colburn and Andreotta were observant Catholics.

Thompson and his crew had flown over the village, intending to aid the infantry fighting Viet Cong. Instead of evidence of a battle, they saw masses of dead civilians. Thompson initially thought that the village was under attack, with Americans protecting villagers, but couldn't figure out where the attack was coming from. He landed the copter amid the chaos and saw one soldier, Sergeant David Mitchell, firing into a mass of injured, wailing civilians in a ditch and another, Captain Ernest Medina, shoot a woman point-blank; Thompson realized who was doing the attacking. He confronted Calley, who was higher ranking than him and told him to mind his damn business.

Thompson saw a group of women and children huddling by a bunker with American soldiers approaching them, preparing to attack. Discussing

Left: *Glenn Andreotta;* Right, right to left: *Hugh Thompson, Lawrence Colburn, and Do Hoa, who they rescued from the ditch as a child, My Lai village, 1998*

what happened next, more than twenty years later, he described his feelings about those soldiers: "It's—they were the enemy at that time, I guess. They were damn sure the enemy to the people on the ground." He did something of dizzying strength and bravery, something that proves every word in this book about how Us/Them categorizations can change in an instant. Hugh Thompson landed his helicopter between the villagers and the soldiers, and with his machine guns oriented toward his fellow Americans, ordered his crew to mow them down if they attempted to further harm the villagers.*,†

Thus we have one person impulsively changing history in twenty countries, another who overcame decades of hatred to catalyze reconcili-

* Thompson radioed fellow helicopter pilots to evacuate survivors to hospitals; Andreotta waded through the dead in the ditch to rescue a miraculously unharmed four-year-old. Thompson reported what he had seen to his commanding officers, who sent word of the events further up the chain. As a result, the senior officer who had commanded the search-and-destroy mission canceled the ones planned for subsequent days in neighboring villages and began the process of covering up what had occurred. Andreotta was dead within three weeks, killed in battle. Colburn and, even more so, Thompson, attempted to inform every military, governmental, and media source available about the events and played key roles in making the My Lai Massacre public. Representative Mendel Rivers, chairman of the House Armed Services Committee, attempted to block the prosecution of Calley and to have Thompson prosecuted instead as a traitor; Thompson testified against Calley at his trial and received death threats for years. It took thirty years for him and Colburn to be honored by the military for their actions. Thompson died in 2006, with Colburn at his bedside.

† I thank two great undergrads, Elena Bridgers and Wyatt Hong, for help with the research throughout this section.

ation, others who overcame every reflex of their training to do the right thing. Time for one last singular person, one who inspires me enormously.

The person was the Anglican cleric John Newton, born in 1725.[54] Well, that doesn't sound too exciting. He's best known for composing the hymn "Amazing Grace." Oh, cool; that, along with Leonard Cohen's "Hallelujah," always move me. Newton also was an abolitionist, a mentor to William Wilberforce in his parliamentary battle to outlaw slavery in the British Empire. Okay, getting better. Now get this—as a young man, Newton had captained a slave ship. Bingo, that's the setup—a man overseeing and profiting from slavery, a flash of religious and moral insight, dramatic recategorization of Us and Them, dramatic expansion of his humanity, dramatic commitment to make amends for the savagery he had done. You can practically see chapter 5's neural plasticity on fire in Newton's brain.

Nothing resembling this occurred.

Newton, the son of a ship captain, goes to sea with his father at age eleven. At eighteen he is pressed into service in the navy, tries to desert, and is flogged. Newton manages to escape and works on a West African slave ship. Get ready for him to see the similarity between the captivity of these people and his own experience, to have a revelation.

No such thing occurs.

He works on the slave ship and is apparently so detested by everyone that they dump him in what is now Sierra Leone with a slaver who gives him to his wife as a slave. He's rescued; the ship he is on, returning to England, is caught in a horrific storm and starts to sink. Newton calls out to God, the ship doesn't sink, and he has a spiritual conversion to evangelical Christianity. He signs up to work on another slave ship. Get ready now—he's found God, has just been a slave himself, and is poised to suddenly recognize the horror that was the slave trade.

Nope.

He professes some sympathy for slaves, grows deeper into his evangelical conversion. He eventually becomes captain of a slave ship and works another six years before stopping. At last he's seen his actions for what they are.

Not that either.

It's because his health was declining from those tough voyages. He works as a tax collector, studies theology, applies to become an Anglican priest. And he *invests his money* in slave-trading ventures. In the parlance of my native Brooklyn, from when it was not yet trendy, can you believe this fuggin' guy?

He becomes a popular preacher, known for his sermons and pastoral concern; he composes hymns, speaks out for the poor and downtrodden. Presumably, somewhere along the way he stops investing in slavery; maybe because of his conscience, maybe because better investments come along. Still, not a word about slavery. Finally he publishes a pamphlet denouncing it, *thirty-four years* after stopping being a slaver. That's a lot of time spent as a blind wretch. Newton's is a rare voice among abolitionists, someone who has witnessed those horrors, let alone inflicted them. He becomes the major abolitionist voice in England and lives to see England ban the slave trade in 1807.

There's no way I could ever be Thompson, Andreotta, or Colburn. I'm not brave; I run away to solitary African field sites instead of confronting difficult things. Maybe, at best, I would have been one of the soldiers standing in confusion, compelled by the inhibitions that Grossman discusses into repeatedly checking my rifle to make sure it was loaded, rather than firing it. I see little indication that as an old man I will achieve the grace and moral stature of a Zenji Abe or a Richard Fiske. Bouazizi's act is incomprehensible to me.

But Newton, Newton is different; Newton is familiar. He takes convenient comfort from the Bible's embrace of slavery, spends decades resisting the possibility of his personal morality moving past its conventions. He shows great empathy but applies it selectively. He expands his circle of who counts as an Us, but only so far. We saw how the person who emerges from the crowd to run into the burning building typically acts before thinking, displaying an ingrained automaticity of doing the harder, better thing. There's no automaticity with Newton. We can practically see his dlPFC laboring with all that rationalizing—"There's nothing I can do," "It's too big for one person to challenge," "Better to be concerned about the needy who are close to home," "I can use the profits from the

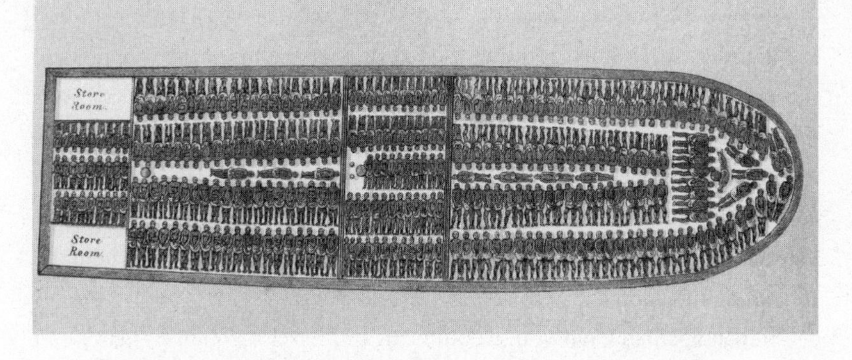

1788 illustration created by abolitionists of the number of slaves (487) a British ship could legally hold during a trans-Atlantic voyage. In actuality, ships transported far more people than that.

investments for good works," "Those people really are so fundamentally different," "I'm tired." Yes, journeys begin with a single step, but with Newton it's ten steps forward, nine self-serving ones back. Thompson's moment of moral perfection feels as unattainable to me as aspiring to be a gazelle or a waterfall or an incandescent sunset. But there's hope for us, with our foibles and inconsistencies and frailties, as we watch Newton slowly lurch his way toward being a moral titan.

Finally—the Potential for Collective Power

There is an anecdote from the Peninsular War of 1807–14, told by Major General George Bell, then an ensign: There was a bridge separating the opposing British and French, with a sentry posted by each side to sound an alarm should the enemy rush across the bridge.[55] A British officer was making rounds and found the British sentry there in an unlikely situation—carrying British and French muskets, one on each shoulder, seemingly guarding the bridge for the two opposing armies, with no French sentry in sight. His explanation? His French counterpart had snuck off to buy some liquor for them to share and, naturally, he was watching the other guy's gun.

Fraternizing between enemy soldiers is remarkably frequent in war. It's most common when they're the same race and major religion and when they are enlisted men rather than officers. It's also more common when individual enemies, rather than groups, encounter each other, when it's the same person day after day (e.g., guarding the bridge opposite you), when someone could have shot you but didn't. Fraternizing rarely involves discussions about life, death, and geopolitics; instead it's things like bartering food (since the other side's rations can't be as bad as yours), cigarettes, or alcohol or complaining about the miserable weather, the miserable officers.[56]

In the Spanish Civil War, Republican and Fascist troops regularly met at night to drink, barter, and exchange newspapers, everyone on the lookout for officers. In the Crimean War there was regular bartering across enemy lines of Russian vodka for French baguettes. One British soldier in the Peninsular War described how in the evenings, British and French troops played cards around campfires. And in the American Civil War, Yankee and Rebel soldiers would fraternize, barter, trade newspapers, and, with piercing poignancy, hold joint baptismal services the evening before a battle that would clearly be a bloodbath.

Thus enemy soldiers have frequently found common ground. A little over a hundred years ago, two such events occurred on a stunning scale.

It must be admitted that some good came of World War I—thanks to

German and British soldiers posing together

the subsequent collapse of three empires, people in the Baltic, the Balkans, and Eastern Europe gained independence. But from anyone else's perspective, it was a pointless slaughter of fifteen million people. The war to end all wars, leading to the ruinous peace to end all peace, turned out to be just another of centuries of examples of Europe devouring its young with meaningless conflict. But amid the quagmire of World War I came two examples of hope that, for want of a better word, seem almost miraculous.

First is the Christmas Truce of 1914, when officers up and down the trenches tentatively shouted, "No shoot," in another language and met opposing officers in no-man's-land. The truce began as an agreement to halt hostilities during Christmas dinner and for retrieval of the dead.

Things spread from there. As extensively documented, soldiers on both sides loaned each other shovels for digging graves. And then helped out. And then held joint burial services. Which led to exchanges of food, drink, and tobacco. Eventually, unarmed soldiers swarmed into no-man's-land, prayed and caroled together, shared dinner, exchanged gifts. Enemy combatants took group portraits; buttons and helmets were exchanged as souvenirs; plans were made to meet when the war was over. Most famously, soccer matches were held with improvised balls, with scores rarely kept.[57]

One historian records a chilling anecdote concerning a German soldier writing home about the truce, mentioning that not everyone participated—there was one soldier who condemned the others as traitors, an obscure corporal named . . . Hitler. But for most of the five hundred miles of trenches, the truce held through Christmas, and often even New Year's. It took officers' threats of court-martial to get everyone back to fighting, soldiers wishing their counterparts a safe war. Stunning, moving, heartbreaking. And with only sporadic exceptions, it never happened again, as even brief Christmas truces to retrieve the dead led to court-martials.

Why did the 1914 truce work? The unique static nature of trench warfare meant that soldiers faced each other day after day. This prompted often-friendly taunting across the lines in the period preceding Christmas, establishing a vague sense of connection. Moreover, the repeated interactions produced a "shadow of the future"—betray the truce, and expect no-holds-barred revenge.

The success was also aided by everyone sharing the same Judeo-Christian tradition and Western European culture; many knew the others' language, had visited the others' country. They were of the same race, and pejoratively calling the enemy "Fritz" completely differs from the pseudospeciation of the Vietnam War's "gooks," "slants," and "dinks."

Additional factors explain why the truce mostly involved British and German troops. While the French fought passionately on their own soil, Brits had no particular animosity toward Germans and typically perceived themselves as fighting to save *les derrières* of the French, their frequent historical enemy. Ironically, during the truce British soldiers would tell Germans that they both should be fighting the French. Meanwhile, by chance, most of the German soldiers were Saxons, who expressed a cousinly affinity for British Anglo-Saxons, suggesting that they should both be fighting the Prussians, the resented dominating group in Germany.

And perhaps most important, the truce was aided by top-down approval. Officers typically negotiated; figures such as the pope called for a truce; it was a holiday that stood for peace and good will toward all men.

Thus we have the Christmas Truce. Remarkably, something even more miraculous occurred during the war. In what has been termed the Live and Let Live phenomenon, soldiers in the trenches repeatedly evolved stable truces without exchanging a word, without a shared religious holiday, without the sanction of officers and leaders.

How did this occur? As documented by the historian Tony Ashworth in *Trench Warfare: 1914–1918*, it would begin passively. Troops on both sides ate around the same time, and guns would go silent then—who wants to interrupt dinner in order to kill someone or be killed? The same would occur during awful weather, when everyone's priority became flooded trenches or avoiding freezing to death.[58]

Mutual restraint also emerged in circumstances shadowed by the future. Wagon trains delivering food were easy artillery targets but were left unharmed, to prevent reciprocal shelling. Similarly, latrines were spared.

These truces emerged when soldiers chose not to do something. But truces were also established by overt action. How? Have your best sniper put a bullet into the wall of an abandoned house near enemy lines. Then have him do it again and again, repeatedly hitting the same spot. What are you communicating? "Look how good our guy is. He could have aimed at you instead but chose not to. What are you going to do about it?" And the other side would reciprocate with their best sniper. An agreement to shoot over each other's heads had been established.

The key was ritualization—shooting repeatedly at the same inconsequential target, renewing the commitment to peace daily at the same time.

Live and Let Live truces could withstand perturbations. Soldiers signaled the other side that they had to shoot for real for a while—officers were coming. The system survived violations. If some gung-ho rookie lobbed a shell into the others' trenches, the most common convention was two shells back, often aimed at important targets. And then the peace would resume. (Ashworth describes such a violation, where Germans unexpectedly fired a shell into British trenches. Soon a German shouted, "We are very sorry about that; we hope no one was hurt. It is not our fault, it is that damned Prussian artillery." And back flew two British shells.)

Live and Let Live truces emerged repeatedly. And repeatedly brass in the rear would intervene, rotating troops, threatening court-martials, ordering savage raids requiring hand-to-hand combat that would shatter any sense of shared interests between enemies.

We see the evolution—initial low-cost overtures with immediate benefits, such as not shooting during dinner, transitioning through gradations of increasingly elaborate restraint and signaling. And we recognize the modified Tit for Tat in dealing with truce violations, with its propensity toward cooperation, punishment for violations, mechanisms for forgiveness, and clear rules.

So, hooray, just like social bacteria, we can evolve cooperation. But one thing that a cooperative bacterium lacks is a psyche. Ashworth thoughtfully explored the psychology of how Live and Let Live participants began to view the enemy.

He described a sequence of steps. First, once any mutual restraint emerged, the enemy had established that they were rational, with incentives to hold fire. This prompted a sense of responsibility in dealing with them; this was initially purely self-serving—don't violate an agreement because they'll violate back. With time, the responsibility developed a moral tinge, tapping into most people's resistance to betraying someone who deals reliably with them. The specific motivations for truces generated insights—"Hey, they don't want dinner disturbed any more than we do; they don't want to fight in this rain either; they also deal with officers who screw up everything." There'd be a creeping sense of camaraderie.

This produced something striking. The war machines in combatant countries spewed the usual pseudospeciating propaganda. But in studying soldiers' diaries and letters, Ashworth observed minimal hostility toward the enemy expressed by trench soldiers; the further from the front, the more hostility. In the words of one frontline soldier, quoted by Ashworth, "At home one abuses the enemy, and draws insulting caricatures. How tired I am of grotesque Kaisers. Out here, one can respect a brave, skillful, and resourceful enemy. They have people they love at home, they too have to endure mud, rain and steel."

Us and Them would be in flux. If someone is shooting at you or your band of brothers, they are certainly a Them. But otherwise Them was

American and German propaganda posters

more likely to be the rats and lice, the mold in the food, the cold. As well as any comfortable officer at headquarters who would be—in the words of another trench soldier—"[an] abstract tactician who from far away disposes of us."

These truces could not persist; the final phases of the war obliterated them, as the British High Command adopted a nightmarish strategy of war by attrition.

In thinking of the Christmas Truce and the Live and Let Live system, I always have the same fantasy, a very different one from the fantasy that began this book. What would have happened if there had been two additional inventions during World War I? The first is modern mass communications—texting, Twitter, Facebook. The second is a mind-set that emerged only among World War I's shattered survivors—the cynicism of modernity. Men up and down hundreds of miles of trenches repeatedly reinvented Live and Let Live, unaware that they were not alone. Imagine texts bouncing along and across the trenches, a million soldiers at death's door saying, "This is bullshit. None of us here want to fight anymore, and we've figured out a way to stop." They could have ended it, tossed down their guns, could have ignored or ridiculed or killed any objecting officer spouting obscenities about God and country, could have gone home to kiss their loved ones and then face the real enemy, the bloated aristocracy who would sacrifice them for their own power.

It is easy to have this fantasy about the Great War, a distant museum piece festooned with twirly mustaches and silly plumed officers' helmets. It behooves us to step back from the grainy black-and-white photos and to consider a hugely difficult thought experiment. Our contemporary adversaries kidnap girls and sell them into slavery, commit atrocities and, instead of concealing them, display the evidence online. When I read the news of the things they've done, I hate them passionately. It's impossible to imagine kicking back, having a group sing-along of "I Saw Mommy Kissing Santa Claus" and exchanging Christmas tchotchkes with Al-Qaeda grunts.

Yet time does interesting things. The hatred between Americans and Japanese during World War II was boundless. American recruiting posters advertised "Jap Hunting Licenses"; one veteran of the Pacific theater described a common event, writing in the *Atlantic* in 1946: [American soldiers] "boiled the flesh off enemy skulls to make table ornaments for sweethearts, or carved their bones into letter openers."[59] And there's the bestial treatment of American POWs by the Japanese. If Richard Fiske had wound up a POW, Zenji Abe might have helped march him to death; if the former had killed the latter in battle, he might have made a souvenir of his skull. And instead, more than fifty years later, one would write a letter of condolence to the other's grandchildren upon the death of Grandpa.

A key point of the previous chapter was that those in the future will look back on us and be appalled at what we did amid our scientific ignorance. A key challenge in this chapter is to recognize how likely we are to eventually look back at our current hatreds and find them mysterious.

Daniel Dennett has pondered a scenario of someone undergoing surgery without anesthesia but with absolute knowledge that afterward

they'd receive a drug that would erase all memory of the event. Would pain be less painful if you knew that it would be forgotten? Would the same happen to hatred, if you knew that with time it would fade and the similarities between Us and Them would outweigh the differences? And that a hundred years ago, in a place that was hell on earth, those with the most temptation to hate often didn't even need the passage of time for that to happen?

The philosopher George Santayana provided us with an aphorism so wise that it has suffered the fate of becoming a cliché—"Those who cannot remember the past are condemned to repeat it." In the context of this final chapter, we must turn Santayana on his head—those who do not remember the extraordinary truces of the World War I trenches, or who do not learn of Thompson, Colburn, and Andreotta, or of the reconciliative distances traveled by Abe and Fiske, Mandela and Viljoen, Hussein and Rabin, or of the stumbling, familiar moral frailties that Newton vanquished, or who do not recognize that science can teach us how to make events like these more likely—those who do not remember these are condemned to be less likely to repeat these reasons to hope.

Epilogue

We've covered lots of ground, and some themes have arisen repeatedly. It's worth reviewing them before considering two final points.

As the single most important of them, virtually every scientific fact presented in this book concerns the *average* of what's being measured. There is always variation, and it's often the most interesting thing about a fact. Not every person activates the amygdala when seeing the face of a Them; not every yeast adheres to another one bearing the same surface protein marker. Instead, *on the average*, both do. Reflecting this, I've just discovered that this book contains variations on "average," "typically," "usually," "often," "tend to," and "generally" more than five hundred times. And I probably should have inserted them even more as reminders. There are individual differences and interesting exceptions everywhere you look in science.

Now, in no particular order:

- It's great if your frontal cortex lets you avoid temptation, allowing you to do the harder, better thing. But it's usually more effective if doing that better thing has become so automatic that it isn't hard.

And it's often easiest to avoid temptation with distraction and reappraisal rather than willpower.

- While it's cool that there's so much plasticity in the brain, it's no surprise—it has to work that way.
- Childhood adversity can scar everything from our DNA to our cultures, and effects can be lifelong, even multigenerational. However, more adverse consequences can be reversed than used to be thought. But the longer you wait to intervene, the harder it will be.
- Brains and cultures coevolve.
- Things that seem morally obvious and intuitive now weren't necessarily so in the past; many started with nonconforming reasoning.
- Repeatedly, biological factors (e.g., hormones) don't so much *cause* a behavior as modulate and sensitize, lowering thresholds for environmental stimuli to cause it.
- Cognition and affect always interact. What's interesting is when one dominates.
- Genes have different effects in different environments; a hormone can make you nicer or crummier, depending on your values; we haven't evolved to be "selfish" or "altruistic" or anything else—we've evolved to be particular ways in particular settings. Context, context, context.
- Biologically, intense love and intense hate aren't opposites. The opposite of each is indifference.
- Adolescence shows us that the most interesting part of the brain evolved to be shaped minimally by genes and maximally by experience; that's how we learn—context, context, context.
- Arbitrary boundaries on continua can be helpful. But never forget that they are arbitrary.
- Often we're more about the anticipation and pursuit of pleasure than about the experience of it.
- You can't understand aggression without understanding fear (and what the amygdala has to do with both).
- Genes aren't about inevitabilities; they're about potentials and vulnerabilities. And they don't determine anything on their own.

Gene/environment interactions are everywhere. Evolution is most consequential when altering *regulation* of genes, rather than genes themselves.

- We implicitly divide the world into Us and Them, and prefer the former. We are easily manipulated, even subliminally and within seconds, as to who counts as each.
- We aren't chimps, and we aren't bonobos. We're not a classic pair-bonding species or a tournament species. We've evolved to be somewhere in between in these and other categories that are clear-cut in other animals. It makes us a much more malleable and resilient species. It also makes our social lives much more confusing and messy, filled with imperfection and wrong turns.
- The homunculus has no clothes.
- While traditional nomadic hunter-gatherer life over hundreds of thousands of years might have been a little on the boring side, it certainly wasn't ceaselessly bloody. In the years since most humans abandoned a hunter-gatherer lifestyle, we've obviously invented many things. One of the most interesting and challenging is social systems where we can be surrounded by strangers and can act anonymously.
- Saying a biological system works "well" is a value-free assessment; it can take discipline, hard work, and willpower to accomplish either something wondrous or something appalling. "Doing the right thing" is always context dependent.
- Many of our best moments of morality and compassion have roots far deeper and older than being mere products of human civilization.
- Be dubious about someone who suggests that other types of people are like little crawly, infectious things.
- When humans invented socioeconomic status, they invented a way to subordinate like nothing that hierarchical primates had ever seen before.
- "Me" versus "us" (being prosocial within your group) is easier than "us" versus "them" (prosociality between groups).

- It's not great if someone believes it's okay for people to do some horrible, damaging act. But more of the world's misery arises from people who, of course, oppose that horrible act . . . but cite some particular circumstances that should make them exceptions. The road to hell is paved with rationalization.
- The certainty with which we act now might seem ghastly not only to future generations but to our future selves as well.
- Neither the capacity for fancy, rarefied moral reasoning nor for feeling great empathy necessarily translates into actually doing something difficult, brave, and compassionate.
- People kill and are willing to be killed for symbolic sacred values. Negotiations can make peace with Them; understanding and respecting the intensity of their sacred values can make lasting peace.
- We are constantly being shaped by seemingly irrelevant stimuli, subliminal information, and internal forces we don't know a thing about.
- Our worst behaviors, ones we condemn and punish, are the products of our biology. But don't forget that the same applies to our best behaviors.
- Individuals no more exceptional than the rest of us provide stunning examples of our finest moments as humans.

Two Last Thoughts

- If you had to boil this book down to a single phrase, it would be "It's complicated." Nothing seems to cause anything; instead everything just modulates something else. Scientists keep saying, "We used to think X, but now we realize that . . ." Fixing one thing often messes up ten more, as the law of unintended consequences reigns. On any big, important issue it seems like 51 percent of the scientific studies conclude one thing, and 49 percent conclude the opposite. And so on. Eventually it can seem hopeless that you can actually fix something, can make things better. But we have no choice but to try. And if you are reading this, you are probably ideally suited to do so. You've amply proven you have intellectual tenacity. You probably also have running water, a home, adequate calories, and low odds of festering with a bad parasitic disease. You probably don't have to worry about Ebola virus, warlords, or being invisible in your world. And you've been educated. In other words, you're one of the lucky humans. So try.

- Finally, you don't have to choose between being scientific and being compassionate.

Acknowledgments

The naturalist Edward O. Wilson, one of the most influential thinkers of our time, has found himself at the center of fiery controversies related to the evolution of human social behavior (discussed in chapter 10). An elegant and graceful man, he has written about those disputes and those who have most strongly opposed him—"Without a trace of irony I can say I have been blessed with brilliant enemies. I owe them a great debt, because they redoubled my energies and drove me in new directions."

When it comes to this book, I count myself even luckier than Wilson, in that I've had the good fortune of brilliant friends, ones who have been enormously helpful and generous with their time in vetting the chapters of this book. They've flagged my errors of omission and commission and my under-, over-, and misinterpretations, and tactfully let me know areas where I was twenty years out of date in my knowledge, or just plain woefully wrong. This book has benefited enormously from their collegial kindness, and I thank them all deeply (while taking credit for any errors remaining). They are:

Ara Norenzayan, University of British Columbia, Canada
Carsten de Dreu, Leiden University/University of Amsterdam, Netherlands
Daniel Weinberger, Johns Hopkins University
David Barash, University of Washington
David Moore, Pitzer College and Claremont Graduate University
Douglas Fry, University of Alabama at Birmingham

Gerd Kempermann, Dresden University of Technology, Germany
James Gross, Stanford University
James Rilling, Emory University
Jeanne Tsai, Stanford University
John Crabbe, Oregon Health and Science University
John Jost, New York University
John Wingfield, University of California at Davis
Joshua Greene, Harvard University
Kenneth Kendler, Virginia Commonwealth University
Lawrence Steinberg, Temple University
Owen Jones, Vanderbilt University
Paul Whalen, Dartmouth College
Randy Nelson, Ohio State University
Robert Seyfarth, University of Pennsylvania
Sarah Hrdy, University of California at Davis
Stephen Manuck, University of Pittsburgh
Steven Cole, University of California at Los Angeles
Susan Fiske, Princeton University

I have also had the fortune to interact with the spectacular students at Stanford University, and a number of them have directly contributed to this book. This has taken the form of their being library assistants, helping out with specific topics, or being members of a small seminar that I taught a few times that focused on the content of this book. They've been wonderful to work with and learn from. They are:

Adam Widman, Alexander Morgan, Ali Maggioncalda, Alice Spurgin, Allison Waters, Anna Chan, Arielle Lasky, Ben Wyler, Bethany Michel, Bilal Mahmood, Carl Cummings, Catherine Le, Christopher Schulze, Davie Yoon, Dawn Maxey, Dylan Alegria, Elena Bridgers, Elizabeth Levey, Ellen Edenberg, Ellora Karmarkar, Erik Lehnert, Ethan Lipka, Felicity Grisham, Gabe Ben-Dor, Gene Lowry, George Capps, Helen McLendon, Helen Shen, Jeffrey Woods, Jonathan Lu, Kaitlin Greene, Katharine Tomalty, Katrina Hui, Kian Eftekhari, Kirsten Hornbeak, Lara Rangel, Lauren Finzer, Lindsay Louie, Lisa Diver, Maisy

Samuelson, Morgan Freret, Nick Hollon, Patrick Wong, Pilar Abascal, Robert Schafer, Sam Bremmer, Sandy Kory, Scott Huckaby, Sean Bruich, Sonia Singh, Stacie Nishimoto, Tom McFadden, Vineet Singhal, Will Peterson, Wyatt Hong, Yun Chu.

I also thank Lisa Pereira of Stanford University, Christopher Richards of Penguin Books, Thea Traff of the *New Yorker*, and Ethan Lipka of the Nueva School for helping tremendously in getting this book into shape during the final stretch. Thanks to Kevin Berger for thinking of the title to chapter 6. Warm, heartfelt thanks to Katinka Matson and Steven Barclay, my publishing and speaking agents, sounding boards, and friends—you both know how long and difficult the gestation of this book was, and thank you for sticking with me throughout. I thank Scott Moyers of Penguin with huge gratitude—you have been a dream of an editor. And for spotting errors in previous printings, I thank John Linderman, Ellis Kirschenbaum, Craig Stephen, Paul Rosenbaum, Dragos Rotaru, Michael Uhl, Amit Sharma, and Robert Moore. And I apologize to anyone whose support I have missed noting here, as I rush frantically to meet the deadline for this book. . . .

Finally, above all else, I thank and madly love those who have given the most support, and who have withstood the most interruptions of board games, while I worked on the book—my family.

Appendix 1

Neuroscience 101

Consider two different scenarios. First:

Think back to when you hit puberty. You'd been primed by a parent or teacher about what to expect. You woke up with a funny feeling, found your jammies alarmingly soiled. You excitedly woke up your parents, who got tearful; they took embarrassing pictures, a sheep was slaughtered in your honor, and you were carried through town in a sedan chair while neighbors chanted in an ancient language. This was a big deal.

But be honest—would your life be so different if those endocrine changes had instead occurred twenty-four hours later?

Second:

Emerging from a store, you are unexpectedly chased by a lion. As part of the stress response, your brain increases your heart rate and blood pressure, dilates blood vessels in your leg muscles, which are now frantically working, and sharpens sensory processing to produce a tunnel vision of concentration.

How would things have turned out if your brain took twenty-four hours to send those commands? You'd be dead meat.

That's what makes the brain special. Hit puberty tomorrow instead of today? So what. Make some antibodies tonight instead of now? Rarely fatal. Same for delaying depositing calcium in your bones. But much of what the nervous system is about is encapsulated in the framing of chapter 2—what happened one second before? Incredible speed.

The nervous system is about contrasts, unambiguous extremes between having something and having nothing to say, maximizing signal-to-noise ratios. And this is demanding and expensive.*

ONE NEURON AT A TIME

The basic cell type of the nervous system, what we typically call a "brain cell," is the neuron. The hundred billion or so in our brains communicate with one another, forming complex circuits. In addition, there are glia cells, which do a lot of gofering—providing structural support and insulation for neurons, storing energy for them, helping to mop up neuronal damage.

Naturally, this neuron/glia comparison is all wrong. There are about ten glial cells for every neuron, coming in various subtypes. They greatly influence how neurons speak to one another, and also form glial networks that communicate completely differently from neurons. So glia are important. Nonetheless, to make this primer more manageable, I'm going to be very neuron-centric.

Part of what makes the nervous system so distinctive is how distinctive neurons are as cells. Cells are usually small, self-contained entities—consider little round red blood cells:

Red Blood Cells

Neurons, in contrast, are highly asymmetrical, elongated beasts, typically with processes sticking out all over the place:

These processes can be elaborated to nutty extents. Consider this single neuron, drawn in

* Which, among other things, is why the nervous system is so vulnerable to injury. Someone has a cardiac arrest. Their heart stops for a few minutes before it is shocked into beating again, and during those few minutes the entire body is deprived of blood, of oxygen and glucose. And at the end of those few minutes of "hypoxia-ischemia," every cell in the body is miserable and queasy. Yet it is preferentially brain cells (and a consistent subset of them) that are now destined to die over the next few days.

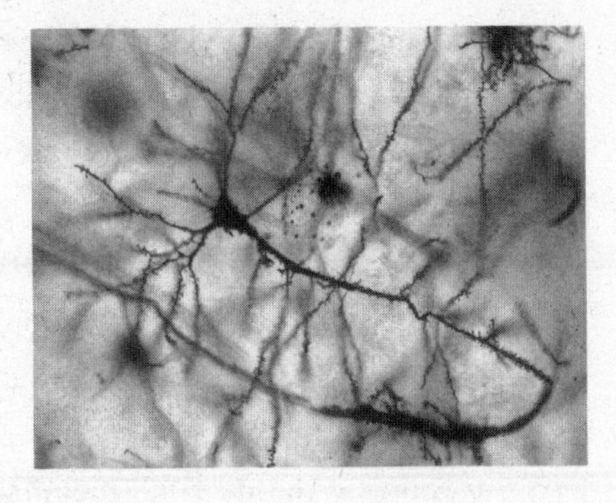

the early twentieth century by one of the gods in the field, Santiago Ramón y Cajal:

It's like the branches of a manic tree, explaining the jargon that this is a highly "arborized" neuron.

Many neurons are also outlandishly large. A zillion red blood cells fit on the proverbial period at the end of this sentence. In contrast, there are single neurons in the spinal cord that send out projection cables many feet long. There are spinal cord neurons in blue whales that are half the length of a basketball court.

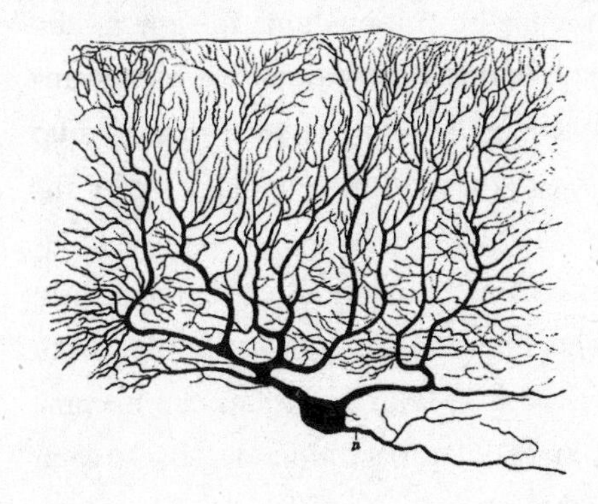

Now for the subparts of a neuron, the key to understanding its function.

What neurons do is talk to one another, cause one another to get excited. At one end of a neuron are its metaphorical ears, specialized processes that receive information from another neuron. At the other end are the processes that are the mouth, that communicate with the next neuron in line.

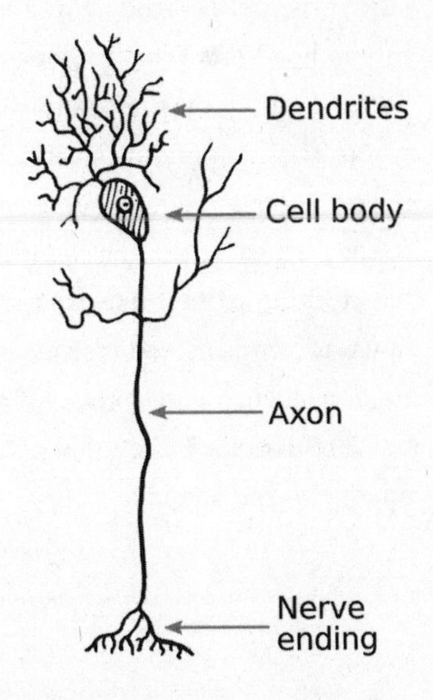

The ears, the inputs, are called dendrites. The output begins with a single long cable called an axon, which then ramifies into axonal endings—these axon terminals are the mouths (ignore the myelin sheath for the moment). Those axon terminals connect to the dendrites of the next neuron in line. Thus a neuron's dendritic ears are informed that the neuron behind it is excited. The flow of information then sweeps from the dendrites to the cell body to the axon to the axon terminals, and is then passed to the next neuron.

Let's translate "flow of information" into quasi chemistry. What actually goes from the dendrites to the axon terminals? A wave of electrical excitation. Inside the neuron are various positively and negatively charged ions. Just outside the neuron's membrane are other positively and negatively charged ions. When a neuron has gotten an exciting signal from the previous neuron at the end of one single dendritic fiber, channels in the membrane in that dendrite open, allowing various ions to flow in and others to flow out, and the net result is that the inside of the end of that dendrite becomes more positively charged. The charge spreads toward the axon terminal, where it is passed to the next neuron. That's it for the chemistry.

Two gigantically important details:

The resting potential. So when a neuron has gotten a hugely excitatory message from the previous neuron in line, its insides can become positively charged relative to the extracellular space around it. Back to our earlier metaphor—the neuron now has something to say and it is screaming its head off. What might things look like then when the neuron has nothing to say, has not been stimulated? Maybe a state of equilibrium, where the inside and outside have equal, neutral charges?* No, never! That's good enough for some cell in your spleen or your big toe. But back to that critical issue, that neurons are all about contrasts. When a neuron has nothing to say, that isn't some passive state of things just trickling down to zero. Instead it's an active process. An active, intentional, forceful, muscular, sweaty process. Instead of the "I have nothing to say" state being one of charge neutrality, the inside of the neuron is *negatively* charged relative to the outside.

* For chemists, in other words, the distribution of charged ions inside and out balance each other.

You couldn't ask for a more dramatic contrast: I have nothing to say = inside of the neuron is negatively charged. I have something to say = inside is positive. No neuron ever confuses the two. The internally negative state is called the "resting potential." The excited state is called the "action potential." And why is generating this dramatic resting potential such an active process? Because neurons have to work like crazy, using various pumps in their membranes, to push some positively charged ions out and to keep some negatively charged ones in, all in order to generate that negative internal resting state. Along comes an excitatory signal; the pumps stop working, channels open, and ions rush this way and that to generate the excitatory positive internal charge. And when that wave of excitation has passed, the channels close and the pumps go back into action, regenerating that negative resting potential. Remarkably, neurons spend nearly half their energy on the pumps that generate the resting potential. It doesn't come cheap to generate dramatic contrasts between having nothing to say and having some exciting news.

Now that we understand resting potentials and action potentials, on to the other gigantically important detail:

That's not what action potentials really are. What I've just outlined is that a single dendritic fibril receives an excitatory signal from the previous neuron (i.e., the previous neuron has had an action potential); this generates an action potential in that dendrite, which propagates toward the cell body, over it, on to the axon, to the axon terminals, and signals the next neuron in line. Not true. Instead:

So the neuron is sitting there with nothing to say, which is to say that it's displaying a resting potential; all of its insides are negatively charged. Along comes an excitatory signal at that one dendritic fibril, emanating from the previous neuron in line. As a result, channels open and ions flow in and out of that one dendrite. But only a little bit. Not enough to make the entire inside of the neuron positively charged, simply a little less negatively charged just inside that dendrite (just to attach some numbers here that don't matter in the slightest, the resting potential charge shifts from around –70 millivolts to around –60 millivolts). Then the channels close. That little hiccup of becoming less negative* spreads farther up the shaft of that dendrite. The pumps have started working, pumping ions back to where they were in the first place. So at the end of that dendritic fibril, the charge went from –70 to –60 mV. But a little bit up the shaft of that fibril, things then go from –70 to –65 mV. Farther up the shaft, –70 to –69. In other words, that excitatory signal dissipates. You've taken a nice smooth calm lake, in its resting state, and tossed a little pebble in. It causes a bit of a ripple right there, which spreads outward, getting smaller in its magnitude, until it dissipates not far from where the pebble hit. And miles away, at the lake's axonal end, that ripple of excitation has had no effect whatsoever.

In other words, if a single dendritic fibril is excited, that's not enough to pass on the excitation down to the axonal end and on to the next neuron. How does a message ever get passed on? Back to that wonderful drawing of a neuron by Cajal on page 681.

That arborized array of bifurcating dendritic branches ends in lots of ends of fibrils (time to introduce the term commonly used: "ends in lots of dendritic *spines*"). And in order to get sufficient excitation to sweep from the dendritic end of the neuron to the axonal end, you have to have summation—the same spine must be stimulated repeatedly and/or, more commonly, a bunch of the spinal neurons must be stimulated at once. You can't get a wave, rather than just a ripple, unless you throw in a lot of pebbles.

At the base of the axon, where it emerges from the cell body, is a specialized part (called the axon "hillock"). If all those summated dendritic inputs produce enough of a ripple to move the resting potential around the hillock from –70 mV to around –40 mV, a threshold is passed. And once that happens, all hell breaks loose. A different class of channels opens in the membrane of the hillock, which allows a massive migration of ions, producing, finally, a positive charge (about 30 mV). In other words, an action potential. Which then opens up those same types of channels in the

* Jargon: that little bit of "depolarization."

next smidgen of axonal membrane, regenerating the action potential there, and then the next, and the next, all the way down to the axon terminals.

From an informational standpoint, a neuron has two different types of signaling systems. From the dendritic spines to the starts of the axon hillock, it's an analogue signal, with gradations of signals that dissipate over space and time. And from the axon hillock to the axon terminals, it's a digital system with all-or-none signaling that regenerates down the length of the axon.

Let's throw in some imaginary numbers. Suppose an average neuron has about one hundred dendritic spines and about one hundred axon terminals. What are the implications of this in the context of the analogue/digital feature of neurons?

Sometimes nothing interesting. Consider neuron A, which, as just introduced, has one hundred axon terminals. Each one of those connects to one of the dendritic spines of the next neuron in line, neuron B. Neuron A has an action potential, which propagates down to all of its one hundred axon terminals, which excites all one hundred dendritic spines in neuron B. The threshold at the axon hillock of neuron B requires fifty of the dendrites to get excited around the same time in order to generate an action potential; thus, with all one hundred of the dendrites firing, neuron B is guaranteed to get an action potential.

Now, instead, neuron A projects half of its axon terminals to neuron B and half to neuron C. It has an action potential; does that guarantee one in neurons B and C? Yes. Each of those neurons' axon hillocks has that threshold of needing a signal from fifty dendritic pebbles at once in order to have an action potential.

Now, instead, neuron A evenly distributes its axon terminals among ten different target neurons, neurons B through K. Is its action potential going to produce action potentials in the target neurons? No way—continuing our example, the ten dendritic spines' worth of pebbles in each target neuron is way below the threshold of fifty pebbles.

So what will ever cause an action potential in, say, neuron K, which has only ten of its dendritic spines getting excitatory signals from neuron A? Well, what's going on with its other ninety dendritic spines? They're

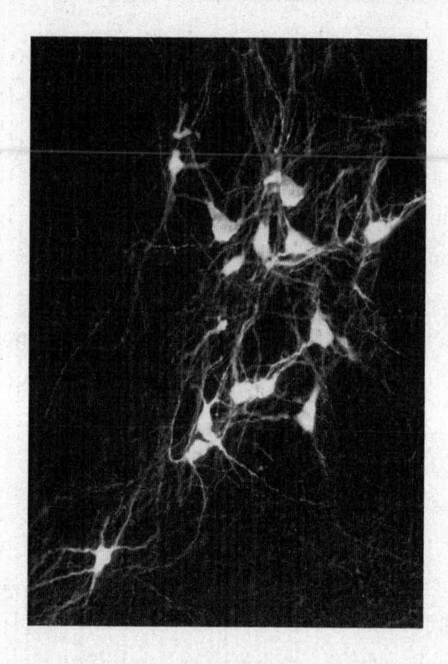

getting inputs from other neurons—nine of them, with ten inputs from each. When will neuron B have an action potential? When at least half of the neurons projecting to it have action potentials. In other words, any given neuron integrates the inputs from all the neurons projecting to it. And out of this comes a rule: *the more neurons that neuron A projects to, by definition, the more neurons it can influence; however, the more neurons it projects to, the smaller its average influence will be at each of those target neurons.* There's a trade-off.

This doesn't matter in the spinal cord, where one neuron typically sends all its projections to the next one in line. But in the brain one neuron will disperse its projections to scads of other ones and receive inputs from scads of other ones, with each neuron's axon hillock determining whether the threshold is reached and an action potential generated. The brain is wired in networks of divergent and convergent signaling.

Now to put in a flabbergasting real number—your average neuron has about *ten thousand* dendritic spines and about the same number of axon terminals. Factor in a hundred billion neurons, and you see why brains, rather than kidneys, write poetry.

Just for completeness, here are a couple of final facts. Neurons have some additional tricks, at the end of an action potential, to enhance the contrast between nothing to say/something to say even more, two means of ending the action potential really fast and dramatically: something called delayed rectification and another thing called the hyperpolarized refractory period. Another minor detail from that diagram above—a type of glial cell wraps around an axon, forming a layer of insulation called a myelin sheath; this "myelination" causes the action potential to shoot down the axon faster.

And one final detail of great future importance: the threshold of the axon hillock can change over time, thus changing the neuron's excitability. What things change thresholds? Hormones, nutritional state, experience, and other factors filling this book's pages.

We've now made it from one end of a neuron to the other. How exactly does a neuron with an action potential communicate its excitation to the next neuron in line?

TWO NEURONS AT A TIME: SYNAPTIC COMMUNICATION

So an action potential has been triggered at the hillock in neuron A and has swept down to all ten thousand axon terminals. How is this excitation passed on to the next neuron(s)?

The Defeat of the Synctitium-ites

For your average nineteenth-century neuroscientist, the answer was easy. Their explanation would be that a fetal brain is made up of huge numbers of separate neurons that slowly grow their dendritic and axonal processes. And eventually the axon terminals of one neuron reach and touch the dendritic spines of the next neuron(s), and they merge, forming a continuous membrane between the two cells. From all those separate fetal neurons, the mature brain forms this continuous, vastly complex net of one single superneuron, which was called a "synctitium." Thus excitation readily flows from one neuron to the next because they aren't really separate neurons.

Late in the nineteenth century an alternative view emerged, namely that each neuron remains an independent unit, and that the axon terminals of one neuron don't actually touch the dendritic spines of the next. Instead there's a tiny gap between the two. This notion was called the "neuron doctrine."

The adherents of the synctitium school thought that the neuron doctrine was asinine. "Show me the gaps between axon terminals and dendritic spines," they demanded of these heretics, "and tell me how excitation jumps from one neuron to the next."

And then in 1873 it all got solved by the Italian neuroscientist Camillo Golgi, who invented a technique for staining brain tissue in a novel fashion. And the aforementioned Cajal used this "Golgi stain" to stain all the processes, all the branches and branchlets and twigs of the dendrites and axon terminals of single neurons. Crucially, the stain didn't spread from one neuron to the next. There wasn't a continuous, merged net of a single superneuron. Individual neurons are discrete entities. The neuron doctrine-ers vanquished the synctitium-ites.*

Hooray, case closed; there are indeed micro-microscopic gaps between axon terminals and dendritic spines; these gaps are called "synapses" (which weren't directly visualized, putting the last nail in the syncytitial coffin, until the invention of electron microscopy in the 1950s). But there's still that problem of how excitation propagates from one neuron to the next, leaping across the synapse.

The answer, whose pursuit dominated neuroscience in the middle half

* Ironic footnote: Cajal was the chief exponent of the neuron doctrine. And the leading voice in favor of synctitiums? Golgi; the technique he invented showed that he was wrong. He apparently moped the entire way to Stockholm to receive his Nobel Prize in 1906—shared with Cajal. The two loathed each other, didn't even speak. In his Nobel address, Cajal managed to muster the good manners to praise Golgi. Golgi, in his, attacked Cajal and the neuron doctrine. Jerk.

of the twentieth century, is that the electrical excitation doesn't leap across the synapse. Instead it gets translated into a different type of signal.

Neurotransmitters

Sitting inside each axon terminal, tethered to the membrane, are little balloons called vesicles, filled with many copies of a chemical messenger. Along comes the action potential that initiated miles away in that neuron's axon hillock. It sweeps over the terminal and triggers the release of those chemical messengers into the synapse. Which they float across, reaching the dendritic spine on the other side, where they excite the neuron. These chemical messengers are called neurotransmitters.

How do neurotransmitters, released from the "presynaptic" side of the synapse, cause excitation in the "postsynaptic" dendritic spine? Sitting on the membrane of the spine are receptors for the neurotransmitter. Time to introduce one of the great clichés of biology. The neurotransmitter molecule has a distinctive shape (with each copy of the molecule having the same). The receptor has a binding pocket of a distinctive shape that is perfectly complementary to the shape of the neurotransmitter. And thus the neurotransmitter—cliché time—fits into the receptor like a key into a lock. No other molecule fits snugly into that receptor; the neurotransmitter molecule won't fit snugly into any other type of receptor. Neurotransmitter binds to receptor, which triggers those channels to open, and the currents of ionic excitation begin in the dendritic spine.

This describes "transsynaptic" communication with neurotransmitters. Except for one detail: what happens to the neurotransmitter molecules after they bind to the receptors? They don't bind forever—remember that action potentials occur on the order of a millisecond. Instead they float off the receptors, at which point the neurotransmitters have to be cleaned up. This occurs in one of two ways. First, for the ecologically minded synapse, there are "reuptake pumps" in the membrane of the axon terminal. They take up the neurotransmitters and recycle them,

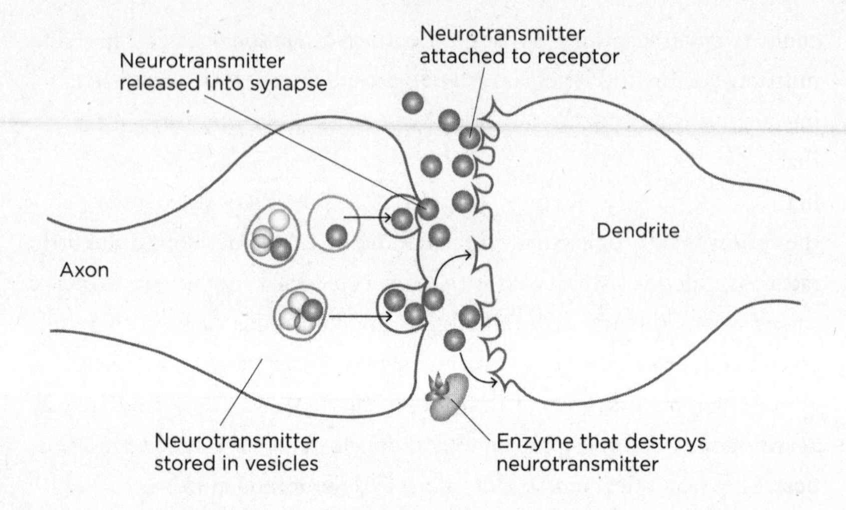

putting them back into those secretory vesicles to be used again.* The second option is for the neurotransmitter to be degraded in the synapse by an enzyme, with the breakdown products flushed out to sea (i.e., the extracellular environment, and from there on to the cerebrospinal fluid, the bloodstream, and eventually the bladder).

These housekeeping steps are hugely important. Suppose you want to increase the amount of neurotransmitter signaling across a synapse. Let's translate that into the excitation terms of the previous section—you want to increase excitability across the synapse, such that an action potential in the presynaptic neuron has more of an oomph in the postsynaptic neuron, which is to say it has an increased likelihood of causing an action potential in that second neuron. You could increase the amount of neurotransmitter released—the presynaptic neuron yells louder. Or you could increase the amount of receptor on the dendritic spine—the postsynaptic neuron is listening more acutely.

But as another possibility, you could decrease the activity of the reuptake pump. As a result, less of the neurotransmitter is removed from the synapse. Thus it sticks around longer and binds to the receptors repeatedly, amplifying the signal. Or, as the conceptual equivalent, you

* More with the keys in locks—the reuptake pumps have a shape that is complementary to the shape of the neurotransmitter, so that the latter is the only thing taken back up into the axon terminal.

could decrease the activity of the degradative enzyme; less neurotransmitter is broken down, so more sticks around longer in the synapse, having an enhanced effect. As we saw, some of the most interesting findings that help explain individual differences in the behaviors that concern us in this book relate to amounts of neurotransmitter made and released, and the amounts and functioning of the receptors, reuptake pumps, and degradative enzymes.

Types of Neurotransmitters

So what is this mythic neurotransmitter molecule, released by action potentials from the axon terminals of all of the hundred billion neurons? Here's where things get complicated, because there is more than one type of neurotransmitter.

Why more than one? The same thing happens in every synapse, which is that the neurotransmitter binds to its key-in-a-lock receptor and triggers the opening of various channels that allow the ions to flow and makes the inside of the spine a bit less negatively charged.

One reason is that different neurotransmitters depolarize to different extents—in other words, some have more excitatory effects than others—and for different durations. This allows for a lot more complexity in information being passed from one neuron to the next.

And now to double the size of our palette, there are some neurotransmitters that don't depolarize, don't increase the likelihood of the next neuron in line having an action potential. They do the opposite—they "hyperpolarize" the dendritic spine, opening different types of channels that make the resting potential even more negative (e.g., shifting from –70 mV to –80 mV). In other words, there are such things as *inhibitory* neurotransmitters. You can see how that has just made things more complicated—a neuron with its ten thousand dendritic spines is getting excitatory inputs of differing magnitudes from various neurons, getting inhibitory ones from other neurons, and integrating all of this at the axon hillock.

Thus there are lots of different classes of neurotransmitters, each binding to a unique receptor site that is complementary to its shape. Are there a bunch of different types of neurotransmitters in each axon terminal, so that an action potential triggers the release of a whole orchestration of signaling? Here is where we invoke Dale's Principle, named for Henry Dale, one of the grand pooh-bahs of the field, who in the 1930s proposed a rule whose veracity forms the very core of each neuroscientist's sense of well-being: an action potential releases the same type of neurotransmitter from all of the axon terminals of a neuron. Therefore there will be a distinctive neurochemical profile to a particular neuron—"Oh, that neuron is a neurotransmitter A–type neuron. And what that also means is that the neurons that it talks to have neurotransmitter A receptors on their dendritic spines."*

There are dozens of neurotransmitters that have been identified. Some of the most renowned: serotonin, norepinephrine, dopamine, acetylcholine, glutamate (the most excitatory neurotransmitter in the brain), and GABA (the most inhibitory). It's at this point that medical students are tortured with all the multisyllabic details of how each neurotransmitter is synthesized—its precursor, the intermediate forms the precursor is converted to until finally arriving at the real thing, the painfully long names of the various enzymes that catalyze the syntheses. Amid that there are some pretty simple rules built around three points:

- **a.** You do not ever want to find yourself running for your life from a lion and, oopsies, the neurons that tell your muscles to run fast go off-line because they've run out of neurotransmitter. Neurotransmitters are therefore made from precursors that are plentiful; often they are simple dietary constituents. Serotonin and dopamine, for example, are made from the dietary amino acids tryptophan and tyrosine, respectively. Acetylcholine is made from dietary choline and lecithin.
- **b.** A neuron can potentially have dozens of action potentials a second. Each involves restocking the vesicles with more neurotransmitter,

* What that also implies is that if a neuron is getting axonal projections to five thousand of its spines from a neurotransmitter A–releasing neuron and five thousand from a neurotransmitter B–releasing one, it expresses different receptors on those two populations of its spines.

releasing them, and mopping up afterward. Given that, you do not want your neurotransmitters to be huge, complex, ornate molecules, each of which requires generations of stonemasons to construct. Instead they are all made in a small number of steps from their precursors. They're cheap and easy to make. For example, it takes only two simple synthetic steps to turn tyrosine into dopamine.

c. Finally, to complete this pattern of neurotransmitter synthesis as cheap and easy, multiple neurotransmitters can be generated from the same precursor. In neurons that use dopamine as the neurotransmitter, for example, there are two enzymes that do those two construction steps. Meanwhile, in norepinephrine-releasing neurons there's an additional enzyme that converts dopamine to norepinephrine.

Cheap, cheap, cheap. Which makes sense. Nothing becomes obsolete faster than a neurotransmitter after it has done its postsynaptic thing. Yesterday's newspaper is useful today only for house-training puppies.

Neuropharmacology

As these neurotransmitterology insights emerged, they allowed scientists to begin to understand how various "neuroactive" and "psychoactive" drugs and medicines work.

Broadly, such drugs fall into two categories: those that increase signaling across a particular type of synapse, and those that decrease it. We already saw some of the strategies for increasing signaling: (a) Stimulate more synthesis of the neurotransmitter (for example, by administering the precursor or using a drug that increases the activity of the enzymes that synthesize the neurotransmitter). As an example, Parkinson's disease involves a loss of dopamine in one brain region, and a staple of treatment is to boost dopamine levels by administering the drug L-DOPA, which is the immediate precursor of dopamine. (b) Administer a synthetic version of the neurotransmitter, or a drug that is structurally close enough to the real thing to fool the receptors. Psilocybin, for example, is structurally similar to serotonin and activates a subtype of its receptors. (c) Stimulate the postsynaptic neuron to make more receptors. Fine in theory, but not easily done. (d) Inhibit degradative enzymes so that more of the neurotransmitter sticks around in the synapse. (e) Inhibit the reuptake of the neurotransmitter, prolonging its effects in the synapse. The modern antidepressant of choice, Prozac, does exactly that in serotonin synapses. Thus it is often referred to as an "SSRI"—a selective serotonin reuptake inhibitor.

Meanwhile, a pharmacopeia of drugs are available to decrease signaling across synapses, and you can see what their underlying mechanisms are going to include—blocking the synthesis of a neurotransmitter, blocking its release, blocking its access to its receptor, and so on. Fun example: Acetylcholine stimulates your diaphragm to contract. Curare, the poison used in darts by Amazonian tribes, blocks acetylcholine receptors. You stop breathing.

A final, very relevant point—just as the threshold of the axon hillock can change over time in response to experience, nearly every facet of the nuts and bolts of neurotransmitterology can be changed by experience as well.

MORE THAN TWO NEURONS AT A TIME

We have now triumphantly reached the point of thinking about three neurons at a time. And within not too many pages, we will have gone wild and considered even more than three. The purpose of this section is to see how circuits of neurons work, the intermediate step before examining what entire regions of the brain have to do with the best and worst of our behaviors. Therefore, the examples here were chosen merely to give a flavor of how things work at this level.

Neuromodulation

Consider the following diagram:

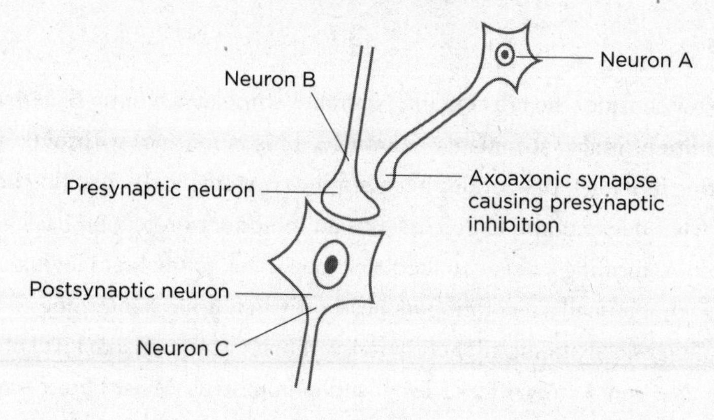

The axon terminal of neuron B forms a synapse with the dendritic spine of the postsynaptic neuron (let's call it neuron C) and releases an excitatory neurotransmitter. The usual. Meanwhile, neuron A sends an axon terminal projection on to neuron B. But not to a normal place, a dendritic spine. Instead its axon terminal synapses onto the axon terminal of neuron B.

What's up with this? Neuron A releases the inhibitory neurotransmitter GABA, which floats across that "axoaxonic" synapse and binds to receptors on that side of neuron B's axon terminal. And such an inhibitory effect (i.e., making that –70 mV resting potential even more negative) snuffs out any action potential hurtling down that branch of the axon, keeps it from getting to the very end and releasing neurotransmitter; in the jargon of the field, neuron A is having a neuromodulatory effect on neuron B.

Sharpening a Signal over Time and Space

Now for a new type of circuitry. To accommodate this, I'm using a simpler way of representing neurons. As diagrammed, neuron A sends all of its axonal projections to neuron B and releases an excitatory neurotransmitter, symbolized by the plus sign. The circle in neuron B represents the cell body plus all the dendritic branches.

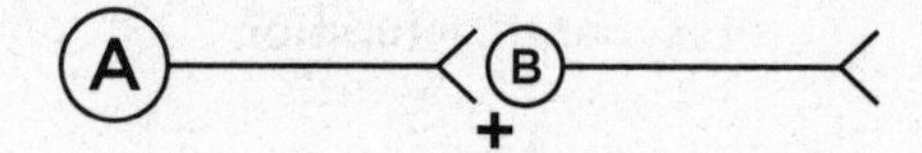

Now consider the next circuit. Neuron A stimulates neuron B, as usual. In addition, it also stimulates neuron C. This is routine, with neuron A splitting its axonal projections between the two target cells, exciting both. And what does neuron C do? It sends an inhibitory projection back onto neuron A, forming a negative feedback loop. Back to the brain loving contrasts, energetically screaming its head off when it has something to say, and energetically being silent otherwise. This is a more macro level of the same. Neuron A fires off a series of action potentials. What better way to energetically communicate that it's all over than to become majorly silent, thanks to the feedback loop? It's a means of sharpening a signal over time.* And note that neuron A can "determine" how powerful that negative feedback signal will be by how many of the ten thousand axon terminals it shunts toward neuron C instead of B.

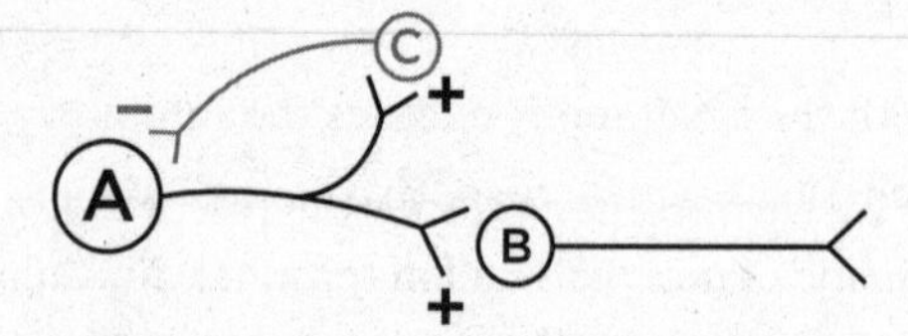

Such temporal sharpening of a signal can be accomplished in another way:

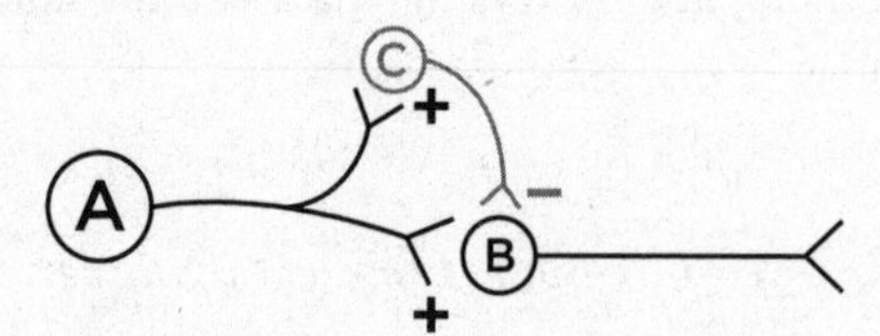

* This makes sense only after introducing an additional fact. Thanks to random, probabilistic hiccups in the ion channels now and then, neurons will occasionally have a random, spontaneous action potential from out of nowhere. So neuron A intentionally fires off ten action potentials, followed soon after by two random ones. That might make it hard to tell if neuron A meant to yell ten, eleven, or twelve times. By calibrating the circuit so that the inhibitory feedback signal shows up right after the tenth action potential, the two random ones afterward are prevented, and it is easier to tell what neuron A meant. The signal has been sharpened by damping the noise.

Neuron A stimulates B and C. Neuron C sends an inhibitory signal on to neuron B that will arrive sometime after B starts getting stimulated (since the A/C/B loop is two synaptic steps, versus one for A/B). Result? Sharpening a signal with "feed-forward inhibition."

Now for another type of sharpening of a signal, of increasing the signal to noise ratio. Consider this six-neuron circuit, where neuron A stimulates B, C stimulates D, and E stimulates F:

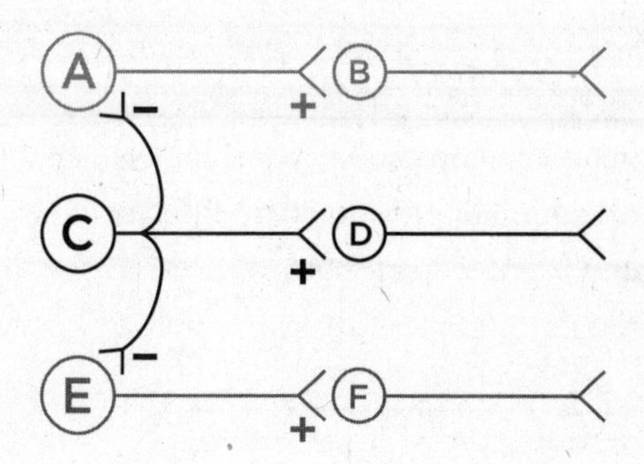

So neuron C sends an excitatory projection on to neuron D. But in addition, neuron C's axon sends collateral inhibitory projections on to neurons A and E.* Thus, if neuron C is stimulated, it both stimulates neuron D *and* silences neurons A and E. With such "lateral inhibition," C screams its head off while A and E become especially silent. It's a means of sharpening a spatial signal (and note that the diagram is simplified, in that I've omitted something obvious—neurons A and E also send inhibitory collateral projection on to neuron C, as well as to the neurons on the other sides of them in this imaginary network).

Lateral inhibition like this is ubiquitous in sensory systems. Shine a tiny dot of light onto an eye. Wait, was that photoreceptor neuron A, C, or E that just got stimulated? Thanks to lateral inhibition, it is clearer that it was C. Ditto in tactile systems, allowing you to tell that it was this

* Thanks to the wisdom of Dale, we know that the same neurotransmitter(s) is coming out of every axon terminal of neuron C. In other words, the same neurotransmitter can be excitatory at some synapses and inhibitory at others. This is determined by what type of ion channel the receptor is coupled to in the dendritic spine.

smidgen of skin that was just touched, not a little this way or that. Or the ears telling you it was definitely an A, not an A-sharp or A-flat.*

Thus what we've seen is another example of contrast enhancement in the nervous system. What is the significance of the fact that the silent state of a neuron is negatively charged, rather than a neutral 0 mV? It's a way of sharpening a signal within a neuron. Feedback, feed-forward, and lateral inhibition? A way of sharpening a signal within a circuit.

Two Different Types of Pain

This next circuit encompasses some of the elements just introduced and explains why there are, broadly, two different types of pain. I love this circuit because it is just so elegant:

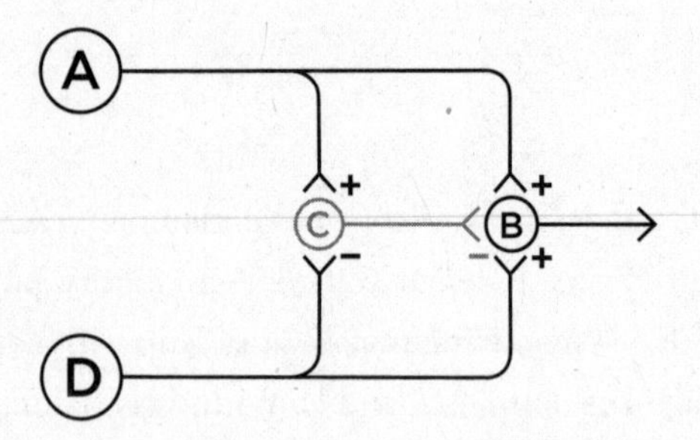

Neuron A's dendrites sit just below the surface of the skin, and the neuron has an action potential in response to a painful stimulus. Neuron A then stimulates neuron B, which projects up the spinal cord, letting you know that something painful just happened. But neuron A also stimulates neuron C, which inhibits B. This is one of our feed-forward inhibitory circuits. Result? Neuron B fires for a while and then is silenced, and you perceive this as a sharp pain—you've been poked with a needle.

Meanwhile, there's neuron D, whose dendrites are in the same general area of the skin and respond to a different type of painful stimulus. As before, neuron D excites neuron B, and the message is sent up to the

* Similar circuitry is also seen in the olfactory system, which has always puzzled me. What's just to the side of the smell of an orange? A tangerine?

brain. But it also sends projections to neuron C, where it *inhibits* it. Result? When neuron D is activated by a pain signal, it inhibits the ability of neuron C to inhibit neuron B. And you perceive it as a throbbing, continuous pain, like a burn or abrasion. Importantly, this is reinforced further by the fact that action potentials travel down the axon of neuron D much slower than in neuron A (having to do with that myelin that I mentioned earlier—details aren't important). So the pain in neuron A's world is not only transient but also fast. Pain in the neuron D branch not only is long-lasting but also has a slower onset.

The two classes of fibers can interact, and we often intentionally force them to. Suppose that you have some sort of continuous, throbbing pain—say, an insect bite. How can you stop the throbbing? Briefly stimulate the fast fiber. This adds to the pain for an instant, but by stimulating neuron C, you shut the system down for a while. And that is precisely what we often do in such circumstances. An insect bite throbs unbearably, we scratch hard right around it to dull the pain, and the slow, chronic pain pathway is shut down for up to a few minutes.

The fact that pain works this way has important clinical implications. For one thing, it has allowed scientists to design treatments for people with severe chronic pain syndromes (for example, someone with a severe back injury). Implanting a little electrode into the fast pain pathway and attaching it to a stimulator on the person's hip enables the patient to buzz that pathway now and then to turn off the chronic pain; it works wonders in many cases.

Thus we have a circuit that encompasses a temporal sharpening mechanism, introduces the double negative of inhibiting inhibitors, and is just all-around cool. And one of the biggest reasons why I love it is that it was first proposed in 1965 by the great neurobiologists Ronald Melzack and Patrick Wall. It was merely proposed as a theoretical model—"No one has ever seen this sort of wiring, but we propose that it's got to look something like this, given how pain works." And subsequent work showed that's exactly how this part of the nervous system is wired.

Which Guy Is It?

One final, completely hypothetical circuit.

Suppose we have a circuit made of two layers of neurons:

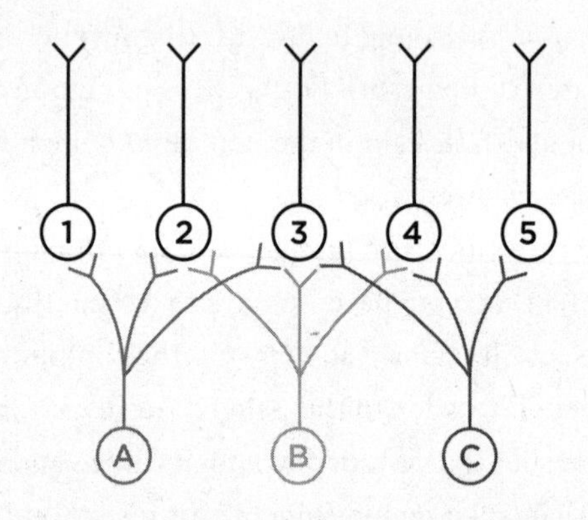

Neuron A projects to neurons 1, 2, and 3; neuron B projects to 2, 3, and 4, etc. Now let's show how hypothetical this circuit is by giving neurons A, B, and C completely imaginary functions. Neuron A responds to the picture of the guy on the left, B to the guy in the middle, C to the one on the right:

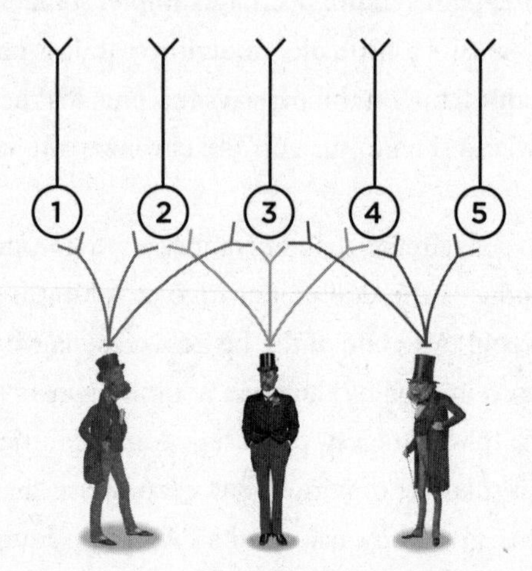

What can neuron 1 learn? How to recognize that particular guy. Neuron 5 is equally specialized. But what can neuron 3 learn? How Victorian gentlemen dressed. It's the neuron that will help you identify the Victorian in the quartet below:

Neuron 3's knowledge is general and comes from the overlap of the first layer's projections. Neurons 2 and 4 are also generalist neurons, but they're less accurate because they have only two exemplars each.

So neuron 3 is at the convergent center of this network. And the fanciest parts of the brain are wired up in a way that resembles this fairy-tale circuit, writ large—at the same time, neuron 3 is a more peripheral element in some other circuit sending projections to it (say, a circuitry that would be drawn perpendicular to this page), neuron 1 is at the very center of some other network in the fourth dimension, and so on. All of these neurons are embedded within multiple networks.

What does this produce? The capacity for association, metaphor, analogy, parable, symbol. To link two disparate things, even from different sensory modalities. To Homerically associate the color of wine with the color of the sea, that both "tomato" and "potato" can be pronounced in two different ways in a song, that a bright red tongue sticking out can remind you of music by the Stones. It's why I associate Stravinsky and Picasso, given that albums (remember those?) of Stravinsky's music always seemed to have a Picasso painting on the cover. And it's why a rectangular piece of cloth with a distinctive pattern of colors on it can stand for an entire nation or people or ideology.

A final point. We differ as to the nature and spread of our associative networks. And extremes of them can produce very interesting things at times. For example, most of us learned early on to associate something resembling the following with the concept of "face":

But then someone comes along whose associate networks of neuronal projections are broader, more idiosyncratic than everyone else's. And they teach the world that this can evoke faceness as well:

What might we call the consequence of some types of atypically wide associative nets of neurons? Creativity.

One More Round of Scaling Up

A neuron, two neurons, a neuronal circuit. We're ready now, as a last step, to scale up to the level of thousands of neurons at once.

Consider the following slice of tissue viewed under a microscope:

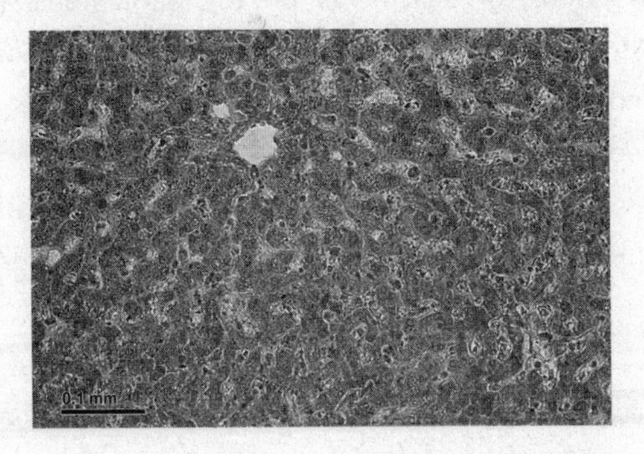

It's a homogeneous field of cells, all organized in roughly the same way. The top left corner and bottom left corner look exactly the same.

This is a liver in cross-section; if you've seen one part, you've seen it all. Boring.

If the brain were this homogeneous and boring, it would be an undifferentiated mass of tissue, with neuronal cell bodies carpeted evenly all over the place, sending out their processes every which way. Instead there's a huge amount of internal organization:

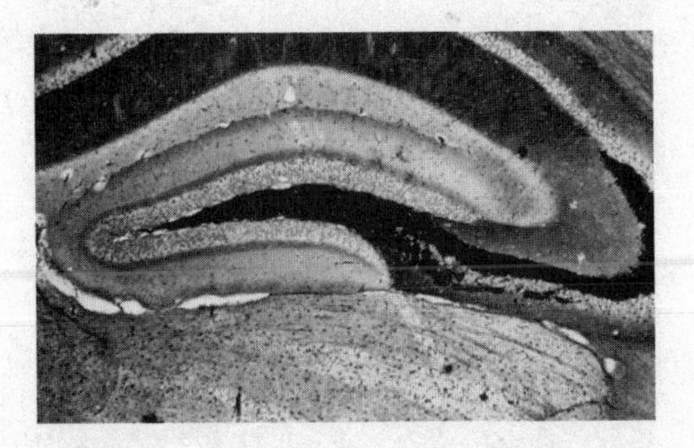

In other words, the cell bodies of neurons that have related functions are clumped together in particular regions of the brain, and the axons that they send to other parts of the brain are organized into projection cables. What all this means, crucially, is that *different parts of the brain do different*

things. All the regions of the brain have names (usually multisyllabic and derived from Greek or Latin), as do the subregions, and the sub-subregions. Moreover, each talks to a consistent collection of other regions (i.e., sends axons to them) and is talked to by a consistent collection (i.e., receives axonal projections from them).

You can go crazy studying all this, as I've seen, tragically, in the case of many a neuroanatomist who relishes all these details. For our purposes there are some key points:

- Each particular region contains millions of neurons. Some familiar names at this level of analysis: hypothalamus, cerebellum, cortex, hippocampus.
- Some regions have very distinct and compact subregions, and each is referred to as a "nucleus." (This is confusing, as the part of every cell that contains the DNA is also called the nucleus. What can you do?) Some probably totally unfamiliar names, just as examples: the basal nucleus of Meynert, the supraoptic nucleus of the hypothalamus, the charmingly named inferior olive nucleus.
- As described, the cell bodies of neurons with related functions are clumped together in their particular region or nucleus and send their axonal projections off in the same direction, merging together into a cable (a "fiber tract"). Here's an example, taken from the hippocampus:

- Back to that myelin wrapping around axons that helps action potentials propagate faster. Myelin tends to be white, sufficiently so that the fiber tract cables in the brain look white. Thus they're generically referred to as "white matter."

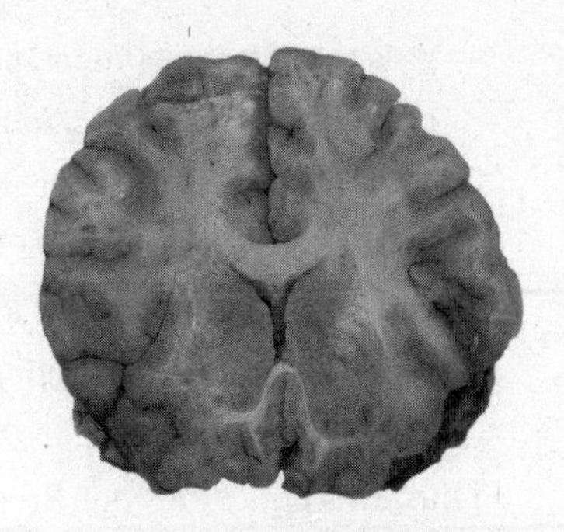

- As can be seen, a lot of the brain is taken up by the fiber tracts—all sorts of regions are talking to one another, often distant parts.*
- Suppose you have someone who has sustained an injury in one particular part of the brain, mysterious spot X. This gives the opportunity to learn something about the brain by now seeing what no longer works right in the person. Neuroscience as a field really got its starts thanks to studies of soldiers who had suffered "missile projection wounds." Viewed in a detached manner, the endless nineteenth-century European military bloodbaths were God's gift to neuroanatomists. The injured individual now does something abnormally. Can you conclude that spot X is the part of the brain responsible for the normal version of that behavior? Only if it's where a cluster of neuronal cell bodies are. If spot X is a fiber tract, you're actually learning something about the region of the brain whose neurons sent axonal projections in that fiber tract, and that region could be at the other end of the brain. So it's important to distinguish between "neuronal nuclei" and "fibers of passage."
- Finally, back to the reference just now about a part of the brain being the center for some behavior. The examples from earlier in the chapter showed how hard it is to make sense of the function of an individual neuron without considering the network that it is part of. Same theme here, writ large. Given that every brain region is

* As an aside, there has been incredibly interesting work concerning emergent properties of the brain that helps explain how the different regions wire up in the developing brain in an optimal way that minimizes the amount (and thus "cost") of axonal projections needed. For aficionados, the things the developing brain does bear some resemblance to some approaches used for the Traveling Salesman Problem.

getting projections from and sending projections to a zillion other places, it is rare that an individual brain region is "the center for" anything. Instead it's all networks where, far more often, a particular region "plays a key role in," "helps mediate," or "influences" a behavior. The function of a particular brain region is embedded in the context of its connections.

Thus, this concludes our Brain 101 primer.

Appendix 2

The Basics of Endocrinology

Endocrinology is the study of hormones, very different sorts of messengers from the neurotransmitters of chapter 2. As a recap, neurotransmitters are released from neurons' axon terminals in response to action potentials. Once released, they travel a microscopic distance across the synapse and bind to receptors on dendrites of the second, postsynaptic neuron, thereby changing that neuron's excitability.

In contrast, a hormone is a chemical messenger released from secretory cells (including neurons) in various glands. Once secreted, it enters the bloodstream, where it can influence any cells throughout the body that possess receptors for it.* So right off the bat we have key differences. First, neurotransmitters directly affect only neurons on the other side of synapses, while a hormone can potentially affect each of the trillions of cells in the body. A second difference is the time course; neurotransmitter signaling across synapses occurs in milliseconds. In contrast, many hormonal effects emerge over hours to days and can persist forever (for example, how often does puberty go away after a while?).

* An implication of these definitions is that the same molecule can serve as either a neurotransmitter or a hormone in different parts of the body. Also (minutia warning), sometimes hormones have "paracrine" effects, influencing cells in the gland in which they were secreted.

Neurotransmitters and hormones also differ in the scale of their effects. A neurotransmitter binds to its postsynaptic receptor, resulting in a local change in the flow of ions across the membrane of that dendritic spine. But depending on the hormone and the target cell being considered, hormones can change the activity of particular proteins, turn certain genes on or off, alter the metabolisms of cells, cause them to grow or atrophy, to divide or to shrivel up and die. Testosterone, for example, increases muscle mass, and progesterone causes the proliferation of cells in the uterus, causing it to thicken during the luteal phase. Conversely, thyroid hormone kills cells in a tadpole's tail as the animal is metamorphosing into a frog, and a class of stress hormones can kill cells in the immune system (helping to explain how stress makes us vulnerable to getting a cold). Hormones are extremely versatile.

Most hormones are part of a "neuroendocrine axis." Recall from chapter 2 how all roads in the limbic system lead to the hypothalamus, with its pivotal role in regulating the autonomic nervous system and hormonal systems. Here's where that second part comes in. Neurons in the hypothalamus secrete a particular hormone that travels in a tiny, local circulatory system connecting to the pituitary, just below the base of the brain. There that hormone stimulates the secretion of a particular pituitary hormone, which enters the general circulation and stimulates the secretion of a third hormone from some peripheral gland. Here's an example involving my three favorite hormones: during stress, hypothalamic neurons secrete CRH (corticotropin-releasing hormone), which stimulates pituitary cells to secrete ACTH (adrenocorticotropic hormone). Once in the general circulation, ACTH gets to the adrenal glands, where it stimulates secretion of steroid stress hormones called glucocorticoids (with the human version being cortisol, aka hydrocortisone). Other hormones (e.g., estrogen, progesterone, testosterone, and thyroid hormone) are released from peripheral glands as the final step of their own "hypothalamic/pituitary/peripheral gland axis."* As a wonderful complication, the secretion of each particular

* Just to make sure we have this sorted out, here's a second example, namely the hypothalamic/pituitary/ovarian axis: the hypothalamus releases GnRH (gonadotropin-releasing hormone), which triggers the pituitary to release LH (lu-

pituitary hormone is often not under the control of only a single hypothalamic releasing hormone. Instead there are multiple types of hormones serving that function, and other hypothalamic hormones that inhibit that particular pituitary hormone's release. For example, an array of hypothalamic hormones in addition to CRH regulates the release of ACTH, where different types of stressors produce different combinations of those hypothalamic hormones.

Not all hormones are regulated in this brain/pituitary/peripheral gland manner. In some cases there's a brain/pituitary two-step, where the pituitary hormone exerts effects throughout the body; growth hormone generally fits this pattern. In other systems the brain sends projections down the spine and to a particular gland, helping to regulate its hormone release; the pancreas and its secretion of insulin are an example (where circulating glucose levels are the main regulator). Then there are weirdo hormones secreted from unlikely places like the heart or gut, where the brain regulates secretion only indirectly.

Hormones, like neurotransmitters, are made cheaply. They are constructed in just a few biosynthetic steps from plentiful precursors—either simple proteins or cholesterol.* Moreover, the body generates multiple types of hormones from the same precursor. For example, the numerous steroid hormones are all generated from cholesterol.

So far we've given short shrift to hormone receptors. They do the same general job as do neurotransmitter receptors; there is a distinctive receptor molecule for each type of hormone,† with a concave binding domain whose shape is complementary to the shape of the hormone. To trot out the same cliché as was used for neurotransmitters, a hormone fits into its receptor like a key fits into a lock. And as with neurotransmitter receptors, there's no free lunch with hormone receptors. The various steroid hormones are structurally similar. Thus, if you're cheap at the production

teinizing hormone), which triggers the ovaries to release estrogen.

* And just to head off a potential misunderstanding at the pass, a zillionth of a percent of the cholesterol in your body is used for hormone synthesis, so changes in levels of cholesterol in the diet won't impact the amount of such steroids made—the body synthesizes enough cholesterol on its own for steroid synthesis.

† Actually typically more than one, but let's not go there.

end, you need subtle, fancy receptors that differentiate among those similar hormones—you do *not* want receptors that confuse, say, estrogen and testosterone.

Hormone/neurotransmitter similarities continue. Like neurotransmitter receptors, a hormone receptor's "avidity" for its hormone can change. This means that the shape of the binding site changes a bit, so that the hormone now fits more or less snugly, thus increasing or decreasing the duration of the hormone's effects. The number of receptors for a particular hormone in a cell can also change, altering the cell's sensitivity to that hormone's effects. The number of receptors in a target cell can be as important as the levels of the hormone itself, and there are endocrine diseases where normal levels of a hormone are secreted but, because of a mutation in the hormone's receptor, no signal gets through. Hormone levels are akin to how loudly someone speaks. Receptor levels are akin to the acuity with which ears detect that voice.

Finally, receptors for a hormone typically occur in only a subset of cells and tissues in the body, meaning that only those are responsive to the hormone. For example, only tail cells contain receptors for thyroid hormone when tadpoles are turning into frogs. Similarly, only some types of breast cancer involve tumors whose cells are "ER positive"—i.e., they contain estrogen receptors and are responsive to the growth-promoting effects of the hormone.

This is our overview of how hormones alter the functions of target cells over the course of hours to days. Hormones were highly pertinent in chapter 7 when considering the effects of hormones in childhood and fetal life. Specifically, hormones can have permanent "organizational" effects during development, shaping how the brain is constructed. In contrast, "activational" effects persist for hours to days. These two domains interact, in that organizational hormone effects on a fetal brain influence what activational effects hormones will have on that brain in adulthood.

Back to the main text to consider specific hormones.

Appendix 3

Protein Basics

Proteins are a class of organic compounds that are the most abundant molecules in living systems. They are hugely important, since numerous hormones, neurotransmitters, and messengers of the immune system are made of protein; ditto for the receptors that respond to those messengers, the enzymes that construct or degrade them,* the scaffolding that shapes a cell, and so on.

A key feature of proteins is their shape, since the shape of a protein determines its function. Proteins that form the scaffolding of a cell have the shape of the different crossbars in scaffolding at construction sites (sort of). A protein hormone will have a distinctive shape that is unique and distinctively different from the shape of a hormone that has different effects.† And a protein receptor must have a shape that is complementary to the shape of the hormone or neurotransmitter that it binds (back to the time-honored cliché of appendix 1, namely that a messenger like a hormone fits into its receptor like a key into a lock).

Some proteins change their shape, typically moving between two

* Naturally, the picture is more complicated than this, as is the case for most everything in this primer. Not all enzymes are made of proteins.

† And as a clarification, there are millions of copies of a particular hormone molecule (e.g., insulin) in the circulation, all sharing that same shape.

conformations. Suppose you have an enzyme (again, a protein) that synthesizes a molecule of sucrose by linking a molecule of glucose to a molecule of fructose. The enzyme must have one conformation that resembles the letter *V*, where one end binds a glucose molecule at a particular angle, the other fructose. The binding of both triggers the enzyme to shift to its other conformation, where the two ends of the *V* move close enough for the glucose and fructose to be linked. The sucrose floats off, and the enzyme flips back to its original conformation.

What determines the shape and function of a protein? Any given protein is made of a string of amino acids. There are about 20 different types of amino acids—including some familiar ones like tryptophan and glutamate. Each protein's string of amino acids is unique—like the string of letters that composes a word. Your typical protein is about 300 amino acids long, and with 20 different amino acid types, there are nearly 10^{400} possible sequences (that's ten followed by four hundred zeros)—more atoms than there are in the universe.* The amino acid sequence of a protein influences the unique shape(s) of that protein. Dogma used to be that amino acid sequence *determines* the shape(s) of that protein, but it turns out that the shape is also subtly altered by things like temperature and acidity—in other words, environmental influences.

And what determines the sequence of amino acids that are strung together to form a particular protein? A particular gene.

DNA AS THE BLUEPRINT FOR CONSTRUCTING PROTEINS

DNA is another class of organic compounds, and just as there are roughly 20 different types of amino acids, there are 4 different "letters" (called nucleotides) that make up DNA. A sequence of 3 nucleotides (called a codon) codes for a single amino acid. If there are 4 different types of nucleotides, and each codon is 3 nucleotides long, there can be a

* Actually, I haven't a clue how many atoms there are in the universe, but you're required to say stuff like this.

total of 64 different codons (4 possibilities in the first place × 4 in the second × 4 in the third = 64). A few of those 64 are reserved to signal the end of a gene, and after eliminating those "stop codons," there are 61 different codons coding for 20 different amino acids. Therefore, there is "redundancy"—almost all amino acids can be specified by more than one unique codon (an average of about 3, i.e., 61/20). Typically the different codons coding for the same amino acid differ by only a single nucleotide. For example, four different codon sequences code for the amino acid alanine: GCA, GCC, GCG, and GCT (A, C, G, and T are the abbreviations for the four types of nucleotides).* Redundancy will be important for understanding gene evolution.

The entire stretch of nucleotides that codes for a single type of protein is called a gene. The entire collection of DNA is called the genome, coding for all of the tens of thousands of genes in an organism; "sequencing" the genome means determining the unique sequence of the billions of nucleotides that make up that organism's genome. That stretch of DNA is so long (containing roughly twenty thousand genes in humans) that it has to be broken into separate volumes, called chromosomes.

This produces a spatial problem. The DNA library is found in the center of the cell, in the nucleus. Proteins, however, occur all over the cell, are constructed all over it (just think of proteins in the axon terminals of a spinal neuron in a blue whale, terminals that are light-years away from that neuron's nucleus). How do you get the DNA information out to where the protein is made? There is an intermediary that completes the picture. The unique nucleotide sequence in DNA that codes for a particular gene is copied into a string of similar nucleotide letters in a related compound called RNA. Any given chromosome contains a staggeringly long stretch of DNA, coding for one gene after another; in contrast, this stretch of RNA is only as long as the particular gene. In other words, a more manageable length. That RNA is then shipped to wherever it is supposed to be in the cell, where it then directs which amino acids are strung together in which sequence to form a protein (and there are amino

* The names of which I'm omitting, to avoid inundating the newcomer.

acids floating around in a cell, ready to be grabbed for the protein-construction project). Think of RNA as a photocopy of a single page out of this vast twenty-thousand-page-long DNA encyclopedia. (And multiple copies of the cognate protein can be made from the instructions in a photocopy page of RNA. This sure helps in circumstances where copies of the protein must wind up in each of the thousands of a single neuron's axon terminals.)

This produces what is termed the "central dogma" of life, a concept first framed in the early 1960s by Francis Crick, half of the renowned Watson and Crick, who discovered the "double helix" structure of DNA (with more than a little purloined help from Rosalind Franklin, but that's another story). Crick's central dogma holds that the nucleotide sequence of DNA that composes a gene determines how a unique stretch of RNA is put together . . . which determines how a unique stretch of amino acids are put together . . . which determines the shape(s) of the resulting protein . . . which determines that protein's function. DNA determines RNA determines protein.* And implicit in that central dogma is another critical point: one type of gene specifies one type of protein.

Just for everyone's sanity, I'm going to mostly ignore RNA. For our purposes, what is interesting is what genes, the starting point, have to do with their end products—proteins and their functions.

MUTATIONS AND POLYMORPHISMS

Genes are inherited from your parents (half the genes from each [not entirely true, as covered in the main text]). Suppose that when someone's DNA genome is being copied for inclusion in their egg or sperm, a mistake is made in the copying of one single nucleotide; with billions of

* The central dogma of "information flows from DNA to RNA to protein" can be wrong. There are circumstances where RNA can determine the sequence of DNA. This has lots to do with how some viruses work but isn't relevant to us. Another bit of revisionism, one that garnered two Nobel Prizes in 2006, is that a huge percentage of RNA does not then specify the construction of some protein. Instead it can target and destroy other sequences of RNA, a phenomenon known as "RNA interference." Still other RNAs are created simply to render some segments of DNA itself "unreadable."

nucleotides, that's bound to happen sometimes. As a result, unless corrected, the gene, now with its nucleotide sequence erroneously differing in one spot, is passed on to an offspring. This is a mutation.

In classical genetics there are three types of mutations that can occur. The first is called a point mutation. One single nucleotide is copied incorrectly. Will this change the amino acid sequence of the protein coded for? It depends. Back to redundancy in the DNA code, from a few paragraphs ago. Suppose there is a codon in a gene with the sequence GCT, coding for alanine. But there has been a mutation, yielding GCA instead. No problem—that still codes for alanine. It's an inconsequential, "neutral" mutation. But suppose the mutation instead was GAT. This codes for a completely different amino acid called asparagine. Uh-oh.

In actuality, though, this may not be a big deal, if the new amino acid looks a lot like the one that was lost. Suppose you have a nucleotide sequence coding for the following metaphorical amino acid sequence:

"I/am/now/going/to/do/the/following"

Thanks to a subtle mutation, there is a change of one amino acid, but one without a ton of consequences:

"I/am/now/going/ta/do/the/following"

This would still be comprehensible to most people; the protein would merely be perceived as coming from New York. Translated into proteinese, the protein has a slightly different shape and does its usual task a bit differently (maybe a little slower or faster). Not the end of the world.

But if the mutation codes for an amino acid that produces a protein with a dramatically different shape, the consequences can be enormous (even fatal).

Back to

"I/am/now/going/to/do/the/following"

What if there is a mutation in a nucleotide helping to code for the first *w*, a mutation with a big consequence?

"I/am/not/going/to/do/the/following"

Trouble.

The next type of classical mutation is called a deletion mutation. In this scenario a copying error is made during the inheritance of a gene. But instead of a nucleotide being miscopied, it is deleted. For example, in a case where the seventh nucleotide is deleted,

"I/am/now/going/to/do/the/following"

becomes

"I/am/now/oingt/od/ot/hef/ollowing"

This can frameshift everything over to generate gibberish, or even a different message (e.g., "For dessert I'd like the mousse" mutating to "For dessert I'd like the mouse").

Deletion mutations can involve the loss of more than a single nucleotide. At an extreme, this can involve the deletion of the entire gene, or even a stretch of genes on a particular chromosome. Definitely not good.

Finally, there are insertion mutations. During copying of the DNA to pass on to the next generation, a nucleotide is inadvertently copied twice, duplicated. Thus:

"I/am/now/going/to/do/the/following"

becomes

"I/am/now/ggoin/gt/od/oth/efollowin"

Gibberish, or perhaps a different message, as in the following case, where an *e* has been inserted near the end of the string of letters: "Mary turned John down for a date because she did not enjoy boweling." In some cases an insertion mutation can involve the insertion of more than a single nucleotide. At an extreme, this can even involve the duplication of an entire gene.

Point, deletion, and insertion mutations are most of what mutations are about.* Deletion and insertion mutations often have major consequences, usually deleterious, but sometimes produce a new, interesting protein.

Back to point mutations. Consider one that results in the substitution of a single amino acid in the protein, one that works a bit differently from the correct amino acid. As noted above, as a result, the protein still does its old job, but maybe does it a bit faster or slower. This could be the grist

* There are other, rarer types of mutations. One class of them, for example, involves the codon coding for an amino acid called glutamine being repeated over and over in the gene, even dozens of times, producing what are called "polyglutamine expansion diseases," the most famous being Huntington's disease. They are extremely rare mutations, though.

for evolutionary change—if the new version is disadvantageous, reducing the reproductive success of anyone who carries it, it will be gradually selected against, removing it from a population. If instead the new version is more advantageous, it will gradually replace the old one in a population. Or if the new version works better than the original in some circumstances but worse in others, it may reach equilibrium in the population with the original version, where a certain percentage of people have the old version, the remainder the new. In this case the particular gene would be described as coming in two different forms or variants, as coming with two different "alleles." Most genes come with multiple alleles. And the result is individual variation in the functioning of genes (this is covered in far more complexity in chapter 8).

Finally, a clarification of the confusion where two sound bites about genetics collide. The first is that, on the average, full (non–identical twin) siblings share 50 percent of their genes.* The other is that we share 98 percent of our genes with chimps. So are we more related to chimps than to our siblings? No. Comparisons between humans and chimps are about *types* of traits—we both have genes coding for traits related to having, for example, eyes, muscle fibers, or dopamine receptors, and both lack genes related to having, for example, gills, antennae, or flower petals. So there's 98 percent overlap at that level of comparison. But comparison between any two humans is about *versions* of those traits—both have a gene that codes for, say, this thing called eye color, but do they share the version that codes for the same particular color? Same for blood type, type of dopamine receptor, and so on. We have 50 percent overlap with siblings at this level of comparison.

* As do a parent and child, while half siblings share 25 percent of their genes, as do grandparents and grandchildren, and so on.

Glossary of Abbreviations

ACC anterior cingulate cortex
ACTH adrenocorticotropic hormone
ADHD attention-deficit/hyperactivity disorder
AIS androgen insensitivity syndrome
APA American Psychological Association
ASD autism spectrum disorders

BDNF brain-derived neurotrophic factor
BLA basolateral amygdala
BMI body mass index
BNST bed nucleus of the stria terminalis

CAH congenital adrenal hyperplasia
CBT cognitive behavioral therapy
COMT catechol-O-methyltransferase
CRH corticotropin-releasing hormone

DAT dopamine transporter
DHEA dehydroepiandrosterone
dlPFC dorsolateral PFC
DZ dizygotic

EEA equal environment assumption
EEG electroencephalographic; EEGs electroencephalograms
ERPS event-related potentials

fMRI functional magnetic resonance imaging
FTD frontotemporal dementia

GABA gamma-aminobutyric acid
GnRH gonadotropin-releasing hormone
GSR galvanic skin resistance
GWAS genomewide association studies

HG hunter-gatherer
HH high-warmth/high-competence
HL high warmth/low competence

IAT Implicit Association Test

LH low warmth/high competence
LH luteinizing hormone
LL low warmth/low competence
LTD long-term depression
LTP long-term potentiation

MAO-A monoamine oxidase-A
MHC major histocompatibility complex
MZ monozygotic

NCAM neural cell adhesion molecule

PAG periaqueductal gray
PD Prisoner's Dilemma
PFC prefrontal cortex
PMC premotor cortex
PMDD premenstrual dysphoric disorder
PMS premenstrual syndrome
PNS parasympathetic nervous system
PTSD post-traumatic stress disorder
PVN paraventricular nucleus

RNA ribonucleic acid
RWA right-wing authoritarianism

SDO social-dominance orientation
SES socioeconomic status
SHRP stress hyporesponsive period

SNPs single-nucleotide polymorphisms
SNS sympathetic nervous system
SPE Stanford Prison Experiment
SSRI selective serotonin reuptake inhibitor
STG superior temporal gyrus
TF transcription factor
TH tryptophan hydroxylase
ToM Theory of Mind
TPJ temporoparietal juncture
TRC truth and reconciliation commission
vlPFC ventrolateral prefrontal cortex
vmPFC ventromedial PFC 54

Abbreviations in the Notes

In order to save forests' worth of paper, references cite only the first one or two authors. The following abbreviations are used for entire journal titles or words within them:

AEL: *Applied Economics Letters.* AGP: *Archives of General Psychiatry.* Am: American. AMFP: *American Journal of Forensic Psychology.* Ann: Annual. ANYAS: *Annals of the New York Academy of Sciences.* Arch: Archives of. ARSR: *Annual Review of Sex Research.* BBR: *Behavioral Brain Research.* BBS: *Behavioral and Brain Sciences.* Behav: Behavior or Behavioral. Biol: Biology or Biological. Biol Lett: *Biology Letters.* BP: *Biological Psychiatry.* Brit: British. Bull: Bulletin. Clin: Clinical. Cog: Cognitive or Cognition. Comp: Comparative. Curr: Current. Dir: Directions in. EHB: *Evolution and Human Behavior.* Endo: Endocrinology. Evol: Evolution. Eur: European. Exp: Experimental. Front: Frontiers in. Horm Behav: *Hormones and Behavior.* Hum: Human. Int: International. J: Journal or Journal of. JAMA: *Journal of the American Medical Association.* JCP: *Journal of Comparative Psychology.* JEP: *Journal of Economic Psychology.* JESP: *Journal of Experimental and Social Psychology.* JPET: *Journal of Pharmacology and Experimental Therapeutics.* JPSP: *Journal of Personality and Social Psychology.* JSS: *Journal of Sports Sciences.* Med: Medical or Medicine. Mol: Molecular. Nat: Nature. NEJM: *New England Journal of Medicine.* Neurobiol: Neurobiology. Neurol: Neurology. Nsci: Neuroscience or Neurosciences. Nsci Biobehav Rev: *Neuroscience and Biobehavioral Reviews.* PLoS: *Public Library of Science.* PNAS: *Proceedings of the National Academy of Science, USA.* PNE: *Psychoneuroendocrinology.* Primat: Primatology. Proc: Proceedings of the. Prog: Progress in. PSPB: *Personality and Social Psychology Bulletin.* PSPR: *Personality and Social Psychology Review.* Psych: Psychology or Psychological. Rep: Report or Reports. Res: Research. Rev: Review or Reviews. SCAN: *Social, Cognitive and Affective Neuroscience.* Sci: Science or Sciences. Sci Am: *Scientific American.* Soc: Society or Social. TICS: *Trends in Cognitive Sciences.* TIEE: *Trends in Ecology and Evolution.* TIGS: *Trends in Genetic Sciences.* TINS: *Trends in Neuroscience.*

Notes

Introduction

1. R. Byrne, "Game 21 Adjourned as Thrust and Parry Give Way to Melee," *New York Times*, December 20, 1990.
2. For reviews of these two "easy" topics, see M. Winklhofer, "An Avian Magnetometer," *Sci* 336 (2012): 991; and L. Kow and D. Pfaff, "Mapping of Neural and Signal Transduction Pathways for Lordosis in the Search for Estrogen Actions on the Central Nervous System," *BBR* 92 (1998): 169.
3. J. Watson, *Behaviorism*, 2nd ed. (New York: Norton, 1930).
4. Footnote: J. Todd and E. Morris, eds., *Modern Perspectives on John B. Watson and Classical Behaviorism* (Westport, CT: Greenwood Press, 1994); H. Link, *The New Psych of Selling and Advertising* (New York: Macmillan, 1932).
5. E. Moniz, quoted in T. Szasz, *Schizophrenia: The Sacred Symbol of Psychiatry* (Syracuse, NY: Syracuse University Press, 1988).
6. K. Lorenz, quoted in R. Learner, *Final Solutions: Biology, Prejudice, and Genocide* (University Park: Penn State Press, 1992).
7. For discussions of Lorenz's activities during the Nazi era, see B. Sax, "What is a 'Jewish Dog'? Konrad Lorenz and the Cult of Wildness," *Soc and Animals* 5 (1997): 3; U. Deichman, *Biologists Under Hitler* (Cambridge MA: Harvard University Press, 1999); and B. Müller-Hill, *Murderous Science: Elimination by Scientific Selection of Jews, Gypsies, and Others, Germany 1933–1945* (Oxford, UK: Oxford University Press).
8. The Wellesley effect was first reported by Martha McClintock of the University of Chicago: M. McClintock, "Menstrual Synchrony and Suppression," *Nat* 229 (1971): 244. While a number of studies have replicated the Wellesley effect, some have not, as summarized in H. Wilson, "A Critical Review of Menstrual Synchrony Research," *PNE* 17 (1992): 565. A critique of that critique can be found in M. McClintock, "Whither Menstrual Synchrony?" *ARSR* 9 (1998): 77.
9. V. S. Naipaul, *Among the Believers: An Islamic Journey* (New York: Vintage Books, 1992). And for the definitive book on this entire field of behavioral biology, see M. Konner, *The Tangled Wing: Biological Constraints on the Human Spirit* (New York: Henry Holt, 2003). This is the finest book in existence on the biology of human social behavior—subtle, nuanced, nondogmatic, and wonderfully written—by the anthropologist/physician Mel Konner. To my vast good fortune, Konner was my academic adviser and mentor when I was an undergraduate, and he has had the greatest intellectual impact on me of anyone in my life. Those who know Mel will recognize his intellectual imprint on every page of this book.

Chapter 1: The Behavior

1. Footnote: F. Gervasi. *The Life and Times of Menachem Begin*. (New York: Putnam, 2009).
2. For good reviews of these distinctions, see K. Miczek et al., "Neurosteroids, GABAA Receptors, and Escalated Aggressive Behavior," *Horm Behav* 44 (2003): 242; and S. Motta et al., "Dissecting the Brain's Fear System Reveals That the Hypothalamus Is Critical for Responding in Subordinate Conspecific Intruders," *PNAS* 106 (2009): 4870.
3. A small, disheartening literature concerns ex–child soldiers and participants in genocides who are able to hold back their symptoms of post-traumatic stress disorder through acts of cruelty: R. Weierstall et al., "When Combat Prevents PTSD Symptoms: Results from a Survey with Former Child Soldiers in Northern Uganda," *BMC Psychiatry* 12 (2012): 41; R. Weierstall et al., "The Thrill of Being Violent as an Antidote to Posttraumatic Stress Disorder in Rwandese Genocide Perpetrators," *Eur J Psychotraumatology* 2 (2011): 6345; V. Nell, "Cruelty's Rewards: The Gratifications of Perpetrators and Spectators," *BBS* 29 (2006): 211; T. Elbert et al., "Fascination Violence: On Mind and Brain of Man Hunters," *Eur Arch Psychiatry and Clin Nsci* 260 (2010): S100.
4. B. Oakley et al., *Pathological Altruism* (Oxford: Oxford University Press, 2011).
5. L. MacFarquhar, "The Kindest Cut," *New Yorker*, July 27, 2009, p. 38.
6. Footnote: For a lengthy overview of Munchausen syndrome by proxy, see R. Sapolsky, "Nursery Crimes," in *Monkeyluv and Other Essays on Our Lives as Animals* (New York: Simon and Schuster/Scribner, 2005).

7. J. King et al., "Doing the Right Thing: A Common Neural Circuit for Appropriate Violent or Compassion Behavior," *NeuroImage* 30 (2006): 1069.

Chapter 2: One Second Before

1. For a summary of MacLean's findings and thinking, see P. MacLean, *The Triune Brain in Evolution* (New York: Springer, 1990).
2. A. Damasio, *Descartes' Error: Emotion, Reason, and the Human Brain* (New York: Putnam, 1994; Penguin, 2005).
3. W. Nauta, "The Problem of the Frontal Lobe: A Reinterpretation," *J Psychiatric Res* 8 (1971): 167; W. Nauta and M. Feirtag, "The Organization of the Brain," *Sci Am* 241 (1979): 88.
4. R. Nelson and B. Trainor, "Neural Mechanisms of Aggression," *Nat Rev Nsci* 8 (2007): 536.
5. For more on the effects of amygdala damage in humans, see A. Young et al., "Face Processing Impairments After Amygdalotomy," *Brain* 118 (1995): 15; H. Narabayashi et al., "Stereotaxic Amygdalotomy for Behavior Disorders," *Arch Neurol* 9 (1963): 1; V. Balasubramaniam and T. Kanaka, "Amygdalotomy and Hypothalamotomy: A Comparative Study," *Confinia Neurologia* 37 (1975): 195; R. Heimburger et al., "Stereotaxic Amygdalotomy for Epilepsy with Aggressive Behavior," *JAMA* 198 (1966): 741; B. Ramamurthi, "Stereotactic Operation in Behavior Disorders: Amygdalotomy and Hypothalamotomy," *Acta Neurochirurgica (Wien)* 44 (1988): 152; G. Lee et al., "Clinical and Physiological Effects of Stereotaxic Bilateral Amygdalotomy for Intractable Aggression," *J Neuropsychiatry and Clin Nsci* 10 (1998): 413; E. Hitchcock and V. Cairns, "Amygdalotomy," *Postgraduate Med J* 49 (1973): 894; and M. Mpakopoulou et al., "Stereotactic Amygdalotomy in the Management of Severe Aggressive Behavioral Disorders," *Neurosurgical Focus* 25 (2008): E6.
6. Some papers touching on the political controversies surrounding amygdalotomies: V. Mark et al., "Role of Brain Disease in Riots and Urban Violence," *JAMA* 201 (1967): 217; P. Breggin, "Psychosurgery for Political Purposes," *Duquesne Law Rev* 13 (1975): 841; E. Valenstein, *Great and Desperate Cures: The Rise and Decline of Psychosurgery and Other Radical Treatments for Mental Illness* (New York: Basic Books 2010).
7. C. Holden, "Fuss over a Terrorist's Brain," *Sci* 298 (2002): 1551
8. D. Eagleman, "The Brain on Trial," *Atlantic*, June 7, 2011; G. Lavergne, *A Sniper in the Tower* (Denton: University of North Texas Press, 1997); H. Hylton, "Texas Sniper's Brother John Whitman Shot," *Palm Beach Post*, July 5, 1973, p. A1.
9. For a great review of the role of aggression in fear, see the superb J. LeDoux, *The Emotional Brain: The Mysterious Underpinnings of Emotional Life* (New York: Simon and Schuster, 1998).
10. N. Kalin et al., "The Role of the Central Nucleus of the Amygdala in Mediating Fear and Anxiety in the Primate," *J Nsci* 24 (2004): 5506; T. Hare et al., "Contributions of Amygdala and Striatal Activity in Emotion Regulation," *BP* 57 (2005): 624; D. Zald, "The Human Amygdala and the Emotional Evaluation of Sensory Stimuli," *Brain Res Rev* 41 (2003): 88.
11. D. Mobbs et al., "When Fear Is Near: Threat Imminence Elicits Prefrontal-Periaqueductal Gray Shifts in Humans," *Sci* 317 (2007): 1079.
12. G. Berns, "Neurobiological Substrates of Dread," *Sci* 312 (2006): 754. Additional papers pertinent to the role of the human amygdala in fear: R. Adolphs et al., "Impaired Recognition of Emotion in Facial Expressions Following Bilateral Damage to the Human Amygdala," *Nat* 372 (1994): 669; A. Young et al., "Face Processing Impairments After Amygdalotomy," *Brain* 118 (1995): 15; J. Feinstein et al., "The Human Amygdala and the Induction and Experience of Fear," *Curr Biol* 21 (2011): 34; A. Bechara et al., "Double Dissociation of Conditioning and Declarative Knowledge Relative to the Amygdala and Hippocampus in Humans," *Sci* 269 (1995): 1115.
13. A. Gilboa et al., "Functional Connectivity of the Prefrontal Cortex and the Amygdala in PTSD," *BP* 55 (2004): 263.
14. M. Hsu et al., "Neural Systems Responding to Degrees of Uncertainty in Human Decision-Making," *Sci* 310 (2006): 1680; J. Rilling et al., "The Neural Correlates of Mate Competition in Dominant Male Rhesus Macaques," *BP* 56 (2004): 364.
15. C. Zink et al., "Know Your Place: Neural Processing of Social Hierarchy in Humans," *Neuron* 58 (2008): 273; M. Freitas-Ferrari et al., "Neuroimaging in Social Anxiety Disorder: A Systematic Review of the Literature," *Prog Neuro-Psychopharmacology and Biol Psychiatry* 34 (2010): 565.
16. G. Berns et al., "Neurobiological Correlates of Social Conformity and Independence During Mental Rotation," *BP* 58 (2005): 245.
17. K. Tye et al., "Amygdala Circuitry Mediating Reversible and Bidirectional Control of Anxiety," *Nat* 471 (2011): 358; S. Kim et al., "Differing Neural Pathways Assemble a Behavioural State from Separable Features in Anxiety," *Nat* 496 (2013): 219.
18. J. Ipser et al., "Meta-analysis of Functional Brain Imaging in Specific Phobia," *Psychiatry and Clin Nsci* 67 (2013): 311; U. Lueken, "Neural Substrates of Defensive Reactivity in Two Subtypes of Specific Phobia," *SCAN* 9 (2013): 11; A. Del Casale et al., "Functional Neuroimaging in Specific Phobia," *Psychiatry Res* 202 (2012): 181; J. Feinstein et al., "Fear and Panic in Humans with Bilateral Amygdala Damage," *Nat Nsci* 16 (2013): 270.
19. M. Cook and S. Mineka, "Selective Associations in the Observational Conditioning of Fear in Rhesus Monkeys," *J Exp Psych and Animal Behav Processes* 16 (1990): 372; S. Mineka and M. Cook, "Immunization Against the Observational Conditioning of Snake Fear in Rhesus Monkeys," *J Abnormal Psych* 95 (1986): 307.
20. S. Rodrigues et al., "Molecular Mechanisms Underlying Emotional Learning and Memory in the Lateral Amygdala," *Neuron* 44 (2004): 75; J. Johansen et al., "Optical Activation of Lateral Amygdala Pyramidal Cells Instructs Associative Fear Learning," *PNAS* 107 (2010): 12692; S. Rodrigues et al., "The Influence of Stress

Hormones on Fear Circuitry," *Ann Rev of Nsci*, 32 (2009): 289; S. Rumpel et al., "Postsynaptic Receptor Trafficking Underlying a Form of Associative Learning," *Sci* 308 (2005): 83.

Other work in this area: C. Herry et al., "Switching On and Off Fear by Distinct Neuronal Circuits," *Nat* 454 (2008): 600; S. Maren and G. Quirk, "Neuronal Signaling of Fear Memory," *Nat Rev Nsci* 5 (2004): 844; S. Wolff et al., "Amygdala Interneuron Subtypes Control Fear Learning Through Disinhibition," *Nat* 509 (2014): 453; R. LaLumiere, "Optogenetic Dissection of Amygdala Functioning," *Front Behav Nsci* 8 (2014): 1.

21. T. Amano et al., "Synaptic Correlates of Fear Extinction in the Amygdala," *Nat Nsci* 13 (2010): 489; M. Milad and G. Quirk, "Neurons in Medial Prefrontal Cortex Signal Memory for Fear Extinction," *Nat* 420 (2002): 70; E. Phelps et al., "Extinction Learning in Humans: Role of the Amygdala and vmPFC," *Neuron* 43 (2004): 897; S. Ciocchi et al., "Encoding of Conditioned Fear in Central Amygdala Inhibitory Circuits," *Nat* 468 (2010): 277; W. Haubensak et al., "Genetic Dissection of an Amygdala Microcircuit That Gates Conditioned Fear," *Nat* 468 (2010): 270.
22. K. Gospic et al., "Limbic Justice: Amygdala Involvement in Immediate Rejections in the Ultimatum Game," *PLoS ONE* 9 (2011): e1001054; B. De Martino et al., "Frames, Biases, and Rational Decision-Making in the Human Brain," *Sci* 313 (2006): 684; A. Bechara et al., "Role of the Amygdala in Decision-Making," *ANYAS* 985 (2003): 356; B. De Martino et al., "Amygdala Damage Eliminates Monetary Loss Aversion," *PNAS* 107 (2010): 3788; J. Van Honk et al., "Generous Economic Investments After Basolateral Amygdala Damage," *PNAS* 110 (2013): 2506.
23. R. Adolphs et al., "The Human Amygdala in Social Judgment," *Nat* 393 (1998): 470
24. D. Zald, "The Human Amygdala and the Emotional Evaluation of Sensory Stimuli," *Brain Res Rev* 41 (2003): 88; C. Saper, "Animal Behavior: The Nexus of Sex and Violence," *Nat* 470 (2011): 179; D. Lin et al., "Functional Identification of an Aggression Locus in Mouse Hypothalamus," *Nat* 470 (2011): 221; M. Baxter and E. Murray, "The Amygdala and Reward," *Nat Rev Nsci* 3 (2002): 563.

    Some other realms where positive stimuli activate the amygdala: S. Aalto et al., "Neuroanatomical Substrate of amusement and Sadness: A PET Activation Study Using Film Stimuli," *Neuroreport* 13 (2002): 67–73; T. Uwano et al., "Neuronal Responsiveness to Various Sensory Stimuli, and Associative Learning in the Rat Amygdala," *Nsci* 68 (1995): 339; K. Tye and P. Janak, "Amygdala Neurons Differentially Encode Motivation and Reinforcement," *J Nsci* 27 (2007): 3937; G. Schoenbaum et al., "Orbitofrontal Cortex and Basolateral Amygdala Encode Expected Outcomes During Learning," *Nat Nsci* 1 (1998): 155; I. Aharon et al., "Beautiful Faces Have Variable Reward Value: fMRI and Behavioral Evidence," *Neuron* 32 (2001): 537.
25. P. Janak and K. Tye, "From Circuits to Behavior in the Amygdala," *Nat* 517 (2015): 284.
26. J. LeDoux, "Coming to Terms with Fear," *PNAS* 111 (2014): 2871; J. LeDoux, "The Amygdala," *Curr Biol* 17 (2007): R868; K. Tully et al., "Norepinephrine Enables the Induction of Associative LTP at Thalamo-Amygdala Synapses," *PNAS* 104 (2007): 14146.
27. T. Rizvi et al., "Connections Between the Central Nucleus of the Amygdala and the Midbrain Periaqueductal Gray: Topography and Reciprocity," *J Comp Neurol* 303 (1991): 121; E. Kim et al., "Dorsal Periaqueductal Gray-Amygdala Pathway Conveys Both Innate and Learned Fear Responses in Rats," *PNAS* 110 (2013): 14795; C. Del-Ben and F. Graeff, "Panic Disorder: Is the PAG Involved?" *Neural Plasticity 2009* (2009): 108135; P. Petrovic et al., "Context Dependent Amygdala Deactivation During Pain," *Neuroimage* 13 (2001): S457; J. Johnson et al., "Neural Substrates for Expectation-Modulated Fear Learning in the Amygdala and Periaqueductal Gray," *Nat Nsci* 13 (2010): 979; W. Yoshida et al., "Uncertainty Increases Pain: Evidence for a Novel Mechanism of Pain Modulation Involving the Periaqueductal Gray," *J Nsci* 33 (2013): 5638.
28. T. Heatherton, "Neuroscience of Self and Self-Regulation," *Ann Rev of Psych* 62 (2011): 363; K. Krendl et al., "The Good, the Bad, and the Ugly: An fMRI Investigation of the Functional Anatomic Correlates of Stigma," *Soc Nsci* 1 (2006): 5; F. Sambataro et al., "Preferential Responses in Amygdala and Insula During Presentation of Facial Contempt and Disgust," *Eur J Nsci* 24, (2006): 2355.
29. X. Liu et al., "Optogenetic Stimulation of a Hippocampal Engram Activates Fear Memory Recall," *Nat* 484 (2012): 381; T. Seidenbecher et al., "Amygdalar and Hippocampal Theta Rhythm Synchronization During Fear Memory Retrieval," *Sci* 301 (2003): 846; R. Redondo et al., "Bidirectional Switch of the Valence Associated with a Hippocampal Contextual Memory Engram," *Nat* 513 (2014): 426; E. Kirby et al., "Basolateral Amygdala Regulation of Adult Hippocampal Neurogenesis and Fear-Related Activation of Newborn Neurons," *Mol Psychiatry* 17 (2012): 527.
30. A. Gozzi, "A Neural Switch for Active and Passive Fear," *Neuron* 67 (2010): 656.
31. G. Aston-Jones and J. Cohen, "Adaptive Gain and the Role of the Locus Coeruleus-Norepinephrine System in Optimal Performance," *J Comp Neurol* 493 (2005): 99; M. Carter et al., "Tuning Arousal with Optogenetic Modulation of Locus Coeruleus Neurons,"*Nat Nsci* 13 (2010): 1526.
32. D. Blanchard et al., "Lesions of Structures Showing FOS Expression to Cat Presentation: Effects on Responsivity to a Cat, Cat Odor, and Nonpredator Threat," *Nsci Biobehav Rev* 29 (2005): 1243.
33. G. Holstege, "Brain Activation During Human Male Ejaculation," *J Nsci* 23 (2003): 9185; H. Lee et al., "Scalable Control of Mounting and Attack by Ersl+ Neurons in the Ventromedial Hypothalamus," *Nat* 509 (2014): 627; D. Anderson, "Optogenetics, Sex, and Violence in the Brain: Implications for Psychiatry," *BP* 71 (2012): 1081.
34. K Blair, "Neuroimaging of Psychopathy and Antisocial Behavior: A Targeted Review," *Curr Psychiatry Rep* 12 (2010): 76; K. Kiehl, *The Psychopath Whisperer: The Nature of Those Without Conscience* (Woodland Hills, CA: Crown Books, 2014); M. Koenigs et al., "Investigating the Neural Correlates of Psychopathy: A Critical Review," *Mol Psychiatry* 16 (2011): 792.

35. A particularly nice consideration of impulsivity and the frontal cortex: J. Dalley et al., "Impulsivity, Compulsivity, and Top-Down Cognitive Control," *Neuron* 69 (2011): 680.
36. J. Rilling and T. Insel, "The Primate Neocortex in Comparative Perspective Using MRI," *J Hum Evol* 37 (1999): 191; R. Barton and C. Venditti, "Human Frontal Lobes Are Not Relatively Large," *PNAS* 110 (2013): 9001; Y. Zhang et al., "Accelerated Recruitment of New Brain Development Genes into the Human Genome," *PLoS Biol* 9 (2011): e1001179; G. Miller, "New Clues About What Makes the Human Brain Special," *Sci* 330 (2010): 1167; K. Semendeferi et al., "Humans and Great Apes Share a Large Frontal Cortex," *Nat Nsci* 5 (2002): 272; P. Schoenemann, "Evolution of the Size and Functional Areas of the Human Brain," *Ann Rev of Anthropology* 35 (2006): 379.
37. J. Allman et al., "The von Economo Neurons in the Frontoinsular and Anterior Cingulate Cortex," *ANYAS* 1225 (2011): 59; C. Butti et al., "Von Economo Neurons: Clinical and Evolutionary Perspectives," *Cortex* 49 (2013): 312; H. Evrard et al., "Von Economo Neurons in the Anterior Insula of the Macaque Monkey," *Neuron* 74 (2012): 482.
38. E. Miller and J. Cohen, "An Integrative Theory of Prefrontal Cortex Function," *Ann Rev of Nsci* 24 (2001): 167.
39. V. Mante et al., "Context-Dependent Computation by Recurrent Dynamics in Prefrontal Cortex," *Nat* 503 (2013): 78. Some more examples of frontal cortical involvement in task switching: S. Bunge, "How We Use Rules to Select Actions: A Review of Evidence from Cognitive Neuroscience," *SCAN* 4 (2004): 564; E. Crone et al., "Evidence for Separable Neural Processes Underlying Flexible Rule Use," *Cerebral Cortex* 16 (2005): 475; R. Passingham et al., "Specialisation Within the Prefrontal Cortex: The Ventral Prefrontal Cortex and Associative Learning," *Exp Brain Res* 133 (2000): 103; D. Liu et al., "Medial Prefrontal Activity During Delay Period Contributes to Learning of a Working Memory Task," *Sci* 346 (2014): 458; 1983, starring Robert De Niro, Diane Keaton, and the young Brad Pitt in his film debut, as the sixth frontocortical neuron from the left.
40. J. Baldo et al., "Memory Performance on the California Verbal Learning Test-II: Findings from Patients with Focal Frontal Lesions," *J the Int Neuropsychological Soc* 8 (2002): 539.
41. D. Freedman, "Categorical Representation of Visual Stimuli in the Primate Prefrontal Cortex," *Sci* 291 (2001): 312. More examples of categorical coding: D. McNamee et al., "Category-Dependent and Category-Independent Goal-Value Codes in Human Ventromedial Prefrontal Cortex," *Nat Nsci* 16 (2013): 479. R. Schmidt et al., "Canceling Actions Involves a Race Between Basal Ganglia Pathways," *Nat Nsci* 16 (2013): 1118.
42. M. Histed et al., "Learning Subtracts in the Primate Prefrontal Cortex and Striatum: Sustained Activity Related to Successful Actions," *Neuron* 63 (2004): 244. For a nice example of the frontal cortex having to keep track of a rule, see D. Crowe et al., "Prefrontal Neurons Transmit Signals to Parietal Neurons That Reflect Executive Control of Cognition," *Nat Nsci* 16 (2013): 1484.
43. M. Rigotti et al., "The Importance of Mixed Selectivity in Complex Cognitive Tasks," *Nat* 497 (2013): 585; J. Cromer et al., "Representation of Multiple, Independent Categories in the Primate Prefrontal Cortex," *Neuron* 66 (2010): 796; M. Cole et al., "Global Connectivity of Prefrontal Cortex Predicts Cognitive Control and Intelligence," *J Nsci* 32 (2012): 8988.
44. L. Grossman et al., "Accelerated Evolution of the Electron Transport Chain in Anthropoid Primates," *Trends in Genetics* 20 (2004): 578.
45. J. W. De Fockert et al., "The Role of Working Memory in Visual Selective Attention," *Sci* 291 (2001): 1803; K. Vohs et al., "Making Choices Impairs Subsequent Self-Control: A Limited-Resource Account of Decision Making, Self-Regulation, and Active Initiative," *JPSP* 94 (2008): 883; K. Watanabe and S. Funahashi, "Neural Mechanisms of Dual-Task Interference and Cognitive Capacity Limitation in the Prefrontal Cortex," *Nat Nsci* 17 (2014): 601.
46. N. Meand et al., "Too Tired to Tell the Truth: Self-Control Resource Depletion and Dishonesty," *JESP* 45 (2009): 594; M. Hagger et al., "Ego Depletion and the Strength Model of Self-Control: A Meta-analysis," *Psych Bull* 136 (2010): 495; C. DeWall et al., "Depletion Makes the Heart Grow Less Helpful: Helping as a Function of Self-Regulatory Energy and Genetic Relatedness," *PSPB* 34 (2008): 1653; W. Hofmann et al., "And Deplete Us Not into Temptation: Automatic Attitudes, Dietary Restraint, and Self-Regulatory Resources as Determinants of Eating Behavior," *JESP* 43 (2007): 497.
47. Footnote: M. Inzlicht and S. Marcora, "The Central Governor Model of Exercise Regulation Teaches Us Precious Little About the Nature of Mental Fatigue and Self-Control Failure," *Front Psych* 7 (2016): 656.
48. J. Fuster, "The Prefrontal Cortex—an Update: Time Is of the Essence," *Neuron* 30 (2001): 319.
49. K. Yoshida et al., "Social Error Monitoring in Macaque Frontal Cortex," *Nat Nsci* 15 (2012): 1307; T. Behrens et al., "Associative Learning of Social Value," *Nat* 456 (2008): 245
50. R. Dunbar, "The Social Brain Meets Neuroimaging," *TICS* 16 (2011): 101; K. Bickart et al., "Intrinsic Amygdala-Cortical Functional Connectivity Predicts Social Network Size in Humans" *J Nsci* 32 (2012): 14729; K. Bickart, "Amygdala Volume and Social Network Size in Humans," *Nat Nsci* 14 (2010): 163; R. Kanai et al., "Online Social Network Size Is Reflected in Human Brain Structure," *Proc Royal Soc B* 279 (2012): 1327; F. Amici et al., "Fission-Fusion Dynamics, Behavioral Flexibility, and Inhibitory Control in Primates," *Curr Biol* 18 (2008): 1415. For a similar finding in corvids, see A. Bond et al., "Serial Reversal Learning and the Evolution of Behavioral Flexibility in Three Species of North American Corvids (*Gymnorhinus cyanocephalus, Nucifraga columbiana, Aphelocoma californica*)," *JCP* 121 (2007): 372.
51. P. Lewis et al., "Ventromedial Prefrontal Volume Predicts Understanding of Others and Social Network Size," *Neuroimage* 57 (2011): 1624; J. Sallet et al., "Social Network Size Affects Neural Circuits in Macaques," *Sci* 334 (2011): 697.
52. J. Harlow, "Recovery from the Passage of an Iron Bar Through the Head," *Publication of the Massachusetts Med Soc* 2 (1868): 327; H. Damasio et al., "The Return of Phineas Gage: Clues About the Brain from the Skull of a

Famous Patient," Sci 264 (1994): 1102; P. Ratiu and I. Talos, "The Tale of Phineas Gage, Digitally Remastered," *NEJM* 351 (2004): e21; J. Van Horn et al., "Mapping Connectivity Damage in the Case of Phineas Gage," *PLoS ONE* 7 (2012): e37454; M. Macmillan, *An Odd Kind of Fame: Stories of Phineas Gage* (Cambridge, MA: MIT Press, 2000); J. Jackson, "Frontis. and Nos. 949–51," in *A Descriptive Catalog of the Warren Anatomical Museum*, reproduced in Macmillan, *An Odd Kind of Fame*. The photographs of Gage come from J. Wilgus and B. Wilgus, "Face to Face with Phineas Gage," *J the History of the Nsci* 18 (2009): 340.

53. W. Seeley et al., "Early Frontotemporal Dementia Targets Neurons Unique to Apes and Humans," *Annals of Neurol* 60 (2006): 660; R. Levenson and B. Miller, "Loss of Cells, Loss of Self: Frontotemporal Lobar Degeneration and Human Emotion," *Curr Dir Psych Sci* 16 (2008): 289.
54. U. Voss et al., "Induction of Self Awareness in Dreams Through Frontal Low Curr Stimulation of Gamma Activity," *Nat Nsci* 17 (2014): 810; J. Georgiadis et al., "Regional Cerebral Blood Flow Changes Associated with Clitorally Induced Orgasm in Healthy Women," *Eur J Nsci* 24 (2006): 3305.
55. A. Glenn et al., "Antisocial Personality Disorder: A Current Review," *Curr Psychiatry Rep* 15 (2013): 427; N. Anderson and K. Kiehl, "The Psychopath Magnetized: Insights from Brain Imaging," *TICS* 16 (2012): 52; L. Mansnerus, "Damaged Brains and the Death Penalty," *New York Times*, July 21, 2001, p. B9; M. Brower and B. Price, "Neuropsychiatry of Frontal Lobe Dysfunction in Violent and Criminal Behaviour: A Critical Review," *J Neurol, Neurosurgery & Psychiatry* 71 (2001): 720.
56. J. Greene et al., "The Neural Bases of Cognitive Conflict and Control in Moral Judgment," *Neuron* 44 (2004): 389; S. McClure et al., "Separate Neural Systems Value Immediate and Delayed Monetary Rewards," *Sci* 306 (2004): 503.
57. A. Barbey et al., "Dorsolateral Prefrontal Contributions to Human Intelligence," *Neuropsychologia* 51 (2013): 1361.
58. D. Knock et al., "Diminishing Reciprocal Fairness by Disrupting the Right Prefrontal Cortex," *Sci* 314 (2006): 829.
59. D. Mobbs et al., "A Key Role for Similarity in Vicarious Reward," *Sci* 324 (2009): 900; P. Janata et al., "The Cortical Topography of Tonal Structures Underlying Western Music," *Sci* 298 (2002): 2167; M. Balter, "Study of Music and the Mind Hits a High Note in Montreal," *Sci* 315 (2007): 758.
60. J. Saver and A. Damasio, "Preserved Access and Processing of Social Knowledge in a Patient with Acquired Sociopathy Due to Ventromedial Frontal Damage," *Neuropsychologia* 29 (1991): 1241; M. Donoso et al., "Foundations of Human Reasoning in the Prefrontal Cortex," *Sci* 344 (2014): 1481; T. Hare, "Exploiting and Exploring the Options," *Sci* 344 (2014): 1446; T. Baumgartner et al., "Dorsolateral and Ventromedial Prefrontal Cortex Orchestrate Normative Choice," *Nat Nsci* 14 (2011): 1468; A. Bechara, "The Role of Emotion in Decision-Making: Evidence from Neurological Patients with Orbitofrontal Damage," *Brain and Cog* 55 (2004): 30.
61. A. Damasio, *The Feeling of What Happens: Body and Emotion in the Making of Consciousness* (Boston: Harcourt, 1999).
62. M. Koenigs et al., "Damage to the Prefrontal Cortex Increases Utilitarian Moral Judgments," *Nat* 446 (2007): 865; B. Thomas et al., "Harming Kin to Save Strangers: Further Evidence for Abnormally Utilitarian Moral Judgments After Ventromedial Prefrontal Damage," *J Cog Nsci* 23 (2011): 2186
63. A. Bechara et al., "Deciding Advantageously Before Knowing the Advantageous Strategy," *Sci* 275 (1997): 1293; A. Bechara et al., "Insensitivity to Future Consequences Following Damage to Human Prefrontal Cortex," *Cog* 50 (1994): 7.
64. L. Young et al., "Damage to Ventromedial Prefrontal Cortex Impairs Judgment of Harmful Intent," *Neuron* 25 (2010): 845.
65. C. Limb and A. Braun, "Neural Substrates of Spontaneous Musical Performance: An fMRI Study of Jazz Improvisation," *PLoS ONE* 3 (2008): e1679; C. Salzman and S. Fusi, "Emotion, Cognition, and Mental State Representation in Amygdala and Prefrontal Cortex," *Ann Rev of Nsci* 33 (2010): 173.
66. J. Greene et al., "An fMRI Investigation of Emotional Engagement in Moral Judgment," *Sci* 293 (2001): 2105; J. Greene et al., "The Neural Bases of Cognitive Conflict and Control in Moral Judgment," *Neuron* 44 (2004): 389–400; J. Greene, *Moral Tribes: Emotion, Reason, and the Gap Between Us and Them* (New York: Penguin, 2013).
67. J. Peters et al., "Induction of Fear Extinction with Hippocampal-Infralimbic BDNF," *Sci* 328 (2010): 1288; M. Milad and G. Quirk, "Neurons in Medial Prefrontal Cortex Signal Memory for Fear Extinction," *Nat* 420 (2002): 70; M. Milad and G. Quirk, "Fear Extinction as a Model for Translational Neuroscience: Ten Years of Progress," *Ann Rev of Psych* 63 (2012): 129; C. Lai et al., "Opposite Effects of Fear Conditioning and Extinction on Dendritic Spine Remodeling," *Nat* 483 (2012): 87. Some recent work suggests involvement of both the ventral mPFC and the basomedial amygdala in this process: A. Adhikari et al., "Basomedial Amygdala Mediates Top-Down Control of Anxiety and Fear," *Nat* 527 (2016): 179.
68. K. Ochsner et al., "Rethinking Feelings: An fMRI Study of the Cognitive Regulation of Emotion," *J Cog Nsci* 14 (2002): 1215; G. Sheppes and J. Gross, "Is Timing Everything? Temporal Considerations in Emotion Regulation," *PSPR* 15 (2011): 319; G. Sheppes and Z. Levin, "Emotion Regulation Choice: Selecting Between Cognitive Regulation Strategies to Control Emotion," *Front Human Neurosci* 7 (2013): 179; J. Gross, "Antecedent- and Response-Focused Emotion Regulation: Divergent Consequences for Experience, Expression, and Physiology," *JPSP* 74 (1998): 224; J. Gross, "Emotion Regulation: Affective, Cognitive, and Social Consequences," *Psychophysiology* 39 (2002): 281; K. Ochsner and J. Gross, "The Cognitive Control of Emotion," *TICS* 9 (2005): 242.
69. M. Lieberman et al., "The Neural Correlates of Placebo Effects: A Disruption Account," *NeuroImage* 22 (2004): 447; P. Petrovic et al., "Placebo and Opioid Analgesia: Imaging a Shared Neuronal Network," *Sci* 295 (2002): 1737.

70. J. Beck, *Cognitive Behavior Therapy*, 2nd edition (New York: Guilford Press, 2011); P. Goldin et al., "Cognitive Reappraisal Self-Efficacy Mediates the Effects of Individual Cognitive-Behavioral Therapy for Social Anxiety Disorder," *J Consulting Clin Psych* 80 (2012): 1034.
71. A. Bechara et al., "Failure to Respond Autonomically to Anticipated Future Outcomes Following Damage to Prefrontal Cortex," *Cerebral Cortex* 6 (1996): 215; C. Martin et al., "The Effects of Vagus Nerve Stimulation on Decision-Making," *Cortex* 40 (2004): 605.
72. G. Bodenhausen et al., "Negative Affect and Social Judgment: The Differential Impact of Anger and Sadness," *Eur J Soc Psych* 24 (1994): 45; A. Sanfey et al., "The Neural Basis of Economic Decision-Making in the Ultimatum Game," *Sci* 300 (2003): 1755; K. Gospic et al., "Limbic Justice: Amygdala Involvement in Immediate Rejections in the Ultimatum Game," *PLoS ONE* 9 (2011): e1001054.
73. D. Wegner, "How to Think, Say, or Do Precisely the Worst Thing on Any Occasion," *Sci* 325 (2009): 58.
74. R. Davidson and S. Begley, *The Emotional Life of Your Brain* (New York: Hudson Street Press, 2011); A. Tomarken and R. Davidson, "Frontal Brain Activation in Repressors and Nonrepressors," *J Abnormal Psych* 103 (1994): 339.
75. A. Ito et al., "The Contribution of the Dorsolateral Prefrontal Cortex to the Preparation for Deception and Truth-Telling," *Brain Res* 1464 (2012): 43; S. Spence et al., "A Cognitive Neurobiological Account of Deception: Evidence from Functional Neuroimaging," *Philosophical Transactions of the Royal Soc London Series B* 359 (2004): 1755; I. Karton and T. Bachmann, "Effect of Prefrontal Transcranial Magnetic Stimulation on Spontaneous Truth-Telling," *BBR* 225 (2011): 209; Y. Yang et al., "Prefrontal White Matter in Pathological Liars," *Brit J Psychiatry* 187 (2005): 320.
76. D. Carr and S. Sesack, "Projections from the Rat Prefrontal Cortex to the Ventral Tegmental Area: Target Specificity in the Synaptic Associations with Mesoaccumbens and Mesocortical Neurons," *J Nsci* 20 (2000): 3864; M. Stefani and B. Moghaddam, "Rule Learning and Reward Contingency Are Associated with Dissociable Patterns of Dopamine Activation in the Rat Prefrontal Cortex, Nucleus Accumbens, and Dorsal Striatum," *J Nsci* 26 (2006): 8810.
77. T. Danjo et al., "Aversive Behavior Induced by Optogenetic Inactivation of Ventral Tegmental Area Dopamine Neurons Is Mediated by Dopamine D2 Receptors in the Nucleus Accumbens," *PNAS* 111 (2014): 6455; N. Schwartz et al., "Decreased Motivation During Chronic Pain Requires Long-Term Depression in the Nucleus Accumbens," *Nat* 345 (2014): 535.
78. J. Cloutier et al., "Are Attractive People Rewarding? Sex Differences in the Neural Substrates of Facial Attractiveness," *J Cog Nsci* 20 (2008): 941; K. Demos et al., "Dietary Restraint Violations Influence Reward Responses in Nucleus Accumbens and Amygdala," *J Cog Nsci* 23 (2011): 1952.
79. Footnote: R. Deaner et al., "Monkeys Pay per View: Adaptive Valuation of Social Images by Rhesus Macaques," *Curr Biol* 15 (2005): 543.
80. V. Salimpoor et al., "Interactions Between the Nucleus Accumbens and Auditory Cortices Predicts Music Reward Value," *Sci* 340 (2013): 216; G. Berns and S. Moore, "A Neural Predictor of Cultural Popularity," *J Consumer Psych* 22 (2012): 154; S. Erk et al., "Cultural Objects Modulate Reward Circuitry," *Neuroreport* 13 (2002): 2499.
81. A. Sanfey et al., "The Neural Basis of Economic Decision-Making in the Ultimatum Game," *Sci* 300 (2003): 1755. Also see J. Moll et al., "Human Front-Mesolimbic Networks Guide Decisions About Charitable Donation," *PNAS* 103 (2006): 15623; W. Harbaugh et al., "Neural Responses to Taxation and Voluntary Giving Reveal Motives for Charitable Donations," *Sci* 316 (2007): 1622.
82. D. De Quervain et al., "The Neural Basis of Altruistic Punishment," *Sci* 305 (2004): 1254; B. Knutson, "Sweet Revenge?" *Sci* 305 (2004): 1246.
83. M. Delgado et al., "Understanding Overbidding: Using the Neural Circuitry of Reward to Design Economic Auctions," *Sci* 321 (2008): 1849; E. Maskin, "Can Neural Data Improve Economics?" *Sci* 321 (2008): 1788.
84. H. Takahasi et al., "When Your Gain Is My Pain and Your Pain Is My Gain: Neural Correlates of Envy and Schadenfreude," *Sci* 323 (2009): 890; K. Fliessbach et al., "Social Comparison Affects Reward-Related Brain Activity in the Human Ventral Striatum," *Sci* 318 (2007): 1305.
85. W. Schultz, "Dopamine Signals for Reward Value and Risk: Basic and Recent Data," *Behav and Brain Functions* 6 (2010): 24.
86. J. Cooper et al., "Available Alternative Incentives Modulate Anticipatory Nucleus Accumbens Activation," *SCAN* 4 (2009): 409; D. Levy and P. Glimcher, "Comparing Apples and Oranges: Using Reward-Specific and Reward-General Subjective Value Representation in the Brain," *J Nsci* 31 (2011): 14693.
87. P. Tobler et al., "Adaptive Coding of Reward Value by Dopamine Neurons," *Sci* 307 (2005): 1642.
88. W. Schultz, "Dopamine Signals for Reward Value and Risk: Basic and Recent Data," *Behav and Brain Functions* 6 (2010): 24; J. Cohen et al., "Neuron-Type-Specific Signals for Reward and Punishment in the Central Tegmental Area," *Nat* 482 (2012): 85; J. Hollerman and W. Schultz, "Dopamine Neurons Report an Error in the Temporal Prediction of Reward During Learning," *Nat Nsci* 1 (1998): 304; A. Brooks et al., "From Bad to Worse: Striatal Coding of the Relative Value of Painful Decisions," *Front Nsci* 4 (2010): 1.
89. B. Knutson et al., "Neural Predictors of Purchases," *Neuron* 53 (2007): 147.
90. P. Sterling, "Principles of Allostasis: Optimal Design, Predictive Regulation, Pathophysiology and Rational Therapeutics," in *Allostasis, Homeostasis, and the Costs of Adaptation*, ed. J. Schulkin (Cambridge, MA: MIT Press, 2004).
91. B. Knutson et al., "Anticipation of Increasing Monetary Reward Selectively Recruits Nucleus Accumbens," *J Nsci* 21 (2001): RC159.
92. G. Stuber et al., "Reward-Predictive Cues Enhance Excitatory Synaptic Strength onto Midbrain Dopamine

Neurons," *Sci* 321 (2008): 1690; A. Luo et al., "Linking Context with Reward: A Functional Circuit from Hippocampal CA3 to Ventral Tegmental Area," *Sci* 33 (2011): 353; J. O'Doherty, "Reward Representations and Reward-Related Learning in the Human Brain: Insights from Neuroimaging," *Curr Opinions in Neurobiol* 14 (2004): 769; M. Cador et al., "Involvement of the Amygdala in Stimulus-Reward Associations: Interaction with the Ventral Striatum," Nsci 30 (1989): 77; J. Britt et al., "Synaptic and Behavioral Profile of Multiple Glutamatergic Inputs to the Nucleus Accumbens," *Neuron* 76 (2012): 790; G. Stuber et al., "Optogenetic Modulation of Neural Circuits That Underlie Reward Seeking," *BP* 71 (2012): 1061; F. Ambroggi et al., "Basolateral Amygdala Neurons Facilitate Reward-Seeking Behavior by Exciting Nucleus Accumbens Neurons," *Neuron* 59 (2008): 648.

93. S. Hyman et al., "Neural Mechanisms of Addiction: The Role of Reward-Related Learning and Memory," *Ann Rev of Nsci* 29 (2006): 565; B. Lee et al., "Maturation of Silent Synapses in Amygdala-Accumbens Projection Contributes to Incubation of Cocaine Craving," *Nat Nsci* 16 (2013): 1644. For a consideration of compulsive behaviors as a sort of addiction: S. Rauch and W. Carlezon, "Illuminating the Neural Circuitry of Compulsive Behaviors," *Sci* 340 (2013): 1174; S. Ahmari et al., "Repeated Cortico-Striatal Stimulation Generates Persistent OCD-like Behavior," *Sci* 340 (2013): 1234; E. Burguiere et al., "Optogenetic Stimulation of Lateral Orbitofronto-Striatal Pathway Suppresses Compulsive Behaviors," *Sci* 340 (2013): 1243.
94. S. Flagel et al., "A Selective Role for Dopamine in Stimulus-Reward Learning," *Nat* 469 (2011): 53; K. Burke et al., "The Role of the Orbitofrontal Cortex in the Pursuit of Happiness and More Specific Rewards," *Nat* 454 (2008): 340.
95. P. Tobler et al., "Adaptive Coding of Reward Value by Dopamine Neurons," *Sci* 307 (2005): 1642; C. Fiorillo et al., "Discrete Coding of Reward Probability and Uncertainty by Dopamine Neurons," *Sci* 299 (2003): 1898.
96. B. Knutson et al., "Distributed Neural Representation of Expected Value," *J Nsci* 25 (2005): 4806; M. Stefani and B. Moghaddam, "Rule Learning and Reward Contingency Are Associated with Dissociable Patterns of Dopamine Activation in the Rat Prefrontal Cortex, Nucleus Accumbens, and Dorsal Striatum," *J Nsci* 26 (2006): 8810.
97. R. Habib and M. Dixon, "Neurobehavioral Evidence for the "Near-Miss" Effect in Pathological Gamblers," *J the Exp Analysis of Behav* 93 (2010): 313; M. Hsu et al., "Neural Systems Responding to Degrees of Uncertainty in Human Decision-Making," *Sci* 310 (2006): 1680.
98. A. Braun et al., "Dorsal Striatal Dopamine Depletion Impairs Both Allocentric and Egocentric Navigation in Rats," *Neurobiol of Learning and Memory* 97 (2012): 402; J. Salamone, "Dopamine, Effort, and Decision Making," *Behavioral Nsci* 123 (2009): 463; I. Whishaw and S. Dunnett, "Dopamine Depletion, Stimulation or Blockade in the Rat Disrupts Spatial Navigation and Locomotion Dependent upon Beacon or Distal Cues," *BBR* 18 (1985): 11; J. Salamone and M. Correa, "The Mysterious Motivational Functions of Mesolimbic Dopamine," *Neuron* 76 (2012): 470; H. Tsai et al., "Phasic Firing in Dopaminergic Neurons Is Sufficient for Behavioral Conditioning," *Sci* 324 (2009): 1080; P. Phillips et al., "Sub-second Dopamine Release Promotes Cocaine Seeking," *Nat* 422 (2003): 614; M. Pessiglione et al., "Dopamine-Dependent Prediction Errors Underpin Reward-Seeking Behavior in Humans," *Nat* 442 (2008): 1042.
99. Footnote: M. Numan and D. Stoltzenberg, "Medial Preoptic Area Interactions with Dopamine Neural systems in the Control of the Onset and Maintenance of Maternal Behavior in Rats," *Front Neuroendo* 30 (2009): 46.
100. S. McClue et al., "Separate Neural Systems Value Immediate and Delayed Monetary Rewards," *Sci* 306 (2004): 503; J. Jennings et al., "Distinct Extended Amygdala Circuits for Divergent Motivational States," *Nat* 496 (2013): 224.
101. M. Howe et al., "Prolonged Dopamine Signaling in Striatum Signals Proximity and Value of Distant Rewards," *Nat* 500 (2013): 575; Y. Niv, "Dopamine Ramps Up," *Nat* 500 (2013): 533.
102. W. Schultz, "Subjective Neuronal Coding of Reward: Temporal Value Discounting and Risk," *Eur J Nsci* 31 (2010): 2124; S. Kobayashi and W. Schultz, "Influence of Reward Delays on Responses of Dopamine Neurons," *J Nsci* 28 (2008): 7837; S. Kim et al., "Prefrontal Coding of Temporally Discounted Values During Intertemporal Choice," *Neuron* 59 (2008): 161; M. Roesch and C. Olson, "Neuronal Activity in Orbitofrontal Cortex Reflects the Value of Time," *J Neurophysiology* 94 (2005): 2457; M. Bermudez and W. Schultz, "Timing in Reward and Decision Processes," *Philosophical Trans of the Royal Soc of London B* 369 (2014): 20120468; B. Figner et al., "Lateral Prefrontal Cortex and Self-Control in Intertemporal Choice," *Nat Nsci* 13 (2010): 538; K. Jimura et al., "Impulsivity and Self-Control During Intertemporal Decision Making Linked to the Neural Dynamics of Reward Value Representation," *J Nsci* 33 (2013): 344; S. McClure et al., "Time Discounting for Primary Rewards," *J Nsci* 27, 5796.
103. K. Ballard and B. Knutson, "Dissociable Neural Representations of Future Reward Magnitude and Delay During Temporal Discounting," *Neuroimage* 45 (2009): 143.
104. A. Lak et al., "Dopamine Prediction Error Responses Integrate Subjective Value from Different Reward Dimensions," *PNAS* 111 (2014): 2343.
105. V. Noreika et al., "Timing Deficits in Attention-Deficit/Hyperactivity Disorder (ADHD): Evidence from Neurocognitive and Neuroimaging Studies," *Neuropsychologia* 51 (2013): 235; A. Pine et al., "Dopamine, Time, and Impulsivity in Humans," *J Nsci* 30 (2010): 8888; W. Schultz, "Potential Vulnerabilities of Neuronal Reward, Risk, and Decision Mechanisms to Addictive Drugs," *Neuron* 69 (2011): 603.
106. G. Brown et al., "Aggression in Humans Correlates with Cerebrospinal Fluid Amine Metabolites," *Psychiatry Res* 1 (1979): 131; M. Linnoila et al., "Low Cerebrospinal Fluid 5-Hydroxyindoleacetic Acid Concentration Differentiates Impulsive from Nonimpulsive Violent Behavior," *Life Sci* 33 (1983): 2609; P. Stevenson and K. Schildberger, "Mechanisms of Experience Dependent Control of Aggression in Crickets," *Curr Opinion in*

Neurobiol 23 (2013): 318; P. Fong and A. Ford, "The Biological Effects of Antidepressants on the Molluscs and Crustaceans: A Review," *Aquatic Toxicology* 151 (2014): 4.

107. M. Linnoila et al., "Low Cerebrospinal Fluid 5-Hydroxyindoleacetic Acid Concentration Differentiates Impulsive from Nonimpulsive Violent Behavior," *Life Sci* 33 (1983): 2609; J. Higley et al., "Excessive Mortality in Young Free-Ranging Male Nonhuman Primates with Low Cerebrospinal Fluid 5-Hydroxyindoleacetic Acid Concentrations," *AGP* 53 (1996): 537; M. Åsberg et al., "5-HIAA in the Cerebrospinal Fluid: A Biochemical Suicide Predictor?" *AGP* 33 (1976): 1193; M. Bortolato et al., "The Role of the Serotonergic System at the Interface of Aggression and Suicide," *Nsci* 236 (2013): 160.
108. H. Clarke et al., "Cognitive Inflexibility After Prefrontal Serotonin Depletion," *Sci* 304 (2004): 878; R. Wood et al., "Effects of Tryptophan Depletion on the Performance of an Iterated PD Game in Healthy Adults," *Neuropsychopharmacology* 1 (2006): 1075.
109. J. Dalley and J. Roiser, "Dopamine, Serotonin and Impulsivity," *Nsci* 215 (2012): 42; P. Redgrave and R. Horrell, "Potentiation of Central Reward by Localized Perfusion of Acetylcholine and 5-Hydroxytryptamine," *Nat* 262 (1976): 305; A. Harrison and A. Markou, "Serotonergic Manipulations Both Potentiate and Reduce Brain Stimulation Reward in Rats: Involvement of Serotonin-1A Receptors," *JPET* 297 (2001): 316.
110. A. Duke, "Revisiting the Serotonin-Aggression Relation in Humans: A Meta-analysis," *Psych Bull* 139 (2013): 1148.
111. A. Gopnik, "The New Neuro-Skeptics," *New Yorker*, September 9, 2013.
112. C. Bukach et al., "Beyond Faces and Modularity: The Power of an Expertise Framework," *TICS* 10 (2006): 159.

Chapter 3: Seconds to Minutes Before

1. Abusive mothering and antibehaviorist results: D. Maestripieri et al., "Neurobiological Characteristics of Rhesus Macaque Abusive Mothers and Their Relation to Social and Maternal Behavior," *Nsci Biobehav Rev* 29 (2005): 51; R. Sullivan et al., "Ontogeny of Infant Fear Learning and the Amygdala," in *Cognitive Neuroscience IV*, ed. M. Gazzaniga (Cambridge, MA: MIT Press, 2009), 889.
2. Pandas' voices: B. Charlton et al., "Vocal Discrimination of Potential Mates by Female Giant Pandas (*Ailuropoda melanoleuca*)," *Biol Lett* 5 (2009): 597. Women's voices: G. Bryant and M. Haselton, "Vocal Cues of Ovulation in Human Females," *Biol Lett* 5 (2009): 12; Footnote: J. Knight, "When Robots Go Wild," *Nat* 434 (2005): 954.
3. Footnote: H. Herzog, *Some We Love, Some We Hate, Some We Eat: Why It's So Hard to Think Straight About Animals* (New York: Harper, 2010).
4. Vibrational communication: P. Hill, *Vibrational Communication in Animals* (Cambridge, MA: Harvard University Press, 2008). Jamming bats: A. Corcoran and W. Conner, "Bats Jamming Bats: Food Competition Through Sonar Interference," *Sci* 346 (2014): 745. Tickling rats: J. Panksepp, "Beyond a Joke: From Animal Laughter to Human Joy?" *Sci* 308 (2005): 62.
5. A review concerning how there is a continuum between subliminal sensory information and information that is sensed but considered to be irrelevant: T. Marteau et al., "Changing Human Behavior to Prevent Disease: The Importance of Targeting Automatic Processes," *Sci* 337 (2012): 1492.
6. Potato chips: M. Zampini and C. Spence, "Assessing the Role of Sound in the Perception of Food and Drink," *Chemical Senses* 3 (2010): 57. K. Edwards, "The Interplay of Affect and Cognition in Attitude Formation and Change," *JPSP* 59 (1990): 212.
7. An excellent review on the subject: J. Kubota et al. "The Neuroscience of Race," *Nat Nsci* 15 (2012): 940; for a good review of the entire subject, see: D. Ariely, *Predictably Irrational: The Hidden Forces That Shape Our Decisions* (New York HarperCollins, 2008).
8. T. Ito and G. J. Urland, "Race and Gender on the Brain: Electrocortical Measures of Attention to the Race and Gender of Multiply Categorizable Individuals," *JPSP* 85 (2003): 616. For a good review of how implicit attitudes are studied, see B. Nosek et al., "Implicit Social Cognition: From Measures to Mechanisms," *TICS* 15 (2011): 152.
9. A. Olsson et al., "The Role of Social Groups in the Persistence of Learned Fear," *Sci* 309 (2005): 785.
10. J. Richeson et al., "An fMRI Investigation of the Impact of Interracial Contact on Executive Function," *Nat Nsci* 6 (2003): 1323; K. Knutson et al., "Why Do Interracial Interactions Impair Executive Function? A Resource Depletion Account," *TICS* 10 (2007): 915; K. Knutson et al., "Neural Correlates of Automatic Beliefs About Gender and Race," *Human Brain Mapping* 28 (2007): 915.
11. N. Kanwisher et al., "The Fusiform Face Area: A Module in Human Extrastriate Cortex Specialized for Face Perception," *J Nsci* 17 (1997): 4302; J. Sergent et al., "Functional Neuroanatomy of Face and Object Processing: A Positron Emission Tomography Study," *Brain* 115 (1992): 15; A. Golby et al., "Differential Responses in the Fusiform Region to Same-Race and Other-Race Faces," *Nat Nsci* 4 (2001): 845; A. J. Hart et al., "Differential Response in the Human Amygdala to Racial Outgroup Versus Ingroup Face Stimuli," *Neuroreport* 11 (2000): 2351.
12. K. Shutts and K. Kinzler, "An Ambiguous-Race Illusion in Children's Face Memory," *Psych Sci* 18 (2007): 763; D. Maner et al., "Functional Projection: How Fundamental Social Motives Can Bias Interpersonal Perception," *JPSP* 88 (2005): 63; K. Hugenberg and G. Bodenhausen, "Facing Prejudice: Implicit Prejudice and the Perception of Facial Threat," *Psych Sci* (2003): 640; J. Van Bavel et al., "The Neural Substrates of In-group Bias: A Functional Magnetic Resonance Imaging Investigation," *Psych Sci* 19 (2008): 1131; J. Van Bavel and W. Cunningham, "Self-Categorization with a Novel Mixed-Race Group Moderates Automatic Social and Racial Biases," *PSPB* 35 (2009): 321.
13. A. Avenanti et al., "Racial Bias Reduces Empathic Sensorimotor Resonance with Other-Race Pain," *Curr Biol* 20 (2010): 1018; V. Mathur et al., "Neural Basis of Extraordinary Empathy and Altruistic Motivation," *Neuroimage* 51 (2010): 1468–75.
14. J. Correll et al., "Event-Related Potentials and the Decision to Shoot: The Role of Threat Perception and Cognitive Control," *JESP* 42 (2006): 120.

15. J. Eberhardt et al., "See Black: Race, Crime, and Visual Processing," *JPSP* 87 (2004): 876; I. Blair et al., "The Influence of Afrocentric Facial Features in Criminal Sentencing," *Psych Sci* 15 (2004): 674; M. Brown et al., "The Effects of Eyeglasses and Race on Juror Decisions Involving a Violent Crime," *AMFP* 26 (2008): 25.
16. J. LeDoux, "Emotion: Clues from the Brain," *Ann Rev of Psych* 46 (1995): 209.
17. T. Ito and G. Urland, "Race and Gender on the Brain: Electrocortical Measures of Attention to the Race and Gender of Multiply Categorizable Individuals," *JPSP* 85 (2003): 616; N. Rule et al., "Perceptions of Dominance Following Glimpses of Faces and Bodies," *Perception* 41 (2012): 687; C. Zink et al., "Know Your Place: Neural Processing of Social Hierarchy in Humans," *Neuron* 58 (2008): 273.
18. T. Tsukiura and R. Cabeza, "Shared Brain Activity for Aesthetic and Moral Judgments: Implications for the Beauty-Is-Good Stereotype," *SCAN* 6 (2011): 138.
19. H. Aviezer et al., "Body Cues, Not Facial Expressions, Discriminate Between Intense Positive and Negative Emotions," *Sci* 338 (2012); 1225; C. Bobst and J. Lobmaier, "Men's Preference for the Ovulating Female Is Triggered by Subtle Face Shape Differences," *Horm Behav* 62 (2012): 413; N. Rule and N. Ambady, "Democrats and Republicans Can Be Differentiated from Their Faces," *PLoS ONE* 5 (2010): e8733; N. Rule et al., "Flustered and Faithful: Embarrassment as a Signal of Prosociality," *JPSP* 102 (2012): 81; N. Rule et al., "On the Perception of Religious Group Membership from Faces," PLoS ONE 5 (2010): e14241.
20. P. Whalen et al., "Human Amygdala Responsivity to Masked Fearful Eye Whites," *Sci* 306 (2004): 2061.
21. Footnote: R. Hill and R. Barton, "Red Enhances Human Performance in Contests," *Nat* 435 (2005): 293; M. Attrill et al., "Red Shirt Colour Is Associated with Long-Term Team Success in English Football," *JSS* 26 (2008): 577; M. Platti et al., "The Red Mist? Red Shirts, Success and Team Sports," *JSS* 15 (2012): 1209; A. Ilie et al., "Better to Be Red Than Blue in Virtual Competition," *CyberPsychology & Behav* 11 (2008): 375; M. Garcia-Rubio et al., "Does a Red Shirt Improve Sporting Performance? Evidence from Spanish Football," *AEL* 18 (2011): 1001; C. Rowe et al., "Sporting Contests: Seeing Red? Putting Sportswear in Context," *Nat* 437 (2005): E10.
22. D. Francey and R. Bergmuller, "Images of Eyes Enhance Investments in a Real-Life Public Good," *PLoS ONE* 7 (2012): e37397; M. Bateson et al., "Cues of Being Watched Enhance Cooperation in a Real-World Setting," *Biol Lett* 2 (2006): 412; K. Haley and D. Fessler, "Nobody's Watching? Subtle Cues Affect Generosity in an Anonymous Economic Game," *EHB* 3 (2005): 245; T. Burnham and B. Hare, "Engineering Human Cooperation," *Hum Nat* 18 (2007): 88; M. Rigdon et al., "Minimal Social Cues in the Dictator Game," *JEP* 30 (2009): 358.
23. C. Forbes et al., "Negative Stereotype Activation Alters Interaction Between Neural Correlates of Arousal, Inhibition and Cognitive Control," *SCAN* 7 (2011): 771.
24. C. Steele, *Whistling Vivaldi and Other Clues to How Stereotypes Affect Us* (New York: Norton, 2010).
25. L. Mujica-Parodi et al., "Chemosensory Cues to Conspecific Emotional Stress Activate Amygdala in Humans," *PLoS ONE* 4 (2009): e6415; W. Zhou and D. Chen, "Fear-Related Chemosignals Modulate Recognition of Fear in Ambiguous Facial Expressions," *Psych Sci* 20 (2009): 177; A. Prehn et al., "Chemosensory Anxiety Signals Augment the Startle Reflex in Humans," *Nsci Letters* 394 (2006): 127.
26. H. Critchley and N. Harrison, "Visceral Influences on Brain and Behavior," *Neuron* 77 (2013): 624; D. Carney et al., "Power Posing Brief Nonverbal Displays Affect Neuroendocrine Levels and Risk Tolerance," *Psych Sci* 21 (2010): 1363. Some related findings: A. Hennenlotter et al., "The Link Between Facial Feedback and Neural Activity Within Central Circuitries of Emotion: New Insights from Botulinum Toxin–Induced Denervation of Frown Muscles," *Cerebral Cortex* 19 (2009): 357; J. Davis, "The Effects of BOTOX Injections on Emotional Experience," Emotion 10 (2010): 433.
27. L. Berkowitz, "Pain and Aggression: Some Findings and Implications," *Motivation and Emotion* 17 (1993): 277.
28. M. Gailliot et al., "Self-Control Relies on Glucose as a Limited Energy Source: Willpower Is More Than a Metaphor," *JPSP* 92 (2007): 325–36; N. Mead et al., "Too Tired to Tell the Truth: Self-Control Resource Depletion and Dishonesty," *JESP* 45 (2009): 594; C. DeWall et al., "Depletion Makes the Heart Grow Less Helpful: Helping as a Function of Self-Regulatory Energy and Genetic Relatedness," *PSPB* 34 (2008): 1653; B. Briers et al., "Hungry for Money: The Desire for Caloric Resources Increases the Desire for Financial Resources and Vice Versa," *Psych Sci* 17 (2006): 939; C. DeWall et al., "Sweetened Blood Cools Hot Tempers: Physiological Self-Control and Aggression," *Aggressive Behav* 37 (2011): 73; D. Benton, "Hypoglycemia and Aggression: A Review," *Int J Nsci* 41 (1988): 163; B. Bushman et al., "Low Glucose Relates to Greater Aggression in Married Couples," *PNAS USA* 111 (2014): 6254. For a reinterpretation of this literature as being about motivation for self-control rather than capacity for it, see M. Inzlicht et al., "Why Self-Control Seems (But May Not Be) Limited," *TICS* 18 (2014): 127.
29. V. Liberman et al., "The Name of the Game: Predictive Power of Reputations Versus Situational Labels in Determining Prisoner's Dilemma Game Moves," *PSPB* 30 (2004): 1175; A. Kay and L. Ross, "The Perceptual Push: The Interplay of Implicit Cues and Explicit Situational Construals on Behavioral Intensions in the Prisoner's Dilemma," *JESP* 39 (2003): 634.
30. Footnote: E. Hall et al., "A Rose by Any Other Name? The Consequences of Subtyping 'African-Americans' from 'Blacks,'" *JESP* 56 (2015): 183.
31. Footnote: K. Jung et al., "Female Hurricanes Are Deadlier Than Male Hurricanes. *PNAS* 111 (2014): 8782.
32. A. Tversky and D. Kahneman, "Rationale Choice and the Framing of Decisions," *J Business* 59 (1986): S251; also see: J. Bargh et al., "Priming In-group Favoritism: The Impact of Normative Scripts in the Minimal Group Paradigm," *JESP* 37 (2001): 316; C. Zogmaister et al., "The Impact of Loyalty and Equality on Implicit Ingroup Favoritism," *Group Processes & Intergroup Relations* 11 (2008): 493.
33. J. Christensen and A. Gomila, "Moral Dilemmas in Cognitive Neuroscience of Moral Decision-Making: A Principled Review," *Nsci Biobehav Rev* 36 (2012): 1249; L. Petrinovich and P. O'Neill, "Influence of Wording and Framing Effects on Moral Intuitions," *Ethology and Sociobiology* 17 (1996): 145; R. O'Hara et al., "Wording

Effects in Moral Judgments," *Judgment and Decision Making* 5 (2010): 547; R. Zahn et al., "The Neural Basis of Human Social Values: Evidence from Functional MRI," *Cerebral Cortex* 19 (2009): 276.

34. D. Butz et al., "Liberty and Justice for All? Implications of Exposure to the U.S. Flag for Intergroup Relations," *PSPB* 33 (2007): 396; M. Levine et al., "Identity and Emergency Intervention: How Social Group Membership and Inclusiveness of Group Boundaries Shape Helping Behavior," *PSPB* 31 (2005): 443; R. Enos, "Causal Effect of Intergroup Contact on Exclusionary Attitudes," *PNAS* 111 (2014): 3699.
35. M. Shih et al., "Stereotype Susceptibility: Identity Salience and Shifts in Quantitative Performance," *Psych Sci* 10 (1999): 80.
36. P. Fischer et al., "The Bystander-Effect: A Meta-analytic Review on Bystander Intervention in Dangerous and Non-dangerous Emergencies," *Psych Bull* 137 (2011): 517.
37. B. Pawlowski et al., "Sex Differences in Everyday Risk-Taking Behavior in Humans," *Evolutionary Psych* 6 (2008): 29; B. Knutson et al., "Nucleus Accumbens Activation Mediates the Influence of Reward Cues on Financial Risk Taking," *Neuroreport* 26 (2008): 509; V. Griskevicius et al., "Blatant Benevolence and Conspicuous Consumption: When Romantic Motives Elicit Strategic Costly Signals," *JPSP* 93 (2007): 85; L. Chang et al., "The Face That Launched a Thousand Ships: The Mating-Warring Association in Men," *PSPB* 37 (2011): 976; S. Ainsworth and J. Maner, "Sex Begets Violence: Mating Motives, Social Dominance, and Physical Aggression in Men," *JPSP* 103 (2012): 819; W. Iredale et al., "Showing Off in Humans: Male Generosity as a Mating Signal," *Evolutionary Psych* 6 (2008): 386; M. Van Vugt and W. Iredale, "Men Behaving Nicely: Public Goods as Peacock Tails," *Brit J Psych* 104 (2013): 3.
38. J. Q. Wilson and G. Kelling, "Broken Windows," *Atlantic Monthly*, March 1982, p. 29.
39. K. Keizer et al., "The Spreading of Disorder," *Sci* 322 (2008): 1681.
40. For some nice examples of how the frontal cortex can direct the nature and focus of sensory processing, see G. Gregoriou et al., "Lesions of Prefrontal Cortex Reduce Attentional Modulation of Neuronal Responses and Synchrony in V4," *Nat Nsci* 17 (2014): 1003; S. Zhang et al., "Long-Range and Local Circuits for Top-Down Modulation of Visual Cortex Processing," *Sci* 345 (2014): 660; and T. Zanto et al., "Causal Role of the Prefrontal Cortex in Top-Down Modulation of Visual Processing and Working Memory," *Nat Nsci* 14 (2011): 656.
41. R. Adolphs et al., "A Mechanism for Impaired Fear Recognition After Amygdala Damage," *Nat* 433 (2005): 68.
42. M. Dadds et al., "Reduced Eye Gaze Explains Fear Blindness in Childhood Psychopathic Traits," *J the Am Academy of Child and Adolescent Psychiatry* 47 (2008): 4; M. Dadds et al., "Attention to the Eyes and Fear-Recognition Deficits in Child Psychopathy," *Brit J Psychiatry* 189 (2006): 280.
43. For an introduction to this cross-cultural literature, see R. Nisbett et al., "Culture and Systems of Thought: Holistic Versus Analytic Cognition," *Psych Rev* 108 (2001): 291; T. Hedden et al., "Cultural Influences on Neural Substrates of Attentional Control," *Psych Sci* 19 (2008): 12; J. Chiao, "Cultural Neuroscience: A Once and Future Discipline," *Prog in Brain Res* 178 (2009): 287; and H. Chua et al., "Cultural Variation in Eye Movements During Scene Perception," *PNAS* 102 (2005): 12629.

Chapter 4: Hours to Days Before

1. Chemical castration as generally effective for obsessive paraphiliacs: F. Berlin, "'Chemical Castration' for Sex Offenders," *NEJM* 336 (1997): 1030. Lack of effectiveness in "hostile" rapists: K. Peters, "Chemical Castration: An Alternative to Incarceration," *Duquesne University Law Rev* 31 (1992): 307. Broad conclusion that it doesn't work particularly well: P. Fagan, "Pedophilia," *JAMA* 288 (2002): 2458. I thank Arielle Lasky for excellent assistance with this research subject.
2. For examples of the lack of correlation in a primate species, see M. Arlet et al., "Social Factors Increase Fecal Testosterone Levels in Wild Male Gray-Cheeked Mangabeys (*Lophocebus albigena*)," *Horm Behav* 59 (2011): 605; J. Archer, "Testosterone and Human Aggression: An Evaluation of the Challenge Hypothesis," *Nsci Biobehav Rev* 30 (2006): 319; the quote is from page 320.
3. J. Oberlander and L. Henderson, "The Sturm und Drang of Anabolic Steroid Use: Angst, Anxiety, and Aggression," *TINS* 35 (2012): 382; R. Agis-Balboa et al., "Enhanced Fear Responses in Mice Treated with Anabolic Androgenic Steroids," *Neuroreport* 22 (2009); 617.
4. E. Hermans, et al., "Testosterone Administration Reduces Empathetic Behavior: A Facial Mimicry Study," *PNE* 31 (2006): 859; J. Honk et al., "Testosterone Administration Impairs Cognitive Empathy in Women Depending on Second-to-Fourth Digit Ratio," *PNAS* 108 (2011): 3448; P. Bos et al., "Testosterone Decreases Trust in Socially Naive Humans," *PNAS* 107 (2010): 9991; P. Bos et al., "The Neural Mechanisms by Which Testosterone Acts on Interpersonal Trust," *Neuroimage* 2 (2012): 730; P. Mehta and J. Beer, "Neural Mechanisms of the Testosterone-Aggression Relation: The Role of the Orbitofrontal Cortex," *J Cog Nsci* 22 (2009): 2357.
5. L. Tsai and R. Sapolsky, "Rapid Stimulatory Effects of Testosterone upon Myotubule Metabolism and Hexose Transport, as Assessed by Silicon Microphysiometry," *Aggressive Behav* 22 (1996): 357; C. Rutte et al., "What Sets the Odds of Winning and Losing?" *TIEE* 21 (2006) 16.

   Confidence and persistence: A. Boissy and M. Bouissou, "Effects of Androgen Treatment on Behavioral and Physiological Responses of Heifers to Fear-Eliciting Situations," *Horm Behav* 28 (1994): 66; R. Andrew and L. Rogers, "Testosterone, Search Behaviour and Persistence," *Nat* 237 (1972): 343; J. Archer, "Testosterone and Persistence in Mice," *Animal Behav* 25 (1977): 479; M. Fuxjager et al., "Winning Territorial Disputes Selectively Enhances Androgen Sensitivity in Neural Pathways Related to Motivation and Social Aggression," *PNAS* 107 (2010): 12393.

   Human sports: M. Elias, "Serum Cortisol, Testosterone, and Testosterone-Binding Globulin Responses to Competitive Fighting in Human Males," *Aggressive Behav* 7 (1981): 215; A. Booth et al., "Testosterone, and Winning and Losing in Human Competition," *Horm Behav* 23 (1989): 556; J. Carré and S. Putnam, "Watching a Previous Victory Produces an Increase in Testosterone Among Elite Hockey Players," *PNE* 35 (2010): 475; A. Mazur et al., "Testosterone and Chess Competition," *Soc Psych Quarterly* 55 (1992): 70; J. Coates and J. Herbert, "Endogenous Steroids and Financial Risk Taking on a London Trading Floor," *PNAS* 105 (2008): 616.

6. N. Wright et al., "Testosterone Disrupts Human Collaboration by Increasing Egocentric Choices," *Proc Royal Soc B* (2012): 2275.
7. P. Mehta and J. Beer, "Neural Mechanisms of the Testosterone-Aggression Relation: The Role of Orbitofrontal Cortex," *J Cog Nsci* 22 (2010): 2357; G. van Wingen et al., "Testosterone Reduces Amygdala–Orbitofrontal Cortex Coupling," *PNE* 35 (2010): 105; P. Bos and E. Hermans et al., "The Neural Mechanisms by Which Testosterone Acts on Interpersonal Trust," *Neuroimage* 2 (2012): 730.
8. Testosterone decreasing fear and anxiety in rodents: C. Eisenegger et al., "The Role of Testosterone in Social Interaction," *TICS* 15 (2011): 263. Testosterone lessens the stress response: V. Viau, "Functional Cross-Talk Between the Hypothalamic- Pituitary-Gonadal and -Adrenal Axes," *J Neuroendocrinology* 14 (2002): 506. Testosterone reduces the startle response in humans: J. van Honk et al., "Testosterone Reduces Unconscious Fear But Not Consciously Experienced Anxiety: Implications for the Disorders of Fear and Anxiety," *BP* 58 (2005): 218; E. J. Hermans et al., "A Single Administration of Testosterone Reduces Fear-Potentiated Startle in Humans," *BP* 59 (2006): 872.
9. General reviews: R. Woods, "Reinforcing Aspects of Androgens," *Physiology & Behav* 83 (2004): 279; A. DiMeo and R. Wood, "Circulating Androgens Enhance Sensitivity to Testosterone Self-Administration in Male Hamsters," *Pharmacology, Biochemistry & Behav* 79 (2004): 383; M. Packard et al., "Rewarding Affective Properties of Intra–Nucleus Accumbens Injections of Testosterone," *Behav Nsci* 111 (1997): 219.
10. A. N. Dimeo and R. I. Wood, "ICV Testosterone Induces Fos in Male Syrian Hamster Brain," *PNE* 31 (2006): 237; M. Packard et al., "Rewarding Affective Properties of Intra–Nucleus Accumbens Injections of Testosterone," *Behav Nsci* 111 (1997): 219; M. Packard et al., "Expression of Testosterone Conditioned Place Preference Is Blocked by Peripheral or Intra-accumbens Injection of Alpha-flupenthixol," *Horm Behav* 34 (1998) 39; M. Fuxjager et al., "Winning Territorial Disputes Selectively Enhances Androgen Sensitivity in Neural Pathways Related to Motivation and Social Aggression," *PNAS* 107 (2010): 12393; A. Lacreuse et al., "Testosterone May Increase Selective Attention to Threat in Young Male Macaques," *Horm Behav* 58 (2010): 854.
11. A. Dixson and J. Herbert, "Testosterone, Aggressive Behavior and Dominance Rank in Captive Adult Male Talapoin Monkeys (*Miopithecus talapoin*)," *Physiology & Behav* 18 (1977): 539.
12. E. Hermans et al., "Exogenous Testosterone Enhances Responsiveness to Social Threat in the Neural Circuitry of Social Aggression in Humans," *BP* 63 (2008): 263; J. van Honk et al., "A Single Administration of Testosterone Induces Cardiac Accelerative Responses to Angry Faces in Healthy Young Women," *Behav Nsci* 115 (2001): 238; R. Ronay and A. Galinsky, "*Lex Talionis*: Testosterone and the Law of Retaliation," *JESP* 47 (2011): 702; P. Mehta and J. Beer, "Neural Mechanisms of the Testosterone-Aggression Relation: The Role of Orbitofrontal Cortex," *J Cog Nsci* 22 (2010): 2357; P. Bos et al., "Testosterone Decreases Trust in Socially Naive Humans," *PNAS* 107 (2010): 9991.
13. K. Kendrick and R. Drewett, "Testosterone Reduces Refractory Period of Stria Terminalis Neurons in the Rat Brain," *Sci* 204 (1979): 877; K. Kendrick, "Inputs to Testosterone-Sensitive Stria Terminalis Neurones in the Rat Brain and the Effects of Castration," *J Physiology* 323 (1982): 437; K. Kendrick, "The Effect of Castration on Stria Terminalis Neurone Absolute Refractory Periods Using Different Antidromic Stimulation Loci," *Brain Res* 248 (1982): 174; K. Kendrick, "Electrophysiological Effects of Testosterone on the Medial Preoptic-Anterior Hypothalamus of the Rat," *J Endo* 96 (1983): 35; E. Hermans et al., "Exogenous Testosterone Enhances Responsiveness to Social Threat in the Neural Circuitry of Social Aggression in Humans," *BP* 63 (2008): 263.
14. J. Wingfield et al., "The 'Challenge Hypothesis': Theoretical Implications for Patterns of Testosterone Secretion, Mating Systems, and Breeding Strategies," *Am Naturalist* 136 (1990): 829.
15. J. Archer, "Sex Differences in Aggression in Real-World Settings: A Meta-analytic Review," *Rev of General Psych* 8 (2004): 291.
16. J. Wingfield, et al., "Avoiding the 'Costs' of Testosterone: Ecological Bases of Hormone-Behavior Interactions," *Brain, Behav and Evolution* 57 (2001): 239; M. Sobolewski et al., "Female Parity, Male Aggression, and the Challenge Hypothesis in Wild Chimpanzees," *Primates* 54 (2013): 81; R. Sapolsky, "The Physiology of Dominance in Stable Versus Unstable Social Hierarchies," in *Primate Social Conflict*, ed. W. Mason and S. Mendoza (New York: SUNY Press, 1993), p. 171. P. Bernhardt et al., "Testosterone Changes During Vicarious Experiences of Winning and Losing Among Fans at Sporting Events," *Physiology & Behav* 65 (1998): 59.
17. M. Muller and R. Wrangham, "Dominance, Aggression and Testosterone in Wild Chimpanzees: A Test of the 'Challenge' Hypothesis," *Animal Behav* 67 (2004): 113; J. Archer, "Testosterone and Human Aggression: An Evaluation of the Challenge Hypothesis," *Nsci Biobehav Rev* 30 (2006): 319.
18. Footnote: L. Gettler et al., "Longitudinal Evidence That Fatherhood Decreases Testosterone in Human Males," *PNAS* 108 (2011): 16194. S. Van Anders et al., "Baby Cries and Nurturance Affect Testosterone in Men," *Horm Behav* 61 (2012): 31. J. Mascaro et al., "Testicular Volume is Inversely Correlated with Nurturing-Related Brain Activity in Human Fathers," *PNAS* 110 (2013): 15746. In some primates, timing is such that males are doing some degree of paternal care of offspring at the same time as doing the male-male competition thing to enhance their future reproductive success. Things get complicated here in that the paternalism and the competition should have opposite effects on testosterone levels. In the one study of this, groin trumped paternalism—testosterone levels were elevated. P. Onyango et al., "Testosterone Positively Associated with Both Male Mating Effort and Paternal Behavior in Savanna Baboons (*Papio cynocephalus*)," *Horm Behav* 63 (2012): 430.
19. J. Higley et al., "CSF Testosterone and 5-HIAA Correlate with Different Types of Aggressive Behaviors," *BP* 40 (1996): 1067.

20. C. Eisenegger et al., "Prejudice and Truth About the Effect of Testosterone on Human Bargaining Behaviour," *Nat* 463 (2010): 356.
21. M. Wibral et al., "Testosterone Administration Reduces Lying in Men," *PLoS ONE* 7 (2012): e46774. Also see: J. Van Honk et al., "New Evidence on Testosterone and Cooperation," *Nat* 485 (2012): E4.
22. Some reviews: O. Bosch and I. Neumann, "Both Oxytocin and Vasopressin Are Mediators of Maternal Care and Aggression in Rodents: From Central Release to Sites of Action," *Horm Behav* 61 (2012): 293; R. Feldman, "Oxytocin and Social Affiliation in Humans," *Horm Behav* 61 (2012): 380; A. Marsh et al., "The Influence of Oxytocin Administration on Responses to Infant Faces and Potential Moderation by OXTR Genotype," *Psychopharmacology* (Berlin) 24 (2012): 469; M. J. Bakermans-Kranenburg and M. H. van Ijzendoorn, "Oxytocin Receptor (OXTR) and Serotonin Transporter (5-HTT) Genes Associated with Observed Parenting," *SCAN* 3 (2008): 128. The hypothalamic pathway that differs by sex: N. Scott et al., "A Sexually Dimorphic Hypothalamic Circuit Controls Maternal Care and Oxytocin Secretion," *Nat* 525 (2016): 519.
23. Footnote: D. Huber et al., "Vasopressin and Oxytocin Excite Distinct Neuronal Populations in the Central Amygdala," *Sci* 308 (2005): 245; D. Viviani and R. Stoop, "Opposite Effects of Oxytocin and Vasopressin on the Emotional Expression of the Fear Response," *Prog Brain Res* 170 (2008): 207.
24. Y. Kozorovitskiy et al., "Fatherhood Affects Dendritic Spines and Vasopressin V1a Receptors in the Primate Prefrontal Cortex," *Nat Nsci* 9 (2006): 1094; Z. Wang et al., "Role of Septal Vasopressin Innervation in Paternal Behavior in Prairie Voles," *PNAS* 91 (1994): 400.
25. A. Smith et al., "Manipulation of the Oxytocin System Alters Social Behavior and Attraction in Pair-Bonding Primates, *Callithrix penicillata*," *Horm Behav* 57 (2010): 255; M. Jarcho et al., "Intransal VP Affects Pair Bonding and Peripheral Gene Expression in Male *Callicebus cupreus*," *Genes, Brain and Behav* 10 (2011): 375; C. Snowdon, "Variation in Oxytocin Is Related to Variation in Affiliative Behavior in Monogamous, Pairbonded Tamarins," *Horm Behav* 58 (2010); 614.
26. Z. Donaldson and L. Young, "Oxytocin, Vasopressin, and the Neurogenetics of Sociality," *Sci* 322 (2008): 900; E. Hammock and L. Young, "Microsatellite Instability Generates Diversity in Brain and Sociobehavioral Traits," *Sci* 308 (2005): 1630; L. Young et al., "Increased Affiliative Response to Vasopressin in Mice Expressing the V1a Receptor from a Monogamous Vole," *Nat* 400 (1999): 766; M. Lim et al., "Enhanced Partner Preference in a Promiscuous Species by Manipulating the Expression of a Single Gene," *Nat* 429 (2004): 754.
27. E. Hammock and L. Young, "Microsatellite Instability Generates Diversity in Brain and Sociobehavioral Traits," *Sci* 308 (2005): 1630.
28. I. Schneiderman et al., "Oxytocin at the First Stages of Romantic Attachment: Relations to Couples' Interactive Reciprocity," *PNE* 37 (2012): 1277.
29. B. Ditzen, et al., "Intranasal Oxytocin Increases Positive Communication and Reduces Cortisol Levels During Couple Conflict," *BP* 65 (2009): 728; D. Scheele et al., "Oxytocin Modulates Social Distance Between Males and Females," *J Nsci* 32 (2012): 16074; H. Walum et al., "Genetic Variation in the Vasopressin Receptor 1a Gene Associates with Pair-Bonding Behavior in Humans," *PNAS* 105 (2008): 14153; H. Walum et al., "Variation in the Oxytocin Receptor Gene Is Associated with Pair-Bonding and Social Behavior," *BP* 71 (2012): 419.
30. M. Nagasawa et al., "Oxytocin-Gaze Positive Loop and the Coevolution of Human-Dog Bonds," *Sci* 348 (2015): 333.
31. M. Yoshida, et al., "Evidence That Oxytocin Exerts Anxiolytic Effects via Oxytocin Receptor Expressed in Serotonergic Neurons in Mice," *J Nsci* 29 (2009): 2259. Oxytocin working in the amygdala: D. Viviani et al., "Oxytocin Selectively Gates Fear Responses Through Distinct Outputs from the Central Nucleus," *Sci* 333 (2011): 104; H. Knobloch et al., "Evoked Axonal Oxytocin Release in the Central Amygdala Attenuates Fear Response," *Neuron* 73 (2012): 553; S. Rodrigues et al., "Oxytocin Receptor Genetic Variation Relates to Empathy and Stress Reactivity in Humans," *PNAS* 106 (2009): 21437; M. Bakermans-Kranenburg and M. van Ijzendoorn, "Oxytocin Receptor (OXTR) and Serotonin Transporter (5-HTT) Genes Associated with Observed Parenting," *SCAN* 3 (2008): 128; G. Domes et al., "Oxytocin Attenuates Amygdala Responses to Emotional Faces Regardless of Valence," *BP* 62 (2007):1187; P. Kirsch, "Oxytocin Modulates Neural Circuitry for Social Cognition and Fear in Humans," *J Nsci* 25 (2005): 11489; I. Labuschagne et al., "Oxytocin Attenuates Amygdala Reactivity to Fear in Generalized Social Anxiety Disorder," *Neuropsychopharmacology* 35 (2010): 2403; M. Heinrichs et al., "Social Support and Oxytocin Interact to Suppress Cortisol and Subjective Responses to Psychosocial Stress," *BP* 54 (2003): 1389; K. Uvnas-Moberg, "Oxytocin May Mediate the Benefits of Positive Social Interaction and Emotions," *PNE* 23 (1998): 819. Carter quoted in P. S. Churchland and P. Winkielman, "Modulating Social Behavior with Oxytocin: How Does It Work? What Does It Mean?" *Horm Behav* 61 (2012): 392.

Oxytocin effects on aggression: M. Dhakar et al., "Heightened Aggressive Behavior in Mice with Lifelong Versus Postweaning Knockout of the Oxytocin Receptor," *Horm Behav* 62 (2012): 86; J. Winslow et al., "Infant Vocalization, Adult Aggression, and Fear Behavior of an Oxytocin Null Mutant Mouse," *Horm Behav* 37 (2005): 145.
32. M. Kosfeld et al., "Oxytocin Increases Trust in Humans," *Nat* 435 (2005): 673; A. Damasio, "Brain Trust," *Nat* 435 (2005): 571; S. Israel et al., "The Oxytocin Receptor (OXTR) Contributes to Prosocial Fund Allocations in the Dictator Game and the Social Value Orientations Task," *PLoS ONE* 4 (2009): e5535; P. Zak et al., "Oxytocin Is Associated with Human Trustworthiness," *Horm Behav* 48 (2005): 522; T. Baumgartner et al., "Oxytocin Shapes the Neural Circuitry of Trust and Trust Adaptation in Humans," *Neuron* 58 (2008): 639; A. Theodoridou et al., "Oxytocin and Social Perception: Oxytocin Increases Perceived Facial Trustworthiness and Attractiveness," *Horm Behav* 56 (2009): 128. A failure of replication: C. Apicella et al., "No Association Between Oxytocin Receptor (OXTR) Gene Polymorphisms and Experimentally Elicited Social Preferences,"

PLoS ONE 5 (2010): e11153. Turning the other cheek: J. Filling et al., "Effects of Intranasal Oxytocin and Vasopressin on Cooperative Behavior and Associated Brain Activity in Men," *PNE* 37 (2012): 447.

33. A. Marsh et al., "Oxytocin Improves Specific Recognition of Positive Facial Expressions," *Psychopharmacology* (Berlin) 209 (2010): 225; C. Unkelbach, et al., "Oxytocin Selectively Facilitates Recognition of Positive Sex and Relationship Words," *Psych Sci* 19 (2008): 102; J. Barraza et al., "Oxytocin Infusion Increases Charitable Donations Regardless of Monetary Resources," *Horm Behav* 60 (2011): 148; A. Kogan et al., "Thin-Slice Study of the Oxytocin Receptor Gene and the Evaluation and Expression of the Prosocial Disposition," *PNAS* 108 (2011): 19189; H. Tost et al., "A Common Allele in the Oxytocin Receptor Gene (OXTR) Impacts Prosocial Temperament and Human Hypothalamic-Limbic Structure and Function," *PNAS* 107 (2010): 13936; R. Hurlemann et al., "Oxytocin Enhances Amygdala-Dependent, Socially Reinforced Learning and Emotional Empathy in Humans," *J Nsci* 30 (2010): 4999.
34. P. Zak et al., "Oxytocin Is Associated with Human Trustworthiness," *Horm Behav* 48 (2005): 522; J. Holt-Lunstad et al., "Influence of a 'Warm Touch' Support Enhancement Intervention Among Married Couples on Ambulatory Blood Pressure, Oxytocin, Alpha Amylase, and Cortisol," *Psychosomatic Med* 70 (2008): 976; V. Morhenn et al., "Monetary Sacrifice Among Strangers Is Mediated by Endogenous Oxytocin Release After Physical Contact," *EHB* 29 (2008): 375; C. Crockford et al., "Urinary Oxytocin and Social Bonding in Related and Unrelated Wild Chimpanzees," *Proc Royal Soc B* 280 (2013): 20122765.
35. Z. Donaldson and L. Young, "Oxytocin, Vasopressin, and the Neurogenetics of Sociality," *Sci* 322 (2008): 900; A. Guastella et al., "Oxytocin Increases Gaze to the Eye Region of Human Faces," *BP* 63 (2008): 3; M. Gamer et al., "Different Amygdala Subregions Mediate Valence-Related and Attentional Effects of Oxytocin in Humans," *PNAS* 107 (2010): 9400; C. Zink et al., "Vasopressin Modulates Social Recognition–Related Activity in the Left Temporoparietal Junction in Humans," *Translational Psychiatry* 1 (2011): e3; G. Domes et al., "Oxytocin Improves 'Mind-Reading' in Humans," *BP* 61 (2007): 731–33; U. Rimmele et al., "Oxytocin Makes a Face in Memory More Familiar," *J Nsci* 29 (2009): 38; M. Fischer-Shofty et al., "Oxytocin Facilitates Accurate Perception of Competition in Men and Kinship in Women," SCAN (2012).
36. C. Sauer et al., "Effects of a Common Variant in the CD38 Gene on Social Processing in an Oxytocin Challenge Study: Possible Links to Autism," *Neuropsychopharmacology* 37 (2012): 1474.
37. E. Hammock and L. Young, "Oxytocin, Vasopressin and Pair Bonding: Implications for Autism," *Philosophical Transactions of the Royal Soc of London B* 361 (2006): 2187; A. Meyer-Lindenberg et al., "Oxytocin and Vasopressin in the Human Brain: Social Neuropeptides for Translational Medicine," *Nat Rev Nsci* 12 (2011): 524; H. Yamasue et al., "Integrative Approaches Utilizing Oxytocin to Enhance Prosocial Behavior: From Animal and Human Social Behavior to Autistic Social Dysfunction," *J Nsci* 32 (2012): 14109.
38. Reviewed in A. Graustella and C. MacLeod, "A Critical Review of the Influence of Oxytocin Nasal Spray on Social Cognition in Humans: Evidence and Future Directions," *Horm Behav* 61 (2012): 410.
39. J. Bartz et al., "Social Effects of Oxytocin in Humans: Context and Person Matter," *TICS* 15 (2011): 301
40. G. Domes et al., "Effects of Intranasal Oxytocin on Emotional Face Processing in Women," *PNE* 35 (2010): 83; G. De Vries, "Sex Differences in Vasopressin and Oxytocin Innervation in the Brain," *Prog Brain Res* 170 (2008): 17; J. Bartz et al., "Effects of Oxytocin on Recollections of Maternal Care and Closeness," *PNAS* 14 (2010): 107.
41. M. Mikolajczak et al., "Oxytocin Not Only Increases Trust When Money Is at Stake, but Also When Confidential Information Is in the Balance," *BP* 85 (2010): 182.
42. H. Kim et al., "Culture, Distress, and Oxytocin Receptor Polymorphism (OXTR) Interact to Influence Emotional Support Seeking," *PNAS* 107 (2010): 15717.
43. O. Bosch and I. Neumann, "Both Oxytocin and Vasopressin Are Mediators of Maternal Care and Aggression in Rodents: From Central Release to Sites of Action," *Horm Behav* 61 (2012): 293.
44. C. Ferris and M. Potegal, "Vasopressin Receptor Blockade in the Anterior Hypothalamus Suppresses Aggression in Hamsters," *Physiology & Behav* 44 (1988): 235; H. Albers, "The Regulation of Social Recognition, Social Communication and Aggression: Vasopressin in the Social Behavior Neural Network," *Horm Behav* 61 (2012): 283; A. Johansson et al., "Alcohol and Aggressive Behavior in Men: Moderating Effects of Oxytocin Receptor Gene (OXTR) Polymorphisms," *Genes, Brain and Behav* 11 (2012): 214; J. Winslow and T. Insel, "Social Status in Pairs of Male Squirrel Monkeys Determines the Behavioral Response to Central Oxytocin Administration," *J Nsci* 11 (1991): 2032; J. Winslow et al., "A Role for Central Vasopressin in Pair Bonding in Monogamous Prairie Voles," *Nat* 365 (1993): 545.
45. T. Baumgartner et al., "Oxytocin Shapes the Neural Circuitry of Trust and Trust Adaptation in Humans," *Neuron* 58 (2008): 639; C. Declerk et al., "Oxytocin and Cooperation Under Conditions of Uncertainty: The Modulating Role of Incentives and Social Information," *Horm Behav* 57 (2010): 368; S. Shamay-Tsoory et al., "Intranasal Administration of Oxytocin Increases Envy and Schadenfreude (Gloating)," *BP* 66 (2009): 864.
46. C. de Dreu, "Oxytocin Modulates Cooperation Within and Competition Between Groups: An Integrative Review and Research Agenda," *Horm Behav* 61 (2012): 419; C. de Dreu et al., "The Neuropeptide Oxytocin Regulates Parochial Altruism in Intergroup Conflict Among Humans," *Sci* 328 (2011): 1408.
47. C. de Dreu et al., "Oxytocin Promotes Human Ethnocentrism," *PNAS* 108 (2011): 1262.
48. Footnote: S. Motta et al., "Ventral Premammillary Nucleus as a Critical Sensory Relay to the Maternal Aggression Network," *PNAS* 110 (2013): 14438.
49. J. Lonstein and S. Gammie, "Sensory, Hormonal, and Neural Control of Maternal Aggression in Laboratory Rodents," *Nsci Biobehav Rev* 26 (2002): 869; S. Parmigiani et al., "Selection, Evolution of Behavior and Animal Models in Behavioral Neuroscience," *Nsci Biobehav Rev* 23 (1999): 957.
50. R. Gandelman and N. Simon, "Postpartum Fighting in the Rat: Nipple Development and the Presence of

Young," *Behav and Neural Biol* 29 (1980): 350; M. Erskine et al., "Intraspecific Fighting During Late Pregnancy and Lactation in Rats and Effects of Litter Removal," *Behav Biol* 23 (1978): 206; K. Flannelly and E. Kemble, "The Effect of Pup Presence and Intruder Behavior on Maternal Aggression in Rats," *Bull of the Psychonomic Soc* 25 (1988): 133.

51. B. Derntl et al., "Association of Menstrual Cycle Phase with the Core Components of Empathy," *Horm Behav* 63 (2013): 97.

For a good review see C. Bodo and E. Rissman, "New Roles for Estrogen Receptor Beta in Behavior and Neuroendocrinology," *Front Neuroendocrinology* 27 (2006): 217.

52. D. Reddy, "Neurosteroids: Endogenous Role in the Human Brain and Therapeutic Potentials," *Prog Brain Res* 186 (2010): 113; F. De Sousa et al., "Progesterone and Maternal Aggressive Behavior in Rats," *Behavioural Brain Res* 212 (2010): 84; G. Pinna et al., "Neurosteroid Biosynthesis Regulates Sexually Dimorphic Fear and Aggressive Behavior in Mice," *Neurochemical Res* 33 (2008): 1990; K. Miczek et al., "Neurosteroids, GABAA Receptors, and Escalated Aggressive Behavior," *Horm Behav* 44 (2003): 242.

53. S. Hrdy, "The 'One Animal in All Creation About Which Man Knows the Least,'" *Philosophical Transactions of the Royal Soc B* 368 (2013): 20130072.

54. The spillover idea is aired in E. Ketterson et al., "Testosterone in Females: Mediator of Adaptive Traits, Constraint on Sexual Dimorphism, or Both?" *Am Naturalist* 166 (2005): 585.

55. C. Voigt and W. Goymann, "Sex-Role Reversal Is Reflected in the Brain of African Black Coucals (*Centropus grillii*)," *Developmental Neurobiol* 67 (2007): 1560; M. Peterson et al., "Testosterone Affects Neural Gene Expression Differently in Male and Female Juncos: A Role for Hormones in Mediating Sexual Dimorphism and Conflict," *PLoS ONE* 8 (2013): e61784.

56. A. Pusey and K. Schroepfer-Walker, "Female Competition in Chimpanzees," *Philosophical Transactions of the Royal Soc B* 368 (2013): 20130077.

57. J. French et al., "The Influence of Androgenic Steroid Hormones on Female Aggression in 'Atypical' Mammals," *Philosophical Transactions of the Royal Soc B* 368 (2013): 20130084; L. Frank et al., "Fatal Sibling Aggression, Precocial Development, and Androgens in Neonatal Spotted Hyenas," *Sci* 252 (1991): 702; S. Glickman et al., "Androstenedione May Organize or Activate Sex-Reversed Traits in Female Spotted Hyenas," *PNAS* 84 (1987): 3444.

58. W. Goymann et al., "Androgens and the Role of Female 'Hyperaggressiveness' in Spotted Hyenas," *Horm Behav* 39 (2001): 83; S. Fenstemaker et al., "A Sex Difference in the Hypothalamus of the Spotted Hyena," *Nat Nsci* 2 (1999): 943; G. Rosen et al., "Distribution of Vasopressin in the Forebrain of Spotted Hyenas," *J Comp Neurol* 498 (2006): 80.

59. P. Chambers and J. Hearn, "Peripheral Plasma Levels of Progesterone, Oestradiol-17β, Oestrone, Testosterone, Androstenedione and Chorionic Gonadotrophin During Pregnancy in the Marmoset Monkey, *Callithrix jacchus*," *J Reproduction Fertility* 56 (1979): 23; C. Drea, "Endocrine Correlates of Pregnancy in the Ring-Tailed Lemur (*Lemur catta*): Implications for the Masculinization of Daughters," *Horm Behav* 59 (2011): 417; M. Holmes et al., "Social Status and Sex Independently Influence Androgen Receptor Expression in the Eusocial Naked Mole-Rat Brain," *Horm Behav* 54 (2008): 278; L. Koren et al., "Elevated Testosterone Levels and Social Ranks in Female Rock Hyrax," *Horm Behav* 49 (2006): 470; C. Kraus et al., "High Maternal Androstenedione Levels During Pregnancy in a Small Precocial Mammal with Female Genital Masculinisation" (Max Planck Institute for Demographic Research Working Paper WP 2008-017, April 2008); C. Kraus et al., "Spacing Behaviour and Its Implications for the Mating System of a Precocial Small Mammal: An Almost Asocial Cavy *Cavia magna*," *Animal Behav* 66 (2003): 225; L. Koren and E. Geffen, "Androgens and Social Status in Female Rock Hyraxes," *Animal Behav* 77 (2009): 233.

60. Footnote: DHEA and local generation of steroids within neurons: K. Soma et al., "Novel Mechanisms for Neuroendocrine Regulation of Aggression," *Front Neuroendocrinology* 29 (2008): 476; K. Schmidt et al., "Neurosteroids, Immunosteroids, and the Balkanization of Endo," *General and Comp Endo* 157 (2008): 266; D. Pradhan et al., "Aggressive Interactions Rapidly Increase Androgen Synthesis in the Brain During the Non-breeding Season," *Horm Behav* 57 (2010): 381.

61. T. Johnson, "Premenstrual Syndrome as a Western Culture-Specific Disorder," *Culture, Med and Psychiatry* 11 (1987): 337; L. Cosgrove and B. Riddle, "Constructions of Femininity and Experiences of Menstrual Distress," *Women & Health* 38 (2003): 37.

62. For the quote in the text, see M. Rodin, "The Social Construction of Premenstrual Syndrome," *Soc Sci & Med* 35 (1992); 49. For the quote in the footnote, see: A. Kleinman, "Depression, Somaticization, and the New 'Cross-Cultural Psychiatry,'" *Social Science Med* 11 (1977): 3.

63. H. Rupp et al., "Neural Activation in the Orbitofrontal Cortex in Response to Male Faces Increases During Follicular Phase," *Horm Behav* 56 (2009): 66. Mareckova K. et al. "Hormonal Contraceptives, Menstrual Cycle and Brain Response to Faces. *SCAN* 9 (2012): 191.

64. A. Rapkin et al., "Menstrual Cycle and Social Behavior in Vervet Monkeys," *PNE* 20 (1995): 289; E. García-Castells et al., "Changes in Social Dynamics Associated to the Menstrual Cycle in the Vervet Monkey (*Cercopithecus aethiops*)," *Boletín de Estudios Médicos y Biológicos* 37 (1989): 11; G. Mallow, "The Relationship Between Aggressive Behavior and Menstrual Cycle Stage in Female Rhesus Monkeys (*Macaca mulatta*)," *Horm Behav* 15 (1981): 259; G. Hausfater and B. Skoblic, "Perimenstrual Behavior Changes Among Female Yellow Baboons: Some Similarities to Premenstrual Syndrome (PMS) in Women," *Animal Behav* 9 (1985): 165.

65. K. Dalton, "School Girls' Behavior and Menstruation," *Brit Med J* 2 (1960): 1647; K. Dalton, "Menstruation and Crime," *Brit Med J* 2 (1961): 1752; K. Dalton, "Cyclical Criminal Acts in Premenstrual Syndrome," *Lancet* 2 (1980): 1070.

66. P. Easteal, "Women and Crime: Premenstrual Issues," *Trends and Issues in Crime and Criminal Justice* 31 (1991): 1–8; J. Chrisler and P. Caplan, "The Strange Case of Dr. Jekyll and Ms. Hyde: How PMS Became a Cultural Phenomenon and a Psychiatric Disorder," *Ann Rev of Sex Res* 13 (2002): 274.
67. For a general review, see R. Sapolsky, *Why Zebras Don't Get Ulcers: A Guide to Stress, Stress-Related Diseases and Coping*, 3rd ed. (New York: Henry Holt, 2004).
68. R. Sapolsky "Stress and the Brain: Individual Variability and the Inverted-U," *Nat Nsci* 25 (2015): 1344.
69. K. Roelofs et al., "The Effects of Social Stress and Cortisol Responses on the Preconscious Selective Attention to Social Threat," *BP* 75 (2007): 1; K. Tully et al., "Norepinephrine Enables the Induction of Associative Long-Term Potentiation at Thalamo-Amygdala Synapses," *PNAS* 104 (2007): 14146; P. Putman et al., "Cortisol Administration Acutely Reduces Threat-Selective Spatial Attention in Healthy Young Men," *Physiology & Behav* 99 (2010): 294; K. Bertsch et al., "Exogenous Cortisol Facilitates Responses to Social Threat Under High Provocation," *Horm Behav* 59 (2011): 428.
70. J. Rosenkranz et al., "Chronic Stress Causes Amygdala Hyperexcitability in Rodents," *BP* 67 (2010): 1128; S. Duvarci and D. Pare, "Glucocorticoids Enhance the Excitability of Principle Basolateral Amygdala Neurons," *J Nsci* 27 (2007): 4482; A. Kavushansky and G. Richter-Levin, "Effects of Stress and Corticosterone on Activity and Plasticity in the Amygdala," *J Nsci Res* 84 (2006): 1580; A. Kavushansky et al., "Activity and Plasticity in the CA1, the Dentate Gyrus, and the Amygdala Following Controllable Versus Uncontrollable Water Stress," *Hippocampus* 16 (2006): 35; P. Rodríguez Manzanares et al., "Previous Stress Facilitates Fear Memory, Attenuates GABAergic Inhibition, and Increases Synaptic Plasticity in the Rat Basolateral Amygdala," *J Nsci* 25 (2005): 8725; H. Lakshminarasimhan and S. Chattarji, "Stress Leads to Contrasting Effects on the Levels of Brain Derived Neurotrophic Factor in the Hippocampus and Amygdala," *PLoS ONE* 7 (2012): e30481; S. Ghosh et al., "Functional Connectivity from the Amygdala to the Hippocampus Grows Stronger After Stress," *J Nsci* 33 (2013): 7234.
71. B. Kolber et al., "Central Amygdala Glucocorticoid Receptor Action Promotes Fear-Associated CRH Activation and Conditioning," *PNAS* 105 (2008): 12004; S. Rodrigues et al., "The Influence of Stress Hormones on Fear Circuitry," *Ann Rev Nsci* 32 (2009): 289; L. Shin and I. Liberzon, "The Neurocircuitry of Fear, Stress, and Anxiety Disorders," *Neuropsychopharmacology* 35, no. 1 (January 2010): 169.
72. M. Milad and G. Quirk, "Neurons in Medial Prefrontal Cortex Signal Memory for Fear Extinction," *Nat* 420 (2002): 70; E. Phelps et al., "Extinction Learning in Humans: Role of the Amygdala and vmPFC," *Neuron* 43 (2004): 897; J. Bremner et al., "Neural Correlates of Exposure to Traumatic Pictures and Sound in Vietnam Combat Veterans With and Without Posttraumatic Stress Disorder: A Positron Emission Tomography Study," *BP* 45 (1999) 806; D. Knox et al., "Single Prolonged Stress Disrupts Retention of Extinguished Fear in Rats," *Learning & Memory* 19 (2012): 43; M. Schmidt et al., "Stress-Induced Metaplasticity: From Synapses to Behavior," *Nsci* 250 (2013): 112; J. Pruessner et al., "Deactivation of the Limbic System During Acute Psychosocial Stress: Evidence from Positron Emission Tomography and Functional Magnetic Resonance Imaging Studies," *BP* 63 (2008): 234.
73. A. Young et al., "The Effects of Chronic Administration of Hydrocortisone on Cognitive Function in Normal Male Volunteers," *Psychopharmacology* (Berlin) 145 (1999): 260; A. Barsegyan et al., "Glucocorticoids in the Prefrontal Cortex Enhance Memory Consolidation and Impair Working Memory by a Common Neural Mechanism," *PNAS* 107 (2010): 16655; A. Arnsten et al., "Neuromodulation of Thought: Flexibilities and Vulnerabilities in Prefrontal Cortical Network Synapses," *Neuron* 76 (2012): 223; B. Roozendaal et al., "The Basolateral Amygdala Interacts with the Medial Prefrontal Cortex in Regulating Glucocorticoid Effects on Working Memory Impairment," *J Nsci* 24 (2004): 1385; C. Liston et al., "Psychosocial Stress Reversibly Disrupts Prefrontal Processing and Attentional Control," *PNAS* 106 (2008): 912.
74. E. Dias-Ferreira et al., "Chronic Stress Causes Frontostriatal Reorganization and Affects Decision-Making," *Sci* 325 (2009): 621; D. Lyons et al., "Stress-Level Cortisol Treatment Impairs Inhibitory Control of Behavior in Monkeys," *J Nsci* 20 (2000): 7816; J. Kim et al., "Amygdala Is Critical for Stress-Induced Modulation of Hippocampal Long-Term Potentiation and Learning," *J Nsci* 21 (2001): 5222; L. Schwabe and O. Wolf, "Stress Prompts Habit Behavior in Humans," *J Nsci* 29 (2009): 7191; L. Schwabe and O. Wolf, "Socially Evaluated Cold Pressor Stress After Instrumental Learning Favors Habits over Goal-Directed Action," *PNE* 35 (2010): 977; L. Schwabe and O. Wolf, "Stress-Induced Modulation of Instrumental Behavior: From Goal-Directed to Habitual Control of Action," *BBR* 219 (2011): 321; L. Schwabe and O. Wolf, "Stress Modulates the Engagement of Multiple Memory Systems in Classification Learning," *J Nsci* 32 (2012): 11042; L. Schwabe et al., "Simultaneous Glucocorticoid and Noradrenergic Activity Disrupts the Neural Basis of Goal-Directed Action in the Human Brain," *J Nsci* 32 (2012): 10146.
75. V. Venkatraman et al., "Sleep Deprivation Biases the Neural Mechanisms Underlying Economic Preferences," *J Nsci* 31 (2011): 3712; M. Brand et al., "Decision-Making Deficits of Korsakoff Patients in a New Gambling Task with Explicit Rules: Associations with Executive Functions," *Neuropsychology* 19 (2005): 267; E. Masicampo and R. Baumeister, "Toward a Physiology of Dual-Process Reasoning and Judgment: Lemonade, Willpower, and Expensive Rule-Based Analysis," *Psych Sci* 19 (2008): 255.
76. S. Preston et al., "Effects of Anticipatory Stress on Decision-Making in a Gambling Task," *Behav Nsci* 121 (2007): 257; R. van den Bos et al., "Stress and Decision-Making in Humans: Performance Is Related to Cortisol Reactivity, Albeit Differently in Men and Women," *PNE* 34 (2009): 1449; N. Lighthall et al., "Acute Stress Increases Sex Differences in Risk Seeking in the Balloon Analogue Risk Task," *PLoS ONE* 4 (2009): e6002; N. Lighthall et al., "Gender Differences in Reward-Related Decision Processing Under Stress," *SCAN* 7, no. 4 (April 2012): 476–84; P. Putman et al., "Exogenous Cortisol Acutely Influences Motivated Decision Making in Healthy Young Men," *Psychopharmacology* 208 (2010): 257; P. Putman et al., "Cortisol Administration Acutely

Reduces Threat-Selective Spatial Attention in Healthy Young Men," *Physiology & Behav* 99 (2010): 294; K. Starcke et al., "Anticipatory Stress Influences Decision Making Under Explicit Risk Conditions," *Behav Nsci* 122 (2008): 1352.

77. E. Mikics et al., "Genomic and Non-genomic Effects of Glucocorticoids on Aggressive Behavior in Male Rats," *PNE* 29 (2004): 618; D. Hayden-Hixson and C. Ferris, "Steroid-Specific Regulation of Agonistic Responding in the Anterior Hypothalamus of Male Hamsters," *Physiology & Behav* 50 (1991): 793; A. Poole and P. Brain, "Effects of Adrenalectomy and Treatments with ACTH and Glucocorticoids on Isolation-Induced Aggressive Behavior in Male Albino Mice," *Prog Brain Res* 41 (1974): 465; E. Mikics et al., "The Effect of Glucocorticoids on Aggressiveness in Established Colonies of Rats," *PNE* 32 (2007): 160; R. Böhnke et al., "Exogenous Cortisol Enhances Aggressive Behavior in Females, but Not in Males," *PNE* 35 (2010): 1034; K. Bertsch et al., "Exogenous Cortisol Facilitates Responses to Social Threat Under High Provocation," *Horm Behav* 59 (2011): 428.
78. S. Levine et al., "The PNE of Stress: A Psychobiological Perspective," in *Psychoneuroendocrinology*, ed. S. Levine and R. Brush (New York: Academic Press, 1988), p. 181; R. Sapolsky and J. Ray, "Styles of Dominance and Their Physiological Correlates Among Wild Baboons," *Am J Primat* 18 (1989): 1; J. C. Ray and R. Sapolsky, "Styles of Male Social Behavior and Their Endocrine Correlates Among High-Ranking Baboons," *Am J Primat* 28 (1992): 231; C. E. Virgin and R. Sapolsky, "Styles of Male Social Behavior and Their Endocrine Correlates Among Low-Ranking Baboons," *Am J Primat* 42 (1997): 25.
79. D. Card and G. Dahl, "Family Violence and Football: The Effect of Unexpected Emotional Cues on Violent Behavior," *Quarterly J Economics* 126 (2011): 103.
80. Footnote: For a study concerning the neurobiology of how stress makes healthy habits harder to maintain, see C. Cifani et al., "Medial Prefrontal Cortex Neuronal Activation and Synaptic Alterations After Stress-Induced Reinstatement of Palatable Food Seeking: A Study Using c-fos-GFP Transgenic Female Rats," *J Nsci* 32 (2012): 8480.
81. K. Starcke et al., "Does Everyday Stress Alter Moral Decision-Making?" *PNE* 36 (2011): 210; F. Youssef et al., "Stress Alters Personal Moral Decision Making," *PNE* 37 (2012): 491.
82. D. Langford et al., "Social Modulation of Pain as Evidence for Empathy in Mice," *Sci* 312 (2006): 1967.
83. S. Taylor et al., "Biobehavioral Responses to Stress in Females: Tend-and-Befriend, Not Fight-or-Flight," *Psych Rev* 107 (2000): 411.
84. B. Bushman, "Human Aggression While Under the Influence of Alcohol and Other Drugs: An Integrative Research Review," *Curr Dir Psych Sci* 2 (1993): 148; L. Zhang et al., "The Nexus Between Alcohol and Violent Crime," *Alcoholism: Clin and Exp Res* 21 (1997): 1264; K. Graham and P. West, "Alcohol and Crime: Examining the Link," in *International Handbook of Alcohol Dependence and Problems*, ed. N. Heather, T. J. Peters, and T. Stockwell (New York: John Wiley & Sons, 2001); I. Quadros et al., "Individual Vulnerability to Escalated Aggressive Behavior by a Low Dose of Alcohol: Decreased Serotonin Receptor mRNA in the Prefrontal Cortex of Male Mice," *Genes, Brain and Behav* 9 (2010): 110; A. Johansson et al., "Alcohol and Aggressive Behavior in Men: Moderating Effects of Oxytocin Receptor Gene (OXTR) Polymorphisms," *Genes, Brain and Behav* 11 (2012): 214.

Chapter 5: Days to Months Before

1. D. O. Hebb, *The Organization of Behaviour* (Hoboken, NJ: John Wiley & Sons, 1949).
2. General reviews: R. Nicoll and K. Roche, "Long-Term Potentiation: Peeling the Onion," *Neuropharmacology* 74 (2013): 18; J. MacDonald et al., "Hippocampal Long-Term Synaptic Plasticity and Signal Amplification of NMDA Receptors," *Critical Rev in Neurobiol* 18 (2006): 71.
3. T. Sigurdsson et al., "Long-Term Potentiation in the Amygdala: A Cellular Mechanism of Fear Learning and Memory," *Neuropharmacology* 52 (2007): 215; J. Kim and M. Jung, "Neural Circuits and Mechanisms Involved in Pavlovian Fear Conditioning: A Critical Review," *Nsci Biobehav Rev* 30 (2006): 188; M. Wolf, "LTP May Trigger Addiction," *Mol Interventions* 3 (2003): 248; M. Wolf et al., "Psychomotor Stimulants and Neuronal Plasticity," *Neuropharmacology* 47, supp. 1 (2004): 61.
4. M. Foy et al., "17beta-estradiol Enhances NMDA Receptor-Mediated EPSPs and Long-Term Potentiation," *J Neurophysiology* 81 (1999): 925; Y. Lin et al., "Oxytocin Promotes Long-Term Potentiation by Enhancing Epidermal Growth Factor Receptor-Mediated Local Translation of Protein Kinase Mζ," *J Nsci* 32 (2012): 15476; K. Tomizawa et al., "Oxytocin Improves Long-Lasting Spatial Memory During Motherhood Through MAP Kinase Cascade," *Nat Nsci* 6 (2003): 384; V. Skucas et al., "Testosterone Depletion in Adult Male Rats Increases Mossy Fiber Transmission, LTP, and Sprouting in Area CA3 of Hippocampus," *J Nsci* 33 (2013): 2338; W. Timmermans et al., "Stress and Excitatory Synapses: From Health to Disease," *Nsci* 248 (2013): 626.
5. S. Rodrigues et al., "The Influence of Stress Hormones on Fear Circuitry," *Ann Rev Nsci* 32 (2009): 289; X. Xu and Z. Zhang, "Effects of Estradiol Benzoate on Learning-Memory Behavior and Synaptic Structure in Ovariectomized Mice," *Life Sci* 79 (2006): 1553; C. Rocher et al., "Acute Stress-Induced Changes in Hippocampal/Prefrontal Circuits in Rats: Effects of Antidepressants," *Cerebral Cortex* 14 (2004): 224.
6. A. Holtmaat and K. Svoboda, "Experience-Dependent Structural Synaptic Plasticity in the Mammalian Brain," *Nat Rev Nsci* 10 (2009): 647; C. Woolley et al., "Naturally Occurring Fluctuation in Dendritic Spine Density on Adult Hippocampal Pyramidal Neurons," *J Nsci* 10 (1990): 4035; W. Kelsch et al., "Watching Synaptogenesis in the Adult Brain," *Ann Rev of Nsci* 33 (2010): 131.
7. B. Leuner and T. Shors, "Stress, Anxiety, and Dendritic Spines: What Are the Connections?" *Nsci* 251 (2013): 108; Y. Chen et al., "Correlated Memory Defects and Hippocampal Dendritic Spine Loss After Acute Stress Involve Corticotropin-Releasing Hormone Signaling," *PNAS* 107 (2010): 13123.
8. J. Cerqueira et al., "Morphological Correlates of Corticosteroid-Induced Changes in Prefrontal Cortex Dependent Behaviours," *J Nsci* 25 (2005): 7792; A. Izquierdo et al., "Brief Uncontrollable Stress Causes Dendritic

Retraction in Infralimbic Cortex and Resistance to Fear Extinction in Mice," *J Nsci* 26 (2006): 5733; C. Liston et al., "Stress-Induced Alterations in Prefrontal Cortical Dendritic Morphology Predict Selective Impairments in Perceptual Attentional Set Shifting," *J Nsci* 26 (2006): 7870; J. Radley, "Repeated Stress Induces Dendritic Spine Loss in the Rat Medial Prefrontal Cortex," *Cerebral Cortex* 16 (2006): 313; A. Arnsten, "Stress Signaling Pathways That Impair Prefrontal Cortex Structure and Function," *Nat Rev Nsci* 10 (2009): 410; C. Sandi and M. Loscertales, "Opposite Effects on NCAM Expression in the Rat Frontal Cortex Induced by Acute vs. Chronic Corticosterone Treatments," *Brain Res* 828 (1999): 127; C. Wellman, "Dendritic Reorganization in Pyramidal Neurons in Medial Prefrontal Cortex After Chronic Corticosterone Administration," *J Neurobiol* 49 (2001): 245; D. Knox et al., "Single Prolonged Stress Decreases Glutamate, Glutamine, and Creatine Concentrations in the Rat Medial Prefrontal Cortex," *Nsci Lett* 480 (2010): 16.

9. E. Dias-Ferreira et al., "Chronic Stress Causes Frontostriatal Reorganization and Affects Decision-Making," *Sci* 325 (2009): 621; M. Fuchikiami et al., "Epigenetic Regulation of BDNF Gene in Response to Stress," *Psychiatry Investigation* 7 (2010): 251.
10. R. Mitra and R. Sapolsky, "Acute Corticosterone Treatment Is Sufficient to Induce Anxiety and Amygdaloid Dendritic Hypertrophy," *PNAS* 105 (2008): 5573; A. Vyas et al., "Chronic Stress Induces Contrasting Patterns of Dendritic Remodeling in Hippocampal and Amygdaloid Neurons," *J Nsci* 22 (2002): 6810; S. Bennur et al., "Stress-Induced Spine Loss in the Medial Amygdala Is Mediated by Tissue-Plasminogen Activator," *Nsci* 144 (2006): 8; A. Govindarajan et al., "Transgenic Brain-Derived Neurotrophic Factor Expression Causes Both Anxiogenic and Antidepressant Effects," *PNAS* 103 (2006): 13208.

    Expansion of the BNST: A. Vyas et al., "Effects of Chronic Stress on Dendritic Arborization in the Central and Extended Amygdala," *Brain Res* 965 (2003): 290; J. Pego et al., "Dissociation of the Morphological Correlates of Stress-Induced Anxiety and Fear," *Eur J Nsci* 27 (2008): 1503.
11. A. Magarinos and B. McEwen, "Stress-Induced Atrophy of Apical Dendrites of Hippocampal CA3c Neurons: Involvement of Glucocorticoid Secretion and Excitatory Amino Acid Receptors," *Nsci* 69 (1995): 89; A. Magarinos et al., "Chronic Psychosocial Stress Causes Apical Dendritic Atrophy of Hippocampal CA3 Pyramidal Neurons in Subordinate Tree Shrews," *J Nsci* 16 (1996): 3534; B. Eadie et al., "Voluntary Exercise Alters the Cytoarchitecture of the Adult Dentate Gyrus by Increasing Cellular Proliferation, Dendritic Complexity, and Spine Density," *J Comp Neurol* 486 (2005): 39.
12. M. Khan et al., "Estrogen Regulation of Spine Density and Excitatory Synapses in Rat Prefrontal and Somatosensory Cerebral Cortex," *Steroids* 78 (2013): 614; B. McEwen, "Estrogen Actions Throughout the Brain," *Recent Prog Hormone Res* 57 (2002): 357; B. Leuner and E. Gould, "Structural Plasticity and Hippocampal Function," *Ann Rev Psych* 61 (2010): 111.
13. R. Hamilton et al., "Alexia for Braille Following Bilateral Occipital Stroke in an Early Blind Woman," *Neuroreport* 11 (2000): 237; E. Striem-Amit et al., "Reading with Sounds: Sensory Substitution Selectively Activates the Visual Word Form Area in the Blind," *Neuron* 76 (2012): 640.
14. S. Florence et al., "Large-Scale Sprouting of Cortical Connections After Peripheral Injury in Adult Macaque Monkeys," *Sci* 282 (1998): 1117; C. Darian-Smith and C. Gilbert, "Axonal Sprouting Accompanies Functional Reorganization in Adult Cat Striate Cortex," *Nat* 368 (1994): 737; M. Kossut and S. Juliano, "Anatomical Correlates of Representational Map Reorganization Induced by Partial Vibrissectomy in the Barrel Cortex of Adult Mice," *Nsci* 92 (1999): 807; L. Merabet and A. Bascual-Leone, "Neural Reorganization Following Sensory Loss: The Opportunity of Change," *Nat Rev Nsci* 11 (2010): 44; A. Pascual-Leone et al., "The Plastic Human Brain Cortex," *Ann Rev Nsci* 28 (2005): 377; B. Becker et al., "Fear Processing and Social Networking in the Absence of a Functional Amygdala," *BP* 72 (2012): 70; L. Colgin, "Understanding Memory Through Hippocampal Remapping," *TINS* 31 (2008): 469; V. Ramirez-Amaya et al., "Spatial Longterm Memory Is Related to Mossy Fiber Synaptogenesis," *J Nsci* 21 (2001): 7340; M. Holahan et al., "Spatial Learning Induces Presynaptic Structural Remodeling in the Hippocampal Mossy Fiber System of Two Rat Strains," *Hippocampus* 16 (2006): 560; I. Galimberti et al., "Long-Term Rearrangements of Hippocampal Mossy Fiber Terminal Connectivity in the Adult Regulated by Experience," *Neuron* 50 (2006): 749; V. De Paola et al., "Cell Type–Specific Structural plasticity of Axonal Branches and Boutons in the Adult Neocortex," *Neuron* 49 (2006): 861; H. Nishiyama et al., "Axonal Motility and Its Modulation by Activity Are Branch-Type Specific in the Intact Adult Cerebellum," *Neuron* 56 (2007): 472.
15. C. Pantev and S. Herholz, "Plasticity of the Human Auditory Cortex Related to Musical Training," *Nsci Biobehav Rev* 35 (2011): 2140.
16. A. Pascual-Leone, "Reorganization of Cortical Motor Outputs in the Acquisition of New Motor Skills," in *Recent Advances in Clin Neurophysiology*, ed. J. Kinura and H. Shibasaki (Amsterdam: Elsevier Science, 1996), pp. 304–8.
17. C. Xerri et al., "Alterations of the Cortical Representation of the Rat Ventrum Induced by Nursing Behavior," *J Nsci* 14 (1994): 171; B. Draganski et al., "Neuroplasticity: Changes in Grey Matter Induced by Training," *Nat* 427 (2004): 311.
18. J. Altman and G. Das, "Autoradiographic and Histological Evidence of Postnatal Hippocampal Neurogenesis in Rats," *J Comp Neurol 124* (1965): 319.
19. M. Kaplan, "Environmental Complexity Stimulates Visual Cortex Neurogenesis: Death of a Dogma and a Research Career," *TINS* 24 (2001): 617.
20. S. Goldman and F. Nottebohm, "Neuronal Production, Migration, and Differentiation in a Vocal Control Nucleus of the Adult Female Canary Brain," *PNAS* 80 (1983): 2390; J. Paton and F. Nottebohm, "Neurons Generated in the Adult Brain Are Recruited into Functional Circuits," *Sci* 225 (1984): 4666; F. Nottebohm, "Neuronal Replacement in Adult Brain," *ANYAS* 457 (1985): 143.

For a great history of the entire neurogenesis saga, see M. Specter, "How the Songs of Canaries Upset a Fundamental Principle of Science," *New Yorker*, July 23, 2001.

21. D. Kornack and P. Rakic, "Continuation of Neurogenesis in the Hippocampus of the Adult Macaque Monkey," *PNAS* 96 (1999): 5768.
22. G. Ming and H. Song, "Adult Neurogenesis in the Mammalian Central Nervous System," *Ann Rev Nsci* 28 (2005): 223. Rate of neuron replacement in the hippocampus: G. Kempermann et al., "More Hippocampal Neurons in Adult Mice Living in an Enriched Environment," *Nat* 386 (1997): 493; H. Cameron and R. McKay, "Adult Neurogenesis Produces a Large Pool of New Granule Cells in the Dentate Gyrus," *J Comp Neurol* 435 (2001): 406. Demonstration in humans: P. Eriksson et al., "Neurogenesis in the Adult Human Hippocampus," *Nat Med* 4 (1998): 1313. Modulators of neurogenesis: C. Mirescu et al., "Sleep Deprivation Inhibits Adult Neurogenesis in the Hippocampus by Elevating Glucocorticoids," *PNAS* 103 (2006): 19170. The role of new neurons in cognition: W. Deng et al., "New Neurons and New Memories: How Does Adult Hippocampal Neurogenesis Affect Learning and Memory?" *Nat Rev Nsci* 11 (2010): 339; T. Shors et al., "Neurogenesis in the Adult Rat Is Involved in the Formation of Trace Memories," *Nat* 410 (2001): 372; T. Shors et al., "Neurogenesis May Relate to Some But Not All Types of Hippocampal-Dependent Learning," *Hippocampus* 12 (2002): 578.
23. Footnote regarding running, glucocorticoids and neurogenesis: S. Droste et al., "Effects of Long-Term Voluntary Exercise on the Mouse Hypothalamic-Pituitary-Adrenocortical Axis," *Endo* 144 (2003): 3012; H. van Praag et al., "Running Enhances Neurogenesis, Learning, and Long-Term Potentiation in Mice," *PNAS* 96 (1999): 13427; G. Kempermann, "New Neurons for 'Survival of the Fittest,'" *Nat Rev Nsci* 13 (2012): 727.
24. L. Santarelli et al., "Requirement of Hippocampal Neurogenesis for the Behavioral Effects of Antidepressants," *Sci* 301 (2003): 80.
25. J. Altmann, "The Discovery of Adult Mammalian Neurogenesis," in *Neurogenesis in the Adult Brain I*, ed. T. Seki, K. Sawamoto, J. Parent, and A. Alvarez-Buylla (New York: Springer-Verlag, 2011).
26. C. Lord et al., "Hippocampal Volumes Are Larger in Postmenopausal Women Using Estrogen Therapy Compared to Past Users, Never Users and Men: A Possible Window of Opportunity Effect," *Neurobiol of Aging* 29 (2008): 95; R. Sapolsky, "Glucocorticoids and Hippocampal Atrophy in Neuropsychiatric Disorders," AGP 57 (2000): 925; A. Mutso et al., "Abnormalities in Hippocampal Functioning with Persistent Pain," *J Nsci* 32 (2012): 5747; J. Pruessner et al., "Stress Regulation in the Central Nervous System: Evidence from Structural and Functional Neuroimaging Studies in Human Populations," *PNE* 35 (2010): 179; J. Kuo et al., "Amygdala Volume in Combat-Exposed Veterans With and Without Posttraumatic Stress Disorder: A Cross-sectional Study," *AGP* 69 (2012): 1080.
27. E. Maguire et al., "Navigation-Related Structural Change in the Hippocampi of Taxi Drivers," *PNAS* 97 (2000): 4398; K. Woollett and E. Maguire, "Acquiring "the Knowledge" of London's Layout Drives Structural Brain Changes," *Curr Biol* 21 (2011): 2109. For an interesting discussion of why you need a bigger hippocampus to become a cab driver in London, revolving around the notoriously difficult licensing exam, see J. Rosen, "The Knowledge, London's Legendary Taxi-Driver Test, Puts Up a Fight in the Age of GPS," *New York Times Magazine*, November 10, 2014.
28. S. Mangiavacchi et al., "Long-Term Behavioral and Neurochemical Effects of Chronic Stress Exposure in Rats," *J Neurochemistry* 79 (2001): 1113; J. van Honk et al., "Baseline Salivary Cortisol Levels and Preconscious Selective Attention for Treat: A Pilot Study," *PNE* 23 (1998): 741; M. Fuxjager et al., "Winning Territorial Disputes Selectively Enhances Androgen Sensitivity in Neural Pathways Related to Motivation and Social Aggression," *PNAS* 107 (2010): 12393; I. McKenzie et al., "Motor Skill Learning Requires Active Central Myelination," *Sci* 346 (2014): 318; M. Bechler and C. ffrench-Constant, "A New Wrap for Neuronal Activity?" *Sci* 344 (2014): 480; E. Gibson et al., "Neuronal Activity Promotes Oligodendrogenesis and Adaptive Myelination in the Mammalian Brain," *Sci* 344 (2014): 487; J. Radley et al., "Reversibility of Apical Dendritic Retraction in the Rat Medial Prefrontal Cortex Following Repeated Stress," *Exp Neurol* 196 (2005): 199; E. Bloss et al., "Interactive Effects of Stress and Aging on Structural Plasticity in the Prefrontal Cortex," *J Nsci* 30 (2010): 6726.
29. N. Doidge, *The Brain That Changes Itself: Stories of Personal Triumph from the Front of Brain Science* (New York: Penguin, 2007); S. Begley, *Train Your Mind, Change Your Brain: How a New Science Reveals Our Extraordinary Potential to Transform Ourselves* (New York: Ballantine Books, 2007); J. Arden, *Rewire Your Brain: Think Your Way to a Better Life* (New York: Wiley, 2010).

Chapter 6: Adolescence; or, Dude, Where's My Frontal Cortex?

1. R. Knickmeyer et al., "A Structural MRI Study of Human Brain Development from Birth to 2 Years," *J Nsci* 28 (2008): 12176.
2. M. Bucholtz, "Youth and Cultural Practice," *Ann Rev Anthropology* 31 (2002): 525; S. Choudhury, "Culturing the Adolescent Brain: What Can Neuroscience Learn from Anthropology?" *SCAN* 5 (2010): 159. Footnote: T. James, "The Age of Majority," *Am J Legal History* 4 (1960): 22; R. Brett, "Contribution for Children and Political Violence," in *Child Soldiering: Questions and Challenges for Health Professionals* (WHO Global Report on Violence), 2000, p. 1; C. MacMullin and M. Loughry, "Investigating Psychosocial Adjustment of Former Child Soldiers in Sierra Leone and Uganda," *J Refugee Studies* 17 (2004): 472.
3. J. Giedd, "The Teen Brain: Insights from Neuroimaging," *J Adolescent Health* 42 (2008): 335. Demonstration of increased intrinsic connectivity of PFC neurons during adolescence in monkeys: X. Zhou et al., "Age-Dependent Changes in Prefrontal Intrinsic Connectivity," *PNAS* 111 (2014): 3853; T. Singer, "The Neuronal Basis and Ontogeny of Empathy and Mind Reading: Review of Literature and Implications for Future Research," *Nsci Biobehav Rev* 30 (2006): 855; P. Shaw et al., "Intellectual Ability and Cortical Development in Children and Adolescents," *Nat* 440 (2006): 676.

4. D. Yurelun-Todd, "Emotional and Cognitive Changes During Adolescence," *Curr Opinion in Neurobiol* 17 (2007): 251; B. Luna et al., "Maturation of Widely Distributed Brain Function Subserves Cognitive Development," *Neuroimage* 13 (2001): 786; B. Schlaggar et al., "Functional Neuroanatomical Differences Between Adults and School-Age Children in the Processing of Single Words," *Sci* 296 (2002): 1476.
5. A. Wang et al., "Developmental Changes in the Neural Basis of Interpreting Communicative Intent," *SCAN* 1 (2006): 107.
6. T. Paus et al., "Maturation of White Matter in the Human Brain: A Review of Magnetic Resonance Studies," *Brain Res Bull* 54 (2001): 255; A. Raznahan et al., "Patterns of Coordinated Anatomical Change in Human Cortical Development: A Longitudinal Neuroimaging Study of Maturational Coupling," *Neuron* 72 (2011): 873; N. Strang et al., "Developmental Changes in Adolescents' Neural Response to Challenge," *Developmental Cog Nsci* 1 (2011): 560.
7. C. Masten et al., "Neural Correlates of Social Exclusion During Adolescence: Understanding the Distress of Peer Rejection," *SCAN* (2009): 143.
8. J. Perrin et al., "Growth of White Matter in the Adolescent Brain: Role of Testosterone and Androgen Receptor," *J Nsci* 28 (2008): 9519; T. Paus et al., "Sexual Dimorphism in the Adolescent Brain: Role of Testosterone and Androgen Receptor in Global and Local Volumes of Grey and White Matter," *Horm Behav* 57 (2010): 63; A. Arnsten and R. Shansky, "Adolescence: Vulnerable Period for Stress-Induced PFC Function?" *ANYAS* 102 (2006): 143; W. Moore et al., "Facing Puberty: Associations Between Pubertal Development and Neural Responses to Affective Facial Displays," *SCAN* 7 (2012): 35; R. Dahl, "Adolescent Brain Development: A Period of Vulnerabilities and Opportunities," *ANYAS* 1021 (2004): 1
9. R. Rosenfield, "Clinical Review: Adolescent Anovulation: Maturational Mechanisms and Implications," *J Clin Endo and Metabolism* 98 (2013): 3572.
10. D. Yurelun-Todd, "Emotional and Cognitive Changes During Adolescence," *Curr Opinion in Neurobiol* 17 (2007): 251; B. Schlaggar et al., "Functional Neuroanatomical Differences Between Adults and School-Age Children in the Processing of Single Words," *Sci* 296 (2002): 1476.
11. W. Moore et al., "Facing Puberty: Associations Between Pubertal Development and Neural Responses to Affective Facial Displays," *SCAN* 7 (2012): 35.
12. D. Gee et al., "A Developmental Shift from Positive to Negative Connectivity in Human Amygdala-Prefrontal Circuitry," *J Nsci* 33 (2013): 4584.
13. K. McRae et al., "Association Between Trait Emotional Awareness and Dorsal Anterior Cingulate Activity During Emotion Is Arousal-Dependent," *Neuroimage* 41 (2008): 648; W. Killgore et al., "Sex-Specific Developmental Changes in Amygdala Responses to Affective Faces," *Neuroreport* 12 (2001): 427; W. Killgore and D. Yurgelun-Todd, "Unconscious Processing of Facial Affect in Children and Adolescents," *Soc Nsci* 2 (2007): 28; T. Hare et al., "Biological Substrates of Emotional Reactivity and Regulation in Adolescence During an Emotional Go-Nogo Task," *BP* 63 (2008): 927; T. Wager et al., "Prefrontal-Subcortical Pathways Mediating Successful Emotion Regulation," *Neuron* 25 (2008): 1037; T. Hare et al., "Self-Control in Decision-Making Involves Modulation of the vmPFC Valuation System," *Sci* 324 (2009): 646; C. Masten et al., "Neural Correlates of Social Exclusion During Adolescence: Understanding the Distress of Peer Rejection," *SCAN* 4 (2009): 143.; Footnote: Shulman et al., "Sex Differences in the Developmental Trajectories of Impulse Control and Sensation-Seeking from Early Adolescence to Early Adulthood," *J Youth and Adolescence* 44 (2013): 1
14. G. Laviola et al., "Risk-Taking Behavior in Adolescent Mice: Psychobiological Determinants and Early Epigenetic Influence," *Nsci Biobehav Rev* 27 (2003): 19; V. Reyna and F. Farley, "Risk and Rationality in Adolescent Decision Making: Implications for Theory, Practice, and Public Policy," *Psych Sci in the Public Interest* 7 (2006): 1; L. Steinberg, "Risk Taking in Adolescence: New Perspectives from Brain and Behavioral Science," *Curr Dir Psych Res* 16 (2007): 55; L. Steinberg, *Age of Opportunity: Lessons from the New Science of Adolescence* (New York: Houghton Mifflin, 2014); C. Moutsiana et al., "Human Development of the Ability to Learn from Bad News," *PNAS* 110 (2013): 16396.
15. Reviewed in A. R. Smith et al., "The Role of the Anterior Insula in Adolescent Decision Making," *Developmental Nsci* 36 (2014): 196.
16. Footnote: Shulman et al., "Sex Differences in the Developmental Trajectories of Impulse Control and Sensation-Seeking from Early Adolescence to Early Adulthood," *J Youth and Adolescence* 44 (2013): 1.
17. R. Sapolsky, "Open Season," *New Yorker*, March 30, 1998, p. 57.
18. D. Rosenberg and D. Lewis, "Changes in the Dopaminergic Innervation of Monkey Prefrontal Cortex During Late Postnatal Development: A Tyrosine Hydroxylase Immunohistochemical Study," *BP* 36 (1994): 272.
19. B. Knutson et al., "FMRI Visualization of Brain Activity During a Monetary Incentive Delay Task," *Neuroimage* 12 (2000): 20; E. Barkley-Levenson and A. Galvan, "Neural Representation of Expected Value in the Adolescent Brain," *PNAS* 111 (2014): 1646; S. Schneider et al., "Risk Taking and the Adolescent Reward System: A Potential Common Link to Substance Abuse," *Am J Psychiatry* 169 (2012): 39; S. Burnett et al., "Development During Adolescence of the Neural Processing of Social Emotion," *J Cog Nsci* 21 (2008): 1; J. Bjork et al., "Developmental Differences in Posterior Mesofrontal Cortex Recruitment by Risky Rewards," *J Nsci* 27 (2007): 4839; J. Bjork et al., "Incentive-Elicited Brain Activation in Adolescents: Similarities and Differences from Young Adults," *J Nsci* 25 (2004): 1793; S. Blakemore et al., "Adolescent Development of the Neural Circuitry for Thinking About Intentions," *SCAN* 2 (2007): 130.
20. A. Galvan et al., "Earlier Development of the Accumbens Relative to Orbitofrontal Cortex Might Underlie Risk-Taking Behavior in Adolescents," *J Nsci* 26 (2006): 6885 (this is also the source of the figure in the text). A demonstration of dopaminergic response to different reward sizes as more linear and accurate in adults: J. Vaidya et al., "Neural Sensitivity to Absolute and Relative Anticipated Reward in Adolescents," *PLoS ONE* 8 (2013): e58708.

21. A. R. Smith et al., "Age Differences in the Impact of Peers on Adolescents' and Adults' Neural Response to Reward," *Developmental Cog Nsci* 11 (2015): 75; J. Chein et al., "Peers Increase Adolescent Risk Taking by Enhancing Activity in the Brain's Reward Circuitry," *Developmental Sci* 14 (2011): F1; M. Gardner and L. Steinberg, "Peer Influence on Risk Taking, Risk Preference, and Risky Decision Making in Adolescence and Adulthood: An Experimental Study," *Developmental Psych* 41 (2005): 625; L. Steinberg, "A Social Neuroscience Perspective on Adolescent Risk-Taking," *Developmental Rev* 28 (2008): 78; M. Grosbras et al., "Neural Mechanisms of Resistance to Peer Influence in Early Adolescence," *J Nsci* 27 (2007): 8040; A. Weigard et al., "Effects of Anonymous Peer Observation on Adolescents' Preference for Immediate Rewards," *Developmental Science* 17 (2014): 71.
22. M. Madden et al., "Teens, Social Media, and Privacy," Pew Research Center, May 23, 2013, www.pewinternet.org/Reports/2013/Teens-Social-Media-And-Privacy/Summary-of-Findings.aspx.
23. A. Guyer et al., "Amygdala and Ventrolateral Prefrontal Cortex Function During Anticipated Peer Evaluation in Pediatric Social Anxiety," *AGP* 65 (2008): 1303; A. Guyer et al., "Probing the Neural Correlates of Anticipated Peer Evaluation in Adolescence," *Child Development* 80 (2009): 1000; B. Gunther Moor et al., "Do You Like Me? Neural Correlates of Social Evaluation and Developmental Trajectories," *Soc Nsci* 5 (2010): 461.
24. N. Eisenberger et al., "Does Rejection Hurt? An fMRI Study of Social Exclusion," *Sci* 302 (2003): 290; N. Eisenberger, "The Pain of Social Disconnection: Examining the Shared Neural Underpinnings of Physical and Social Pain," *Nat Rev Nsci* 3 (2012): 421.
25. C. Sebastian et al., "Development Influences on the Neural Bases of Responses to Social Rejection: Implications of Social Neuroscience for Education," *NeuroImage* 57 (2011): 686; C. Masten et al., "Neural Correlates of Social Exclusion During Adolescence: Understanding the Distress of Peer Rejection," *SCAN* 4 (2009): 143; J. Pfeifer and S. Blakemore, "Adolescent Social Cognitive and Affective Neuroscience: Past, Present, and Future," *SCAN* 7 (2012): 1.
26. J. Pfeifer et al., "Entering Adolescence: Resistance to Peer Influence, Risky Behavior, and Neural Changes in Emotion Reactivity," *Neuron* 69 (2011): 1029; L. Steinberg and K. Monahan, "Age Differences in Resistance to Peer Influence," *Developmental Psych* 43 (2007): 1531; M. Grosbras et al., "Neural Mechanisms of Resistance to Peer Influence in Early Adolescence," *J Nsci* 27 (2007): 8040.
27. I. Almas et al., "Fairness and the Development of Inequality Acceptance," *Sci* 328 (2010): 1176.
28. J. Decety and K. Michalska, "Neurodevelopmental Changes in the Circuits Underlying Empathy and Sympathy from Childhood to Adulthood," *Developmental Sci* 13 (2010): 886.
29. N. Eisenberg et al., "The Relations of Emotionality and Regulation to Dispositional and Situational Empathy-Related Responding," *JPSP* 66 (1994): 776; J. Decety et al., "The Developmental Neuroscience of Moral Sensitivity," *Emotion Rev* 3 (2011): 305.
30. E. Finger et al., "Disrupted Reinforcement Signaling in the Orbitofrontal Cortex and Caudate in Youths with Conduct Disorder or Oppositional Defiant Disorder and a High Level of Psychopathic Traits," *Am J Psychiatry* 168 (2011): 152; A. Marsh et al., "Reduced Amygdala-Orbitofrontal Connectivity During Moral Judgments in Youths with Disruptive Behavior Disorders and Psychopathic Traits," *Psychiatry Res* 194 (2011): 279.
31. L. Steinberg, "The Influence of Neuroscience on US Supreme Court Decisions About Adolescents' Criminal Culpability," *Nat Rev Nsci* 14 (2013): 513.
32. Roper v. Simmons, 543 U.S. 551 (2005).
33. J. Sallet et al, "Social Network Size Affects Neural Circuits in Macaques," *Sci* 334 (2011): 697.

Chapter 7: Back to the Crib, Back to the Womb

1. P. Yakovlev and A. Lecours, "The Myelogenetic Cycles of Regional Maturation of the Brain," in *Regional Development of the Brain in Early Life*, ed. A. Minkowski (Oxford: Blackwell, 1967); H. Kinney et al., "Sequence of Central Nervous System Myelination in Human Infancy: II. Patterns of Myelination in Autopsied Infants," *J Neuropathology & Exp Neurol* 47 (1988): 217; S. Deoni et al., "Mapping Infant Brain Myelination with MRI," *J Nsci* 31 (2011): 784; N. Baumann and D. Pham-Dinh, "Biology of Oligodendrocyte and Myelin in the Mammalian CNS," *Physiological Rev* 81 (2001): 871.
2. Demonstration of the predictive power of degree of connectivity: N. Dosenbach et al., "Prediction of Individual Brain Maturity Using fMRI," *Sci* 329 (2010): 1358.
3. N. Uesaka et al., "Retrograde Semaphorin Signaling Regulates Synapse Elimination in the Developing Mouse Brain," *Sci* 344 (2014): 1020; R. C. Paolicelli et al., "Synaptic Pruning by Microglia Is Necessary for Normal Brain Development," *Sci* 333 (2011): 1456; R. Buss et al., "Adaptive Roles of Programmed Cell Death During Nervous System Development," *Ann Rev of Nsci* 29 (2006): 1; D. Nijhawan et al., "Apoptosis in Neural Development and Disease," *Ann Rev of Nsci* 23 (2000): 73; C. Kuan et al., "Mechanisms of Programmed Cell Death in the Developing Brain," *TINS* 23 (2000): 291.
4. J. Piaget, *Main Trends in Psychology* (London: George Allen & Unwin, 1973); J. Piaget, *The Language and Thought of the Child* (New York: Psychology Press, 1979).
5. Other realms of stage development: R. Selman et al., "Interpersonal Awareness in Children: Toward an Integration of Developmental and Clinical Child Psychology," *Am J Orthopsychiatry* 47 (1977): 264; T. Singer, "The Neuronal Basis and Ontogeny of Empathy and Mind Reading: Review of Literature and Implications for Future Research," *Nsci Biobehav Rev* 30 (2006): 855.
6. S. Baron-Cohen, "Precursors to a Theory of Mind: Understanding Attention in Others," in *Natural Theories of Mind: Evolution, Development and Simulation of Everyday Mindreading*, ed. A. Whiten (Oxford: Basil Blackwell, 1991); J. Topal et al., "Differential Sensitivity to Human Communication in Dogs, Wolves, and Human Infants," *Sci* 325 (2009): 1269; G. Lakatos et al., "A Comparative Approach to Dogs' (*Canis familiaris*) and Human

Infants' Comprehension of Various Forms of Pointing Gestures," *Animal Cog* 12 (2009): 621 J. Kaminski et al., "Domestic Dogs are Sensitive to a Human's Perspective," *Behaviour* 146 (2009): 979.

7. S. Baron-Cohen et al., "Does the Autistic Child Have a 'Theory of Mind'?" *Cog* 21 (2985): 37.
8. L. Young et al., "Disruption of the Right Temporal Lobe Function with TMS Reduces the Role of Beliefs in Moral Judgments," *PNAS* 107 (2009): 6753; Y. Moriguchi et al., "Changes of Brain Activity in the Neural Substrates for Theory of Mind During Childhood and Adolescence," *Psychiatry and Clin Nsci* 61 (2007): 355; A. Saitovitch et al., "Social Cognition and the Superior Temporal Sulcus: Implications in Autism," *Rev of Neurol* (Paris) 168 (2012): 762; P. Shaw et al., "The Impact of Early and Late Damage to the Human Amygdala on 'Theory of Mind' Reasoning," *Brain* 127 (2004): 1535.
9. B. Sodian and S. Kristen, "Theory of Mind During Infancy and Early Childhood Across Cultures, Development of," *Int Encyclopedia of the Soc & Behav Sci* (Amsterdam: Elsevier, 2015), p. 268.
10. S. Nichols, "Experimental Philosophy and the Problem of Free Will," *Sci* 331 (2011): 1401.
11. D. Premack and G. Woodruff, "Does the Chimpanzee Have a Theory of Mind?" *BBS* 1 (1978): 515. Evidence against: D. Povinelli and J. Vonk, "Chimpanzee Minds: Suspiciously Human?" *TICS* 7 (2003): 157. Evidence for: B. Hare et al., "Do Chimpanzees Know What Conspecifics Know and Do Not Know?" *Animal Behav* 61 (2001): 139. Footnote: L. Santo Let al., "Rhesus Monkeys (*Macaca mulatta*) Know What Others Can and Cannot Hear," *Animal Behav* 71 (2006): 1175.
12. J. Decety et al., "The Contribution of Emotion and Cognition to Moral Sensitivity: A Neurodevelopmental Study," *Cerebral Cortex* 22 (2011): 209.
13. J. Decety et al., "Who Caused the Pain? An fMRI Investigation of Empathy and Intentionality in Children," *Neuropsychologia* 46 (2008): 2607; J. Decety et al., "The Contribution of Emotion and Cognition to Moral Sensitivity: A Neurodevelopmental Study," *Cerebral Cortex* 22 (2012): 209; J. Decety and K. Michalska, "Neurodevelopmental Changes in the Circuits Underlying Empathy and Sympathy from Childhood to Adulthood," *Developmental Sci* 13 (2010): 886.
14. J. Decety et al., "The Contribution of Emotion and Cognition to Moral Sensitivity: A Neurodevelopmental Study," *Cerebral Cortex* 22 (2012): 209; N. Eisenberg et al., "The Relations of Emotionality and Regulation to Dispositional and Situational Empathy-Related Responding," *JPSP* 66 (1994): 776.
15. P. Blake et al., "The Ontogeny of Fairness in Seven Societies," *Nat* 528 (2016): 258.
16. I. Almas et al., "Fairness and the Development of Inequality Acceptance," *Sci* 328 (2010): 1176; E. Fehr et al., "Egalitarianism in Young Children," *Nat* 454 (2008): 1079; K. Olson et al., "Children's Responses to Group-Based Inequalities: Perpetuation and Rectification," *Soc Cog* 29 (2011): 270; M. Killen, "Children's Social and Moral Reasoning About Exclusion," *Curr Dir Psych Sci* 16 (2007): 32.
17. D. Garz, *Lawrence Kohlberg: An Introduction* (Cologne, Germany: Barbara Budrich, 2009).
18. C. Gilligan, *In a Different Voice: Psychological Theory and Women's Development* (Cambridge, MA: Harvard University Press, 1982).
19. N. Eisenberg, "Emotion, Regulation, and Moral Development," *Ann Rev of Psych* 51 (2000): 665; J. Hamlin et al., "Social Evaluation by Preverbal Infants," *Nat* 450 (2007): 557; M. Hoffman, *Empathy and Moral Development: Implications for Caring and Justice* (Cambridge: Cambridge University Press, 2001).
20. W. Mischel et al., "Cognitive and Attentional Mechanisms in Delay of Gratification," *JPSP* 21 (1972): 204; W. Mischel, *The Marshmallow Test: Understanding Self-Control and How to Master It* (New York: Bantam Books, 2014); K. McRae et al., "The Development of Emotion Regulation: An fMRI Study of Cognitive Reappraisal in Children, Adolescents and Young Adults," *SCAN* 7 (2012): 11; H. Palmeri and R. N. Aslin, "Rational Snacking: Young Children's Decision-Making on the Marshmallow Task is Moderated by Beliefs About Environmental Reliability," *Cog* 126 (2013): 109.
21. B. J. Casey et al., "From the Cover: Behavioral and Neural Correlates of Delay of Gratification 40 Years Later," *PNAS* 108 (2011): 14998; N. Eisenberg et al., "Contemporaneous and Longitudinal Prediction of Children's Social Functioning from Regulation and Emotionality," *Child Development* 68 (1997): 642; N. Eisenberg et al., "The Relations of Regulation and Emotionality to Resiliency and Competent Social Functioning in Elementary School Children," *Child Development* 68 (1997): 295.
22. L. Holt, *The Care and Feeding of Children* (NY: Appleton-Century, 1894). This book went through fifteen editions between 1894 and 1915.
23. For a history of hospitalism, see R. Sapolsky, "How the Other Half Heals," *Discover*, April 1998, p. 46.
24. J. Bowlby *Attachment and Loss*, vol. 1, *Attachment* (New York: Basic Books, 1969); J. Bowlby, *Attachment and Loss*, vol. 2, *Separation* (London: Hogarth Press, 1973); J. Bowlby, *Attachment and Loss*, vol. 3, *Loss: Sadness & Depression* (London: Hogarth Press, 1980).
25. D. Blum, *Love at Goon Park: Harry Harlow and the Science of Affection* (New York: Perseus, 2002). This is the source of the Harlow quote.
26. R. Rosenfeld, "The Case of the Unsolved Crime Decline," *Sci Am*, February 2004, p. 82; J. Donohue III and S. Levitt, "The Impact of Legalized Abortion on Crime," *Quarterly J Economics* 116 (2001): 379. Raine et al., "Birth Complications Combined with Early Maternal Rejection at Age 1 Year Predispose to Violent Crime at Age 18 Years," *AGP* 51 (1994): 984; Footnote: J. Bowlby, "Forty-four Juvenile Thieves: Their Characters and Home-Life," *Int J Psychoanalysis* 25 (1944): 107.
27. G. Barr et al., "Transitions in Infant Learning Are Modulated by Dopamine in the Amygdala," *Nat Nsci* 12 (2009): 1367; R. Sullivan et al., "Good Memories of Bad Events," *Nat* 407 (2000): 38; S. Moriceau et al., "Dual Circuitry for Odor-Shock Conditioning During Infancy: Corticosterone Switches Between Fear and Attraction via Amygdala," *J Nsci* 26 (2006): 6737; R. Sapolsky, "Any Kind of Mother in a Storm," *Nat Nsci* 12 (2009): 1355.

28. R. Sapolsky and M. Meaney, "Maturation of the Adrenocortical Stress Response: Neuroendocrine Control Mechanisms and the Stress Hyporesponsive Period," *Brain Res Rev* 11 (1986): 65.
29. L. M. Renner and K. S. Slack, "Intimate Partner Violence and Child Maltreatment: Understanding Intra- and Intergenerational Connections," *Child Abuse & Neglect* 30 (2006): 599.
30. D. Maestripieri, "Early Experience Affects the Intergenerational Transmission of Infant Abuse in Rhesus Monkeys," *PNAS* 102 (2005): 9726.
31. C. Hammen et al., "Depression and Sensitization to Stressors Among Young Women as a Function of Childhood Adversity," *J Consulting Clin Psych* 68 (2000): 782; E. McCrory et al., "The Link Between Child Abuse and Psychopathology: A Review of Neurobiological and Genetic Research," *J the Royal Soc of Med* 105 (2012): 151; K. Lalor and R. McElvaney, "Child Sexual Abuse, Links to Later Sexual Exploitation/High-Risk Sexual Behavior, and Prevention/Treatment Programs," *Trauma Violence & Abuse* 11 (2010): 159; Y. Dvir et al., "Childhood Maltreatment, Emotional Dysregulation, and Psychiatric Comorbidities," *Harvard Rev of Psychiatry* 22 (2014): 149; E. Mezzacappa et al., "Child Abuse and Performance Task Assessments of Executive Functions in Boys," *J Child Psych and Psychiatry* 42 (2001): 1041; M. Wichers et al., "Transition from Stress Sensitivity to a Depressive State: Longitudinal Twin Study," *Brit J Psychiatry* 195 (2009): 498.
32. C. Heim et al., "Pituitary-Adrenal and Autonomic Responses to Stress in Women After Sexual and Physical Abuse in Childhood," *JAMA* 284 (2000): 592; E. Binder et al., "Association of FKBP5 Polymorphisms and Childhood Abuse with Risk of Posttraumatic Stress Disorder Symptoms in Adults," *JAMA* 299 (2008): 1291; C. Heim et al., "The Dexamethasone/Corticotropin-Releasing Factor Test in Men with Major Depression: Role of Childhood Trauma," *BP* 63 (2008): 398; R. Lee et al., "Childhood Trauma and Personality Disorder: Positive Correlation with Adult CSF Corticotropin-Releasing Factor Concentrations," *Am J Psychiatry* 162 (2005): 995; R. J. Lee et al., "CSF Corticotropin-Releasing Factor in Personality Disorder: Relationship with Self-Reported Parental Care," *Neuropsychopharmacology* 31: (2006): 2289; L. Carpenter et al., "Cerebrospinal Fluid Corticotropin-Releasing Factor and Perceived Early-Life Stress in Depressed Patients and Healthy Control Subjects," *Neuropsychopharmacology* 29 (2004): 777; T. Rinne et al., "Hyperresponsiveness of Hypothalamic-Pituitary-Adrenal Axis to Combined Dexamethasone/Corticotropin-Releasing Hormone Challenge in Female Borderline Personality Disorder Subjects with a History of Sustained Childhood Abuse," *BP* 52 (2002): 1102; P. McGowan et al., "Epigenetic Regulation of the Glucocorticoid Receptor in Human Brain Associates with Childhood Abuse," *Nat Nsci* 12 (2009): 342; M. Toth et al., "Post-weaning Social Isolation Induces Abnormal Forms of Aggression in Conjunction with Increased Glucocorticoid and Autonomic Stress Responses," *Horm Behav* 60 (2011): 28.
33. S. Lupien et al., "Effects of Stress Throughout the Lifespan on the Brain, Behaviour and Cognition," *Nat Rev Nsci* 10 (2009): 434; V. Carrion et al., "Stress Predicts Brain Changes in Children: A Pilot Longitudinal Study on Youth Stress, Posttraumatic Stress Disorder, and the Hippocampus," *Pediatrics* 119 (2007): 509; F. L. Woon and D. W. Hedges, "Hippocampal and Amygdala Volumes in Children and Adults with Childhood Maltreatment–Related Posttraumatic Stress Disorder: A Meta-analysis," *Hippocampus* 18 (2008): 729.
34. S. J. Lupien et al., "Effects of Stress Throughout the Lifespan on the Brain, Behaviour and Cognition," *Nat Rev Nsci* 10 (2009): 434; D. Hackman et al., "Socioeconomic Status and the Brain: Mechanistic Insights from Human and Animal Research," *Nat Rev Nsci* 11 (2010): 651; M. Sheridan et al., "The Impact of Social Disparity on Prefrontal Function in Childhood," *PLoS ONE* 7 (2012): e35744; J. L. Hanson et al., "Structural Variations in Prefrontal Cortex Mediate the Relationship Between Early Childhood Stress and Spatial Working Memory," *J Nsci* 32 (2012): 7917; M. Sweitzer et al., "Polymorphic Variation in the Dopamine D4 Receptor Predicts Delay Discounting as a Function of Childhood Socioeconomic Status: Evidence for Differential Susceptibility," *SCAN* 8 (2013): 499; E. Tucker-Drob et al., "Emergence of a Gene X Socioeconomic Status Interaction on Infant Mental Ability Between 10 Months and 2 Years," *Psych Sci* 22 (2011): 125; I. Liberzon et al., "Childhood Poverty Alters Emotional Regulation in Adulthood," *SCAN* 10 (2015): 1596; K. G. Noble et al., "Family Income, Parental Education and Brain Structure in Children and Adolescents," *Nat Nsci* 18 (2015): 773.
35. Footnote: R. Nevin, "Understanding International Crime Trends: The Legacy of Preschool Lead Exposure," *Environmental Res* 104 (2007): 315.
36. Reviewed in R. Sapolsky, *Why Zebras Don't Get Ulcers: A Guide to Stress, Stress-Related Diseases and Coping*, 3rd ed. (New York: Holt, 2004). Baboon equivalent: P. O. Onyango et al., "Persistence of Maternal Effects in Baboons: Mother's Dominance Rank at Son's Conception Predicts Stress Hormone Levels in Subadult Males," *Horm Behav* 54 (2008): 319.
37. F. L. Woon and D. W. Hedges, "Hippocampal and Amygdala Volumes in Children and Adults with Childhood Maltreatment–Related Posttraumatic Stress Disorder: A Meta-analysis," *Hippocampus* 18 (2008): 729; D. Gee et al., "Early Developmental Emergence of Human Amygdala-PFC Connectivity After Maternal Deprivation," *PNAS* 110 (2013): 15638; A. K. Olsavsky et al., "Indiscriminate Amygdala Response to Mothers and Strangers After Early Maternal Deprivation," *BP* 74 (2013): 853.
38. L. M. Oswald et al., "History of Childhood Adversity Is Positively Associated with Ventral Striatal Dopamine Responses to Amphetamine," *Psychopharmacology* (Berlin) 23 (2014): 2417; E. Hensleigh and L. M. Pritchard, "Maternal Separation Increases Methamphetamine-Induced Damage in the Striatum in Male, But Not Female Rats," *BBS* 295 (2014): 3; A. N. Karkhanis et al., "Social Isolation Rearing Increases Nucleus Accumbens Dopamine and Norepinephrine Responses to Acute Ethanol in Adulthood," *Alcohol: Clin Exp Res* 38 (2014): 2770.
39. C. Anacker et al., "Early Life Adversity and the Epigenetic Programming of Hypothalamic-Pituitary-Adrenal Function," *Dialogues in Clin Nsci* 16 (2014): 321.
40. S. L. Buka et al., "Youth Exposure to Violence: Prevalence, Risks, and Consequences," *Am J Orthopsychiatry* 71

(2001): 298; M. B. Selner-O'Hagan et al., "Assessing Exposure to Violence in Urban Youth," *J Child Psych and Psychiatry* 39 (1998): 215; P. T. Sharkey et al., "The Effect of Local Violence on Children's Attention and Impulse Control," *Am J Public Health* 102 (2012): 2287; J. B. Bingenheimer et al., "Firearm Violence Exposure and Serious Violent Behavior," *Sci* 308 (2005): 1323. Footnote: I. Shaley et al., "Exposure to Violence During Childhood Is Associated with Telomere Erosion from 5 to 10 Years of Age: A Longitudinal Study," *Mol Psychiatry* 18 (2013): 576.

41. For a particularly good review, see L. Huesmann and L. Taylor, "The Role of Media Violence in Violent Behavior," *Ann Rev of Public Health* 27 (2006): 393. See also J. D. Johnson et al., "Differential Gender Effects of Exposure to Rap Music on African American Adolescents' Acceptance of Teen Dating Violence," *Sex Roles* 33 (1995): 597; J. Johnson et al., "Television Viewing and Aggressive Behavior During Adolescence and Adulthood," *Sci* 295 (2002): 2468; J. Savage and C. Yancey, "The Effects of Media Violence Exposure on Criminal Aggression: A Meta-analysis," *Criminal Justice and Behav* 35 (2008): 772; C. Anderson et al., "Violent Video Game Effects on Aggression, Empathy, and Prosocial Behavior in Eastern and Western Countries: A Meta-analytic Review," *Psych Bull* 136, 151; C. J. Ferguson, "Evidence for Publication Bias in Video Game Violence Effects Literature: A Meta-analytic Review," *Aggression and Violent Behavior* 12 (2007): 470; C. Ferguson, "The Good, the Bad and the Ugly: A Meta-analytic Review of Positive and Negative Effects of Violent Video Games," *Psychiatric Quarterly* 78 (2007): 309.
42. W. Copeland et al., "Adult Psychiatric Outcomes of Bullying and Being Bullied by Peers in Childhood and Adolescence," *JAMA Psychiatry* 70 (2013): 419; S. Woods and E. White, "The Association Between Bullying Behaviour, Arousal Levels and Behaviour Problems," *J Adolescence* 28 (2005): 381; D. Jolliffe and D. P. Farrington, "Examining the Relationship Between Low Empathy and Bullying," *Aggressive Behav* 32 (2006): 540; G. Gini, "Social Cognition and Moral Cognition in Bullying: What's Wrong?" *Aggressive Behav* 32 (2006): 528; S. Shakoor et al., "A Prospective Longitudinal Study of Children's Theory of Mind and Adolescent Involvement in Bullying," *J Child Psych and Psychiatry* 53 (2012): 254.
43. J. D. Unenever, "Bullies, Aggressive Victims, and Victims: Are They Distinct Groups?" *Aggressive Behav* 31 (2005): 153; D. P. Farrington and M. M. Tofi, "Bullying as a Predictor of Offending, Violence and Later Life Outcomes," *Criminal Behaviour and Mental Health* 21 (2011): 90; M. Tofi et al., "The Predictive Efficiency of School Bullying Versus Later Offending: A Systematic/Meta-analytic Review of Longitudinal Studies," *Criminal Behaviour and Mental Health* 21 (2011): 80; T. R. Nansel et al., "Cross-National Consistency in the Relationship Between Bullying Behaviors and Psychosocial Adjustment," *Arch Pediatrics & Adolescent Med* 158 (2004): 730; J. A. Stein et al., "Adolescent Male Bullies, Victims, and Bully-Victims: A Comparison of Psychosocial and Behavioral Characteristics," *J Pediatric Psych* 32 (2007): 273; P. W. Jansen et al., "Prevalence of Bullying and Victimization Among Children in Early Elementary School: Do Family and School Neighbourhood Socioeconomic Status Matter?" *BMC Public Health* 12 (2012): 494; A. Sourander et al., "What Is the Early Adulthood Outcome of Boys Who Bully or Are Bullied in Childhood? The Finnish 'From a Boy to a Man' Study," *Pediatrics* 120 (August 2007): 397; A. Sourander et al., "Childhood Bullies and Victims and Their Risk of Criminality in Late Adolescence," *Arch Pediatrics & Adolescent Med* 161 (2007): 546; C. Winsper et al., "Involvement in Bullying and Suicide-Related Behavior at 11 Years: A Prospective Birth Cohort Study," *J the Am Academy of Child and Adolescent Psychiatry* 51 (2012): 271; F. Elgar et al., "Income Inequality and School Bullying: Multilevel Study of Adolescents in 37 Countries," *J Adolescent Health* 45 (2009): 351.
44. G. M. Glew et al., "Bullying, Psychosocial Adjustment, and Academic Performance in Elementary School," *Arch Pediatrics & Adolescent Med* 159 (2005): 1026.
45. K. Appleyard et al., "When More Is Not Better: The Role of Cumulative Risk in Child Behavior Outcomes," *J Child Psych and Psychiatry* 46 (2005): 235.
46. M. Sheridan et al., "Variation in Neural Development as a Result of Exposure to Institutionalization Early in Childhood," *PNAS* 109 (2012): 12927; M. Carlson and F. Earis, "Psychological and Neuroendocrinological Sequelae of Early Social Deprivation in Institutionalized Children in Romania," *ANYAS* 15 (1997): 419; N. Tottenham, "Human Amygdala Development in the Absence of Species-Expected Caregiving," *Developmental Psychobiology* 54 (2012): 598; M. A. Mehta et al., "Amygdala, Hippocampal and Corpus Callosum Size Following Severe Early Institutional Deprivation: The English and Romanian Adoptees Study Pilot," *J Child Psych and Psychiatry* 50 (2009): 943; N. Tottenham et al., "Prolonged Institutional Rearing Is Associated with Atypically Large Amygdala Volume and Difficulties in Emotion Regulation," *Developmental Sci* 13 (2010): 46; M. M. Loman et al., "The Effect of Early Deprivation on Executive Attention in Middle Childhood," *J Child Psych and Psychiatry* 54 (2012): 37; T. Eluvathingal et al., "Abnormal Brain Connectivity in Children After Early Severe Socioemotional Deprivation: A Diffusion Tensor Imaging Study," *Pediatrics* 117 (2006): 2093; H. T. Chugani et al., "Local Brain Functional Activity Following Early Deprivation: A Study of Postinstitutionalized Romanian Orphans," *Neuroimage* 14 (2001): 1290.
47. Her idea is nicely summarized in M. Small, *Our Babies, Ourselves* (New York: Anchor Books, 1999).
48. H. Arendt, *The Origins of Totalitarianism* (New York: Harcourt 1951); T. Adorno et al., *The Authoritarian Personality* (New York: Harper & Row, 1950).
49. D. Baumrind, "Child Care Practices Anteceding Three Patterns of Preschool Behavior," *Genetic Psych Monographs* 75 (1967): 43.
50. E. E. Maccoby and J. A. Martin, "Socialization in the Context of the Family: Parent-Child Interaction," in *Handbook of Child Psychology*, ed. P. Mussen (New York: Wiley, 1983).
51. J. R. Harris, *The Nurture Assumption: Why Children Turn Out the Way They Do* (New York: Simon & Schuster, 1998).
52. J. Huizinga, *Homo Ludens: A Study of the Play-Element in Culture* (London: Routledge & Kegan Paul, 1938); A.

Berghänel et al., "Locomotor Play Drives Motor Skill Acquisition at the Expense of Growth: A Life History Trade-off," *Sci Advances* 1 (2015): 1; J. Panksepp and W. W. Beatty, "Social Deprivation and Play in Rats," *Behav and Neural Biol* 39 (1980): 197; M. Bekoff and J. A. Byers, *Animal Play: Evolutionary, Comparative, and Ecological Perspectives* (Cambridge: Cambridge University Press, 1998); M. Spinka et al., "Mammalian Play: Training for the Unexpected," *Quarterly Rev of Biol* 76 (2001): 141.

53. S. M. Pellis, "Sex Differences in Play Fighting Revisited: Traditional and Nontraditional Mechanisms of Sexual Differentiation in Rats," *Arch Sexual Behav* 31 (2002): 17; B. Knutson et al., "Ultrasonic Vocalizations as Indices of Affective States in Rats," *Psych Bull* 128 (2002): 961; Y. Delville et al., "Development of Aggression," in *Biology of Aggression*, ed. R. Nelson (Oxford: Oxford University Press, 2005).
54. J. Tsai, "Ideal Affect: Cultural Causes and Behavioral Consequences," *Perspectives on Psych Sci* 2 (2007): 242; S. Kitayama and A. Uskul, "Culture, Mind, and the Brain: Current Evidence and Future Directions," *Ann Rev of Psych* 62 (2011): 419.
55. C. Kobayashi et al., "Cultural and Linguistic Influence on Neural Bases of 'Theory of Mind': An fMRI Study with Japanese Bilinguals," *Brain and Language* 98 (2006): 210; C. Lewis et al., "Social Influences on False Belief Access: Specific Sibling Influences or General Apprenticeship?" *Child Development* 67 (1996): 2930; J. Perner et al., "Theory of Mind Is Contagious: You Catch It from Your Sibs," *Child Development* 65 (1994): 1228; D. Liu et al., "Theory of Mind Development in Chinese Children: A Meta-analysis of False-Belief Understanding Across Cultures and Languages," *Developmental Psych* 44 (2008): 523.
56. C. Anderson et al., "Violent Video Game Effects on Aggression, Empathy, and Prosocial Behavior in Eastern and Western Countries: A Meta-analytic Review," *Psych Bull* 136 (2010): 151.
57. R. E. Nisbett and D. Cohen, *Culture of Honor: The Psychology of Violence in the South* (Boulder, CO: Westview Press, 1996).
58. A. Kusserow, "De-homogenizing American Individualism: Socializing Hard and Soft Individualism in Manhattan and Queens," *Ethos* 27 (1999): 210.
59. S. Ullal-Gupta et al., "Linking Prenatal Experience to the Emerging Musical Mind," *Front Systems Nsci* 3 (2013): 48.
60. A. DeCasper and W. Fifer, "Of Human Bonding: Newborns Prefer Their Mothers' Voices," *Sci* 6 (1980): 208; A. J. DeCasper and P. A. Prescott, "Human Newborns' Perception of Male Voices: Preference, Discrimination, and Reinforcing Value," *Developmental Psychobiology* 17 (1984): 481; B. Mampe et al., "Newborns' Cry Melody Is Shaped by Their Native Language," *Curr Biol* 19 (2009): 1994; A. DeCasper and M. Spence, "Prenatal Maternal Speech Influences Newborns' Perception of Speech Sounds," *Infant Behav and Development* 9 (1986): 133.
61. J. P. Lecanuet et al., "Fetal Perception and Discrimination of Speech Stimuli: Demonstration by Cardiac Reactivity: Preliminary Results," *Comptes rendus de l'Académie des sciences III* 305 (1987): 161; J. P. Lecanuet et al., "Fetal Discrimination of Low-Pitched Musical Notes," *Developmental Psychobiology* 36 (2000): 29; C. Granier-Deferre et al., "A Melodic Contour Repeatedly Experienced by Human Near-Term Fetuses Elicits a Profound Cardiac Reaction One Month After Birth," *PLoS ONE* 23 (2011): e17304.
62. G. Kolata, "Studying Learning in the Womb," *Sci* 225 (1984): 302; A. J. DeCasper and M. J. Spence, "Prenatal Maternal Speech Influences Newborns' Perception of Speech Sounds," *Infant Behav and Development* 9 (1986): 133.
63. P. Y. Wang et al., "Müllerian Inhibiting Substance Contributes to Sex-Linked Biases in the Brain and Behavior," *PNAS* 106 (2009): 7203; S. Baron-Cohen et al., "Sex Differences in the Brain: Implications for Explaining Autism," *Sci* 310 (2005): 819.
64. R. Goy and B. McEwen, *Sexual Differentiation of the Brain* (Cambridge, MA: MIT Press, 1980).
65. J. Money, "Sex Hormones and Other Variables in Human Eroticism," in *Sex and Internal Secretions*, ed. W. C. Young, 3rd ed. (Baltimore: Williams and Wilkins, 1963), p. 138.
66. G. M. Alexander and M. Hines, "Sex Differences in Response to Children's Toys in Nonhuman Primates (*Cercopithecus aethiops sabaeus*)," *EHB* 23 (2002): 467. (This is the source of the figure in the text). J. M. Hassett et al., "Sex Differences in Rhesus Monkey Toy Preferences Parallel Those of Children," *Horm Behav* 54 (2008): 359.
67. K. Wallen and J. M. Hassett, "Sexual Differentiation of Behavior in Monkeys: Role of Prenatal Hormones," *J Neuroendocrinology* 21 (2009): 421; J. Thornton et al., "Effects of Prenatal Androgens on Rhesus Monkeys: A Model System to Explore the Organizational Hypothesis in Primates," *Horm Behav* 55 (2009): 633.
68. M. Hines, *Brain Gender* (New York: Oxford University Press, 2004); G. A. Mathews et al., "Personality and Congenital Adrenal Hyperplasia: Possible Effects of Prenatal Androgen Exposure," *Horm Behav* 55 (2009): 285; R. W. Dittmann et al., "Congenital Adrenal Hyperplasia. I: Gender-Related Behavior and Attitudes in Female Patients and Sisters," *PNE* 15 (1990): 401; A. Nordenstrom et al., "Sex-Typed Toy Play Correlates with the Degree of Prenatal Androgen Exposure Assessed by CYP21 Genotype in Girls with Congenital Adrenal Hyperplasia," *J Clin Endo and Metabolism* 87 (2002): 5119; V. L. Pasterski et al., "Increased Aggression and Activity Level in 3- to 11-Year-Old Girls with Congenital Adrenal Hyperplasia," *Horm Behav* 52 (2007): 368.
69. C. A. Quigley et al., "Androgen Receptor Defects: Historical, Clinical, and Molecular Perspectives," *Endocrine Rev* 16 (1995): 271; N. P. Mongan et al., "Androgen Insensitivity Syndrome," *Best Practice & Res: Clin Endo & Metabolism* 29 (2015): 569.
70. F. Brunner et al., "Body and Gender Experience in Persons with Complete Androgen Insensitivity Syndrome," *Zeitschrift für Sexualforschung* 25 (2012): 26; F. Brunner et al., "Gender Role, Gender Identity and Sexual Orientation in CAIS ('XY-Women') Compared with Subfertile and Infertile 46,XX Women," *J Sex Res* 2 (2015): 1; D. G. Zuloaga et al., "The Role of Androgen Receptors in the Masculinization of Brain and Behavior: What We've Learned from the Testicular Feminization Mutation," *Horm Behav* 53 (2008): 613; H. F. L. Meyer-Bahlburg,

"Gender Outcome in 46,XY Complete Androgen Insensitivity Syndrome: Comment on T'Sjoen et al.," *Arch Sexual Behav* 39 (2010): 1221; G. T'Sjoen et al., "Male Gender Identity in Complete Androgen Insensitivity Syndrome," *Arch Sexual Behav* 40 (2011): 635.

71. J. Hönekopp et al., "2nd to 4th Digit Length Ratio (2D:4D) and Adult Sex Hormone Levels: New Data and a Meta-analytic Review," *PNE* 32 (2007): 313.

72. Findings from males regarding aggression and assertiveness: C. Joyce et al., "2nd to 4th Digit Ratio Confirms Aggressive Tendencies in Patients with Boxers Fractures," *Injury* 44 (2013): 1636; M. Butovskaya et al., "Digit Ratio (2D:4D), Aggression, and Dominance in the Hadza and the Datoga of Tanzania," *Am J Human Biology* 27 (2015): 620;

ADHD and autism: D. McFadden et al., "Physiological Evidence of Hypermasculination in Boys with the Inattentive Subtype of ADHD," *Clinical Neurosci Res* 5 (2005): 233; M. Martel et al., "Masculinized Finger-Length Ratios of Boys, but Not Girls, Are Associated with Attention-Deficit/Hyperactivity Disorder," *Behavioral Neuroscience* 122 (2008): 273; J. Manning et al., "The 2nd to 4th Digit Ratio and Autism," *Development Medicine Child Neurology* 43 (2001): 160.

Depression and anxiety: A. Bailey et al., "Depression in Men Is Associated with More Feminine Finger Length Ratios," *Pers Individ Diff* 39 (2005): 829; M. Evardone et al., "Anxiety, Sex-linked Behavior, and Digit Ratios," *Arch Sex Behav.* 38 (2009): 442–55.

Dominance: N. Neave et al., "Second to Fourth Digit Ratio, Testosterone and Perceived Male Dominance," *Proc Royal Society B* 270 (2003): 2167.

Handwriting: J. Beech et al., "Do Differences in Sex Hormones Affect Handwriting Style? Evidence from Digit Ratio and Sex Role Identity as Determinants of the Sex of Handwriting," *Pers Individ Diff* 39 (2005): 459.

Sexual orientation: K. Hirashi et al., "The Second to Fourth Digit Ratio in a Japanese Twin Sample: Heritability, Prenatal Hormone Transfer, and Association with Sexual Orientation," *Arch Sex Behav* 41 (2012): 711; A. Churchill et al., "The Effects of Sex, Ethnicity, and Sexual Orientation on Self-Measured Digit Ratio," *Arch Sex Behav* 36 (2007): 251.

Findings from females regarding autism: J. Manning et al., "The 2nd to 4th Digit Ratio and Autism," *Dev Med Child Neurol* 43 (2001): 160.

Anorexia: S. Quinton et al., "The 2nd to 4th Digit Ratio and Eating Disorder Diagnosis in Women," *Pers Individ Diff* 51 (2011): 402.

Handedness: B. Fink et al., "2nd to 4th Digit Ratio and Hand Skill in Austrian Children," *Biol Psychology* 67 (2004): 375.

Sexual orientation and sexual behavior: T. Grimbos et al., "Sexual Orientation and the 2nd to 4th Finger Length Ratio: A Meta-Analysis in Men and Women," *Behav Neurosci* 124 (2010): 278; W. Brown et al., "Differences in Finger Length Ratios Between Self-Identified 'Butch' and 'Femme' Lesbians," *Arch Sex Behav* 31 (2002): 123.

73. Footnote: A. Lamminmaki et al., "Testosterone Measured in Infancy Predicts Subsequent Sex-Typed Behavior in Boys and in Girls," *Horm Behav* 61 (2012): 611; G. Alexander and J. Saenz, "Early Androgens, Activity Levels and Toy Choices of Children in the Second Year of Life," *Horm Behav* 62 (2012): 500.

74. B. Heijmans et al., "Persistent Epigenetic Differences Associated with Prenatal Exposure to Famine in Humans," *PNAS* 105 (2008): 17046.

75. For a great review, see D. Moore, *The Developing Genome: An Introduction to Behavioral Genetics.* (Oxford: Oxford University Press, 2015).

76. Weaver et al., "Epigenetic Programming by Maternal Behavior," *Nature Neurosci* 7 (2004): 847; R. Sapolsky, "Mothering Style and Methylation," *Nature Neurosci* 7 (2004): 791; D. Francis et al., "Nongenomic Transmission Across Generations of Maternal Behavior and Stress Response in the Rat," *Science* 286 (2004): 1155.

77. N. Provencal et al., "The Signature of Maternal Rearing in the Methylome in Rhesus Macaque Prefrontal Cortex and T Cells," *J Neurosci* 32 (2012): 15626; T. L. Roth et al., "Lasting Epigenetic Influence of Early-Life Adversity on the BDNF Gene," *BP* 65 (2009): 760; E. C. Braithwaite et al., "Maternal Prenatal Depressive Symptoms Predict Infant NR3C1 1F and BDNF IV DNA Methylation," *Epigenetics* 10 (2015): 408; C. Murgatroyd et al., "Dynamic DNA Methylation Programs Persistent Adverse Effects of Early-Life Stress," *Nat Nsci* 12 (2009): 1559; M. J. Meaney and M. Szyf, "Environmental Programming of Stress Responses Through DNA Methylation: Life at the Interface Between a Dynamic Environment and a Fixed Genome," *Dialogues in Clin Neuroscience* 7 (2005): 103; P. O. McGowan et al., "Broad Epigenetic Signature of Maternal Care in the Brain of Adult Rats," *PLoS ONE* 6 (2011): e14739; D. Liu et al., "Maternal Care, Hippocampal Glucocorticoid Receptors, and Hypothalamic-Pituitary-Adrenal Responses to Stress," *Sci* 277 (1997): 1659; T. Oberlander et al., "Prenatal Exposure to Maternal Depression, Neonatal Methylation of Human Glucocorticoid Receptor Gene (NR3C1) and Infant Cortisol Stress Responses," *Epigenetics* 3 (2008): 97; F. A. Champagne, "Epigenetic Mechanisms and the Transgenerational Effects of Maternal Care," *Front Neuroendocrinology* 29 (2008): 386; J. P. Curley et al., "Transgenerational Effects of Impaired Maternal Care on Behaviour of Offspring and Grandoffspring," *Animal Behav* 75 (2008): 1551; J. P. Curley et al., "Social Enrichment During Postnatal Development Induces Transgenerational Effects on Emotional and Reproductive Behavior in Mice," *Front Behav Nsci* 3 (2009): 1; F. A. Champagne, "Maternal Imprints and the Origins of Variation," *Horm Behav* 60 (2011): 4; F. A. Champagne and J. P. Curley, "Epigenetic Mechanisms Mediating the Long-Term Effects of Maternal Care on Development," *Nsci Biobehav Rev* 33 (2009): 593; F. A. Champagne et al., "Maternal Care Associated with Methylation of the Estrogen Receptor-alpha1b Promoter and Estrogen Receptor-Alpha Expression in the Medial Preoptic Area of Female Offspring," *Endo* 147 (2006): 2909; F. A. Champagne and J. P. Curley, "How Social Experiences Influence the Brain," *Curr Opinion in Neurobiol* 15 (2005): 704.

Chapter 8: Back to When You Were Just a Fertilized Egg

1. Footnote: E. Suhay and T. Jayaratne, "Does Biology Justify Ideology? The Politics of Genetic Attribution," *Public Opinion Quarterly* (2012): doi:10.1093/poq/nfs049. See also M. Katz, "The Biological Inferiority of the Undeserving Poor," *Social Work and Soc* 11 (2013): 1.
2. E. Uhlmann et al., "Blood Is Thicker: Moral Spillover Effects Based on Kinship," *Cog* 124 (2012): 239.
3. E. Pennisi, "ENCODE Project Writes Eulogy for Junk DNA," *Sci* 337 (2012): 1159.
4. M. Bastepe, "The GNAS Locus: Quintessential Complex Gene Encoding Gsa, XLas, and Other Imprinted Transcripts," *Curr Genomics* 8 (2007): 398.
5. Y. Gilad et al., "Expression Profiling in Primates Reveals a Rapid Evolution of Human Transcription Factors," *Nat* 440 (2006): 242.
6. D. Moore, *The Developing Genome: An Introduction to Behavioral Genetics* (Oxford: Oxford University Press, 2015); H. Wang et al., "Histone Deacetylase Inhibitors Facilitate Partner Preference Formation in Female Prairie Voles," *Nat Nsci* 16 (2013): 919.
7. I. Weaver et al., "Epigenetic Programming by Maternal Behavior," *Nat Nsci* 7 (2004): 847.
8. Y. Wei et al., "Paternally Induced Transgenerational Inheritance of Susceptibility to Diabetes in Mammals," *PNAS* 111 (2014): 1873; M. Anway et al., "Epigenetic Transgenerational Actions of Endocrine Disruptors and Male Fertility," *Sci* 308 (2005): 1466; K. Siklenka et al., "Disruption of Histone Methylation in Developing Sperm Impairs Offspring Health Transgenerationally," *Sci* 350 (2016): 651. For the controversy, see J. Kaiser, "The Epigenetics Heretic," *Sci* 343 (2014): 361.
9. E. Jablonka and M. Lamb, *Epigenetic Inheritance and Evolution: The Lamarckian Dimension* (Oxford: Oxford University Press, 1995).
10. E. T. Wang et al., "Alternative Isoform Regulation in Human Tissue Transcriptomes," *Nat* 456 (2008): 470; Q. Pan et al., "Deep Surveying of Alternative Splicing Complexity in the Human Transcriptome by High-Throughput Sequencing," *Nat* Gen, 40 (2008): 1413.
11. A. Muotri et al., "Somatic Mosaicism in Neuronal Precursor Cells Mediated by L1 Retrotransposition," *Nat* 435 (2005): 903; P. Perrat et al., "Transposition-Driven Genomic Heterogeneity in the *Drosophila* Brain," Sci 340 (2013): 91; G. Vogel, "Do Jumping Genes Spawn Diversity?" *Sci* 332 (2011): 300; J. Baillie et al., "Somatic Retrotransposition Alters the Genetic Landscape of the Human Brain," *Nat* 479 (2011): 534.
12. A. Eldar and M. Elowitz, "Functional Roles for Noise in Genetic Circuits," *Nat* 467 (2010): 167; C. Finch and T. Kirkwood, *Chance, Development, and Aging* (Oxford: Oxford University Press, 2000).
13. Some of the early, classic adoption studies: L. L. Heston, "Psychiatric Disorders in Foster Home Reared Children of Schizophrenic Mothers," *Brit J Psychiatry* 112 (1966): 819; S. Kety et al., "Mental Illness in the Biological and Adoptive Families of Adopted Schizophrenics," *Am J Psychiatry* 128 (1971): 302; D. Rosenthal et al., "The Adopted-Away Offspring of Schizophrenics," *Am J Psychiatry* 128 (1971): 307.
14. For an extraordinary example of a mix-up of babies shortly after birth, and the implications, see S. Dominus, "The Mixed-Up Brothers of Bogotá," *New York Times Magazine*, July 9, 2015, www.nytimes.com/2015/07/12/magazine/the-mixed-up-brothers-of-bogota.html.
15. R. Ebstein et al., "Genetics of Human Social Behavior," *Neuron* 65 (2008): 831; S. Eisen et al., "Familial Influence on Gambling Behavior: An Analysis of 3359 Twin Pairs," *Addiction* 93 (1988): 1375. Footnote: W. Hopkins et al., "Chimpanzee Intelligence Is Heritable," *Curr Biol* 24 (2014): 1649.
16. T. Bouchard and M. McGue, "Genetic and Environmental Influences on Human Psychological Differences," *J Neurobiol* 54 (2003): 4; D. Cesarini et al., "Heritability of Cooperative Behavior in the Trust Game," *PNAS* 105 (2008): 3721; S. Zhong et al., "The Heritability of Attitude Toward Economic Risk," *Twin Res and Hum Genetics* 12 (2009): 103; D. Cesarini et al., "Genetic Variation in Financial Decision-Making," *J the Eur Economic Association* 7 (2010): 617.
17. K. Verweij et al., "Shared Aetiology of Risky Sexual Behaviour and Adolescent Misconduct: Genetic and Environmental Influences," *Genes, Brain and Behav* 8 (2009): 107; K. Verweij et al., "Genetic and Environmental Influences on Individual Differences in Attitudes Toward Homosexuality: An Australian Twin Study," *Behav Genetics* 38 (2008): 257.
18. K. Verweij et al., "Evidence for Genetic Variation in Human Mate Preferences for Sexually Dimorphic Physical Traits. *PLoS ONE* 7 (2012): e49294; K. Smith et al., "Biology, Ideology and Epistemology: How Do We Know Political Attitudes Are Inherited and Why Should We Care?" *Am J Political Sci* 56 (2012): 17; K. Arceneaux et al., "The Genetic Basis of Political Sophistication," *Twin Res and Hum Genetics* 15 (2012): 34; J. Fowler and D. Schreiber, "Biology, Politics, and the Emerging Science of Human Nature," *Sci* 322 (2008): 912.
19. J. Ray et al., "Heritability of Dental Fear," *J Dental Res* 89 (2010): 297; G. Miller et al., "The Heritability and Genetic Correlates of Mobile Phone Use: Twin Study of Consumer Behavior," *Twin Res and Hum Genetics* 15 (2012): 97.
20. L. Littvay et al., "Sense of Control and Voting: A Genetically-Driven Relationship," *Soc Sci Quarterly* 92 (2011): 1236; J. Harris, *The Nurture Assumption: Why Children Turn Out the Way They Do* (NY: Free Press, 2009); A. Seroczynski et al., "Etiology of the Impulsivity/Aggression Relationship: Genes or Environment?" *Psychiatry Res* 86 (1999): 41; E. Coccaro et al., "Heritability of Aggression and Irritability: A Twin Study of the Buss-Durkee Aggression Scales in Adult Male Subjects," *BP* 41 (1997): 273.
21. E. Hayden, "Taboo Genetics," *Nat* 502 (2013): 26.
22. Some strong criticisms of twin and adoption approaches: R. Rose, "Genes and Human Behavior," *Ann Rev Psych* 467 (1995): 625; J. Joseph, "Twin Studies in Psychiatry and Psychology: Science or Pseudoscience?" *Psychiatric Quarterly* 73 (2002): 71; K. Richardson and S. Norgate, "The Equal Environments Assumption of Classical Twin Studies May Not Hold," *Brit J Educational Psych* 75 (2005): 339; R. Fosse et al., "A Critical Assessment of

the Equal-Environment Assumption of the Twin Method for Schizophrenia," *Front Psychiatry* 6 (2015): 62; A. V. Horwitz et al., "Rethinking Twins and Environments: Possible Social Sources for Assumed Genetic Influences in Twin Research," *J Health and Soc Behav* 44 (2003): 111.

23. Work of some of the most prominent defenders of the approaches:

Kenneth Kendler: K. S. Kendler, "Twin Studies of Psychiatric Illness: An Update," *AGP* 58 (2001): 1005; K. S. Kendler et al., "A Test of the Equal-Environment Assumption in Twin Studies of Psychiatric Illness," *Behav Genetics* 23 (1993): 21; K. S. Kendler and C. O. Gardner Jr., "Twin Studies of Adult Psychiatric and Substance Dependence Disorders: Are They Biased by Differences in the Environmental Experiences of Monozygotic and Dizygotic Twins in Childhood and Adolescence?" *Psych Med* 8 (1998): 625; K. S. Kendler et al., "A Novel Sibling-Based Design to Quantify Genetic and Shared Environmental Effects: Application to Drug Abuse, Alcohol Use Disorder and Criminal Behavior," *Psych Med* 46 (2016): 1639; K. S. Kendler et al., "Genetic and Familial Environmental Influences on the Risk for Drug Abuse: A National Swedish Adoption Study," *AGP* 69 (2012): 690; K. S. Kendler et al., "Tobacco Consumption in Swedish Twins Reared Apart and Reared Together," *AGP* 57 (2000): 886.

Thomas Bouchard: Y. Hur and T. Bouchard, "Genetic Influences on Perceptions of Childhood Family Environment: A Reared Apart Twin Study," *Child Development* 66 (1995): 330; M. McGue and T. J. Bouchard, "Genetic and Environmental Determinants of Information Processing and Special Mental Abilities: A Twin Analysis," in *Advances in the Psychology of Hum Intelligence*, ed. R. J. Sternberg, vol. 5 (Hillsdale, NJ: Erlbaum, 1989), pp. 7–45; T. J. Bouchard et al., "Sources of Human Psychological Differences: The Minnesota Study of Twins Reared Apart," *Sci* 250 (1990): 223.

Robert Plomin: R. Plomin et al., *Behavioral Genetics*, 5th ed. (New York: Worth, 2008); K. Hardy-Brown et al., "Selective Placement of Adopted Children: Prevalence and Effects," *J Child Psych and Psychiatry* 21 (1980) 143; N. L. Pedersen et al., "Genetic and Environmental Influences for Type A–Like Measures and Related Traits: A Study of Twins Reared Apart and Twins Reared Together," *Psychosomatic Med* 51 (1989): 428; N. L. Pedersen et al., "Neuroticism, Extraversion, and Related Traits in Adult Twins Reared Apart and Reared Together," *JPSP* 55 (1988): 950.

Also: E. Coccaro et al., "Heritability of Aggression and Irritability: A Twin Study of the Buss-Durkee Aggression Scales in Adult Male Subjects," *BP* 41 (1997): 273; A. Bjorklund et al., "The Origins of Intergenerational Associations: Lessons from Swedish Adoption Data," *Quarterly J Economics* 121 (2006): 999; E. P. Gunderson et al., "Twins of Mistaken Zygosity (TOMZ): Evidence for Genetic Contributions to Dietary Patterns and Physiologic Traits," *Twin Res and Hum Genetics* 9 (2006): 540; B. N. Sánchez et al., "A Latent Variable Approach to Study Gene-Environment Interactions in the Presence of Multiple Correlated Exposures," *Biometrics* 68 (2012): 466.

24. Evidence that chorionic status is a meaningful variable: M. Melnick et al., "The Effects of Chorion Type on Variation in IQ in the NCPP Twin Population," *Am J Hum Genetics* 30 (1978): 425; N. Jacobs et al., "Heritability Estimates of Intelligence in Twins: Effect of Chorion Type," *Behav Genetics* 31 (2001): 209; M. Melnick et al., "The Effects of Chorion Type on Variation in IQ in the NCPP Twin Population," *Am J Hum Genetics* 30 (1978): 425; R. J. Rose et al., "Placentation Effects on Cognitive Resemblance of Adult Monozygotes," in *Twin Research 3: Epidemiological and Clinical Studies*, ed. L. Gedda et al. (New York: Alan R. Liss, 1981), p. 35; K. Beekmans et al., "Relating Type of Placentation to Later Intellectual Development in Monozygotic (MZ) Twins (Abstract)," *Behav Genetics* 23 (1993): 547; M. Carlier et al., "Manual Performance and Laterality in Twins of Known Chorion Type," *Behav Genetics* 26 (1996): 409.

Mixed findings: L. Gutknecht et al. "Long-Term Effect of Placental Type on Anthropometrical and Psychological Traits Among Monozygotic Twins: A Follow Up Study," *Twin Res* 2 (1999): 212; D. K. Sokol et al., "Intrapair Differences in Personality and Cognitive Ability Among Young Monozygotic Twins Distinguished by Chorion Type," *Behav Genetics* 25 (1996): 457; A. C. Bogle et al., "Replication of Asymmetry of a-b Ridge Count and Behavioral Discordance in Monozygotic Twins," *Behav Genetics* 24 (1994): 65; J. O. Davis et al., "Prenatal Development of Monozygotic Twins and Concordance for Schizophrenia," *Schizophrenia Bull* 21 (1995): 357.

Evidence against: Y. M. Hur, "Effects of the Chorion Type on Prosocial Behavior in Young South Korean Twins," Twin Res and Hum Genetics 10 (2007): 773; M. C. Wichers et al., "Chorion Type and Twin Similarity for Child Psychiatric Symptoms," *AGP* 59 (2002): 562; P. Welch et al., "Placental Type and Bayley Mental Development Scores in 18 Month Old Twins," in *Twin Research: Psychology and Methodology*, ed. L. Gedda et al. (New York: Alan R Liss, 1978), pp. 34–41.

Quote from: C. A. Prescott et al., "Chorion Type as a Possible Influence on the Results and Interpretation of Twin Study Data," *Twin Res* 2 (1999): 244.

25. R. Simon and H. Alstein, *Adoption, Race and Identity: From Infancy to Young Adulthood* (New Brunswick, NJ: Transaction Publishers, 2002); Child Welfare League of America, *Standards of Excellence: Standards of Excellence for Adoption Services*, rev. ed. (Washington, DC: Child Welfare League of America, 2000); M. Bohman, *Adopted Children and Their Families: A Follow-up Study of Adopted Children, Their Background, Environment and Adjustment* (Stockholm: Proprius, 1970).

26. L. J. Kamin and A. S. Goldberger, "Twin Studies in Behavioral Research: A Skeptical View," *Theoretical Population Biol* 61 (2002): 83.

27. M. Stoolmiller, "Correcting Estimates of Shared Environmental Variance for Range Restriction in Adoption Studies Using a Truncated Multivariate Normal Model," *Behav Gen* 28 (1998) 429; M. Stoolmiller, "Implications of Restricted Range of Family Environments for Estimates of Heritability and Nonshared Environment in Behavior-Genetic Adoption Studies," *Psych Bull* 125 (1999): 392; M. McGue et al., "The Environments of Adopted and Non-adopted Youth: Evidence on Range Restriction from the Sibling Interaction and Behavior Study (SIBS)," *Behav Gen* 37 (2007): 449.

28. R. Ebstein et al., "Genetics of Human Social Behavior," *Neuron* 65 (2008): 831.
29. This example comes from N. Block, "How Heritability Misleads About Race," *Cog* 56 (1995): 99–128.
30. D. Moore, *The Dependent Gene: The Fallacy of "Nature Versus Nurture"* (NY: Holt, 2001); M. Ridley, *Nature via Nurture* (New York: HarperCollins, 2003); A. Tenesa and C. Haley, "The Heritability of Human Disease: Estimation, Uses and Abuses," *Nat Rev Genetics* 14 (2013): 139; P. Schonemann, "On Models and Muddles of Heritability," *Genetica* 99 (1997): 97.
31. T. Bouchard and M. McGue, "Genetic and Environmental Influences on Human Psychological Differences," *J Neurobiol* 54 (2003): 4.
32. L. E. Duncan and M. C. Keller, "A Critical Review of the First 10 Years of Candidate Gene-by-Environment Interaction Research in Psychiatry," *Am J Psychiatry* 168 (2011): 1041; S. Manuck and J. McCaffery, "Gene-Environment Interaction," *Ann Rev of Psych* 65 (2014): 41.
33. A. Caspi et al., "Influence of Life Stress on Depression: Moderation by a Polymorphism in the 5-HTT Gene," *Sci* 297 (2002): 851.
34. A. Caspi et al., "Moderation of Breastfeeding Effects on the IQ by Genetic Variation in Fatty Acid Metabolism," *PNAS* 104 (2007): 18860; B. K. Lipska and D. R. Weinberger, "Genetic Variation in Vulnerability to the Behavioral Effects of Neonatal Hippocampal Damage in Rats," *PNAS* 92 (1995): 8906.
35. J. Crabbe et al., "Genetics of Mouse Behavior: Interactions with Laboratory Environment," *Sci* 284 (1999): 1670.
36. A nice example of a dual environment hit: N. P. Daskalakis et al., "The Three-Hit Concept of Vulnerability and Resilience: Toward Understanding Adaptation to Early-Life Adversity Outcome," *PNE* 38 (2013): 1858.
37. E. Turkheimer et al., "Socioeconomic Status Modifies Heritability of IQ in Young Children," *Psych Sci* 14 (2003): 623; E. M. Tucker-Drob et al., "Emergence of a Gene x Socioeconomic Status Interaction on Infant Mental Ability Between 10 Months and 2 Years," *Psych Sci* 22 (2010): 125; M. Rhemtulla and E. M. Tucker-Drob, "Gene-by-Socioeconomic Status Interaction on School Readiness," *Behav Genetics* 42 (2012): 549; D. Reiss et al., "How Genes and the Social Environment Moderate Each Other," *Am J Public Health* 103 (2013): S111; S. A. Hart et al., "Expanding the Environment: Gene × School-Level SES Interaction on Reading Comprehension," *J Child Psych and Psychiatry* 54 (2013): 1047; J. R. Koopmans et al., "The Influence of Religion on Alcohol Use Initiation: Evidence for Genotype × Environment Interaction," *Behav Genetics* 29 (1999): 445.
38. S. Nielsen et al., "Prevalence of Alcohol Problems Among Adult Somatic Inpatients of a Copenhagen Hospital," *Alcohol and Alcoholism* 29 (1994): 583; S. Manuck et al., "Aggression and Anger-Related Traits Associated with a Polymorphism of the Tryptophan Hydroxylase Gene," *BP* 45 (1999): 603; J. Hennig et al., "Two Types of Aggression Are Differentially Related to Serotonergic Activity and the A779C TPH Polymorphism," *Behav Nsci* 119 (2005): 16; A. Strobel et al., "Allelic Variation in 5-HT1A Receptor Expression Is Associated with Anxiety- and Depression-Related Personality Traits," *J Neural Transmission* 110 (2003): 1445; R. Parsey et al., "Effects of Sex, Age, and Aggressive Traits in Man on Brain Serotonin 5-HT1A Receptor Binding Potential Measured by PET Using [C-11]WAY-100635," *Brain Res* 954 (2002): 173; A. Benko et al., "Significant Association Between the C(-1019)G Functional Polymorphism of the HTR1A Gene and Impulsivity," *Am J Med Genetics, Part B, Neuropsychiatric Genetics* 153 (2010): 592, M. Soyka et al., "Association of 5-HT1B Receptor Gene and Antisocial Behavior and Alcoholism," *J Neural Transmission* 111 (2004): 101; L. Bevilacqua et al., "A Population-Specific HTR2B Stop Codon Predisposes to Severe Impulsivity," *Nat* 468 (2010): 1061; C. A. Ficks and I. D. Waldman, "Candidate Genes for Aggression and Antisocial Behavior: A Meta-analysis of Association Studies of the 5HTTLPR and MAOA-uVNTR," *Behav Genetics* 44 (2014): 427; I. Craig and K. Halton, "Genetics of Human Aggressive Behavior," *Hum Genetics* 126 (2009): 101.
39. H. Brunner et al., "Abnormal Behavior Associated with a Point Mutation in the Structural Gene for Monoamine Oxidase A," *Sci* 262 (1993): 578; H. G. Brunner et al., "X-Linked Borderline Mental Retardation with Prominent Behavioral Disturbance: Phenotype, Genetic Localization, and Evidence for Disturbed Monoamine Metabolism," *Am J Hum Genetics* 52 (1993): 1032.
40. O. Cases et al., "Aggressive Behavior and Altered Amounts of Brain Serotonin and Norepinephrine in Mice Lacking MAOA," *Sci* 268 (1995): 1763; J. J. Kim et al., "Selective Enhancement of Emotional, but Not Motor, Learning in Monoamine Oxidase A–Deficient Mice," *PNAS* 94 (1997): 5929.
41. J. Buckholtz and A. Meyer-Lindenberg, "MAOA and the Neurogenetic Architecture of Human Aggression," *TINS* 31 (2008): 120; A. Meyer-Lindenberg et al., "Neural Mechanisms of Genetic Risk for Impulsivity and Violence in Humans," *PNAS* 103 (2006): 6269; J. Fan et al., "Mapping the Genetic Variation of Executive Attention onto Brain Activity," *PNAS* 100 (2003): 7406; L. Passamonti et al., "Monoamine Oxidase-A Genetic Variations Influence Brain Activity Associated with Inhibitory Control: New Insight into the Neural Correlates of Impulsivity," *BP* 59 (2006): 334; N. Eisenberger et al., "Understanding Genetic Risk for Aggression: Clues from the Brain's Response to Social Exclusion," *BP* 61 (2007): 1100.
42. O. Cases et al., "Aggressive Behaviour and Altered Amounts of Brain Serotonin and Norepinephrine in Mice Lacking MAOA," *Sci* 268 (1995): 1763; J. S. Fowler et al., "Evidence That Brain MAO A Activity Does Not Correspond to MAO A Genotype in Healthy Male Subjects," *BP* 62 (2007): 355.
43. The "warrior gene" in the science literature: C. Holden, "Parsing the Genetics of Behavior," *Sci* 322 (2008): 892; D. Eccles et al., "A Unique Demographic History Exists for the MAO-A Gene in Polynesians," *J Hum Genetics* 57 (2012): 294; E. Feresin, "Lighter Sentence for Murder with 'Bad Genes,'" *Nat News* (30 October, 2009); P. Hunter, "The Psycho Gene," *EMBO Rep* 11 (2010): 667.

Criticism of the scientists in the Maori study for overselling the significance of their finding: D. Wensley and M. King, "Scientific Responsibility for the Dissemination and Interpretation of Genetic Research: Lessons from the 'Warrior Gene' Controversy," *J Med Ethics* 34 (2008): 507; S. Halwani and D. Krupp, "The Genetic Defense: The Impact of Genetics on the Concept of Criminal Responsibility," *Health Law J* 12 (2004): 35.

44. A. Caspi et al., "Influence of Life Stress on Depression: Moderation by a Polymorphism in the 5-HTT Gene," *Sci* 297 (2002): 851.
45. J. Buckholtz and A. Meyer-Lindenberg, "MAOA and the Neurogenetic Architecture of Human Aggression," *TINS* 31 (2008): 120.
46. J. Kim-Cohen et al., "MAOA, Maltreatment, and Gene Environment Interaction Predicting Children's Mental Health: New Evidence and a Meta-analysis," *Mol Psychiatry* 11 (2006): 903; A. Byrd and S. Manuck, "MAOA, Childhood Maltreatment and Antisocial Behavior: Meta-analysis of a Gene-Environment Interaction," *BP* 75 (2013): 9; G. Frazzetto et al., "Early Trauma and Increased Risk for Physical Aggression During Adulthood: The Moderating Role of MAOA Genotype," *PLoS ONE* 2 (2007): e486; C. Widom and L. Brzustowicz, "MAOA and the 'Cycle of Violence': Childhood Abuse and Neglect, MAOA Genotype, and Risk for Violent and Antisocial Behavior," *BP* 60 (2006): 684; R. McDermott et al., "MAOA and Aggression: A Gene-Environment Interaction in Two Populations," *J Conflict Resolution* 1 (2013): 1043; T. Newman et al., "Monoamine Oxidase A Gene Promoter Variation and Rearing Experience Influences Aggressive Behavior in Rhesus Monkeys," *BP* 57 (2005): 167; X. Ou et al., "Glucocorticoid and Androgen Activation of Monoamine Oxidase A Is Regulated Differently by R1 and Sp1," *J Biol Chemistry* 281 (2006): 21512.

    Replication: D. L. Foley et al., "Childhood Adversity, Monoamine Oxidase A Genotype, and Risk for Conduct Disorder," *AGP* 61 (2004): 738; D. M. Fergusson et al., "MAOA, Abuse Exposure and Antisocial Behaviour: 30-Year Longitudinal Study," *Brit J Psychiatry* 198 (2011): 457.

    Weaker effect in girls: E. C. Prom-Wormley et al., "Monoamine Oxidase A and Childhood Adversity as Risk Factors for Conduct Disorder in Females," *Psych Med* 39 (2009): 579.

    Replicates for whites but not blacks: C. S. Widom and L. M. Brzustowicz, "MAOA and the 'Cycle of Violence': Childhood Abuse and Neglect, MAOA Genotype, and Risk for Violent and Antisocial Behavior," *BP* 60 (2006): 684.

    Failure of replication: D. Huizinga et al., "Childhood Maltreatment, Subsequent Antisocial Behavior, and the Role of Monoamine Oxidase A Genotype," *BP* 60 (2006): 677; S. Young et al., "Interaction Between MAO-A Genotype and Maltreatment in the Risk for Conduct Disorder: Failure to Confirm in Adolescent Patients," *Am J Psychiatry* 163 (2006): 1019.
47. R. Sjoberg et al., "A Non-additive Interaction of a Functional MAO-A VNTR and Testosterone Predicts Antisocial Behavior," *Neuropsychopharmacology* 33 (2008): 425; R. McDermott et al., "Monoamine Oxidase A Gene (MAOA) Predicts Behavioral Aggression Following Provocation," *PNAS* 106 (2009): 2118; D. Gallardo-Pujol et al., "MAOA Genotype, Social Exclusion and Aggression: An Experimental Test of a Gene-Environment Interaction," *Genes, Brain and Behav* 12 (2013): 140; A. Reif et al., "Nature and Nurture Predispose to Violent Behavior: Serotonergic Genes and Adverse Childhood Environment," *Neuropsychopharmacology* 32 (2007): 2375.
48. A. Rivera et al., "Cellular Localization and Distribution of Dopamine D4 Receptors in the Rat Cerebral Cortex and Their Relationship with the Cortical Dopaminergic and Noradrenergic Nerve Terminal Networks," *Nsci* 155 (2008): 997; O. Schoots and H. Van Tol, "The Human Dopamine D4 Receptor Repeat Sequences Modulate Expression," *Pharmacogenomics J* 3 (2003): 343; C. Broeckhoven and S. Gestel, "Genetics of Personality: Are We Making Progress? *Mol Psychiatry* 8 (2003): 840; M. R. Munafò et al., "Association of the Dopamine D4 Receptor (DRD4) Gene and Approach-Related Personality Traits: Meta-analysis and New Data," *BP* 63 (2007): 197; R. Ebstein et al., "Dopamine D4 Receptor (D4DR) Exon III Polymorphism Associated with the Human Personality Trait of Novelty Seeking," *Nat Genetics* 12 (1996): 78; J. Carpenter et al., "Dopamine Receptor Genes Predict Risk Preferences, Time Preferences, and Related Economic Choices," *J Risk and Uncertainty* 42 (2011): 233; J. Garcia et al., "Associations Between Dopamine D4 Receptor Gene Variation with Both Infidelity and Sexual Promiscuity," *PLoS ONE* 5 (2010): e14162; D. Li et al., "Meta-analysis Shows Significant Association Between Dopamine System Genes and Attention Deficit Hyperactivity Disorder (ADHD)," *Human Mol Genetics* 15 (2006): 2276; L. Ray et al., "The Dopamine D4 Receptor (DRD4) Gene Exon III Polymorphism, Problematic Alcohol Use and Novelty Seeking: Direct and Mediated Genetic Effects," *Addiction Biol* 14 (2008): 238; A. Dreber et al., "The 7R Polymorphism in the Dopamine Receptor D4 Gene (DRD4) Is Associated with Financial Risk-Taking in Men," *EHB* 30 (2009): 85; D. Eisenberg et al., "Polymorphisms in the Dopamine D4 and D2 Receptor Genes and Reproductive and Sexual Behaviors," *Evolutionary Psych* 5 (2007): 696; A. N. Kluger et al., "A Meta-analysis of the Association Between DRD4 Polymorphism and Novelty Seeking," *Mol Psychiatry* 7 (2002): 712; S. Zhong et al., "Dopamine D4 Receptor Gene Associated with Fairness Preference in Ultimatum Game," *PLoS ONE* 5 (2010): e13765.
49. M. Bakermans-Kranenburg and M. van Ijzendoorn, "Differential Susceptibility to Rearing Environment Depending on Dopamine-Related Genes: New Evidence and a Meta-analysis," *Development Psychopathology* 23 (2011): 39; J. Sasaki et al., "Religion Priming Differentially Increases Prosocial Behavior Among Variants of the Dopamine D4 Receptor (DRD4) Gene," *SCAN* 8 (2013): 209; M. Sweitzer et al., "Polymorphic Variation in the Dopamine D4 Receptor Predicts Delay Discounting as a Function of Childhood Socioeconomic Status: Evidence for Differential Susceptibility," *SCAN* 8 (2013): 499.
50. F. Chang et al., "The World-wide Distribution of Allele Frequencies at the Human Dopamine D4 Receptor Locus," *Hum Genetics* 98 (1996): 91; C. Chen et al., "Population Migration and the Variation of Dopamine D4 Receptor (DRD4) Allele Frequencies Around the Globe," *EHB* 20 (1999): 309.
51. M. Reuter and J. Hennig, "Association of the Functional Catechol-O-Methyltransferase VAL158MET Polymorphism with the Personality Trait of Extraversion," *Neuroreport* 16 (2005): 1135; T. Lancaster et al., "COMT

val158met Predicts Reward Responsiveness in Humans," *Genes, Brain and Behav* 11 (2012): 986; A. Caspi et al., "A Replicated Molecular-Genetic Basis for Subtyping Antisocial Behavior in ADHD," *AGP* 65 (2007): 203; N. Perroud et al., "COMT but Not Serotonin-Related Genes Modulates the Influence of Childhood Abuse on Anger Traits," *Genes, Brain and Behav* 9 (2010): 193.

COMT variants also associated with cognitive end points: F. Papaleo et al., "Genetic Dissection of the Role of Catechol-O-Methyltransferase in Cognition and Stress Reactivity in Mice," *J Nsci* 28 (2008): 8709; F. Papaleo et al., "Effects of Sex and COMT Genotype on Environmentally Modulated Cognitive Control in Mice," *PNAS* 109 (2012): 20160; F. Papaleo et al., "Epistatic Interaction of COMT and DTNBP1 Modulates Prefrontal Function in Mice and in Humans," *Mol Psychiatry* 19 (2013): 311.

52. D. Enter et al., "Dopamine Transporter Polymorphisms Affect Social Approach-Avoidance Tendencies," *Genes, Brain and Behav* 11 (2012): 671; G. Guo et al., "Dopamine Transporter, Gender, and Number of Sexual Partners Among Young Adults," *Eur J Hum Genetics* 15 (2007): 279; S. Lee et al., "Association of Maternal Dopamine Transporter Genotype with Negative Parenting: Evidence for Gene X Environment Interaction with Child Disruptive Behavior," *Mol Psychiatry* 15 (2010): 548 M. van Ijzendoorn et al., "Dopamine System Genes Associated with Parenting in the Context of Daily Hassles," *Genes, Brain and Behav* 7 (2008): 403.
53. D. Gothelf et al., "Biological Effects of Catechol-O-Methyltransferase Haplotypes and Psychosis Risk in 22q11.2 Deletion Syndrome," *BP* 75 (2013): 406.
54. M. Dadds et al., "Polymorphisms in the Oxytocin Receptor Gene Are Associated with the Development of Psychopathy," *Development Psychopathology* 26 (2014): 21; A. Malik et al., "The Role of Oxytocin and Oxytocin Receptor Gene Variants in Childhood-Onset Aggression," *Genes, Brain and Behav* 11 (2012): 545; H. Walum et al., "Variation in the Oxytocin Receptor Gene Is Associated with Pair-Bonding and Social Behavior," *BP* 71 (2012): 419.
55. S. Rajender et al., "Reduced CAG Repeats Length in Androgen Receptor Gene Is Associated with Violent Criminal Behavior," *Int J Legal Med* 122 (2008): 367; D. Cheng et al., "Association Study of Androgen Receptor CAG Repeat Polymorphism and Male Violent Criminal Activity," *PNE* 31 (2006): 548; A. Raznahan et al., "Longitudinally Mapping the Influence of Sex and Androgen Signaling on the Dynamics of Human Cortical Maturation in Adolescence," *PNAS* 107 (2010): 16988; H. Vermeersch et al., "Testosterone, Androgen Receptor Gene CAG Repeat Length, Mood and Behaviour in Adolescent Males," *Eur J Endo* 163 (2010): 319; S. Manuck et al., "Salivary Testosterone and a Trinucleotide (CAG) Length Polymorphism in the Androgen Receptor Gene Predict Amygdala Reactivity in Men," *PNE* 35 (2010): 94; J. Roney et al., "Androgen Receptor Gene Sequence and Basal Cortisol Concentrations Predict Men's Hormonal Responses to Potential Mates," *Proc Royal Soc B* 277 (2010): 57.
56. D. Comings et al., "Multivariate Analysis of Associations of 42 Genes in ADHD, ODD and Conduct Disorder," *Clin Genetics* 58 (2000): 31; Z. Prichard et al., "Association of Polymorphisms of the Estrogen Receptor Gene with Anxiety-Related Traits in Children and Adolescents: A Longitudinal Study," *Am J Med Genetics* 114 (2002): 169; H. Tiemeier et al., "Estrogen Receptor Alpha Gene Polymorphisms and Anxiety Disorder in an Elderly Population," *Mol Psychiatry* 10 (2005): 806; D. Crews et al., "Litter Environment Affects Behavior and Brain Metabolic Activity of Adult Knockout Mice," *Front Behav Nsci* 3 (2009): 1.
57. R. Bogdan et al., "Mineralocorticoid Receptor Iso/Val (rs5522) Genotype Moderates the Association Between Previous Childhood Emotional Neglect and Amygdala Reactivity," *Am J Psychiatry* 169 (2012): 515; L. Bevilacqua et al., "Interaction Between FKBP5 and Childhood Trauma and Risk of Aggressive Behavior," *AGP* 69 (2012): 62; E. Binder et al., *JAMA* 299 (2008): 1291; M. White et al., "FKBP5 and Emotional Neglect Interact to Predict Individual Differences in Amygdala Reactivity," *Genes, Brain and Behav* 11 (2012): 869.
58. L. Schmidt et al., "Evidence for a Gene-Gene Interaction in Predicting Children's Behavior Problems: Association of Serotonin Transporter Short and Dopamine Receptor D4 Long Genotypes with Internalizing and Externalizing Behaviors in Typically Developing 7-Year-Olds," *Developmental Psychopathology* 19 (2007): 1105; M. Nobile et al., "Socioeconomic Status Mediates the Genetic Contribution of the Dopamine Receptor D4 and Serotonin Transporter Linked Promoter Region Repeat Polymorphisms to Externalization in Preadolescence," *Developmental Psychopathology* 19 (2007): 1147.
59. M. J. Arranz et al., "Meta-analysis of Studies on Genetic Variation in 5-HT2A Receptors and Clozapine Response," *Schizophrenia Res* 32 (1998): 93.
60. H. Lango Allen, et al., "Hundreds of Variants Clustered in Genomic Loci and Biological Pathways Affect Human Height," *Nat* 467 (2010): 832.
61. E. Speliotes et al., "Association Analyses of 249,796 Individuals Reveal 18 New Loci Associated with Body Mass Index," *Nat Genetics* 42 (2010): 937; J. Perry et al., "Parent-of-Origin-Specific Allelic Associations Among 106 Genomic Loci for Age at Menarche," *Nat* 514 (2014): 92; S. Ripke et al., "Biological Insights from 108 Schizophrenia-Associated Genetic Loci," *Nat* 511 (2014): 421; F. Flint and M. Munafo, "Genesis of a Complex Disease," *Nat* 511 (2014): 412; J. Tennessen et al., "Evolution and Functional Impact of Rare Coding Variation from Deep Sequencing of Human Exomes," *Sci* 337 (2012): 64; F. Casals and J. Bertranpetit, "Human Genetic Variation, Shared and Private," *Sci* 337 (2012): 39.
62. C. Rietveld et al., "GWAS of 126,559 Individuals Identifies Genetic Variants Associated with Educational Attainment," *Sci* 340 (2013): 1467; J. Flint and M. Munafo, "Herit-Ability," *Sci* 340 (2013): 1416.
63. S. Cole et al., "Social Regulation of Gene Expression in Human Leukocytes," *Genome Biol* 8 (2007): R189.
64. C. Chabris et al., "The Fourth Law of Behavior Genetics," *Curr Dir Psych Sci* 24 (2015): 304; K. Haddley et al., "Behavioral Genetics of the Serotonin Transporter," *Curr Topics in Behav Nsci* 503 (2012): 503; F. S. Neves et al., "Is the Serotonin Transporter Polymorphism (5-HTTLPR) a Potential Marker for Suicidal Behavior in Bipolar

Disorder Patients?" *J Affective Disorders* 125 (2010): 98; T. Y. Wang et al., "Bipolar: Gender-Specific Association of the SLC6A4 and DRD2 Gene Variants in Bipolar Disorder," *Int J Neuropsychopharmacology* 17 (2014): 211; P. R. Moya et al., "Common and Rare Alleles of the Serotonin Transporter Gene, SLC6A4, Associated with Tourette's Disorder," *Movement Disorders* 28 (2013): 1263.

65. E. Turkheimer, "Three Laws of Behavior Genetics and What They Mean," *Curr Dir Psych Sci* 9 (2000): 160.

Chapter 9: Centuries to Millennia Before

1. L. Guiso et al., "Culture, Gender, and Math," *Sci* 320 (2008): 1164.
2. R. Fisman and E. Miguel, "Corruption, Norms, and Legal Enforcement: Evidence from Diplomatic Parking Tickets," *J Political Economics* 115 (2007): 1020; M. Gelfand et al., "Differences Between Tight and Loose Cultures: A 33-Nation Study," *Sci* 332 (2011): 1100; A. Alesina et al., "On the Origins of Gender Roles: Women and the Plough," *Quarterly J Economics* 128 (2013): 469.
3. For a good discussion of this, see A. Norenzayan, "Explaining Human Behavioral Diversity," *Sci* 332 (2011): 1041.
4. E. Tylor. *Primitive Culture* (1871; repr. New York: J. P. Putnam's Sons, 1920).
5. A. Whitten "Incipient Tradition in Wild Chimpanzees," *Nat* 514 (2014): 178; R. O'Malley et al., "The Cultured Chimpanzee: Nonsense or Breakthrough?" *J Curr Anthropology* 53 (2012): 650; J. Mercador et al., "4,300-Year-Old Chimpanzee Sites and the Origins of Percussive Stone Technology," *PNAS* 104 (2007): 3043; E. van Leeuwen et al., "A Group-Specific Arbitrary Tradition in Chimpanzees (*Pan troglodytes*)," *Animal Cog* 17 (2014): 1421.
6. J. Mann et al., "Why Do Dolphins Carry Sponges?" *PLoS ONE* 3 (2008): e3868; M. Krutzen et al., "Cultural Transmission of Tool Use in Bottlenose Dolphins," *PNAS* 102 (2005): 8939; M. Möglich and G. Alpert, "Stone Dropping by *Conomyrma bicolor* (Hymenoptera: Formicidae): A New Technique of Interference Competition," *Behav Ecology and Sociobiology* 2 (1979): 105.
7. M. Pagel, "Adapted to Culture," *Nat* 482 (2012): 297; C. Kluckhohn et al., *Culture: A Critical Review of Concepts and Definitions* (Chicago: University of Chicago Press, 1952); C. Geertz, *The Interpretation of Cultures* (New York: Basic Books, 1973).
8. D. Brown, *Human Universals* (New York: McGraw-Hill, 1991); D. Smail, *On Deep History and the Brain* (Oakland: University of California Press, 2008).
9. U.S. Central Intelligence Agency, "Life Expectancy at Birth," in *The World Factbook*, https://cia.gov/library/publications/the-world-factbook/rankorder/2102rank.html; W. Lutz and S. Scherbov, *Global Age-Specific Literacy Projections Model (GALP): Rationale, Methodology and Software* (Montreal: UNESCO Institute for Statistics Adult Education and Literacy Statistics Programme, 2006), www.uis.unesco.org/Library/Documents/GALP2006_en.pdf; U.S. Central Intelligence Agency, "Infant Mortality Rate," in *The World Factbook*, https://cia.gov/library/publications/the-world-factbook/rankorder/2091rank.html; International Monetary Fund, *World Economic Outlook Database*, October 2015.
10. Homicide: United Nations Office on Drugs and Crime, *Global Study on Homicide 2013* (April 2014); K. Devries, "The Global Prevalence of Intimate Partner Violence Against Women," *Sci* 340 (2013): 1527.

    Rape data: NationMaster, "Rape Rate: Countries Compared," www.nationmaster.com/country-info/stats/Crime/Rape-rate; L. Melhado, "Rates of Sexual Violence are High in Democratic Republic of the Congo," *Int Perspectives on Sexual and Reproductive Health* 36 (2010): 210; K. Johnson et al., "Association of Sexual Violence and Human Rights Violations with Physical and Mental Health in Territories of the Eastern Democratic Republic of the Congo," *JAMA* 304 (2010): 553. Bullying data: F. Elgar et al., "Income Inequality and School Bullying: Multilevel Study of Adolescents in 37 Countries," *J Adolescent Health* 45 (2009): 351.
11. B. Snyder, "The Ten Best Countries for Women," *Fortune*, October 27, 2014, http://fortune.com/2014/10/27/best-countries-for-women/. The Global Gender Gap Report was first published in 2006 by the World Economic Forum. Inter-Parliamentary Union, "Women in National Parliaments," IPU.org, August 1, 2016, www.ipu.org/wmn-e/classif.htm; U.S. Central Intelligence Agency, "Maternal Mortality Rate," in *The World Factbook*, https://cia.gov/library/publications/the-world-factbook/rankorder/2223rank.html.
12. Gallup Poll International, "Do You Feel Loved?" February 2013; J. Henrich et al., "The Weirdest People in the World? *BBS* 33 (2010): 61; M. Morris et al. "Culture, Norms and Obligations: Cross-National Differences in Patterns of Interpersonal Norms and Felt Olibgations Toward Coworkers," *The Practice of Social Influence in Multiple Cultures* 84107 (2001).
13. H. Markus and S. Kitayama, "Culture and Self: Implications for Cognition, Emotion, and Motivation," *Psych Rev* 98 (1991): 224; S. Kitayama and A. Uskul, "Culture, Mind, and the Brain: Current Evidence and Future Directions," *Ann Rev of Psych* 62 (2011): 419; J. Sui and S. Han, "Self-Construal Priming Modulates Neural Substrates of Self-Awareness," *Psych Sci* 18 (2007): 861; B. Park et al., "Neural Evidence for Cultural Differences in the Valuation of Positive Facial Expressions," *SCAN* 11 (2016): 243.
14. H. Katchadourian, *Guilt: The Bite of Conscience* (Palo Alto, CA: Stanford General Books, 2011); J. Jacquet, *Is Shame Necessary? New Uses for an Old Tool* (New York: Pantheon, 2015); B. Cheon et al., "Cultural Influences on Neural Basis of Intergroup Empathy," *Neuroimage* 57 (2011): 642; A. Cuddy et al., "Stereotype Content Model Across Cultures: Towards Universal Similarities and Some Differences," *Brit J Soc Psych* 48 (2009): 1.
15. R. Nisbett, *The Geography of Thought: How Asians and Westerners Think Differently . . . And Why* (New York: Free Press, 2003).
16. T. Hedden et al., "Cultural Influences on Neural Substrates of Attentional Control," *Psych Sci* 19 (2008): 12; S. Han and G. Northoff, "Culture-Sensitive Neural Substrates of Human Cognition: A Transcultural Neuroimaging Approach," *Nat Rev Nsci* 9 (2008): 646; T. Masuda and R. E. Nisbett, "Attending Holistically vs. Analytically: Comparing the Context Sensitivity of Japanese and Americans," *JPSP* 81 (2001): 922.

17. J. Chiao, "Cultural Neuroscience: A Once and Future Discipline," *Prog Brain Res* 178 (2009): 287.
18. Nisbett, *The Geography of Thought*; Y. Ogihara et al., "Are Common Names Becoming Less Common? The Rise in Uniqueness and Individualism in Japan," *Front Psych* 6 (2015): 1490.
19. A. Mesoudi et al., "How Do People Become W.E.I.R.D.? Migration Reveals the Cultural Transmission Mechanisms Underlying Variation in Psychological Processes," *PLoS ONE* 11 (2016): e0147162.
20. A. Terrazas and J. Batalova, *Frequently Requested Statistics on Immigrants in the United States* (Migration Policy Institute, 2009); J. DeParle, "Global Migration: A World Ever More on the Move," *New York Times*, June 25, 2010; Pew Research Center, "Second-Generation Americans: A Portrait of the Adult Children of Immigrants," February 7, 2013, www.pewsocialtrends.org/2013/02/07/second-generation-americans/.
21. J. Lansing, "Balinese 'Water temples' and the Management of Irrigation," *Am Anthropology* 89 (1987): 326.
22. T. Talhelm et al., "Large-Scale Psychological Differences Within China Explained by Rice Versus Wheat Agriculture," *Sci* 344 (2014): 603.
23. A. Uskul et al., "Ecocultural Basis of Cognition: Farmers and Fishermen Are More Holistic than Herders," *PNAS* 105 (2008): 8552.
24. Z. Dershowitz, "Jewish Subcultural Patterns and Psychological Differentiation," *Int J Psych* 6 (1971): 223.
25. H. Harpending and G. Cochran, "In Our Genes," *PNAS* 99 (2002): 10; F. Chang et al., "The World-wide Distribution of Allele Frequencies at the Human Dopamine D4 Receptor Locus," *Hum Genetics* 98 (1996): 891; K. Kidd et al., "An Historical Perspective on 'The World-wide Distribution of Allele Frequencies at the Human Dopamine D4 Receptor Locus,'" *Hum Genetics* 133 (2014): 431; C. Chen et al., "Population Migration and the Variation of Dopamine D4 Receptor (DRD4) Allele Frequencies Around the Globe," *EHB* 20 (1999): 309.
26. C. Ember and M. Ember, "Warfare, Aggression, and Resource Problems: Cross-Cultural Codes," *Behav Sci Res* 26 (1992): 169; R. Textor, "Cross Cultural Summary: Human Relations Area Files" (1967); H. People and F. Marlowe, "Subsistence and the Evolution of Religion," *Hum Nat* 23 (2012): 253.
27. R. McMahon, *Homicide in Pre-famine and Famine Ireland* (Liverpool, UK: Liverpool University Press, 2013).
28. R. Nisbett and D. Cohen, *Culture of Honor: The Psychology of Violence in the South* (Boulder, CO: Westview Press, 1996).
29. W. Borneman, *Polk: The Man Who Transformed the Presidency and America* (New York: Random House, 2008); B. Wyatt-Brown, *Southern Honor: Ethics and Behavior in the Old South* (Oxford: Oxford University Press, 1982).
30. F. Stewart, *Honor* (Chicago: University of Chicago Press, 1994).
31. D. Fischer, *Albion's Seed* (Oxford: Oxford University Press, 1989).
32. P. Chesler, "Are Honor Killings Simply Domestic Violence?" *Middle East Quarterly*, Spring 2009, pp. 61–69, www.meforum.org/2067/are-honor-killings-simply-domestic-violence.
33. M. Borgerhoff Mulder et al., "Intergenerational Wealth Transmission and the Dynamics of Inequality in Small-Scale Societies," *Sci* 326 (2009): 682.
34. P. Turchin, *War and Peace and War: The Rise and Fall of Empires* (NY: Penguin Press, 2006); D. Rogers et al., "The Spread of Inequality," *PLoS ONE* 6 (2011): e24683.
35. R. Wilkinson, *Mind the Gap: Hierarchies, Health and Human Evolution* (London: Weidenfeld and Nicolson, 2000).
36. F. Elgar et al., "Income Inequality, Trust and Homicide in 33 Countries," *Eur J Public Health* 21, 241; F. Elgar et al., "Income Inequality and School Bullying: Multilevel Study of Adolescents in 37 Countries," *J Adolescent Health* 45 (2009): 351; B. Herrmann et al., "Antisocial Punishment Across Societies," *Sci* 319 (2008): 1362.
37. F. Durante et al., "Nations' Income Inequality Predicts Ambivalence in Stereotype Content: How Societies Mind the Gap," *Brit J Soc Psych* 52 (2012): 726.
38. N. Adler et al., "Relationship of Subjective and Objective Social Status with Psychological and Physiological Functioning: Preliminary Data in Healthy White Women," *Health Psych* 19 (2000): 586; N. Adler and J. Ostrove, "SES and Health: What We Know and What We Don't," *ANYAS* 896 (1999): 3; I. Kawachi et al., "Crime: Social Disorganization and Relative Deprivation," *Soc Sci and Med* 48 (1999): 719; I. Kawachi and B. Kennedy, *The Health of Nations: Why Inequality Is Harmful to Your Health* (New York: New Press, 2002); J. Lynch et al., "Income Inequality, the Psychosocial Environment, and Health: Comparisons of Wealthy Nations," *Lancet* 358 (2001): 194; G. A. Kaplan et al., "Inequality in Income and Mortality in the United States: Analysis of Mortality and Potential Pathways," *Brit Med J* 312 (1996): 999; J. R. Dunn et al., "Income Distribution, Public Services Expenditures, and All Cause Mortality in US States," *J Epidemiology and Community Health* 59 (2005): 768; C. R. Ronzio et al., "The Politics of Preventable Deaths: Local Spending, Income Inequality, and Premature Mortality in US Cities," *J Epidemiology and Community Health* 58 (2004): 175.
39. R. Evans et al., *Why Are Some People Healthy and Others Not? The Determinants of Health of Populations* (New York: Aldine de Gruyter, 1994).
40. D. Chon, "The Impact of Population Heterogeneity and Income Inequality on Homicide Rates: A Cross-National Assessment," *Int J Offender Therapy and Comp Criminology* 56 (2012): 730; F. J. Elgar and N. Aitken, "Income Inequality, Trust and Homicide in 33 Countries," *Eur J Public Health* 21 (2010): 241; C. Hsieh and M. Pugh, "Poverty, Income Inequality, and Violent Crime: A Meta-analysis of Recent Aggregate Data Studies," *Criminal Justice Rev* 18 (1993): 182; M. Daly et al., "Income Inequality and Homicide Rates in Canada and the United States," *Canadian J Criminology* 32 (2001): 219.
41. K. A. DeCellesa and M. I. Norton, "Physical and Situational Inequality on Airplanes Predicts Air Rage," *PNAS* 113 (2016): 5588.
42. M. Balter, "Why Settle Down? The Mystery of Communities," *Sci* 282 (1998): 1442; P. Richerson, "Group Size Determines Cultural Complexity," *Nat* 503 (2013): 351; M. Derex et al., "Experimental Evidence for the Influence of Group Size on Cultural Complexity," *Nat* 503 (2013): 389; A. Gibbons, "How We Tamed Ourselves—and Became Modern," *Sci* 346 (2014): 405.

43. F. Lederbogen et al., "City Living and Urban Upbringing Affect Neural Social Stress Processing in Humans," *Nat* 474 (2011): 498; D. P. Kennedy and R. Adolphs, "Stress and the City," *Nat* 474 (2011): 452; A. Abbott, "City Living Marks the Brain," *Nat* 474 (2011): 429.
44. J. Henrich et al., "Markets, Religion, Community Size, and the Evolution of Fairness and Punishment," *Sci* 327 (2010): 1480; Footnote: B. Maheer, "Good Gaming," *Nat* 531 (2016): 568.
45. A. Norenzayan, *Big Gods: How Religions Transformed Cooperation and Conflict* (Princeton, NJ: Princeton University Press, 2015).
46. L. R. Florizno et al., "Differences Between Tight and Loose Cultures: A 33-Nation Study," *Sci* 332 (2011): 1100.
47. J. B. Calhoun, "Population Density and Social Pathology," *Sci Am* 306 (1962): 139; E. Ramsden, "From Rodent Utopia to Urban Hell: Population, Pathology, and the Crowded Rats of NIMH," *Isis* 102 (2011): 659; J. L. Freedman et al., "Environmental Determinants of Behavioral Contagion," *Basic and Applied Soc Psych* 1 (1980): 155; O. Galle et al., "Population Density and Pathology: What Are the Relations for Man?" *Sci* 176 (1972): 23.
48. A. Parkes, "The Future of Fertility Control," in J. Meade, ed., *Biological Aspects of Social Problems* (NY: Springer, 1965).
49. M. Lim et al., "Global Pattern Formation and Ethnic/Cultural Violence," *Sci* 317 (2007): 1540; A. Rutherford et al., "Good Fences: The Importance of Setting Boundaries for Peaceful Coexistence," *PLoS ONE* 9 (2014): e95660.
50. Florizno et al., "Differences Between Tight and Loose."
51. The following papers examine the effects of normal weather fluctuations, extremes of weather, and global warming on a variety of social end points: J. Brashares et al., "Wildlife Decline and Social Conflict," *Sci* 345 (2014): 376; S. M. Hsiang et al., "Civil Conflicts Are Associated with the Global Climate," *Nat* 476 (2011): 438; A. Solow, "Climate for Conflict," *Nat* 476 (2011): 406; S. Schiermeier, "Climate Cycles Drive Civil War," *Nat* 476 (2011): 406; E. Miguel et al., "Economic Shocks and Civil Conflict: An Instrumental Variables Approach," *J Political Economy* 112 (2004): 725; M. Burke et al., "Warming Increases Risk of Civil War in Africa," *PNAS* 106 (2009): 20670; J. P. Sandholt and K. S. Gleditsch, "Rain, Growth, and Civil War: The Importance of Location," *Defence and Peace Economics* 20 (2009): 359; H. Buhaug, "Climate Not to Blame for African Civil Wars," *PNAS* 107 (2010): 16477; D. D. Zhang et al., "Global Climate Change, War and Population Decline in Recent Human History," *PNAS* 104 (2007): 19214; R. S. J. Tol and S. Wagner, "Climate Change and Violent Conflict in Europe over the Last Millennium," *Climatic Change* 99 (2009): 65; A. Solow, "A Call for Peace on Climate and Conflict," *Nat* 497 (2013): 179; J. Bohannon, "Study Links Climate Change and Violence, Battle Ensues," *Sci* 341 (2013): 444; S. M. Hsiang et al., "Quantifying the Influence of Climate on Human Conflict," *Sci* 341 (2013): 1212.
52. R. Sapolsky, "Endocrine and Behavioral Correlates of Drought in the Wild Baboon," *Am J Primat* 11 (1986): 217.
53. J. Bohannon, "Study Links Climate Change and Violence, Battle Ensues," *Sci* 341 (2013): 444.
54. E. Culotta, "On the Origins of Religion," *Sci* 326 (2009): 784 (this is the source of the quote); C. A. Botero et al., "The Ecology of Religious Beliefs," *PNAS* 111 (2014): 16784; A. Shariff and A. Norenzayan, "God Is Watching You: Priming God Concepts Increases Prosocial Behavior in an Anonymous Economic Game," *Psych Science* 18 (2007): 803; R. Wright, *The Evolution of God* (Boston, MA: Little, Brown, 2009).
55. L. Keeley, *War Before Civilization: The Myth of the Peaceful Savage* (Oxford: Oxford University Press, 1996).
56. S. Pinker, *The Better Angels of Our Nature: Why Violence Has Declined* (New York: Penguin, 2011).
57. G. Milner, "Nineteenth-Century Arrow Wounds and Perceptions of Prehistoric Warfare," *Am Antiquity* 70 (2005): 144.
58. See this entire volume: D. Fry, *War, Peace, and Human Nature: The Convergence of Evolutionary and Cultural Views* (Oxford: Oxford University Press, 2015). In particular, see these chapters in it: R. Ferguson, "Pinker's List: Exaggerating Prehistoric War Mortality," p. 112; R. Sussman "Why the Legend of the Killer Ape Never Dies: The Enduring Power of Cultural Beliefs to Distort Our View of Human Nature," p. 92; and R. Kelly, "From the Peaceful to the Warlike: Ethnographic and Archeological Insights into Hunter-Gatherer Warfare and Homicide," p. 151.
59. F. Wendorf, *The Prehistory of Nubia* (Dallas: Southern Methodist University Press, 1968).
60. R. A. Marlar et al., "Biochemical Evidence of Cannibalism at a Prehistoric Puebloan Site in Southwestern Colorado," *Nat* 407 (2000): 74; M. Balter, "Did Neandertals Dine In?" *Sci* 326 (2009): 1057.
61. N. Chagnon, *Yanomamo: The Fierce People* (NY: Holt McDougal, 1984); N. A. Chagnon, "Life Histories, Blood Revenge, and Warfare in a Tribal Population," *Sci* 239 (1988): 985.
62. A. Lawler, "The Battle over Violence," *Sci* 336 (2012): 829.
63. G. Benjamin et al., "Violence: Finding Peace," *Sci* 338 (2012): 327; S. Pinker, "Violence: Clarified," *Sci* 338 (2012): 327.
64. A. R. Ramos, "Reflecting on the Yanomami: Ethnographic Images and the Pursuit of the Exotic," *Cultural Anthropology* 2 (1987): 284; R. Ferguson, *Yanomami Warfare: A Political History*, a School for Advanced Research Resident Scholar Book (1995); E. Eakin, "How Napoleon Chagnon Became Our Most Controversial Anthropologist," *New York Times Magazine*, 2013, p. 13; D. Fry, *Beyond War: The Human Potential for Peace* (Oxford: Oxford University Press, 2009).
65. L. Glowacki and R. Wrangham, "Warfare and Reproductive Success in a Tribal Population," *PNAS* 112 (2015): 348. For related findings, see: J. Moore, "The Reproductive Success of Cheyenne War Chiefs: A Contrary Case to Chagnon's Yanomamo," *Curr Anthropology* 31 (1990): 322; S. Beckerman et al., "Life Histories, Blood Revenge and Reproductive Success Among the Waorani of Ecuador," *PNAS* 106 (2009): 8134.
66. The original research cited by Pinker and Fry: K. Hill and A. Hurtado, *Ache Life History: The Ecology and Demography of a Foraging People* (New York: Aldine de Gruyter, 1996).

67. S. Corry, "The Case of the 'Brutal Savage': Poirot or Clouseau? Why Steven Pinker, Like Jared Diamond, Is Wrong," London: Survival International website, 2013.
68. K. Lorenz, *On Aggression* (MFJ Books, 1997); R. Ardrey, *The Territorial Imperative: A Personal Inquiry into the Animal Origins of Property and Nations* (Delta Books, 1966); R. Wrangham and D. Peterson, *Demonic Males: Apes and the Origin of Human Violence* (Boston: Houghton Mifflin, 1996).
69. C. H. Boehm, *Hierarchy in the Forest: The Evolution of Egalitarian Behavior* (Cambridge, MA: Harvard University Press, 1999); K. Hawkes et al., "Hunting Income Patterns Among the Hadza: Big Game, Common Goods, Foraging Goals, and the Evolution of the Human Diet," *Philosophical Transactions of the Royal Soc of London B* 334 (1991): 243; B. Chapais, "The Deep Social Structure of Humankind," *Sci* 331 (2011): 1276; K. Hill et al., "Co-residence Patterns in Hunter-Gatherer Societies Show Unique Human Social Structure," *Sci* 331 (2011): 1286; K. Endicott, "Peace Foragers: The Significance of the Batek and Moriori for the Question of Innate Human Violence," in Fry, *War, Peace, and Human Nature*, p. 243; M. Butovskaya, "Aggression and Conflict Resolution Among the Nomadic Hadza of Tanzania as Compared with Their Pastoralist Neighbors," in Fry, *War, Peace, and Human Nature*, p. 278.
70. C. Apicella et al., "Social Networks and Cooperation in Hunter-Gatherers," *Nat* 481 (2012): 497; J. Henrich, "Hunter-Gatherer Cooperation," *Nat* 481 (2012): 449.
71. E. Thomas, *The Harmless People* (New York: Vintage Books, 1959); M. Shostak *Nisa: The Life and Words of a !Kung Woman* (Cambridge, MA: Harvard University Press, 2006); R. Lee, *The !Kung San: Men, Women and Work in a Foraging Society* (Cambridge: Cambridge University Press, 1979).
72. C. Ember, "Myths About Hunter-Gatherers," *Ethnology* 17 (1978): 439.
73. Ferguson 1995, op cit; Fry 2009, op cit; R. B. Lee, "Hunter-Gatherers on the Best-Seller List: Steven Pinker and the 'Bellicose School's' Treatment of Forager Violence," *J Aggression, Conflict and Peace Res* 6 (2014): 216; M. Guenther, "War and Peace Among Kalahari San," *J Aggression, Conflict and Peace Res* 6 (2014): 229; D. P. Fry and P. Soderberg, "Myths About Hunter-Gatherers Redux: Nomadic Forager War and Peace," *J Aggression, Conflict and Peace Res* 6 (2014): 255; R. Kelley, *Warless Societies and the Evolution of War* (Ann Arbor: University of Michigan Press, 2000).
74. M. M. Lahr et al., "Inter-group Violence Among Early Holocene Hunter-Gatherers of West Turkana, Kenya," *Nat* 529 (2016): 394.
75. C. Boehm, *Moral Origins: The Evolution of Virtue, Altruism, and Shame* (New York: Basic Books, 2012).
76. M. C. Stiner et al., "Cooperative Hunting and Meat Sharing 400–200 kya at Qesem Cave, Israel," *PNAS* 106 (2009): 13207.
77. P. Wiessner, "The Embers of Society: Firelight Talk Among the Ju/'hoansi Bushmen," *PNAS* 111 (2014): 14013; P. Wiessner, "Norm Enforcement Among the Ju/'hoansi Bushmen: A Case of Strong Reciprocity?" *Hum Nat* 16 (2004): 115.

Chapter 10: The Evolution of Behavior

1. T. Dobzhansky, "Nothing in Biology Makes Sense Except in the Light of Evolution," *Am Biol Teacher* 35 (1973): 125.
2. A. J. Carter and A. Q. Nguyen, "Antagonistic Pleiotropy as a Widespread Mechanism for the Maintenance of Polymorphic Disease Alleles," *BMC Med Genetics* 12 (2011): 160.
3. J. Gratten et al., "Life History Trade-offs at a Single Locus Maintain Sexually Selected Genetic Variation," *Nat* 502 (2013): 93.
4. A. Brown, *The Darwin Wars: The Scientific Battle for the Soul of Man* (New York: Touchstone/Simon and Schuster, 1999).
5. V. C. Wynne-Edwards, *Evolution Through Group Selection* (London: Blackwell Science, 1986).
6. W. D. Hamilton, "The Genetical Evolution of Social Behavior," *J Theoretical Biol* 7 (1964): 1; G. C. Williams, *Adaptation and Natural Selection* (Princeton, NJ: Princeton University Press, 1966). See also: E. O. Wilson, *Sociobiology: The New Synthesis* (Cambridge, MA: Harvard University Press, 1975); and R. Dawkins, *The Selfish Gene* (Oxford: Oxford University Press, 1976).
7. S. B. Hrdy, *The Langurs of Abu: Female and Male Strategies of Reproduction* (Cambridge, MA: Harvard University Press, 1977).
8. Pathology argument: P. Dolhinow, "Normal Monkeys?" *Am Scientist* 65 (1977): 266. Just overflow of male aggression: R. Sussman et al., "Infant Killing as an Evolutionary Strategy: Reality or Myth?" *Evolutionary Anthropology* 3 (1995): 149.
9. Primates: G. Hausfater and S. Hrdy, *Infanticide: Comparative and Evolutionary Perspectives* (New York: Aldine, 1984); M. Hiraiwa-Hasegawa, "Infanticide in Primates and a Possible Case of Male-Biased Infanticide in Chimpanzees," in *Animal Societies: Theories and Facts*, ed. J. L. Brown and J. Kikkawa (Tokyo: Japan Scientific Societies Press, 1988), pp. 125–39; S. Hrdy, "Infanticide Among Mammals: A Review, Classification, and Examination of the Implications for the Reproductive Strategies of Females," *Ethology and Sociobiology* 1 (1979): 13. Rodents, lions: G. Perrigo et al., "Social Inhibition of Infanticide in Male House Mice," *Ecology Ethology and Evolution* 5 (1993): 181; A. Pusey and C. Packer, 1984, "Infanticide in Carnivores," in Hausfater and Hrdy, *Infanticide*; S. Gursky-Doyen, "Infanticide by a Male Spectral Tarsier (*Tarsius spectrum*)," *Primates* 52 (2011): 385. See also: D. Lukas and E. Huchard, "The Evolution of Infanticide by Males in Mammalian Societies," *Sci* 346 (2014): 841.
10. J. Berger, "Induced Abortion and Social Factors in Wild Horses," *Nat* 303 (1983): 59; E. Roberts et al., "A Bruce Effect in Wild Geladas," Sci 335 (2012): 1222; H. Bruce, "An Exteroceptive Block to Pregnancy in the Mouse," *Nat* 184 (1959): 105.

11. A. Pusey and K. Schroepfer-Walker, "Female Competition in Chimpanzees," *Philosophical Transactions of the Royal Soc of London B* 368 (2013): 1471.
12. D. Fossey, "Infanticide in Mountain Gorillas (*Gorilla gorilla beringei*) with Comparative Notes on Chimpanzees," in Hausfater and Hrdy, *Infanticide.*
13. L. Fairbanks, "Reciprocal Benefits of Allomothering for Female Vervet Monkeys," *Animal Behav* 40 (1990): 553.
14. V. Baglione et al., "Kin Selection in Cooperative Alliances of Carrion Crows," *Sci* 300 (2003): 1947.
15. J. Buchan et al., "True Paternal Care in a Multi-male Primate Society," *Nat* 425 (2003): 179.
16. D. Cheney and R. Seyfarth, *How Monkeys See the World: Inside the Mind of Another Species* (Chicago: University of Chicago Press, 1992).
17. D. Cheney and R. Seyfarth, "Recognition of Other Individuals' Social Relationships by Female Baboons," *Animal Behav* 58 (1999): 67; R. Wittig et al., "Kin-Mediated Reconciliation Substitutes for Direct Reconciliation in Female Baboons," *Proc Royal Soc B* 274 (2007): 1109.
18. T. Bergman et al., "Hierarchical Classification by Rank and Kinship in Baboons," *Sci* 203 (2003): 1234.
19. H. Fisher and H. Hoekstra, "Competition Drives Cooperation Among Closely Related Sperm of Deer Mice," *Nat* 463 (2010): 801.
20. J. Hoogland, "Nepotism and Alarm Calling in the Black-Tailed Prairie Dog (*Cynomys ludovicianus*)," *Animal Behav* 31 (1983): 472; G. Schaller, *The Serengeti Lion: A Study of Predator-Prey Relations* (Chicago: University of Chicago Press, 1972); P. Sherman, "Recognition Systems," in *Behavioural Ecology*, ed. J. R. Krebs and N. B. Davies (Oxford: Blackwell Scientific, 1997); C. Packer et al., "A Molecular Genetic Analysis of Kinship and Cooperation in African Lions," *Nat* 351 (1991): 6327; A. Pusey and C. Packer, "Non-offspring Nursing in Social Carnivores: Minimizing the Costs," *Behav Ecology* 5 (1994): 362.
21. Footnote: G. Alvarez et al., "The Role of Inbreeding in the Extinction of a European Royal Dynasty," *PLoS ONE* 4 (2009): e5174.
22. Theoretical model: B. Bengtsson, "Avoiding Inbreeding: At What Cost?" *J Theoretical Biol* 73 (1978): 439.
23. Insects: S. Robinson et al., "Preference for Related Mates in the Fruit Fly, *Drosophila melanogaster*," *Animal Behav* 84 (2012): 1169. Lizards: M. Richard et al., "Optimal Level of Inbreeding in the Common Lizard," *Proc Royal Soc of London B* 276 (2009): 2779. Fish, and related parents invested more in rearing: T. Thünken et al., "Active Inbreeding in a Cichlid Fish and Its Adaptive Significance," *Curr Biol* 17 (2007): 225. Numerous birds: P. Bateson, "Preferences for Cousins in Japanese Quail," *Nat* 295 (1982): 236; L. Cohen and D. Dearborn, "Great Frigatebirds, *Fregata minor*, Choose Mates That Are Genetically Similar," *Animal Behav* 68 (2004): 1129; N. Burley et al., "Social Preference of Zebra Finches for Siblings, Cousins and Non-kin," *Animal Behav* 39 (1990): 775. Birds sneaking outside monogamy: O. Kleven et al., "Extrapair Mating Between Relatives in the Barn Swallow: A Role for Kin Selection?" *Biol Lett* 1 (2005): 389; C. Wang and X. Lu, "Female Ground Tits Prefer Relatives as Extra-pair Partners: Driven by Kin-Selection?" *Mol Ecology* 20 (2011): 2851. I assume that no one on earth is ever going to read this sentence, so if you do, I'd love to hear from you, in order to congratulate you on your extraordinarily thorough reading habits—sapolsky@stanford.edu. Rodents: S. Sommer, "Major Histocompatibility Complex and Mate Choice in a Monogamous Rodent," *Behav Ecology and Sociobiology* 58 (2005): 181; C. Barnard and J. Fitzsimons, "Kin Recognition and Mate Choice in Mice: The Effects of Kinship, Familiarity and Interference on Intersexual Selection," *Animal Behav* 36 (1988): 1078; M. Peacock and A. Smith, "Nonrandom Mating in Pikas *Ochotona princeps*: Evidence for Inbreeding Between Individuals of Intermediate Relatedness," *Mol Ecology* 6 (1997): 801.
24. A. Helgason et al., "An Association Between the Kinship and Fertility of Human Couples," *Sci* 319 (2008): 813: S. Jacob et al., "Paternally Inherited HLA Alleles Are Associated with Women's Choice of Male Odor," *Nat Genetics* 30 (2002): 175.
25. T. Shingo et al., "Pregnancy-Stimulated Neurogenesis in the Adult Female Forebrain Mediated by Prolactin," *Sci* 299 (2003): 117; C. Larsen and D. Grattan, "Prolactin, Neurogenesis, and Maternal Behaviors," *Brain, Behav and Immunity* 26 (2012): 201.
26. W. D. Hamilton, "The Genetical Evolution of Social Behaviour," *J Theoretical Biol* 7 (1964): 1.
27. S. West and A. Gardner, "Altruism, Spite and Greenbeards," *Sci* 327 (2010): 1341.
28. S. Smukalla et al., "FLO1 Is a Variable Green Beard Gene That Drives Biofilm-like Cooperation in Budding Yeast," *Cell* 135 (2008): 726; E. Queller et al., "Single-Gene Greenbeard Effects in the Social Amoeba *Dictyostelium discoideum*," *Sci* 299 (2003): 105.
29. B. Kerr et al., "Local Dispersal Promotes Biodiversity in a Real-Life Game of Rock-Paper-Scissors," *Nat* 418 (2002): 171; J. Nahum et al., "Evolution of Restraint in a Structured Rock-Paper-Scissors Community," *PNAS* 108 (2011): 10831.
30. G. Wilkinson, "Reciprocal Altruism in Bats and Other Mammals," *Ethology and Sociobiology* 9 (1988): 85; G. Wilkinson, "Reciprocal Food Sharing in the Vampire Bat," *Nat* 308 (1984): 181.
31. W. D. Hamilton, "Geometry for the Selfish Herd," *J Theoretical Biol* 31 (1971): 295.
32. R. Trivers, "The Evolution of Reciprocal Altruism," *Quarterly Rev of Biol* 46 (1971): 35.
33. R. Seyfarth and D. Cheney, "Grooming, Alliances and Reciprocal Altruism in Vervet Monkeys," *Nat* 308 (1984): 541.
34. R. Axelrod and W. D. Hamilton, "The Evolution of Cooperation," *Sci* 211 (1981): 1390.
35. M. Nowak and K. Sigmund, "Tit for Tat in Heterogeneous Populations," *Nat* 355 (1992): 250; R. Boyd, "Mistakes Allow Evolutionary Stability in the Repeated Prisoner's Dilemma Game," *J Theoretical Biol* 136 (1989): 4756.
36. Nowak and R. Highfield, *SuperCooperators: Altruism, Evolution, and Why We Need Each Other to Succeed* (New York: Simon & Schuster, 2012). Footnote: Nowak and K. Sigmund, "A Strategy of Win-Stay, Lose-Shift that Outperforms Tit-for-Tat in the Prisoner's Dilemma Game," *Nat* 364 (1993): 56.

37. E. Fischer, "The Relationship Between Mating System and Simultaneous Hermaphroditism in the Coral Reef Fish, *Hypoplectrus nigricans* (Serranidae)," *Animal Behav* 28 (1980): 620.
38. M. Milinski, "Tit for Tat in Sticklebacks and the Evolution of Cooperation," *Nat* 325 (1987): 433.
39. C. Packer et al., "Egalitarianism in Female African Lions," *Sci* 293 (2001): 690; M. Scantlebury et al., "Energetics Reveals Physiologically Distinct Castes in a Eusocial Mammal," *Nat* 440 (2006): 795; R. Heinsohn and C. Packer, "Complex Cooperative Strategies in Group-Territorial African Lions," *Sci* 269 (1995): 1260.
40. R. Trivers, "Parent-Offspring Conflict," *Am Zoologist* 14 (1974): 249.
41. D. Maestripieri, "Parent-Offspring Conflict in Primates," *Int J Primat* 23 (2002): 923.
42. D. Haig, "Genetic Conflicts in Human Pregnancy," *Quartery Rev of Biol* 68 (1993): 495; R. Sapolsky, "The War Between Men and Women," *Discover*, May 1999, p. 56.
43. S. J. Gould, "Caring Groups and Selfish Genes," in *The Panda's Thumb: More Reflections in Natural History* (London: Penguin Books, 1990), p. 72.
44. S. Okasha, *Evolution and the Levels of Selection* (Oxford: Clarendon Press, 2006).
45. P. Bijma et al., "Multilevel Selection 1: Quantitative Genetics of Inheritance and Response to Selection," *Genetics* 175 (2007): 277. A similar example to the chickens, in spiders: J. Pruitt and C. Goodnight, "Site-Specific Group Selection Drives Locally Adapted Group Compositions," *Nat* 514 (2014): 359.
46. S. Bowles, "Conflict: Altruism's Midwife," *Nat* 456 (2008): 326.
47. D. S. Wilson and E. O. Wilson, "Rethinking the Theoretical Foundation of Sociobiology," *Quarterly Rev of Biol* 82 (2008): 327.
48. F. de Waal, *Our Inner Ape* (NY: Penguin, 2005); I. Parker, "Swingers: Bonobos Are Celebrated as Peace-Loving, Matriarchal, and Sexually Liberated. Are They?" *New Yorker*, July 30, 2007, p. 48; R. Wrangham and D. Peterson, *Demonic Males: Apes and the Origins of Human Violence* (NY: Houghton Mifflin, 1996); R. Wrangham et al., "Comparative Rates of Violence in Chimpanzees and Humans," *Primates* 47 (2006): 14.
49. D. Falk et al., "Brain Shape in Human Microcephalics and *Homo floresiensis*," *PNAS* 104 (2007): 2513. The opposite view: M. Henneberg et al., "Evolved Developmental Homeostasis Disturbed in LB1 from Flores, Indonesia, Denotes Down Syndrome and Not Diagnostic Traits of the Invalid Species *Homo floresiensis*," *PNAS* 111 (2014): 11967.
50. K. Prufer et al., "The Bonobo Genome Compared with the Chimpanzee and Human Genomes," *Nat* 486 (2012): 527; W. Enard et al., "Intra- and Interspecific Variation in Primate Gene Expression Patterns," *Sci* 296 (2002): 340.
51. D. Barash and J. Lipton, *The Myth of Monogamy: Fidelity and Infidelity in Animals and People* (New York: Henry Holt, 2002); B. Chapais, *Primeval Kinship: How Pair-Bonding Gave Birth to Human Society* (Cambridge, MA: Harvard University Press).
52. T. Zerjal et al., "The Genetic Legacy of the Mongols," *Am J Hum Genetics* 72 (2003): 713.
53. M. Daly and M. Wilson, "Evolutionary Social Psychology and Family Homicide," Sci 242 (1988): 519. Replication: V. Weekes-Shackelford and T. K. Shackelford, "Methods of Filicide: Stepparents and Genetic Parents Kill Differently," *Violence and Victims* 19 (2004): 75. Swedish failures of replication: H. Temrin et al., "Step-Parents and Infanticide: New Data Contradict Evolutionary Predictions," *Proc Royal Soc B* 267 (2000): 943; M. Van Ijzendoorn et al., "Elevated Risk of Child Maltreatment in Families with Stepparents but Not with Adoptive Parents," *Child Maltreatment* 14 (2009): 369.; J. Nordlund and H. Temrin, "Do Characteristics of Parental Child Homicide in Sweden Fit Evolutionary Predictions?" *Ethology* 113 (2007): 1029.
54. K. Hill et al., "Co-residence Patterns in Hunter-Gatherer Societies Show Unique Human Social Structure," *Sci* 331 (2011): 1286.
55. R. Topolski et al., "Choosing Between the Emotional Dog and the Rational Pal: A Moral Dilemma with a Tail," *Anthrozoös* 26 (2013): 253.
56. B. Thomas et al., "Harming Kin to Save Strangers: Further Evidence for Abnormally Utilitarian Moral Judgments After Ventromedial Prefrontal Damage," *J Cog Nsci* 23 (2011): 2186.
57. R. Sapolsky, "Would You Break That Law for Your Family?" *Los Angeles Times*, November 17, 2013.
58. J. Persico, *My Enemy, My Brother: Men and Days of Gettysburg* (Cambridge, MA: Da Capo Press, 1996).
59. R. MacMahon, *Homicide in Pre-famine and Famine Ireland* (Liverpool, UK: Liverpool University Press, 2014). Cheeseburger murder: J. Berlinger and T. Marco, "Man Kills Brother in Argument over Cheeseburger, Police Say," CNN.com, May 9, 2016, www.cnn.com/2016/05/08/us/man-allegedly-kills-brother-over-cheeseburger/index.html.
60. Footnote: "MP Comes to the Aid of 5 Year Old Girl at Risk of Being Sold," *Kenya Daily Nation*, October 13, 2014, www.nation.co.ke/video/-/1951480/2484684/-/gditgq/-/index.html.
61. S. Friedman and P. Resnick, "Child Murder by Mothers: Patterns and Prevention," *World Psychiatry* 6 (2007): 137; S. West, et al., "Fathers Who Kill Their Children: An Analysis of the Literature," *J Forensic Sci* 54 (2009): 463; S. B. Hrdy, *Mother Nature: A History of Mothers, Infants and Natural Selection* (New York: Pantheon, 1999).
62. J. Shepher, "Mate Selection Among Second Generation Kibbutz Adolescents and Adults: Incest Avoidance and Negative Imprinting," *Arch Sexual Behav* 1 (1971): 293; A. Wolf, *Sexual Attraction and Childhood Association: A Chinese Brief for Edward Westermarck* (Palo Alto, CA: Stanford University Press, 1995).
63. K. Hill et al., "Co-residence Patterns in Hunter-Gatherer Societies Show Unique Human Social Structure," *Sci* 331 (2011): 1286.
64. N. Eldredge and S. J. Gould, "Punctuated Equilibria: An Alternative to Phyletic Gradualism," in *Models in Paleobiology*, ed. T. J. M. Schopf (San Francisco: Freeman Cooper, 1972), p. 82.
65. J. Goldman, "Man's New Best Friend? A Forgotten Russian Experiment in Fox Domestication," *Sci Am*, September 2010; D. Belyaev and L. Trut, "Behaviour and Reproductive Function of Animals. II: Correlated Changes Under Breeding for Tameness," *Bull Moscow Soc of Naturalists B Series* (in Russian) 69 (1964): 5.

66. S. Sternthal, "Moscow's Stray Dogs," *Financial Times*, January 16, 2010.
67. Footnote: M. Carneiro et al., "Rabbit Genome Analysis Reveals a Polygenic Basis for Phenotypic Change During Domestication," *Sci* 345 (2014): 1074.
68. S. Fisher and M. Ridley, "Culture, Genes, and the Human Revolution," *Sci* 340 (2013) 929; D. Swallow, "Genetics of Lactase Persistence and Lactose Intolerance," *Ann Rev of Genetics* 37 (2003): 197; J. Troelsen, "Adult-Type Hypolactasia and Regulation of Lactase Expression," *Biochimica et Biophysica Acta* 1723 (2005): 19.
69. N. Mekel-Bobrov et al., "Ongoing Adaptive Evolution of ASPM, a Brain Size Determinant in *Homo sapiens*," *Sci* 309 (2005): 1720.
70. J. Weiner, *The Beak of the Finch: A Story of Evolution in Our Time* (New York: Knopf, 1994); J. Neel, "Diabetes Mellitus: A 'Thrifty' Genotype Rendered Detrimental by 'Progress'?" *Am J Hum Genetics* 14 (1962): 353; J. Diamond, "Sweet Death," *Natural History*, February 1992. American versus Mexican Pimas: P. Kopelman, "Obesity as a Medical Problem," *Nat* 404 (2000): 635. Genes identified: C. Ezzell, "Fat Times for Obesity Research," *J NIH Research* 7 (1995): 39; C. Holden, "Race and Medicine," *Sci* 302 (2003): 594; J. Diamond, "The Double Puzzle of Diabetes," *Nat* 423 (2003): 599.
71. E. Pennisi, "The Man Who Bottled Evolution," *Sci* 342 (2013): 790.
72. S. J. Gould and N. Eldredge, "Punctuated Equilibria: The Tempo and Mode of Evolution Reconsidered," *Paleobiology* 3 (1977): 115.
73. P. W. Andrews et al., "Adaptationism—How to Carry Out an Exaptationist Program," *BBS* 25 (2002): 489; S. J. Gould and E. S. Vrba, "Exaptation—a Missing Term in the Science of Form," *Paleobiology* 8 (1982): 4; A. Figueredo and S. Berry, "'Just Not So Stories': Exaptations, Spandrels, and Constraints," *BBS* 25 (2002): 517; J. Roney and D. Maestripieri, "The Importance of Comparative and Phylogenetic Analyses in the Study of Adaptation," *BBS* 25 (2002): 525.
74. A. Brown, *The Darwin Wars: The Scientific Battle for the Soul of Man* (New York: Touchstone/Simon and Schuster, 1999).
75. S. J. Gould and R. Lewontin, "The Spandrels of San Marco and the Panglossian Paradigm: A Critique of the Adaptationist Programme," *Proc Royal Soc of London B* 205 (1979): 581.
76. D. Barash and J. Lipton, "How the Scientist Got His Ideas," *Chronicle of Higher Education*, January 3, 2010.

Chapter 11: Us Versus Them

1. D. Hofstede, *Planet of the Apes: An Unofficial Companion* (Toronto: ECW Press, 2001).
2. T. A. Ito and G. R. Urland, "Race and Gender on the Brain: Electrocortical Measures of Attention to the Race and Gender of Multiply Categorizable Individuals," *JPSP* 85 (2003): 616; T. Ito and B. Bartholow, "The Neural Correlates of Race," *TICS* 13 (2009): 524.
3. A. Greenwald et al., "Measuring Individual Differences in Implicit Cognition: The Implicit Association Test," *JPSP* 74 (1998): 1464.
4. N. Mahajan et al., "The Evolution of Intergroup Bias: Perceptions and Attitudes in Rhesus Macaques," *JPSP* 100 (2011): 387.
5. H. Tajfel, "Social Psychology of Intergroup Relations," *Ann Rev of Psych* 33 (1982): 1; H. Tajfel, "Experiments in Intergroup Discrimination," *Sci Am* 223 (1970): 96.
6. E. Losin et al., "Own-Gender Imitation Activates the Brain's Reward Circuitry," *SCAN* 7 (2012): 804; B. C. Müller et al., "Prosocial Consequences of Imitation," *Psych Rep* 110 (2012): 891.
7. S. B. Flagel et al., "A Selective Role for Dopamine in Stimulus-Reward Learning," *Nat* 469 (2011): 53–57.
8. A. S. Baron and M. R. Banaji, "The Development of Implicit Attitudes: Evidence of Race Evaluations from Ages 6, 10, and Adulthood," *Psych Sci* 17 (2006): 53; F. E. Aboud, *Children and Prejudice* (New York: Blackwell, 1988); R. S. Bigler et al., "Social Categorization and the Formation of Intergroup Attitudes in Children," *Child Development* 68 (1997): 530; L. A. Hirschfeld, "Natural Assumptions: Race, Essence and Taxonomies of Human Kinds," *Soc Res* 65 (1998): 331; R. S. Bigler et al., "Developmental Intergroup Theory: Explaining and Reducing Children's Social Stereotyping and Prejudice," *Curr Dir Psych Sci* 16 (2007): 162; P. Bronson and A. Merryman, "See Baby Discriminate," *Newsweek*, September 14, 2009, p. 53 (from their book, *Nurture Shock*).
9. K. D. Kinzler et al., "The Native Language of Social Cognition," *PNAS* 104 (2007); 12577; S. Sangrigoli and S. De Schonen, "Recognition of Own-Race and Other-Race Faces by Three-Month-Old Infants," *J Child Psych and Psychiatry* 45 (2004): 1219.
10. S. Sangrigoli et al., "Reversibility of the Other-Race Effect in Face Recognition During Childhood," *Psych Sci* 16 (2005): 440.
11. R. Bigler and L. Liben, "Developmental Intergroup Theory: Explaining and Reducing Children's Social Stereotyping and Prejudice," *Curr Dir Psych Sci* 16 (2007): 162.
12. A. J. Cuddy et al., "Stereotype Content Model Across Cultures: Towards Universal Similarities and Some Differences," *Brit J Soc Psych* 48 (2009): 1; H. Bernhard et al., "Parochial Altruism in Humans," *Nat* 442 (2006): 912.
13. M. Levine et al., "Self-Categorization and Bystander Non-intervention: Two Experimental Studies," *J Applied Soc Psych* 32 (2002): 1452; J. M. Engelmann and E. Hermann, "Chimpanzees Trust Their Friends," *Curr Biol* 26 (2016): 252.
14. M. Levine et al., "Identity and Emergency Intervention: How Social Group Membership and Inclusiveness of Group Boundaries Shape Helping Behavior," *PSPB* 31 (2005): 443.
15. H. A. Hornstein et al., "Effects of Sentiment and Completion of a Helping Act on Observer Helping: A Case for Socially Mediated Zeigarnik Effects," *JPSP* 17 (1971): 107.
16. L. Gaertner and C. Insko, "Intergroup Discrimination in the Minimal Group Paradigm: Categorization, Reciprocation, or Fear?" *JPSP* 79 (2000): 77; T. Wildschut et al., "Intragroup Social Influence and Intergroup

Competition," *JPSP* 82 (2002): 975; C. A. Insko et al., "Interindividual-Intergroup Discontinuity as a Function of Trust and Categorization: The Paradox of Expected Cooperation," *JPSP* 88 (2005): 365.

17. M. Cikara et al., "Us Versus Them: Social Identity Shapes Neural Responses to Intergroup Competition and Harm," *Psych Sci* 22 (2011): 306; E. R. de Bruijn et al., "When Errors Are Rewarding," *J Nsci* 29 (2009): 12183; J. J. Van Bavel et al., "Modulation of the Fusiform Face Area Following Minimal Exposure to Motivationally Relevant Faces: Evidence of In-group Enhancement (Not Out-group Disregard)," *J Cog Nsci* 223 (2011): 3343. Footnote: M. Cikar et al., "Their Pain Gives Us Pleasure: How Intergroup Dynamics Shape Empathic Failures and Counter-empathic Responses," *JESP* 55 (2014) 110.
18. T. Singer et al., "Empathic Neural Responses Are Modulated by the Perceived Fairness of Others," *Nat* 439 (2006): 466; H. Takahashi et al., "When Your Gain Is My Pain and Your Pain Is My Gain: Neural Correlates of Envy and Schadenfreude," *Sci* 323 (2009): 937.
19. G. Hertel and N. L. Kerr, "Priming In-group Favoritism: The Impact of Normative Scripts in the Minimal Group Paradigm," *JESP* 37 (2001): 316.
20. J. N. Gutsell and M. Inzlicht, "Intergroup Differences in the Sharing of Emotive States: Neural Evidence of an Empathy Gap," *SCAN* 7 (2012): 596; J. Y. Chiao et al., "Cultural Specificity in Amygdala Response to Fear Faces," *J Cog Nsci* 20 (2008): 2167.
21. P. K. Piff et al., "Me Against We: In-group Transgression, Collective Shame, and In-group-Directed Hostility," *Cog & Emotion* 26 (2012): 634.
22. W. Barrett, "Thug Life: The Shocking Secret History of Harold Giuliani, the Mayor's Ex-Convict Dad," *Village Voice*, 5 July, 2000; D. Strober and G. Strober, *Giuliani: Flawed or Flawless?* (New York: Wiley, 2007).
23. Footnote: J. A. Lukas, "Judge Hoffman Is Taunted at Trial of the Chicago 7 After Silencing Defense Counsel," *New York Times*, February 6, 1970.
24. S. Svonkin, *Jews Against Prejudice: American Jews and the Fight for Civil Liberties* (New York: Columbia University Press, 1997). Footnote: A. Zahr, "I Refuse to Condemn," *Civil Arab*, January 9, 2015, www.civilarab.com/i-refuse-to-condemn/.
25. D. A. Stanley et al., "Implicit Race Attitudes Predict Trustworthiness Judgments and Economic Trust Decisions," *PNAS* 108 (2011): 7710; Y. Dunham, "An Angry = Outgroup Effect," *JESP* 47 (2011): 668; D. Maner et al., "Functional Projection: How Fundamental Social Motives Can Bias Interpersonal Perception," *JPSP* 88 (2005): 63; K. Hugenberg and G. Bodenhausen, "Facing Prejudice: Implicit Prejudice and the Perception of Facial Threat," *Psych Sci* 14 (2003): 640; A. Rattan et al., "Race and the Fragility of the Legal Distinction Between Juveniles and Adults," *PLoS ONE* 7 (2012): e36680; Y. J. Xiao and J. J. Van Bavel, "See Your Friends Close and Your Enemies Closer: Social Identity and Identity Threat Shape the Representation of Physical Distance," *PSPB* 38 (2012): 959; B. Reiek et al., "Intergroup Threat and Outgroup Attitudes: A Meta-analytic Review," *PSPR* 10 (2006): 336; H. A. Korn, et al., "Neurolaw: Differential Brain Activity for Black and White Faces Predicts Damage Awards in Hypothetical Employment Discrimination Cases," *Soc Nsci* 7 (2012): 398. Activation of insula when interacting with outgroup in game: J. Rilling et al., "Social Cognitive Neural Networks During In-group and Out-group Interactions," *NeuroImage* 41 (2008): 1447.
26. P. Rozin et al., "From Oral to Moral," Science 323 (2009): 1179.
27. G. Hodson and K. Costello, "Interpersonal Disgust, Ideological Orientations, and Dehumanization as Predictors of Intergroup Attitudes," *Psych Sci* 18 (2007):691.
28. G. Hodson et al., "A Joke Is Just a Joke (Except When It Isn't): Cavalier Humor Beliefs Facilitate the Expression of Group Dominance Motives," *JPSP* 99 (2010): 460.
29. D. Berreby, *Us and Them: The Science of Identity* (Chicago: University of Chicago Press, 2008).
30. Leyens et al., "The Emotional Side of Prejudice: The Attribution of Secondary Emotions to Ingroups and Outgroups," *PSPR* 4 (2000): 186; K. Wailoo, *Pain: A Political History* (Baltimore: Johns Hopkins University Press, 2014).
31. J. T. Jost and O. Hunyad, "Antecedents and Consequences of System-Justifying Ideologies," *Curr Dir Psych Sci* 14 (2005): 260; G. E. Newman and P. Bloom, "Physical Contact Influences How Much People Pay at Celebrity Auctions," *PNAS* 111 (2013): 3705.
32. J. Greenberg et al., "Evidence for Terror Management II: The Effects of Mortality Salience on Reactions to Those Who Threaten or Bolster the Cultural Worldview," *JPSP* 58 (1990): 308.
33. J. Haidt, "The Emotional Dog and Its Rational Tail: A Social Intuitionist Approach to Moral Judgment," *Psych Rev* 108 (2001): 814; J. Haidt, *The Righteous Mind: Why Good People Are Divided by Politics and Religion* (New York: Pantheon Books, 2012).
34. Berreby, *Us and Them*.
35. W. Cunningham et al., "Implicit and Explicit Ethnocentrism: Revisiting the Ideologies of Prejudice," *PSPB* 30 (2004): 1332.
36. Footnote: M. J. Wood et al., "Dead and Alive: Beliefs in Contradictory Conspiracy Theories," *Social Psych and Personality Sci* 3 (2012): 767.
37. C. Zogmaister et al., "The Impact of Loyalty and Equality on Implicit Ingroup Favoritism," *Group Processes & Intergroup Relations* 11 (2008): 493.
38. C. D. Navarrete et al., "Race Bias Tracks Conception Risk Across the Menstrual Cycle," *Psych Sci* 20 (2009): 661. C. Navarrete et al., "Fertility and Race Perception Predict Voter Preference for Barack Obama," *EHB* 31 (2010): 391.
39. G. E. Newman and P. Bloom, "Physical Contact Influences How Much People Pay at Celebrity Auctions," *PNAS* 111 (2013): 3705; R. Sapolsky, "Magical Thinking and the Stain of Madoff's Sweater," *Wall Street Journal*, July 12, 2014.

40. Footnote: S. Boria, *Animals in the Third Reich: Pets, Scapegoats, and the Holocaust* (Providence, RI: Yogh and Thorn Books, 2000).
41. A. Rutland and R. Brown, "Stereotypes as Justification for Prior Intergroup Discrimination: Studies of Scottish National Stereotyping," *Eur J Soc Psych* 31 (2001): 127.
42. C. S. Crandall et al., "Stereotypes as Justifications of Prejudice," *PSPB* 37 (2011): 1488.
43. R. Niebuhr, *The Nature and Destiny of Man*, vol. 1 (London: Nisbet, 1941); B. P. Meier and V. B. Hinsz, "A Comparison of Human Aggression Committed by Groups and Individuals: An Interindividual Intergroup Discontinuity," *JESP* 40 (2004): 551; T. Wildschut et al., "Beyond the Group Mind: A Quantitative Review of the Interindividual-Intergroup Discontinuity Effect," *Psych Bull* 129 (2003): 698.
44. T. Cohen et al., "Group morality and Intergroup Relation: Cross-Cultural and Experimental Evidence," *PSPB* 32 (2006): 1559; T. Wildschut et al., "Intragroup Social Influence and Intergroup Competition," *JPSP* 82 (2002): 975.
45. S. Bowles, "Conflict: Altruism's Midwife," *Nat* 456 (2008): 326.
46. M. Shih et al., "Stereotype Susceptibility: Identity Salience and Shifts in Quantitative Performance," *Psych Sci* 10 (1999): 80; T. Harada et al., "Dynamic Social Power Modulates Neural Basis of Math Calculation," *Front Hum Nsci* 6 (2012): 350; J. Van Bavel and W. Cunningham, "Self-Categorization with a Novel Mixed-Race Group Moderates Automatic Social and Racial Biases," *PSPB* 35 (2009): 321; G. Bohner et al., "Situational Flexibility of In-group-Related Attitudes: A Single Category IAT Study of People with Dual National Identity," *Group Processes & Intergroup Relations* 11 (2008): 301.
47. N. Jablonski, *Skin: A Natural History* (Oakland, CA: University of California Press, 2006): A. Gibbons, "Shedding Light on Skin Color," *Sci* 346 (2014): 934.
48. R. Hahn, "Why Race Is Differentially Classified on U.S. Birth and Infant Death Certificates: An Examination of Two Hypotheses," *Epidemiology* 10 (1999): 108.
49. C. D. Navarrete et al., "Fear Extinction to an Out-group Face: The Role of Target Gender," *Psych Sci* 20 (2009): 155; J. P. Mitchell et al., "Contextual Variations in Implicit Evaluation," *J Exp Psych: General* 132 (2003): 455; this latter paper is the one involving politicians versus athletes.
50. R. Kurzban et al., "Can Race Be Erased? Coalitional Computation and Social Categorization," *PNAS* 98 (2001): 15387.
51. M. E. Wheeler and S. T. Fiske, "Controlling Racial Prejudice: Social-Cognitive Goals Affect Amygdala and Stereotype Activation," *Psych Sci* 16 (2005): 56; J. P. Mitchell et al., "The Link Between Social Cognition and Self-Referential Thought in the Medial Prefrontal Cortex," *J Cog Nsci* 17 (2005): 1306.
52. M. A. Halleran, *The Better Angels of Our Nature: Freemasonry in the American Civil War* (Tuscaloosa AL: : University of Alabama Press, 2010).
53. T. Kennealy, *The Great Shame: And the Triumph of the Irish in the English-Speaking World* (New York: Anchor Books, 2000).
54. Patrick Leigh Fermor obituary, *Daily Telegraph* (London), June, 11, 2011. For footage of the reunion with Kreipe, see "Η ΑΠΑΓΩΓΗ ΤΟΥ ΣΤΡΑΤΗΓΟΥ ΚΡΑΙΠΕ," uploaded by Idomeneas Kanakakis on October 21, 2010, www.youtube.com/watch?v=8zlUhJwddFU. For a documentary about the kidnapping and journey, see "The Abduction of Gengeral Kreipe.avi," uploaded by Nico Mastorakis on February 25, 2012, www.youtube.com/watch?v=vN1qrghgCqI.
55. E. Krusemark and W. Li, "Do All Threats Work the Same Way? Divergent Effects of Fear and Disgust on Sensory Perception and Attention," *J Nsci* 31 (2011): 3429.
56. Footnote: M. Plitt et al., "Are Corporations People Too? The Neural Correlates of Moral Judgments About Companies and Individuals," *Social Nsci* 10 (2015): 113.
57. S. Fiske et al., "A Model of (Often Mixed) Stereotype Content: Competence and Warmth Respectively Follow from Perceived Status and Competition," *JPSP* 82 (2002): 878; L. T. Harris and S. T. Fiske, "Dehumanizing the Lowest of the Low: Neuroimaging Responses to Extreme Out-groups," *Psych Sci* 17 (2006): 847; L. T. Harris and S. T. Fiske, "Social Groups That Elicit Disgust Are Differentially Processed in mPFC," *SCAN* 2 (2007): 45. Also see: S. Morrison et al., "The Neuroscience of Group Membership," *Neuropsychologia* 50 (2012): 2114.
58. T. Ashworth, *Trench Warfare: 1914–1918* (London: Pan Books, 1980).
59. K. B. Clark and M. P. Clark, "Racial Identification and Preference Among Negro Children," in *Readings in Social Psychology*, ed. E. L. Hartley (New York: Holt, Rinehart, and Winston, 1947); K. Clark and C. Mamie, "The Negro Child in the American Social Order," *J Negro Education* 19 (1950): 341; J. Jost et al., "A Decade of System Justification Theory: Accumulated Evidence of Conscious and Unconscious Bolstering of the Status Quo," *Political Psych* 25 (2004): 881; J. Jost et al., "Non-conscious Forms of System Justification: Implicit and Behavioral Preferences for Higher Status Groups," *JESP* 38 (2002): 586.
60. S. Lehrman, "The Implicit Prejudice," *Sci Am* 294 (2006): 32.
61. K. Kawakami et al., "Mispredicting Affective and Behavioral Responses to Racism," *Sci* 323 (2009): 276; B. Nosek, "Implicit-Explicit Relations," *Curr Dir Psych Sci* 16 (2007): 65; L. Rudman and R. Ashmore, "Discrimination and the Implicit Association Test," *Group Processes & Intergroup Relations* 10 (2007): 359; J. Dovidio et al., "Implicit and Explicit Prejudice and Interracial Interaction," *JPSP* 82 (2002): 62. For an additional approach to uncovering implicit biases, see I. Blair, "The Malleability of Automatic Stereotypes and Prejudice," *PSPR* 6 (2002): 242.
62. W. Cunningham et al., "Separable Neural Components in the Processing of Black and White Faces," *Psych Sci* 15 (2004): 806; W. A. Cunningham et al., "Neural Correlates of Evaluation Associated with Promotion and Prevention Regulatory Focus," *Cog, Affective & Behav Nsci* 5 (2005): 202; K. M. Knutso et al., "Neural Correlates of Automatic Beliefs About Gender and Race," *Hum Brain Mapping* 28 (2007): 915.

63. B. K. Payne, "Conceptualizing Control in Social Cognition: How Executive Functioning Modulates the Expression of Automatic Stereotyping," *JPSP* 89 (2005): 488.
64. J. Dovidio et al., "Why Can't We Just Get Along? Interpersonal Biases and Interracial Distrust," *Cultural Diversity & Ethnic Minority Psych* 8 (2002): 88.
65. J. Richeson et al., "An fMRI Investigation of the Impact of Interracial Contact on Executive Function," *Nat Nsci* 12 (2003): 1323; J. Richeson and J. Shelton, "Negotiating Interracial Interactions: Cost, Consequences, and Possibilities," *Curr Dir Psych Sci* 16 (2007): 316.
66. J. N. Shelton et al., "Expecting to Be the Target of Prejudice: Implications for Interethnic Interactions," *PSPB* 31 (2005): 1189.
67. P. M. Herr, "Consequences of Priming: Judgment and Behavior," *JPSP* 51 (1986): 1106; N. Dasgupta and A. Greenwald, "On the Malleability of Automatic Attitudes: Combating Automatic Prejudice with Images of Admired and Disliked Individuals," *JPSP* 81 (2001): 800.
68. W. A. Cunningham et al., "Rapid Social Perception Is Flexible: Approach and Avoidance Motivational States Shape P100 Responses to Other-Race Faces," *Front Hum Nsci* 6 (2012): 140.
69. A. D. Galinsky and G. B. Moskowitz, "Perspective-Taking: Decreasing Stereotype Expression, Stereotype Accessibility, and In-group Favoritism," *JPSP* 78 (2000): 708; I. Blair et al., "Imagining Stereotypes Away: The Moderation of Implicit Stereotypes Through Mental Imagery," *JPSP* 81 (2001): 828; T. J. Allen et al., "Social Context and the Self-Regulation of Implicit Bias," *Group Processes & Intergroup Relations* 13 (2010): 137; J. Fehr and K. Sassenberg, "Willing and Able: How Internal Motivation and Failure Help to Overcome Prejudice," *Group Processes & Intergroup Relations* 13 (2010): 167.
70. C. Macrae et al., "The Dissection of Selection in Person Perception: Inhibitory Processes in Social Stereotyping," *JPSP* 69 (1995): 397.
71. T. Pettigrew and L. A. Tropp, "A Meta-analytic Test of Intergroup Contact Theory," *JPSP* 90 (2006): 751.
72. A. Rutherford et al., "Good Fences: The Importance of Setting Boundaries for Peaceful Coexistence," *PLoS ONE* 9 (2014): e95660; L. G. Babbitt and S. R. Sommers, "Framing Matters: Contextual Influences on Interracial Interaction Outcomes," *PSPB* 37 (2011): 1233.
73. M. J. Williams and J. L. Eberhardt, "Biological Conceptions of Race and the Motivation to Cross Racial Boundaries," *JPSP* 94 (2008): 1033.
74. G. Hodson et al., "A Joke Is Just a Joke (Except When It Isn't): Cavalier Humor Beliefs Facilitate the Expression of Group Dominance Motives," *JPSP* 99 (2010): 460; F. Pratto and M. Shih, "Social Dominance Orientation and Group Context in Implicit Group Prejudice," *Psych Sci* 11 (2000): 515; F. Pratto et al., "Social Dominance Orientation and the Legitimization of Inequality Across Cultures," *J Cross-Cultural Psych* 31 (2000); 369; F. Durante et al., "Nations' Income Inequality Predicts Ambivalence in Stereotype Content: How Societies Mind the Gap," *Brit J Soc Psych* 52 (2012): 726; A. C. Kay and J. T. Jost, "Complementary Justice: Effects of 'Poor but Happy' and 'Poor but Honest' Stereotype Exemplars on System Justification and Implicit Activation of the Justice Motive," *JPSP* 85 (2003): 823; A Kay, et al., "Victim Derogation and Victim Enhancement as Alternate Routes to System Justification," *Psych Sci* 16 (2005): 240.
75. C. Sibley and J. Duckitt, "Personality and Prejudice: A Meta-analysis and Theoretical Review," *PSPR* 12 (2008): 248.
76. J. Dovidio et al., "Commonality and the Complexity of 'We': Social Attitudes and Social Change.," *PSPR* 13 (2013): 3; E. Hehman et al., "Group Status Drives Majority and Minority Integration Preferences," *Psych Sci* 23 (2011): 46.
77. A demonstration that a reward shared with an in-group member activates dopaminergic reward pathways more than does the same reward shared with a stranger: J. B. Freeman and D. Fareri et al., "Social Network Modulation of Reward-Related Signals," *J Nsci* 32 (2012): 9045.

Chapter 12: Hierarchy, Obedience, and Resistance

1. J. Freeman et al., "The Part: Social Status Cues Shape Race Perception," *PLoS ONE* 6 (2011): e25107.
2. Footnote: George, "Faith and Toilets," *Sci Am*, November 19, 2015.
3. R. I. Dunbar and S. Shultz, "Evolution in the Social Brain," *Sci* 317 (2007): 1344; R. I. Dunbar, "The Social Brain Hypothesis and Its Implications for Social Evolution," *Ann Hum Biol* 36 (2009): 562; F. J. Pérez-Barbería et al. "Evidence for Coevolution of Sociality and Relative Brain Size in Three Orders of Mammals," *Evolution* 61 (2007): 2811; J. Powell et al., "Orbital Prefrontal Cortex Volume Predicts Social Network Size: An Imaging Study of Individual Differences in Humans," *Proc Royal Soc B: Biol Sci* 279 (2012): 2157; P. A. Lewis et al., "Ventromedial Prefrontal Volume Predicts Understanding of Others and Social Network Size," *Neuroimage* 57 (2011): 1624; J. L. Powell et al., "Orbital Prefrontal Cortex Volume Correlates with Social Cognitive Competence," *Neuropsychologia* 48 (2010): 3554; J. Lehmann and R. I. Dunbar, "Network Cohesion, Group Size and Neocortex Size in Female-Bonded Old World Primates," *Proc Royal Soc B: Biol Sci* 276 (2009): 4417; J. Sallet et al., "Social Network Size Affects Neural Circuits in Macaques," *Sci* 334 (2011): 697.
4. F. Amici et al., "Fission-Fusion Dynamics, Behavioral Flexibility, and Inhibitory Control in Primates," *Curr Biol* 18 (2008): 1415; A. B. Bond et al., "Serial Reversal Learning and the Evolution of Behavioral Flexibility in Three Species of North American Corvids (*Gymnorhinus cyanocephalus, Nucifraga columbiana, Aphelocoma californica*)," *JCP* 121 (2007): 372; A. Bond et al., "Social Complexity and Transitive Inference in Corvids," *Animal Behav* 65 (2003): 479.
5. J. Lehmann and R. I. Dunbar, "Network Cohesion, Group Size and Neocortex Size in Female-Bonded Old World Primates," *Proc Royal Soc B: Biol Sci* 276 (2009): 4417.
6. J. Powell et al., "Orbital Prefrontal Cortex Volume Predicts Social Network Size: An Imaging Study of

Individual Differences in Humans," *Proc Royal Soc B: Biol Sci* 279 (2012): 2157; P. A. Lewis et al., "Ventromedial Prefrontal Volume Predicts Understanding of Others and Social Network Size," *Neuroimage* 57 (2011): 1624; J. L. Powell et al., "Orbital Prefrontal Cortex Volume Correlates with Social Cognitive Competence," *Neuropsychologia* 48 (2010): 3554; K. C. Bickart et al., "Amygdala Volume and Social Network Size in Humans," *Nat Nsci* 14 (2011): 163; R. Kanai et al., "Online Social Network Size Is Reflected in Human Brain Structure," *Proc Royal Soc B: Biol Sci* 279 (2012): 1327.

7. F. Elgar et al., "Income Inequality and School Bullying: Multilevel Study of Adolescents in 37 Countries," *J Adolescent Health* 45 (2009): 351.
8. E. González-Bono et al., "Testosterone, Cortisol and Mood in a Sports Team Competition," *Horm Behav* 35 (2009): 55; E. González-Bono et al., "Testosterone and Attribution of Successful Competition," *Aggressive Behav* 26 (2000): 235.
9. N. O. Rule et al., "Perceptions of Dominance Following Glimpses of Faces and Bodies," *Perception* 41 (2012): 687.
10. L. Thomsen et al., "Big and Mighty: Preverbal Infants Mentally Represent Social Dominance," *Sci* 331 (2011): 477.
11. S. V. Shepherd et al., "Social Status Gates Social Attention in Monkeys," *Curr Biol* 16 (2006): R119; J. Massen et al., "Ravens Notice Dominance Reversals Among Conspecifics Within and Outside Their Social Group," *Nat Communications* 5 (2013); 3679.
12. M. Karafin et al., "Dominance Attributions Following Damage to the Ventromedial Prefrontal Cortex," *J Cog Nsci* 16 (2004): 1796; L. Mah et al., "Impairment of Social Perception Associated with Lesions of the Prefrontal Cortex," *Am J Psychiatry* 161 (2004): 1247; T. Farrow et al., "Higher or Lower? The Functional Anatomy of Perceived Allocentric Social Hierarchies," *Neuroimage* 57 (2011): 1552; C. F. Zink et al., "Know Your Place: Neural Processing of Social Hierarchy in Humans," *Neuron* 58 (2008): 273.
13. A. A. Marsh et al., "Dominance and Submission: The Ventrolateral Prefrontal Cortex and Responses to Status Cues," *J Cog Nsci* 21 (2009): 713; T. Allison et al., "Social Perception from Visual Cues: Role of the STS Region," *TICS* 4 (2000): 267; J. B. Freeman et al., "Culture Shapes a Mesolimbic Response to Signals of Dominance and Subordination That Associates with Behavior," *Neuroimage* 47 (2009): 353.
14. M. Nader et al., "Social Dominance in Female Monkeys: Dopamine Receptor Function and Cocaine Reinforcement," *BP* 72 (2012): 414; M. P. Noonan et al., "A Neural Circuit Covarying with Social Hierarchy in Macaques," *PLoS Biol* 12 (2014): e1001940; F. Wang et al., "Bidirectional Control of Social Hierarchy by Synaptic Efficacy in Medial Prefrontal Cortex," *Sci* 334 (2011): 693.
15. M. Rushworth et al., "Are There Specialized Circuits for Social Cognition and Are They Unique to Humans?" *PNAS* 110 (2013): 10806.
16. For example: J. C. Beehner et al., "Testosterone Related to Age and Life-History Stages in Male Baboons and Geladas," *Horm Behav* 56 (2009): 472.
17. J. Brady et al., "Avoidance Behavior and the Development of Duodenal Ulcers," *J the Exp Analysis of Behav* 1 (1958): 69; J. Weiss, "Effects of Coping Responses on Stress," *J Comp Physiological Psych* 65 (1968): 251.
18. R. Sapolsky, "The Influence of Social Hierarchy on Primate Health," *Sci* 308 (2005): 648; H. Uno et al., "Hippocampal Damage Associated with Prolonged and Fatal Stress in Primates," *J Nsci* 9 (1989): 1705; R. Sapolsky et al., "Hippocampal Damage Associated with Prolonged Glucocorticoid Exposure in Primates," *J Nsci* 10 (1990): 2897; See also E. Archie et al., "Social Status Predicts Wound Healing in Wild Baboons," *PNAS* 109 (2012): 9017.
19. R. Sapolsky, "The Physiology of Dominance in Stable Versus Unstable Social Hierarchies," in *Primate Social Conflict*, ed. W. Mason and S. Mendoza (New York: SUNY Press, 1993).
20. L. R. Gesquiere et al., "Life at the Top: Rank and Stress in Wild Baboons," *Sci* 333 (2011): 357.
21. D. Abbott et al., "Are Subordinates Always Stressed? A Comparative Analysis of Rank Differences in Cortisol Levels Among Primates," *Horm Behav* 43 (2003): 67.
22. R. Sapolsky and J. Ray, "Styles of Dominance and Their Physiological Correlates Among Wild Baboons," *Am J Primat* 18 (1989) 1; J. C. Ray and R. Sapolsky, "Styles of Male Social Behavior and Their Endocrine Correlates Among High-Ranking Baboons," *Am J Primat* 28 (1992): 231; C. E. Virgin and R. Sapolsky, "Styles of Male Social Behavior and Their Endocrine Correlates Among Low-Ranking Baboons," *Am J Primat* 42 (1997): 25.
23. J. Chiao et al., "Neural Basis of Preference for Human Social Hierarchy Versus Egalitarianism," *ANYAS* 1167 (2009): 174; J. Sidanius et al., "You're Inferior and Not Worth Our Concern: The Interface Between Empathy and Social Dominance Orientation," *J Personality* 81 (2012): 313.
24. G. Sherman et al., "Leadership Is Associated with Lower Levels of Stress," *PNAS* 109 (2012): 17903; R. Sapolsky, "Importance of a Sense of Control and the Physiological Benefits of Leadership," *PNAS* 109 (2012): 17730.
25. N. Adler and J. Ostrove, "SES and Health: What We Know and What We Don't," *ANYAS* 896 (1999): 3; R. Wilkinson, *Mind the Gap: Hierarchies, Health and Human Evolution* (London: Weidenfeld and Nicolson, 2000); I. Kawachi and B. Kennedy, *The Health of Nations: Why Inequality Is Harmful to Your Health* (New York: New Press, 2002); M. Marmot, *The Status Syndrome: How Social Standing Affects Our Health and Longevity* (New York: Bloomsbury, 2015).
26. A. Todorov et al., "Inferences of Competence from Faces Predict Election Outcomes," *Sci* 308 (2005): 1623.
27. T. Tsukiura and R. Cabeza, "Shared Brain Activity for Aesthetic and Moral Judgments: Implications for the Beauty-Is-Good Stereotype," *SCAN* 6 (2011): 138.
28. K. Dion et al., "What Is Beautiful Is Good," *JPSP* 24 (1972): 285.
29. N. K. Steffens and S. A. Haslam, "Power Through 'Us': Leaders' Use of We-Referencing Language Predicts Election Victory," *PLoS ONE* 8 (2013): e77952.
30. B. R. Spisak et al., "Warriors and Peacekeepers: Testing a Biosocial Implicit Leadership Hypothesis of Intergroup Relations Using Masculine and Feminine Faces," *PLoS ONE* 7 (2012): e30399; B. R. Spisak, "The General Age of Leadership: Older-Looking Presidential Candidates Win Elections During War," *PLoS ONE* 7

(2012): e36945; B. R. Spisak et al., "A Face for All Seasons: Searching for Context-Specific Leadership Traits and Discovering a General Preference for Perceived Health," *Front Hum Nsci* 8 (2014): 792.

31. J. Antonakis and O. Dalgas, "Predicting Elections: Child's Play!" *Sci* 323 (2009): 1183.
32. K. Smith et al., "Linking Genetics and Political Attitudes: Reconceptualizing Political Ideology," *Political Psych* 32 (2011): 369.
33. G. Hodson and M. Busseri, "Bright Minds and Dark Attitudes: Lower Cognitive Ability Predicts Greater Prejudice Through Right-Wing Ideology and Low Intergroup Contact," *Psych Sci* 32 (2012): 187; C. Sibley and J. Duckitt, "Personality and Prejudice: A Meta-analysis and Theoretical Review," *PSPR* 12 (2008): 248.
34. L. Skitka et al., "Dispositions, Ideological Scripts, or Motivated Correction? Understanding Ideological Differences in Attributions for Social Problems," *JPSP* 83 (2002): 470; L. J. Skitka, "Ideological and Attributional Boundaries on Public Compassion: Reactions to Individuals and Communities Affected by a Natural Disaster," *PSPB* 25 (1999): 793; L. J. Skitka and P. E. Tetlock, "Providing Public Assistance: Cognitive and Motivational Processes Underlying Liberal and Conservative Policy Preferences," *JPSP* (1993): 65, 1205; G. S. Morgan et al., "When Values and Attributions Collide: Liberals' and Conservatives' Values Motivate Attributions for Alleged Misdeeds," *PSPB* 36 (2010): 1241; J. T. Jost and M. Krochik, "Ideological Differences in Epistemic Motivation: Implications for Attitude Structure, Depth of Information Processing, Susceptibility to Persuasion, and Stereotyping," *Advances in Motivation Sci* 1 (2014): 181.
35. S. Eidelman et al., "Low-Effort Thought Promotes Political Conservatism," *PSPB* 38 (2012): 808; H. Thórisdóttir and J. T. Jost, "Motivated Closed-Mindedness Mediates the Effect of Threat on Political Conservatism," *Political Psych* 32 (2011): 785.
36. B. Briers et al., "Hungry for Money: The Desire for Caloric Resources Increases the Desire for Financial Resources and Vice Versa," *Psych Sci* 17 (2006): 939; S. Danziger et al., "Extraneous Factors in Judicial Decisions," *PNAS* 108 (2011): 6889. The preceding is the source of the figure in the text. C. Schein and K. Gray, "The Unifying Moral Dyad," *PSPB* 41 (2015): 1147.
37. S. J. Thoma, "Estimating Gender Differences in the Comprehension and Preference of Moral Issues," *Developmental Rev* 6 (1986): 165; S. J. Thoma, "Research on the Defining Issues Test," in *Handbook of Moral Development*, ed. M. Killen and J. Smetana (New York: Psychology Press 2006), p. 67; N. Mahwa et al., "The Distinctiveness of Moral Judgment," *Educational Psych Rev* 11 (1999): 361; E. Turiel, *The Development of Social Knowledge: Morality and Convention* (Cambridge: Cambridge University Press, 1983); N. Kuyel and R. J. Clover, "Moral Reasoning and Moral Orientation of U.S. and Turkish University Students," *Psych Rep* 107 (2010): 463.
38. J. Haidt, "The New Synthesis in Moral Psychology," *Sci* 316 (2007): 998; G. L. Baril and J. C. Wright, "Different Types of Moral Cognition: Moral Stages Versus Moral Foundations," *Personality and Individual Differences* 53 (2012): 468.
39. N. Shook and R. Fazio, "Political Ideology, Exploration of Novel Stimuli, and Attitude Formation," *JESP* 45 (2009): 995; M. D. Dodd et al., "The Political Left Rolls with the Good and the Political Right Confronts the Bad: Connecting Physiology and Cognition to Preferences," *Philosophical Transactions of the Royal Soc B* 640 (2012) 640; K. Bulkeley, "Dream Content and Political Ideology," *Dreaming* 12 (2002): 61; J. Vigil, "Political Leanings Vary with Facial Expression Processing and Psychosocial Functioning," *Group Processes & Intergroup Relations* 13 (2011): 547; J. Jost et al., "Political Conservatism as Motivated Social Cognition," *Psych Bull* 129 (2003): 339; L. Castelli and L. Carraro, "Ideology Is Related to Basic Cognitive Processes Involved in Attitude Formation," *JESP* 47 (2011): 1013; L. Carraro et al., "Implicit and Explicit Illusory Correlation as a Function of Political Ideology," *PLoS ONE* 9 (2014): e96312; J. R. Hibbing et al., "Differences in Negativity Bias Underlie Variations in Political Ideology," *BBS* 37 (2014): 297.
40. For an interesting analysis of the relationships among rank, stability, and risk aversion, see J. Jordan et al., "Something to Lose and Nothing to Gain: The Role of Stress in the Interactive Effect of Power and Stability on Risk Taking," *Administrative Sci Quarterly* 56 (2011): 530. Discussed in: J. Jost et al., "Political Conservatism as Motivated Social Cognition," *Psych Bull* 129 (2003): 339.
41. P. Nail et al., "Threat Causes Liberals to Think Like Conservatives," *JESP* 45 (2009): 901; J. Greenberg et al., "The Causes and Consequences of the Need for Self-Esteem: A Terror Management Theory," in *Public Self and Private Self*, ed. R. Baumeister (New York: Springer, 1986); T. Verlag Pyszczynski et al., "A Dual Process Model of Defense Against Conscious and Unconscious Death-Related Thoughts: An Extension of Terror Management Theory," *Psych Rev* 106 (1999): 835.
42. J. L. Napier and J. T. Jost, "Why Are Conservatives Happier Than Liberals?" *Psych Sci* 19 (2008): 565.
43. J. Block and J. Block, "Nursery School Personality and Political Orientation Two Decades Later," *J Res in Personality* 40 (2006): 734. Also see: M. R. Tagar et al., "Heralding the Authoritarian? Orientation Toward Authority in Early Childhood," *Psych Sci* 25 (2014): 883; R. C. Fraley et al., "Developmental Antecedents of Political Ideology: A Longitudinal Investigation from Birth to Age 18 Years," *Psych Sci* 23 (2012): 1425.
44. Y. Inbar et al., "Disgusting Smells Cause Decreased Liking of Gay Men," *Emotion* 12 (2012): 23; T. Adams et al., "Disgust and the Politics of Sex: Exposure to a Disgusting Odorant Increases Politically Conservative Views on Sex and Decreases Support for Gay Marriage," *PLoS ONE* 9 (2014): e95572; H. A. Chapman and A. K. Anderson, "Things Rank and Gross in Nature: A Review and Synthesis of Moral Disgust," *Psych Bull* 139 (2013): 300.
45. G. Hodson and K. Costello, "Interpersonal Disgust, Ideological Orientations, and Dehumanization as Predictors of Intergroup Attitudes," *Psych Sci* 18 (2007): 691; K. Smith et al., "Disgust Sensitivity and the Neurophysiology of Left-Right Political Orientations," *PLoS ONE* 6 (2011): e2552.
46. J. Lee et al., "Emotion Regulation as the Foundation of Political Attitudes: Does Reappraisal Decrease Support

for Conservative Policies?" *PLoS ONE* 8 (2013): e83143; M. Feinberg et al., "Gut Check: Reappraisal of Disgust Helps Explain Liberal-Conservative Differences on Issues of Purity," *Emotion* 14 (2014): 513.

47. J. Haidt, *The Righteous Mind: Why Good People Are Divided by Politics and Religion* (New York: Pantheon, 2012); L. Kass, "The Wisdom of Repugnance: Why We Should Ban the Cloning of Human Beings," *New Republic*, June 2, 1997.
48. R. Kanai et al., "Political Orientations Are Correlated with Brain Structure in Young Adults," *Curr Biol* 21 (2011): 677; D. Schreiber et al., "Red Brain, Blue Brain: Evaluative Processes Differ in Democrats and Republicans," *PLoS ONE* 8 (2013): e52970; W. Ahn et al., "Nonpolitical Images Evoke Neural Predictors of Political Ideology," *Curr Biol* 24 (2014): 2693. For a general review, see J. Hibbing et al., "The Deeper Source of Political Conflict: Evidence from the Psychological, Cognitive, and Neurosciences," *TICS* 18 (2014): 111.
49. J. Settle et al., "Friendships Moderate an Association Between a Dopamine Gene Variant and Political Ideology," *J Politics* 72 (2010): 1189; K. Smith et al., "Linking Genetics and Political Attitudes: Reconceptualizing Political Ideology," *Political Psych* 32 (2011): 369; L. Buchen, "The Anatomy of Politics," *Nat* 490 (2012): 466.

    Some papers on the genetics of political orientation and involvement:

    Twin studies: N. G. Martin et al., "Transmission of Social Attitudes," *PNAS* 83 (1986): 4364; R. I. Lake et al., "Further Evidence Against the Environmental Transmission of Individual Differences in Neuroticism from a Collaborative Study of 45,850 Twins and Relatives on Two Continents," *Behav Genetics* 30 (2000): 223; J. R. Alford et al., "Are Political Orientations Genetically Transmitted," *Am Political Sci Rev* 99 (2005): 153.

    Genomewide linkage: P. Hatemi et al., "A Genome-wide Analysis of Liberal and Conservative Political Attitudes," *J Politics* 73 (2011): 1; D. Amodio et al., "Neurocognitive Correlates of Liberalism and Conservatism," *Nat Nsci* 10 (2007): 1246.
50. T. Kameda and R Hastie, "Herd Behavior: Its Biological, Neural, Cognitive and Social Underpinnings," in *Emerging Trends in the Social and Behavioral Sciences*, ed. R. Scott and S. Kosslyn (Hoboken, NJ: Wiley and Sons, 2015); H. Kelman, "Compliance, Identification, and Internalization: Three Processes of Attitude Change," *J Conflict Resolution* 2 (1958): 51.
51. Footnote: B. O. McGonigle and M. Chalmers, "Are Monkeys Logical?" *Nat* 267 (1977): 694; D. J. Gillian, "Reasoning in the Chimpanzee: II. Transitive Inference," *J Exp Psych: Animal Behav Processes* 7 (1981): 87; H. Davis, "Transitive Inference in Rats (*Rattus norvegicus*)," *J Comparative Psych* 106 (1992): 342; W. Roberts and M. Phelps, "Transitive Inference in Rats: A Test of the Spatial Coding Hypothesis," *Psych Sci* 5 (1994): 368; L. von Fersen et al., "Transitive Inference Formation in Pigeons," *J Exp Psych: Animal Behav Processes* 17 (1991): 334; J. Stern et al., "Transitive Inference in Pigeons: Simplified Procedures and a Test of Value Transfer Theory," *Animal Learning & Behav* 23 (1995): 76; A. B. Bond et al., "Social Complexity and Transitive Inference in Corvids," *Animal Behav* 65 (2003): 479; L. Grosenick et al., "Fish Can Infer Social Rank by Observation Alone," *Nat* 445 (2007): 429.
52. C. Watson and C. Caldwell, "Neighbor Effects in Marmosets: Social Contagion of Agonism and Affiliation in Captive *Callithrix jacchus*," *Am J Primat* 72 (2010): 549; K. Baker and F. Aureli, "The Neighbor Effect: Other Groups Influence Intragroup Agonistic Behavior in Captive Chimpanzees," *Am J Primat* 40 (1996): 283.
53. L. A. Dugatkin, "Animals Imitate, Too," *Sci Am* 283 (2000): 67.
54. K. Bonnie et al., "Spread of Arbitrary Conventions Among Chimpanzees: A Controlled Experiment," *Proc Royal Soc of London B* 274 (2007): 367; M. Dindo et al., "In-group Conformity Sustains Different Foraging Traditions in Capuchin Monkeys (*Cebus apella*)," *PLoS ONE* 4 (2009): e7858; D. Fragaszy and E. Visalberghi, "Socially Biased Learning in Monkeys," *Learning Behav* 32 (2004): 24; L. Aplin et al., "Experimentally-Induced Innovations Lead to Persistent Culture via Conformity in Wild Birds," *Nat* 518 (2014): 538. One study that failed to replicate the basic de Waal finding: E. Van Leeuwen et al., "Chimpanzees (*Pan troglodytes*) Flexibly Adjust Their Behaviour in Order to Maximize Payoffs, Not to Conform to Majorities," *PLoS ONE* 8 (2013): e80945.
55. E. van de Waal et al., "Potent Social Learning and Conformity Shape a Wild Primate's Foraging Decisions," *Sci* 340 (2013): 483.
56. A. Shestakova et al., "Electrophysiological Precursors of Social Conformity," *SCAN* 8 (2013): 756
57. H. Tajfel and J. C. Turner, "The Social Identity Theory of Intergroup Behaviour," in Psychology of Intergroup Relations, ed. S. Worchel and W. G. Austin (Chicago IL: Nelson-Hall, 1986), pp. 7–24; E. A. Losin et al., "Own-Gender Imitation Activates the Brain's Reward Circuitry," *SCAN* 7 (2012): 804; R. Yu and S. Sun, "To Conform or Not to Conform: Spontaneous Conformity Diminishes the Sensitivity to Monetary Outcomes," *PLoS ONE* 28 (2013): e64530.
58. R. Huber et al., "Neural Correlates of Informational Cascades: Brain Mechanisms of Social Influence on Belief Updating," *Neuroimage* 249 (2010): 2687; G. Berns et al., "Neural Mechanisms of the Influence of Popularity on Adolescent Ratings of Music," *BP* 58 (2005): 245; M. Edelson et al., "Following the Crowd: Brain Substrates of Long-Term Memory Conformity," *Sci* 333 (2011): 108; H. L. Roediger and K. B. McDermott, "Remember When?" *Sci* 333 (2011): 47; J. Chen et al., "ERP Correlates of Social Conformity in a Line Judgment Task," *BMC Nsci* 13 (2012): 43; K. Izuma, "The Neural Basis of Social Influence and Attitude Change," *Curr Opinion in Neurobiol* 23 (2013): 456.
59. J. Zaki et al., "Social Influence Modulates the Neural Computation of Value," *Psych Sci* 22 (2011): 894.
60. V. Klucharev et al., "Downregulation of the Posterior Medial Frontal Cortex Prevents Social Conformity," *J Nsci* 31 (2011): 11934; See also: A. Shestakova et al., "Electrophysiological Precursors of Social Conformity," *SCAN* 8 (2013): 756; V. Klucharev et al., "Reinforcement Learning Signal Predicts Social Conformity," *Neuron* 61 (2009): 140.
61. G. Berns et al., "Neurobiological Correlates of Social Conformity and Independence During Mental Rotation," *BP* 58 (2005): 245.

62. S. Asch, "Opinions and Social Pressure," *Sci Am* 193 (1955): 35; S. Asch, "Studies of Independence and Conformity: A Minority of One Against a Unanimous Majority," *Psych Monographs* 70 (1956): 1.
63. S. Milgram, *Obedience to Authority: An Experimental View* (New York: HarperCollins, 1974).
64. C. Haney et al., "Study of Prisoners and Guards in a Simulated Prison," *Naval Research Rev* 9 (1973): 1; C. Haney et al., "Interpersonal Dynamics in a Simulated Prison," *Int J Criminology and Penology* 1 (1973): 69.
65. M. Banaji, "Ordinary Prejudice," *Psych Sci Agenda* 8 (2001): 8.
66. Footnote: C. Hofling et al., "An Experimental Study of Nurse-Physician Relationships," *J Nervous and Mental Disease* 141 (1966): 171.
67. S. Fiske et al., "Why Ordinary People Torture Enemy Prisoners," *Sci* 306 (2004): 1482.
68. P. Zimbardo, *The Lucifer Effect: Understanding How Good People Turn Evil* (New York: Random House, 2007). This is also one source of the Solzhenitsyn quote.
69. Ibid.
70. G. Perry, *Behind the Shock Machine: The Untold Story of the Notorious Milgram Psych Experiments* (New York: New Press, 2013).
71. T. Carnahan and S. McFarland, "Revisiting the Stanford Prison Experiment: Could Participant Self-Selection Have Led to the Cruelty?" *PSPB* 33 (2007): 603; S. H. Lovibond et al., "Effects of Three Experimental Prison Environments on the Behavior of Non-convict Volunteer Subjects," *Psychologist* 14 (1979): 273.
72. S. Reiche and S. A. Haslam, "Rethinking the Psychology of Tyranny: The BBC Prison Study," *Brit J Soc Psych* 45 (2006): 1; S. A. Haslam and S. D. Reicher, "When Prisoners Take Over the Prison: A Social Psychology of Resistance," *PSPR* 16 (2012): 154.
73. P. Zimbardo, "On Rethinking the Psychology of Tyranny: The BBC Prison Study," *Brit J Soc Psych* 45 (2006): 47.
74. A. Abbott, "How the Brain Responds to Orders," *Nat* 530 (2016): 394.
75. B. Müller-Hill, *Murderous Science: Elimination by Scientific Selection of Jews, Gypsies, and Others, Germany 1933–1945* (Oxford: Oxford University Press, 1988).
76. S. Asch, "Opinions and Social Pressure," *Sci Am*, 193 (1955): 35.
77. R. Sapolsky, "Measures of Life," *Sciences*, March/April 1994, p. 10.
78. R. Watson, "Investigation into Deindividuation Using a Cross-Cultural Survey Technique," *JPSP* 25 (1973): 342.
79. A. Bandura et al., "Disinhibition of Aggression Through Diffusion of Responsibility and Dehumanization of Victims," *J Res in Personality* 9 (1975): 253.
80. L. Bègue et al., "Personality Predicts Obedience in a Milgram Paradigm," *J Personality* 83 (2015): 299; V. Zeigler-Hill, et al., "Neuroticism and Negative Affect Influence the Reluctance to Engage in Destructive Obedience in the Milgram Paradigm," *J Soc Psych* 153 (2013): 161; T. Blass, "Right-Wing Authoritarianism and Role as Predictors of Attributions About Obedience to Authority," *Personality and Individual Differences* 1 (1995): 99; P. Burley and J. McGuinnes, "Effects of Social Intelligence on the Milgram Paradigm," *Psych Rep* 40 (1977): 767.
81. A. H. Eagly and L. L. Carli, "Sex of Researchers and Sex-Typed Communications as Determinants of Sex Differences in Influenceability: A Meta-analysis of Social Influence Studies," *Psych Bull* 90 (1981): 1; S. Ainsworth and J. Maner, "Sex Begets Violence: Mating Motives, Social Dominance, and Physical Aggression in Men," *JPSP* 103 (2012): 819; H. Reitan and M. Shaw, "Group Membership, Sex-Composition of the Group, and Conformity Behavior," *J Soc Psych* 64 (1964): 45.
82. S. Milgram, "Nationality and Conformity," *Sci Am* 205 (1961): 45.

Chapter 13: Morality and Doing the Right Thing, Once You've Figured Out What That Is

1. A. Shenhav and J. D. Greene, "Moral Judgments Recruit Domain-General Valuation Mechanisms to Integrate Representations of Probability and Magnitude," *Neuron* 67 (2010): 667; P. N. Tobler et al., "The Role of Moral Utility in Decision Making: An Interdisciplinary Framework," *Cog, Affective & Behav Nsci* 8 (2008): 390; B. Harrison et al., "Neural Correlates of Moral Sensitivity in OCD," *AGP* 69 (2012): 741.
2. L. Young et al., "The Neural Basis of the Interaction Between Theory of Mind and Moral Judgment," *PNAS* 104 (2007): 8235; L. Young and R. Saxe, "Innocent Intentions: A Correlation Between Forgiveness for Accidental Harm and Neural Activity," *Neuropsychologia* 47 (2009): 2065; L. Young et al., "Disruption of the Right Temporoparietal Junction with TMS Reduces the Role of Beliefs in Moral Judgments," *PNAS* 107 (2009): 6753; L. Young and R. Saxe, "An fMRI Investigation of Spontaneous Mental State Inference for Moral Judgment," *J Cog Nsci* 21 (2009): 1396.
3. J. Knobe, "Intentional Action and Side Effects in Ordinary Language Analysis," 63 (2003): 190; J. Knobe, "Theory of Mind and Moral Cognition: Exploring the Connections," *TICS* 9 (2005): 357.
4. J. Knobe, "Theory of Mind and Moral Cognition: Exploring the Connections," *TICS* 9 (2005): 357.
5. P. Singer, "Sidgwick and Reflective Equilibrium," *Monist* 58 (1974), reprinted in *Unsatisfying Human Life*, ed. H. Kulse (Oxford: Blackwell, 2002).
6. J. Haidt, "The Emotional Dog and Its Rational Tail: A Social Intuitionist Approach to Moral Judgment," *Psych Rev* 108 (2001): 814–34; J. Haidt, "The New Synthesis in Moral Psychology," *Sci* 316 (2007): 996.
7. J. S. Borg et al., "Infection, Incest, and Iniquity: Investigating the Neural Correlates of Disgust and Morality," *J Cog Nsci* 20 (2008): 1529.
8. M. Haruno and C. D. Frith, "Activity in the Amygdala Elicited by Unfair Divisions Predicts Social Value Orientation," *Nat Nsci* 13 (2010): 160; C. D. Batson, "Prosocial Motivation: Is It Ever Truly Altruistic?" *Advances in*

Exp. Soc Psych 20 (1987): 65; A. G. Sanfey et al., "The Neural Basis of Economic Decision-Making in the Ultimatum Game," *Sci* 300 (2003): 1755.

9. J. Van Bavel et al., "The Importance of Moral Construal: Moral Versus Non-moral Construal Elicits Faster, More Extreme, Universal Evaluations of the Same Actions," *PLoS ONE* 7 (2012): e48693.
10. G. Miller, "The Roots of Morality," *Sci* 320 (2008): 734.
11. For this entire section on rudiments of morality in young children, see the excellent P. Bloom, *Just Babies: The Origins of Good and Evil* (Portland, OR: Broadway Books, 2014). This source applies to the subsequent half dozen paragraphs.
12. S. F. Brosnan and F. B. M. de Waal, "Monkeys Reject Unequal Pay," *Nat* 425 (2003): 297.
13. F. Range et al., "The Absence of Reward Induces Inequity Aversion in Dogs," *PNAS* 106 (2009): 340; C. Wynne "Fair Refusal by Capuchin Monkeys," *Nat* 428 (2004): 140; D. Dubreuil et al., "Are Capuchin Monkeys (*Cebus apella*) Inequity Averse?" *Proc Royal Soc of London B* 273 (2006): 1223.
14. S. F. Brosnan and F. B. M. de Waal, "Evolution of Responses to (un)Fairness," *Sci* 346 (2014): 1251776; S. F. Brosnan et al., "Mechanisms Underlying Responses to Inequitable Outcomes in Chimpanzees, *Pan troglodytes*," *Animal Behav* 79 (2010): 1229; M. Wolkenten et al., "Inequity Responses of Monkeys Modified by Effort," *PNAS* 104 (2007): 18854.
15. K. Jensen et al., "Chimpanzees Are Rational Maximizers in an Ultimatum Game," *Sci* 318 (2007): 107; D. Proctor et al., "Chimpanzees Play the Ultimatum Game," *PNAS* 110 (2013): 2070.
16. V. R. Lakshminarayanan and L. R. Santos, "Capuchin Monkeys Are Sensitive to Others' Welfare," *Curr Biol* 17 (2008): 21; J. M. Burkart et al., "Other-Regarding Preferences in a Non-human Primate: Common Marmosets Provision Food Altruistically," *PNAS* 104 (2007): 19762; J. B. Silk et al., "Chimpanzees Are Indifferent to the Welfare of Unrelated Group Members," *Nat* 437 (2005); 1357; K. Jensen et al., "What's in It for Me? Self-Regard Precludes Altruism and Spite in Chimpanzees," *Proc Royal Soc B* 273 (2006): 1013; J. Vonk et al., "Chimpanzees Do Not Take Advantage of Very Low Cost Opportunities to Deliver Food to Unrelated Group Members," *Animal Behav* 75 (2008): 1757.
17. F. De Waal and S. Macedo, *Primates and Philosophers: How Morality Evolved* (Princeton, NJ: Princeton Science Library, 2009).
18. B. Thomas et al., "Harming Kin to Save Strangers: Further Evidence for Abnormally Utilitarian Moral Judgments After Ventromedial Prefrontal Damage," *J Cog Nsci* 23 (2011): 2186.
19. J. Greene et al., "An fMRI Investigation of Emotional Engagement in Moral Judgment," *Sci* 293 (2001): 2105; J. Greene et al., "The Neural Bases of Cognitive Conflict and Control in Moral Judgment," *Neuron* 44 (2004): 389; J. Greene, *Moral Tribes: Emotion, Reason and the Gap Between Us and Them* (New York: Penguin, 2014).
20. D. Ariely, *Predictably Irrational: The Hidden Forces That Shape Our Decisions* (New York: Harper Perennial, 2010).
21. P. Singer, "Famine, Affluence, and Morality," *Philosophy and Public Affairs* 1 (1972) 229.
22. D. A. Smalia et al., "Sympathy and Callousness: The Impact of Deliberative Thought on Donations to Identifiable and Statistical Victims," *Organizational Behav and Hum Decision Processes* 102 (2007): 143; L. Petrinovich and P. O'Neill, "Influence of Wording and Framing Effects on Moral Intuitions," *Ethology and Sociobiology* 17 (1996): 145; L. Petrinovich et al., "An Empirical Study of Moral Intuitions: Toward an Evolutionary Ethics," *JPSP* 64 (1993): 467; R. E. O'Hara et al., "Wording Effects in Moral Judgments," *Judgment and Decision Making* 5 (2010): 547.
23. A. Cohn et al., "Business Culture and Dishonesty in the Banking Industry," *Nat* 516 (2014): 86. See also M. Villeval, "Professional Identity Can Increase Dishonesty," *Nat* 516 (2014): 48.
24. R. Zahn et al., "The Neural Basis of Human Social Values: Evidence from Functional MRI," *Cerebral Cortex* 19 (2009): 276.
25. K. Starcke et al., "Does Stress Alter Everyday Moral Decision-Making?" *PNE* 36 (2011): 210; F. Youssef et al., "Stress Alters Personal Moral Decision Making," *PNE* 37 (2012): 491.
26. E. Pronin, "How We See Ourselves and How We See Others," *Sci* 320 (2008): 1177.
27. R. M. N. Shweder et al., "The 'Big Three' of Morality and the 'Big Three' Explanations of Suffering," in *Morality and Health*, ed. A. M. B. P. Rozin (Oxford: Routledge, 1997).
28. M. Shermer, *The Science of Good and Evil* (New York: Holt, 2004).
29. F. W. Marlowe et al., "More 'Altruistic' Punishment in Larger Societies," *Sci* 23 (2006): 1767; J. Henrich et al., "'Economic Man' in Cross-Cultural Perspective: Behavioral Experiments in 15 Small-Scale Societies," *BBS* 28 (2005): 795.
30. R. Benedict, *The Chrysanthemum and the Sword* (Nanjing, China: Yilin Press1946); H. Katchadourian, *Guilt: The Bite of Conscience* (Palo Alto, CA: Stanford General Books, 2011); J. Jacquet, *Is Shame Necessary? New Uses for an Old Tool* (New York: Pantheon, 2015).
31. C. Berthelsen, "College Football: 9 Enter Pleas in U.C.L.A. Parking Case," *New York Times*, July 29, 1999, www.nytimes.com/1999/07/29/sports/college-football-9-enter-pleas-in-ucla-parking-case.html.
32. J. Bakan, *The Corporation: The Pathological Pursuit of Profit and Power* (New York: Simon & Schuster 2005).
33. Greene, *Moral Tribes.*
34. D. G. Rand et al., "Spontaneous Giving and Calculated Greed," *Nat* 489 (2012): 427.
35. S. Bowles, "Policies Designed to Self-Interested Citizens May Undermine 'The Moral Sentiments': Evidence from Economic Experiments," *Sci* 320 (2008): 1605; E. Fehr and B. Rockenbach, "Detrimental Effects of Sanctions on Human Altruism," *Nat* 422 (2003): 137.
36. M. M. Littlefield et al., "Being Asked to Tell an Unpleasant Truth About Another Person Activates Anterior

Insula and Medial Prefrontal Cortex," *Front Hum Nsci* 9 (2015): 553; Footnote: S. Harris, *Lying*. Four Elephants Press, 2013. e-book.

37. For a tour of animal deception, see the following: B. C. Wheeler, "Monkeys Crying Wolf? Tufted Capuchin Monkeys Use Anti-predator Calls to Usurp Resources from Conspecifics," *Proc Royal Soc B Biol Sci* 276 (2009): 3013; F. Amici et al., "Variation in Withholding of Information in Three Monkey Species," *Proc Royal Soc B Biol Sci* 276 (2009): 3311; A. le Roux et al., "Evidence for Tactical Concealment in a Wild Primate," *Nat Communications* 4 (2013): 1462; A. Whiten and R. W. Byrne, "Tactical Deception in Primates," *BBS* 11 (1988): 233; F. de Waal, *Chimpanzee Politics: Power and Sex Among Apes* (Baltimore, MD: Johns Hopkins University Press, 1982); G. Woodruff and D. Premack, "Intentional Communication in the Chimpanzee: The Development of Deception," *Cog* 7 (1979): 333; R. W. Byrne and N. Corp, "Neocortex Size Predicts Deception Rate in Primates," *Proc Royal Soc B Biol Sci* 271 (2004): 693; C. A. Ristau, "Language, Cognition, and Awareness in Animals?" *ANYAS* 406 (1983): 170; T. Bugnyar and K. Kotrschal, "Observational Learning and the Raiding of Food Caches in Ravens, *Corvus corax*: Is It 'Tactical' Deception?" *Animal Behav* 64 (2002): 185; J. Bro-Jorgensen and W. M. Pangle, "Male Topi Antelopes Alarm Snort Deceptively to Retain Females for Mating," *Am Nat* 176 (2010): E33; C. Brown et al., "It Pays to Cheat: Tactical Deception in a Cephalopod Social Signalling System," *Biol Lett* 8 (2012): 729; T. Flower, "Fork-Tailed Drongos Use Deceptive Mimicked Alarm Calls to Steal Food," *Proc Royal Soc B Biol Sci* 278 (2011): 1548.
38. K. G. Volz et al., "The Neural Basis of Deception in Strategic Interactions," *Front Behav Nsci* 9 (2015): 27.
39. Y. Yang et al., "Prefrontal White Matter in Pathological Liars," *Br J Psychiatry* 187 (2005): 325; Y. Yang et al., "Localisation of Increased Prefrontal White Matter in Pathological Liars," *Br J Psychiatry* 190 (2007):174.
40. D. D. Langleben et al., "Telling Truth from Lie in Individual Subjects with Fast Event-Related fMRI," *Hum Brain Mapping* 26 (2005): 262; J. M. Nunez et al., "Intentional False Responding Shares Neural Substrates with Response Conflict and Cognitive Control," *Neuroimage* 25 (2005): 267; G. Ganis et al., "Neural Correlates of Different Types of Deception: An fMRI Investigation," *Cerebral Cortex* 13 (2003): 830; K. L. Phan et al., "Neural Correlates of Telling Lies: A Functional Magnetic Resonance Imaging Study at 4 Tesla," *Academic Radiology* 12 (2005): 164; N. Abe et al., "Dissociable Roles of Prefrontal and Anterior Cingulate Cortices in Deception," *Cerebral Cortex* 16 (2006): 192; N. Abe, "How the Brain Shapes Deception: An Integrated Review of the Literature," *Neuroscientist* 17 (2011): 560.
41. A. Priori et al., "Lie-Specific Involvement of Dorsolateral Prefrontal Cortex in Deception," *Cerebral Cortex* 18 (2008): 451; L. Zhu et al., "Damage to Dorsolateral Prefrontal Cortex Affects Tradeoffs Between Honesty and Self-Interest," *Nat Nsci* 17 (2014): 1319.
42. T. Baumgartner et al., "The Neural Circuitry of a Broken Promise," *Neuron* 64 (2009): 756.
43. Footnote: F. Sellal et al., "'Pinocchio Syndrome': A Peculiar Form of Reflex Epilepsy?" *J Neurol, Neurosurgery and Psychiatry* 56 (1993): 936.
44. J. D. Greene and J. M. Paxton, "Patterns of Neural Activity Associated with Honest and Dishonest Moral Decisions," *PNAS* 106 (2009): 12506.
45. L. Pascual et al., "How Does Morality Work in the Brain? A Functional and Structural Perspective of Moral Behavior," *Front Integrative Nsci* 7 (2013): 65.
46. D. G. Rand and Z. G., Epstein, "Risking Your Life Without a Second Thought: Intuitive Decision-Making and Extreme Altruism," *PLoS ONE* 9, no. 10 (2014): e109687; R. W. Emerson, *Essays, First Series: Heroism* (1841).

Chapter 14: Feeling Someone's Pain, Understanding Someone's Pain, Alleviating Someone's Pain

1. Great reads on this general topic by leading scientists in the field: D. Keltner et al., *The Compassionate Instinct: The Science of Human Goodness* (New York: W. W. Norton, 2010); R. Davidson and S. Begley, *The Emotional Life of Your Brain* (New York: Plume, 2012).
2. G. Hein et al., "The Brain's Functional Network Architecture Reveals Human Motives," *Sci* 351 (2016): 1074. Also see: S. Gluth and L. Fontanesi, "Wiring the Altruistic Brain," *Sci* 351 (2016): 1028.
3. A. Whiten et al., "Imitative Learning of Artificial Fruit Processing in Children (*Homo sapiens*) and Chimpanzees (*Pan troglodytes*)," *JCP* 110 (1996): 3; V. Horner and A. Whiten, "Causal Knowledge and Imitation/Emulation Switching in Chimpanzees (*Pan troglodytes*) and Children (*Homo sapiens*)," *Animal Cog* 8 (2005): 164.
4. D. Jeon et al., "Observational Fear Learning Involves Affective Pain System and Ca1.2. CA2 Channels in ACC," *Nat Nsci* 13 (2010): 482.
5. B. L. Warren et al., "Neurobiological Sequelae of Witnessing Stressful Events in Adult Mice," *BP* 73 (2012): 7.
6. D. J. Langford et al., "Social Modulation of Pain as Evidence for Empathy in Mice," *Sci* 312 (2006): 1967.
7. M. Tomasello and V. Amrisha, "Origins of Human Cooperation and Morality," *Ann Rev Psych* 64 (2013): 231; D. Povinelli et al., review of *Reaching into Thought: The Minds of the Great Apes*, ed. A. E. Russon et al., *TICS* 2 (1998): 158.
8. F. de Waal and A. van Roosmalen, "Reconciliation and Consolation Among Chimpanzees," *Behav Ecology and Sociobiology* 5 (1979): 55; E. Palagi and G. Cordoni, "Postconflict Third-Party Affiliation in *Canis lupus*: Do Wolves Share Similarities with the Great Apes?" *Animal Behav* 78 (2009): 979; A. Cools et al., "Canine Reconciliation and Third-Party-Initiated Postconflict Affiliation: Do Peacemaking Social Mechanisms in Dogs Rival Those of Higher Primates?" *Ethology* 14 (2008): 53; O. Fraser and T. Bugnyar, "Do Ravens Show Consolation? Responses to Distressed Others," *PLoS ONE* 5, no. 5 (2010), doi:10.1371/journal.pone.0010605; A. Seed et al., "Postconflict Third-Party Affiliation in Rooks, *Corvus frugilegus*," *Curr Biol* 2 (2006): 152; J. Plotnik and F. de Waal, "Asian Elephants (*Elephas maximus*) Reassure Others in Distress," *PeerJ* 2 (2014), doi:10.7717/peerj.278; Z. Clay and F. de Waal, "Bonobos Respond to Distress in Others: Consolation Across the Age Spectrum," *PLoS ONE* 8 (2013): e55206.

9. J. P. Burkett et al., "Oxytocin-Dependent Consolation Behavior in Rodents," *Sci* 351 (2016): 375.
10. G. E. Rice and P. Gainer, "'Altruism' in the Albino Rat," *J Comp and Physiological Psych* 55 (1962): 123; J. S. Mogil, "The Surprising Empathic Abilities of Rodents," *TICS* 16 (2012): 143; I. Ben-Ami Bartal et al., "Empathy and Pro-social Behavior in Rats," *Sci* 334 (2011): 1427–30.
11. I. B. A. Bartal et al., "Pro-social Behavior in Rats is Modulated by Social Experience," *eLife* 3 (2014): e01385.
12. C. Lamm et al., "Meta-analytic Evidence for Common and Distinct Neural Networks Associated with Directly Experienced Pain and Empathy for Pain," *Neuroimage* 54 (2011): 2492; B. C. Bernhardt and T. Singer, "The Neural Basis of Empathy," *Ann Rev Nsci* 35 (2012): 1.
13. A. Craig, "How Do You Feel? Interoception: The Sense of the Physiological Condition of the Body," *Nat Rev Nsci* 3 (2002): 655; J. Kong et al., "A Functional Magnetic Resonance Imaging Study on the Neural Mechanisms of Hyperalgesic Nocebo Effect," *J Nsci* 28 (2008): 13354.
14. B. Vogt, "Pain and Emotion Interactions in Subregions of the Cingulate Gyrus," *Nat Rev Nsci* 6 (2005): 533; K. Ochsner et al., "Your Pain or Mine? Common and Distinct Neural Systems Supporting the Perception of Pain in Self and Other," *SCAN* 3 (2008): 144; this is the source of the Ochsner quote.
15. N. Eisenberger et al., "Does Rejection Hurt? An fMRI Study of Social Exclusion," *Sci* 302 (2003): 290; D. Pizzagalli, "Frontocingulate Dysfunction in Depression: Toward Biomarkers of Treatment Response," *Neurophyschopharmacology* 36 (2011): 183.
16. C. Lamm et al., "The Neural Substrate of Human Empathy: Effects of Perspective-Taking and Cognitive Appraisal," *J Cog Nsci* 19 (2007): 42; P. Jackson et al., "Empathy Examined Through the Neural Mechanisms Involved in Imagining How I Feel Versus How You Feel Pain," *Neuropsycholo´gia* 44 (2006): 752; M. Saarela et al., "The Compassionate Brain: Humans Detect Intensity of Pain from Another's Face," *Cerebral Cortex* 17 (2007): 230; N. Eisenberg et al., "The Relations of Emotionality and Regulation to Dispositional and Situational Empathy-Related Responding," *JPSP* 66 (1994): 776; J. Burkett et al., "Oxytocin-Dependent Consolation Behavior in Rodents," *Sci* 351 (2016): 6271; M. Botvinick et al., "Viewing Facial Expressions of Pain Engages Cortical Areas Involved in the Direct Experience of Pain," *Neuroimage* 25 (2005): 312; C. Lamm et al., "The Neural Substrate of Human Empathy: Effects of Perspective-Taking and Cognitive Appraisal," *J Cog Nsci* 19 (2007): 42; C. Lamm et al., "What Are You Feeling? Using Functional Magnetic Resonance Imaging to Assess the Modulation of Sensory and Affective Responses During Empathy for Pain," *PLoS ONE* 2 (2007): e1292.
17. D. Jeon et al., "Observational Fear Learning Involves Affective Pain System and Ca(v)1.2 Ca2+ Channels in ACC," *Nat Nsci* 13 (2010): 482.
18. A. Craig, "How Do You Feel—Now? The Anterior Insula and Human Awareness," *Nat Rev Nsci* 10 (2009): 59; B. King-Casas et al., "The Rupture and Repair of Cooperation in Borderline Personality Disorder," *Sci* 321 (2008): 806; M. H. Immordino-Yang et al., "Neural Correlates of Admiration and Compassion," *PNAS* 106 (2009): 8021.
19. J. Decety and K. Michalska, "Neurodevelopmental Changes in the Circuits Underlying Empathy and Sympathy from Childhood to Adulthood," *Developmental Sci* 13 (2009): 886; J. Decety, "The Neuroevolution of Empathy," *ANYAS* 1231 (2011): 35; this second reference is the source of the quote.
20. E. Brueau et al., "Distinct Roles of the 'Shared Pain' and 'Theory of Mind' Networks in Processing Others' Emotional Suffering," *Neuropsychologia* 50 (2012): 219; C. Lamm et al., "How Do We Empathize with Someone Who Is Not Like Us? A Functional Magnetic Resonance Imaging Study," *J Cog Nsci* 22 (2010): 362; C. Keysers et al., "Somatosensation in Social Perception," *Nat Rev Nsci* 11 (2010): 417.
21. L. Harris and S. Fiske, "Dehumanizing the Lowest of the Low: Neuroimaging Responses to Extreme Outgroups," *Psych Sci* 17 (2006): 847.
22. I. Konvalinka et al., "Synchronized Arousal Between Performers and Related Spectators in a Fire-Walking Ritual," *PNAS* 108 (2011): 8514; Y. Cheng et al., "Love Hurts: An fMRI Study," *NeuroImage* 51 (2010): 923.
23. A. Avenanti et al., "Transcranial Magnetic Stimulation Highlights the Sensorimotor Side of Empathy for Pain," *Nat Nsci* 8 (2005): 955; X. Xu et al., "Do You Feel My Pain? Racial Group Membership Modulates Empathic Neural Responses," *J Nsci* 29 (2009): 8525; V. Mathur et al., "Neural Basis of Extraordinary Empathy and Altruistic Motivation," *NeuroImage* 51 (2010): 1468; G. Hein et al., "Neural Responses to Ingroup and Outgroup Members' Suffering Predict Individual Differences in Costly Helping," *Neuron* 68 (2010): 149; E. Bruneau et al., "Social Cognition in Members of Conflict Groups: Behavioural and Neural Responses in Arabs, Israelis and South Americans to Each Other's Misfortunes," *Philosophical Transactions of the Royal Soc B* 367 (2012): 717; E. Bruneau and R. Saxe, "Attitudes Towards the Outgroup are Predicted by Activity in the Precuneus in Arabs and Israelis," *NeuroImage* 52 (2010): 1704; J. Gutsell and M. Inzlicht, "Intergroup Differences in the Sharing of Emotive States: Neural Evidence of an Empathy Gap," *SCAN* 10 (2011): 1093; J. Freeman et al., "The Neural Origins of Superficial and Individuated Judgments About Ingroup and Outgroup Members," *Hum Brain Mapping* 31 (2010): 150.
24. Footnote: K. Wailoo, *Pain: A Political History* (Baltimore, MD: Johns Hopkins University Press, 2014).
25. C. Oveis et al., "Compassion, Pride, and Social Intuitions of Self-Other Similarity," *JPSP* 98 (2010): 618; M. W. Kraus et al., "Social Class, Contextualism, and Empathic Accuracy," *Psych Sci* 21 (2012): 1716; J. Stellar et al., "Class and Compassion: Socioeconomic Factors Predict Responses to Suffering," *Emotion* 12 (2012): 449; P. Piff et al., "Higher Social Class Predicts Increased Unethical Behavior," *PNAS* 109 (2012): 4086.
26. J. Gutsell and M. Inzlicht, "Intergroup Differences in the Sharing of Emotive States: Neural Evidence of an Empathy Gap," *SCAN* 10 (2011): 1093; H. Takahasi et al., "When Your Gain Is My Pain and Your Pain Is My Gain: Neural Correlates of Envy and Schadenfreude," *Sci* 323 (2009): 890; T. Singer et al., "Empathic Neural Responses Are Modulated by the Perceived Fairness of Others," *Nat* 439 (2006): 466; S. Preston and F. de Waal, "Empathy: Its Ultimate and Proximate Bases," *BBS* 25 (2002): 1.
27. C. N. Dewall et al., "Depletion Makes the Heart Grow Less Helpful: Helping as a Function of Self-Regulatory

Energy and Genetic Relatedness," *PSPB* 34 (2008): 1653. Mother Theresa is quoted in: P. Slovic, "'If I Look At the Mass, I Will Never Act': Psychic Numbing and Genocide," *Judgment and Decision Making,* 2 (2007): 1. The quote has been attributed to Stalin in many places, including: L Lyons, "Looseleaf Notebook," *Washington Post,* January 30, 1947.

28. A. Jenkins and J. Mitchell, "Medial Prefrontal Cortex Subserves Diverse Forms of Self-Reflection," *Soc Nsci* 6 (2011): 211.
29. G. Di Pellegrino et al., "Understanding Motor Events: A Neurophysiological Study," *Exp Brain Res* 91 (1992): 176; G. Rizzolatti et al., "Premotor Cortex and the Recognition of Motor Actions," *Cog Brain Res* 3 (1996): 131; also see: P. Ferrari et al., "Mirror Neurons Responding to the Observation of Ingestive and Communicative Mouth Actions in the Ventral Premotor Cortex," *Eur J Nsci* 17 (2003): 1703; G. Rizzolatti and L. Craighero, "The Mirror-Neuron System," *Ann Rev Nsci* 27 (2004): 169.
30. Footnote: P. Molenberghs et al., "Is the Mirror Neuron System Involved in Imitation? A Short Review and Meta-analysis," *Nsci and Biobehavioral Reviews* 33 (2009): 975.
31. Human MRI studies: V. Gazzola and C. Keysers, "The Observation and Execution of Actions Share Motor and Somatosensory Voxels in All Tested Subjects: Single-Subject Analyses of Unsmoothed fMRI Data," *Cerebral Cortex* 19 (2009): 1239; M. Iacoboni et al., "Cortical Mechanisms of Human Imitation," *Sci* 286 (1999): 2526. Single neuron recordings in humans: C. Keysers and V. Gazzola, "Social Neuroscience: Mirror Neurons Recorded in Humans," *Curr Biol* 20 (2010): R353; J. Kilner and A. Neal, "Evidence of Mirror Neurons in Human Inferior Frontal Gyrus," *J Nsci* 29 (2009): 10153.
32. M. Rochat et al., "The Evolution of Social Cognition: Goal Familiarity Shapes Monkeys' Action Understanding," *Curr Biol* 18 (2008): 227; M. Lacoboni, "Grasping the Intentions of Others with One's Own Mirror Neuron System," *PLoS Biol* 3 (2005): e79.
33. C. Catmur et al., "Sensorimotor Learning Configures the Human Mirror System," *Curr Biol* 17 (2007): 1527.
34. G. Hickok, "Eight Problems for the Mirror Neuron Theory of Action Understanding in Monkeys and Humans," *J Cog Nsci* 7 (2009): 1229.
35. V. Gallese and A. Goldman, "Mirror Neurons and the Simulation Theory," *TICS* 2 (1998): 493.
36. V. Caggiano et al., "Mirror Neurons Differentially Encode the Peripersonal and Extrapersonal Space of Monkeys," *Sci* 324 (2009): 403.
37. V. Gallese et al., "Mirror Neurons," *Perspectives on Psych Sci* 6 (2011): 369.
38. A sampling of some relevant papers: L. Oberman et al., "EEG Evidence for Mirror Neuron Dysfunction in Autism Spectrum Disorders," *Brain Res: Cog Brain Res* 24 (2005): 190; M. Dapretto et al., "Understanding Emotions in Others: Mirror Neuron Dysfunction in Children with Autism Spectrum Disorders," *Nat Nsci* 9 (2006): 28; I. Dinstein et al., "A Mirror Up to Nature," *Curr Biol* 19 (2008): R13; A. Hamilton, "Reflecting on the Mirror Neuron System in Autism: A Systematic Review of Current Theories," *Developmental Cog Nsci* 3 (2013): 91.
39. G. Hickok, *The Myth of Mirror Neurons: The Real Neuroscience of Communication and Cognition* (New York: Norton, 2014).
40. D. Freedberg and V. Gallese, "Motion, Emotion and Empathy in Esthetic Experience," *TICS* 11 (2007): 197; S. Preston and F. de Waal, "Empathy: Its Ultimate and Proximate Bases," *BBS* 25 (2002); 1; J. Decety and P. Jackson, "The Functional Architecture of Human Empathy," *Behav and Cog Nsci Rev* 3 (2004): 71.
41. J. Pfeifer et al., "Mirroring Others' Emotions Relates to Empathy and Interpersonal Competence in Children," *NeuroImage* 39 (2008): 2076; V. Gallese, "The 'Shared Manifold' Hypothesis: From Mirror Neurons to Empathy," *J Consciousness Studies* 8 (2001): 33.
42. J. Kaplan and M. Iacoboni, "Getting a Grip on Other Minds: Mirror Neurons, Intention Understanding, and Cognitive Empathy," *Soc Nsci* 1 (2006): 175.
43. Center for Building a Culture of Empathy, "Mirror Neurons," http://cultureofempathy.com/, no date, http://cultureofempathy.com/References/Mirror-Neurons.htm; J. Marsh, "Do Mirror Neurons Give Us Empathy?" *Greater Good Newsletter,* March 29, 2012; V. Ramachandran, "Mirror Neurons and Imitation Learning as the Driving Force Behind 'the Great Leap Forward' in Human Evolution," *Edge,* May 31, 2000.
44. Grayling is quoted in C. Jarrett, "Mirror Neurons: The Most Hyped Concept in Neuroscience?" *Psychology Today,* December 10, 2012, www.psychologytoday.com/blog/brain-myths/201212/mirror-neurons-the-most-hyped-concept-in-neuroscience; C. Buckley, "Why Our Hero Leapt onto the Tracks and We Might Not," *New York Times,* January 7, 2007.
45. All quotes are from Hickok, 2014, op cit. For some more analysis of the skepticism, see C. Jarrett, "A Calm Look at the Most Hyped Concept in Neuroscience: Mirror Neurons," *Wired,* December 13, 2013; D. Dobbs, "Mirror Neurons: Rock Stars or Backup Singers?" *News Blog,* ScientificAmerican.com, December 18, 2007; B. Thomas, "What's So Special About Mirror Neurons?" *Guest Blog,* ScientificAmerican.com, November 6, 2012; A. Gopnik, "Cells That Read Minds?" *Slate,* April 26, 2007; and "A Mirror to the World," *Economist,* May 12, 2005, www.economist.com/node/3960516.
46. L. Jamison, "Forum: Against Empathy," *Boston Review,* September 10, 2014.
47. C. Lamm et al., "The Neural Substrate of Human Empathy: Effects of Perspective-Taking and Cognitive Appraisal," *J Cog Nsci* 19 (2007): 42.
48. N. Eisenberg et al., "The Relations of Emotionality and Regulation to Dispositional and Situational Empathy-Related Responding," *JPSP* 66 (1994): 776; G. Carlo et al., "The Altruistic Personality: In What Contexts Is It Apparent?" *JPSP* 61 (1991): 450.
49. B. Briers et al., "Hungry for Money: The Desire for Caloric Resources Increases the Desire for Financial Resources and Vice Versa?" *Psych Sci* 17 (2006): 939; J. Twenge et al., "Social Exclusion Decreases Prosocial

Behavior," *JPSP* 92 (2007): 56; L. Martin et al., "Reducing Social Stress Elicits Emotional Contagion of Pain in Mouse and Human Strangers," *Curr Biol* 25 (2015): 326.

50. R. Davidson and S. Begley, *The Emotional Life of Your Brain* (NY: Avery, 2012); M. Ricard et al., "Mind of the Meditator," *Sci Am* 311 (2014): 39.
51. A. Lutz et al., "Long-Term Meditators Self-Induce High-Amplitude Gamma Synchrony During Mental Practice," *PNAS* 101 (2004): 16369; T. Singer and M. Ricard, eds., *Caring Economics: Conversations on Altruism and Compassion, Between Scientists, Economists, and the Dalai Lama* (New York: St Martin's Press, 2015); O. Klimecki et al., "Functional Neural Plasticity and Associated Changes in Positive Affect After Compassion Training," *Cerebral Cortex* 23 (2013): 1552.
52. P. Bloom, "Against Empathy," *Boston Review*, September 10, 2014; B. Oakley, *Cold-Blooded Kindness* (Amherst, NY: Prometheus Books, 2011); Y. Cheng et al., "Expertise Modulates the Perception of Pain in Others," *Curr Biol* 17 (2007): 1708; Davidson and Begley, op cit.; this is the source of the quote.
53. K. Izuma et al., "Processing of the Incentive for Social Approval in the Ventral Striatum During Charitable Donation," *J Cog Nsci* 22 (2010): 621; K. Izuma et al., "Processing of Social and Monetary Rewards in the Human Striatum," *Neuron* 58 (2008): 284; E. Dunn et al., "Spending Money on Others Promotes Happiness," *Sci* 319 (2008): 1687.
54. B. Purzycki et al., "Moralistic Gods, Supernatural Punishment and the Expansion of Human Sociality," *Nat* 530 (2016): 327.
55. L. Penner et al., "Prosocial Behavior: Multilevel Perspectives," *Ann Rev Psych* 56 (2005): 365.
56. W. Harbaugh et al., "Neural Responses to Taxation and Voluntary Giving Reveal Motives for Charitable Donations," *Sci* 316 (2007): 1622.
57. E. Tricomi et al., "Neural Evidence for Inequality-Averse Social Preferences," *Nat* 463 (2010): 1089.

Chapter 15: Metaphors We Kill By

1. "Fighting and Dying for the Colors at Gettysburg," HistoryNet.com, June 7, 2007, www.historynet.com/fighting-and-dying-for-the-colors-at-gettysburg.htm.
2. The killing of Tavin Price: Brainuser1, "Mentally Challenged Teen Shot Dead for Wearing Wrong Color Shoes," EurThisNThat.com, September 22, 2016, www.eurthisnthat.com/2015/06/03/mentally-challenged-teen-shot-dead-for-wearing-wrong-color-shoes/comment-page-1/. Irish hunger strikers: "1981 Irish Hunger Strike," Wikipedia.com, https://en.wikipedia.org/wiki/1981_Irish_hunger_strike#First_hunger_strike. "My Way" killings: N. Onishi, "Sinatra Song Often Strikes Deadly Chord," *New York Times*, February 7, 2010.
3. Footnote: T. Appenzeller, "Old Masters," *Nat* 497 (2013): 302.
4. R. Hughes, *The Shock of the New* (New York: Knopf, 1991). The following reference is included in the hopes that it will make it seem like I actually read this book: M. Foucault, *This Is Not a Pipe* (Oakland: University of California Press, 1983).
5. T. Deacon, *The Symbolic Species: The Coevolution of Language and the Brain* (New York: Norton, 1997).
6. Footnote: L. Boroditsky, "How Language Shapes Thought," *Sci Am*, February, 2011.
7. G. Lakoff and M. Johnson, *Metaphors We Live By* (Chicago: University of Chicago Press, 1980); G. Lakoff, *Moral Politics: What Conservatives Know That Liberals Don*'t (Chicago: University of Chicago Press, 1996).
8. T. Singer and C. Frith, "The Painful Side of Empathy," *Nat Nsci* 8 (2005): 845.
9. M. Kramer et al., "Distinct Mechanism for Antidepressant Activity by Blockade of Central Substance P Receptors," *Sci* 281 (1998): 1640; B. Bondy et al., "Substance P Serum Levels are Increased in Major Depression: Preliminary Results," *BP* 53 (2003): 538; G. S. Berns et al., "Neurobiological Substrates of Dread," *Sci* 312 (2006): 754.
10. H. Takahasi et al., "When Your Gain Is My Pain and Your Pain Is My Gain: Neural Correlates of Envy and Schadenfreude," *Sci* 323 (2009): 890.
11. P. Ekman and W. Friesen, *Unmasking the Face: A Guide to Recognizing Emotions from Facial Cues* (Upper Saddle River, NJ: Prentice Hall, 1975).
12. M. Hsu et al., "The Right and the Good: Distributive Justice and Neural Encoding of Equity and Efficiency," *Sci* 320 (2008): 1092; F. Sambataro et al., "Preferential Responses in Amygdala and Insula During Presentation of Facial Contempt and Disgust," *Eur J Nsci* 24 (2006): 2355; P. S. Russell and R. Giner-Sorolla, "Bodily Moral Disgust: What It Is, How It Is Different from Anger, and Why It Is an Unreasoned Emotion," *Psych Bull* 139 (2013): 328; H. A. Chapman and A. K. Anderson, "Things Rank and Gross in Nature: A Review and Synthesis of Moral Disgust," *Psych Bull* 139 (2013): 300; H. Chapman et al., "In Bad Taste: Evidence for the Oral Origins of Moral Disgust," Sci 323 (2009): 1222; P. Rozin et al., "From Oral to Moral," *Sci* 323 (2009): 1179.
13. C. Chan et al., "Moral Violations Reduce Oral Consumption," *J Consumer Psych* 24 (2014): 381; K. J. Eskine et al., "The Bitter Truth About Morality: Virtue, Not Vice, Makes a Bland Beverage Taste Nice," *PLoS ONE* 7 (2012): e41159.
14. E. J. Horberg et al., "Disgust and the Moralization of Purity," *JPSP* 97 (2009): 963.
15. K. Smith et al., "Disgust Sensitivity and the Neurophysiology of Left-Right Political Orientations," *PLoS ONE* 6 (2011): e2552; G. Hodson and K. Costello, "Interpersonal Disgust, Ideological Orientations, and Dehumanization as Predictors of Intergroup Attitudes," *Psych Sci* 18 (2007): 691; M. Landau et al., "Evidence That Self-Relevant Motives and Metaphoric Framing Interact to Influence Political and Social Attitudes," *Psych Sci* 20 (2009): 1421.
16. A. Sanfey et al., "The Neural Basis of Economic Decision-Making in the Ultimatum Game," *Sci* 300 (2003): 1755.
17. T. Wang et al., "Is Moral Beauty Different from Facial Beauty? Evidence from an fMRI Study," *SCAN* 10 (2015): 814.

18. S. Lee and N. Schwarz, "Washing Away Postdecisional Dissonance," *Sci* 328 (2010): 709.
19. S. Schnall et al., "With a Clean Conscience: Cleanliness Reduces the Severity of Moral Judgments," *Psych Sci* 19 (2008): 1219; K. Kaspar et al., "Hand Washing Induces a Clean Slate Effect in Moral Judgments: A Pupillometry and Eye-Tracking Study," *Sci Rep* 5 (2015): 10471.
20. C. B. Zhong and K. Liljenquist, "Washing Away Your Sins: Threatened Morality and Physical Cleansing," *Sci* 313 (2006): 1451; L. N. Harkrider et al., "Threats to Moral Identity: Testing the Effects of Incentives and Consequences of One's Actions on Moral Cleansing," *Ethics & Behav* 23 (2013): 133.
21. M. Schaefer et al., "Dirty Deeds and Dirty Bodies: Embodiment of the Macbeth Effect Is Mapped Topographically onto the Somatosensory Cortex," *Sci Rep* 5 (2015): 18051. See also C. Denke et al., "Lying and the Subsequent Desire for Toothpaste: Activity in the Somatosensory Cortex Predicts Embodiment of the Moral-Purity Metaphor," *Cerebral Cortex* 26 (2016): 477. A debate about these findings: D. Johnson et al., "Does Cleanliness Influence Moral Judgments? A Direct Replication of Schnall, Benton, and Harvey (2008)," *Soc Psych* 45 (2014): 209; J. L. Huang, "Does Cleanliness Influence Moral Judgments? Response Effort Moderates the Effect of Cleanliness Priming on Moral Judgments," *Front Psych* 5 (2014): 1276.
22. S. W. Lee et al., "A Cultural Look at Moral Purity: Wiping the Face Clean," *Front Psych* 6 (2015): 577.
23. H. Xu et al., "Washing the Guilt Away: Effects of Personal Versus Vicarious Cleansing on Guilty Feelings and Prosocial Behavior," *Front Hum Nsci* 8 (2014): 97.
24. J. Ackerman et al., "Incidental Haptic Sensations Influence Social Judgments and Decisions," *Sci* 328 (2010): 1712; also see: M. V. Day and D. R. Bobocel, "The Weight of a Guilty Conscience: Subjective Body Weight as an Embodiment of Guilt," *PLoS ONE* 8 (2013): e69546.
25. L. Williams and J. Bargh, "Experiencing Physical Warmth Promotes Interpersonal Warmth," *Sci* 322 (2008): 606; Y. Kang et al., "Physical Temperature Effects on Trust Behavior: The Role of Insula," *SCAN* 6 (2010): 507.
26. B. Briers et al., "Hungry for Money: The Desire for Caloric Resources Increases the Desire for Financial Resources and Vice Versa," *Psych Sci* 17 (2006): 939; X. Wang and R. Dvorak, "Sweet Future: Fluctuating Blood Glucose Levels Affect Future Discounting," *Psych Sci* 21 (2010): 183.
27. M. Anderson, "Neural Reuse: A Fundamental Organizational Principle of the Brain," *BBS* 245 (2014); 245; G. Lakoff, "Mapping the Brain's Metaphor Circuitry: Metaphorical Thought in Everyday Reason," *Front Hum Nsci* (2014), doi:10.3389/fnhum.2014.00958.
28. P. Gourevitch, *We Wish to Inform You That Tomorrow We Will Be Killed with Our Families* (New York: Farrar, Straus and Giroux 2000); R. Guest, *The Shackled Continent* (Washington, DC: Smithsonian Books, 2004); G. Stanton, "The Rwandan Genocide: Why Early Warning Failed," *J African Conflicts and Peace Studies* 1 (2009) 6; R. Lemarchand, "The 1994 Rwandan Genocide," in *Century of Genocide*, ed. S. Totten and W. Parsons, 3rd ed. (Abingdon, UK: Routledge, 2009), p. 407.
29. S. Atran et al., "Sacred Barriers to Conflict Resolution," *Sci* 317 (2007): 1039.
30. Hussein quote from CNN, Nov 6, 1995.
31. D. Thornton, "Peter Robinson and Martin McGuinness Shake Hands for the First Time," *Irish Central*, January 18, 2010, www.irishcentral.com/news/peter-robinson-and-martin-mcguinness-shake-hands-for-the-first-time-81957747-237681071.html.
32. J. Carlin, *Playing the Enemy: Nelson Mandela and the Game That Made a Nation* (New York: Penguin Press, 2008); D. Cruywagen, *Brothers in War and Peace: Constand and Abraham Viljoen and the Birth of the New South Africa* (Cape Town, South Africa: Zebra Press, 2014).

Chapter 16: Biology, the Criminal Justice System, and (Oh, Why Not?) Free Will

1. Innocence Project, "DNA Exonerations in the United States," www.innocenceproject.org/dna-exonerations-in-the-united-states/.
2. N. Schweitzer and M. Saks, "Neuroimage Evidence and the Insanity Defense," *Behav Sci & the Law* 29 (2011): 4; A. Roskies et al., "Neuroimages in Court: Less Biasing Than Feared," *TICS* 17 (2013): 99.
3. J. Marks, "A Neuroskeptic's Guide to Neuroethics and National Security," *Am J Bioethics: Nsci* 1 (2010): 4; A. Giridharadas, "India's Use of Brain Scans in Courts Dismays Critics," *New York Times*, September 15, 2008; A. Madrigal, "MRI Lie Detection to Get First Day in Court," *Wired*, March 16, 2009.
4. S. Reardon, "Smart Enough to Die?" *Nat* 506 (2014): 284.
5. J. Monterosso et al., "Explaining Away Responsibility: Effects of Scientific Explanation on Perceived Culpability," *Ethics & Behav* 15 (2005): 139; S. Aamodt, "Rise of the Neurocrats," *Nat* 498 (2013): 298.
6. J. Rosen, "The Brain on the Stand," *New York Times Magazine*, March 11, 2007.
7. Footnote: S. Lucas, "Free Will and the Anders Breivik Trial," *Humanist*, Sept/Oct 2012, p. 36; J. Greene and J. Cohen, "For the Law, Neuroscience Changes Nothing and Everything," *Philosophical Transactions of the Royal Soc B, Biol Sci* 359 (2004): 1775.
8. D. Robinson, *Wild Beasts and Idle Humours: The Insanity Defense from Antiquity to the Present* (Cambridge, MA: Harvard University Press, 1996).
9. S. Kadri, *The Trial: Four Thousand Years of Courtroom Drama* (New York: Random House, 2006).
10. J. Quen, "An Historical View of the M'Naghten Trial," *Bull of the History of Med* 42 (1968): 43.
11. Both O'Connor and Scalia are quoted from their dissenting opinions in *Roper v. Simmons*, 545 U.S. 551 (2005).
12. L. Buchen, "Arrested Development," *Nat* 484 (2012): 304.
13. Rosen, "Brain on the Stand."
14. L. Mansnerus, "Damaged Brains and the Death Penalty," *New York Times*, July 21, 2001, p. B9; M. Brower and B. Price, "Neuropsychiatry of Frontal Lobe Dysfunction in Violent and Criminal Behaviour: A Critical Review," *J Neurol, Neurosurgery and Psychiatry* 71 (2001): 720.

15. M. Gazzaniga, "Free Will Is an Illusion, but You're Still Responsible for Your Actions," *Chronicle of Higher Education*, March 18, 2012; M. Gazzaniga, *Who's in Charge? Free Will and the Science of the Brain* (New York: Ecco, 2012).
16. L. Steinberg et al., "Are Adolescents Less Mature Than Adults? Minors' Access to Abortion, the Juvenile Death Penalty, and the Alleged APA 'Flip-flop,'" *Am Psychologist* 64 (2009): 583.
17. S. Morse, "Brain and Blame," *Georgetown Law J* 84 (1996): 527.
18. B. Libet, "Can Conscious Experience Affect Brain Activity?" *J Consciousness Studies* 10 (2003): 24; B. Libet et al., "Time of Conscious Intention to Act in Relation to Onset of Cerebral Activity (Readiness-Potential)," *Brain* 106 (1983): 623.
19. V. Ramachandran, *The Tell-Tale Brain: A Neuroscientist's Quest for What Makes Us Human* (NY: Norton, 2012).
20. C. Dweck, *Mindset: How You Can Fulfill Your Potential* (London, UK: Constable & Robinson, 2012); C. Dweck, "Motivational Processes Affecting Learning," *Am Psychologist* 41 (1986): 1040; S. Levy and C. Dweck, "Trait-Focused and Process-Focused Social Judgment," *Soc Cog* (1998); 151; C. Mueller and C. Dweck, "Intelligence Praise Can Undermine Motivation and Performance," *JPSP* 75 (1998): 33–52.
21. J. Cantor, "Do Pedophiles Deserve Sympathy?" CNN.com, June 21, 2012.
22. S. Morse, "Neuroscience and the Future of Personhood and Responsibility," in *Constitution 3.0: Freedom and Technological Change*, ed. J. Rosen and B. Wittes (Washington, DC: Brookings Institution Press, 2011); J. Rosen, "Brain on the Stand" *New York Times*, March 11, 2007; S. Morse, "Brain Overclaim Syndrome and Criminal Responsibility: A Diagnostic Note," *Ohio State J Criminal Law* 397 (2006): 397; this is the source of the Morse quotes in the subsequent paragraphs.
23. H. Bok, "Want to Understand Free Will? Don't Look to Neuroscience," *Chronicle Review*, March 23, 2012.
24. Morse, "Neuroscience and the Future of Personhood"; S. Nichols, "Experimental Philosophy and the Problem of Free Will," *Sci* 331 (2011): 1401.
25. Morse, 2011, op cit.
26. Marvin Minsky, quoted in J. Coyne, "You Don't Have Free Will," *Chronicle Review*, March 23, 2012.
27. Footnote: J. Kaufman et al., "Brain-Derived Neurotrophic Factor–5-HTTLPR Gene Interactions and Environmental Modifiers of Depression in Children," *BP* 59 (2006): 673.
28. J. Russell, *Witchcraft in the Middle Ages* (Ithaca, NY: Cornell University Press, 1972).
29. D. Dennett, *Elbow Room: The Varieties of Free Will Worth Wanting* (Cambridge, MA: MIT Press, 1984).
30. Greene and Cohen, "For the Law, Neuroscience Changes Nothing."
31. M. Hoffman, *The Punisher's Brain: The Evolution of Judge and Jury* (Cambridge, MA: Cambridge University Press, 2014)
32. K. Gospic et al., "Limbic Justice: Amygdala Involvement in Immediate Rejections in the Ultimatum Game," *PLoS ONE* 9 (2011): e1001054; Buckholtz, "Neural Correlates of Third-Party Punishment."
33. D. de Quervain et al., "The Neural Basis of Altruistic Punishment," *Sci* 305 (2004): 1254; B. Knutson, "Sweet Revenge?" *Sci* 305 (2004): 1246.
34. Footnote: J. Bonnefon et al., "The Social Dilemma of Autonomous Vehicles," *Sci* 352 (2016): 1573; J. Greene, "Our Driverless Dilemma," *Sci* 352 (2016): 1514.

Chapter 17: War and Peace

1. M. Fisher, "The Country Where Slavery Is Still Normal," *Atlantic*, June 28, 2011; C. Welzel, *Freedom Rising: Human Empowerment and the Quest for Emancipa*tion (Cambridge: Cambridge University Press, 2013).
2. S. Pinker, *The Better Angels of Our Nature: Why Violence Has Declined* (New York: Penguin, 2011).
3. N. Elias, *The Civilizing Process: Sociogenetic and Psychogenetic Investigations*, rev. ed. (Malden, MA: Blackwell, 2000); W. Yang, "Nasty, Brutish, and Long," *New York*, October 16, 2011.
4. S. Herman and D. Peterson, "Steven Pinker on the Alleged Decline of Violence," *Int Socialist Rev*, November/December, 2012.
5. R. Douthat, "Steven Pinker's History of Violence," *New York Times*, October 17, 2011; J. Gray, "Delusions of Peace," *Prospect*, October 2011; E. Kolbert, "Peace in Our Time: Steven Pinker's History of Violence," *New Yorker*, October 3, 2011; T. Cowen, "Steven Pinker on Violence," *Marginal Revolution*, October 7, 2011.
6. C. Apicella et al., "Social Networks and Cooperation in Hunter-Gatherers," *Nat* 481 (2012): 497.
7. S. Huntington, "Democracy for the Long Haul," *J Democracy* 7 (1996): 3; T. Friedman, *The Lexus and the Olive Tree* (New York: Anchor Books, 1999).
8. L. Rhue and A. Sundararajan, "Digital Access, Political Networks and the Diffusion of Democracy," *Soc Networks* 36 (2014): 40.
9. M. Inzlicht et al., "Neural Markers of Religious Conviction," *Psych Sci* 20 (2009): 385; M. Anastasi and A. Newberg, "A Preliminary Study of the Acute Effects of Religious Ritual on Anxiety," *J Alternative and Complementary Med* 14 (2008): 163.
10. U. Schjoedt et al., "Reward Prayers," *Nsci Letters* 433 (2008): 165; N. P. Azari et al., "Neural Correlates of Religious Experience," *Eur J Nsci* 13 (2001): 1649; U. Schjoedt et al., "Highly Religious Participants Recruit Areas of Social Cognition in Personal Prayer," *SCAN* 4 (2009): 199; A. Norenzayan and W. Gervais, "The Origins of Religious Disbelief," *TICS* 17 (2013): 20; U. Schjoedt et al., "The Power of Charisma: Perceived Charisma Inhibits the Frontal Executive Network of Believers in Intercessory Prayer," *SCAN* 6 (2011): 119.
11. L. Galen, "Does Religious Belief Promote Prosociality? A Critical Examination," *Psych Bull* 138 (2012): 876; S. Georgianna, "Is a Religious Neighbor a Good Neighbor?" *Humboldt J Soc Relations* 11 (1994): 1; J. Darley and C. Batson, "From Jerusalem to Jericho: A Study of Situational and Dispositional Variables in Helping Behavior," *JPSP* 27 (1973): 100; L. Penner et al., "Prosocial Behavior: Multilevel Perspectives," *Ann Rev Psych* 56 (2005): 365.

12. C. Batson et al., *Religion and the Individual: A Social-Psychological Perspective* (Oxford: Oxford University Press, 1993); D. Malhotra, "(When) Are Religious People Nicer? Religious Salience and the 'Sunday Effect' on Prosocial Behavior," *Judgment and Decision Making* 5 (2010): 138.
13. A. Norenzayan and A. Shariff, "The Origin and Evolution of Religious Prosociality," *Sci* 422 (2008): 58.
14. A. Shariff and A. Norenzayan, "God Is Watching You: Priming God Concepts Increases Prosocial Behavior in an Anonymous Economic Game," *Psych Sci* 18 (2007): 803; W. Gervais, "Like a Camera in the Sky? Thinking About God Increases Public Self-Awareness and Socially Desirable Responding," *JESP* 48 (2012): 298. See also: I. Pichon et al., "Nonconscious Influences of Religion on Prosociality: A Priming Study," *Eur J Soc Psych* 37 (2007): 1032; M. Bateson et al., "Cues of Being Watched Enhance Cooperation in Real-World Setting," *Biol Lett* 2 (2006): 412.
15. S. Jones, "Defeating Terrorist Groups," RAND Corporation, CT-314 (testimony presented before the House Armed Services Committee, Subcommittee on Terrorism and Unconventional Threats and Capabilities), September 18, 2008; P. Shadbolt, "Karma Chameleons: What Happens When Buddhists Go to War," CNN.com, April 22, 2013.
16. J. LaBouff et al., "Differences in Attitudes Toward Outgroups in Religious and Nonreligious Contexts in a Multinational Sample: A Situational Context Priming Study," *Int J for the Psych of Religion* 22 (2011): 1; B. J. Bushman et al., "When God Sanctions Killing: Effect of Scriptural Violence on Aggression," *Psych Sci* 18 (2007): 204. This is the source of the figure in the text. H. Ledford, "Scriptural Violence Can Foster Aggression," *Nat* 446 (2007): 114.
17. J. Ginges et al., "Religion and Support for Suicide Attacks," *Psych Sci* 20 (2009): 224.
18. G. Allport, *The Nature of Prejudice* (Boston: Addison-Wesley, 1954).
19. T. Pettigrew and L. Tropp, "A Meta-analytic Test of Intergroup Contact Theory," *JPSP* 90 (2006): 751.
20. A. Al Ramiah and M. Hewstone, "Intergroup Contact as a Tool for Reducing, Resolving, and Preventing Intergroup Conflict: Evidence, Limitations, and Potential," *Am Psychologist* 68 (2013): 527; Y. Yablon and Y. Katz, "Internet-Based Group Relations: A High School Peace Education Project in Israel," *Educational Media Int* 38 (2001): 175; L. Goette and S. Meier, "Can Integration Tame Conflicts?" *Sci* 334 (2011): 1356; M. Alexander and F. Christia, "Context Modularity of Human Altruism," *Sci* 334 (2011): 1392; M. Kalman, "Israeli/Palestinian Camps Don't Work," *San Francisco Chronicle*, October 19, 2008.
21. I. Beah, *A Long Way Gone* (New York: Sarah Crichton Books, 2007).
22. R. Weierstall et al., "Relations Among Appetitive Aggression, Post-traumatic Stress and Motives for Demobilization: A Study in Former Colombian Combatants," *Conflict and Health* 7 (2012): 9; N. Boothby, "What Happens When Child Soldiers Grow Up? The Mozambique Case Study," *Intervention* 4 (2006): 244.
23. J. Arthur, "Remember Nayirah, Witness for Kuwait?" *New York Times*, January 6, 1992; J. Macarthur, "Kuwaiti Gave Consistent Account of Atrocities; Retracted Testimony," *New York Times*, January 24, 1992; "Deception on Capitol Hill" (editorial), *New York Times*, January 15, 1992; T. Regan, "When Contemplating War, Beware of Babies in Incubators," Christian Science Monitor, September 6, 2002; R. Sapolsky, "'Pseudokinship' and Real War," *San Francisco Chronicle*, March 2, 2003. For Nayirah's actual testimony, see. www.youtube.com/watch?v=LmfVs3WaE9Y.
24. E. Queller et al., "Single-Gene Greenbeard Effects in the Social Amoeba *Dictyostelium discoideum*," *Sci* 299 (2003): 105; M. Nowak, "Five Rules for the Evolution of Cooperation," *Sci* 314 (2006): 1560.
25. C. Camerer and E. Fehr, "When Does Economic Man Dominate Social Behavior?" *Sci* 311 (2006): 47; J. McNamara et al., "Variation in Behaviour Promotes Cooperation in the Prisoner's Dilemma Game," *Nat* 428 (2004): 745; C. Hauert and M. Doebeli, "Spatial Structure Often Inhibits the Evolution of Cooperation in the Snowdrift Game," *Nat* 428 (2004): 643.
26. M. Milinski et al., "Reputation Helps Solve the 'Tragedy of the Commons,'" *Nat* 415 (2002): 424.
27. M. Nowak et al., "Fairness Versus Reason in the Ultimatum Game," *Sci* 289 (2000: 1773; G. Vogel, "The Evolution of the Golden Rule," *Sci* 303 (2004): 1128.
28. J. Henrich et al., "Costly Punishment Across Human Societies," *Sci* 312 (2006): 1767; B. Vollan and E. Olstrom, "Cooperation and the Commons," *Sci* 330 (2010): 923; D. Rustagi et al., "Conditional Cooperation and Costly Monitoring Explain Success in Forest Commons Management," *Sci* 330 (2010): 961.
29. S. Gachter et al., "The Long-Run Benefits of Punishment," *Sci* 322 (2008): 1510.
30. B. Knutson, "Sweet Revenge?" *Sci* 305 (2004): 1246; D. de Quervain et al., "The Neural Basis of Altruistic Punishment," *Sci* 305 (2004): 1254; E. Fehr and S. Gachter, "Altruistic Punishment in Humans," *Nat* 415 (2002): 137; E. Fehr and B. Rockenbach, "Detrimental Effects of Sanctions on Human Altruism," *Nat* 422 (2003): 137; C. T. Dawes et al., "Egalitarian Motives in Humans," *Nat* 446 (2007): 794
31. E. Fehr and U. Fischbacher, "The Nature of Human Altruism," *Nat* 425 (2003): 785; M. Janssen et al., "Lab Experiments for the Study of Social-Ecological Systems," *Sci* 328 (2010): 613; R. Boyd et al., "Coordinated Punishment of Defectors Sustains Cooperation and Can Proliferate When Rare," *Sci* 328 (2010): 617.
32. J. Jordan et al., "Third-Party Punishment as a Costly Signal of Trustworthiness," *Nat* 530 (2016): 473.
33. A. Gneezy et al., "Shared Social Responsibility: A Field Experiment in Pay-What-You-Want Pricing and Charitable Giving," *Sci* 329 (2010): 325; S. DellaVigna, "Consumers Who Care," *Sci* 329 (2010): 287.
34. J. McNamara et al., "The Coevolution of Choosiness and Cooperation," *Nat* 451 (2008): 189.
35. IDASA, *National Elections Survey, August 1994* (Cape Town: Institute for Democracy in South Africa, 1994); Human Science Research Council, *Omnibus, May 1995* (Pretoria, South Africa: HSRC/Mark Data, 1995); B. Hamber et al., "'Telling It Like It Is . . .': Understanding the Truth and Reconciliation Commission from the Perspective of Survivors," *Psych in Soc* 26 (2000): 18.
36. D. Filkins, "Atonement: A Troubled Iraq Veteran Seeks Out the Family He Harmed," *New Yorker*, October 29,

2012; D. Margolick, *Elizabeth and Hazel: Two Women of Little Rock* (New Haven, CT: Yale University Press, 2011).

37. R. Fehr and M. Gelfand, "When Apologies Work: How Matching Apology Components to Victims' Self-Construals Facilitates Forgiveness," *Organizational Behav and Hum Decision Processes* 113 (2010): 37.
38. M. McCullough, *Beyond Revenge: The Evolution of the Forgiveness Instinct* (Hoboken, New Jersy: Jossey-Bass, 2008).
39. M. Berman, "'I Forgive You.' Relatives of Charleston Church Shooting Victims Address Dylann Roof," *Washington Post*, June 19, 2015.
40. J. Thompson-Cannino et al., *Picking Cotton: Our Memoir of Injustice and Redemption* (New York: St. Martin's Griff, 2010).
41. L. Toussaint et al., "Effects of Lifetime Stress Exposure on Mental and Physical Health in Young Adulthood: How Stress Degrades and Forgiveness Protects Health," *J Health Psych* 21 (2014): 1004; K. A. Lawler et al., "A Change of Heart: Cardiovascular Correlates of Forgiveness in Response to Interpersonal Conflict," *J Behav Med* 26 (2003): 373; M. C. Whited et al., "The Influence of Forgiveness and Apology on Cardiovascular Reactivity and Recovery in Response to Mental Stress," *J Behav Med* 33 (2010): 293; C. vanOyen Witvliet et al., "Granting Forgiveness or Harboring Grudges: Implications for Emotion, Physiology, and Health," *Psych Sci* 12 (2001): 117; P. A. Hannon et al., "The Soothing Effects of Forgiveness on Victims' and Perpetrators' Blood Pressure," *Personal Relationships* 19 (2011): 27; G. L. Reed and R. D. Enright, "The Effects of Forgiveness Therapy on Depression, Anxiety, and Posttraumatic Stress for Women After Spousal Emotional Abuse," *J Consulting Clin Psych* 74 (2006): 920.
42. D. Kahneman and J. Renshon, "Why Hawks Win," *Foreign Policy*, January/February 2007.
43. D. Laitin, "Confronting Violence Face to Face," *Sci* 320 (2008): 51.
44. D. Grossman, *On Killing: The Psychological Costs of Learning to Kill in War and Society* (New York: Back Bay Books, 1995).
45. M. Power, "Confessions of a Drone Warrior," *GQ*, October 22, 2013; J. L. Otto and B. J. Webber, "Mental Health Diagnoses and Counseling Among Pilots of Remotely Piloted Aircraft in the United States Air Force," *MSMR* 20 (2013): 3; J. Dao, "Drone Pilots Are Found to Get Stress Disorders Much as Those in Combat Do," *New York Times*, February 22, 2013.
46. J. Altmann et al., "Body Size and Fatness of Free-Living Baboons Reflect Food availability and Activity Level," *Am J Primat* 30 (1993): 149; J. Kemnitz et al., "Effects of Food Availability on Insulin and Lipid Levels in Free-Ranging Baboons," *Am J Primat* 57 (2002): 13; W. Banks et al., "Serum Leptin Levels as a Marker for a Syndrome X-Like Condition in Wild Baboons," *J Clin Endo and Metabolism* 88 (2003): 1234.
47. R. Tarara et al., "Tuberculosis in Wild Baboon (*Papio cynocephalus*) in Kenya," *J Wildlife Diseases* 21 (1985): 137; R. Sapolsky and J. Else, "Bovine Tuberculosis in a Wild Baboon Population: Epidemiological Aspects," *J Med Primat* 16 (1987): 229.
48. R. Sapolsky and L. Share, "A Pacific Culture Among Wild Baboons, Its Emergence and Transmission," *PLoS Biol* 2 (2004): E106; R. Sapolsky, "Culture in Animals, and a Case of a Non-human Primate Culture of Low Aggression and High Affiliation," *Soc Forces* 85 (2006): 217; R. Sapolsky, "Social Cultures in Non-human Primates," *Curr Anthropology* 47 (2006): 641; R. Sapolsky, "A Natural History of Peace," *Foreign Affairs* 85 (2006): 104.
49. I. DeVore, *Primate Behavior: Field Studies of Monkeys and Apes* (New York: Holt, 1965).
50. A. McAvoy, "Pearl Harbor Vets Reconcile in Hawaii," Associated Press, December 6, 2006; R. Ohira, "Zenji Abe, the Enemy Who Became a Friend," *Honolulu Advertiser*, April 12, 2007.
51. N. Rhee, "Why US Veterans Are Returning to Vietnam," *Christian Science Monitor*, November 10, 2013.
52. K. Sim and M. Bilton, *Remember My Lai*, (PBS Video, 1989); G. Eckhardt, *My Lai: An American Tragedy* (Kansas City: University of Missouri—Kansas City Law Review, Summer 2000); M. Bilton and K. Sim, *Four Hours in My Lai* (New York: Penguin, 1993); this is the source of the Varnado Simpson quote; T. Angers, *The Forgotten Hero of My Lai: The Hugh Thompson Story* (Lafayette, LA: Acadian House, 1999); this is the source of the Hugh Thompson quote.
53. Footnote: M. Bilton and K. Sim, *Four Hours in My Lai* (NY: Penguin, 1993).
54. A. Hochschild, *Bury the Chains: The British Struggle to Abolish Slavery* (Basingstoke, UK: Pan Macmillan, 2005); E. Metaxas, *Amazing Grace: William Wilberforce and the Heroic Campaign to End Slavery* (New York: HarperOne, 2007).
55. G. Bell, *Rough Notes by an Old Soldier: During Fifty Years' Service, from Ensign G. B. to Major-General C. B.* (London: Day, 1867).
56. M. Seidman, "Quiet Fronts in the Spanish Civil War," libcom.org, Summer 1999; F. Robinson, *Diary of the Crimean War* (1856); E. Costello, *The Adventures of a Soldier* (1841); BiblioLife, 2013; J. Persico *My Enemy, My Brother: Men and Days of Gettysburg* (Cambridge, MA: Da Capo Press, 1996).
57. S. Weintraub, *Silent Night: The Story of the World War I Christmas Truce* (New York: Plume Press, 2002).
58. T. Ashworth, *Trench Warfare, 1914–1918: The Live and Let Live System* (London: Pan Books, 1980). Live and Let Live is also analyzed in R. Axelrod, *The Evolution of Cooperation* (New York: Basic Books, 2006).
59. E. Jones, "One War Is Enough," *Atlantic*, February 1946.

Illustration Credits

Pages:

30 Courtesy Chickensaresocute/CC BY-SA 3.0.
190 Photo Researchers, Inc./Science Source.
201 Courtesy Angela Catlin.
268 AFP/Getty Images.
268 Zoonar GmbH/Alamy.
270 Katherine Cronin and Edwin van Leeuwen/Chimfunshi Wildlife Orphanage Trust.
278 Courtesy Yulin Jia/Dale Bumpers National Rice Research Center/U.S. Department of Agriculture/CC BY 2.0.
307 (Right) Augustin Ochsenreiter/South Tyrol Museum of Archaeology.
307 (Left) Eurac/Samadelli/Staschitz/South Tyrol Museum of Archaeology.
320 Courtesy Mopane Game Safaris/CC BY-SA 4.0.
355 (Bottom) SD Dirk/Wikimedia Commons.
361 Courtesy Liz Schulze.
378 Vincent J. Musi/National Geographic Creative.
387 ZUMA Press, Inc./Alamy.
472 (Top right) Jacob Halls/Alamy.
472 (Bottom) Walt Disney Studios Motion Pictures/Lucasfilm Ltd.
556 (Top) Dennis Hallinan/Alamy.
556 (Bottom) Courtesy © 2016 C. Herscovici/Artists Rights Society (ARS), New York.
577 Moshe Milner/Israel's Government Press Office/Flickr.
631 (Top) Courtesy © 2013 Marcus Bleasdale/VII for Human Rights Watch.
631 (Bottom left) Courtesy Pierre Holtz/UNICEF CAR/CC BY-SA 2.0.
631 (Bottom right) Bjorn Svensson/Alamy.
653 Chris Belsten/Flickr.
655 (Left) NA (Public Domain).
655 (Right) Courtesy Maureen Monte.
658 (Left) Courtesy Jim Andreotta.
658 (Right) Courtesy Susan T. Kummer.
669 (Top) Ralph Crane/The LIFE Images Collection/Getty Images.
669 (Bottom) War posters/Alamy.
680 Courtesy BruceBlaus/CC BY 3.0.
681 (Top) Courtesy MethoxyRoxy/CC BY-SA 2.5.
686 Deco Images/Alamy.
702 Keystone Pictures USA/Alamy.
703 (Top) Courtesy Doc. RNDr. Josef Reischig, CSc./CC BY SA 3.0.
703 (Bottom) Courtesy Blacknick and Nataliia Skrypnyk, Taras Shevchenko National University of Kyiv/CC BY-SA 4.0.
704 Courtesy Livet, Sanes, and Lichtman/Harvard University.
705 Courtesy Rajalakshmi L. Nair et al./CC BY 2.0.

Index

Page numbers in *italics* refer to illustrations.